Energy, Environment, and Sustainability

Series Editor

Avinash Kumar Agarwal, Department of Mechanical Engineering, Indian Institute of Technology Kanpur, Kanpur, Uttar Pradesh, India

AIMS AND SCOPE

This books series publishes cutting edge monographs and professional books focused on all aspects of energy and environmental sustainability, especially as it relates to energy concerns. The Series is published in partnership with the International Society for Energy, Environment, and Sustainability. The books in these series are edited or authored by top researchers and professional across the globe. The series aims at publishing state-of-the-art research and development in areas including, but not limited to:

- Renewable Energy
- Alternative Fuels
- Engines and Locomotives
- Combustion and Propulsion
- Fossil Fuels
- Carbon Capture
- Control and Automation for Energy
- Environmental Pollution
- Waste Management
- Transportation Sustainability

Review Process

The proposal for each volume is reviewed by the main editor and/or the advisory board. The chapters in each volume are individually reviewed single blind by expert reviewers (at least four reviews per chapter) and the main editor.

Ethics Statement for this series can be found in the Springer standard guidelines here https://www.springer.com/us/authors-editors/journal-author/journal-author-helpdesk/before-you-start/before-you-start/1330#c14214

More information about this series at https://link.springer.com/bookseries/15901

Swatantra P. Singh · Avinash Kumar Agarwal ·
Tarun Gupta · Shihabudheen M. Maliyekkal
Editors

New Trends in Emerging Environmental Contaminants

Springer

Editors
Swatantra P. Singh
Department of Environmental Science and Engineering (ESED)
Indian Institute of Technology Bombay
Mumbai, Maharashtra, India

Tarun Gupta
Department of Civil Engineering
Indian Institute of Technology Kanpur
Kanpur, Uttar Pradesh, India

Avinash Kumar Agarwal
Department of Mechanical Engineering
Indian Institute of Technology Kanpur
Kanpur, Uttar Pradesh, India

Shihabudheen M. Maliyekkal
Department of Civil and Environmental Engineering
Indian Institute of Technology Tirupati
Tirupati, Andhra Pradesh, India

ISSN 2522-8366 ISSN 2522-8374 (electronic)
Energy, Environment, and Sustainability
ISBN 978-981-16-8369-5 ISBN 978-981-16-8367-1 (eBook)
https://doi.org/10.1007/978-981-16-8367-1

This Springer imprint is published by the registered company Springer Nature Singapore Pte Ltd.
The registered company address is: 152 Beach Road, #21-01/04 Gateway East, Singapore 189721, Singapore

Preface

Water, soil and air pollution have been receiving remarkable consideration due to increasing population, industrialization and urbanization. Moreover, extensive usage of various compounds in modernized pocesses has also led to an increasing in various toxic pollutants in water as well as in the atmosphere which are harmful at very low concentration. These pollutants term as emerging pollutants. Water, soil and air pollution have showed wide range of adverse effects to both human beings and the environment.

The International Society for Energy, Environment and Sustainability (ISEES) was founded at the Indian Institute of Technology Kanpur (IIT Kanpur), India, in January 2014 to spread knowledge/awareness and catalyze research activities in the fields of Energy, Environment, Sustainability, and Combustion. Society's goal is to contribute to the development of clean, affordable and secure energy resources and a sustainable environment for society and spread knowledge in the areas mentioned above and create awareness about the environmental challenges the world is facing today. The unique way adopted by ISEES was to break the conventional silos of specializations (engineering, science, environment, agriculture, biotechnology, materials, fuels, etc.) to tackle the problems related to energy, environment, and sustainability in a holistic manner. This is quite evident by the participation of experts from all fields to resolve these issues. The ISEES is involved in various activities such as conducting workshops, seminars and conferences in the domains of its interests. Society also recognizes the outstanding works of young scientists, professionals and engineers for their contributions in these fields by conferring them awards under various categories.

Fifth International Conference on 'Sustainable Energy and Environmental Challenges' (V–SEEC) was organized under the auspices of ISEES from December 19–21, 2020, in virtual mode due to restrictions on travel because of the ongoing COVID-19 pandemic situation. This conference provided a platform for discussions between eminent scientists and engineers from various countries, including India, Spain, Austria, Bangladesh, Mexico, USA, Malaysia, China, UK, Netherlands, Germany, Israel and Saudi Arabia. At this conference, eminent international speakers presented their views on energy, combustion, emissions and alternative energy resources for sustainable development and a cleaner environment. The conference presented two high voltage plenary talks by Dr. VK Saraswat, Honorable Member, NITI Ayog, on 'Technologies for Energy Security and Sustainability,' and Prof. Sandeep Verma, Secretary, SERB, on 'New and Equitable R&D Funding Opportunities at SERB.'

The conference included nine technical sessions on topics related to energy and environmental sustainability. Each session had 6–7 eminent scientists from all over the world, who shared their opinion and discussed the trends for the future. The technical sessions in the conference included Emerging Contaminants: Monitoring and Degradation Challenges; Advanced Engine Technologies and Alternative Transportation Fuels; Future Fuels for Sustainable Transport; Sustainable Bioprocessing for Biofuel/ Non-biofuel Production by Carbon Emission Reduction; Future of Solar Energy; Desalination and Wastewater Treatment by Membrane Technology; Biotechnology in Sustainable Development; Emerging Solutions for Environmental Applications' and Challenges and Opportunities for Electric Vehicle Adoption. 500+ participants and speakers from all over the world attended this three-day conference. The conference concluded with a high voltage panel discussion on 'Challenges and Opportunities for Electric Vehicle Adoption,' where the panelists were Prof. Gautam Kalghatgi (University of Oxford), Prof. Ashok Jhunjhunwala (IIT Madras), Dr. Kelly Senecal (Convergent Science), Dr. Amir Abdul Manan (Saudi Aramco) and Dr. Sayan Biswas (University of Minnesota, USA). Prof. Avinash K Agarwal, ISEES, moderated the panel discussion. This conference laid out the roadmap for technology development, opportunities and challenges in energy, environment and sustainability domain. All these topics are very relevant for the country and the world in the present context. We acknowledge the support received from various agencies and organizations for the successful conduct of the Fifth ISEES conference V-SEEC, where these books germinated. We want to acknowledge SERB (Special thanks to Dr. Sandeep Verma, Secretary) and our publishing partner Springer (Special thanks to Ms. Swati Meherishi).

The editors would like to express their sincere gratitude to large number of authors from all over the world for submitting their high-quality work in a timely manner and revising it appropriately at a short notice. We would like express our special thanks to Dr. Amritanshu Shriwastav, Dr. Manoj Kumar Tiwari, Dr. Alok Sinha, Dr. Bhaskar Bahaduri, Dr. Shreeprasad P. S., Dr. Tabish Nawaz, Dr. Vamsi V. B. and Dr. Seetha N. who reviewed various chapters of this monograph and provided their valuable suggestions to improve the manuscripts.

This book is based on recent trends for the research in emerging environmental contaminants in different compartment of the environment. Emerging environmental contaminants such as pharmaceutical and personal care products (PPCPs) and microplastics have started creating a lot of problems in the ecosystem. The current understanding and treatment technologies are not equipped to remove these pollutants leading to bioaccumulation and magnification in the environment. This book aims to provide a recent understanding for the fate, transport and degradation of emerging contaminants in different environmental sectors, including water, air and soil. In this book, several chapters have been incorporated to discuss the fate and transport of microplastics, PPCPs, along with the method of detection and degradation. This book includes removal of variety of pollutants including microplastics, phamaceuticals and personal care products from the water using adsorption technique, electrooxidation, membrane technology and other advance oxidation methods. We are confident that the book would be of great attention to the professionals and post-graduate students involved in environmental science and engineering research.

Mumbai, India — Swatantra P. Singh
Tirupati, India — Shihabudheen M. Maliyekkal
Kanpur, India — Tarun Gupta
Kanpur, India — Avinash Kumar Agarwal

Contents

Editors and Contributors

About the Editors

Dr. Swatantra P. Singh is an environmental engineer with experience in membrane fabrication, environmental nanotechnology, fate and transport of pollutants and emerging contaminants in the environment. Currently, he is an assistant professor in the Environmental science and Engineering Department (ESED) at IIT Bombay. He has developed a key technology to fabricate the membranes for water purification and print graphene in situ in a single step. He has four US patents (two granted and two provisional) on membrane and laser-based graphene fabrication techniques. He has authored 19 journal articles, one book and three book chapters. He has recently won INAE Young Engineer Award (2020) and ISEES Young Scientists Award (2020).

Prof. Avinash Kumar Agarwal joined IIT Kanpur in 2001. He worked at the Engine Research Center, UW@Madison, the USA as a Post-Doctoral Fellow (1999–2001). His interests are IC engines, combustion, alternate and conventional fuels, lubricating oil tribology, optical diagnostics, laser ignition, HCCI, emissions, and particulate control, 1D and 3D Simulations of engine processes, and large-bore engines. Prof. Agarwal has published 435+ peer-reviewed international journal and conference papers, 70 edited books, 92 books chapters, and 12200+ Scopus and 19000+

Google Scholar citations. He is the associate principal editor of FUEL. He has edited "Handbook of Combustion" (5 Volumes; 3168 pages), published by Wiley VCH, Germany. Prof. Agarwal is a Fellow of SAE (2012), Fellow of ASME (2013), Fellow of ISEES (2015), Fellow of INAE (2015), Fellow of NASI (2018), Fellow of Royal Society of Chemistry (2018), and a Fellow of American Association of Advancement in Science (2020). He is the recipient of several prestigious awards such as Clarivate Analytics India Citation Award-2017 in Engineering and Technology, NASI-Reliance Industries Platinum Jubilee Award-2012; INAE Silver Jubilee Young Engineer Award-2012; Dr. C. V. Raman Young Teachers Award: 2011; SAE Ralph R. Teetor Educational Award-2008; INSA Young Scientist Award-2007; UICT Young Scientist Award-2007; INAE Young Engineer Award-2005. Prof. Agarwal received Prestigious CSIR Shanti Swarup Bhatnagar Award-2016 in Engineering Sciences. Prof. Agarwal is conferred upon Sir J C Bose National Fellowship (2019) by SERB for his outstanding contributions. Prof. Agarwal was a highly cited researcher (2018) and was in the top ten HCR from India among 4000 HCR researchers globally in 22 fields of inquiry.

Prof. Tarun Gupta is a doctorate from Harvard University and M.Tech. from IIT Bombay. He has authored more than 125 ISI indexed journal publications, 3 books, 10 book chapters, 4 patents and has been reviewer of more than 36 journals. He has guided 6 Ph.D. and 34 M.Tech. theses. A submicron aerosol sampler designed, developed and evaluated by him at IIT Kanpur has been commercialized by Envirotech (Delhi). He has developed a high volume fine PM sampler and transferred technology to BARC. He is currently P. K. Kelkar research fellow and selected member of INYAS and INAE Associate. He has recently won INAE Innovator and Entrepreneur Award (2018), VNMM award (2017), NASI-SCOPUS Award (2015), INSA Medal for Young Scientist (2011), INAE Young Engineer Award (2009) and IEI Young Engineer Award (2008). He is N. C. Nigam Chair Professor and serving as the head of department Civil Engineering at IIT Kanpur.

Dr. Shihabudheen M. Maliyekkal is an Assistant Professor in the Civil and Environmental Engineering Department at the Indian Institute of Technology Tirupati. He earned his master's and Ph.D. from the Indian Institute of Technology Madras in Environmental Engineering. The focus of his research is to address the challenging task of providing safe water for all. One of his arsenic removal technologies has been implemented in the field, and more than one million affected people benefit from the invention. He has extensive experience in the synthesis and characterization of nanoscale materials and their application in water purification. Besides water and wastewater treatment, his interests include developing green and smart materials for energy and infrastructure applications. Dr. Maliyekkal has authored over 70 articles in reputed journals and conference proceedings, and he is an inventor/co-inventor in 16 patents or patent applications. He is one of the founding directors of an IIT Tirupati incubated start-up NGEN Water Solutions Pvt. Ltd.

Contributors

Avinash Kumar Agarwal Department of Mechanical Engineering, IIT Kanpur, Kanpur, UP, India

Mohit Aggarwal School of Infrastructure, Indian Institute of Technology Bhubaneswar, Argul, Odisha, India

Shaik Mahamad Allabakshi Department of Civil and Environmental Engineering, Indian Institute of Technology Tirupati, Tirupati, India

Najmul Haque Barbhuiya Environmental Science and Engineering Department, IIT Bombay, Mumbai, India

Sadia Basri School of Environmental Sciences (SES), Jawaharlal Nehru University, New Delhi, India

Bhaskar Bhaduri Department of Chemical Engineering, Indian Institute of Technology, Kharagpur, West Bengal, India

Shefali Chauhan School of Environmental Sciences (SES), Jawaharlal Nehru University, New Delhi, India

Venkata S. N. Chava Department of Chemistry and Biochemistry, The University of Texas at El Paso, El Paso, TX, USA

Swathi Desireddy School of Civil Engineering, Vellore Institute of Technology (VIT), Chennai, India

Ashutosh Kumar Dixit Department of Electronics and Communication Engineering, VFSTR University, Vadlamudi, Guntur, Andhra Pradesh, India

Nandini Dixit Environmental Science and Engineering Department (ESED), Indian Institute of Technology Bombay, Mumbai, India

Shibil Abdul Gafoor Department of Civil and Environmental Engineering, Indian Institute of Technology Tirupati, Tirupati, India

Reetesh Kumar Gangwar Department of Physics & CAMOST, Indian Institute of Technology Tirupati, Tirupati, India

Tarun Gupta Department of Civil Engineering, IIT Kanpur, Kanpur, UP, India

Hansa Department of Materials Science and Engineering, Gachon University, Seongnam-si, Gyeonggi-do, South Korea

Afrah Harafan Department of Civil and Environmental Engineering, Indian Institute of Technology Tirupati, Tirupati, India

Mahima John Horta Department of Civil Engineering, Indian Institute of Technology Hyderabad, Hyderabad, Telangana, India

Angel Joseph Department of Civil Engineering, Indian Institute of Technology Delhi, New Delhi, India

Pradip Kalbar Centre for Urban Science & Engineering, Indian Institute of Technology Bombay, Mumbai, India

Bhavana Kanwar Environmental Science and Engineering Department, IIT Bombay, Mumbai, India

Ansaf V. Karim Environmental Science and Engineering Department, Indian Institute of Technology Bombay, Mumbai, India

Indrajeet Vijay Kharade Environmental Science and Engineering Department, IIT Bombay, Mumbai, India

Tae Young Kim Department of Materials Science and Engineering, Gachon University, Seongnam-si, Gyeonggi-do, South Korea

Sheetal S. Kothawale Centre for Research in Nanotechnology and Science, Indian Institute of Technology Bombay, Mumbai, India

Yerramilli Sai Rama Krishna Department of Civil Engineering, Indian Institute of Technology Hyderabad, Hyderabad, Telangana, India

Sukanya Krishnan Environmental Science and Engineering Department, Indian Institute of Technology Bombay, Mumbai, India

Ashish Kumar Environmental Science and Engineering Department (ESED), Indian Institute of Technology Bombay, Mumbai, India

Pramod Kumar Environmental Science and Engineering Department (ESED), Indian Institute of Technology Bombay, Mumbai, India

Tadimeti Divya Kusuma Department of Civil and Environmental Engineering, Indian Institute of Technology Tirupati, Tirupati, India

Chinthalapudi Naga Lakshmi Department of Chemical Engineering, Indian Institute of Technology Tirupati, Tirupati, Andhra Pradesh, India

Shihabudheen M. Maliyekkal Department of Civil and Environmental Engineering, Indian Institute of Technology Tirupati, Tirupati, India

Bikram Mishra School of Infrastructure, Indian Institute of Technology Bhubaneswar, Argul, Odisha, India

Utkarsh Misra Centre for Research in Nanotechnology & Science (CRNTS), Indian Institute of Technology Bombay, Mumbai, India

Monisha Monachan Environmental Science and Engineering Department (ESED), Indian Institute of Technology Bombay, Mumbai, India

Azmat Naseem Department of Civil Engineering, Indian Institute of Technology Delhi, New Delhi, India

Tabish Nawaz Environmental Science and Engineering Department (ESED), Indian Institute of Technology Bombay, Powai, Mumbai, Maharashtra, India

Darakhshan Nayyar Environmental Science and Engineering Department, Indian Institute of Technology Bombay, Powai, Mumbai, Maharashtra, India

Vartika Nishad Environmental Science and Engineering Department (ESED), Indian Institute of Technology Bombay, Mumbai, India

Swatantra P. Singh Interdisciplinary Program in Climate Studies, Environmental Science and Engineering Department, Centre for Research in Nanotechnology & Science (CRNTS), Indian Institute of Technology Bombay, Mumbai, India

Aishwarya Rastogi School of Environmental Science and Engineering, Indian Institute of Technology, Kharagpur, India

M. K. Ravi Varma Department of Physics, National Institute of Technology Calicut, Calicut, Kerala, India

Sudharshan Reddy Environmental Science and Engineering Department, IIT Bombay, Mumbai, India

Neelancherry Remya School of Infrastructure, Indian Institute of Technology Bhubaneswar, Argul, Odisha, India

P. C. Sabumon School of Civil Engineering, Vellore Institute of Technology (VIT), Chennai, India

Shalini Sahani Department of Materials Science and Engineering, Gachon University, Seongnam-si, Gyeonggi-do, South Korea

N. Seetha Department of Civil Engineering, Indian Institute of Technology Hyderabad, Hyderabad, Telangana, India

Abinaya Sekar Department of Civil Engineering, National Institute of Technology Calicut, Calicut, Kerala, India

S. Sethulekshmi Environmental Science and Engineering Department, Indian Institute of Technology Bombay, Mumbai, India

Mohd Ahmed Naim Shaikh Environmental Science and Engineering Department, Indian Institute of Technology Bombay, Powai, Mumbai, Maharashtra, India

Yogesh Chandra Sharma Department of Chemistry, Indian Institute of Technology (Banaras Hindu University), Varanasi, India

Amritanshu Shriwastav Environmental Science and Engineering Department, Indian Institute of Technology Bombay, Mumbai, India

Narendra Singh Department of Chemical Engineering, Indian Institute of Technology Tirupati, Tirupati, Andhra Pradesh, India

Alok Sinha Indian Institute of Technology (Indian School of Mines), Dhanbad, India

Charulata Sivodia Indian Institute of Technology (Indian School of Mines), Dhanbad, India

Sreeprasad T. Sreenivasan Department of Chemistry and Biochemistry, The University of Texas at El Paso, El Paso, TX, USA

P. S. N. S. R. Srikar Department of Physics & CAMOST, Indian Institute of Technology Tirupati, Tirupati, India

Manoj Kumar Tiwari School of Water Resources, Indian Institute of Technology, Kharagpur, India

Kumud Malika Tripathi Department of Chemistry, Indian Institute of Petroleum and Energy, Visakhapatnam, Andhra Pradesh, India

George K. Varghese Department of Civil Engineering, National Institute of Technology Calicut, Calicut, Kerala, India

Arya Vijayanandan Department of Civil Engineering, Indian Institute of Technology Delhi, New Delhi, India

Part I
Emerging Contaminants: Sources, Occurrence, and Their Fate

Chapter 1
Introduction of New Trends in Emerging Environmental Contaminants

Swatantra P. Singh, Shihabudheen M. Maliyekkal, Tarun Gupta, and Avinash Kumar Agarwal

Abstract Clean water and unpolluted air and soil are very critical for sustaining human life and the ecosystem. The importance of these resources has long been recognized. Nevertheless, a significant portion of these essential resources is contaminated, affecting millions of human lives and the productivity of the ecosystems globally. Among the various contaminants detected in the environment, new and emerging pollutants pose a great challenge due to their increasing presence and unknown toxicity. This book is a compendium of chapters by different authors, reviewing various aspects of emerging contaminants in different environmental compartments. The current understanding of fate, transport, and degradation of emerging pollutants in different environmental sectors is discussed. The method of detection and the challenges associated with the degradation of these contaminants are also discussed. The book also reviews both conventional and advanced treatment technologies, including adsorption, electrooxidation, membrane filtration, advanced oxidation process, and application of nanomaterials in removing these pollutants from the environment. The authors believe that the book will be of great use to the professionals and students involved in environmental science and engineering research.

Keywords Emerging contaminants · Photocatalysts · Desalination · Membrane technology · Microplastics

---

S. P. Singh (✉)
Environmental Science and Engineering Department, IIT Bombay, Mumbai 400076, India
e-mail: swatantra@iitb.ac.in

S. M. Maliyekkal
Department of Civil and Environmental Engineering, Indian Institute of Technology Tirupati, Tirupati 517506, India

T. Gupta
Department of Civil Engineering, IIT Kanpur, Kanpur, UP 208016, India

A. K. Agarwal
Department of Mechanical Engineering, IIT Kanpur, Kanpur, UP 208016, India

S. P. Singh et al. (eds.), *New Trends in Emerging Environmental Contaminants*, Energy, Environment, and Sustainability,
https://doi.org/10.1007/978-981-16-8367-1_1

1.1 Introduction

Rapid industrialization, modern lifestyle, and indiscriminate use of chemicals in various industrial, domestic, and agricultural sectors have paved the way for the ubiquitous presence of emerging contaminants (ECs) in the environment. The ECs include a wide range of chemical compounds such as pesticides, active pharmaceutical compounds, disinfection by-products, personal care products (PCPs), food additives, artificial sweeteners, flame retardants, endocrine-disrupting compounds, nanomaterials, microplastics, and surfactants. These new and emerging contaminants are of grave concern due to their toxicity, persistence, and bioaccumulative nature. Many emerging contaminants can sustain after primary and secondary wastewater treatment, and there is only limited information available regarding their toxicity and long-term impact on human health and the environment. Being an involving field, understanding the renewed knowledge in the area is of prime importance. The edited book comprises 22 chapters describing various aspects of emerging contaminants, including sources, occurrence, fate and transport, treatment, and detection. The authors across the globe contribute to the book, giving a global perspective of emerging contaminants in the environment and associated issues. The section below presents a summary of various topics discussed in the book.

We have divided the book into two parts; part one has seven chapters and is based on new emerging contaminants sources, occurrence, fate, and transport; the second section has 15 chapters based on the treatment and detection of emerging pollutants with advanced technologies such as plasma, microbial fuel cell, and membrane technologies. The first chapter of this book has summarized the book's overall content with the basic introduction of ECs. The second chapter of the book provides an introduction of EC in the aquatic systems, and these EC includes the trace compounds such as pharmaceuticals (PhACs), personal care products (PCPs), endocrine disrupting compounds (EDCs). This chapter also provided the overall problem with potential solutions.

Microplastics are an important class of emerging contaminants and are omnipresent. They are plastic particles of size <5 mm that are either manufactured or formed from the disintegration of large plastics in the environment. Microplastics occur in various forms and are extensively used for a multitude of applications. They are potentially toxic to humans and the environment. Chapter 3 presents an overview of the microplastics' source, occurrence, fate, and transport in various environmental compartments, including air, water, and soil. Microplastics' detection and extraction methods from different environmental matrices and microplastics' capability to act as a carrier for organic contaminants are also discussed. The chapter also highlights various exposure mechanisms and ecotoxicological effects of microplastics. Chapter 4 also deals with microplastic pollution, emphasizing its origin and occurrence in terrestrial soils such as floodplain soils. The chapter provides strategies for the chemical extraction, quantification, and analysis of microplastics from terrestrial soil. Furthermore, the chapter emphasizes the microplastic control and remediation techniques in these natural ecosystems for scientific inquiry.

Wastewater treatment plants act as sinks for ECs to accumulate and undergo transformations. They are recognized as the primary source of contamination of ECs in aquatic and terrestrial environments. Hence, there is a requirement to understand the fate and behavior of ECs in such systems. Chapter 5 discusses microplastics' occurrence and fate during various treatment processes such as biodegradation, adsorption, membrane processes, filtration, electrocoagulation, and advanced oxidation processes. In addition, the mechanism of removal and the challenges involved in removing microplastics from each treatment unit are discussed in detail. Finally, a comparative analysis of removal efficiency between conventional treatment systems and recent technologies is also presented.

Volatile organic compounds (VOCs) are an essential class of ECs. Industrial accidents are the preliminary cause of the discharge of significant VOCs into the air environment, causing many deaths and damaging the ecosystem. Among the VOCs, chloroform causes serious health effects and is found abundantly in the environment. These compounds are not regulated due to their short atmospheric lifetime. However, chlorocarbons pose a significant threat to the ozone layer. Chapter 6 summarizes the properties of chloroform and gives a broad spectrum of information on its sources, metabolic pathways, health effects, and occurrence in the air medium.

Phthalate esters (PAE) are widely used as industrial additives, primary plasticizers for polyvinyl chloride, cellulosic, elastomers, and adhesives. They are extensively used in the plastic industry owing to their ability to improve the flexibility and durability of polymers. Their presence in water bodies has received increased attention due to their endocrine-disrupting effects. Chapter 7 provides an overview of various types and sources of PAEs, and their fate and transport in the environment. In addition, concise information regarding their extraction and detection is also discussed. The chapter also highlights the health risks associated with acute and chronic exposure of these pollutants to aquatic and terrestrial organisms and microbes. Furthermore, the chapter also reviews various treatment technologies employed for the removal of PAEs from the environment.

Pharmaceuticals are widely used across the globe to treat diseases and promote growth in animals and plants. According to statistics, India is one of the top five emerging markets for pharmaceuticals. In recent years, pharmaceuticals as ECs have gained increased attention in the research community because of their high persistence, low biodegradable nature, bioaccumulation, relative stability, and tendency to develop antibiotic-resistant genes. In this context, Chapter 8 deals with the sources and occurrence of pharmaceuticals, their toxic effects on biotic organisms, and their removal by various technologies. A brief outlook on the detection and analysis of pharmaceuticals is also included.

Several technologies are developed to remove ECs from various matrices and protect the environment. These include both traditional and advanced technologies. Although advance technologies exhibit excellent capability in removing ECs, their energy-intensive nature and large ecological footprint limit their usage. This urges the need for sustainable treatment technologies. Among various techniques, microbial degradation can be viewed as a preferential method to eradicate the ECs. Numerous bacterial, fungal, and algal strains have shown substantial efficacy in degrading ECs, using them as the secondary carbon source. However, it is necessary to understand their degradation potential, pathways, and mechanisms for different classes of ECs. The traditional technologies such as biodegradation and adsorption are covered in the Chaps. 9–14. The Chap. 9 provides a detailed outlook on microbial degradation of various pharmaceuticals and their mechanism of action. The chapter also analyzes the prevalent hindrances in the degradation of drugs in field-scale wastewater treatment plants. The adsorption process appears promising treatment technologies due to its low cost, simplicity in operation, and excellent removal efficiency. However, the selectiveness of adsorbents towards various ECs depends upon the surface, chemical, and textural properties of adsorbents and types and the chemical nature of contaminants. Conventionally, biochar, obtained from waste biomass by thermal reduction, has been extensively used as an adsorbent in removing ECs. The presence of surface functional groups like hydroxyl, carboxyl, carbonyl, and amines on the biochar enhances its potential to remove contaminants. Typically, they are modified to increase their removal efficiency. Chapter 10 discusses the status quo of functionalization and characterization of the biochar. The recent advancements in using as derived or modified biochar for ECs. Chapter 11 also discusses the adsorptive removal of ECs using natural waste-derived adsorbents, the factors affecting and removal mechanisms.

Nanotechnology-based techniques have gained momentum recently due to their exciting properties such as increased surface area, high reactivity, distinctive quantum effects, tunable pore size, high catalytic activity, and superior electrical, chemical, and magnetic properties promising in the treatment of ECs. Chapter 12 reviews the recent developments in nanotechnology and novel nanomaterial-based systems for removing ECs from water and wastewater. Magnetic nanoparticles (MNP) serve as promising adsorbents for the removal of ECs due to their excellent catalytic and adsorption performance and ease of separation via the application of an external magnetic field. Among the various MNPs, iron oxide-based magnetic nanoparticles are more popular. Chapter 13 evaluates the recent progress in the development of MNPs and their hybrid material-based techniques for the remediation of ECs from wastewater. The chapter also surveys the potential development, technical challenges, and engineering concerns and future directions to overcome them and improve water security. Another important class of iron-based nanomaterial is zero-valent iron (ZVI). Although ZVI is an excellent choice for treating ECs in water, they are usually immobilized on a supporting matrix due to their tendency to agglomerate. Clay minerals have been identified as an excellent supporting material for nano ZVI due to their ability to tune the immobilization and form functionalized nanocomposite with distinct physicochemical properties. Chapter 14 defines various strategies employed to synthesize nano-ZVI and the effective utilization of clay as supporting material. Finally, the material's applicability as an adsorbent and a catalyst in removing ECs is discussed in detail.

The limitation of conventional treatment technologies in EC remediation led to the development of advanced treatment technologies. The advanced oxidation processes (AOPs) provide a promising route to degrade and mineralize ECs by generating reactive chemical species via an in-situ process. It is a potential technique and can degrade and mineralize complex organic molecules. The processes rely on the in-situ generation of reactive chemical species (RCS). Most of the RCS with higher oxidizing potential is short-lived, and hence the effective production of these compounds is crucial for this technology's success. Various AOPs include the Fenton and photo-Fenton processes, photocatalysis, sonolysis, plasma-based oxidation, catalytic wet air oxidation, and combinations thereof. Chapter 15 focuses on the recent developments in plasma-based reactors and their application in the degradation of ECs. The chapter also highlights plasma-based technology's current challenges and prospects in treating ECs and various operating parameters influencing the process.

Photocatalysis is an essential class of APOs due to their enhanced ability to remove ECs from various environments. Typically, semiconductor oxides are preferred as photocatalysts. However, due to the quick recombination of electron–hole pairs and the inability to effectively harness solar energy, their use is limited. Recently, graphene nanomaterials have been extensively used as efficient photocatalysts due to their improved reactivity under light irradiation. These are usually coupled or modified with transition/noble metals to function as heterogeneous photocatalysts. Chapter 16 reviews the application of various graphene-modified photocatalysts in the treatment of ECs in water. The chapter also discusses the working principle, factors influencing the removal, a brief history, challenges, and prospects in the development of photocatalysis. Plasma-assisted AOP is an emerging technology capable of simultaneously generating various reactive chemical species at a controlled rate in an ambient environment without any additional chemical/catalyst.

Membrane-based technologies are classified as ultrafiltration, microfiltration, forward osmosis, reverse osmosis (RO) and nanofiltration (NF). Among them, NF, and RO have broad applicability in the efficient reclamation of wastewater. However, their practical use is limited owing to their intense energy use. Chapter 17 focuses mainly on the recent studies in the removal of ECs using NF and RO. The chapter also reviews the integration of recent technologies such as membrane bioreactor and AOPs with RO/NF process for ECs removal. Finally, an outline of current knowledge gaps and future research scope related to the application of RO and NF for wastewater treatment is also discussed. Another widely preferred technology is membrane distillation due to its added advantages, such as low-pressure requirements, less susceptibility to fouling, low-temperature requirements, and complete retention of non-volatile compounds. Chapter 18 elaborates on various membrane distillation technologies, the possible materials, configurations, operating parameters, and energy requirements. The chapter also highlights the future research trends and challenges for the sustainable and commercial application of the technology.

Most existing centralized water treatment facilities are inadequate to meet the growing demand for safe drinking water. Establishing large-scale new water treatment systems is not feasible due to high capital investment, lack of skilled labor, poor governance, water scarcity, poor track record of operation and maintenance. Point-of-use (POU) water treatment systems contribute significantly to achieving one of the UNs sustainable development goals, i.e., clean water and sanitation. Many studies have reported the capability of POU systems in reducing waterborne diseases by improving the microbial quality of water. Numerous physical, chemical, light, and heat-based methods are available for POU water treatment. Chapter 19 reviews various technologies available for POU water treatment focusing mainly on membrane-based POU water filters. The chapter also discusses in detail the performance of household POU water filters in removing emerging contaminants. Electrocoagulation is a quick and effective method in the removal of ECs. The process utilizes a combination of electricity and sacrificial metal electrodes for the remediation of pollutants. Chapter 20 mainly focuses on removing ECs via the electrocoagulation process, its mechanism, and the operational parameters. Finally, the chapter presents the future challenges in the treatment of ECs from its various sources.

Due to substantial scientific and technological advancements, there has been a significant rise in energy consumption derived from fossil fuels. This has, in turn, contributed negatively to climate change. Hence, there is a need for a fast transition towards renewable and clean energy sources for addressing climate change problems. Such transition is possible only through evolving technologies, socioeconomics, and government policies. Chapter 21 talks about utilizing microbial fuel cells for ECs removals with power generation. Furthermore, energy production using solar photovoltaics has been considered the primary choice for a clean and renewable energy economy due to its ability to harness solar energy. However, the health and safety concerns arising due to the generation and consumption of solar cells are often overlooked. Chapter 22 presents the various types of photovoltaic cells and the methodology involved in their fabrication process. The chapter also aims to provide a comprehensive overview of the safety and health concerns of various materials, processes used to manufacture these cells, and their practical use.

The establishment of sustainable water treatment systems must not be limited to potable water production but must also refrain from contributing to secondary contamination. Colloidal contaminants such as engineered nanoparticles, mineral precipitates of iron, aluminum, and manganese, natural organic matter, bacteria, viruses, and clays have been frequently detected in the subsurface. These may enter from various sources such as land application of wastewater, reuse of untreated sewage for irrigation and sanitary landfills, or improper disposal of treatment materials. Hence, understanding colloidal transport in the subsurface is essential for assessing the safe distance of drinking water wells from the source of contamination, bioremediation of contaminated sites, and degree of treatment required for land application of wastewater. Chapter 23 discusses the mechanisms of colloid retention in porous media at various scales, the effect of heterogeneity on colloid transport at each scale, and upscaling of transport processes. Overall, this book includes detection, fate, transport, and removal of various pollutants, including microplastics, pharmaceuticals, and personal care products from the water using adsorption technique, electrooxidation, membrane technology, and other advanced oxidation methods. We are confident that the book would be of great attention to the professionals, post-graduate students involved in environmental science and engineering research.

Chapter 2
Emerging Contaminants in Wastewater and Surface Water

Shalini Sahani, Hansa, Yogesh Chandra Sharma, and Tae Young Kim

Abstract Currently, severe contamination of water bodies has led to the scarcity of usable water for drinking and other purposes. The heavy emergence of pollutants in the aquatic environment ranging from ng/l to mg/L is just a consequence of drastic and speedy anthropogenic activities. As compared to developed countries, developing countries, in general, face a more serious impact of alleviating contaminants in water bodies owing to the lack of adequate studies on the current status of water contamination as well as the fate and impact of ECs in wastewater and surface water. Further, the occurrence of pollutants (generally, organics) in wastewater is increasing non-regulated manner above the safe limits. These are not only harmful to aquatic flora and fauna but also for terrestrial lives and recently have been observed by advanced analytical detectors. These trace compounds are termed emerging contaminants (ECs) derived from pharmaceuticals (PhACs), personal care products (PCPs), endocrine disrupting compounds (EDCs). In this book chapter, we would thoroughly discuss the above-mentioned individual types of emerging contaminants in detail and also present the survey on their local distribution aqueous matrix in India. Finally, we would elaborate on the possible solution to tackle this problematic situation of ECs dumping into wastewater and surface water.

Keywords Emerging contaminants · Pharmaceuticals · Personal care products · Nanomaterials

---

S. Sahani · Hansa · T. Y. Kim (✉)
Department of Materials Science and Engineering, Gachon University, Seongnam-daero, Sujeong-gu, 1342, Seongnam-si, Gyeonggi-do 13120, South Korea
e-mail: taeykim@gachon.ac.krl

Y. C. Sharma (✉)
Department of Chemistry, Indian Institute of Technology (Banaras Hindu University), Varanasi, India
e-mail: ysharma.apc@itbhu.ac.in

S. P. Singh et al. (eds.), *New Trends in Emerging Environmental Contaminants*, Energy, Environment, and Sustainability,
https://doi.org/10.1007/978-981-16-8367-1_2

2.1 Introduction

Nowadays, fresh and clean water has become scanty across the globe as surface water is no more able to be leveraged directly for various domestic or commercial purposes. Heavy pollution of surface water has made it absolutely of no use. The major contribution to water pollution is the accumulation of emerging contaminants (ECs). ECs are predominantly organic and inorganic compounds for examples chemical ingredients of pharmaceuticals, cosmetic products, taste and odor compounds, surfactants, X-ray contrast media, and steroid hormones. Their direct consumption has a fatal impact on human health in form of lethal and incurable diseases like cancer. Excessive presence in the aquatic environment ranging from ng/l to mg/L is just a consequence of drastic and speedy anthropogenic activities (Verlicchi and Zambello 2015). For the last few decades, several kinds of ECs are getting dumped in the aquatic bodies of our ecosystem through various outputs such as waste sludges from wastewater treatment plants, domestic outlets, industrial units, sewer leakage, agriculture, and surface runoffs from rural and urban zones, and manure waste. Even, groundwater is also severely contaminated across the globe (Bu et al. 2013; Zhang et al. 2015). Predominantly, wastewater treatment plants are of major concern because they directly dump the emerging ECs into water bodies just by overlooking their proper treatment. This serious issue has alarmed the research fraternity engineers and civil authorities to look into the matter to take some quick action as the level of these hazardous pollutants has trespassed their permissible limit in wastewater and surface waters. Therefore, the supply of clean and safe water to urban as well as rural populations has become an extremely challenging task. Even, thorough treatment of wastewater has been proven a tedious task for its further recycling to reuse.'

Ironically, conventional water treatment plants are only sensitive towards water nutrients, bacteria, heavy metals, pesticides, petroleum hydrocarbons, and other primary pollutants (Pal et al. 2014). Overall, the present water treatment plant only focuses on primary pollutants whose impact on human health and animal is known.

However, as compared to developed countries, developing countries, in general, lack adequate studies on issues related to ECs (Rehman et al. 2015). India is one among them. In contrast, the production and consumption rate of compounds in ECs categories in India is growing day by day.

The occurrence of pollutants (generally, organics) in wastewater is increasing in a non-regulated manner above the safe limits. These are not only harmful to aquatic flora and fauna but also to terrestrial lives as recently observed by advanced analytical detectors. These trace compounds are termed as ECs.

Generally, the organic and inorganic trace pollutants found in wastewater have the following broad categories they are derived from.

(i) Pharmaceuticals (PhACs)
(ii) Personal Care Products (PCPs)
(iii) Endocrine Disrupting Compounds (EDCs)
(iv) Per- and polyfluoroalkyl substances (PFASs)
(v) Artificial sweeteners (ASs)
(vi) Flame retardants (FRs)
(vii) Nanomaterials (NMs)

Out of all the above-mentioned categories of ECs, PhACs, PCPs, and EDCs are found in very high concentrations in the aquatic system throughout the globe. This emerging pollutant disturbs the bacterial biomass and hence the biological activity in wastewater and surface water. Especially, nanomaterials consisting of inorganic metal oxides cause malfunctioning in bacterial activity hindering the removal of emerging pollutants in wastewater treatment plants (Wang et al. 2012). It is to be noted that though there are specific groups of ECs as discussed above still there is no clear discrimination amongst the groups of ECs. The compositions of ECs subgroups are overlapping each other.

In this book chapter, a thorough discussion is dedicated to the major ECs i.e. PhACs, PCPs, and EDCs in detail in subsequent sections along with their carcinogenic impact on animal and human health.

2.2 Emerging Contaminants (ECs) in Wastewater and Surface Water

As a consequence of population explosion, ECs are easily circulated over surface water and even reach the groundwater from open resources of water. Disposal of municipal sludge is also a very challenging task as it is considered as the principal source of the liberation of emerging pollutants from open aquatic ecosystem where industries and domestic outlets (as point sources) are directly connected to water bodies without significant treatment of wastewater to make it free of emerging pollutants. The emerging pollutants are categorized into three above said major groups of chemicals. Firstly, PhACs are further subdivided into different categories viz. antibiotics, analgesics, steroids, and nonsteroids, etc. Secondly, Personal care products (PCPs) are identified as cosmetic and health care chemicals available in the market, for example, galaxolide, tonalide, etc. Endocrine-disrupting compounds (EDCs) are groups of compounds or chemicals that show an adverse impact on human endocrine systems by causing androgenic or estrogenic activities even at low concentrations, for example, estradiol, phthalates, etc.

After executing a thorough review of reports on emerging contaminants found in surface water and wastewater, the following major classes falling in the above discussed categories are formed as per their abundant quantity in the aquatic system (Luo et al. 2014; Subedi et al. 2015; Semblante et al. 2017).

(i) Antibiotics
(ii) Antifungal/antimicrobial agents
(iii) Nonsteroidal anti-inflammatory drugs (NSAIDs)
(iv) Anticonvulsants/antidepressants
(v) Artificial sweeteners
(vi) Beta-adrenoceptor blocking agents
(vii) Lipid regulating drugs
(viii) Steroidal hormones
(ix) X-ray contrast media
(x) UV filters
(xi) Stimulants
(xii) Anti-itching drugs
(xiii) Insect repellents
(xiv) Plasticizers
(xv) Pesticides
(xvi) Metal oxides and composites
(xvii) Perfluorochemicals (PFCs)

2.3 Pharmaceuticals (PhACs)

PhACs are a set of developing ecological contaminants that are broadly and progressively being utilized as a part of human and veterinary medication. They include compounds of environmental concern like antibiotics, legal and illicit drugs, analgesics, steroids, β-blockers, anticonvulsants, antihypertensive, etc. (Luo et al. 2014).

The formation of the above subcategories is based on a mass number of observations by researchers. They have examined their pros and cons over human health and the environment. This classification was performed while keeping four main points into consideration i.e. (i) consumption across the globe (ii) observation frequencies (iii) effect on surroundings (iv) quantity of analytical data.

Keeping all the critical points in view, Table 2.1 has been prepared which ensemble the major classes of pharmaceutical chemicals depicting potential risks on human health and other organisms (Minh et al. 2009; Kurunthachalam 2012; Bu et al. 2013; Luo et al. 2014; Stefanakis and Becker 2015; aus der Beek et al. 2016; Singh et al. 2016; Tiwari et al. 2016; Pavithra et al. 2017; Bai et al. 2018; Gogoi et al. 2018; Nag et al. 2018; Brown and Winterstein 2019).

PhACs include organic compounds like antibiotics, legal and illicit drugs, analgesics, steroids, beta-blockers, anticonvulsants, antihypertensive, etc. (Luo et al. 2014; Pal et al. 2014; Zhang et al. 2015; Taheran et al. 2018). Pharmaceuticals have been defined as substances that are primarily being used for therapeutic, preventive, and diagnostic purposes. Recreational drugs, such as cocaine and caffeine, have been excluded from the present analysis. Other substances, such as homeopathic, minerals, proteins, and immunologic substances, also were not included. If not otherwise indicated, the terms "pharmaceuticals" and "pharmaceutical substances" refer only to the actual active pharmaceutical ingredients. These pharmaceutical pollutants are found to be bio-accumulating in nature as they are very much active in the biological system. Their persistence in the body occurs due to their specific mode of action. Therefore, they are traced in wastewater, surface water, and drinking water which ultimately cause bio-resistant properties in bacteria present anywhere. Even the metabolites of pharmaceutically active compounds have been causing a lethal impact on the ecosystem gradually. Due to their very minute concentration, they have been unnoticed and never been prioritized by Governmental guideline makers to treat wastewater for their removal. That's why they have been termed pseudo-persistent pollutants which accumulate into the environment at trace level (Archana et al. 2017). This situation has been continuing for few decades and no assessment has been executed seriously except a few years back, few reports got surfaced and were noticed for the risk assessment of pharmaceutical ingredients and their derivatives. Recently, more than 160 different pharmaceutical compounds have been detected in water bodies supremely, wastewater treatment plant effluent in the concentration range of ng L^{-1} to low $\mu g\ L^{-1}$ (Archana et al. 2017). These pharmaceutical drugs are not only affecting human beings but also wildlife tremendously (Archana et al. 2017). So far, extremely inadequate information is collected about the toxicological effect of pharmaceutical drugs on terrestrial and aquatic lives, especially in the Indian subcontinent. The exact fate of bioaccumulation of PhACs and their impact on water-bodies and surroundings are therefore quite unpredictable. It is only assured once the consolidated database of PhACs distribution throughout the country is available.

2.3.1 Sources of PhACs

There are various inlets of PhACs through which PhACs are dumped into surface water. Amongst all, sewage effluents are realized as a prominent source of PhACs and their metabolites. Generally, sewage effluents consist of the domestic pharmaceutical left out, discharge from hospitals, and even sometimes from pharmaceutical manufacturing plants. According to a report, Patancheru Enviro Tech Limited (PETL) WTP near Hyderabad received 1.5 MLD effluents from ~90 bulk drug manufacturers in the vicinity in Patancheru. This was found to record the highest levels of PhACs ever reported in wastewater from anywhere else across the globe (Balakrishna et al. 2017). Sometimes the overflow from WTPs also contributes to the contamination of surface water. Even the untreated sewage discharged from cruise and boats also contributes to PhACs contamination in the riverine system. For example, sewage effluents received from typhoon shelters were found highly contaminated with several kinds of antibiotics like β-lactams, fluoroquinolones, macrolides, sulfonamides, tetracyclines, trimethoprim, and amphenicols (Minh et al. 2009). Large scaled animal husbandry and horticulture have been also noticed to pollute the surface water due to excessive use of antibiotics to cure animals (Kaval et al. 2015). Despite the ban on the use of antibiotics at several places like Europe, there are still some regions where a low dose of antibiotics in feed as growth promoters is given to animals. Besides, there are many reports on the use of antibiotics including oxytetracycline and streptomycin on horticultural crops in some countries (Kaval et al. 2015). Leachates from landfills might be the smoother route for pharmaceuticals disposed of household and clinical wastes entering the surface water and even groundwater.

2.3.2 Occurrence of PhACs in Indian Environment

Analgesics, anticonvulsants, antidepressants, antiepileptics, antihypertensives, and β-blockers are commonly observed in water systems owing to higher solubility in water and slower metabolic rate (Brown and Winterstein 2019). The antiepileptic gabapentin has been recorded with the highest median concentration in the aquatic system followed by metformin, lamotrigine, desmethylvenlafaxine, hydrochlorothiazide, sulfamethoxazole, and hydroxycarbamazepine with (>100 ng/L) concentration.

In developing countries, more monitoring results have recently become available, but a concise picture of measured environmental concentrations (MECs) is still elusive. Through a comprehensive literature review of 1016 original publications and 150 review articles (aus der Beek et al. 2016), the authors collected MECs for human and veterinary pharmaceutical substances reported worldwide in surface water, groundwater, tap/drinking water, manure, soil, and other environmental matrices in a comprehensive database. Approximately, 631 different pharmaceutical substances were found at MECs above the detection limit of the respective analytical methods employed, revealing distinct regional patterns. Sixteen pharmaceutical substances i.e. diclofenac, carbamazepine, ibuprofen, sulfamethoxazole, naproxen, estrone, estradiol, ethinylestradiol, trimethoprim, paracetamol, clofibric acid, ciprofloxacin, ofloxacin, estriol, norfloxacin, and acetylsalicylic acid were detected in each of the 5 United Nations (UN) regional groups (Africa Group, Asia–Pacific Group, Eastern Europe Group, group of Latin American and the Caribbean States, and Western Europe and Others Group, which also includes North America, Australia, and New Zealand). For example, the anti-inflammatory drug diclofenac has been detected in environmental matrices in 50 countries, and concentrations found in several locations exceeded predicted no-effect concentrations. aus der Beek et al. (2016) studied the distribution of various PhACs in different regions of the world as shown in Fig. 2.1.

Recently, Philip et al. (2018) analyze the data derived from the monitoring of the presence of PhACs across the country. Unfortunately, it is found that data are scarce regarding the presence of ECs primarily from northeast, north, central parts of the

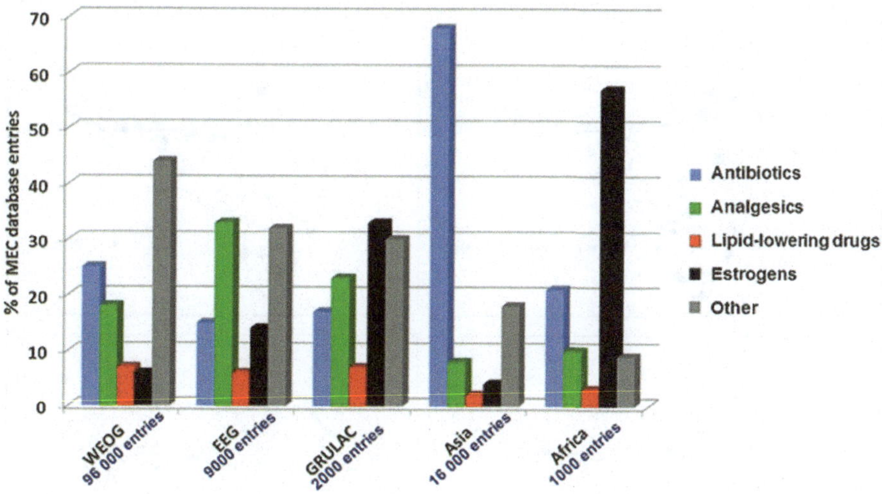

Fig. 2.1 Regional patterns of pharmaceutical therapeutic groups analyzed in each United Nations region. MEC = measured environmental concentration; EEG = Eastern Europe Group; GRULAC = Latin American and Caribbean States; WEOG, Western Europe and Others Group (aus der Beek et al. 2016)

country as visible from Fig. 2.2. Only southern parts have been almost monitored for the presence of ECs in comparison to other parts of the country. However, whatever data are reflected they show the presence of ECs way more than permissible limit almost every part of the country. The drastic accumulation of ECs in water bodies and wastewater systems is attributable to the boom in the application of medicines due to the increase in population across the globe. India stands in the second position in the queue of highest populated countries after China (Kalotra 2014). India is now the largest exporter of generic drugs nearly (20% of gross global exports) as generics rule over more than 70% of the total market share in the Indian pharmaceutical sector which is further scaled to be enhanced in near future. Moreover, multinational companies have been attracted to this giant market of pharmaceutics in the greed of cheap-cost production and flexible environment tribunal guidelines. This is further contributed by efficient and fast techniques, quality research, the abundance of educated personnel, and skilled laborers making the Indian market a fascinating choice for the foreign pharmaceutical majors for their drug production and outsourcing (Kalotra 2014; Bai et al. 2018). Today, the Indian pharmaceutical industry has become one of the leading industries of the nation which has the very high caliber to make right from analgesics to medicine for any fatal disease like cancer and cardiac treatment. Abnormally high concentrations of PhACs are the result of heavy usages and also inefficient wastewater treatment plants (WTPs) to decontaminate the water from these pollutants. Recently, the occurrence of explicitly used PhACs and drugs either in their original or metabolite form raises the concern of the scientific fraternity. Recent reports suggested the significant occurrence of PhACs in

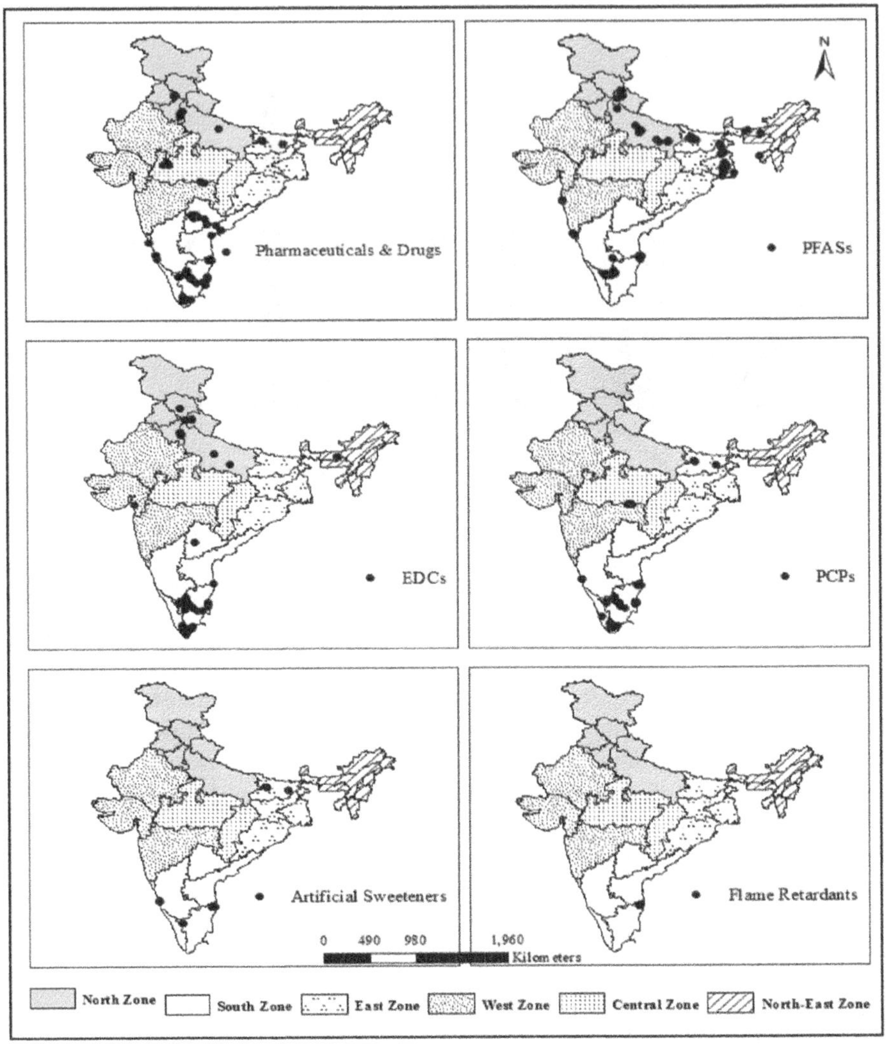

Fig. 2.2 Distribution of different classes of ECs in various Indian environmental matrices reprinted with permission from Philip et al. (2018). Copyright 2018, Elsevier

wastewater and surface water bodies in India. Carbamazepine, atenolol, triclocarban, triclosan, trimethoprim, and sulfamethoxazole, ibuprofen, and acetaminophen are the most commonly found at extremely high concentrations in WTPs and rivers. The concentration of most of these PhACs such as ciprofloxacin, sulfamethoxazole, amoxicillin, norfloxacin, and ofloxacin in WTPs across India might be 40 times higher than that in other countries in Europe, Australia, Asia, and North America (Balakrishna et al. 2017). The pattern of PhACs occurrence in the riverine system is analogous to that of China. This might be due to similar pharmaceutical infrastructure and population compatibility.

Notably, according to a study on antibiotic consumption, India was found to be the largest consumer even more than 75% during the decade (2000–2010). The excessive application of PhACs especially antibiotics has caused the accumulation of antibiotic-resistant genes (ARGs) at an alarming rate almost everywhere across the country. Coming to riverine systems, due to ARGs, the most fatal damage has been caused to holy rivers of the country named Ganga, Gomati, Narmada, Yamuna, Kshipra, etc. (Singh et al. 2016; Gogoi et al. 2018; Nag et al. 2018; Philip et al. 2018). This has caused a distressing situation for riverine life and human health as well as other organisms as flora and fauna belonging to the riverside use the river water for food consumption and various other purposes. The water from these rivers is widely supplied to the urban area of countries for domestic and commercial use. This may be the reason for spreading several kinds of incurable diseases such as cancer, and respiratory malfunctioning. According to a report (Singh et al. 2016), heavy contamination of β-lactam resistance determinant gene (bla_{TEM}) has been observed in the Indo-Gangetic region. It was seen that surface water and sediments of the river Ganga and Gomti Rivers exhibited high numbers of bla_{TEM} gene copies and varied significantly ($p < 0.05$) among the sampling locations. This heavy accumulation of this resistance gene is the exorbitant consumption of antimicrobial of the β-lactam group by humans due to its high efficiency and low cost. The water sample taken from drinking water facility and clinical settings had a considerable number of bla_{TEM} gene copies (13 $\pm$ 0.44–10,200 $\pm$ 316 gene copies/100 mL). Recently, Biswas and Vellanki (2021) reported the presence of 16 different types of PhACs and PCPs in water samples collected from the 13 locations spanning 575 km along the river Yamuna. Spatiotemporal variations in the occurrence of these ECs were detected during summer and post-monsoon. The appreciably high concentration of ECs was detected during the post-monsoon in comparison to the summer season. This is due to the assimilation of rain runoff from the catchment areas along river Yamuna. The ECs which were found in high concentrations include trimethoprim (8807.6 ng/L), caffeine (6489.9 ng/L), and gemfibrozil (2991 ng/L). The water samples with very high concentrations of ECs were mainly from Okhla barrage (ponding of water from drains traversing Delhi), the confluence of Yamuna with Shahadara drain (industrial and poultry cluster, and Ghazipur dumping yard), and Agra city (industrial clusters). According to a survey (Hanna et al. 2020), there has been a severe accumulation of antibiotic residues and antibiotic-resistant *Escherichia coli* (*E. coli*) in the water and sediment of the Kshipra river. This report analyzed the water samples collected from seven different sites over a long span of 2014–2016 during different seasons. Sulfamethoxazole was detected with the highest concentration of 4.66 μg/L. Even, ARGs were found against various antibiotics including sulfamethoxazole, norfloxacin, ciprofloxacin, cefotaxime, co-trimoxazole, ceftazidime, meropenem, ampicillin, amikacin, metronidazole, tetracycline, and tigecycline. It was found that the presence of antibiotics and anti-resistant genes was directly associated with water quality parameters. In another study (Archana et al. 2017), acetaminophen and ciprofloxacin, caffeine, irgasan, and benzophenone were found present in the influent and the effluent of the sewage treatment plant (STP) and surface water bodies (six major lakes) in and around Nagpur. These ECs were present in the range of 1–174 $\mu g\ L^{-1}$ in collected surface water, 12– 373 $\mu g\ L^{-1}$ in the influent, and 11–233 $\mu g\ L^{-1}$ in the effluent of the STP. They also carried out the ecotoxicity test on the targeted organism. Though they found the hazard quotient (HQ) less than 1 reflecting no hazardous effect of these analytes on the living organism still there is the uncertainty of the extent of effect over long-term exposure as ECs get inhabited in the water bodies for a quite long period.

Unfortunately, the relationship of antibiotics with aquatic dwellers has not been thoroughly explored in the country. Only a few research groups are involved in such studies. The monitoring efficiency must be improved by leaps and bounds and also except exploring the impact in sewage effluents from commercials units, domestics, and surface water systems should also be carefully monitored.

Fortunately, after several reports on the presence of lethal ARGs, the Government of India (GOI) has taken some necessary action but still, their proper implementation is underway. This might be due to improper health infrastructure, inaccessible medical facilities in remote areas, and lack of laboratory facilities for national-level monitoring of antimicrobial resistance.

2.3.3 Effects of PhACs on the Animal and Human Health

Toxicity caused by pharmaceutical substances in the Indian environment is undoubtedly a concerning issue (Kurunthachalam 2012; Philip et al. 2018). Several studies reported that the presence of PhACs in the aquatic system may cause complex biological effects on dwelling flora and fauna by significant alterations in biochemical parameters, hematological indices, enzymological activity, ionoregulatory responses, gene responses, plasma phosphate levels, and hormonal levels (Philip et al. 2018). The extinction of the vulture population from the Indian subcontinent is one of the finest examples of the environmental impact caused by PhACs residues. The exemplary mass extinction of mainly three species of vulture (*Gyps bengalensis*, *Gyps indicus*, and *Gyps tenuirostris*) is due to the consumption of livestock remains that were medicated with diclofenac (NSAID). Generally, high adult mortality and hence lives loss have been related to renal failure and kidney malfunction. These are attributed to in vivo assimilation of commonly used PhACs in high concentrations over a period. According to a study (Stefanakis and Becker 2015), drugs associated with the central nervous system, cardiovascular and anti-infective classes can turn out to be extremely lethal to aquatic life as well as the outer environment. Anti-infectants are also found very hazardous. Moreover, the long-term effect of these ECs is unknown especially at low concentrations in surface water. However, researchers are addressing only the specific classes of PhACs found in high concentrations in water bodies rather than non-specified. However, the effect of these non-prescribed PhACs can be more concerning. Hence there should be proper monitoring of every class of PhACs which can inhabit the wastewater and so surface water. The contamination of groundwater is of primary concern as still, a huge population uses the groundwater as drinking water directly for oral consumption in India. Furthermore, most of the respective scientific studies focus on certain groups such as antibiotics, hormones, or antineoplastics, but other groups such as antiulcerants and non-prescribed drugs also possess a potential negative impact on the environment.

According to reports on the high concentration of PhACs and ARGs in wastewater and surface water, it is concluded that these residues are extremely toxic and lethal to aquatic life too. Devarajan et al. (2015) analyzed the water samples collected from hospital outlet pipes (HOP) and the Cauvery River Basin (CRB) in Tiruchirappalli, Tamil Nadu, India. They analyzed these samples for organic matter, toxic metals, and their ecotoxicity. The analysis report manifested the high concentration of toxic metals in HOP, reaching values (mg kg^{-1}) of 1851 (Cr), 210 (Cu), 986 (Zn), 82 (Pb), and 17 (Hg) whereas in CRB concentrations were lower than that of in the HOP (except for Cu, Pb), with maximum values (mg kg^{-1}) of 75 (Cr), 906 (Cu), 649 (Zn), 111 (Pb), and 0.99 (Hg). As a result of the analysis report, the metal concentrations in both sampling sites were much ahead of the permissible limit recommended by Sediment Quality Guidelines (SQGs) and the Probable Effect Concentration (PEC) for the Protection of Aquatic Life. In this study, the ecotoxicity test with ostracods exposed to the sediment samples presented a mortality rate ranging from 22 to 100% (in sediments from HOP) and 18–87% (in sediments from CRB). The effluents from hospitals and pharmaceuticals outlets are extremely metal and PhACs loaded, and hence it is a major source of pollution of water resources and may place aquatic organisms and human health at risk.

2.4 Personal Care Products

Personal care products (PCPs) consist of generally organic compounds employed in cosmetic and healthcare products which include deodorants, perfumes, shampoos, sun scream, facial cream, sanitizers, cleaning products, and other products with estrogenic activity (Pal et al. 2014; Gogoi et al. 2018).

The following Table 2.2 consists of all possible types of PCPs composed of carcinogenic materials (Bu et al. 2013; Noguera-Oviedo and Aga 2016; Gogoi et al. 2018; Philip et al. 2018).

A list of emerging PCPs in surface water and wastewater (Verlicchi and Zambello 2015; Gogoi et al. 2018; Philip et al. 2018; Wang et al. 2021).

PCPs are dumped from domestic sources and cosmetic industries into surface water and wastewater in their native and derivative forms. The major lethal effect causing concern is the lipophilic nature of their constituents which hinders the metabolism and the lifecycle of flora and fauna in the water ecosystem.

2.4.1 Sources of PCPs

PCPs are most commonly dumped from domestic sources and cosmetic industries into surface water and wastewater in their native and derivative forms. Primarily, three categories of PCPs i.e. musk fragrances, disinfectants, antimicrobials, and preservatives are monitored in water and surface water across the world (Bu et al. 2013). Several reports are demonstrating the finding of these three categories of PCPs in surface water of the Haihe River in China, Michigan Lake in the USA, the North Sea in Germany, etc. (Bu et al. 2013). Notably, PhACs and PCPs can be accumulated in sediments which further act as a source of ECs in the aquatic atmosphere (Bu et al. 2013). The expired or unused PCPs, as well as drugs, even are generally dumped from households into the water which ultimately turns into fatal for animal or humans which consumes the untreated water in any forms (Stefanakis and Becker 2015). Besides Triclosan (TCS) and trichlorocarbon (TCC), paraben is drastically used in PCPs. Their adverse health effects have become a public concern. According to a study Wang et al. (2021), around six types of parabens, TCS and TCC were found in 129 indoor dust samples. They are also observed in 203 urine samples from two different cities in China along with four paraben metabolites. These trace level PCPs like phenols, paraben, and phthalates are almost found in the aquatic system of every part of the world (Santos et al. 2016; Sangeetha et al. 2021).

2.4.2 Occurrence of PCPs in Indian Environment

Primarily, three categories of PCPs i.e. musk fragrances, disinfectants, and preservatives are monitored in water and surface water across the world. Nag et al. (2018) monitored the occurrence of (TCS), the antibacterial agent commonly used in PCPs in water, sediment, and fish samples collected from a stretch of about 450 km of River Gomti, a major tributary of River Ganga, in India. TCS was detected in the range of 1.1–9.65 μg/l while in sediments and fish samples, the level was 5.11–50.36 μg/kg and 13–1040 μg/kg (wet weight basis), respectively. According to this evaluation, TCS uptake via contaminated fish appeared below the permissible limit (50 μg/kg body wt/day) and so there is no health hazard over a certain period. The high concentration of irgasan was monitored at various places in the country varying in seasons throughout the year. An extreme level of PCPs is dumped into the riverine system especially on the occasion of religious gatherings like Kumbh Mela where millions of pilgrims take a bath in holy rivers. Nowadays varieties of UV filters are used in PCPs. Among UV filters, Benzophenone (BPs) and 4-Methylbenzylidene Camphor (4-MBC) are often used to protect the skin and hair from UV radiations. BPs and 4-MBC are directly discharged to water bodies while cleaning and personal hygiene process. Since the conventional treatment plants are incapable to remove the BPs and 4-MBC thoroughly. Therefore, BPs and 4-MBC have been consistently observed in wastewater and off-course surface water in various parts of the country as well as in other parts of the world (Zhang et al. 2021a, b).

2.4.3 *Effects of PCPs on the Animal and Human Health*

Although the risk assessment is still underway for several components due to the absence of extensive research still several forms of allergy and mutation are observed in animals and humans as well. Hence an extensive form of research is required to analyze the long-term impact of these ECs. PCPs like triclosan commonly used as an antibacterial agent, shows highly toxic to aquatic flora and fauna such as algae, zooplankton, and fish. It is highly bio-accumulative and has endocrine disruptive properties. According to a report, UV filters used in PCPs are extremely ecotoxic *C. vulgaris* (algae), *D. Magna* (zooplankton), and *Brachydanio rerio* (fish) (Zhang et al. 2021a).

2.5 Endocrine Disruptors (EDCs)

Endocrine-disrupting compounds (EDCs) includes hormones, steroids, perfluorinated compounds, surfactants, and surfactant metabolites, flame retardants, industrial additives and agents, gasoline additives, antiseptics, plastic precursor, food additives, herbicides, insecticide, artificial sweeteners, and nanomaterials that manifest carcinogenic effects on the human endocrine system due to their androgenic or estrogenic activities even at trace level (Table 2.3).

A list of emerging EDCs in surface water and wastewater (Pal et al. 2014; aus der Beek et al. 2016; Tiwari et al. 2016; Kasonga et al. 2021).

These chemicals may disrupt the functioning of vital organs in human as well as animal bodies adversely. United States Environmental Protection Agency (EPA) has declared as external chemicals mess with the metabolism and anabolism of the body's natural hormones which functions in homeostasis, development, reproduction, and behavior. There are three main classes of EDCs; estrogenic mocks the functioning of the body's natural estrogens; androgenic duplicates natural testosterone and last thyroidal disrupts the functioning of the thyroid.

2.5.1 *Sources of EDCs*

A majority of ECs are not produced naturally and their occurrence in the aquatic environment is entirely due to anthropogenic activities. The emergence of EDCs is predominantly accelerated by above mentioned domestic and industrial waste products dumped directly into the water which finally reaches surface water or groundwater because wastewater is further dumped into water bodies without EDCs specific treatment in water treatment plants, unfortunately. The presence of EDCs in the aquatic environment was observed earlier in the 1970s but has received more attention in the 1990s with the use of advanced analytical methods for their detection at very low concentrations. Most of PhACs and sometimes their metabolites were found to exhibit EDC properties. Though they are found in immensely low concentrations (ng L^{-1} or μg L^{-1}) in wastewater they can be very hazardous in long-term exposure to human beings and other organisms.

Phthalates like DBP and DEHP are mainly utilized plasticizers to give flexibility to rigid polymers. These materials are components of industrial solvents, adhesive, wax, ink, pharmaceutical products, insecticide materials, and cosmetics. The direct human exposure of EDCs is mainly through the diet while using canned materials as eateries. The phthalate is not covalently bound to PVC so it can leach, migrate or evaporate into the body and environment and becomes ubiquitous (Tiwari et al. 2016).

2.5.2 *Occurrence of EDCs in Indian Aquatic Environment*

There are several reports on the occurrence of EDCs in Indian water reservoirs, sediments as well soil (Tiwari et al. 2016; Gogoi et al. 2018; Philip et al. 2018; Kasonga et al. 2021). According to a report (Tiwari et al. 2016), phthalates (PAEs), bisphenol A (BPA), Di-n-butyl phthalate (DBP), and other EDCs were found significantly in surface sediments at different stations across Thane Creek, India. BPA and DBP were detected in the range 0.13 and 0.4 mg kg^{-1} and 16.3 to 35.79 μg kg^{-1}, respectively. According to a study of the androgenic potential of WTPs influents and effluents in the Northern part of the county, the concentration of four EDCs; nonylphenol, hexachlorobenzene, and two testosterone was very appreciable, and hence despite severe treatment, final treated effluent from WTP still had enough androgenic and ecotoxic impact to general public health (Kumar et al. 2008).

Philip et al. (2018) reported the presence of EDCs in all the environmental matrices including air, water, sediment, wastewater, and sludge collected from the North zone, south zone, west zone, and north zone. The ubiquitous presence of five phthalates such as DEHP, DMP, DBP, DOP, and DEP has been reported in bed sediments of Gomti River (Srivastava et al. 2010). The presence of phthalates was also detected in indoor air, outdoor air, indoor dust, outdoor dust, and drinking water samples collected from Jawaharlal Nehru University campus, India.

2.5.3 Effects of EDCs on the Animal and Human Health

Their emergence in aquatic bodies can cause disruptive physiological processes, sexual impairment, lesser fertility, and cancer in the human and animal body if directly consumed even in low concentrations. Therefore, even a minute concentration of EDCs can increase the aquatic toxicity manifold and make the water system inhabitable for its dwellers. EDCs are even reported to mock or alter the functioning of hormones and distort the body fluid in the living organism. DEHP was observed to cause a drastic reduction in sperm production, motility, and velocity in goldfish over a monthly exposure, and also there was a noteworthy decrease in 11-ketotestosterone and luteinizing hormone levels over 15–30 days of exposure (Golshan et al. 2015). Zhang et al. (2021b) have demonstrated the impact of DEHP and BPA throughout exposure on rats and found that it can elevate the risk of breast cancer in females. The data manifested that BPA even individually or in combination with DEHP may induce hyperplasia of mammary glands, including the proliferation of ductal epithelial cells and an increase in the number of lobules and acinus after a 30-week exposure. In another study Zhou et al. (2020), the combined exposure to PFOS and BPA can provoke the increment in collagen and interventricular septal thickness (IVS) and damage heart development.

2.6 Monitoring and Regulation of ECs in Wastewater and Surface Water

2.6.1 Regulations and Guidelines

Since the conventional WTPs can not remove ECs completely as demonstrated by their presence in drinking water, this is a potential threat to public health. The detection of ECs across the globe makes them ubiquitous. Moreover, several ECs are degraded over the period but their constant use and abuse release them into the environment and hence in water. That's why they can be called "pseudo-persistent" (Ebele et al. 2017). Pseudo-persistent PhACs are more fatal to the environment than other organic ECs as they are getting replenished even after decomposition processes such as biodegradation and photodegradation.

Attributing to awareness about the adverse impact of ECs on human and animal health, several countries have imposed strict rules and regulations to monitor and tackle the panic situation throughout the globe. Indian governments have also been establishing several policies and rules over this issue of severe contamination of water bodies and groundwater. Concerning the abrupt usage of antibiotics, GOI has formulated a national policy on the application of antimicrobial drugs. Unfortunately, the implementation was not practiced up to the mark due to various reasons, for example, unorganized health infrastructure, poor medical facilities in remote areas, lack of coordination national level monitoring of antimicrobial resistance.

Later, to pose more restrictions on counter sales of PhACs and drugs, the Drugs and Cosmetic Rule, 1945 underwent amendation realizing Schedule H1 in the year 2013 which incorporated 46 more drugs including 3rd and 4th generation antibiotics, anti TB drugs, and narcotic drugs (Philip et al. 2018). After 1 year this Schedule H1 was implemented in the country in March 2014 by Central Drugs Standard Control Organization (CDSCO). Afterward, authorities took the initiative of 'Chennai declaration' in 2012 to tackle the issue of antimicrobial resistance in the country and made a 5-year strategy to control antibiotic resistance. Onwards, National Action Plan on Antimicrobial Resistance (NAP-AMR) 2017–2021, was actuated in the country along with six strategic rules to combat antimicrobial resistance and contribute towards the global efforts to tackle this public health threat (Oviedo and Aga 2016). Despite all of this, the country is facing the problem of emerging contaminants in surface water and groundwater bodies. This is basically because of inefficient treatment of pharmaceutical industrial effluents, hospital effluents, urban runoff, etc. (as presented in Fig. 2.3) as well as extremely poor sanitation facilities in both rural and urban areas. The abrupt use of PhACs and PCPs is going to result in a panic situation for highly populated countries such as India. The distribution and fate of ECs in the environmental matrices are still unclear due to the lack of comprehensive studies. A few studies have reported the presence of PhACs and PCPs in the rivers but are limited to few sampling locations and target compounds. In India, the capacity of WTPs is considerably less than the sewage produced leading to the discharge of partially treated or untreated sewage being discharged into the surface water bodies.

In nutshell, to tackle this problem, there should be strict rules on the implementation of existing national drug policies. Moreover, an awareness program on the application of PhACs, PCPs, EDCs, and other abusive drugs should be run across the

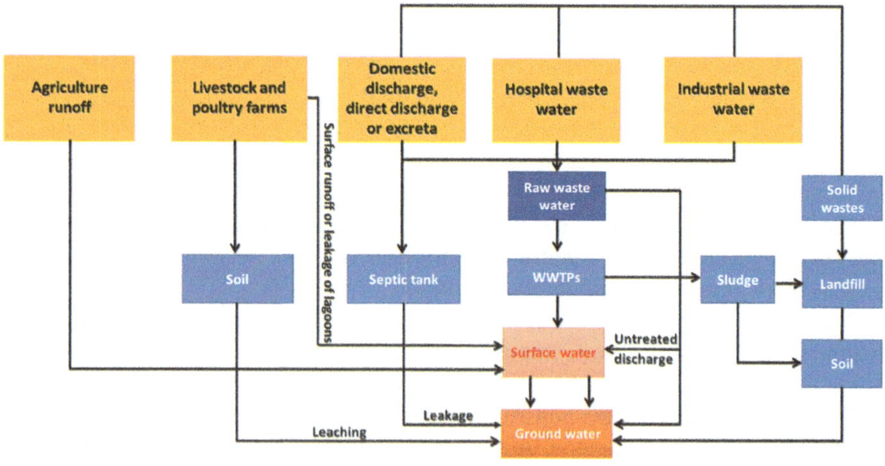

Fig. 2.3 The main sources and possible pollution pathways of ECs accumulation in surface water and groundwater

country. The proper handling of expired and unused drugs is the utmost requirement today through establishing a new regulation over its disposal and counter-sell.

2.6.2 Analytical Methods for Monitoring of ECs

There has been great improvement in analytical tools to achieve better chemical identification and quantification of ECs as well as their metabolites. Conventionally, ultra-violet visible (UV–Visible) spectrophotometry has been employed for the determination of PhACs and other drugs in various types of water matrices. But over the period, there have been advancements in varieties of ECs so new techniques were developed which can even trace the minute concentrations of ECs i.e. ppm or sub-ppm level in surface water, groundwater, and wastewater. The analytical techniques like gas chromatography with mass spectrometry (GC–MS) or GC with tandem MS (GC–MS/MS) and liquid chromatography with mass spectrometry (LC–MS) or LC with tandem MS (LC–MS/MS) are employed for trace level detection of ECs. Amongst, LC–MS/MS is the most preferred analytical technique in ECs detection. LC–MS/MS technique performs the separation and exact detection of co-eluted ECs having similar molecular mass, but different productions. Additionally, MS/MS combination allows better analytical selectivity and sensitivity in complex water samples (Luo et al. 2014; Pal et al. 2014; Rehman et al. 2015; Stefanakis and Becker 2015; Verlicchi and Zambello 2015; Pavithra et al. 2017; Philip et al. 2018; Kasonga et al. 2021).

Besides the trace level detection of ECs, it is also very important to monitor the health impact of these ECs on macroinvertebrates and other organisms in the water. Hershberger assays are generally used to monitor the androgenicity caused by EDCs (Kumar et al. 2008). To make the traditional WTPs capable to treat ECs, they have to be equipped with modern technologies including adsorption, biodegradation, membrane technology, and advanced oxidation processes (AOPs). Further X-rays radiation can be employed for the removal of parabens and EDCs in water. Moreover, the heterogeneous photocatalytic process appears to be more favored for the removal of parabens due to its ability to mineralize parabens in water. However, more work is needed to improve this ability of heterogeneous photocatalysts.

2.7 Conclusions

The emergence of pollutants in wastewater has affected the physiological and reproduction processes of the organism by impairing the functioning biological system inside the body. This has also increased the probability of cancer in human beings and other organisms and developed antibiotic resistance in bacteria. As a result of excessive drugs and their metabolite loading in biota, ARGs have been recorded widely and this can be fatal for the ecosystem which may not be further cured.

So far, any serious research to dive deep over emerging pollutants and their possible effect has not yet been checked thoroughly. Furthermore, the pharmaceutical industry is all overspread in the country but still, the major clusters are concentrated in Maharashtra, Gujrat, West Bengal, Andhra Pradesh, Tamil Nadu Madhya Pradesh, and Goa, but unfortunately, there are very few assessments surfaced to reflect the status of ECs in wastewater and surface water. Hence the authorities and academia responsible for the monitoring of ECs should be more active in fieldwork rather than doing paperwork. Further, time-wise risk assessment and status of all possible ECs in wastewater or surface water present in every province must be updated earnestly on record.

Additionally, WTPs are still not well equipped to treat the wastewater and make it free of ECs. Resultantly, water effluents carrying an abundance of ECs have been carried out to rivers and other open water bodies causing a threat for water bodies as well as human beings.

Therefore, wastewater effluents after treatment from WTPs must be checked for the presence of ECs. WTPs must be equipped with effective detection and removal techniques to ensure the absence of ECs in wastewater effluents. After removal of ECs and mitigating much below the permissible limit wastewater effluents must be discharged to the aquatic system. Additionally, the long-term effect of ECs exposure to the ecosystem must be carefully analyzed to avoid any kind of casualties or incidence of aquatic pollution.

Noteworthy, most of the reported studies reflecting the presence of ECs in diverse environmental matrices including biological samples in India are executed by scientific groups overseas. Various types of possible samples such as influent, effluent, sludge, water, sediment, soil, air, and biological samples were collected from different locations in India and analyzed at laboratories situated abroad. Therefore, a monitoring committee should be established to monitor the time-wise accumulation of ECs into the aquatic system of the country as soon as possible.

Regarding more strict regulation on monitoring of ECs across the country, GOI must make more stringent rules as well as the national authorities responsible to give time-wise report on the risk analysis and status of ECs in Indian aquatic life both surface water and groundwater. Hence there is a suggestion given inspired by the European Union recommendation that each Member of Parliament can submit a monitoring and measuring program to the European Commission, to achieve a good chemical status of the surface water.

Table 2.1 A list of major classes of PhACs depicting potential risks on human health and other organisms

Emerging pharmaceuticals contaminant groups	Contaminants
Human antibiotics and veterinary	Trimethoprim, streptomycin, amoxicillin, lincomycin, sulfamethoxazole, chloramphenicol, Triclosan
Analgesics, anti-inflammatory drugs	Ibuprofene, diclofenac, codein, acetaminophen, acetylsalicilyc acid, fenoprofen, tramadol
Psychiatric drugs (Anticonvulsant)	Diazepam, carbamazepine, primidone, salbutamol
β-blockers	Metoprolol, propanolol, timolol, atenolol, sotalol
Lipid regulators	Bezafibrate, clofibric acid, fenofibric acid, etofibrate, gemfibrozil
X-ray contrasts	Iopromide, iopamidol, diatrizoate
Stimulants	Caffeine, nicotine
Antihypertensive	Triamterene, valsartan, hydrochlorothiazide
Antidepressant	Desmethyl-venlafaxine, antihypertensive, temazepam, venlafaxine hydroxybupropion
Antiepileptic	Phenytoin, pregabalin, lamotrigine, gabapentin
Antihyperlipidemic	Gemfibrozil
Antiarrhythmic	Lidocaine
Drug precursor	Acetophenone

Table 2.2 A list of all possible types of PCPs composed of carcinogenic materials

Emerging PCPs Contaminant groups	Contaminants
Fragrances	Nitro, polycyclic and macrocyclic musks, phthalates (galaxolide, celestolide, tonalide, phantolide, traseolide, musk ketone, musk xylene Cashmeran)
Sun-screen agents	Benzophenone, methylbenzylidene camphor
Shampoos and facial creams	Dimethicone, sodium laureth sulfate, sodium laureth sulfate, cocamidopropyl betaine
Cleansing products	Perchloroethylene, bleach
Toners	Iopamidol, diatrizoic acid
Soaps and detergents	Triclocarban, triclosan

Table 2.3 A list of emerging EDCs in surface water and wastewater

Emerging EDCs contaminant groups	Contaminants
Hormones and steroids	Estradiol, estrone, estriol, diethylstilbestrol (DES), 17-α-Ethinylestradiol
Perfluorinated compounds	Perfluorotoctane sulfonates (PFOS), perfluoroctanoic acid (PFOA)
Surfactants and surfactant metabolites	Alkylphenol ethoxylates, 4-nonylphnol, 4-octylphenol, alkylphenol carboxylates, tributyl phosphate, octylphenols, nonylphenols, di(2-ethylhexyl) phthalate (DEHP)
Flame retardants	Polybrominated diphenyl ethers (PBDEs): polybrominated biphenyls (PBBs), polybrominated dibenzo-*p*-dioxins (PBDDs), polybrominated dibenzofurans (PBDFs), tetrabromo bisphenol A, C10-C13 chloroalkanes, tris (2-chloroethyl)phosphate, hexabromocyclododecane (HBCDs)
Industrial additives and agents	Chelating agents (EDTA), aromatic sulfonates
Gasoline additives	Dialkyl ethers, methyl-t-butyl ether (MTBE)
Antiseptics	Triclosan, chloroprene
Plastic precursor	Bisphenol A, phenol
Food additives	Butylated hydroxyanisole, triethyl citrate
Herbicides	2,4-D, atrazine, bromacil, dDiuron, MCPP, metolachlor, metolachlor ESA, triclopyr
Insecticide	Imidacloprid, carbaryl
Artifical sweeteners	Acesulfame, cyclamate, saccharin, sucralose
Nanomaterials	Inorganic metal oxides
UV-filters	Octocrylene, oxybenzone
Insect repellents	*N,N*-diethyltoluamide

References

Archana G, Dhodapkar R, Kumar A (2017) Ecotoxicological risk assessment and seasonal variation of some pharmaceuticals and personal care products in the sewage treatment plant and surface water bodies (lakes). Environ Monit Assess 189(9). https://doi.org/10.1007/s10661-017-6148-3

aus der Beek T et al (2016) Pharmaceuticals in the environment-Global occurrences and perspectives. Environ Toxicol Chem 35(4):823–835. https://doi.org/10.1002/etc.3339

Bai X et al (2018) Occurrence, distribution, and seasonality of emerging contaminants in urban watersheds. Chemosphere 200:133–142. https://doi.org/10.1016/j.chemosphere.2018.02.106

Balakrishna K et al (2017) A review of the occurrence of pharmaceuticals and personal care products in Indian water bodies. Ecotoxicol Environ Saf 137:113–120. https://doi.org/10.1016/j.ecoenv.2016.11.014

Biswas P, Vellanki BP (2021) Occurrence of emerging contaminants in highly anthropogenically influenced river Yamuna in India. Sci Total Environ 782:146741. https://doi.org/10.1016/j.scitotenv.2021.146741

Brown J, Winterstein A (2019) Potential adverse drug events and drug-drug interactions with medical and consumer cannabidiol (CBD) use. J Clin Med 8(7):989. https://doi.org/10.3390/jcm8070989

Bu Q et al (2013) Pharmaceuticals and personal care products in the aquatic environment in China: a review. J Hazard Mater 262:189–211. https://doi.org/10.1016/j.jhazmat.2013.08.040

Devarajan N et al (2015) Hospital and urban effluent waters as a source of accumulation of toxic metals in the sediment receiving system of the Cauvery River, Tiruchirappalli, Tamil Nadu, India. Environ Sci Pollut Res 22(17):12941–12950. https://doi.org/10.1007/s11356-015-4457-z

dos Santos MM et al (2016) Occurrence and risk assessment of parabens and triclosan in surface waters of southern Brazil: a problem of emerging compounds in an emerging country. Rbrh 21(3):603–617. https://doi.org/10.1590/2318-0331.011616018

Ebele AJ, Abou-Elwafa Abdallah M, Harrad S (2017) Pharmaceuticals and personal care products (PPCPs) in the freshwater aquatic environment. Emerg Contam 3(1):1–16. https://doi.org/10.1016/j.emcon.2016.12.004

Gogoi A et al (2018) Occurrence and fate of emerging contaminants in water environment: a review. Groundwater Sustain Dev 6:169–180. https://doi.org/10.1016/j.gsd.2017.12.009

Golshan M et al (2015) Di-(2-ethylhexyl)-phthalate disrupts pituitary and testicular hormonal functions to reduce sperm quality in mature goldfish. Aquat Toxicol 163:16–26. https://doi.org/10.1016/j.aquatox.2015.03.017

Hanna N et al (2020) Monitoring of water quality, antibiotic residues, and antibiotic-resistant escherichia coli in the kshipra river in india over a 3-year period. Int J Environ Res Public Health 17(21):1–22. https://doi.org/10.3390/ijerph17217706

Kalotra A (2014) Marketing strategies of different pharmaceutical companies. J Drug Delivery Ther 4(2):64–71. https://doi.org/10.22270/jddt.v4i2.771

Kasonga TK et al (2021) Endocrine-disruptive chemicals as contaminants of emerging concern in wastewater and surface water: a review. J Environ Manage 277:111485. https://doi.org/10.1016/j.jenvman.2020.111485

Kaval İ, Behçet L, Çakilcioğlu U (2015) Survey of wild food plants for human consumption in Geçitli (Hakkari, Turkey). Indian J Tradit Knowl 14(2):183–190

Kumar V et al (2008) Androgenic endocrine disruptors in wastewater treatment plant effluents in India: their influence on reproductive processes and systemic toxicity in male rats. Toxicol Appl Pharmacol 226(1):60–73. https://doi.org/10.1016/j.taap.2007.08.023

Kurunthachalam SK (2012) Pharmaceutical substances in India are a point of great concern? J Waste Water Treat Anal 03(05):3–5. https://doi.org/10.4172/2157-7587.1000e103

Luo Y et al (2014) A review on the occurrence of micropollutants in the aquatic environment and their fate and removal during wastewater treatment. Sci Total Environ 473–474:619–641. https://doi.org/10.1016/j.scitotenv.2013.12.065

Minh TB et al (2009) Antibiotics in the Hong Kong metropolitan area: ubiquitous distribution and fate in Victoria Harbour. Mar Pollut Bull 58(7):1052–1062. https://doi.org/10.1016/j.marpolbul.2009.02.004

Nag SK, Das Sarkar S, Manna SK (2018) Triclosan–an antibacterial compound in water, sediment and fish of River Gomti, India. Int J Environ Health Res 28(5):461–470. https://doi.org/10.1080/09603123.2018.1487044

Noguera-Oviedo K, Aga DS (2016) Lessons learned from more than two decades of research on emerging contaminants in the environment. J Hazard Mater 316:242–251. https://doi.org/10.1016/j.jhazmat.2016.04.058

Pal A et al (2014) Emerging contaminants of public health significance as water quality indicator compounds in the urban water cycle. Environ Int 71:46–62. https://doi.org/10.1016/j.envint.2014.05.025

Pavithra KG et al (2017) Sources and impacts of pharmaceutical components in wastewater and its treatment process: a review. Korean J Chem Eng 34(11):2787–2805. https://doi.org/10.1007/s11814-017-0255-2

Philip JM, Aravind UK, Aravindakumar CT (2018) Emerging contaminants in Indian environmental matrices—a review. Chemosphere 190:307–326. https://doi.org/10.1016/j.chemosphere.2017.09.120
Rehman MSU et al (2015) Global risk of pharmaceutical contamination from highly populated developing countries. Chemosphere 138:1045–1055. https://doi.org/10.1016/j.chemosphere.2013.02.036
Sangeetha S, Vimalkumar K, Loganathan BG (2021) Environmental contamination and human exposure to select endocrine-disrupting chemicals: a review. Sustain Chem 2(2):343–380. https://doi.org/10.3390/suschem2020020
Semblante GU et al (2017) Fate of trace organic contaminants in oxic-settling-anoxic (OSA) process applied for biosolids reduction during wastewater treatment. Biores Technol 240:181–191. https://doi.org/10.1016/j.biortech.2017.02.053
Singh G et al (2016) Exploring the potential reservoirs of non specific TEM beta lactamase (blaTEM) gene in the Indo-Gangetic region: a risk assessment approach to predict health hazards. J Hazard Mater 314:121–128. https://doi.org/10.1016/j.jhazmat.2016.04.036
Srivastava A et al (2010) Occurrence of phthalic acid esters in Gomti River Sediment, India. Environ Monit Assess 169(1–4):397–406. https://doi.org/10.1007/s10661-009-1182-4
Stefanakis AI, Becker JA (2015) A review of emerging contaminants in water: Classification, sources, and potential risks. Impact Water Pollut Human Health Environ Sustain 55–80. https://doi.org/10.4018/978-1-4666-9559-7.ch003
Subedi B et al (2015) Mass loading and removal of pharmaceuticals and personal care products, including psychoactive and illicit drugs and artificial sweeteners, in five sewage treatment plants in India. J Environ Chem Eng 3(Part 4A):2882–2891. https://doi.org/10.1016/j.jece.2015.09.031
Taheran M et al (2018) Emerging contaminants: here today, there tomorrow! Environ Nanotechnol Monit Manage 10:122–126. https://doi.org/10.1016/j.enmm.2018.05.010
Tiwari M, Sahu SK, Pandit GG (2016) Distribution and estrogenic potential of endocrine disrupting chemicals (EDCs) in estuarine sediments from Mumbai, India. Environ Sci Pollut Res 23(18):18789–18799. https://doi.org/10.1007/s11356-016-7070-x
Verlicchi P, Zambello E (2015) Pharmaceuticals and personal care products in untreated and treated sewage sludge: Occurrence and environmental risk in the case of application on soil—a critical review. Sci Total Environ 538:750–767. https://doi.org/10.1016/j.scitotenv.2015.08.108
Wang Y, Westerhoff P, Hristovski KD (2012) Fate and biological effects of silver, titanium dioxide, and C 60 (fullerene) nanomaterials during simulated wastewater treatment processes. J Hazard Mater 201–202:16–22. https://doi.org/10.1016/j.jhazmat.2011.10.086
Wang Y et al (2021) Occurrence of parabens, triclosan and triclocarban in paired human urine and indoor dust from two typical cities in China and its implications for human exposure. Sci Total Environ 786:147485. https://doi.org/10.1016/j.scitotenv.2021.147485
Zhang QQ et al (2015) Comprehensive evaluation of antibiotics emission and fate in the river basins of China: source analysis, multimedia modeling, and linkage to bacterial resistance. Environ Sci Technol 49(11):6772–6782. https://doi.org/10.1021/acs.est.5b00729
Zhang J et al (2021a) Mutagenicity evaluation to UV filters of benzophenone-6, benzophenone-8, and 4-methylbenzylidene camphor by Ames test. Plos One 16(9):e0255504. https://doi.org/10.1371/journal.pone.0255504
Zhang X et al (2021b) Co-exposure to BPA and DEHP enhances susceptibility of mammary tumors via up-regulating Esr1/HDAC6 pathway in female rats. Ecotoxicol Environ Safety 221:112453. https://doi.org/10.1016/j.ecoenv.2021.112453
Zhou R et al (2020) Combined effects of BPA and PFOS on fetal cardiac development: in vitro and in vivo experiments. Environ Toxicol Pharmacol 80(639). https://doi.org/10.1016/j.etap.2020.103434

Chapter 3
Occurrence, Fate, and Health Hazards of Microplastics Pollution

S. Sethulekshmi, Sheetal S. Kothawale, Sukanya Krishnan, Ansaf V. Karim, Pradip Kalbar, Swatantra P. Singh, and Amritanshu Shriwastav

Abstract An emerging class of pervasive pollutants, microplastics are being increasingly detected in almost all the environmental matrices. Microplastics are plastic particles having a size of less than 5 mm, either formed from the disintegration of large plastics in the environment due to natural and anthropogenic factors, named as secondary microplastics, or manufactured that way for industrial and commercial uses, namely primary microplastics. They occur in a variety of shapes such as beads, fibres, fragments, foams, etc. in nature. As an emerging pollutant, they are gaining popularity among researchers due to massive abundance and yet to be identified potential impacts on different tiers of the ecosystem. This chapter reviews the occurrence and abundance of microplastics in all the environmental compartments, various detection methods, different sources of generation, and their fate and transport in air, water, and soil. A discussion on current trends of their exposure levels on biota and ecotoxicological effects is also included. Further, the potential of microplastics as the carriers of other organic pollutants and leaching out of toxic chemicals from microplastics are also reviewed.

Keywords Microplastics · Occurrence · Sources · Fate · Exposure · Ecotoxicology · Detection

---

S. Sethulekshmi · S. Krishnan · A. V. Karim · S. P. Singh · A. Shriwastav (✉)
Environmental Science and Engineering Department, Indian Institute of Technology Bombay, Mumbai 400 076, India
e-mail: amritan@iitb.ac.in

S. S. Kothawale
Centre for Research in Nanotechnology and Science, Indian Institute of Technology Bombay, Mumbai 400 076, India

P. Kalbar
Centre for Urban Science & Engineering, Indian Institute of Technology Bombay, Mumbai 400 076, India

S. P. Singh et al. (eds.), *New Trends in Emerging Environmental Contaminants*, Energy, Environment, and Sustainability,
https://doi.org/10.1007/978-981-16-8367-1_3

3.1 Introduction

The invention of plastics has made our everyday lives more comfortable and simpler such that plastic commodities mark their presence omnipresent. Easy availability, convenience to handle, clubbed together with cost-effectiveness resulted in mismanagement of plastics. This carelessness or inefficiency in the system attribute to their loss from the global value chain. A notable portion of the total manufactured plastics is getting lost into the environmental compartments during their use and/or disposal phase (Ryberg et al. 2019). Apart from the plastic debris loss into the natural system (macroplastics), small plastic particles, named microplastics (MPs), are also getting released into the environment. Lately, researchers have recognized MPs as one of the important classes of emerging pollutants due to their perseverance in the environment and possible environmental toxicity (Seth and Shriwastav 2018).

MPs are characterized as plastic particles having a size <5 mm (Browne et al. 2007; Qiu et al. 2015; Zhang et al. 2018; Wu et al. 2019a). The lower size limit in the classification is usually taken as 1 μm and therefore plastic particles smaller than 1 μm in length are named nano plastics (WHO 2019). Based on their shapes, MPs are mainly classified as microbeads, foams, fibers, films, and fragments (Andrady 2011; UNEP 2018). They are broadly classified as primary and secondary MPs based on origin. Primary MPs are produced less than 5 mm in size mostly in the form of beads and fibres for various industrial and commercial applications. Secondary MPs are generated by the disintegration of larger plastic debris upon the action of various physicochemical factors. These generally occur mostly as fibers, foams, and fragments (Wu et al. 2019a). These plastic particles occur in nature in a variety of polymer types viz. polystyrene (PS), polypropylene (PP), polyethylene (PE), polyurethane (PU), polyvinyl chloride (PVC), etc.

This chapter discusses the basic information about the occurrence and abundance of this emerging pollutant. Different methods used at present for their identification are discussed. Further, various sources of generation of MPs and their fate and transport in different environmental matrices are discussed. A special focus on exposure levels on different biota and reported results of ecotoxicological effects studies are presented. Apart from being an important class of pollutants on their own, MPs are also carriers of persistent organic pollutants (POPs) in the environment. Therefore, a brief discussion on the potentiality of MPs as the carriers of other organic pollutants is discussed. Adding on, the leaching of different chemical additives from MPs is also an issue of concern. Thus, a review of the leaching of plastic additives and adjacent risks are discussed. Altogether, the chapter provides an overview of MPs as an emerging contaminant in the environment.

3.2 Occurrence and Abundance in Various Matrices

The data published by the researchers and scientists on the MP occurrence in the ecosystem in recent times have increased the awareness among people about microplastic pollution. (Picó and Barceló 2019; Tang et al. 2020b; Shruti et al. 2020a; Patti et al. 2020; Selvam et al. 2021). The analytical data indicates that there are broadly five shape types are present in the environment viz. foams, pellets(microbeads), fragments, fibres, and films. These are made up of commercially available polymers such as polyethylene, polypropylene, polystyrene, polyvinyl chloride (PVC), and sometimes a composite of any two polymers (Tiwari et al. 2019; Shruti et al. 2020b; Selvam et al. 2021). Out of these, around 60% of MPs have a size range of less than 1 mm. Scientist discovered that MPs comprised large quantity of secondary MPs out of which fragments and fiber has a share of 48% and 24%, respectively (Karthik et al. 2018; Tiwari et al. 2019).

MP contamination differs geographically. The occurrence and abundance of MP are primarily governed by environmental (Imhof et al. 2013; Kim et al. 2015; Veerasingam et al. 2016) as well as anthropogenic factors (Sarafraz et al. 2016). Nevertheless, environmental parameters possibly play a major role in the dispersal of MP than anthropogenic ones (Zhang et al. 2019; Herrera et al. 2020). Environmental factors such as the direction of the wind, wave motion, hurricanes, and river hydrodynamics outline the occurrence of microplastic (Liubartseva et al. 2016; Besseling et al. 2017; Cai et al. 2017). The abundance of MP is directly proportional to the intensity of these environmental factors (Besseling et al. 2017; Cai et al. 2017). On the other side, anthropogenic factors include population density and human activities that lead to the build-up of these minute plastic particles in the milieu (Sarafraz et al. 2016). The level of microplastic pollution could be closely correlated to specific human activities than the global plastic production and its utilization (Anbumani and Kakkar 2018). Taking into account the hotspots of MPs, aquatic ecosystems are acting as a sink for MP. But, the recent studies in the area of MP pollution direct their existence on every corner of an ecosystem (Anbumani and Kakkar 2018; Guzzetti et al. 2018; Zhang et al. 2020a; Pham et al. 2021; Wang et al. 2021).

3.2.1 Aquatic Environment

Studies have stated the presence of MPs in aquatic environments around the globe and are projected to increase two folds by 2030 (Wang and Wang 2018; Isobe et al. 2019; Tang et al. 2020a). MP has not only been detected in beaches, surface waters, sediments, basically throughout the water column but also within the biota (Duis and Coors 2016; Anbumani and Kakkar 2018; Klein et al. 2018; Wang and Wang 2018). However, research in the marine environment dominates freshwater and other aquatic expanses. For a fact, less than 4% of MP-related research is allied with freshwater habitat (Wagner and Lambert 2018). Nonetheless, existing studies revealed that the abundance of MPs in freshwaters is analogous to that of the saltwater ecosystem but the distribution is extremely diverse (Anbumani and Kakkar 2018; Li et al. 2018; Sridharan et al. 2021). A range of 38–234 items/m^3 of polyamide (PA), polypropylene (PP), polystyrene (PS), and polyethylene (PE) types of MPs were observed from the ice cores in the Arctic (Obbard et al. 2014). On the other hand, in the Atlantic Ocean, Kanhai et al. (2017) quantified the average microplastic as 1.15 ± 1.45 items/m^3. In 2015, about 3000 microplastic particles were gathered from six beaches along the Goa coast (Veerasingam et al. 2016), likewise 932 items/m^3 in Southern Ocean (Isobe et al. 2019) and 0.13 items/m^3 in Northern South China Sea (Sun et al. 2018). Research findings revealed the occurrence of MPs in lakes, rivers, and estuaries (Imhof et al. 2013; Klein et al. 2018). Considering Indian scenario, freshwater MP particles in the quantity of 28 items/m^3 and 5900 items/m^3 were detected from the Veeranam lake and Red Hills lake of Tamil Nadu respectively (Gopinath et al.; Bharath et al. 2021). The source of freshwater MPs could be a fragmentation of plastic particles in the water column or from leaching of MP particles from terrestrial or from direct effluent discharge from wastewater treatment plants and households (Li et al. 2018, 2020a; Boyle and Örmeci 2020).

3.2.2 Terrestrial Environment

The origin of a greater number of aquatic MPs pollution is from in-land sources, therefore the presence of MPs in the terrestrial environment cannot be ignored (Huang et al. 2020; Boyle and Örmeci 2020; Wang et al. 2021; Sridharan et al. 2021). Though the occurrence of MPs in the terrestrial environment is apparent due to littering and application of sewage sludge to land and also soils have been assumed to be a sink for MPs, there is very little information on the concentration of MPs in the terrestrial ecosystem (Rillig and Bonkowski 2018; Yang et al. 2021a; Sridharan et al. 2021). So far, research on terrestrial MP s has been conducted on a global scale, including in Asia, Europe, North America, and Australia (Fuller and Gautam 2016; Zhou et al. 2018; Huang et al. 2020; Li et al. 2020c). Since China accounts for 31% of the world's plastic production, most of the MP research is pursued in China (URL-1). The study of MPs contamination in seashore sand of Bohai sea and Yellow sea, China, accounted for an average of 14,712.5 items/kg (Zhou et al. 2018). One of the major sources behind the terrestrial contamination is the improper disposal of agricultural plastic films that indirectly resulted in the accumulation of MPs and film-like MPs (Huang et al. 2020). A recent study on agricultural soil in Xinjiang province indicates that the average concentration of 40.35 mg/kg of MP of size range 0.9–2.0 mm has been covered with plastic film for 30 years (Li et al. 2020c). In Southern America's Chile's Regi6n Metropolitana, the average concentrations of MPs present in the soil collected from farmlands and meadows are 306 ± 306 and 184 ± 266 items/kg, respectively (Corradini et al. 2021). Comparing with agricultural soil, the soil of an industrial area in Sydney has shown a high concentration of MP contamination, it varies from 300 to 6900 items/kg (Fuller and Gautam 2016). In Washington, USA, Helcoski et al. (2020) reported a parallel abundance of MPs ($1,270 \pm 150$ items/kg) in a tidal freshwater wetland.

3.2.3 Atmosphere

Besides studies of MPs in aquatic and terrestrial environments, recent research has detected MPs from atmospheric fallout in cities of China, France, and Germany (Cai et al. 2017; Dris et al. 2017; Klein and Fischer 2019). The high-density urban populations of these cities generate a huge amount of plastic waste. (Dris et al. 2015; Cai et al. 2017; Klein and Fischer 2019). Though the presence of MP in the atmosphere has become prominent, research regarding atmospheric MPs pollution is in its initial stages. Studies examining atmospheric microplastic primarily detected natural as well as synthetic fibers in the air (Can-Güven; Cai et al. 2017; Dris et al. 2017; Wright et al. 2020; Zhang et al. 2020a). Among the various polymer types, polyethylene terephthalate (PET) and polyethylene (PE) were reported at high concentrations, along with innumerable colours of MPs (Can-Güven; Dris et al. 2017; Allen et al. 2019; Szewc et al. 2021). Initiating research carried out on atmospheric fallout samples in Paris, France, reported 29–280 particles of MPs, of which 90% were fibers (Dris et al. 2015). Another study in the same city reported a higher number of MP fibers in indoor air (5.4 fibers m^{-3}) than outdoor air (0.9 fibers m^{-3}) of two private apartments and one office (Dris et al. 2017). Research conducted in Dongguan, China, determined 175–313 particles m^{-2} day^{-1} of fibers, foams, fragments, and films shaped MPs among which fibers were prevailing (Cai et al. 2017). Prime sources for atmospheric MP pollution were specified as fabrics, various anthropogenic activities, and break down of large plastics and the wind was the principal transport system (Wright et al. 2020; Zhang et al. 2020a, b).

Allen et al. (2019) studied the abundance of atmospheric MPs in an isolated area of the Pyrenees mountains in southwestern Europe and reported the MP value as 366 particles $m^{-2}d^{-1.}$ Similar research also indicated that tiny plastic particles (300um) were able to get easily carried away over long spaces by the air (Allen et al. 2019). In the shoreline of the southern Baltic area, Europe, atmospheric MP were deposited with an abundance from 0 m^{-2} day^{-1} to 30 m^{-2} day^{-1} (Szewc et al. 2021). Among the total MPs, fibers were the prevailing type with 60% concentration, whereas films and fragments individually were founded at 26% and 14% concentration. Some of the reported concentrations of MPs in different environmental matrices are shown in Table 3.1.

Currently, one of the major limitations of perception on occurrence and abundance of MP pollution research is the dearth of standardization of data and approaches that are extensively used by the researchers. The relationship between the detection techniques and sample period representation is a further important study area that

Table 3.1 Occurrence and abundance of MPs in the environment

Matrices	Study Area	MP type	Size	Abundance	References
Freshwater	The western river downstream, China Elbe river, Germany	PE, PP, PVC, PET PS	≤0.5 mm k 0.125–5 mm	Sediment: 2560–10,240 items/kg and Surface water: 2.99–9.87 items/L Sediment: 3,350,000 particles/m^3 and surface water: 5.57 particles/m^3	Huang et al. (2021) Scherer et al. (2020)
Marine	Monterey Bay National Marine Sanctuary, California	PE and PP	≤5 mm	Average concentration of 1.32 ± 0.70 particles/m^3	Kashiwabara et al. (2021)
Wastewater Treatment Plant	Sydney, Australia Cadiz, Spain	PS 12 Different types of MPs	≤5 mm 12 0.1–5 mm	3.6 × 10^6 MPs/day after tertiary treatment Urban WWTP—16.40 ± 7.85 MPs/L. Industrial WWTP—131.35 ± 95.36 MPs/L	Ziajahromi et al. (2021) Franco et al. (2021)
Soil	Southeast China Qinghai-Tibet plateau, China	PE, PP, PES PE, PP	0.5–5 mm ≤5 mm	1250 ± 640 particles/kg in pig manure Shallow soil: 53.2 ± 29.7 particles/kg Deep Soil: 43.9 ± 22.3 particles/kg	Yang et al. (2021a, b) Feng et al. (2021)
Atmospheric deposition	Hamburg, Germany Kusatsu (Japan), Da Nang (Vietnam), and Kathmandu (Nepal)	PE Different types of MPs	0.063–0.3 mm 0.1–5 mm	275 particles/m^2/Day[1] Road dust from three cities: 0.10–39.6 pieces/m^2	Klein and Fischer (2019) Yukioka et al. (2020)

requires taking into account to retrieve more accurate data on the abundance in future microplastic research (Wang and Wang 2018; Allen et al. 2019; Tang et al. 2020a; Zhang et al. 2021; Sridharan et al. 2021).

3.3 Detection Methods of MPs

Prior to detection and quantification of MPs in environmental systems, these pollutant particles must be extracted from the background media. After sampling the respective media, the extraction process of MPs begins. The first and foremost step is the volume reduction of the medium. This is usually done by sieving, elutriation, filtration, etc. After volume reduction, the background is to be eliminated by removing all the inorganic and organic contamination present. This step is followed by the identification and quantification of MPs (Wagner and Lambert 2018). The series of these steps may vary with the compartment under study and the nature of contamination. Literature-reported steps in the extraction and detection methods of MPs in various matrices are drafted below.

3.3.1 Sampling

Representative sampling is the most important step, details of which vary with the research question it addresses. For water sampling, various types of equipment like neuston or manta nets, plankton nets, sieves, pumps, and ex-situ filtration are commonly used for sample collection (Prata et al. 2019). The choice of equipment depends on the quantity of sample, location, the budget of study, availability, etc. Any choice of grab sampling or composite sampling is preferred for sampling, however, to account for the extreme variability in shape, size, and abundance of MPs within a specified sampling region, composite samples are highly recommended (Möller et al. 2020). The size of samples varies with different types of samples, accounting for the nature and extent of pollution The MP concentration in water samples is usually expressed in the number of particles per litre of sample or mass of MPs per litre of samples.

Usually, sediments samples which are collected from the shoreline and bottom region of aquatic bodies are analyzed for the presence of MPs. In Sediments and soil samples, samples are collected mainly from the top 2–5 cm layer since a majority of MPs tend to be found in that depth (Wagner and Lambert 2018). Grid sampling or bulk sampling is preferred in shoreline sampling and for the waterbed sampling, core cutters or grid samplers are preferred (Prata et al. 2019). MP concentration in sediments is generally reported as the number of particles per kg of sediments.

For air sampling of MPs, two kinds of samplers are used: passive atmospheric deposition sampling and active pumped samplers (Chen et al. 2020b). In passive sampling, the atmospheric fallout is collected in a water column and analysed for the presence of MPs (Dris et al. 2016; Cai et al. 2017). Other passive methods include the collection of dust with vacuum cleaners and brushes and further analysis (Dris et al. 2017). Pumped air samplers like PM 10 and PM 2.5 (Particulate matter with aerodynamic diameter less than 10 μm and 2.5 μm respectively) are often used for active air sampling of MPs. The setup consists of pump and filter system for the collection of air-borne particles (Chen et al. 2020b). The MP pollution in air is expressed as the number of particles/m^2 day.

3.3.2 Extraction

Reducing the volume of background media is the first step to be followed in the MP extraction. Mostly, this is done by sieving and/or filtration of the collected samples. Further volume reduction is achieved by using various separation techniques like density separation, elutriation, magnetic separation, electrostatic techniques, etc. In density separation, the difference in densities of MPs and bulk medium is employed for the separation of MPs from bulk media. Salt solutions of sodium chloride, sodium iodide, and zinc chloride are commonly used in this process (Wagner and Lambert 2018). Density separation is also achieved with another method called elutriation. In the elutriation method, water is injected from the bottom of the column to make the MPs get floated on the top, which is further subjected to density separation using salt solutions. This method consumes a lesser quantity of salts but is time-consuming (Kedzierski et al. 2016). Another method utilises hydrophobic properties of plastics for their separation using oils (Prata et al. 2019). Other efficient techniques include particle impact electro-chemistry using electrostatic behavior of plastic particles (Felsing et al. 2018), oil extraction protocol which utilises oleophilic properties of MPs (Crichton et al. 2017), magnetic separation (Helm et al. 2020), froth floatation (Marques and Tenório 2000), etc.

Environmental samples contain organic matter associated with MPs and need to be removed before analysis, particularly with the visual identification method of MPs. Acid digestion, alkali digestion, oxidative digestion, and enzymatic digestion are the popular ways to remove organic debris from environmental samples (Prata et al. 2019). Acid digestion employs the heating of a mixture of acid (mostly HNO_3) and sample at 60 °C or above, however has its demerit to degrade some polymers (like nylon, PET, etc.) and therefore not highly recommended (Rocha-Santos and Duarte 2017). Alkali digestion commonly employs the use of NaOH and KOH as the digestion solution. Though this method is efficient enough to achieve the removal of organics from the matrix, it owes to the disadvantage of time-consuming and destruction/decoloration of MPs (Foekema et al. 2013; Karami et al. 2017). Oxidizing agent hydrogen peroxide (H_2O_2 30%) is recommended over acid and alkali digestion with little effect on MPs for their extraction from the organic media (Prata et al. 2019; Möller et al. 2020). NOAA (Marine Debris Program 2015) recommends the use of wet peroxide oxidation method in which a mixture of 0.05 M Fe(II) solution and 30% H_2O_2 heated up to 75 °C were effectively used for the extraction of MPs from sediment media. In the enzymatic digestion methods, the different enzymes like corolase, trypsin, papain, pepsin, etc. are used with different operational conditions. The enzymatic method is promising on the extraction of MPs as very less effect on MPs is observed with the enzyme digestion. However, it is rarely employed due to the high cost of enzymes. Even though a series of steps were discussed above for the sample extraction, it is to be noted that some studies employ organics removal prior to inorganics, and therefore the series of processes depend on the properties of samples.

3.3.3 Identification Methods of MPs

Visual identification with the help of a microscope is one of the most commonly adopted methods for identifying extracted MPs. Fluorescence microscopy is recommended if the particles are colourless or transparent in appearance. However, the visual identification with the help of microscopic methods is inefficient and less accurate as polymer type identification is not possible. The polymer type identification is often achieved with spectroscopic or spectrometric methods. Nevertheless, visual methods should not be used separately from spectroscopic methods, the latter can be used to confirm the results. Lately, a novel method of visual identification of MPs in environmental samples is developed based on specific fluorescent staining with Nile Red dye (Shim et al. 2016; Maes et al. 2017). This approach is based on the property of dye adsorbed plastic particles emitting green fluorescence in blue light. Staining followed by fluorescence microscopy helps in the detection of MPs in the samples. The microscopic images are further analysed to quantify the MPs particles.

Sophisticated methods of identification are spectroscopic methods like Fourier transform infrared spectroscopy, Raman spectroscopy, and scanning electron microscopy. Other methods include a combination of density separation and C:H:N ratio analysis (Morét-Ferguson et al. 2010) and pyrolysis GC/MS (Fries et al. 2013). In the density separation method, different polymers are identified based on their difference in estimated density using experimentation and confirmation based on C:H:N ratio analysis. This method provides a rough estimate and is time-consuming. In pyrolysis GC/MS, the characteristic pyrograms of combustion products are compared with the pyrograms of polymer types for identification. Identification of the chemical composition of MPs is possible with pyrolysis GC/MS, but owes its demerit in identifying smaller-sized particles. Table 3.2 describes the advantages and disadvantages of different identification methods of MPs. With the existing limitations of currently used techniques, there is in need for developing more techniques that are simple, easy, and sensitive in analysing MPs samples.

3.4 Sources of Generation of MPs

The inventory of plastics has been a significant logical accomplishment that has influenced practically every part of human life. A world without the presence of plastics is difficult to envision now, however, their large-scale manufacturing and use go back just to 1950 (UNEP 2018). Plastics' biggest use is in the packaging sector where the reusable items are supplanted with single-use plastics and as a result of this, the worldwide generation of plastic wastes in MSW expanded from 1% in 1960 to over 10% (by mass) in 2005 (Geyer et al. 2017). However, a significant share of this produced plastics is lost into the environment during the use phase or end of life phase due to the inefficient management of wastes. This plastic loss is

Table 3.2 Identification methods of MPs (Shim et al. 2017; Chen et al. 2020b)

Method	Principle	Merits	Demerits
Visual identification	Observation under microscope	Cost-effective	Cannot be used for smaller particles Time-consuming Polymer identification is not possible
Fourier transform infrared spectroscopy	Polymer-specific excited molecular vibrations are detected when samples are exposed to IR radiation	Non-destructive technique, Smaller particles can be detected, Reveals chemical composition	Expensive Experienced operators are needed Sample pretreatment needed
Raman Spectroscopy	Excited light is detected after a laser irradiated onto target sample. The frequency shift between two light is used to identify the molecular structure and chemical components of samples	Works for smaller particles, Polymer type can be identified	Time consuming Expensive Experienced operators are needed Sample pretreatment required
Scanning Electron Microscopy	The images are obtained by the interaction of a high-intensity electron beam with the sample	Surface Morphology can be obtained	Time consuming Labour intensive
Pyrolysis–Gas Chromatography-Mass	Thermal degradation products are analysed for polymer types	Polymer type identification	Time consuming Sample destruction Polymer identification is limited

polluting the environment as MPs which is resulted from various anthropogenic and natural factors.

Household-level generation of plastic waste is an important contributor to plastic pollution in the environment. Different plastic items used in households end up in the generation of plastic litter and further generation of secondary MPs. In a study conducted by Qu et al. (2009), about 10% of total waste generated in the household is comprised of plastics. Another study by Phuc and Yasuhiro (2011) in Vietnam reported an average plastic waste generation of 17.24 g/cap/day. This plastic debris lost from households and industries eventually results in the formation of secondary MPs. Household sources of primary MPs are personal care products (van Wezel et al. 2016), laundry (Henry et al. 2019), paints, and furnishings (Gaylarde et al. 2021), abrasion of plastic items (Sobhani et al. 2020), etc. This section describes the most significantly identified inland sources of MPs in the three environmental compartments: air, water, and soil.

3.4.1 Aquatic Bodies

About 80% of plastic contamination in aquatic bodies is accounted to the land-based sources and the remaining 20% to sea-based activities (Galafassi et al. 2019). The identified inland sources of microplastic pollution are WWTPs, reuse of sewage sludge, surface runoffs, municipal solid waste, plastic litter, and industrial wastes/discharges (Galafassi et al. 2019; Ryberg et al. 2019; Wu et al. 2019a).

MPs in the wastewater are mostly primary MPs having an average size of about 250 μm, originating mainly from the cosmetic and personal care products like facial cleansers, shampoos, toothpaste (Praveena et al. 2018), etc. along with synthetic textile fibers which are resulted from laundry washing (Sillanpää and Sainio 2017). Other important ways are abrasion or breaking or opening of plastic containers and bags in houses, plastic utensils/sieves/appliances (Sobhani et al. 2020), MPs from tyre abrasion, road markings (Siegfried et al. 2017), and paints (Gaylarde et al. 2021) reaching the urban stormwater runoff, industrial applications like blasting and molding operations, Landfill leachate, wet sedimentation process/ atmospheric fallouts (Ngo et al. 2019). Wastewater treatment plants (WWTPs) are identified as the significant contributor to anthropogenic emission of MPs in aquatic bodies and terrestrial systems though they also act as barriers to MPs (Prata 2018a). Though WWTPs are capable of removing a majority of the MPs from effluent, the removed particles persist in sewage sludge and directly reach the agricultural land by land application of sludge (Rolsky et al. 2020). Also, the inability of existing WWTPs to remove very small MPs along with the high volume of wastewater treated makes WWTP effluent a significant contributor of MPs in aquatic bodies (Magni et al. 2019). Therefore, WWTPs act as both collectors of MPs from anthropogenic use as well as a source for MP pollution in the natural environment (Carr et al. 2016; Magni et al. 2019; Hatinoğlu and Sanin 2021).

Pathway of generation of MPs in surface runoffs is attributed to the road dust, road markings and paints, tyre abrasion, wet sedimentation/precipitation (Horton et al. 2017; Galafassi et al. 2019; Treilles et al. 2021), etc. Another important contributor of MPs in aquatic bodies is the mismanaged MSW and consequently unauthorised plastic litter (Jambeck et al. 2015; UNEP 2018). About 4.1 Mt of plastics are reaching the oceans from mismanaged MSW according to a recent estimate (Ryberg et al. 2019) which accounts for the majority of the share of plastic pollution in the aquatic environment. Apart from these sources, blast abrasion media and other industrial applications are also identified as the contributors of MP pollution in aquatic media (Galafassi et al. 2019).

3.4.2 Air

MP pollution in the air has resulted from synthetic textiles, abrasion of plastic products, road dust, road markings, and tyre abrasion. Synthetic textile fibres constitute one of the major contributors to atmospheric MP pollution (Dris et al. 2016). According to Gasperi et al. (2018), the two-thirds portion of total textile fibers produced in the year 2016 were synthetic fibers and these plastic fibers may lead to MP fibre pollution in the atmosphere (Gasperi et al. 2018). Other significant sources of MPs in the air are emissions from traffic, road paintings, tyre abrasion, road dust, incineration of solid waste, industrial emissions, etc. (Chen et al. 2020a).

3.4.3 Soil

With the close evaluation of sources in all the three matrices, it is inferred that all sources of MP generation are not mutually exclusive for any compartment and one matrix can also act as a potential MP source to another. The sources such as WWTP, land application of sludge, landfill leachate, litter, mismanaged MSW, road dust, paints and markings, artificial turfs etc. which are mentioned for aquatic system and air, are also applicable to terrestrial systems (Karbalaei et al. 2018; Prata et al. 2020). Apart from these regular sources, various plastic items being used for medical purposes for our convenience and safety are often discarded without proper treatment and act as a source of secondary MPs in the environment. A recent study reported that the current usage manner of disposable face asks during the COVID-19 pandemic time could be a source of MPs soon (Aragaw 2020). An estimate of the release of MPs to air, water, and soil reported in some studies is shown in Table 3.3.

The sources of generation of MPs in the natural environment are plenty and recent research focussed only on a very few sources. Tyre abrasion, paints, and furnishings, sea activities, etc. need to be examined closely for knowing the exact generation rates.

3.5 Fate and Transport of MPs

Air, water, and soil are the common pathways of transport of MPs in the environment. Figure 3.1 describes the common routes for the fate and transport of MPs in air, water, and soil compartments. The densities of MP particles are an important factor that affects their fate in the environment (Wu et al. 2019a). In aquatic systems, heavier particles are subjected to sinking and rising and lighter particles are likely to be carried away by water and wind currents. However, biofouling and weathering forces acting on these particles in the natural environment alter their properties and subsequently their fate. Air-borne MPs are mostly smaller in size and are transported

Table 3.3 Sources of generation of MPs

Sl. No	Source	MPs quantity	Matrix considered	Region	Reference
1	WWTP	16 × 10^7 MPs/day	Water	Italy	Magni et al. (2019)
2	Surface runoffs	29 MPs/L	Water	Paris	Treilles et al. (2021)
3	Landfill leachate	24 MPs/L	Water	China	He et al. (2019)
4	Synthetic textile	2–355 MPs/m^2/day	Air	Paris	Dris et al. (2016)
5	Incineration of MSW	1.9–565 MPs/kg of bottom ash	Air Water Soil	China	Yang et al. (2021)
6	Road dust	0.10–39.6 pieces/m^2	Air	Japan, Vietnam, Nepal	Yukioka et al. (2020)
	Road dust	50–120 MPs/day	Air	India	Sathish et al. (2019)
7	Street dust	227.94 ± 91.37	Street dust	India	Patchaiyappan et al. (2021)

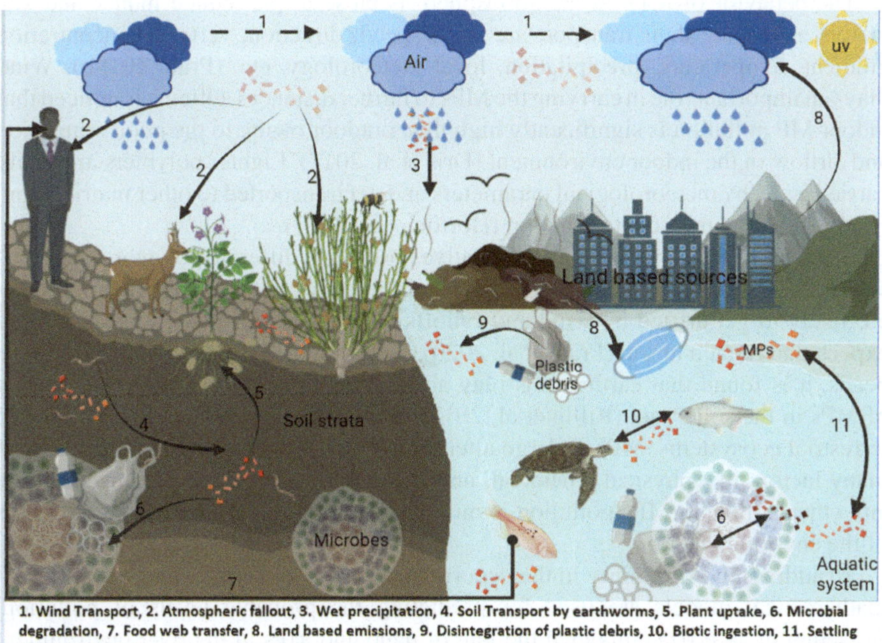

Fig. 3.1 Fate and transport of MPs in the natural environment

in the atmosphere by prevailing wind currents (Zhang et al. 2020b). In soil, soil organisms play a key role in the transport of MPs along the soil strata. Some of the important processes that are involved in the fate and transport of MPs in air, water, and soil are wind transport, atmospheric fallout, wet precipitation, transport by earthworms, uptake by plants, microbial degradation, food web transfer, the disintegration of plastic debris and settling and resuspension of MPs in aquatic bodies (Rillig et al. 2017; Wagner and Lambert 2018; Zhang et al. 2020b). In natural systems, these particles are subjected to physical, chemical, and biological degradation factors which lead to the alteration of their properties and aging.

The important factors affecting the transport and presence of MPs in aquatic systems are depth of water body, surface currents, prevailing wind, physicochemical conditions of plastic particles. In aquatic bodies, settling and resuspension of MPs are controlled by the density of particles, prevailing surface currents, biofouling conditions, water hydraulics, etc. Though lakes act as sinks of MPs, most of the plastic debris in freshwater bodies is expected to reach marine waters eventually with the flow of water (Wu et al. 2019a). Upon reaching the marine environment, it is estimated that about 70% of plastic debris settles at the marine benthic zones and the remaining is distributed as floating MPs and at the shorelines (UNEP 2018). In the aquatic environment, apart from being a pollutant on their own, MP particles can act as the carriers of various other emerging contaminants (Carr et al. 2016).

The behavior of MPs in the atmosphere is close to particulate matter and key factors governing their transport are wind speed, direction, vertical concentration gradient, temperature, precipitation, local meteorology, etc. (Prata 2018b). Wind plays an important role in carrying the MPs to farther distances. Often it is noticed that indoor MP pollution is significantly high than outdoor owing to the poor ventilation and airflow in the indoor environment (Dris et al. 2017). Lighter polymers are easily carried away by meteorological parameters and get transported to other matrices and results in subsequent contamination (Horton et al. 2017).

In the terrestrial environment, the relative position of different polymers in the soil strata depends on the density of polymers and sources of pollution. Lighter polymers are likely to get drifted by wind and runoffs whereas higher density polymers are expected to remain in soil (Prata et al. 2020). In a study conducted with polyethylene beads, it is found that earthworms play an important role in the vertical transport of MPs in the soil strata (Rillig et al. 2017). Similar to the aquatic environment, in terrestrial ecosystems, MPs undergo alteration in their properties upon the action of many factors like physical, chemical, and biological degradation parameters. Like any other pollutant, MPs pollution is mutually inclusive to different compartments of the environment.

To address the variability in the occurrence of MPs in the environment, different mathematical models are used for predicting the probable levels of pollution. Besseling et al. (2017) modeled the fate of nano and MPs in the river environment considering factors like advection, dispersion, biofouling, aggregation, degradation, settling, resuspension, and burial. Siegfried et al. (2017) modeled the approximate quantity of land-based point source MPs released into the sea. In a study by van Wezel et al. (2016), a modeling approach was used to estimate the primary MPs release into the domestic sewage. The development of fate models is a challenging process since many processes need to be accounted for to predict the fate of MPs in the environment.

3.6 Exposure Levels and Ecotoxicological Effects on Biota

The data on the MPs contamination in the various environmental sectors is rising exponentially, as a consequence, it has become the subject of rising concern. Apart from the research on marine MP pollution, recent studies also indicate that MP particles are also present in flora and fauna (Lu et al. 2016; Provencher et al. 2017; Mao et al. 2018; Cox et al. 2019; Picó and Barceló 2019; Jiang et al. 2020; Yu et al. 2021). Briefly, MP pollution turns out to be a worldwide problem, and the ultimate fate of MPs intensifying the concern globally.

The vast presence of MPs in seawaters has received the global attention of many researchers as the presence of MP is causing detrimental effects on marine life (Jambeck et al. 2015; Provencher et al. 2017; Guzzetti et al. 2018; Tang et al. 2020b; Yu et al. 2021). Along with large plastics, MPs have also been stated to be ingested by around 700 genera of aquatic wildlife (Provencher et al. 2017). It has been reported that the growth of microalgae can be significantly reduced by MPs exposure in both freshwater and seawater (González-Pleiter et al. 2021). A study conducted by Phillips and Bonner (2015) reported the occurrence of a higher concentration (29%) of MPs in freshwater fish in urban areas than in non-urbanized areas. The intake of MPs by fish in urban waterways has been reported in the range of 6.8–29% (Sanchez et al. 2014). Aquatic creatures become susceptible to captivating floating MPs both directly and indirectly via the ingestion of infested prey (Phillips and Bonner 2015; Lopes et al. 2020). The ingestion of MPs by aquatic animals causes false repletion which leads to their starvation and at last death (Tang et al. 2020b; Cormier et al. 2021). Moreover, a recent study found out that MPs in the aquatic environment can play a role of a carrier of heavy metals which ultimately end up forming intricate pollutants (Selvam et al. 2021). It has been determined that atmospheric MPs direct to inhalation exposure. As soon as airborne MPs come in contact with the human body, they enter the lungs and thus can cause serious health issues (Wright and Kelly 2017).

The occurrence of MPs on inland systems affects soil properties, which adversely impacts process rates and net primary production. This can also have an indirect effect on the atmosphere, comprising greenhouse gases (Cai et al. 2017; Klein and Fischer 2019; Zhang et al. 2020a; Huang et al. 2021b). Scholars found that the association of MPs and plants would not only responsible for the physical damage to the plants but also affect their metabolism and gene expression (Wang et al. 2020b; Yu et al. 2021). In the case of new era environment-friendly plastic i.e. bioplastics, a study shows that polylactic acid (PLA) which is the most common form of bioplastic either remain fossilized for more than 3 years or go through anaerobic biodegradation and can emit toxic gas, methane (Selvamurugan Muthusamy and Pramasivam 2019).

MPs have lately been discovered not only in the metropolitan but also remote and rural regions which are distant from source areas of MP, implying the possible long-distance atmospheric transportation of MPs (Gasperi et al. 2018; Zhang et al. 2019). MPs exist in the air as airborne pollutants originating from plastic textile fiber production (Dris et al. 2017; Gasperi et al. 2018). MPs amass several contaminants and tenacious organic pollutants and also act as carriers for distant transportation. Likewise, it can also supplement numerous antibiotics and their transportation which can become the basis of many severe effects (Caruso 2019; Guo and Wang 2019; Fred-Ahmadu et al. 2020). Based on the requirement for various modifications, particular organics and minerals are further incorporated into the synthetic polymer processing, for example, biological stabilizers, antioxidants, lubricants, etc. Among those, phthalates and bisphenols act as triggers for endocrine diseases (Jambeck et al. 2015; Hong et al. 2018). In the future, humans may have to face the terrible consequences of such pollutants.

Human consumption of microplastic is a hot topic in the research area (Cox et al. 2019; Shruti et al. 2020a, b). The results revealed that the contamination of beverages by MPs is not only limited to beers but also span across soft drinks, cold tea, and energy drinks (Shruti et al. 2020a). Out of the various types, fiber-shaped and small-sized MPs dominated the samples (Shruti et al. 2020a, b). Study findings show that free public drinking water facilities are also a budding hotspot of MP and deliver valuable references for qualification procedures (Shruti et al. 2020b). MPs have also been found in edible fruits and vegetables (Oliveri Conti et al. 2020). The new method established by the research scholars of Arizona State University helps to find out the infection of MPs in the human tissues and also has detected chemical traces of plastic in tissue (URL-2). The analysis carried out by Senathirajah et al. (2021) displays that globally, on average, humans could potentially be ingesting 0.1–5 g of MPs per week. In recent times, researchers investigated the potential exposure of infants to MPs from consuming formula prepared in infant feeding bottles made of polypropylene (PP) (Li et al. 2020b). These bottles were found to release up to 16,200,000 particles per liter (Li et al. 2020b). Therefore, it would be naïve to believe that there is plastic everywhere but just not in us. Human exposure routes of MPs and associated risks are displayed in Fig. 3.2. Table 3.4 is the summary of various studies exploring the ecotoxicological effects of MPs.

3.7 MPs as the Carriers of Other Organic Pollutants

The persistence of MPs in the environmental matrices, particularly in the aqueous medium can result in their interaction with the variety of pollutants present in water. The longer stability of MPs in the water column may result in accumulating organic compounds or microbes that come in contact with their surface. This accumulation of active communities of microbes or organic pollutants can easily transfer into the food chain and substantially contribute to potential health impacts on the ecosystem (Ziajahromi et al. 2016). Owing to the lower density of MPs, they tend to accumulate

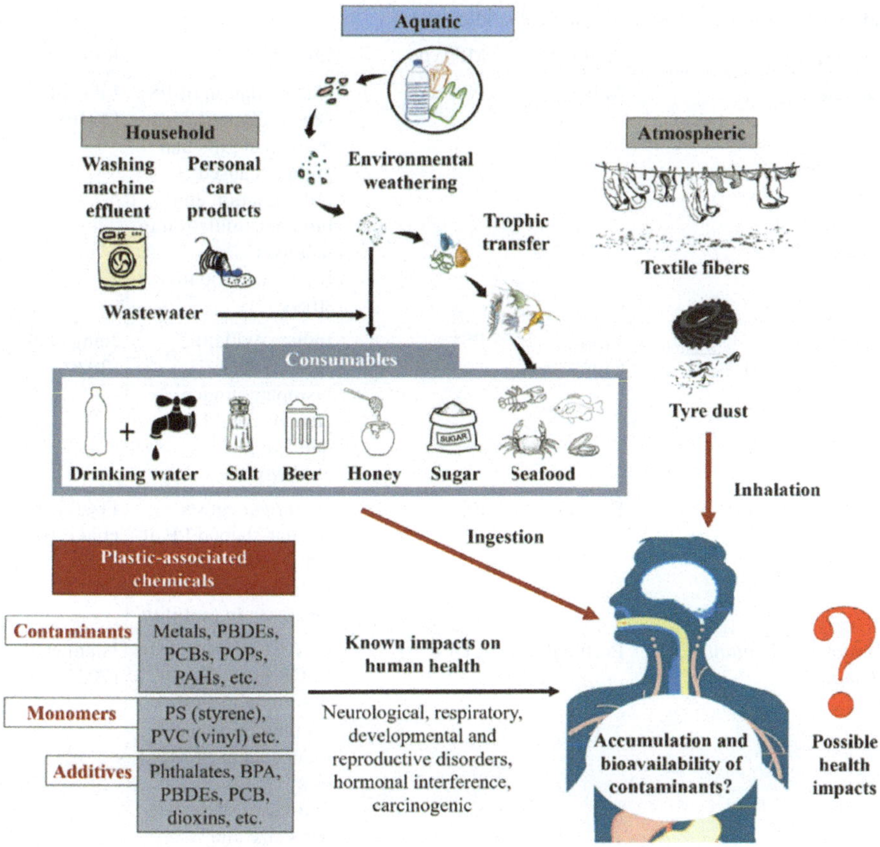

Fig. 3.2 Human exposure to MPs and associated risks. Reprinted with permission from Senathirajah et al. (2021). Copyright © 2021 Elsevier Ltd

in the surface microlayer of the aquatic organism and may subsequently cause toxic effects (Frydkjær et al. 2017; Adeogun et al. 2020).

The sorption behavior of contaminants and microbes depends upon the characteristic properties of MPs such as size, chemical composition, type of plastics, concentrations, hydrophobicity, etc. (Frydkjær et al. 2017; Wang et al. 2018). They are also dependent on the specific surface and porosity of MPs. The adsorption studies of metals, Zn, Cu, Co, Ni, Cd, and Cr on five different metals in different water sources were studied and the results concluded that among the different MPs, PP and PS had the highest specific surface, but their adsorption properties were moderate when compared to that of PVC and PE on the metals (Godoy et al. 2019). They also observed that the water matrix and pH of the solution altered the adsorption of metals on the MPs' surface. Similar to the above study, the level of adsorption of fungicides, hexaconazole, myclobutanil, and triadimenol on PS was primarily due to the pH, MP particle size, and ionic strength of the aqueous solution (Fang et al.

Table 3.4 Summary of ecotoxicological studies of MPs

Biota species	Study area	MP type	Effects	References
Zebrafish	China	PS beads	Accumulation of PS beads (20 μm) seen in the fish gills and gut and caused inflammation and lipid accumulation in fish liver Increased oxidative stress	Lu et al. (2016)
Earthworm (*Eisenia fetida*)	Jurong, China	PS	Induce oxidative stress, histopathological changes, and DNA damage in earthworms	Jiang et al. (2020)
Daphnia Magna	Belgium	PE	*D. magna* egests regular-shaped PE at a higher rate than irregular ones Decrease in mobility	Frydkjær et al.(2017)
Scrobicularia plana (Clam)	Portugal	LDPE	PEs act as a vector of PFOS in *S. plana* tissues and PFOS accumulation was independent of MPs size Oxidative stress was MPs size and tissue-dependent	Islam et al. (2021)
Chlorella pyrenoidosa and Microcystis flosaquae	Wuhan, China	PP, PVC	Decrease photosynthesis efficiency by reducing the chlorophyll level in cells	Wu et al. (2019a, b)
Vibrio fischeri, Caenorhabditis elegans and Lactuca sativa	Madrid, Spain	PE with two contaminants, ibuprofen and simazine	PE acted as a carrier for both contaminants and affected survival in *V. fischeri* and development in *L. sativa* whereas *C. elegans* was least affected	Martín et al. (2021)
Salvinia cucullate (Plant)	China	PS	PS MPs and glyphosate caused single and synergistic effects on leaf growth Oxidative stress accumulation	Yu et al. (2021)

(continued)

Table 3.4 (continued)

Biota species	Study area	MP type	Effects	References
Chlorella pyrenoidosa (microalgae)	Wuhan, China	PS	Hampered algal cell growth by reducing photosynthetic activity and cell membrane damage	Mao et al. (2018)
Sea urchins, jellyfish ephyrae	Guadeloupe archipelago, France	PE,PP	Leachates (Cu, Zn, Cd, Pb) induced sublethal effects on the growth of sea urchins and the pulsation frequency of jellyfish ephyrae	Cormier et al. (2021)
Dicentrarchus labrax (Fish)	Messina, Italy	PVC	Intestinal functions were affected due to the saturation of PVC pellets	Pedà et al. (2016)
Tegillarca granosa (Clams)	China	Different MPs associated with POPs	MPs and POPs (benzo-a-pyrene and 17β-estradiol) exposure led to immunotoxicity and inhibition of gene expressions in blood clam	Yu Tang et al. (2020)
Human	Italy	PP and other pigmented MPs	12 MP fragments(5-10 μm) were found in all placental portions of a woman with physiological pregnancy	Ragusa et al. (2021)

2019). These understandings on the sorption behavior of various MPs for contaminants help in explaining the distribution and risk assessment of various contaminants in the ecosystem.

Due to the hydrophobic nature of MPs and higher surface, they can interact with organic compounds in the water column and can potentially accumulate these toxic compounds (Frydkjær et al. 2017; Sheng et al. 2021). Also, the aging of these MPs could improve the interactions with pollutants. For instance, the aging of MPs in aqueous solutions may disintegrate resulting in increasing their surface area. The longer residence time of these disintegrated MPs in aqueous solution may allow the easy diffusion of hydrophobic pollutants into the particle surface (Wang et al. 2018). However, the sorption of organic compounds onto MPs is not limited to their hydrophobicity of organic compounds, can also occur with hydrophilic compounds. MPs undergo deformations and structural changes with longer residence time in aquatic environments. In a study conducted by Liu et al. (2019), concluded that aged PVC and PS MPs were able to adsorb ciprofloxacin owing to the generation of functional groups on MP surface with aging. The sorption behavior of hydrophilic compounds is significantly affected by the matrix, pH, and ionic strength which would either promote or suppress the transport of pollutants on MPs surface (Atugoda et al. 2021).

The interactions of MPs with organic pollutants may vary with the nature and type of polymer material. Further, the stronger sorption of contaminants on MPs surface even at trace concentration can reduce the availability of these compounds in sediments also (Hüffer and Hofmann 2016). In a study conducted by Guo et al. (2019), the sorption behavior of an antibiotic pollutant sulfamethazine (SMT) was studied on six types of MPs and concluded that the sorption capacity was dependent on the interaction energies between MPs and SMT. Further, the sorption also varied with the salinity and pH of the solution which will have a significant role in understanding the possibility of MPs as a vector for pollutants. The irregularity in the shape of MPs also affects their interaction with organic compounds and microbes. In another study conducted by Frydkjær et al. (2017) concluded that irregular-shaped polyethylene MPs adsorbed less amount of phenanthrene compared to different plankton organisms in a water column experiment.

The interactions of MPs with heavy metals in the environment can results in their adsorption onto their surface and can easily be transported to food resources (Liao and Yang 2020; Liu et al. 2021). These adsorbed metals may get easily desorbed depending upon the environmental conditions in aqueous conditions causing ecological threats to macroinvertebrates. In a study conducted by Zhou et al. (2020) evaluated the adsorption–desorption processes of Cd(II) on five different MPs under sediment and gut environments. Their findings implied that the higher ecological risk associated with the desorption of Cd(II) is found in simulated gut environments and long-term studies are required to understand more about the bioavailability of ingested MPs. In another study conducted by Brennecke et al. (2016) evaluated the interaction between polystyrene and polyvinylchloride (PVC) on the heavy metals Cu and Zn leaching from the antifouling paints. They found that the higher surface area and polarity of PVC resulted in higher adsorption of Cu when compared to PS. Another study on the adsorption of Cd on HDPE under laboratory conditions inferred that based on the particle size and solution constituents, the adsorption capacity varied. The outcomes of the study provided evidence on the possible threat to organisms that can be caused by metal particles adsorbed on MPs (Wang et al. 2019).

The adhesion of microbes to the plastic debris can consequently act as a vector for the spread of microbial contamination. However, in some cases, the attached microbes can easily degrade the MPs into smaller particles and results in increasing their surface for more adsorption. The interactions between plastic debris and microbes are needed to be studied to understand the potential of MPs as carriers for pathogens. Also, the various microbes may colonize on the MP surface creating a "plastisphere" which is comprised of autotrophic as well as heterotrophic microorganisms (Caruso et al. 2018). The significance of biofilms on MPs surfaces to enhance the adsorption of trace metals was studied by Guan et al. (2020). They have observed that the trace metal adsorption capacity of expanded polystyrene was enhanced due to the development of biofilm on their surface. The studies concluded that the intraparticle diffusion mechanism enhanced the adsorption of trace metals to virgin PS particles whereas film diffusion governed them onto biofilm-covered MPs. In another study, Viršek et al. (2017) examined the bacterial species colonized on the MPs surface in the North Adriatic Sea. They were able to identify 28 bacterial species from MP particles and confirmed the presence of *Aeromonas salmonicida* species that are responsible for diseases in fish. Recent studies also reported that MPs can act as an artificial substrate for biofilm formation in aquatic environments and these microbial communities are different from natural free-living microbial communities (Yang et al. 2020). MPs characteristics such as polymer type, surface roughness, hydrophobicity, and environmental conditions such as temperature, salinity, nutrient conditions have a significant effect on these biofilm formations.

The research was also conducted to identify the unintended toxicity caused by MPs in the marine ecosystem by acting as carriers of micropollutants. Apart from the aqueous phase, MPs are potential carriers of toxic organic compounds in soil. Hodson et al. (2017) conducted a study analysing the interaction of MPs and Zn metal in soil and their possible risk on terrestrial organisms. The experiments were carried out with fragmented HDPE bags in synthetic soil solution with Zn metal and earthworms. The studies on earthworms confirmed that MPs only acted as a vector for transporting Zn metals into the guts of earthworms causing exposure and were thus not retained in their guts. They also concluded that smaller particles with a larger surface area had more impact on the relative adsorption of Zn metals owing to their partitioning in soil. In another study, the effect of adsorbed antibiotics azithromycin and clarithromycin on freshwater microalgae when exposed in contact with 5 five different MPs was evaluated in batch studies (González-Pleiter et al. 2021). The studies concluded that antibiotic-loaded MPs were toxic and inhibited the growth and chlorophyll content of the cyanobacterium. The effect of ingested MPs on the bioaccumulation of hydrophobic organic compounds in earthworms was studied (Wang et al. 2020a). They found that the polluted MPs in clean soil were more bioavailable for earthworms when compared to the clean MPs in polluted soil and higher bioaccumulation was observed in the former case. Further, the smaller-sized MPs showed greater potential for desorbing organics into worm tissues posing a greater risk to the overall bioaccumulation. A study conducted by Beiras and Tato (2019) analysed the toxicity to marine plankton caused by the ingestion of polyethylene MPs carrying hydrophobic compound 4-n-nonylphenol. However, Contrary to the toxic effects caused by MPs on living organisms, the findings from the study concluded that the active ingestion of MPs containing 4-n-nonylphenol does not increase the toxicity to planktonic larvae. Though several studies have identified the toxic effects of MPs acting as a vector for POPs and microbes, a deeper understanding of the structural and morphological changes on target organisms is required to understand their toxicity.

3.8 Leaching of Compounds from Plastics and MPs

The leaching of plastic additives used in the manufacturing stage of various plastic consumer products is of great concern these days due to their environmental toxicity (Ye et al. 2020). The major plastic additives include phthalates (Keresztes et al. 2013; Kastner et al. 2012), bisphenol-A (BPA) (Romera-Castillo et al. 2018; Andaluri et al. 2018), and brominated flame retarders (Sun et al. 2021a; Choi et al. 2009), etc. which are introduced during the manufacturing stages and are easily released into the environment. The important factors influencing the leaching of additives from various plastic products and MPs include temperature, ultraviolet radiation, aging, weathering, chemical oxidation, mechanical wear, and tear and biological forces, etc. (Sun et al. 2021b). The major studies on the leaching of different plastic additives from various sources of MPs and their maximum detected concentrations are shown in Table 3.5.

There are several studies on the leaching of BPA from different daily used products especially infant feeding bottles (Johnson et al. 2015), drinking water bottles (Cooper et al. 2011), water distribution pipe networks (Cantoni et. al. 2021; Rajasärkkä et al. 2016), etc. Their migration is mainly caused by the diffusion of unbounded BPA present in polycarbonate and hydrolysis reactions (Johnson et al. 2015). The study by Johnson et al. (2015) investigated the leaching behavior of BPA from feeding bottles filled with milk, water, and apple juice under different temperatures in 30 min exposure time. It was observed that the leaching concentration of BPA depends upon the temperature of water-filled. A maximum concentration of around 46.05 ppb was detected in the sterilized water for the feeding bottles whereas no leaching of BPA was observed with apple juice filled under room temperature. Besides, BPA was found in the range of 7.06 ppb for bottles filled with boiling water with milk powder at 95 °C due to the hydrolysis of the polymer at elevated temperatures (Johnson et al. 2015). Similar higher leaching behavior of BPA was observed by Cooper et al. (2011) at elevated temperature for water stored in epoxy lined reusable plastic, steel, and aluminum water bottles. They have determined a maximum of 0.3 mg L^{-1} and 1.9 mg L^{-1} of BPA leaching from polycarbonate and aluminium bottles lined with epoxy respectively. Also, there was no leaching of BPA from uncoated aluminium and steel bottles.

Another important source for leaching of BPA was found to be the drinking water pipe networks. For instance, the study by Cantoni et. al. (2021) monitored the leaching of BPA as a function of water conditions (deionized water, tap water, and tap water with 1 $mgCl_2$ L^{-1} of chlorine concentration) and residual chlorine concentration from three different epoxy resin pipes for 170 days of test time. The results indicate that as the chlorine concentration increases, the BPA concentration decreases due to the formation of chlorination by-products with BPA. Also, the BPA leaching was found to be lower in resin with a higher cross-linking degree where the free BPA bonded well and trapped inside the lattice. Similarly, another study by Rajasärkkä et al. (2016) examined the drinking water contamination with BPA with epoxy resin coated by employing old and recent technology. The maximum leaching

Table 3.5 Major studies on leaching of various plastic additives from different sources

Source	Plastic additives leached	Analyzed samples for additives	Detected concentration	Study area	Reference
Children soft toys	Bisphenol A and Phthalates	Wetted wipes from toys	13–280 ng cm^{-2}	USA	Andaluri et al. (2018)
Packed juice containers	Diethyl phthalate (DEP) and Diethyl hexyl phthalate (DEHP)	Fruit juice	0.26–0.59 $\mu g\ L^{-1}$	Iran	Rastkari et al. (2018)
Take-out food containers	di-n-butyl phthalate (DnBP), and bis(2-ethylhexyl) phthalate (DEHP)	Food samples	0.12–86.1 mg g^{-1}	China	Han et al. (2021)
Thermal receipts	Bisphenol-A	Thermal receipts papers	13.9 mg g^{-1}	Japan, Korea and Vietnam	Liao and Kannan (2011)
Landfill leachates	Polybrominated diphenyl ethers (PBDEs)	Landfill leachate sample	11–620 ng L^{-1}	Japan	Osako et al. (2004)
Infant feeding bottle	Bisphenol-A	Milk and water samples	46.05 ppb	India	Johnson et al. (2015)
Plastic bags	Di(2-ethylhexyl) phthalate, di-n-butyl phthalate, and di-isobutyl phthalate	Shredded plastic bag pieces	181.4–5320.6 ng g^{-1}	China	Xu et al. (2020)
Polyethylene terephthalate (PET) water bottles	benzyl-butyl phthalate and di(2-ethyl-hexyl) phthalate (DEHP)	Mineral water	0.1–1.7 $\mu g\ L^{-1}$	Hungary	Keresztes et al. (2013)

of BPA around 0.25 mg L^{-1} was observed for pipelines with older technology and lesser concentration of around 10 ng L^{-1} (cold water) and 23.5 μgL^{-1} (hot water) for recent technology pipelines in 8–10 h of exposure time. They have also identified that the aging of epoxy lining plays an important role in the leaching behavior. The amount of BPA leached for old pipelines (8–9 yrs) was 4–20 times more as compared with the 2 years old.

Phthalates used as softening agents, are commonly leached from various soft PVC products such as soft toys for children (Andaluri et al. 2018), plastic bags (Xu et al. 2020), PET bottles (Keresztes et al. 2013), are not tightly bound to the plastic, hence they will leach out easily from the plastic matrix (Stringer et al. 2000). For instance, the study by Keresztes et al. (2013) investigated the phthalates leaching from carbonated and non-carbonated mineral water stored in PET bottles. The highest concentration of around 0.1 and 1.7 μg L^{-1} was observed for benzyl-butyl phthalate and di(2-ethyl-hexyl) phthalate (DEHP) respectively when it is stored for 44 days at 22 °C. Also, it was found that 0.5 L bottles have higher phthalates concentration as compared with 1.5 and 2 L due to higher surface-to-volume ratio. Similarly, plastic bags were also found to be the potential source for phthalates leaching. The study by Xu et al. (2020) monitored the leaching of phthalates from plastic express bags in China and detected around 181.4–5320.6 ng g^{-1} of most dominant types such as di(2-ethyl hexyl) phthalate, di-n-butyl phthalate, and di-isobutyl phthalate after 48 h of exposure time.

Additionally, brominated flame retarders are another class of compounds used as plasticizers, defoaming, and stabilizing agents, which tend to leach out mainly from E-waste such as electronic equipment and printed circuit boards, etc. These have been reported mainly in landfill leachates, and soil (Osako et al. 2004; Zhou et al. 2013; Liao et al. 2020). The study by Zhou et al. (2013) investigated the leaching of polybrominated diphenyl ethers (PBDEs) and tetra bromo-bisphenol A (TBBPA) from plastic circuit boards and observed the highest concentration in landfill leachate tests of around 30.4 and 12.27 μg L^{-1} respectively. The reports suggest that polybrominated diphenyl ethers (PBDEs) are hydrophobic hence their leachability could be greatly affected by the other constituents in leachates especially dissolved organic matter (Osako et al. 2004). Whereas tetra bromo-bisphenol A (TBBPA) is relatively hydrophilic and thus readily dissolved in rainwater and easily get distributed in the environment (Liao et al. 2020). Besides, the study by Kim et al. (2006) examined the leaching behavior of PBDEs in simulated leachate solution with 20% methanol and 1000 mg L^{-1} dissolved humic solution for a contact period of 5 days at 30 °C. A higher leaching concentration of around 1200–23,000 ng L^{-1} was detected in the presence of methanol and dissolve humic solution as compared with blank distilled water (220 ng L^{-1}), this is mainly attributed to the higher solubility effect, and formation of a complex with PBDEs and dissolved humic matter.

There are many studies on the leaching of plastic additives from plastic containers and other daily used products as a whole, but limited studies are available on the leaching of additives from microplastic particles. For instance, the study by Wei et al. (2019) observed the leaching of BPA from PVC microplastics in activated sludge during wastewater treatment. As these PVC microplastics contain major additives such as BPA, DEHP, DiNP, etc. which are employed in the manufacturing process, these may leach out by solubilization, acidification, and hydrolysis reactions during the treatment process. They have observed a decrease in the production of methane in the anaerobic sludge digestor due to the presence of the PVC MPs and the leaching of additives. Another study by Luo et al. (2019) investigated the leaching of additives from polyurethane microplastics in acidic, alkaline, saline, and natural waters. It was observed that compounds similar to 3,3′-diaminobenzidine were found in the leachates after 12–24 h of exposure time. In a recent study by Schrank et al. (2019), investigated the impact of *Daphnia Magna* (planktonic crustacean) lifetime due to the exposure of microplastics and leaching out of additives from MPs. It was observed that a maximum concentration of around 2.67 mg L^{-1} of diisononylphthalate (DiNP) was found to be leaching out from flexible PVC to the surrounding medium, thereby affecting the body length and relative tail spine length of *D. magna*.

3.9 Summary and Conclusion

The chapter discussed the overview of MP pollution in the natural environment. MPs are widely distributed all over the environment. They are identified as an emerging contaminant and their identification, fate, and transport, remediation studies are in the growing stage. Studies are still focussing on the inventory and quantification of MPs. Existing sample preparation and identification methods have their own merits and demerits and therefore newer technologies are likely to be developed. Though many sources that contribute to MP pollution are being identified, data is lacking about the actual generation of MPs with varying use patterns of these sources. There exists a large variation in the spatial and temporal distribution of MPs across the world. Fate and transport studies of MPs in the natural environment often employ modelling methods for knowing the extent of the contamination. Current studies are mostly concentrated on the occurrence and distribution of this ubiquitous pollutant, however, there is an actual need of developing remediation measures to control MP pollution. Much recent research is focussing on the ecotoxicological effects of MPs on biota including humans. However, research is in the progressing stage and health effects on humans are not yet understood completely. Apart from being a pollutant of concern on their own, MPs also acts as carriers of other organic pollutants in the environment. Further, the leaching of chemical additives projects another major risk associated with MPs. Altogether, MPs represent a class of important emerging pollutant that demands major concern to all levels of the ecosystem.

References

Adeogun AO, Ibor OR, Khan EA et al (2020) Detection and occurrence of microplastics in the stomach of commercial fish species from a municipal water supply lake in southwestern Nigeria. Environ Sci Pollut Res 27:31035–31045. https://doi.org/10.1007/s11356-020-09031-5

Allen S, Allen D, Phoenix VR et al (2019) Atmospheric transport and deposition of microplastics in a remote mountain catchment. Nat Geosci 12:339–344. https://doi.org/10.1038/s41561-019-0335-5

Anbumani S, Kakkar P (2018) Ecotoxicological effects of microplastics on biota: a review. Environ Sci Pollut Res 25:14373–14396. https://doi.org/10.1007/s11356-018-1999-x

Andaluri G, Manickavachagam M, Suri R (2018) Plastic toys as a source of exposure to bisphenol-A and phthalates at childcare facilities. Environ Monit Assess 190. https://doi.org/10.1007/s10661-017-6438-9

Andrady AL (2011) Microplastics in the marine environment. Mar Pollut Bull 62:1596–1605. https://doi.org/10.1016/j.marpolbul.2011.05.030

Aragaw TA (2020) Surgical face masks as a potential source for microplastic pollution in the COVID-19 scenario. Mar Pollut Bull 159:111517. https://doi.org/10.1016/j.marpolbul.2020.111517

Atugoda T, Vithanage M, Wijesekara H et al (2021) Interactions between microplastics, pharmaceuticals and personal care products: implications for vector transport. Environ Int 149:106367. https://doi.org/10.1016/j.envint.2020.106367

Beiras R, Tato T (2019) Microplastics do not increase toxicity of a hydrophobic organic chemical to marine plankton. Mar Pollut Bull 138:58–62. https://doi.org/10.1016/j.marpolbul.2018.11.029

Besseling E, Quik JTK, Sun M, Koelmans AA (2017) Fate of nano- and microplastic in freshwater systems: A modeling study. Environ Pollut 220:540–548. https://doi.org/10.1016/j.envpol.2016.10.001

Bharath KM, Srinivasalu S, Natesan U et al (2021) Microplastics as an emerging threat to the freshwater ecosystems of Veeranam lake in south India: a multidimensional approach. Chemosphere 264. https://doi.org/10.1016/j.chemosphere.2020.128502

Boyle K, Örmeci B (2020) Microplastics and nanoplastics in the freshwater and terrestrial environment: a review. Water (Switzerland) 12

Brennecke D, Duarte B, Paiva F et al (2016) Microplastics as vector for heavy metal contamination from the marine environment. Estuar Coast Shelf Sci 178:189–195. https://doi.org/10.1016/j.ecss.2015.12.003

Browne MA, Galloway T, Thompson R (2007) MicropLastic—an Emerging contaminant of potential concern? Integr Environ Assess Manag 3:559–566. https://doi.org/10.1001/archderm.98.3.219

Cai L, Wang J, Peng J et al (2017) Characteristic of microplastics in the atmospheric fallout from Dongguan city, China: preliminary research and first evidence. Environ Sci Pollut Res 24:24928–24935. https://doi.org/10.1007/s11356-017-0116-x

Can-Güven E Microplastics as emerging atmospheric pollutants: a review and bibliometric analysis. https://doi.org/10.1007/s11869-020-00926-3/Published

Cantoni B, Riguzzi AC, Turolla MA (2021) Bisphenol A leaching fromepoxy resins in the drinking-water distribution networks as human health risk determinant. Sci Total Environ J 783:146908. https://doi.org/10.1080/09593330.2010.53517

Carr SA, Liu J, Tesoro AG (2016) Transport and fate of microplastic particles in wastewater treatment plants. Water Res 91:174–182. https://doi.org/10.1016/j.watres.2016.01.002

Caruso G, Pedà C, Cappello S et al (2018) Effects of microplastics on trophic parameters, abundance and metabolic activities of seawater and fish gut bacteria in mesocosm conditions. Environ Sci Pollut Res 25:30067–30083. https://doi.org/10.1007/s11356-018-2926-x

Caruso G (2019) Microplastics as vectors of contaminants. Mar Pollut Bull 146:921–924. https://doi.org/10.1016/j.marpolbul.2019.07.052

Chen G, Feng Q, Wang J (2020a) Mini-review of microplastics in the atmosphere and their risks to humans. Sci Total Environ 703:135504. https://doi.org/10.1016/j.scitotenv.2019.135504

Chen G, Fu Z, Yang H, Wang J (2020b) An overview of analytical methods for detecting microplastics in the atmosphere. TrAC—Trends Anal Chem 130:115981. https://doi.org/10.1016/j.trac.2020.115981

Choi KI, Lee SH, Osako M (2009) Leaching of brominated flame retardants from TV housing plastics in the presence of dissolved humic matter. Chemosphere 74:460–466. https://doi.org/10.1016/j.chemosphere.2008.08.030

Cooper JE, Kendig EL, Belcher SM (2011) Assessment of bisphenol A released from reusable plastic, aluminium and stainless steel water bottles. Chemosphere 85:943–947. https://doi.org/10.1016/j.chemosphere.2011.06.060

Cormier B, Gambardella C, Tato T et al (2021) Chemicals sorbed to environmental microplastics are toxic to early life stages of aquatic organisms. Ecotoxicol Environ Saf 208. https://doi.org/10.1016/j.ecoenv.2020.111665

Corradini F, Casado F, Leiva V et al (2021) Microplastics occurrence and frequency in soils under different land uses on a regional scale. Sci Total Environ 752. https://doi.org/10.1016/j.scitotenv.2020.141917
Cox KD, Covernton GA, Davies HL et al (2019) Human consumption of microplastics. Environ Sci Technol 53:7068–7074. https://doi.org/10.1021/acs.est.9b01517
Crichton EM, Noël M, Gies EA, Ross PS (2017) A novel, density-independent and FTIR-compatible approach for the rapid extraction of microplastics from aquatic sediments. Anal Methods 9:1419–1428. https://doi.org/10.1039/c6ay02733d
Dris R, Gasperi J, Rocher V et al (2015) Microplastic contamination in an urban area: a case study in Greater Paris. Environ Chem. https://doi.org/10.1071/EN14167
Dris R, Gasperi J, Saad M et al (2016) Synthetic fibers in atmospheric fallout: a source of microplastics in the environment? Mar Pollut Bull 104:290–293. https://doi.org/10.1016/j.marpolbul.2016.01.006
Dris R, Gasperi J, Mirande C et al (2017) A first overview of textile fibers, including microplastics, in indoor and outdoor environments. Environ Pollut 221:453–458. https://doi.org/10.1016/j.envpol.2016.12.013
Duis K, Coors A (2016) Microplastics in the aquatic and terrestrial environment: sources (with a specific focus on personal care products), fate and effects. Environ Sci Eur 28:1–25. https://doi.org/10.1186/s12302-015-0069-y
Fang S, Yu W, Li C et al (2019) Adsorption behavior of three triazole fungicides on polystyrene microplastics. Sci Total Environ 691:1119–1126. https://doi.org/10.1016/j.scitotenv.2019.07.176
Felsing S, Kochleus C, Buchinger S et al (2018) A new approach in separating microplastics from environmental samples based on their electrostatic behavior. Environ Pollut 234:20–28. https://doi.org/10.1016/j.envpol.2017.11.013
Feng S, Lu H, Liu Y (2021) The occurrence of microplastics in farmland and grassland soils in the Qinghai-Tibet plateau: different land use and mulching time in facility agriculture. Environ Pollut 279. https://doi.org/10.1016/j.envpol.2021.116939
Foekema EM, De Gruijter C, Mergia MT et al (2013) Plastic in north sea fish. Environ Sci Technol 47:8818–8824. https://doi.org/10.1021/es400931b
Franco AA, Arellano JM, Albendín G et al (2021) Microplastic pollution in wastewater treatment plants in the city of Cádiz: Abundance, removal efficiency and presence in receiving water body. Sci Total Environ 776. https://doi.org/10.1016/j.scitotenv.2021.145795
Fred-Ahmadu OH, Bhagwat G, Oluyoye I et al (2020) Interaction of chemical contaminants with microplastics: principles and perspectives. Sci Total Environ 706
Fries E, Dekiff JH, Willmeyer J et al (2013) Identification of polymer types and additives in marine microplastic particles using pyrolysis-GC/MS and scanning electron microscopy. Environ Sci Process Impacts 15:1949–1956. https://doi.org/10.1039/c3em00214d
Frydkjær CK, Iversen N, Roslev P (2017) Ingestion and egestion of microplastics by the Cladoceran Daphnia magna: effects of regular and irregular shaped plastic and sorbed phenanthrene. Bull Environ Contam Toxicol 99:655–661. https://doi.org/10.1007/s00128-017-2186-3
Fuller S, Gautam A (2016) A procedure for measuring microplastics using pressurized fluid extraction. Environ Sci Technol 50:5774–5780. https://doi.org/10.1021/acs.est.6b00816
Galafassi S, Nizzetto L, Volta P (2019) Plastic sources: a survey across scientific and grey literature for their inventory and relative contribution to microplastics pollution in natural environments, with an emphasis on surface water. Sci Total Environ 693:133499. https://doi.org/10.1016/j.scitotenv.2019.07.305
Gasperi J, Wright SL, Dris R et al (2018) ScienceDirect microplastics in air: are we breathing it in ? Curr Opin Environ Sci Heal 1:1–5. https://doi.org/10.1016/j.coesh.2017.10.002
Gaylarde CC, Neto JAB, da Fonseca EM (2021) Paint fragments as polluting microplastics: a brief review. Mar Pollut Bull 162. https://doi.org/10.1016/j.marpolbul.2020.111847
Geyer R, Jambeck JR, Law KL (2017) Production, use, and fate of all plastics ever made. Sci Adv 3:25–29. https://doi.org/10.1126/sciadv.1700782

Godoy V, Blázquez G, Calero M et al (2019) The potential of microplastics as carriers of metals. Environ Pollut 255. https://doi.org/10.1016/j.envpol.2019.113363

González-Pleiter M, Pedrouzo-Rodríguez A, Verdú I et al (2021) Microplastics as vectors of the antibiotics azithromycin and clarithromycin: effects towards freshwater microalgae. Chemosphere 268. https://doi.org/10.1016/j.chemosphere.2020.128824

Gopinath K, Seshachalam S, Neelavannan K et al, Quantification of microplastic in Red Hills Lake of Chennai city, Tamil Nadu, India. https://doi.org/10.1007/s11356-020-09622-2/Published

Guan J, Qi K, Wang J et al (2020) Microplastics as an emerging anthropogenic vector of trace metals in freshwater: Significance of biofilms and comparison with natural substrates. Water Res 184:116205. https://doi.org/10.1016/j.watres.2020.116205

Guo X, Liu Y, Wang J (2019) Sorption of sulfamethazine onto different types of microplastics: a combined experimental and molecular dynamics simulation study. Mar Pollut Bull 145:547–554. https://doi.org/10.1016/j.marpolbul.2019.06.063

Guo X, Wang J (2019) Sorption of antibiotics onto aged microplastics in freshwater and seawater. Mar Pollut Bull 149. https://doi.org/10.1016/j.marpolbul.2019.110511

Guzzetti E, Sureda A, Tejada S, Faggio C (2018) Microplastic in marine organism: environmental and toxicological effects. Environ Toxicol Pharmacol 64:164–171

Han Y, Cheng J, Tang Z et al (2021) Widespread occurrence of phthalates in popular take-out food containers from China and the implications for human exposure. J Clean Prod 290:125851. https://doi.org/10.1016/j.jclepro.2021.125851

Hatinoğlu MD, Sanin FD (2021) Sewage sludge as a source of microplastics in the environment: a review of occurrence and fate during sludge treatment. J Environ Manage 295:113028. https://doi.org/10.1016/j.jenvman.2021.113028

He P, Chen L, Shao L et al (2019) Municipal solid waste (MSW)landfill: a source of microplastics? Evidence of microplastics in landfill leachate. Water Res 159:38–45. https://doi.org/10.1016/j.watres.2019.04.060

Helcoski R, Yonkos LT, Sanchez A, Baldwin AH (2020) Wetland soil microplastics are negatively related to vegetation cover and stem density. Environ Pollut 256. https://doi.org/10.1016/j.envpol.2019.113391

Helm P, Athey S, Rochman CM (2020) Microplastics entering northwestern Lake Ontario are diverse and linked to urban sources Jelena Grbi 174. https://doi.org/10.1016/j.watres.2020.115623

Henry B, Laitala K, Klepp IG (2019) Microfibres from apparel and home textiles: prospects for including microplastics in environmental sustainability assessment. Sci Total Environ 652:483–494. https://doi.org/10.1016/j.scitotenv.2018.10.166

Herrera A, Raymond E, Martínez I et al (2020) First evaluation of neustonic microplastics in the Macaronesian region, NE Atlantic. Mar Pollut Bull 153. https://doi.org/10.1016/j.marpolbul.2020.110999

Hodson ME, Duffus-Hodson CA, Clark A et al (2017) Plastic bag derived-microplastics as a vector for metal exposure in terrestrial invertebrates. Environ Sci Technol 51:4714–4721. https://doi.org/10.1021/acs.est.7b00635

Hong SH, Shim WJ, Jang M (2018) Chemicals associated with marine plastic debris and microplastics: Analyses and contaminant levels. In: Microplastic contamination in aquatic environments: an emerging matter of environmental urgency. Elsevier, pp 271–315

Horton AA, Svendsen C, Williams RJ et al (2017) Large microplastic particles in sediments of tributaries of the River Thames, UK—abundance, sources and methods for effective quantification. Mar Pollut Bull 114:218–226. https://doi.org/10.1016/j.marpolbul.2016.09.004

Huang Y, Liu Q, Jia W et al (2020) Agricultural plastic mulching as a source of microplastics in the terrestrial environment. Environ Pollut 260. https://doi.org/10.1016/j.envpol.2020.114096

Huang D, Li X, Ouyang Z et al (2021a) The occurrence and abundance of microplastics in surface water and sediment of the West River downstream, in the south of China. Sci Total Environ 756. https://doi.org/10.1016/j.scitotenv.2020.143857

Huang Y, He T, Yan M, et al (2021b) Atmospheric transport and deposition of microplastics in a subtropical urban environment. J Hazard Mater 416. https://doi.org/10.1016/j.jhazmat.2021.126168

Hüffer T, Hofmann T (2016) Sorption of non-polar organic compounds by micro-sized plastic particles in aqueous solution. Environ Pollut 214:194–201. https://doi.org/10.1016/j.envpol.2016.04.018

Imhof HK, Ivleva NP, Schmid J et al (2013) Contamination of beach sediments of a subalpine lake with microplastic particles. Curr Biol 23:R867–R868. https://doi.org/10.1016/j.cub.2013.09.001

Islam N, Garcia da Fonseca T, Vilke J et al (2021) Perfluorooctane sulfonic acid (PFOS) adsorbed to polyethylene microplastics: accumulation and ecotoxicological effects in the clam Scrobicularia plana. Mar Environ Res 164. https://doi.org/10.1016/j.marenvres.2020.105249

Isobe A, Iwasaki S, Uchida K, Tokai T (2019) Abundance of non-conservative microplastics in the upper ocean from 1957 to 2066. Nat Commun 10. https://doi.org/10.1038/s41467-019-08316-9

Jambeck JR, Geyer R, Wilcox C et al (2015) Plastic waste inputs from land into the ocean. Science (80-) 347:768–771. https://doi.org/10.1126/science.1260352

Jiang X, Chang Y, Zhang T et al (2020) Toxicological effects of polystyrene microplastics on earthworm (Eisenia fetida). Environ Pollut 259. https://doi.org/10.1016/j.envpol.2019.113896

Johnson S, Saxena P, Sahu R (2015) Leaching of Bisphenol A from Baby Bottles. Proc Natl Acad Sci India Sect B—Biol Sci 85:131–135. https://doi.org/10.1007/s40011-013-0246-y

Kanhai LDK, Officer R, Lyashevska O et al (2017) Microplastic abundance, distribution and composition along a latitudinal gradient in the Atlantic Ocean. Mar Pollut Bull 115:307–314. https://doi.org/10.1016/j.marpolbul.2016.12.025

Karami A, Golieskardi A, Choo CK et al (2017) A high-performance protocol for extraction of microplastics in fish. Sci Total Environ 578:485–494. https://doi.org/10.1016/j.scitotenv.2016.10.213

Karbalaei S, Hanachi P, Walker TR, Cole M (2018) Occurrence, sources, human health impacts and mitigation of microplastic pollution. Environ Sci Pollut Res 25:36046–36063. https://doi.org/10.1007/s11356-018-3508-7

Kashiwabara LM, Kahane-Rapport SR, King C et al (2021) Microplastics and microfibers in surface waters of Monterey Bay National Marine Sanctuary, California. Mar Pollut Bull 165. https://doi.org/10.1016/j.marpolbul.2021.112148

Kastner J, Cooper DG, Marić M et al (2012) Aqueous leaching of di-2-ethylhexyl phthalate and "green" plasticizers from poly(vinyl chloride). Sci Total Environ 432:357–364. https://doi.org/10.1016/j.scitotenv.2012.06.014

Kedzierski M, Le V, Bourseau P et al (2016) Microplastics elutriation from sandy sediments: a granulometric approach. MPB 107:315–323. https://doi.org/10.1016/j.marpolbul.2016.03.041

Keresztes S, Tatár E, Czégény Z et al (2013) Study on the leaching of phthalates from polyethylene terephthalate bottles into mineral water. Sci Total Environ 458–460:451–458. https://doi.org/10.1016/j.scitotenv.2013.04.056

Kim YJ, Osako M, Sakai S, ichi (2006) Leaching characteristics of polybrominated diphenyl ethers (PBDEs) from flame-retardant plastics. Chemosphere 65:506–513. https://doi.org/10.1016/j.chemosphere.2006.01.019

Kim IS, Chae DH, Kim SK et al (2015) Factors influencing the spatial variation of microplastics on high-tidal coastal beaches in Korea. Arch Environ Contam Toxicol 69. https://doi.org/10.1007/s00244-015-0155-6

Klein S, Dimzon IK, Eubeler J, Knepper TP (2018) Analysis, occurrence, and degradation of microplastics in the aqueous environment. In: Handbook of environmental chemistry. Springer, pp 51–67

Klein M, Fischer EK (2019) Microplastic abundance in atmospheric deposition within the Metropolitan area of Hamburg, Germany. Sci Total Environ 685:96–103. https://doi.org/10.1016/j.scitotenv.2019.05.405

Li J, Liu H, Paul Chen J (2018) Microplastics in freshwater systems: a review on occurrence, environmental effects, and methods for microplastics detection. Water Res 137:362–374. https://doi.org/10.1016/j.watres.2017.12.056

Li D, Shi Y, Yang L et al (2020) Microplastic release from the degradation of polypropylene feeding bottles during infant formula preparation. Nat Food 1:746–754. https://doi.org/10.1038/s43016-020-00171-y

Li C, Busquets R, Campos LC (2020a) Assessment of microplastics in freshwater systems: a review. Sci Total Environ 707:135578. https://doi.org/10.1016/j.scitotenv.2019.135578

Li W, Wufuer R, Duo J et al (2020c) Microplastics in agricultural soils: Extraction and characterization after different periods of polythene film mulching in an arid region. Sci Total Environ 749. https://doi.org/10.1016/j.scitotenv.2020.141420

Liao C, Kannan K (2011) Widespread occurrence of bisphenol A in paper and paper products: implications for human exposure. Environ Sci Technol 45:9372–9379. https://doi.org/10.1021/es202507f

Liao R, Jiang J, Li Y et al (2020) Distribution and leaching behavior of organophosphorus and brominated flame retardants in soil in Chengdu. Environ Sci Process Impacts 22:1295–1305. https://doi.org/10.1039/d0em00106f

Liao YL, Yang JY (2020) Microplastic serves as a potential vector for Cr in an in-vitro human digestive model. Sci Total Environ 703:134805. https://doi.org/10.1016/j.scitotenv.2019.134805

Liu G, Zhu Z, Yang Y et al (2019) Sorption behavior and mechanism of hydrophilic organic chemicals to virgin and aged microplastics in freshwater and seawater. Environ Pollut 246:26–33. https://doi.org/10.1016/j.envpol.2018.11.100

Liu S, Shi J, Wang J et al (2021) Interactions Between Microplastics and Heavy Metals in Aquatic Environments: A Review. Front Microbiol 12:1–14. https://doi.org/10.3389/fmicb.2021.652520

Liubartseva S, Coppini G, Lecci R, Creti S (2016) Regional approach to modeling the transport of floating plastic debris in the Adriatic Sea. Mar Pollut Bull 103:115–127. https://doi.org/10.1016/j.marpolbul.2015.12.031

Lopes C, Raimundo J, Caetano M, Garrido S (2020) Microplastic ingestion and diet composition of planktivorous fish. Limnol Oceanogr Lett 5:103–112. https://doi.org/10.1002/lol2.10144

Lu Y, Zhang Y, Deng Y et al (2016) Uptake and Accumulation of Polystyrene Microplastics in Zebrafish (Danio rerio) and Toxic Effects in Liver. Environ Sci Technol 50:4054–4060. https://doi.org/10.1021/acs.est.6b00183

Luo H, Xiang Y, He D et al (2019) Leaching behavior of fluorescent additives from microplastics and the toxicity of leachate to Chlorella vulgaris. Sci Total Environ 678:1–9. https://doi.org/10.1016/j.scitotenv.2019.04.401

Maes T, Jessop R, Wellner N et al (2017) A rapid-screening approach to detect and quantify microplastics based on fluorescent tagging with Nile Red. Sci Rep 7. https://doi.org/10.1038/srep44501

Magni S, Binelli A, Pittura L et al (2019) The fate of microplastics in an Italian Wastewater Treatment Plant. Sci Total Environ 652:602–610. https://doi.org/10.1016/j.scitotenv.2018.10.269

Mao Y, Ai H, Chen Y et al (2018) Phytoplankton response to polystyrene microplastics: Perspective from an entire growth period. Chemosphere 208:59–68. https://doi.org/10.1016/j.chemosphere.2018.05.170

Marine Debris Program N (2015) Laboratory Methods for the Analysis of Microplastics in the Marine Environment: Recommendations for quantifying synthetic particles in waters and sediments

Marques GA, Tenório JAS (2000) Use of froth flotation to separate PVC/PET mixtures. Waste Manag 20:265–269. https://doi.org/10.1016/S0956-053X(99)00333-5

WHO (2019) Marsden P, Koelmans AA, Bourdon-Lacombe J, Gouin T, Anglada LD', Cunliffe D, Jarvis P, Fawell J, France J De (2019) Microplastics in drinking water

Martín C, Fajardo C, Costa G et al (2021) Bioassays to assess the ecotoxicological impact of polyethylene microplastics and two organic pollutants, simazine and ibuprofen. Chemosphere 274. https://doi.org/10.1016/j.chemosphere.2021.129704

Möller JN, Löder MGJ, Laforsch C (2020) Finding Microplastics in Soils: A Review of Analytical Methods. Environ Sci Technol 54:2078–2090. https://doi.org/10.1021/acs.est.9b04618

Morét-Ferguson S, Law KL, Proskurowski G et al (2010) The size, mass, and composition of plastic debris in the western North Atlantic Ocean. Mar Pollut Bull 60:1873–1878. https://doi.org/10.1016/j.marpolbul.2010.07.020

Ngo PL, Pramanik BK, Shah K, Roychand R (2019) Pathway, classification and removal efficiency of microplastics in wastewater treatment plants. Environ Pollut 255. https://doi.org/10.1016/j.envpol.2019.113326

Obbard RW, Sadri S, Wong YQ et al (2014) Global warming releases microplastic legacy frozen in Arctic Sea ice. Earth's Futur 2:315–320. https://doi.org/10.1002/2014ef000240

Oliveri Conti G, Ferrante M, Banni M, et al (2020) Micro- and nano-plastics in edible fruit and vegetables. The first diet risks ass]essment for the general population. Environ Res 187. https://doi.org/10.1016/j.envres.2020.109677

Osako M, Kim YJ, Sakai SI (2004) Leaching of brominated flame retardants in leachate from landfills in Japan. Chemosphere 57:1571–1579. https://doi.org/10.1016/j.chemosphere.2004.08.076

Patchaiyappan A, Dowarah K, Zaki Ahmed S, et al (2021) Prevalence and characteristics of microplastics present in the street dust collected from Chennai metropolitan city, India. Chemosphere 269:128757. https://doi.org/10.1016/j.chemosphere.2020.128757

Patti TB, Fobert EK, Reeves SE, Burke da Silva K (2020) Spatial distribution of microplastics around an inhabited coral island in the Maldives, Indian Ocean. Sci Total Environ 748. https://doi.org/10.1016/j.scitotenv.2020.141263

Pedà C, Caccamo L, Fossi MC et al (2016) Intestinal alterations in European sea bass Dicentrarchus labrax (Linnaeus, 1758) exposed to microplastics: Preliminary results. Environ Pollut 212:251–256. https://doi.org/10.1016/j.envpol.2016.01.083

Pham TH, Do HT, Phan Thi LA et al (2021) Global challenges in microplastics: from fundamental understanding to advanced degradations toward sustainable strategies. Chemosphere 267. https://doi.org/10.1016/j.chemosphere.2020.129275

Phillips MB, Bonner TH (2015) Occurrence and amount of microplastic ingested by fishes in watersheds of the Gulf of Mexico. Mar Pollut Bull 100:264–269. https://doi.org/10.1016/j.marpolbul.2015.08.041

Phuc N, Yasuhiro T (2011) Assessment of plastic waste generation and its potential recycling of household solid waste in Can Tho City, Vietnam, pp 23–35. https://doi.org/10.1007/s10661-010-1490-8

Picó Y, Barceló D (2019) Analysis and prevention of microplastics pollution in water: current perspectives and future directions. ACS Omega 4:6709–6719. https://doi.org/10.1021/acsomega.9b00222

Prata JC (2018) Microplastics in wastewater: State of the knowledge on sources, fate and solutions. Mar Pollut Bull 129:262–265. https://doi.org/10.1016/j.marpolbul.2018.02.046

Prata JC (2018) Airborne microplastic: consequences to human health? Environ Pollut 234:115–126. https://doi.org/10.1016/j.envpol.2017.11.043

Prata JC, da Costa JP, Duarte AC, Rocha-Santos T (2019) Methods for sampling and detection of microplastics in water and sediment: A critical review. TrAC - Trends Anal Chem 110:150–159. https://doi.org/10.1016/j.trac.2018.10.029

Prata JC, da Costa JP, Lopes I et al (2020) Environmental exposure to microplastics: an overview on possible human health effects. Sci Total Environ 702:134455. https://doi.org/10.1016/j.scitotenv.2019.134455

Praveena SM, Shaifuddin SNM, Akizuki S (2018) Exploration of microplastics from personal care and cosmetic products and its estimated emissions to marine environment: an evidence from Malaysia. Mar Pollut Bull 136:135–140. https://doi.org/10.1016/j.marpolbul.2018.09.012

Provencher JF, Bond AL, Avery-Gomm S et al (2017) Quantifying ingested debris in marine megafauna: A review and recommendations for standardization. Anal Methods 9:1454–1469

Qiu Q, Peng J, Yu X et al (2015) Occurrence of microplastics in the coastal marine environment: First observation on sediment of China. Mar Pollut Bull 98:274–280. https://doi.org/10.1016/j.marpolbul.2015.07.028

Qu X, yan, Li Z shan, Xie X yuan, et al (2009) Survey of composition and generation rate of household wastes in Beijing, China. Waste Manag 29:2618–2624. https://doi.org/10.1016/j.wasman.2009.05.014

Ragusa A, Svelato A, Santacroce C, et al (2021) Plasticenta: First evidence of microplastics in human placenta. Environ. Int. 146

Rajasärkkä J, Pernica M, Kuta J et al (2016) Drinking water contaminants from epoxy resin-coated pipes: a field study. Water Res 103:133–140. https://doi.org/10.1016/j.watres.2016.07.027

Rastkari N, Jeddi MZ, Yunesian M, Ahmadkhaniha R (2018) Effect of sunlight exposure on phthalates migration from plastic containers to packaged juices. J Environ Heal Sci Eng 16:27–33. https://doi.org/10.1007/s40201-018-0292-8

Rillig MC, Bonkowski M (2018) Microplastic and soil protists: a call for research. Environ Pollut 241:1128–1131

Rillig MC, Ziersch L, Hempel S (2017) Microplastic transport in soil by earthworms. Sci Rep 7:1–6. https://doi.org/10.1038/s41598-017-01594-7

Rocha-Santos T, Duarte AC (2017) Characterization and analysis of microplastics. Elsevier

Rolsky C, Kelkar V, Driver E, Halden RU (2020) Science direct municipal sewage sludge as a source of microplastics in the environment. Curr Opin Environ Sci Heal 14:16–22. https://doi.org/10.1016/j.coesh.2019.12.001

Romera-Castillo C, Pinto M, Langer TM et al (2018) Leaching of bisphenol A (BPA) from polycarbonate plastic to water containing amino acids and its degradation by radical oxygen species. Nat Commun 55:861–867. https://doi.org/10.1016/j.chemosphere.2003.11.065

Ryberg MW, Hauschild MZ, Wang F et al (2019) Global environmental losses of plastics across their value chains. Resour Conserv Recycl 151:104459. https://doi.org/10.1016/j.resconrec.2019.104459

Sanchez W, Bender C, Porcher JM (2014) Wild gudgeons (Gobio gobio) from French rivers are contaminated by microplastics: preliminary study and first evidence. Environ Res 128:98–100. https://doi.org/10.1016/j.envres.2013.11.004

Sarafraz J, Rajabizadeh M, Kamrani E (2016) The preliminary assessment of abundance and composition of marine beach debris in the northern Persian Gulf, Bandar Abbas City. Iran. J Mar Biol Assoc United Kingdom 96:131–135. https://doi.org/10.1017/S0025315415002076

Sathish N, Jeyasanta KI, Patterson J (2019) Abundance, characteristics and surface degradation features of microplastics in beach sediments of five coastal areas in Tamil Nadu, India. Mar Pollut Bull 142:112–118. https://doi.org/10.1016/j.marpolbul.2019.03.037

Scherer C, Weber A, Stock F, et al (2020) Comparative assessment of microplastics in water and sediment of a large European river. Sci Total Environ 738. https://doi.org/10.1016/j.scitotenv.2020.139866

Schrank I, Trotter B, Dummert J et al (2019) Effects of microplastic particles and leaching additive on the life history and morphology of Daphnia magna. Environ Pollut 255. https://doi.org/10.1016/j.envpol.2019.113233

Selvam S, Jesuraja K, Venkatramanan S et al (2021) Hazardous microplastic characteristics and its role as a vector of heavy metal in groundwater and surface water of coastal south India. J Hazard Mater 402. https://doi.org/10.1016/j.jhazmat.2020.123786

Selvamurugan Muthusamy M, Pramasivam S (2019) Bioplastics—an eco-friendly alternative to petrochemical plastics. Curr World Environ 14:49–59. https://doi.org/10.12944/cwe.14.1.07

Senathirajah K, Attwood S, Bhagwat G et al (2021) Estimation of the mass of microplastics ingested—a pivotal first step towards human health risk assessment. J Hazard Mater 404. https://doi.org/10.1016/j.jhazmat.2020.124004

Seth CK, Shriwastav A (2018) Contamination of Indian sea salts with microplastics and a potential prevention strategy. Environ Sci Pollut Res 25:30122–30131. https://doi.org/10.1007/s11356-018-3028-5

Sheng Y, Ye X, Zhou Y, Li R (2021) Microplastics (MPs) Act as sources and vector of pollutants-impact hazards and preventive measures. Bull Environ Contam Toxicol. https://doi.org/10.1007/s00128-021-03226-3

Shim WJ, Song YK, Hong SH, Jang M (2016) Identi fi cation and quanti fi cation of microplastics using Nile Red staining. https://doi.org/10.1016/j.marpolbul.2016.10.049

Shim WJ, Hong H, Eo S (2017) Analytical Methods Identi fi cation methods in microplastic analysis: 1384–1391. https://doi.org/10.1039/c6ay02558g

Shruti VC, Pérez-Guevara F, Elizalde-Martínez I, Kutralam-Muniasamy G (2020a) First study of its kind on the microplastic contamination of soft drinks, cold tea and energy drinks—Future research and environmental considerations. Sci Total Environ 726. https://doi.org/10.1016/j.scitotenv.2020.138580

Shruti VC, Pérez-Guevara F, Kutralam-Muniasamy G (2020b) Metro station free drinking water fountain—a potential "microplastics hotspot" for human consumption. Environ Pollut 261. https://doi.org/10.1016/j.envpol.2020.114227

Siegfried M, Koelmans AA, Besseling E, Kroeze C (2017) Export of microplastics from land to sea. A modelling approach. Water Res 127:249–257. https://doi.org/10.1016/j.watres.2017.10.011

Sillanpää M, Sainio P (2017) Release of polyester and cotton fibers from textiles in machine washings. Environ Sci Pollut Res 24:19313–19321. https://doi.org/10.1007/s11356-017-9621-1

Sobhani Z, Lei Y, Tang Y et al (2020) Microplastics generated when opening plastic packaging. Sci Rep 10:4841. https://doi.org/10.1038/s41598-020-61146-4

Sridharan S, Kumar M, Bolan NS, et al (2021) Are microplastics destabilizing the global network of terrestrial and aquatic ecosystem services? Environ. Res. 198

Stringer R, Labunska I, Santillo D et al (2000) Concentrations of phthalate esters and identification of other additives in PVC children's toys. Environ Sci Pollut Res 7:27–36. https://doi.org/10.1065/espr199910.007

Sun X, Liang J, Zhu M et al (2018) Microplastics in seawater and zooplankton from the Yellow Sea. Environ Pollut 242:585–595. https://doi.org/10.1016/j.envpol.2018.07.014

Sun B, Liu J, Zhang YQ et al (2021a) Leaching of polybrominated diphenyl ethers from microplastics in fish oil: kinetics and bioaccumulation. J Hazard Mater 406:124726. https://doi.org/10.1016/j.jhazmat.2020.124726

Sun P, Liu X, Zhang M et al (2021b) Sorption and leaching behaviors between aged MPs and BPA in water: the role of BPA binding modes within plastic matrix. Water Res 195:116956. https://doi.org/10.1016/j.watres.2021.116956

Szewc K, Graca B, Dołęga A (2021) Atmospheric deposition of microplastics in the coastal zone: characteristics and relationship with meteorological factors. Sci Total Environ 761. https://doi.org/10.1016/j.scitotenv.2020.143272

Tang Y, Liu Y, Chen Y, et al (2020a) A review: Research progress on microplastic pollutants in aquatic environments. Sci Total Environ

Tang Y, Rong J, Guan X et al (2020b) Immunotoxicity of microplastics and two persistent organic pollutants alone or in combination to a bivalve species. Environ Pollut 258. https://doi.org/10.1016/j.envpol.2019.113845

Tiwari M, Rathod TD, Ajmal PY et al (2019) Distribution and characterization of microplastics in beach sand from three different Indian coastal environments. Mar Pollut Bull 140:262–273. https://doi.org/10.1016/j.marpolbul.2019.01.055

Treilles R, Gasperi J, Gallard A et al (2021) Microplastics and microfibers in urban runoff from a suburban catchment of Greater Paris. Environ Pollut 287:117352. https://doi.org/10.1016/j.envpol.2021.117352

UNEP (2018) Mapping of global plastics value chain and plastics losses to the environment (with a particular focus on marine environment). 1–99

URL-1 https://www.plasticseurope.org/en/resources/publications/4312-plastics-facts-2020 last accessed on 09/07/2021

URL-2 https://www.theguardian.com/environment/2020/aug/17/microplastic-particles-discovered-in-human-organs last accessed on 09/07/2021

van Wezel A, Caris I, Kools SAE (2016) Release of primary microplastics from consumer products to wastewater in the Netherlands. Environ Toxicol Chem 35:1627–1631. https://doi.org/10.1002/etc.3316
Veerasingam S, Saha M, Suneel V et al (2016) Characteristics, seasonal distribution and surface degradation features of microplastic pellets along the Goa coast, India. Chemosphere 159:496–505. https://doi.org/10.1016/j.chemosphere.2016.06.056
Viršek MK, Lovšin MN, Koren Š et al (2017) Microplastics as a vector for the transport of the bacterial fish pathogen species Aeromonas salmonicida. Mar Pollut Bull 125:301–309. https://doi.org/10.1016/j.marpolbul.2017.08.024
Wagner M, Lambert S (2018) Freshwater microplastics—the handbook of environmental chemistry 58
Wang W, Wang J (2018) Investigation of microplastics in aquatic environments: an overview of the methods used, from field sampling to laboratory analysis. TrAC—Trends Anal Chem 108:195–202
Wang F, Wong CS, Chen D et al (2018) Interaction of toxic chemicals with microplastics: a critical review. Water Res 139:208–219. https://doi.org/10.1016/j.watres.2018.04.003
Wang F, Yang W, Cheng P et al (2019) Adsorption characteristics of cadmium onto microplastics from aqueous solutions. Chemosphere 235:1073–1080. https://doi.org/10.1016/j.chemosphere.2019.06.196
Wang J, Li J, Wang Q, Sun Y (2020) Microplastics as a vector for HOC bioaccumulation in earthworm eisenia fetida in soil: importance of chemical diffusion and particle size. Environ Sci Technol 54:12154–12163. https://doi.org/10.1021/acs.est.0c03712
Wang T, Wang L, Chen Q et al (2020b) Interactions between microplastics and organic pollutants: Effects on toxicity, bioaccumulation, degradation, and transport. Sci Total Environ 748
Wang C, Zhao J, Xing B (2021) Environmental source, fate, and toxicity of microplastics. J Hazard Mater 407
Wei W, Huang QS, Sun J et al (2019) Polyvinyl Chloride Microplastics Affect Methane Production from the Anaerobic digestion of waste activated sludge through leaching toxic bisphenol-A. Environ Sci Technol 53:2509–2517. https://doi.org/10.1021/acs.est.8b07069
Wright SL, Ulke J, Font A et al (2020) Atmospheric microplastic deposition in an urban environment and an evaluation of transport. Environ Int 136:105411. https://doi.org/10.1016/j.envint.2019.105411
Wright SL, Kelly FJ (2017) Plastic and human health: a micro issue? Environ Sci Technol 51:6634–6647. https://doi.org/10.1021/acs.est.7b00423
Wu Y, Guo P, Zhang X et al (2019) Effect of microplastics exposure on the photosynthesis system of freshwater algae. J Hazard Mater 374:219–227. https://doi.org/10.1016/j.jhazmat.2019.04.039
Wu P, Huang J, Zheng Y et al (2019a) Ecotoxicology and Environmental Safety Environmental occurrences , fate , and impacts of microplastics. Ecotoxicol Environ Saf 184:109612. https://doi.org/10.1016/j.ecoenv.2019.109612
Xu Z, Xiong X, Zhao Y, et al (2020) Pollutants delivered every day: phthalates in plastic express packaging bags and their leaching potential. J Hazard Mater 384:121282. https://doi.org/10.1016/j.jhazmat.2019.121282
Yang Y, Liu W, Zhang Z et al (2020) Microplastics provide new microbial niches in aquatic environments. Appl Microbiol Biotechnol 104:6501–6511. https://doi.org/10.1007/s00253-020-10704-x
Yang J, Li R, Zhou Q, et al (2021a) Abundance and morphology of microplastics in an agricultural soil following long-term repeated application of pig manure. Environ Pollut 272. https://doi.org/10.1016/j.envpol.2020.116028
Yang Z, Lü F, Zhang H, et al (2021b) Is incineration the terminator of plastics and microplastics? J Hazard Mater 401:123429. https://doi.org/10.1016/j.jhazmat.2020.123429
Ye X, Wang P, Wu Y et al (2020) Microplastic acts as a vector for contaminants: the release behavior of dibutyl phthalate from polyvinyl chloride pipe fragments in water phase. Environ Sci Pollut Res 27:42082–42091. https://doi.org/10.1007/s11356-020-10136-0

Yu H, Peng J, Cao X et al (2021) Effects of microplastics and glyphosate on growth rate, morphological plasticity, photosynthesis, and oxidative stress in the aquatic species Salvinia cucullata. Environ Pollut 279. https://doi.org/10.1016/j.envpol.2021.116900

Yukioka S, Tanaka S, Nabetani Y et al (2020) Occurrence and characteristics of microplastics in surface road dust in Kusatsu (Japan), Da Nang (Vietnam), and Kathmandu (Nepal). Environ Pollut 256:113447. https://doi.org/10.1016/j.envpol.2019.113447

Zhang S, Yang X, Gertsen H et al (2018) A simple method for the extraction and identification of light density microplastics from soil. Sci Total Environ 616–617:1056–1065. https://doi.org/10.1016/j.scitotenv.2017.10.213

Zhang Y, Gao T, Kang S, Sillanpää M (2019) Importance of atmospheric transport for microplastics deposited in remote areas. Environ Pollut 254:1–4. https://doi.org/10.1016/j.envpol.2019.07.121

Zhang Y, Kang S, Allen S, et al (2020a) Atmospheric microplastics: A review on current status and perspectives. Earth-Science Rev. 203

Zhang Y, Kang S, Allen S et al (2020b) Atmospheric microplastics: a review on current status and perspectives. Earth-Science Rev 203:103118. https://doi.org/10.1016/j.earscirev.2020.103118

Zhang K, Hamidian AH, Tubić A, et al (2021) Understanding plastic degradation and microplastic formation in the environment: A review. Environ. Pollut. 274

Zhou X, Guo J, Lin K et al (2013) Leaching characteristics of heavy metals and brominated flame retardants from waste printed circuit boards. J Hazard Mater 246–247:96–102. https://doi.org/10.1016/j.jhazmat.2012.11.065

Zhou Q, Zhang H, Fu C et al (2018) The distribution and morphology of microplastics in coastal soils adjacent to the Bohai Sea and the Yellow Sea. Geoderma 322:201–208. https://doi.org/10.1016/j.geoderma.2018.02.015

Zhou Y, Yang Y, Liu G, et al (2020) Adsorption mechanism of cadmium on microplastics and their desorption behavior in sediment and gut environments: The roles of water pH, lead ions, natural organic matter and phenanthrene. Water Res 184:116209. https://doi.org/10.1016/j.watres.2020.116209

Ziajahromi S, Neale PA, Leusch FDL (2016) Wastewater treatment plant effluent as a source of microplastics: review of the fate, chemical interactions and potential risks to aquatic organisms. Water Sci Technol 74:2253–2269. https://doi.org/10.2166/wst.2016.414

Ziajahromi S, Neale PA, Telles Silveira I et al (2021) An audit of microplastic abundance throughout three Australian wastewater treatment plants. Chemosphere 263:128294. https://doi.org/10.1016/j.chemosphere.2020.128294

Chapter 4
Microplastics in Terrestrial Soils: Occurrence, Analysis, and Remediation

Shefali Chauhan and Sadia Basri

Abstract The production of synthetic polymers has increased in manifolds since the advancement of petroleum engineering. These cosmopolitan artificial materials are now persistent in almost every ecosystem on planet earth. This could be attributed to the higher tendency of polymer disintegration and lower polymer degradation in the ambient environments. The processes of polymer disintegration lead to the formation of microscopic plastic particles, referred to as 'microplastics'. Significant research in emerging environmental contaminants focuses on microplastics, their occurrence in terrestrial and aquatic ecosystems, and adverse effects on biota, humans and other environmental compartments. Therefore, taking account of the background, this chapter summarizes the different strategies for the chemical extraction of microplastics from soil samples and further qualitative and quantitative analysis via distinct spectrophotometric techniques. All over the world, rivers, lakes, and associated floodplains are some of the most polluted ecosystems. The Indian rivers, including Meghna, Brahmaputra and Ganges are polluted with 72,845 tonnes of plastic loads per annum which eventually is transported into the oceans. The existing literature also presents a picture that microplastic pollution studies are usually focused on aquatic ecosystems such as oceans, lakes and rivers. However, microplastic pollution in floodplain soils is a major neglected research question. Therefore, this chapter aims to focus on the occurrence of microplastics in terrestrial soils such as floodplains, their origin and sources, and how to extract and quantify microplastic presence in the geological samples. Furthermore, microplastics are non-uniform in terms of their physicochemical characteristics and degradation. Numerous forms of microplastics pollute aquatic and terrestrial ecosystems. Therefore, an emphasis is also laid down on the microplastic control and remediation techniques in these natural ecosystems for scientific inquiry.

Keywords Synthetic polymers · Microplastics · Floodplains · Sediments

S. Chauhan (✉) · S. Basri
School of Environmental Sciences (SES), Jawaharlal Nehru University, New Delhi 110067, India

S. P. Singh et al. (eds.), *New Trends in Emerging Environmental Contaminants*, Energy, Environment, and Sustainability,
https://doi.org/10.1007/978-981-16-8367-1_4

4.1 Introduction

The global annual plastic production has escalated to 368 million metric tonnes in 2019, with almost an increase of 9 million metric tonnes as of 2018 (Tiseo 2021; Klemeš et al. 2021). These simple to complex synthetic polymers have become emerging environmental pollutants in various terrestrial and aquatic ecosystems posing a serious threat to the sustainability and viability of these natural biosystems. The large plastic debris is today's major concern of environmental pollution throughout the world, however there is an increasing interest of environmental researchers about the environmental and toxicological impacts of tiny plastic particles called **'Microplastics'**. Microplastics can be found in different environmental domains of sediment, water, biota, air/dust and salt (Ranjani et al. 2021). Microplastics occur in the heterogeneous forms of fragment, fiber, pellet, film, granules, bead or foam, with sizes ranging from 100 nm to 5 mm (Sarkar et al. 2019). The major sources of microplastics are polymer nanoparticles manufactured for cosmetic products and printer inks known as primary microplastics. Secondary microplastics are generated via fragmentation of macro-plastics by mechanical abrasion, photo-dissociation and microbial degradation, and discharges from wastewater treatment plant (Karthik et al. 2018; Pico et al. 2019). Plastic fragmentation in terrestrial compartments can be caused by physical or chemical actions such as UV exposure, higher temperatures and physical forces, whereas fragmentation in aquatic ecosystems is attributed to the bacterial degradation (Ricciardi et al. 2021) (Fig. 4.1).

Microplastics have been considered as cosmopolitan particles by few researchers as they have been detected in some very remote habitats. Microplastics presence

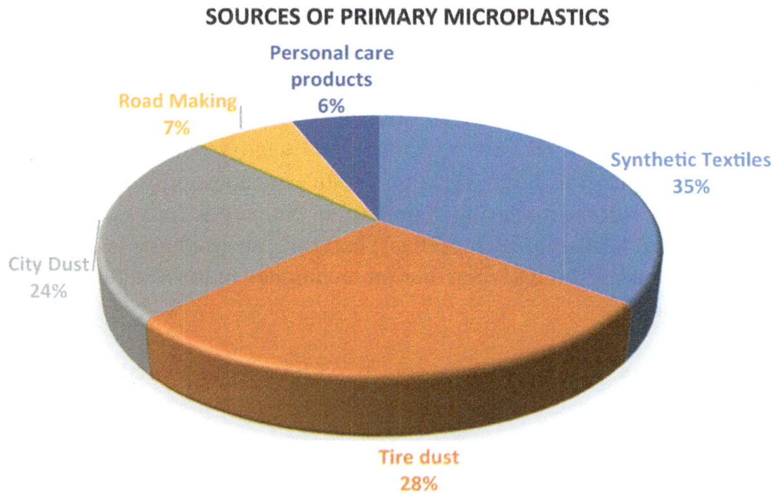

Fig. 4.1 Global evaluation of sources of primary microplastics in the oceans (European Commission 2018)

in different environments is based on human activities, inherent natural conditions, sampling locations and approaches. Wastewater treatment plants are one of the most dominant sources of microplastics in terrestrial ecosystems (Li et al. 2018), however, river and coastal plastic pollution is the main source for microplastics in oceans. When biological interactions are concerned, microplastics can be bio-accumulated in organisms through different food chains in terrestrial as well as in aquatic ecosystems, leading to toxicological effects in organisms and humans (Yang et al. 2021). Furthermore, these particles are also reported for their adsorbing potential for Persistent Organic Pollutants (POP's), Polyaromatic Hydrocarbons (PAH's), and Polycyclic Biphenyls (PCB's) in their polymer structures (Sarkar et al. 2020), further enhances their bio-toxicity. The adsorption behaviour is attributed to their small size, large surface area to volume ratio, and their chemical composition and subsequent desorption of these chemicals from microplastics can cause severe harm to biota (Cole et al. 2011). According to Pandey et al. (2021), microplastics can interact chemically with the toxic contaminants like Arsenic and other heavy metals, leading to an increase in their concentrations in the sensitive North-Indian floodplains and develop a potential risk for a large population. Meanwhile, the first successful attempt for the assessment of microplastics in floodplain soils is worth mentioning in this chapter. Scheurer and Bigalke (2018) assessed the presence of microplastics in the Swiss floodplain soils, and their results suggested that 90% of Swiss floodplain soils contained microplastics. Therefore, it can be vividly said that microplastic presence in floodplain soils is an emerging but prevalent environmental contamination concern.

Microplastics are ubiquitous, persistent and bio-toxic and therefore it is essential to understand their accumulation, sources and occurrence in different aquatic and terrestrial ecosystems. A vast literature exists for freshwater and marine ecosystems, however, terrestrial ecosystems such as agro-ecosystems and floodplains are still to be explored for microplastic occurrence and impacts, especially the densely populated areas of fast-developing countries like the Indo-Gangetic plains of India. Presently, India considerably lacks the documentation of plastic pollution data for the Ganges River and its associated floodplains (Sarkar et al. 2019). The Indo-Gangetic plains are the most densely populated north-Indian region comprising of the plains of river Ganges, Indus River plains and the short plain of Brahmaputra River. These rivers hold enormous importance related to cultural, religious and industrial significance in northern India. Accordingly, it is evident that rapid urbanization, industrialization, and population explosion in these plains are the major cause of microplastic pollution associated with rivers and subsequently, floodplains. The existing literature suggests that field-based studies on microplastics were done for Ganges, Indus, and Brahmaputra River up to some extent (Sarkar et al. 2019; Tsering et al. 2021), however, there are very less significant data or research orientation on the microplastic pollution associated with the floodplains of these rivers. In this light, this chapter aims to outline the spatial occurrence of microplastics in the soils of terrestrial floodplains, analysis of microplastics from geogenic samples using different chemical and physical techniques, and the control measures and remediation for microplastic pollution.

4.2 Occurrence of Microplastics in the Floodplain Soils: An IGP Perspective

The Indo-Gangetic Plains are the extensive alluvial plains of the Ganges, Indus and Brahmaputra rivers and their tributaries, which separate the Himalayan Ranges from Peninsular India. These extensive fluvial floodplains cover the northern, eastern and western part of India, eastern Pakistan, Bangladesh and southern Nepal, covering approximately 43.7 million hectares of land in India and comprising of eight agro-ecological regions (AER) of India (Pal et al. 2009). The soil deposited by the Indus, Ganges and Brahmaputra River in the Indo-Gangetic plains is alluvial soil that consists of silt, sand and clay in different proportions. Furthermore, the soil texture in the Indo-Gangetic plains vary as per the climatic conditions, such as sandy loam in the western parts, silty loam in the central plains, silty clayey loam in the eastern parts, and silty clay texture in the southern parts of the plains (Srivastava et al. 2015). The Indo-Gangetic plains are among the most fertile lands in the world in terms of agricultural production, industrial significance and economic growth. According to Pal et al. (2009), the Indo-Gangetic plains contribute to approximately 50% of agricultural production of India and feeds about 40% population of the country. Therefore, the key drivers of environmental pollution in these floodplains are industrial invasion, deforestation, migration and human population growth. The major environmental concern in these overpopulated floodplains includes plastic waste management, and mitigation of air and water pollution. According to an Annual Report (2018–2019) on Plastic Waste Management Rules by Central Pollution Control Board (CPCB, New Delhi), the estimated annual plastic waste generation was 33,60,043 tons/annum during the year 2018–2019 and a total of 15,03,556 tons/annum generated solely in the Indo-Gangetic plains which account for approximately 45% of the total annual plastic waste generation in India (CPCB 2019). Therefore, it is an essential research question to address the occurrence of microplastics in the soils of the Indo-Gangetic plains.

Microplastics are introduced in the floodplain soils or any other soil environment via multiple routes and are mainly related to the anthropogenic activities (Sarkar et al. 2019). Several studies have reported varied concentrations of microplastics in the non-agricultural soils such as floodplains, with higher concentrations near urban or industrial areas (Dioses-Salinas et al. 2020). There are several sources which can be attributed for the microplastic accumulation in the floodplain soils, including continuing land application of sludge and organic fertilizers which have been detected with the presence of microplastic fibres (Zubris and Richards 2005), atmospheric deposition and atmospheric fallout of synthetic fibres on the soil surfaces (Dris et al. 2016), agricultural plastic films subjected to physical, chemical and biological degradation, and other anthropogenic sources, such as plastic littering, wastewater irrigation and surface runoff (Zhu et al. 2019). The floodplains are considered as a potential temporal sink for microplastics, therefore, in an attempt to understand the behaviour of microplastics in floodplain soils, Weber et al. (2021) presented the spatial heterogeneity within the floodplain soils of Lahn River, Germany, elaborating the deposition and in-situ vertical transfer of mesoplastics and coarse microplastics within these soils (Weber and Opp 2020).

The quantification of microplastic pollution in floodplains of Virginia was conducted by Christensen et al. (2020) in a view of analysing microplastic accumulation and transportation in the floodplain soils. The study concluded that microplastic abundance in floodplains was much higher as compared to the river channels and soil sampling at six different sites revealed that the size of microplastics in floodplain soils was bigger than in the stream channel. This study conducted by Christensen et al. (2020) proposes that floodplains are more vulnerable and abundant in microplastic concentrations, therefore, risk assessment and microplastics concentration quantification in the context of the Indo-Gangetic plains is an indispensable concern. The Ganges River is one of the most important catchments in the Indo-Gangetic plains, which has the highest plastic loads among Indian rivers. Despite this fact, microplastic contamination in the river Ganges is still a very limited issue in scientific studies. However, some research has contributed to the sediment analysis of the Ganges River to detect microplastics. The investigation carried out by Sarkar et al. (2019) highlighted the importance of the estimation of microplastics in terrestrial and river ecosystems. The study included the analysis of sediments of Ganges River on the lower stretch for the detection of meso and microplastics, using Fourier Transform Infrared Spectroscopy. Further, the conclusion relates that all the sediment samples contained microplastics and a total of six different morphotypes of microplastics were extracted from the sediment samples. The abundance of meso and microplastics in the sediments was attributed to the direct inflow of plastic debris through sewage and littering.

The Indo-Gangetic plains harbour a large population in the Indian sub-continent, whereas it is developing as an extremely sensitive floodplain due to its natural geogenic contaminants such as Arsenic and Fluoride, and on the other hand, the increasing presence of microplastics makes it more vulnerable due to anthropogenic activities. The natural tropical environment further impacts positively on the weathering ability of microplastics because sediments and temperature are both suitable for microplastic degradation processes (Pandey et al. 2021). The southwest monsoon is attributed for the nutrient and enrichment dynamics in the Indian sub-continent therefore, it is also the biggest driver for the transportation and mixing of microplastics in the soil strata of the Indo-Gangetic floodplains (Pandey et al. 2021). Therefore, it is undeniable that there is a potential threat to biological diversity, environment and human population due to microplastic pollution in the Indo-Gangetic plains, also owing to the favourable environmental conditions of the Indian sub-continent.

4.3 Extraction of Microplastics from Soils

4.3.1 Sample Collection Methods

The collection of soil samples is the most crucial step in quantifying environmental microplastic pollution. The sampling design should be distinctively strategic for heterogeneous medium such as floodplain soils containing microplastics to accomplish reliable results. Thereby, the objective of the research, the sampling site, as well as the sampling depths must be precisely defined in advance of the study (Moller et al. 2020). Based on the research objectives, numerous diverse sampling approaches might be used. In the case of microplastics in soils along the floodplain, transect sampling approach can be used to identify the contamination extent along with the linear features like roads, floodplains, drainage ditches (Moller et al. 2020). Therefore, the selection of sampling site would have a substantial impact on the amounts and types of microplastics present in the floodplain soils. Since microplastic particles are particulate in nature and vary strongly in their size, it is probable that the distribution of microplastics within the flood plain soil also varies significantly. To deal with this heterogeneity, composite samples (several discrete point samples of the same size within a spatial unit are combined and homogenized to a single one) taken from defined subunits within a sampling site may be helpful to obtain a more representative image of the degree of microplastic contamination and it minimizes the issues of transporting, storing and analysing numerous soil samples (Moller et al. 2020). Zhang and Liu 2018 applied this approach in their study of the distribution of microplastics in soil aggregate fraction in southwestern China by defining five plots at each sampling site, and taking six-point samples randomly from each plot and thereby, merging them into a single composite sample (Zhang and Liu 2018; Moller et al. 2020). Soil is defined as a solid medium; therefore, sampling depth is of supreme importance and should always be well-defined. The depth of soil sampling is considered based on the sample of concern like undisturbed soil, vertically disturbed soil by ploughing or microplastics at varying depth of soil. For the study of the distribution of microplastics on the surface of undisturbed soil, bulk sampling of the first few centimetres might be enough. And if the vertical distribution of microplastics after ploughing is the focus of the study, then sampling depth must be adapted accordingly. However, if the contamination of microplastics at different soil layers is to be estimated, then core sampling is considered for the study. Usually, the soil surface within the upper 10 cm is the usual range which contains a bulk of microplastic particles (Moller et al. 2020). Generally, 0–5 cm is preferred for sampling microplastics in the floodplain areas (Scheurer et al. 2018). The amount of sample (mass/volume) of each sampling point should also be determined. Due to the heterogeneous nature of microplastic distribution, the larger quantity of samples is more representative rather than the small quantity in the field and these samples can be reduced to smaller samples in the laboratory by homogenization and splitting processes. After considering all the points sincerely, required samples can be collected in the zip lock bags using steel tools efficiently. Precautions should be taken to avoid the contamination of samples from tools, clothing, or the ambient air of laboratories while carrying out the experimental process.

4.3.2 Extraction of Microplastics

Extraction of microplastics from the soil samples is a crucial step for the analysis and quantification of microplastic particles. The soil samples are firstly sun dried and sieved through a mesh (<5 mm), thereafter the sieved samples are further processed for the extraction of microplastics using methods such as density separation, manual extraction, oil extraction etc. The presence of organic materials on the surface of MPs can interfere in their counting and identification process. After extraction, it is necessary to remove the organic matter adhered to microplastics using 30% H_2O_2 solution. Some of the methods for the extraction of microplastics from soil samples are:

(a) **Manual extraction**: The simplest method for microplastic isolation is sieving and manual sorting, further using a stereo microscope to eliminate the mineral or biogenic matter. The most commonly applied technique for isolating microplastics from soil samples is on-site sieving and this technique has been successfully applied for the larger microplastic particles (>1 mm) (Moller et al. 2020).

(b) **Density separation**: Another technique for the extraction of microplastics from soil samples is the density separation method which utilizes high-density salt solutions as extraction media combined with the filtration process (Hanvey et al. 2017; Moller et al. 2020). The density separation method is based on the difference in the densities of different polymers. Van Cauwenberghe et al. (2015) suggested that the optimum density of the solution should be 1.6–1.8 g/cm^3, which could be attained by using $ZnCl_2$ or NaI salts (He et al. 2018). The frequently used $ZnCl_2$ (1.6–1.7 g/cm^3) salt solution is as an aqueous solvent for high-density polymers like polyethylene terephthalate (PET), polyvinyl chloride (PVC). Furthermore, stirring the solution containing high-density salt solution in the sample is an important step to ensure that the polymers are separated from each other. Stirring of the sample can either be done mechanically with a centrifuge or mechanical shaker or by vigorous manual shaking. Afterwards, the vigorous stirring is followed by the filtration process. The dry sorting process involves the sieving of the whole sample, whereas vacuum filtration is generally used for the wet sorting process. Wet sorting is done after density separation for the isolation of the floating microplastics by the vacuum filtration method. The filtrate consists of interfering with organic matter along with the concerned microplastics. Therefore, the final step after filtration is the digestion of organic matter using 30% H_2O_2 solution. Hydrogen peroxide is preferred for the digestion process as it does not alter the shape and size of the microplastic polymers (Qiu et al. 2016). One important aspect to mention is that all the organic matter is digested before the analysis of the sample.

(c) **Oil Extraction:** Crichton et al. developed a method for the extraction of microplastics from solid samples using oil due to the lipophilic surface characteristics of most of the plastics (Moller et al. 2020). Dried samples are mixed with water and canola oil, and stirred thoroughly and left to settle down until the oil, water, and mineral layers are completely separated. Microplastic particles in contact with the oil will be carried by the oil layer and can be transferred to the filter paper efficiently. The procedure is very simple, safe, cheap, and less time consuming but it requires a supplementary step for the removal of organic substances from the sample (Moller et al. 2020). After the extraction process, further analysis of microplastics is carried out which consists of physical and chemical characterization.

4.4 Microplastics Analysis in Soils

4.4.1 Quantitative Analysis of Microplastics

The microplastic analysis consists of two processes, namely (a) physical characterization of microplastics (microscopy) and (b) chemical characterization (spectroscopy) for the confirmation of plastics (Tiwari et al. 2019). The quantitative abundance of microplastics is estimated using microscopes, including binocular microscope, scanning electron microscope (SEM), dissected microscope, stereomicroscope and fluorescent microscope (Qiu et al. 2016). The microplastics can be of several types, such as Styrofoam, fibres, films, granules and pellets. Determination of microplastics visually should meet the mandatory criteria which include, (1) no cellular or organic structures are visible in the microplastics, (2) fibres should be equally thick throughout their entire length, (3) microplastics must show clear and homogeneous colours, and (4) if particles are transparent or white, they shall be examined with a microscope at high magnification, or using a fluorescence microscope to exclude the organic origin (Qiu et al. 2016). All these criteria should be fulfilled while analysing the samples using different microscopes.

Fluorescence Microscopy: The combination of fluorescence staining method and density separation provides a simple and sensitive approach for highlighting the most common polymer fragments. The magnified fluorescence images can be recorded via this technique up to particles of size in micrometres. This method also rules out the non-plastic items. The solvatochromic nature of Nile Red (the ability of Nile Red to change its colour due to change in solvent polarity) also offers the possibility of plastic categorization based on the polarity of surfaces of identified particles (Tiwari et al. 2019). Usually, the staining of microplastic samples is carried out using Nile Red (NR) dye with the concentration of 10 $\mu g\ mL^{-1}$ (in Acetone) and an exposure time of 30 min is optimized. Each NR-stained microplastic is then tested with an excitation wavelength of blue light and emission wavelengths are processed using

4′,6 diamidino-2-phenylindole (DAPI), fluorescein isothiocyanate (FITC) and Texas red filter with a fluorescent microscope (for example, Carl Zeiss AXIO, Imager M1). Microplastics were categorized manually by Tiwari et al. (2019) in granules, fibres, and films shape by observing their fluorescence images.

4.4.2 Qualitative Analysis of Microplastics

Some of the most commonly used analytical methods for the identification of microplastics are Fourier transform infrared spectroscopy (FTIR) and Raman spectroscopy. Both are based on vibrational spectroscopic techniques known for the analysis and characterisation of synthetic organic polymers and their products. In vibrational spectroscopy, molecular vibrations of a sample are excited and detected, which finally leads to characteristic spectral fingerprints in the FTIR or Raman spectra. Therefore, characterisation on the basis of the polymeric chemical structure and then identification by comparing with the known reference spectra becomes possible. However, there is a basic physical difference between both the vibrational techniques, FTIR and Raman spectroscopy. In short, IR absorption depends on the change of the permanent dipole moment of a chemical bond leading to the fact that polar functional groups (like carbonyl groups) are very well detectable. On the contrary, Raman spectroscopy depends on a change in the polarizability of a chemical bond, due to which bonds like aromatic bonds, C–H and C = C double bonds can easily be excited. Consequently, FTIR and Raman spectroscopy are complementary vibrational techniques, which means that, bands which have strong Raman intensities in most cases have weak IR intensities and vice versa. These physical fundamentals should be well thought-out when FTIR and Raman spectroscopy are applied for the microplastics analysis (Kappler et al. 2016).

(a) **Fourier Transform Infrared spectroscopy (FTIR)**: The FTIR spectroscopy is the most common technique used for the qualitative analysis of microplastics (Qiu et al. 2016). For the analysis of microplastics with the help of FTIR spectroscopy, the extracted sample can be irradiated with infra-red light of wavenumber range 400–4000 cm^{-1} for Mid-IR. Some part of the IR radiation can be absorbed depending on the molecular structure of the sample and finally can be measured in transmission or reflection mode (Kappler et al. 2016). The FTIR spectroscopy offers information on the chemical bonds present in the molecules. Carbon 39 containing polymers are straightforwardly recognized by FTIR and different bond configurations produce unique IR spectra that distinguishes plastics from other organic and inorganic interferences (Tiwari et al. 2019). Most common FTIR methods used are micro-FTIR and ATR-FTIR (Attenuated Total Reflection- Fourier Transform Infrared Spectroscopy) (Tiwari et al. 2019). The microplastic polymer types could be identified quickly and directly by comparing the resulting spectra with those of the known plastic polymers. In the ATR-FTIR, there are two feasible measuring modes, i.e., reflectance and transmittance modes. Spectra in the reflectance mode have shown the potential for spectral distortion for refractive error that occurs when irregularly shaped microplastics are analyzed, therefore the infrared transparent filters would be required for the analysis. However, it is essential that samples must be dried prior to spectral analysis (Qiu et al. 2016).

(b) **Scanning electron microscopy (SEM–EDS)**: Scanning electron microscopy (SEM) gives extremely perfect and highly magnified images of microplastics. SEM is used to define the elemental composition of the microplastics using real-time analysis with energy-dispersive X-ray spectroscopy (EDS). The elemental composition of particles is utilized for identifying carbon-dominant plastics over other interfering material. This information can be used for the monitoring of the likely microplastic particles, thus ruling out the non-plastic materials. Samples can be fixed on double-sided adhesive carbon tabs on aluminium SEM stubs (Tiwari et al. 2019). Then, samples are sputter coated (physical vapour deposition method) with a thin film in a MCM-100P gold coating unit with an average 10 nm coating thickness at the vacuum level of 8–10 Pa. Chemical and morphological 40 characterization of the particles is performed using a scanning electron microscope (SEM) equipped with an energy-dispersive X-ray microanalyzer (Tiwari et al. 2019).

(c) **Raman spectroscopy**: Raman spectroscopy has been also applied in the detection of microplastics in several studies. A technique was developed called as a bioimaging technique based on coherent anti-Stokes Raman Scattering (CARS) microscopy, which can visualize 2–3 μm microplastics in plankton (Qiu et al. 2016). Raman spectroscopy can identify extremely small microplastics and perform better in wet samples, and generate spatial chemical images of microplastics. This is a scattering method in which monochromatic light (laser) is used as a source. The light radiation interacts with the extracted sample. A part of the scattered photons experiences an energy shift and gives information about molecule vibrations in the sample. The resulting Raman spectrum provides analogous information as an IR spectrum (Kappler et al. 2016). Furthermore, Raman spectroscopy can detect some molecular vibrations which are IR inactive or inactive. Nevertheless, the technique of Raman spectroscopy would be interrupted by the presence of colour, additives, and attached contaminant microplastics (Qiu et al. 2016).

4.5 Microplastics Control and Remediation

The origin of microplastic pollution in any ecosystem is anthropogenic such as industrial, agricultural, and individual activities. Therefore, the first step to control microplastic pollution is source reduction. The replacement of plastic polymers with biodegradable polymers in the industrial and manufacturing sector could be a promising approach to reduce plastic waste generation and thereby, microplastic pollution. Replacement or reduction of agricultural plastic films and plastic mulching films is another source reduction strategy for mitigating soil microplastic pollution (Qi et al. 2020). Another reduction strategy is removal of plastic microbeads from personal care products and enhancing usage of biodegradable materials such as Polylactatide (PLA) and Polyhydroxyalkanoates (PHA) (Wu et al. 2017).

In addition to these source reduction measures, several technologies can also be harnessed for on-site microplastic remediation. Technological advancement of wastewater treatment plants could be a promising solution for the reduction of microplastic inputs in the environment, especially from the point sources (Pico et al. 2019). Another approach could be the utilization of biotechnology-based solutions such as microorganisms (bacteria and fungi) capable of degrading small-sized plastic polymers. A potential bioremediation solution for microplastics could be the marine fungus '*Zalerion maritimum*', which has been tested for the biodegradation of polyethylene pellets (Paço et al. 2017). Furthermore, recent research in biodegradation of plastics indicates that bacterial strains of *Bacillus* and *Enterobacter* are capable of degradation of Polyethylene (PE) and Polystyrene (PS), respectively. The strains of *Ideonella sakaiensis* have been discovered which degrade Polyethylene (PET) in PET contaminated soils (Wu et al. 2017). The physicochemical methods for microplastic removal from drinking waters include coagulation, sedimentation, sand filtration, and clarification (Bhatt et al. 2021). If scientifically implemented, these solutions can also prove compelling in the context of microplastics in floodplain soils. Therefore, the strategies for microplastic remediation in floodplains such as Indo-Gangetic floodplains need to be explored, since microplastic pollution is expected to increase exponentially in the coming years (Fig. 4.2).

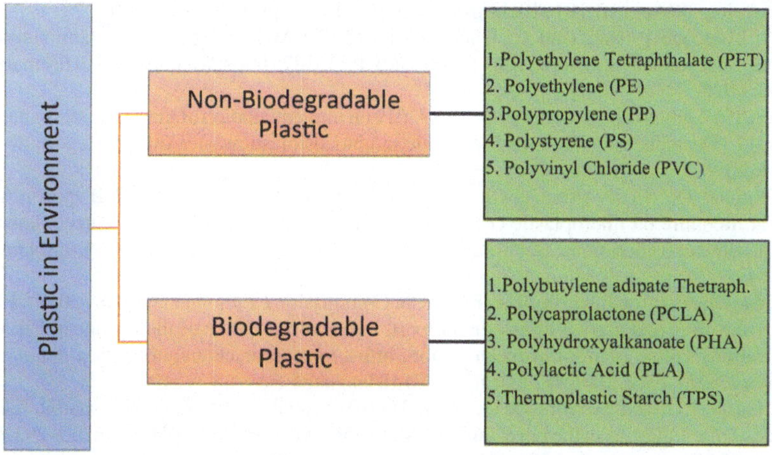

Fig. 4.2 Biodegradable and non-biodegradable plastic/microplastic polymers

4.6 Conclusion

This chapter is an attempt at exploring the unfolding vulnerable situation in the terrestrial floodplains with respect to microplastics pollution. Plastic littering, agriculture, plastic inflow, and fragmentation in river systems are some of the primary activities which contribute towards microplastics in floodplains. Furthermore, it is likely to increase in microplastic abundance in the IGP plains for the foreseeable future, therefore it is necessary to prohibit the microplastic sources by source reduction or minimizing its utilization through alternatives. There should be some mechanism to check the microplastics emission at the source itself as it is evident that an increased presence of microplastics in the ecologically sensitive floodplains may develop into a potential risk for a large population. Another perspective for combating microplastic pollution could be the awareness for microplastics existence in the floodplain soils, thus sufficient data and research should be accomplished in this area to ensure sustainable development and restoration of soil health.

References

Bhatt P, Pathak VM, Bagheri AR, Bilal M (2021) Microplastic contaminants in the aqueous environment, fate, toxicity consequences, and remediation strategies. Environ Res 200:111762. https://doi.org/10.1016/j.envres.2021.111762

Christensen ND, Wisinger CE, Maynard LA, Chauhan N, Schubert JT, Czuba JA, Barone JR (2020) Transport and characterization of microplastics in inland waterways. Journal of Water Process Engineering 38(August):101640. https://doi.org/10.1016/j.jwpe.2020.101640

Cole M, Lindeque P, Halsband C, Galloway TS (2011) Microplastics as contaminants in the marine environment: a review. Mar Pollut Bull 19:33–37. https://doi.org/10.1016/j.marpolbul.2011.09.025

CPCB (2019) Annual report for the year 2018–2019 on implementation of plastic waste management rules. 17 (December), 1–18. https://cpcb.nic.in/uploads/plasticwaste/Annual_Report_2018-19_PWM.pdf

Dioses-Salinas DC, Pizarro-Ortega CI, De-la-Torre GE (2020) A methodological approach of the current literature on microplastic contamination in terrestrial environments: current knowledge and baseline considerations. Sci Total Environ 730:13. https://doi.org/10.1016/j.scitotenv.2020.139164

Directorate-General for Environment (European Commission), Eunomia and ICF (2018). Plastics: reuse, recycling and marine litter : final report. Publications Office of the European Union. LU: Publications Office of the European Union. https://op.europa.eu/en/publication-detail/-/publication/3cdca2d1-c5f2-11e8-9424-01aa75ed71a1/language-en

Dris R, Gasperi J, Saad M, Mirande C, Tassin B (2016) Synthetic fibers in atmospheric fallout: A source of microplastics in the environment? Marine Pollution Bulletin 104(1–2):290–293. https://doi.org/10.1016/j.marpolbul.2016.01.006

Hanvey JS, Lewis PJ, Lavers JL, Crosbie ND, Pozo K, Clarke BO (2017) A review of analytical techniques for quantifying microplastics in sediments. Anal Methods 9(9):1369–1383. https://doi.org/10.1039/c6ay02707e

He D, Luo Y, Lu S, Liu M, Song Y, Lei L (2018) Microplastics in soils: analytical methods, pollution characteristics and ecological risks. Trends Anal Chem 109:163–172. https://doi.org/10.1016/j.trac.2018.10.006

Käppler A, Fischer D, Oberbeckmann S, Schernewski G, Labrenz M, Eichhorn KJ, Voit B (2016) Analysis of environmental microplastics by vibrational microspectroscopy: FTIR, Raman or both? Anal Bioanal Chem 408(29):8377–8391. https://doi.org/10.1007/s00216-016-9956-3

Karthik R, Robin RS, Purvaja R, Ganguly D, Anandavelu I, Raghuraman R, Hariharan G, Ramakrishna A, Ramesh R (2018) Microplastics along the beaches of southeast coast of India. Sci Total Environ 645:1388–1399. https://doi.org/10.1016/j.scitotenv.2018.07.242

Klemeš JJ, Fan YV, Jiang P (2021) Plastics: friends or foes? The circularity and plastic waste footprint. Energy Sources Part a: Recovery Utilization Environ Effects 43(13):1549–1565. https://doi.org/10.1080/15567036.2020.1801906

Li J, Liu H, Paul Chen J (2018) Microplastics in freshwater systems: a review on occurrence, environmental effects, and methods for microplastics detection. Water Res 137:362–374. https://doi.org/10.1016/j.watres.2017.12.056

Moller JN, Loder MGJ, Laforsch C (2020) Finding microplastics in soils: a review of analytical methods. Environ Sci Technol 54(4):2078–2090. https://doi.org/10.1021/acs.est.9b04618

Paço A, Duarte K, da Costa JP, Santos SM, Pereira R, Pereira ME, Freitas AC, Duarte AC, Rocha-Santos TAP (2017) Biodegradation of polyethylene microplastics by the marine fungus Zalerion maritimum. Sci Total Environ 586:10–15. https://doi.org/10.1016/j.scitotenv.2017.02.017

Pal DK, Bhattacharyya T, Srivastava P, Chandran P, Ray SK (2009) Soils of the Indo-Gangetic plains: their historical perspective and management. Current Sci 96(9):1193–1202. https://www.researchgate.net/publication/228648568

Pandey D, Singh A, Ramanathan A, Kumar M (2021) The combined exposure of microplastics and toxic contaminants in the floodplains of North India: a review. J Environ Manage 279:111557. https://doi.org/10.1016/j.jenvman.2020.111557

Pico Y, Alfarhan A, Barcelo D (2019) Nano- and microplastic analysis: focus on their occurrence in freshwater ecosystems and remediation technologies. TrAC Trends Anal Chem 113:409–425. https://doi.org/10.1016/j.trac.2018.08.022

Piehl S, Leibner A, Löder MGJ, Dris R, Bogner C, Laforsch C (2018) Identification and quantification of macro- and microplastics on an agricultural farmland. Sci Rep 8(1):1–9. https://doi.org/10.1038/s41598-018-36172-y

Qi R, Jones DL, Li Z, Liu Q, Yan C (2020) Behavior of microplastics and plastic film residues in the soil environment: a critical review. Sci Total Environ 703(2017):134722. https://doi.org/10.1016/j.scitotenv.2019.134722

Qiu Q, Tan Z, Wang J, Peng J, Li M, Zhan Z (2016) Extraction, enumeration and identification methods for monitoring microplastics in the environment. Estuar Coast Shelf Sci 176:102–109. https://doi.org/10.1016/j.ecss.2016.04.012

Ranjani M, Veerasingam S, Venkatachalapathy R, Mugilarasan M, Bagaev A, Mukhanov V, Vethamony P (2021) Assessment of potential ecological risk of microplastics in the coastal sediments of India: a meta-analysis. Mar Pollut Bull 163. https://doi.org/10.1016/j.marpolbul.2021.111969

Ricciardi M, Pironti C, Motta O, Miele Y, Proto A, Montano L (2021) Microplastics in the aquatic environment: occurrence, persistence, analysis, and human exposure. Water 2021(13):973. https://doi.org/10.3390/w13070973

Sarkar DJ, Das Sarkar S, Das BK, Manna RK, Behera BK, Samanta S (2019) Spatial distribution of meso and microplastics in the sediments of river Ganga at Eastern India. Sci Total Environ 694:1–7. https://doi.org/10.1016/j.scitotenv.2019.133712

Sarkar DJ, Sarkar SD, Manna RK, Samanta S, Das BK (2020) Microplastics pollution: an emerging threat to freshwater aquatic ecosystem of India. J Inland Fisheries Soc India 52(1):005. https://doi.org/10.47780/jifsi.52.1.2020.106513

Scheurer M, Bigalke M (2018) Microplastics in Swiss floodplain soils. Environ Sci Technol 52(6):3591–3598. https://doi.org/10.1021/acs.est.7b06003

Srivastava P, Pal DK, Aruche KM, Wani SP, Sahrawat KL (2015) Soils of the Indo-Gangetic plains: a pedogenic response to landscape stability, climatic variability and anthropogenic activity during the Holocene. Earth Sci Rev 140:54–71. https://doi.org/10.1016/j.earscirev.2014.10.010

Tiseo I (2021) Global plastic production from 1950–2019. https://www.statista.com/statistics/282732/global-production-of-plastics-since-1950/
Tiwari M, Rathod TD, Ajmal PY, Bhangare RC, Sahu SK (2019) Distribution and characterization of microplastics in beach sand from three different Indian coastal environments. Mar Pollut Bull 140:262–273. https://doi.org/10.1016/j.marpolbul.2019.01.055
Tsering T, Sillanpää M, Sillanpää M, Viitala M, Reinikainen S-P (2021) Microplastics pollution in the Brahmaputra River and the Indus River of the Indian Himalaya. Sci Total Environ 789:147968. https://doi.org/10.1016/j.scitotenv.2021.147968
Van Cauwenberghe L, Devriese L, Galgani F, Robbens J, Janssen CR (2015) Microplastics in sediments: a review of techniques, occurrence and effects. Mar Environ Res 111:5–17. https://doi.org/10.1016/j.marenvres.2015.06.007
Weber CJ, Opp C (2020) Spatial patterns of mesoplastics and coarse microplastics in floodplain soils as resulting from land use and fluvial processes. Environ Pollut 267:115390. https://doi.org/10.1016/j.envpol.2020.115390
Weber CJ, Opp C, Prume JA, Koch M, Andersen TJ, Chifflard P (2021) Deposition and in-situ translocation of microplastics in floodplain soils. May. Manuscript submitted to Earth's Future
Wu WM, Yang J, Criddle CS (2017) Microplastics pollution and reduction strategies. Front Environ Sci Eng 11(1):1–4. https://doi.org/10.1007/s11783-017-0897-7
Yang L, Zhang Y, Kang S, Wang Z, Wu C (2021) Microplastics in freshwater sediment: a review on methods, occurrence, and sources. Sci Total Environ 754:141948. https://doi.org/10.1016/j.scitotenv.2020.141948
Zhang GS, Liu YF (2018) The distribution of microplastics in soil aggregate fractions in southwestern China. Sci Total Environ 642:12–20. https://doi.org/10.1016/j.scitotenv.2018.06.004
Zhu F, Zhu C, Wang C, Gu C (2019) Occurrence and ecological impacts of microplastics in soil systems: a review. Bull Environ Contam Toxicol 102(6):741–749. https://doi.org/10.1007/s00128-019-02623-z
Zubris KAV, Richards BK (2005) Synthetic fibers as an indicator of land application of sludge. Environ Pollut 138(2):201–211. https://doi.org/10.1016/j.envpol.2005.04.013

Chapter 5
Microplastics in Wastewater Treatment Plants: Occurrence, Fate and Mitigation Strategies

Angel Joseph, Azmat Naseem, and Arya Vijayanandan

Abstract Environmental contamination caused by microplastics (MPs) is an issue of grave concern which is pervasive in global water, air, and soil. Wastewater treatment plants (WWTPs) are recognized as the vital source of MPs into the environment. Hence understanding the fate and behavior of MPs in WWTPs is of utmost importance. Studies have also reported the presence of MPs in the treated effluent. Challenges in MPs removal in different treatment units need to be identified, and advancements of existing treatment technologies are to be explored for enhancing the removal of MPs in WWTPs. This chapter presents the occurrence and fate of MPs in different treatment processes such as biodegradation, adsorption, membrane processes, filtration, electrocoagulation, and advanced oxidation processes. The removal efficiency of MPs in different treatment units, the mechanism of removal, and the challenges involved in removing MPs in each treatment unit are discussed in detail. The efficiency of the existing treatment technologies in WWTPs are compared, and the modifications suggested by recent studies to improve the removal of MPs are presented in this chapter.

Keywords Microplastics · Occurrence · Mitigation · Advanced oxidation processes · Adsorption · Bioreactor · Electrocoagulation · Filtration · Wastewater treatment

5.1 Introduction

Plastics are considered the modern marvel due to their benefits across all the fields ranging from health to food sectors saving countless lives (Golwala et al. 2021; Wang et al. 2021). The property that makes the plastics suitable also renders their ubiquitous presence in the environment (Guo et al. 2020; Yang et al. 2021). Depending on the usage pattern of plastic, the shelf life of the plastic items varies from 1 to 50 years or beyond before they are dumped as trash from where it can be recycled (9%),

A. Joseph · A. Naseem · A. Vijayanandan (✉)
Department of Civil Engineering, Indian Institute of Technology Delhi, Hauz Khas, Delhi 110016, India
e-mail: aryav@civil.iitd.ac.in

S. P. Singh et al. (eds.), *New Trends in Emerging Environmental Contaminants*, Energy, Environment, and Sustainability,
https://doi.org/10.1007/978-981-16-8367-1_5

abandoned in the landfill (8%), utilized for energy recovery (12%), or dissipated in the environment (71%) (Dris et al. 2015a; Gregory 2009; Foerster 2018). Microplastics (MPs) are of significant concern de to their persistence and capability to act as carriers for other toxic pollutants such as dioxins, aldrin, perfluorooctane sulfonic acid, triclosan, tonalite chlordane, hexachlorobenzene, etc. (Mammo et al. 2020). Once MPs enter the food chain, it would result in bioaccumulation and will impart adverse health effects on living organisms and human beings as well (Miller et al. 2020). It is imperative to highlight that MPs are not a single type of contaminant but a suite of contaminants and are considered among the emerging contaminants (Jain et al. 2021). Owing to this serious concern, the European Parliament recently framed a resolution proposal (TA/2019/0071) emphasizing the urgent need for addressing MPs contamination related to wastewater treatment (Edo et al. 2020). MPs are small fragments ($\leq$5 mm), synthetic polymer fragments which have either been deliberately manufactured in smaller sizes known as primary MPs or developed due to the disintegration of bigger plastic known as secondary MPs (Frias and Nash 2019; Hartmann et al. 2019).

MPs are reported to be extensively present in the soil (Guo et al. 2020), oceans (Wang et al. 2020c), atmosphere (Abbasi et al. 2019), freshwater systems (Han et al. 2020), and are also detected in the sediments (González-Pleiter et al. 2019). It is essential to consider that WWTPs majorly act as the source for the entry of MPs into the environment as single treatment plant effluent can release up to 10^{10} MPs per day (Mintenig et al. 2017). These fragments of plastics, whether in larger or smaller forms, would cause serious adverse repercussions for biota, ecosystems, and the environment (Hartline et al. 2016; He et al. 2020; Magnusson and Norén 2014; Michielssen et al. 2016; Ou and Zeng 2018; Rummel et al. 2017; Wang et al. 2020b; Wright et al. 2020). In the natural hydrological cycle, an appreciable proportion of MPs is detected in lakes, oceans, rivers, atmospheric precipitation, and wastewater and treated water (Alia et al. 2016; Andrady, 2011; Dris et al. 2015b). In the case of the anthropogenic hydrological cycle, micro-sized litter finds its way in sewers (Tagg et al. 2015), which are ultimately sent to the influent load of the WWTPs (Carr et al. 2016).

Previous studies focused on the source, composition, and concentration of MPs have reported that the MPs in the influent of WWTPs are majorly composed of small-sized fibers discharged from laundries (Baldwin et al. 2011), microbeads used in personal grooming products (Ezel et al. 2016). Pollutants such as heavy metals (Yazdani Foshtomi et al. 2019), polycyclic aromatic hydrocarbons (Sørensen et al. 2020), pharmaceutical compounds and cosmetic products (Liu et al. 2019a; Ma et al. 2019), and polybrominated biphenyl ethers (Singla et al. 2020) gets adsorbed from surrounding media to MPs due to their large surface area and low volume (Thompson et al. 2004). High concentrations of micro-sized plastic litter in water affect the efficiency of water purification processes like filtration units that get choked or malfunctioned due to clogging (Chesters et al. 2013; Guo et al. 2012). There is also a possibility that MPs experience shear forces generated by pumping or mixing, which further disintegrates MPs into much smaller fragments leading to the formation of nano plastics released in the water matrix (Lee et al. 2019).

The research focusing on the sources, occurrence, and fate of MPs in WWTPs has gained a lot of attention in recent years (Bakaraki Turan et al. 2021; Edo et al. 2020; Habib et al. 2020; Liu et al. 2021; Ngo et al. 2019; Raju et al. 2018; Sun et al. 2019;

Xu et al. 2021). Various technologies are being adopted for the removal of MPs at different treatment stages. A detailed review of the occurrence and the fate of MPs in WWTPs are presented in this chapter. In addition, this article aims to provide a comprehensive review of the mechanism of removal of MPs in specific treatment units for a better understanding of research gaps that need to be addressed in the future. Occurrence of MPs in wastewater treatment plants.

Contamination of MPs in municipal sewage is reported in several parts of the world (Iyare et al. 2020). WWTPs are considered the prime source of MPs in the water bodies (Tang et al. 2020a, 2020b) MPs are mainly introduced into wastewater through stormwater discharge and outlets from residential structures and multiple production units (Salmi et al. 2021). The release of MPs can occur throughout the life cycle of the plastic products from the production and formulation stage to transport, utilization, and final disposal (Wang et al. 2021). These MPs are being transported to different environmental matrices through runoff, sewage, landfills, industries, and weathering (Lambert et al. 2014).

The origin and composition of MPs found in WWTPs are challenging to analyze due to their smaller particle size and wide range of unknown sources (Kaliszewicz et al. 2020; Ziajahromi et al. 2016, 2017). Primary MPs detected in WWTPs include tiny beads that are used in several cosmetic products such as scrubs, face wash, toothpaste, etc. (Chang 2015; Fendall and Sewell 2009). The other sources can be plastic pellets, fishing gear, and building paints which yield small-sized MPs (0.5–6 mm) released into water systems (Kärrman et al. 2016). However, the most considerable contribution is from the automobile sector. The tires release 50,000–135,000 tons of MPs of range > 0.01 mm, and the smallest recorded micro fragments of size > 5 μm originate from fertilizers and detergents (Viveknand et al. 2021). A common source of MPs in the WWTPs is microbeads (Kärrman et al. 2016), and it is estimated that nearly 94,500 microbeads are discharged on single wash from cosmetic products (Napper et al. 2015).

Secondary MPs directly discharged into the sewers include tiny fiber litter (of size 12–16 μm) from clothing, and it is estimated that every wash of a single garment using a household machine would generate > 1900 fibers (Browne et al. 2011). The formation of secondary MPs is due to regular disintegration of larger plastics under nominal atmospheric conditions due to exposure to the sunlight or by mechanical disintegration during pumping and mixing in WWTPs (Eriksen et al. 2014). Larger plastic items are broken down by different mechanisms such as hydrolysis, photo-oxidation, mechanical breakage caused by turbulent water or sand abrasion, or microorganisms-assisted bio-assimilation (Gewert et al. 2015; Niaounakis 2017). Determining particle shape, size, amount, and surface chemistry of MPs is challenging, especially in the case of secondary MPs (Andrady 2017). In a recent study, the samples obtained from the WWTPs reported MPs' presence that can be categorized into six shapes: pellets, granules, fibers, films, fragments, and foams (Yang et al. 2021). Fibers and fragments are most abundantly found in WWTPs, and these fibers possessing a greater length to width ratio, are difficult to remove in the treatment units (Sol et al. 2020).

Physical, chemical, and biological treatment techniques could be responsible for changing the configuration of primary MPs to secondary MPs (González-Pleiter et al. 2019), rendering WWTPs the key contributor of secondary MPs in the terrestrial as well as in aquatic media (Liu et al. 2021). Depending upon the level of treatment, WWTPs have a high propensity of removing MPs; however, they are critical contributors to contamination as water remains the prime carrier for the diffusion of MPs in the environment (Bayo et al. 2020; Bretas Alvim et al. 2020; Edo et al. 2020; Sol et al. 2020; Ziajahromi et al. 2016). At various treatment levels involved in WWTPs, such as preliminary followed by primary, secondary, and terminating at the tertiary, approximately 87–99% of plastic litter is removed from the sewage (Viveknand et al. 2021). A schematic representation of the sources and pathway of MPs during in WWTPs is shown in Fig. 5.1.

Despite the high rates of removal of MPs in WWTP, still, a large proportion of this debris-loaded wastewater is regularly released to the water streams (Carr et al. 2016). A Scotland-based treatment plant was reported to release about 6.5 $\times$ 107 MPs daily into its neighborhood (Murphy et al. 2016), and another WWTP in Italy was found to discharge around 1.6×10^8 MPs per day (Magni et al. 2019). Murphy et al. 2016 reported that the influent loading of MPs in a WWTP in England was close to 15.70 ($\pm$ 5.20) particles/L and was lowered to 0.25 ($\pm$ 0.04) particles/L in the discharged effluent. A study conducted in a WWTP in China reported that the MPs with mass < 10 kg were released into the water system per day, and the proportion of fragments is abundant (Lv et al. 2019). Gies et al. (2018) estimated the dominance of different types of MPs such as fibers (65.6%), fragments (28.1%), pellets (5.4%), granules (0.45%), foam (0.22%), and sheets (0.20%) in samples collected from a WWTP in Canada.

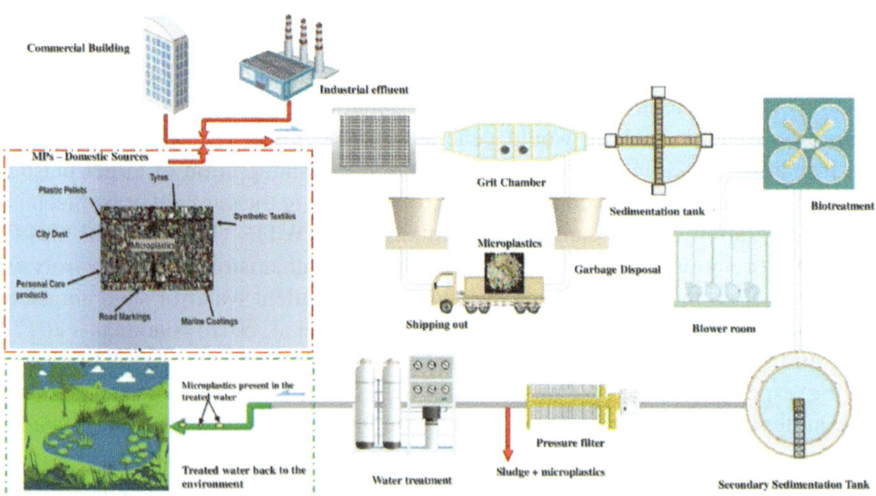

Fig. 5.1 Different sources, occurrence and pathways of MPs in WWTPs (figure modified from Zhang et al.(2020))

5.2 Treatment Methods and the Mechanism Involved in Removing MPs from WWTPs

Existing WWTPs are not carefully designed for removing MPs from wastewater. Literature suggests that a significant quantity of MPs was released to the environment in each treatment option adopted in the WWTPs (Joo et al. 2021). Depending upon the treatment methods and physicochemical characteristics of the polymer, such as particle density, size, charge, hydrophobicity, etc., variable removal efficiencies can be expected in different treatment technologies (Bond et al. 2018). The efficacy of any removal treatment method is highly dependent on the quality of influent and the characteristics of MPs in it.

Primary treatments involving coarse screening remove most of the MPs, while grit sedimentation, screening, degreasing, and primary sedimentation contribute to further removal. The biological processes followed by sedimentation in secondary treatment accounts for approximately 20% removal efficiency (Xu et al. 2021). About 69–79% of MPs introduced into WWTPs are eliminated at the primary screening stage with grit removal methods (Ziajahromi et al. 2021). Low-weight floating MPs could bypass the grease skimming method (Sun et al. 2019). Primary sedimentation could remove particles with a spherical shape having a diameter > 27–149 μm (Iyare et al. 2020). In combination with the grit and grease removal units, the primary settling unit potentially improves the efficacy of removing MPs. A recent study reported that treatment units such as sedimentation and skimming units have the potential to remove almost 72% of MPs present in sewage influent (Xu et al. 2021).

Another study reported higher removal efficiency for MPs having size (106–300 μm) than MPs with range (>300 μm) (Lee and Kim, 2018). Smaller MPs can be trapped easily in the grit and grease removal stage and adhere to the biofilm or flocs. A study conducted by Liu et al. (2019) suggested that MPs removal is closely associated with turbidity removal from primary sedimentation as well as secondary treatment. In WWTPs, the major fraction of microbeads ends up in the sludge, with the remaining microbeads released into receiving streams (Mason et al. 2016; Murphy et al. 2016). Many of the available treatment techniques eliminate MPs by entrapping this litter in sludge; however, dedicated techniques for removing MPs are lacking. There is still the persistence of MPs in the final effluent discharged via WWTPs, signifying the inability of existing treatment methods in completely eliminating MPs in wastewater (Ziajahromi et al. 2017). Secondary and tertiary treatments in WWTPs remove 88–94% of MPs loaded in the sewage influent, and about 72% removal efficacy was reported during preliminary and primary treatment (Iyare et al. 2020). The following section discusses the detailed removal mechanisms of MPs in specific treatment units in WWTPs.

5.2.1 *Biological Processes*

Bioreactors are adopted as a secondary treatment option for the removal of MPs in WWTPs. The MPs are hardly degraded within a short period. The extracellular polymer substances secreted by the microorganisms facilitate the initial trapping of MPs (Zhang et al. 2020). The degradation of MPs by the microbes followed by the formation of sludge aggregates causes the effective removal of MPs in bioreactors. The accumulated MPs in the sludge were then removed in a secondary clarifier (Jeong et al. 2016). Recent studies have reported the biological degradation of MPs by various organisms (Yuan et al. 2020; Miri et al. 2021). Polymer digestion by the microbes present in the sludge is the plausible mechanism of MPs removal (Miri et al. 2022).

Biodegradation of MPs takes place in four successive steps: colonization, bio-fragmentation, assimilation, and mineralization (Dussud and Ghiglione 2014). The biodegradation of MPs initiates with the formation of microbial colonies over the surface of the polymer (Oberbeckmann et al. 2015). This colonization causes macro-molecular modifications on the MPs by physical (biofilms forcing cracks and pits on the polymer surface) and chemical (the secretion of acids by the microbial colonies weakens the polymer structure) pathways (Yoshida et al. 2016). Moreover, the enzymes secreted on the MPs aids in disintegrating the polymer to a low molecular weight oligomer during bio-fragmentation (Skariyachan et al. 2018). The bio-fragmentation stage is continued with subsequent assimilation of fragmented oligomers by the microbial colony. This stage is called assimilation, and in the final stage of mineralization, bacteria consume all useful energy out of oligomers and release oxidized waste into the surrounding environment through multiple biochemical pathways (Hou et al. 2021; Miri et al. 2022). However, in most of the biological treatment processes, degradation of MPs is minimum; rather removal occurs due to adsorption and entrapment (Zhang et al. 2020).

The activated sludge process (ASP) has been effective in mitigating wastewater MPs after preliminary treatments (Bretas Alvim et al. 2021). The mineralization of MPs in ASP involves two processes (a) sorption and (b) degradation (Fig. 5.2). Most

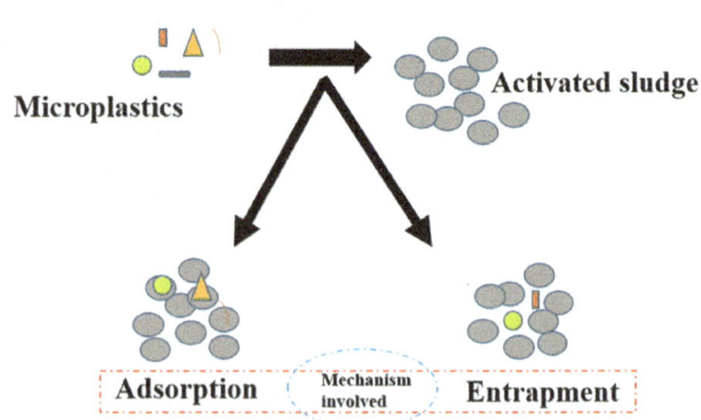

Fig. 5.2 Interactions of MPs with biomass in ASP (Figure modified from Zhang et al. 2020)

of the secondary treatment facilities in WWTPs combine biological processes with a clarification unit for the effective removal of biomass which also aids in the removal of MPs (Tang et al. 2020a, 2020b). The biological flocs promote growth on MPs' surface, which causes accumulation of MPs debris in the system, and subsequently, MPs may get settled in the clarifier (Hurley et al. 2018). Experimental results from a lab-scale SBR detected the presence of smaller fragments in the sludge and larger MPs in the effluent with a removal efficiency of 52% (Kalčíková et al. 2017).

The introduction of biofilters to remove MPs is considered a very simple and innovative concept. Liu et al. (2020) tested the biofilter consisting of different layers of stone wool as an advanced refining step for treated wastewater. The secondary effluent entered the filter through the top portion, and treated effluent was discharged from the bottom of the filter (Liu et al. 2020). The study reported that the filter could reduce MPs by 79–89% in effluents of the wastewater treatment plant. However, it is also observed that the biofilters are more recommended for the removal of larger plastic fragments, and hence a complete MPs removal could not be achieved by this method (Kuoppamäki et al. 2021). Maximum removal occurred at the top of the biofilter with the lower layers allowing limited surplus treatment efficacy (Thuptimdang et al. 2021). Large-sized MPs were retained by biofilter with no particles greater than 100 μm from the filter in the final effluent (Kuoppamäki et al. 2021).

Lee and Kim (2018) compared the removal efficiencies of 3 different configurations of the activated sludge process for MPs removal: sequencing batch reactor (SBR), anaerobic–anoxic–oxic (A^2O), and media process (A^2O basin with filled carrier). It was observed that all processes attained removal efficiencies > 98%, and SBR reported the highest rate of 99.2%. Jiang et al.2020 reported that the anoxic–oxic process (A/O) removed approximately 17% of MPs in wastewater. MPs show higher retention potential in most of the biological processes, and the treatment of MPs fraction with the sludge is of great concern (Liu et al. 2019c).

A significant portion of the MPs is reported to be accumulated in the waste activated sludge (WAS). WAS produced in the specific treatment units is further treated with an anaerobic digester (AD) (Wei et al. 2019). The treatment of WAS enriched with MPs has been challenging in recent times. The size and composition of the MPs are the predominant factors that determine the efficacy of anaerobic digestion (Chen et al. 2021). The composition of WAS enriched with larger-sized fractions of MPs has shown a reduction in methane production than in smaller fragments (Wei et al. 2019; Zhang et al. 2020). It has been reported that the presence of MPs in the WAS reduces the efficiency of the anaerobic digestion as the MPs also act as a carrier for the other antibiotics and other toxic pollutants (Wei et al. 2019). A recent study has investigated the co-effect of MPs and antibiotics in the anaerobic digestion and observed adsorption of antibiotics to MPs due to its hydrophobicity and larger specific area (Wang et al. 2020a). This causes a reduction in the anaerobic efficiency and the prolonged residence of antibiotics in the anaerobic digester, which causes inhibition in antibiotic degradation (Zhu et al. 2018). Higher content of MPs in WAS causes inhibition of the hydrolysis stage of anaerobic digestion that results in the increased formation of dissolved organic matter (Chen et al. 2021). This further causes the inhibition of the acidogenesis stage that leads to lower methane production (Lin et al. 2020). In general, biodegradation of MPs in anaerobic digestion processes is difficult, and understanding the fate of MPs in the system is essential. The leaching of monomers during anaerobic digestion is another matter of concern, and the mechanisms involved in the anaerobic digestion of MPs are still not clearly understood.

5.2.2 Filtration

Rapid sand filtration (RSF) is the commonly adopted low-cost treatment method for the removal of MPs. The removal of MPs has been enhanced by incorporating RSF units as the last stage option in tertiary treatment (Padervand et al. 2020). After the systematic investigation of WWTPs in Finland, it was observed that RSF could remove 97% of MPs present in the effluent from the secondary treatment stage (Talvitie et al. 2017). Although higher removal efficiencies are being reported, the main demerit of such systems is that it causes fragmentation of MPs. Hence the magnitude as well the surface properties of the MPs are of great importance in this context, as these factors determine the extent of MPs and filter media interaction (Bhattacharjee 2016). A recent study in South Korea reported that RSF and coagulation units could be coupled together to achieve improved removal efficiency (Hidayaturrahman and Lee 2019). Investigation on the transport behavior of smaller fragments of MPs (0.02 μm and 2 μm) through sand and biochar medium has been evaluated in a recent study (Tong et al. 2020). The study also concluded that improvement in the surface roughness of the filter media has contributed to enhanced interaction between the polymer and caused better deposition of MPs in the filter media. In addition, the development of composite filter media such as biochar/Fe_3O_4-biochar having magnetic attraction can act as a permeable barrier and has great potential to immobilize the MPs fragments (Faisal et al. 2018). A study focused on the elucidation of the mechanism behind the immobilization of MPs (Fig. 5.3) in biochar media concluded that stuck, trapped, and entangled are the three factors that control MPs fate in biochar media (Wang et al. 2020d). The development of such cost-effective filter media can improve the overall performance of tertiary treatment systems.

5.2.3 Adsorption

Several studies have targeted the development of adsorbent media that can be used suitably to adsorb the MPs present in wastewater (Li et al. 2019; Nolte et al. 2017). In a recent study, the application of graphene oxide and chitin for the adsorption of MPs had been explored and identified to be an effective method (Sun et al. 2020). The adsorption of MPs to graphene-supported adsorbents is primarily due to the pie bond, H-bond, and electrostatic attraction. Three-dimensional graphene-based adsorbents can be used for the separation of MPs. The adherence and high sorption

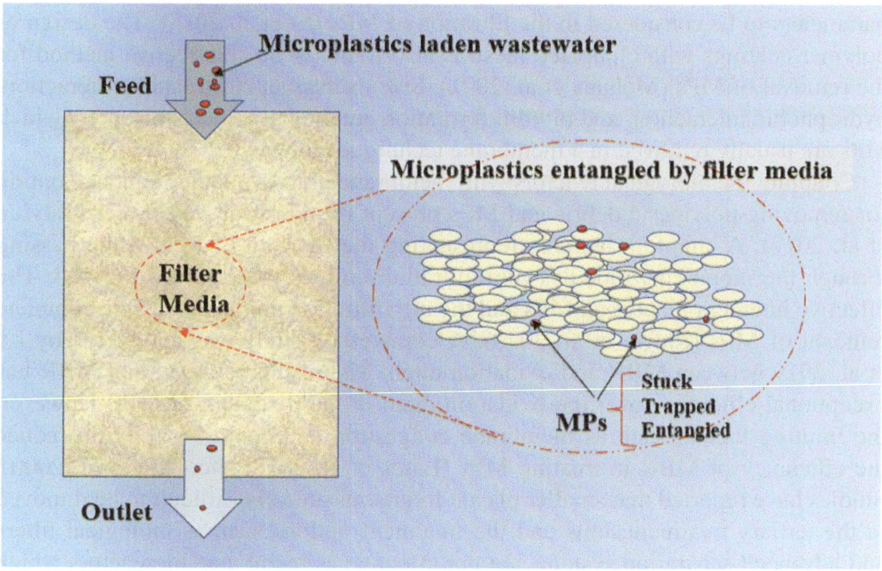

Fig. 5.3 Mechanisms involved in the removal of MPs using filtration (figure modified from Wang et al.(2020d))

of MPs to the surface of the marine microalgae and seaweed have been reported in a recent study (Sundbæk et al. 2018). The alginate released from the seaweed surfaces enhances the sorption capacity, and about 94% of MPs have been removed in that manner (Padervand et al. 2020). The effect of surface charges on MPs and microalgae in the adsorption process has been analyzed in a study conducted by Nolte et al. (2017). The study also concluded that the rate of adsorption of positively charged polystyrene is comparatively higher than the negatively charged MPs. Anionic polysaccharides present in the algal chemical structure enhanced the adsorption of MPs on the surface of algae (Dey et al. 2021). Hence the surface charge of the MPs is an important parameter that determines the rate of adsorption. A recent study shows effective MPs removal using granular activated carbon (GAC) adsorption with thermal regeneration. This study proposes to retrofit the existing STP with GAC combined with thermal regeneration as the process ensures more than 92% removal of MPs ranging from 20 to 50 μm (Kim & Park 2021).

5.2.4 Membrane Technologies

The application of membranes is highly effective for the treatment of MPs present in wastewater. Such membrane filtration units are reported to remove MPs with higher efficiency and maintain stable effluent quality and ease of treatment (Poerio et al. 2019). It is reported that the influent load and particle concentration are the critical parameters to be considered in the filtration of MPs (Li et al. 2018). The design of polymer coatings with elongated mesh is also found to be an effective method for the removal of MPs (Mohana et al. 2021). Size exclusion, electrostatic interaction, hydrophobic interaction, and biofilm formation are the basic mechanisms by which MPs are usually removed in a membrane technique (Joo et al. 2021).

Coupling the biological reactors with membrane units is another effective option for removing polymeric debris and MPs present in industrial wastewater (Talvitie et al. 2017). A simple schematic representing the removal of MPs while passing through the membrane modules (ultrafiltration unit) is presented in Fig. 5.4. The effective interaction between MPs and the ultrafiltration unit results in the complete removal of MPs (Enfrin et al. 2020). A comparison study was conducted by Lv et al. 2019 between MBR and oxidation ditch. The results revealed that MBR has exceptional efficiency over the oxidation ditch in the treatment of MPs. However, the limiting factors such as membrane congestion and control of biofilm reduce the efficiency of MBR in treating MPs (Lares et al. 2018; Nicolella et al. 2000). Studies have reported that smaller plastic fragments are very difficult to be removed in the tertiary treatment units and the treatment options such as biological filters and advanced separation systems are not effective as membrane bioreactors which ensures almost 99.9% removal efficiency (Mason et al. 2016; Talvitie et al. 2017). However, optimization of different parameters, including membrane surface charge, fouling phenomena, transmembrane pressure, membrane material, pore size, and hydraulic retention time are important for the effective removal of MPs.

A case study has showcased the capability of removing MPs in effluent at a WWTP using membrane process in Mikkeli, Finland (Hermabessiere et al. 2017). Dynamic membrane (DM) is considered an appealing membrane-based method for wastewater that works on the principle of cake layer-based formation (Poerio et al. 2019). In wastewater treatment, DM performs the function of a secondary membrane filter; however, compared to MF/UF, membrane performance is weakened due to dense fouling and thick layers (Baresel et al. 2019). The reverse osmosis (RO) process can also be effectively implemented in WWTPs to remove MPs. Tertiary treatment involving ultrafiltration and RO are reported to be efficient for removing MPs (Ziajahromi et al. 2017).

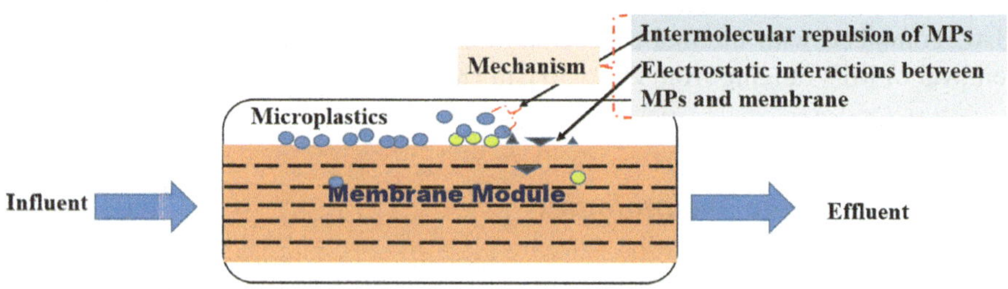

Fig. 5.4 Removal of MPs using ultrafiltration unit (figure modified from Enfrin et al. 2020)

5.2.5 *Electrocoagulation*

The electrocoagulation processes are considered to be an effective method for the removal of MPs as the coagulants produced from the metal electrodes are more likely to encounter the MPs present in wastewater with several advantages, including lower sludge production and minimal operating cost (Akarsu & Deniz, 2021; Elkhatib et al. 2021). The performance of electrocoagulation processes has been evaluated using a bench-scale electrocoagulation unit (Fig. 5.5) and found to be most effective in removing MPs at neutral pH as the rate of coagulant formation is higher at neutral pH (Perren et al. 2018).

In a recent study conducted by Ariza-Tarazona et al. (2019), the efficiency of iron and aluminum salt coagulants in removing polyethylene has been compared and concluded that aluminum-based coagulants had shown better removal efficiency than iron salt. This study also noted a reduction in the removal efficiency with the rise in pH, and the smaller plastic fragments below 0.5 mm are found to be hardly removed at a higher pH range. The study also inferred that an enhancing coagulation agent such as polyacrylamide had improved the removal efficiency of smaller MPs. Electrocoagulation processes have been successfully employed for the removal of microbeads using aluminum as the electrode material. Perren et al. (2018) investigated different parameters, including the effect of pH and electrode to electrode distance that affects the MPs removal rate. Higher removal efficiency (more than 90%) was reported over the lower pH values (pH 3). Kim et al. (2021) reported that about 90% of MPs were converted to separable flocs in the electrocoagulation unit, which would later be removed in any tertiary treatment unit.

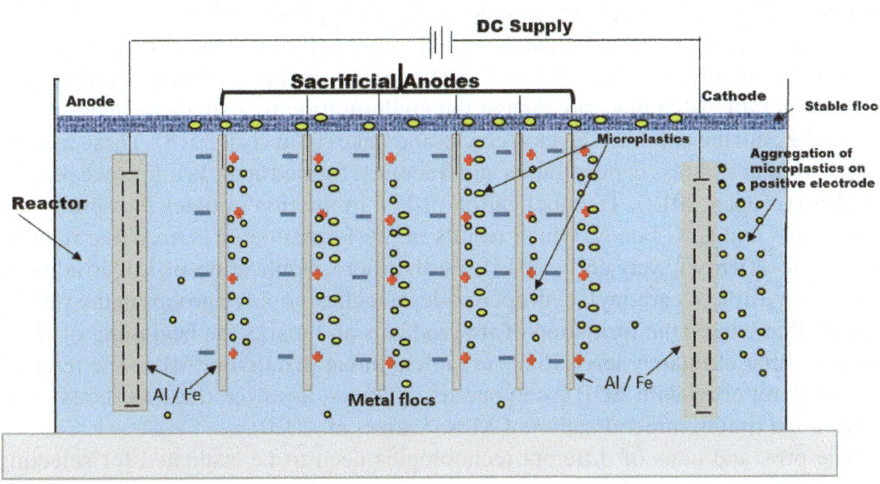

Fig. 5.5 Electrocoagulation unit adopted for the removal of MPs (figure modified from Perren et al. 2018)

5.2.6 Advanced Oxidation Processes

The applicability of the advanced oxidation process (AOP) has been explored as a tertiary treatment option for removing MPs. Recently, the combined influence of Fenton and heat-activated persulfate methods in the long-term degradation of polyethylene and polystyrene MPs have been studied (Liu et al. 2019c). The influence of the average size and oxygen to carbon ratio (O/C) ratio of MPs in the surface properties and the adsorption capacity was analyzed in this study, and concluded that these parameters have significantly changed the oxidation rate of MPs. The alterations in the surface of MPs in an advanced oxidation process are usually evaluated for the O/C ratio (Liu et al. 2019b). The morphological and structural variations of polyethylene MPs during the dark and UV light degradation have been evaluated in a recent study (Da Costa et al. 2018). It is observed that salinity increased the rate of degradation of MPs compared to UV irradiation (Suhrhoff et al. 2016). Hence the salinity of the surrounding media plays an important factor in the photodegradation of MPs as salts enhance the formation of oxidized sites (Vasile and Pascu 2005).

The disintegration of MPs by the photocatalytic method has been recently explored and reported as a potential treatment option for the mineralization of polymer (Tofa et al. 2019). The light source is chosen depending upon the type of photocatalyst and the amount of energy required for the extraction of electrons from the valence band of the photocatalyst. Many of the recent studies have effectively utilized photocatalysis for the effective degradation of MPs. The applicability of TiO_2 (Ariza-Tarazona et al. 2020; Luo et al. 2021) and ZnO (Uheida et al. 2021) photocatalysts for the photocatalytic degradation of MPs have been majorly explored in this field. Liang et al. (2013) reported improved photocatalytic degradation of low-density polyethylene by grafting hydrophilic polyacrylamide on the TiO_2 photocatalyst.

The apparent conditions of the degradation for polypropylene (PP), polystyrene (PS), and polyethylene (PE) are cracks/flakes and granular oxidation. However, the most vulnerable MPs that can change the configuration to a smaller range or even nano range are the ones that develop cracks and flakes (Cai et al. 2018). These fissures are referred to as stress concentrators and fractures, formed at the brittle or weak spots of MPs (Cooper, 2012). The application of UV irradiation initiates the splitting of C-H bonds and C–C bonds, which results in the formation of peroxy free radicals (Cai et al. 2018; Gewert et al. 2015). Furthermore, application of UV irradiation activates hydroxyl, carbonyl, hydroperoxide, and chromophore groups in the surface of MPs that causes the formation of free radicals and marks the beginning of chain reactions that ultimately leads to the complete mineralization of MPs. The removal efficiency reported with AOP is comparatively lower; however, these methods could achieve maximum mineralization of MPs (Lam et al. 2020).

The pros and cons of different technologies need to be evaluated for selecting the suitable treatment scheme for removing MPs. High energy demand, membrane fouling, sludge accumulation, microbial aggregation on the membrane surface are some of the inherent demerits encountered with conventional treatments and membrane processes that can be avoided by the extended application of photocatalytic processes with efficient utilization of solar energy. Filtration systems are a comparatively low-cost option for MPs mitigation. However, the lower removal rate of MPs has been reported with filtration units, and this might be due to the longitudinal movement of MPs fragments through the filter media (Sun et al. 2019). Hence additional final stage treatment units should be supported with filtration systems in WWTPs to remove smaller fragments of MPs. In most cases, during the backwashing of the filters, the solids removed may again be sent back to the beginning stage, adding the loading rate of MPs in the WWTPs.

The retention of a significant fraction of MPs in the sewage sludge limits the visibility of biological processes in the treatment of MPs. However, research focusing on the leaching of monomers from MPs during the biodegradation and studies focusing on the transformation of MPs in specific units of WWTPs are scanty. The interactions of MPs with the surface of the membrane are not clearly understood, and further studies need to be conducted in this area. The effect of other pollutants and the operating parameters on the fate of MPs in different treatment units should be investigated, and the underlying mechanisms of removal need to be substantiated. Future studies should also focus on the reaction intermediates generated during photocatalysis and other AOPs during the treatment of MPs.

5.3 Conclusions

Plastic polymers are inevitable in human lives due to their continued demand in all sectors. Though it seems a considerable removal, still WWTPs are majorly contributing to MPs contamination to the environment. The MPs-loaded effluent released from the WWTPs ultimately merges with waterbodies. Being an emerging contaminant, MPs serve as carriers of metals and organic contaminants. Bioaccumulation of MPs is inevitable considering the cumulative emission of MPs into the water bodies. Therefore, maximizing the removal of MPs in WWTPs would eliminate a possible pathway of MPs into the environment. However, there are uncertainties in comparing the removal efficiencies of MPs in WWTPs due to the multiple possible removal mechanisms and lack of standardized sampling/identification methods. Proper standardization and the periodic assessment of treatment units are required for a greater degree of evaluation of the fate of the MPs in WWTPs. Different technologies are being explored for the removal of MPs, such as adsorption, filtration, membrane technologies, electrocoagulation, and advanced oxidation processes such as photocatalysis. The coupling of different secondary and tertiary schemes would significantly reduce the MPs concentration in the treated effluent. Membrane filtration is reported to be a promising technology; however, further research is required to address the challenges related to membrane fouling issues. Ample attention should be given to the development of treatment methods that focus on the complete mineralization of the MPs to curb the emission, and further studies are needed in this direction.

References

Abbasi S, Keshavarzi B, Moore F, Turner A, Kelly FJ, Dominguez AO, Jaafarzadeh N (2019) Distribution and potential health impacts of microplastics and micro rubbers in air and street dusts from Asaluyeh County, Iran. Environ Pollut 244:153–164. https://doi.org/10.1016/j.envpol.2018.10.039

Akarsu C, Deniz F (2021) Electrocoagulation/electroflotation process for removal of organics and microplastics in laundry wastewater. Clean: Soil, Air, Water. https://doi.org/10.1002/clen.202000146

Alia JM, Herrmann KMK, Rakestrawb MJ, Alan S, Coady K (2016) Learned discourses: timely scientific opinions learned discourse: timely scientific opinions 12(2), 397–406. https://doi.org/10.1002/ieam.1732

Andrady AL (2011) Microplastics in the marine environment. Mar Pollut Bull 62(8):1596–1605. https://doi.org/10.1016/j.marpolbul.2011.05.030

Andrady AL (2017) The plastic in microplastics: a review. Mar Pollut Bull 119(1):12–22

Ariza-Tarazona MC, Villarreal-Chiu JF, Barbieri V, Siligardi C, Cedillo-González EI (2019) New strategy for microplastic degradation: green photocatalysis using a protein-based porous N-TiO_2 semiconductor. Ceram Int 45(7):9618–9624

Ariza-Tarazona MC, Villarreal-Chiu JF, Hernández-López JM, De la Rosa JR, Barbieri V, Siligardi C, Cedillo-González EI (2020) Microplastic pollution reduction by a carbon and nitrogen-doped TiO_2: effect of pH and temperature in the photocatalytic degradation process. J Hazard Mater 395:122632

Bakaraki Turan N, Sari Erkan H, Onkal Engin G (2021) Microplastics in wastewater treatment plants: occurrence, fate and identification. Process Saf Environ Prot. https://doi.org/10.1016/j.psep.2020.08.039

Baldwin V, Bhatia M, Luckey M (2011) Folding studies of purified LamB protein, the maltoporin from the Escherichia coli outer membrane: trimer dissociation can be separated from unfolding. Biochim Biophys Acta (BBA)—Biomembranes 1808(9):2206–2213. https://doi.org/10.1016/j.bbamem.2011.05.013

Baresel C, Harding M, Fang J (2019) Ultrafiltration/granulated active carbon-biofilter: efficient removal of a broad range of micropollutants. Appl Sci (Switzerland). https://doi.org/10.3390/app9040710

Bayo J, Olmos S, López-Castellanos J (2020) Microplastics in an urban wastewater treatment plant: the influence of physicochemical parameters and environmental factors. Chemosphere 238:124593

Bhattacharjee S (2016) DLS and zeta potential—what they are and what they are not? J Control Release. https://doi.org/10.1016/j.jconrel.2016.06.017

Bond T, Ferrandiz-Mas V, Felipe-Sotelo M, Van Sebille E (2018) The occurrence and degradation of aquatic plastic litter based on polymer physicochemical properties: a review. Crit Rev Environ Sci Technol 48(7–9):685–722

Bretas Alvim C, Castelluccio S, Ferrer-Polonio E, Bes-Piá MA, Mendoza-Roca JA, Fernández-Navarro J, Alonso JL, Amorós I (2021) Effect of polyethylene microplastics on activated sludge process—accumulation in the sludge and influence on the process and on biomass characteristics. Process Saf Environ Prot. https://doi.org/10.1016/j.psep.2020.10.014

Bretas Alvim C, Mendoza-Roca JA, Bes-Piá A (2020) Wastewater treatment plant as microplastics release source—quantification and identification techniques. J Environ Manage 255, 109739. https://doi.org/10.1016/j.jenvman.2019.109739

Browne MA, Crump P, Niven SJ, Teuten E, Tonkin A, Galloway T, Thompson R (2011) Accumulation of microplastic on shorelines woldwide: sources and sinks. Environ Sci Technol 45(21):9175–9179. https://doi.org/10.1021/es201811s

Cai L, Wang J, Peng J, Wu Z, Tan X (2018) Observation of the degradation of three types of plastic pellets exposed to UV irradiation in three different environments. Sci Total Environ 628:740–747

Carr SA, Liu J, Tesoro AG (2016) Transport and fate of microplastic particles in wastewater treatment plants. Water Res. https://doi.org/10.1016/j.watres.2016.01.002

Chang M (2015) Reducing microplastics from facial exfoliating cleansers in wastewater through treatment versus consumer product decisions. Mar Pollut Bull 101(1):330–333. https://doi.org/10.1016/j.marpolbul.2015.10.074

Chen H, Tang M, Yang X, Tsang YF, Wu Y, Wang D, Zhou Y (2021) Polyamide 6 microplastics facilitate methane production during anaerobic digestion of waste activated sludge. Chem Eng J. https://doi.org/10.1016/j.cej.2020.127251

Chesters SP, Pena N, Gallego S, Fazel M, Armstrong MW, del Vigo F (2013) Results from 99 seawater RO membrane autopsies. IDA J Desalin Water Reuse 5(1):40–47

Cooper DA (2012) Effects of chemical and mechanical weathering processes on the degradation of plastic debris on marine beaches

Dey TK, Uddin ME, Jamal M (2021) Detection and removal of microplastics in wastewater: evolution and impact. Environ Sci Pollut Res 1–23

Dris R, Gasperi J, Rocher V, Saad M, Renault N, Tassin B (2015a) Microplastic contamination in an urban area: a case study in Greater Paris. Environ Chem 12(5):592–599

Dris R, Imhof H, Sanchez W, Gasperi CJ (2015) Special front issue beyond the ocean: contamination of freshwater ecosystems with (micro-) plastic particles, 539–550

Dussud C, Ghiglione JF (2014) Bacterial degradation of synthetic plastics. In: CIESM workshop monogr 46:49–54

Edo C, González-Pleiter M, Leganés F, Fernández-Piñas F, Rosal R (2020) Fate of microplastics in wastewater treatment plants and their environmental dispersion with effluent and sludge. Environ Pollut 259. https://doi.org/10.1016/j.envpol.2019.113837

Elkhatib D, Oyanedel-Craver V, Carissimi E (2021) Electrocoagulation applied for the removal of microplastics from wastewater treatment facilities. Sep Purif Technol. https://doi.org/10.1016/j.seppur.2021.118877

Enfrin M, Lee J, Le-Clech P, Dumée LF (2020) Kinetic and mechanistic aspects of ultrafiltration membrane fouling by nano- and microplastics. J Membr Sci. https://doi.org/10.1016/j.memsci.2020.117890

Eriksen M, Lebreton LCM, Carson HS, Thiel M, Moore CJ, Borerro JC, Galgani F, Ryan PG, Reisser J (2014) Plastic pollution in the world's oceans: more than 5 trillion plastic pieces weighing over 250,000 tons afloat at sea. PLoS ONE. https://doi.org/10.1371/journal.pone.0111913

Ezel ANVANW, Aris INEZC, Ools STAEK (2016) Plastic debris in the aquatic environment release of primary microplastics from consumer products to wastewater in The Netherlands. Environ Toxicol 35(7):1627–1631. https://doi.org/10.1002/etc.3316

Faisal AAH, Sulaymon AH, Khaliefa QM (2018) A review of permeable reactive barrier as passive sustainable technology for groundwater remediation. Int J Environ Sci Technol. https://doi.org/10.1007/s13762-017-1466-0

Fendall LS, Sewell MA (2009) Contributing to marine pollution by washing your face: microplastics in facial cleansers. Mar Pollut Bull 58(8):1225–1228

Foerster KH (2018) Plastics-the facts 2017. Association of Plastic Manufacturers, Brussels

Frias JPGL, Nash R (2019) Microplastics: finding a consensus on the definition. Mar Pollut Bull 138:145–147. https://doi.org/10.1016/j.marpolbul.2018.11.022

Gewert B, Plassmann MM, Macleod M (2015) Pathways for degradation of plastic polymers floating in the marine environment. Environ Sci Process Impacts 17(9):1513–1521. https://doi.org/10.1039/c5em00207a

Gies EA, LeNoble JL, Noël M, Etemadifar A, Bishay F, Hall ER, Ross PS (2018) Retention of microplastics in a major secondary wastewater treatment plant in Vancouver, Canada. Mar Pollut Bull 133:553–561

Golwala H, Zhang X, Iskander SM, Smith AL (2021) Solid waste: An overlooked source of microplastics to the environment. Sci Total Environ. https://doi.org/10.1016/j.scitotenv.2020.144581

González-Pleiter M, Tamayo-Belda M, Pulido-Reyes G, Amariei G, Leganés F, Rosal R, Fernández-Piñas F (2019) Secondary nanoplastics released from a biodegradable microplastic severely impact freshwater environments. Environ Sci Nano 6(5):1382–1392. https://doi.org/10.1039/c8en01427b

Gregory MR (2009) Environmental implications of plastic debris in marine settings—entanglement, ingestion, smothering, hangers-on, hitch-hiking and alien invasions. Philos Trans R Soc B Biol Sci 364(1526):2013–2025

Guo J-J, Huang X-P, Xiang L, Wang Y-Z, Li Y-W, Li H, Cai Q-Y, Mo C-H, Wong M-H (2020) Source, migration and toxicology of microplastics in soil. Environ Int 137:105263. https://doi.org/10.1016/j.envint.2019.105263

Guo W, Ngo H-H, Li J (2012) A mini-review on membrane fouling. Bioresour Technol 122:27–34. https://doi.org/10.1016/j.biortech.2012.04.089

Habib RZ, Thiemann T, Al Kendi R (2020) Microplastics and wastewater treatment plants—a review. J Water Resour Prot. https://doi.org/10.4236/jwarp.2020.121001

Han M, Niu X, Tang M, Zhang B-T, Wang G, Yue W, Kong X, Zhu J (2020) Distribution of microplastics in surface water of the lower Yellow River near estuary. Sci Total Environ 707:135601. https://doi.org/10.1016/j.scitotenv.2019.135601

Hartline NL, Bruce NJ, Karba SN, Ruff EO, Sonar SU, Holden PA (2016) Microfiber masses recovered from conventional machine washing of new or aged garments. Environ Sci Technol 50(21):11532–11538

Hartmann NB, Hüffer T, Thompson RC, Hassellöv M, Verschoor A, Daugaard AE, Rist S, Karlsson T, Brennholt N, Cole M, Herrling MP, Hess MC, Ivleva NP, Lusher AL, Wagner M (2019) Are we speaking the same language? Recommendations for a definition and categorization framework for plastic debris. Environ Sci Technol 53(3):1039–1047. https://doi.org/10.1021/acs.est.8b05297

He D, Bristow K, Filipović V, Lv J, He H (2020) Microplastics in terrestrial ecosystems: a scientometric analysis. Sustainability (Switzerland). https://doi.org/10.3390/su12208739

Hermabessiere L, Dehaut A, Paul-Pont I, Lacroix C, Jezequel R, Soudant P, Duflos G (2017) Occurrence and effects of plastic additives on marine environments and organisms: a review. Chemosphere 182:781–793

Hidayaturrahman H, Lee TG (2019) A study on characteristics of microplastic in wastewater of South Korea: identification, quantification, and fate of microplastics during treatment process. Mar Pollut Bull. https://doi.org/10.1016/j.marpolbul.2019.06.071

Hou L, Kumar D, Yoo CG, Gitsov I, Majumder ELW (2021) Conversion and removal strategies for microplastics in wastewater treatment plants and landfills. Chem Eng J. https://doi.org/10.1016/j.cej.2020.126715

Hurley RR, Lusher AL, Olsen M, Nizzetto L (2018) Validation of a method for extracting microplastics from complex, organic-rich, environmental matrices. Environ Sci Technol. https://doi.org/10.1021/acs.est.8b01517

Iyare PU, Ouki SK, Bond T (2020) Microplastics removal in wastewater treatment plants: a critical review. Environ Sci: Water Res Technol 6(10):2664–2675. https://doi.org/10.1039/d0ew00397b

Jain S, Mishra D, Khare P (2021) Microplastics as an emerging contaminant in environment: occurrence, distribution, and management strategy. In: Management of contaminants of emerging concern (CEC) in environment. https://doi.org/10.1016/b978-0-12-822263-8.00010-5

Jeong CB, Won EJ, Kang HM, Lee MC, Hwang DS, Hwang UK, Lee JS (2016) Microplastic size-dependent toxicity, oxidative stress induction, and p-JNK and p-p38 activation in the monogonont rotifer (Brachionus koreanus). Environ Sci Technol 50(16):8849–8857

Jiang J, Wang X, Ren H, Cao G, Xie G, Xing D, Liu B (2020) Investigation and fate of microplastics in wastewater and sludge filter cake from a wastewater treatment plant in China. Science Total Environ 746:141378. https://doi.org/10.1016/j.scitotenv.2020.141378

Joo SH, Liang Y, Kim M, Byun J, Choi H (2021) Microplastics with adsorbed contaminants: mechanisms and treatment. Environ Challenges 100042

Kalčíková G, Alič B, Skalar T, Bundschuh M, Gotvajn AŽ (2017) Wastewater treatment plant effluents as source of cosmetic polyethylene microbeads to freshwater. Chemosphere. https://doi.org/10.1016/j.chemosphere.2017.08.131

Kaliszewicz A, Winczek M, Karaban K, Kurzydłowski D, Górska M, Koselak W, Romanowski J (2020) The contamination of inland waters by microplastic fibres under different anthropogenic pressure: preliminary study in Central Europe (Poland) 1–8. https://doi.org/10.1177/0734242X20938448

Kärrman A, Schönlau C, Engwall M (2016) Exposure and effects of microplastics on wildlife: a review of existing data

Kim KT, Park S (2021) Enhancing microplastics removal from wastewater using electrocoagulation and granule-activated carbon with thermal regeneration. Processes. https://doi.org/10.3390/pr9040617

Kuoppamäki K, Pflugmacher Lima S, Scopetani C, Setälä H (2021) The ability of selected filter materials in removing nutrients, metals, and microplastics from stormwater in biofilter structures. J Environ Qual. https://doi.org/10.1002/jeq2.20201

Lambert S, Sinclair C, Boxall A (2014) Occurrence, degradation, and effect of polymer-based materials in the environment. Rev Environ Contam Toxicol 227:1–53

Lares M, Ncibi MC, Sillanpää M, Sillanpää M (2018) Occurrence, identification and removal of microplastic particles and fibers in conventional activated sludge process and advanced MBR technology. Water Res 133:236–246

Lee H, Kim Y (2018) Treatment characteristics of microplastics at biological sewage treatment facilities in Korea. Mar Pollut Bull 137:1–8. https://doi.org/10.1016/j.marpolbul.2018.09.050

Lee J, Enfrin M, Dum LF (2019) Nano/microplastics in water and wastewater treatment processes—origin, impact and potential solutions. Water Res 161. https://doi.org/10.1016/j.watres.2019.06.049

Li L, Xu G, Yu H, Xing J (2018) Dynamic membrane for micro-particle removal in wastewater treatment: performance and influencing factors. Sci Total Environ 627:332–340

Li X, Mei Q, Chen L, Zhang H, Dong B, Dai X, Zhou J (2019) Enhancement in adsorption potential of microplastics in sewage sludge for metal pollutants after the wastewater treatment process. Water Res 157:228–237

Liang W, Luo Y, Song S, Dong X, Yu X (2013) High photocatalytic degradation activity of polyethylene containing polyacrylamide grafted TiO_2. Polym Degrad Stab 98(9):1754–1761

Lin X, Su C, Deng X, Wu S, Tang L, Li X, Liu J, Huang X (2020) Influence of polyether sulfone microplastics and bisphenol A on anaerobic granular sludge: performance evaluation and microbial community characterization. Ecotoxicol Environ Saf. https://doi.org/10.1016/j.ecoenv.2020.111318

Liu G, Zhu Z, Yang Y, Sun Y, Yu F, Ma J (2019a) Sorption behavior and mechanism of hydrophilic organic chemicals to virgin and aged microplastics in freshwater and seawater. Environ Pollut 246:26–33. https://doi.org/10.1016/j.envpol.2018.11.100

Liu P, Qian L, Wang H, Zhan X, Lu K, Gu C, Gao S (2019b) New Insights into the aging behavior of microplastics accelerated by advanced oxidation processes. Environ Sci Technol 53(7):3579–3588. https://doi.org/10.1021/acs.est.9b00493

Liu X, Yuan W, Di M, Li Z, Wang J (2019c) Transfer and fate of microplastics during the conventional activated sludge process in one wastewater treatment plant of China. Chem Eng J 362:176–182

Liu F, Nord NB, Bester K, Vollertsen J (2020) Microplastics removal from treated wastewater by a biofilter. Water (switzerland) 12(4):1–11. https://doi.org/10.3390/W12041085

Liu W, Zhang J, Liu H, Guo X, Zhang X, Yao X, Cao Z, Zhang T (2021) A review of the removal of microplastics in global wastewater treatment plants: characteristics and mechanisms. Environ Int 146:106277

Luo H, Xiang Y, Li Y, Zhao Y, Pan X (2021) Photocatalytic aging process of Nano-TiO_2 coated polypropylene microplastics: combining atomic force microscopy and infrared spectroscopy (AFM-IR) for nanoscale chemical characterization. J Hazard Mater 404:124159

Lv X, Dong Q, Zuo Z, Liu Y, Huang X, Wu WM (2019) Microplastics in a municipal wastewater treatment plant: fate, dynamic distribution, removal efficiencies, and control strategies. J Clean Prod 225:579–586

Ma J, Zhao J, Zhu Z, Li L, Yu F (2019) Effect of microplastic size on the adsorption behavior and mechanism of triclosan on polyvinyl chloride. Environ Pollut 254:113104. https://doi.org/10.1016/j.envpol.2019.113104

Magni S, Binelli A, Pittura L, Avio CG, Della Torre C, Parenti CC, Gorbi S, Regoli F (2019) The fate of microplastics in an Italian Wastewater Treatment Plant. Sci Total Environ 652:602–610. https://doi.org/10.1016/j.scitotenv.2018.10.269

Magnusson K, Norén F (2014) Screening of microplastic particles in and down-stream a wastewater treatment plant

Mammo FK, Amoah ID, Gani KM, Pillay L, Ratha SK, Bux F, Kumari S (2020) Microplastics in the environment: interactions with microbes and chemical contaminants. Sci Tot Environ 140518

Mason SA, Garneau D, Sutton R, Chu Y, Ehmann K, Barnes J, Fink P, Papazissimos D, Rogers DL (2016) Microplastic pollution is widely detected in US municipal wastewater treatment plant effluent. Environ Pollut 218:1045–1054. https://doi.org/10.1016/j.envpol.2016.08.056

Michielssen MR, Michielssen ER, Ni J, Duhaime MB (2016) Fate of microplastics and other small anthropogenic litter (SAL) in wastewater treatment plants depends on unit processes employed. Environ Sci Water Res Technol 2(6):1064–1073

Miller ME, Hamann M, Kroon FJ (2020) Bioaccumulation and biomagnification of microplastics in marine organisms: a review and meta-analysis of current data. PLoS ONE. https://doi.org/10.1371/journal.pone.0240792

Mintenig SM, Int-Veen I, Löder MGJ, Primpke S, Gerdts G (2017) Identification of microplastic in effluents of waste water treatment plants using focal plane array-based micro-Fourier-transform infrared imaging. Water Res 108:365–372

Miri S, Saini R, Davoodi SM, Pulicharla R, Brar SK, Magdouli S (2021) Biodegradation of microplastics: better late than never. Chemosphere 131670

Miri S, Saini R, Davoodi SM, Pulicharla R, Brar SK, Magdouli S (2022) Biodegradation of microplastics: better late than never. Chemosphere. https://doi.org/10.1016/j.chemosphere.2021.131670

Mohana AA, Farhad SM, Haque N, Pramanik BK (2021) Understanding the fate of nano-plastics in wastewater treatment plants and their removal using membrane processes. Chemosphere 284:131430

Murphy F, Ewins C, Carbonnier F, Quinn B (2016) Wastewater treatment works (WwTW) as a source of microplastics in the aquatic environment. Environ Sci Technol 50(11):5800–5808

Napper IE, Bakir A, Rowland SJ, Thompson RC (2015) Characterisation, quantity and sorptive properties of microplastics extracted from cosmetics. Mar Pollut Bull 99(1–2):178–185

Ngo PL, Pramanik BK, Shah K, Roychand R (2019) Pathway, classification and removal efficiency of microplastics in wastewater treatment plants. Environ Pollut. https://doi.org/10.1016/j.envpol.2019.113326

Niaounakis M (2017) Management of marine plastic debris. William Andrew

Nicolella C, Van Loosdrecht MCM, Heijnen JJ (2000) Wastewater treatment with particulate biofilm reactors. J Biotechnol 80(1):1–33

Nolte TM, Hartmann NB, Kleijn JM, Garnæs J, Van De Meent D, Hendriks AJ, Baun A (2017) The toxicity of plastic nanoparticles to green algae as influenced by surface modification, medium hardness and cellular adsorption. Aquat Toxicol 183:11–20

Oberbeckmann S, Löder MG, Labrenz M (2015) Marine microplastic-associated biofilms–a review. Environ Chem 12(5):551–562

Ou H, Zeng EY (2018) Occurrence and fate of microplastics in wastewater treatment plants. In: Microplastic contamination in aquatic environments: an emerging matter of environmental urgency. Elsevier Inc. https://doi.org/10.1016/B978-0-12-813747-5.00010-2

Padervand M, Lichtfouse E, Robert D, Wang C (2020) Removal of microplastics from the environment. A review. Environ Chem Lett 1–22

Perren W, Wojtasik A, Cai Q (2018) Removal of microbeads from wastewater using electrocoagulation. ACS Omega 3(3):3357–3364

Poerio T, Piacentini E, Mazzei R (2019) Membrane processes for microplastic removal. Molecules 24(22):4148

Raju S, Carbery M, Kuttykattil A, Senathirajah K, Subashchandrabose SR, Evans G, Thavamani P (2018) Transport and fate of microplastics in wastewater treatment plants: implications to environmental health. Rev Environ Sci Biotechnol. https://doi.org/10.1007/s11157-018-9480-3

Rummel CD, Jahnke A, Gorokhova E, Kühnel D, Schmitt-Jansen M (2017) Impacts of biofilm formation on the fate and potential effects of microplastic in the aquatic environment. Environ Sci Technol Lett 4(7):258–267

Salmi P, Ryymin K, Karjalainen AK, Mikola A, Uurasjärvi E (2021) Uncorrected proof particle balance and return loops for microplastics in a tertiary-level wastewater treatment plant. Uncorrected Proof 1–12. https://doi.org/10.2166/wst.2021.209

Singla M, Díaz J, Broto-Puig F, Borrós S (2020) Sorption and release process of polybrominated diphenyl ethers (PBDEs) from different composition microplastics in aqueous medium: solubility parameter approach. Environ Pollut 262:114377. https://doi.org/10.1016/j.envpol.2020.114377

Skariyachan S, Patil AA, Shankar A, Manjunath M, Bachappanavar N, Kiran S (2018) Enhanced polymer degradation of polyethylene and polypropylene by novel thermophilic consortia of Brevibacillus sps. and Aneurinibacillus sp. screened from waste management landfills and sewage treatment plants. Polym Degrad Stab. https://doi.org/10.1016/j.polymdegradstab.2018.01.018

Sol D, Laca A, Laca A, Díaz M (2020) Approaching the environmental problem of microplastics: Importance of WWTP treatments. Sci Total Environ 740:140016. https://doi.org/10.1016/j.scitotenv.2020.140016

Sørensen L, Rogers E, Altin D, Salaberria I, Booth AM (2020) Sorption of PAHs to microplastic and their bioavailability and toxicity to marine copepods under co-exposure conditions. Environ Pollut 258:113844. https://doi.org/10.1016/j.envpol.2019.113844

Suhrhoff TJ, Scholz-Böttcher BM (2016) Qualitative impact of salinity, UV radiation and turbulence on leaching of organic plastic additives from four common plastics—a lab experiment. Mar Pollut Bull 102(1):84–94

Sun J, Dai X, Wang Q, Van Loosdrecht MCM, Ni B (2019) Microplastics in wastewater treatment plants: detection, occurrence and removal. Water Res 152:21–37. https://doi.org/10.1016/j.watres.2018.12.050

Sun Q, Ren SY, Ni HG (2020) Incidence of microplastics in personal care products: an appreciable part of plastic pollution. Sci Total Environ 742:140218

Sundbæk KB, Koch IDW, Villaro CG, Rasmussen NS, Holdt SL, Hartmann NB (2018) Sorption of fluorescent polystyrene microplastic particles to edible seaweed Fucus vesiculosus. J Appl Phycol 30(5):2923–2927

Tagg AS, Sapp M, Harrison JP, Ojeda JJ (2015) Identification and quantification of microplastics in wastewater using focal plane array-based reflectance micro-FT-IR imaging. Anal Chem 87(12):6032–6040. https://doi.org/10.1021/acs.analchem.5b00495

Talvitie J, Mikola A, Koistinen A, Setälä O (2017) Solutions to microplastic pollution–removal of microplastics from wastewater effluent with advanced wastewater treatment technologies. Water Res 123:401–407

Tang N, Liu X, Xing W (2020a) Microplastics in wastewater treatment plants of Wuhan, Central China: abundance, removal, and potential source in household wastewater. Sci Total Environ 745:141026

Tang W, Li X, Liu H, Wu S, Zhou Q, Du C, Teng Q, Zhong Y, Yang C (2020b) Sequential vertical flow trickling filter and horizontal flow multi-soil-layering reactor for treatment of decentralized domestic wastewater with sodium dodecyl benzene sulfonate. Biores Technol. https://doi.org/10.1016/j.biortech.2019.122634

Thompson RC, Olsen Y, Mitchell RP, Davis A, Rowland SJ, John AWG, McGonigle D, Russell AE (2004) Lost at sea: where is all the plastic? Science(Washington) 304(5672):838

Thuptimdang P, Siripattanakul-Ratpukdi S, Ratpukdi T, Youngwilai A, Khan E (2021) Biofiltration for treatment of recent emerging contaminants in water: current and future perspectives. Water Environ Res. https://doi.org/10.1002/wer.1493

Tofa TS, Kunjali KL, Paul S, Dutta J (2019) Visible light photocatalytic degradation of microplastic residues with zinc oxide nanorods. Environ Chem Lett. https://doi.org/10.1007/s10311-019-00859-z

Tong M, He L, Rong H, Li M, Kim H (2020) Transport behaviors of plastic particles in saturated quartz sand without and with biochar/Fe_3O_4-biochar amendment. Water Res. https://doi.org/10.1016/j.watres.2019.115284

Uheida A, Mejía HG, Abdel-Rehim M, Hamd W, Dutta J (2021) Visible light photocatalytic degradation of polypropylene microplastics in a continuous water flow system. J Hazard Mater 406:124299

Vasile C, Pascu M (2005) Practical guide to polyethylene. iSmithers Rapra Publishing

Viveknand AC, Mohapatra S, Tyagi VK (2021) Microplastics in aquatic environment: challenges and perspectives. ECSN 131151. https://doi.org/10.1016/j.chemosphere.2021.131151

Wang J, Liu X, Dai Y, Ren J, Li Y, Wang X, Zhang P, Peng C (2020a) Effects of co-loading of polyethylene microplastics and ciprofloxacin on the antibiotic degradation efficiency and microbial community structure in soil. Sci Total Environ. https://doi.org/10.1016/j.scitotenv.2020.140463

Wang Q, Zhang Y, Wangjin X, Wang Y, Meng G, Chen Y (2020b) The adsorption behavior of metals in aqueous solution by microplastics effected by UV radiation. J Environ Sci 87:272–280

Wang S, Chen H, Zhou X, Tian Y, Lin C, Wang W, Zhou K, Zhang Y, Lin H (2020c) Microplastic abundance, distribution and composition in the mid-west Pacific Ocean. Environ Pollut 264:114125. https://doi.org/10.1016/j.envpol.2020.114125

Wang Z, Sedighi M, Lea-Langton A (2020d) Filtration of microplastic spheres by biochar: removal efficiency and immobilisation mechanisms. Water Res. https://doi.org/10.1016/j.watres.2020.116165

Wang C, Zhao J, Xing B (2021) Environmental source, fate, and toxicity of microplastics. J Hazard Mater. https://doi.org/10.1016/j.jhazmat.2020.124357

Wei W, Huang QS, Sun J, Dai X, Ni BJ (2019) Revealing the mechanisms of polyethylene microplastics affecting anaerobic digestion of waste activated sludge. Environ Sci Technol. https://doi.org/10.1021/acs.est.9b02971

Wright SL, Ulke J, Font A, Chan KLA, Kelly FJ (2020) Atmospheric microplastic deposition in an urban environment and an evaluation of transport. Environ Int 136:105411. https://doi.org/10.1016/j.envint.2019.105411

Xu Z, Bai X, Ye Z (2021) Removal and generation of microplastics in wastewater treatment plants: a review. J Clean Prod. https://doi.org/10.1016/j.jclepro.2021.125982

Yang L, Zhang Y, Kang S, Wang Z, Wu C (2021) Microplastics in soil: a review on methods, occurrence, sources, and potential risk. Sci Total Environ. https://doi.org/10.1016/j.scitotenv.2021.146546

Yazdani Foshtomi M, Oryan S, Taheri M, Darvish Bastami K, Zahed MA (2019) Composition and abundance of microplastics in surface sediments and their interaction with sedimentary heavy metals, PAHs and TPH (total petroleum hydrocarbons). Mar Pollut Bull 149:110655. https://doi.org/10.1016/j.marpolbul.2019.110655

Yoshida S, Hiraga K, Takehana T, Taniguchi I, Yamaji H, Maeda Y, Toyohara K, Miyamoto K, Kimura Y, Oda K (2016) A bacterium that degrades and assimilates poly(ethylene terephthalate). Science. https://doi.org/10.1126/science.aad6359

Yuan J, Ma J, Sun Y, Zhou T, Zhao Y, Yu F (2020) Microbial degradation and other environmental aspects of microplastics/plastics. Sci Total Environ 715:136968

Zhang J, Ma L (2020) Environmental sustainability assessment of a new sewage treatment plant in China based on infrastructure construction and operation phases emergy analysis. Water 12(2):484

Zhang J, Zhao M, Li C, Miao H, Huang Z, Dai X, Ruan W (2020) Evaluation the impact of polystyrene micro and nanoplastics on the methane generation by anaerobic digestion. Ecotoxicol Environ Saf. https://doi.org/10.1016/j.ecoenv.2020.111095

Zhu D, Chen QL, An XL, Yang XR, Christie P, Ke X, Wu LH, Zhu YG (2018) Exposure of soil collembolans to microplastics perturbs their gut microbiota and alters their isotopic composition. Soil Biol Biochem. https://doi.org/10.1016/j.soilbio.2017.10.027

Ziajahromi S, Neale PA, Leusch FDL (2016). Wastewater treatment plant ef fl uent as a source of microplastics: review of the fate, chemical interactions and potential risks to aquatic organisms. Water Sci Technol. https://doi.org/10.2166/wst.2016.414

Ziajahromi S, Neale PA, Rintoul L, Leusch FD (2017) Wastewater treatment plants as a pathway for microplastics: development of a new approach to sample wastewater-based microplastics. Water Res 112:93–99

Ziajahromi S, Neale PA, Telles Silveira I, Chua A, Leusch FDL (2021) An audit of microplastic abundance throughout three Australian wastewater treatment plants. Chemosphere 263:128294. https://doi.org/10.1016/j.chemosphere.2020.128294

Chapter 6
Chloroform—An Emerging Pollutant in the Air

Abinaya Sekar, George K. Varghese, and M. K. Ravi Varma

Abstract Around the world, about seven million people die each year from air pollution. World Health Organization (WHO) has reported that 9 out of 10 people breathe air that exceeds the WHO guideline values for the concentration of pollutants. The highest exposure is observed in low and middle-income countries. Urbanization and industrialization have led to the formation of many new chemical pollutants. The list of such emerging pollutants (EP) in the atmosphere is ever increasing. The sources and health impacts caused by EPs are still not well understood due to various reasons such as the paucity of monitoring data and lack of efficient detection methodology. Consequently, regulatory standards are missing for many of the EPs. The detection of such pollutants is difficult due to their complex behavior and sources in the atmosphere. Chloroform ($CHCl_3$) is one such EP in air, which is categorized as 2B carcinogen (probable human carcinogen) by the United States Environmental Protection Agency (US-EPA). $CHCl_3$ gets readily absorbed in the human body, where 60–80% of the inhaled quantity is absorbed and gets distributed throughout the body. The concentration of $CHCl_3$ found in indoor air is 10 folds higher than the outdoor air. Despite serious health effects caused by $CHCl_3$, it is not frequently monitored and included in National Ambient Air Quality Standards (NAAQs) of different countries. To address the knowledge gap, an in-depth analysis of literature was conducted. The toxicity profile of the compound, its spatial and temporal distribution, fate in the atmosphere, monitoring methods and health effects caused by it are discussed. The information presented will help policymakers to add/modify the regulatory standards on Chloroform in future amendments.

Keywords Air pollution · Chloroform · Carcinogen · Emerging pollutant

A. Sekar · G. K. Varghese (✉)
Department of Civil Engineering, National Institute of Technology Calicut, Calicut, Kerala, India
e-mail: gkv@nitc.ac.in

M. K. R. Varma
Department of Physics, National Institute of Technology Calicut, Calicut, Kerala, India

S. P. Singh et al. (eds.), *New Trends in Emerging Environmental Contaminants*, Energy, Environment, and Sustainability,
https://doi.org/10.1007/978-981-16-8367-1_6

6.1 Introduction

6.1.1 General

Air pollution is an issue of global concern posing a significant threat to the environment and driving climate change (Li 2019). Air pollution is considered one of the major risk factors for mortality and remains one of the major causes of the global burden of disease. The global burden of disease includes both years lost due to premature death and years living in poor health (Ritchie and Roser 2017). Air pollution is a serious issue impacting the economy and normal life in many developing countries. For example, the Indian cities are continuously getting featured in the list of most polluted cities of the world. Besides the National Capital-Delhi, there are twenty-one other cities of India among the top thirty polluted cities in the world (IQAir 2020). In a recent study, it was found that air pollution solely is responsible for the loss of 7 lakh crore rupees annually in India due to reduction in productivity, absence from work and premature deaths. The amount lost is equal to 3% of 2019 GDP, twice the amount used for annual public health expenditure and 40% of the cost involved in tackling the COVID-19 pandemic in India (Dalberg and CAF 2021).

Monitoring air quality is one of the important steps in combating air pollution (Dalberg and CAF 2021). Due to this, the Central Pollution Control Board of India had launched a National air quality monitoring Programme (NAMP). Under this scheme, four air pollutants (PM10, PM2.5, NO_2 and SO_2) are monitored on a regular basis (CPCB 2020). Similarly, in the recently launched National Clean Air Programme (NCAP) Continuous Ambient Air Quality Monitoring Stations (CAAQMS) were installed in 71 cities in 12 states to monitor eight pollutants (PM10, PM2.5, NO_x, SO_2, NH_3, CO, O_3 and benzene) (MoEF 2019). But, certain other pollutants that are more harmful to human health are left out from the monitoring scheme (Sekar et al. 2019). These unregulated contaminants are the Emerging pollutants (EP) which can be future candidates for regulations as and when more information becomes available on their occurrence in the environment and health effects (BarcelóE et al. 2005). Classification of EPs in the air with examples is provided in Fig. 6.1.

VOCs are an important class of emerging pollutants. There have been multiple instances where industrial accidants have discharged significant quantities of VOCs into the air environment causing deaths of many and damaging the ecosystem. One of the latest in the line was the styrene gas leak that occurred in the Indian state of Andhra Pradesh in 2020 that resulted in 12 human casualities (Bhattacharjee 2020). IARC has classified styrene in category 2B (Probably carcinogenic to humans) and it is also an emerging pollutant of concern (IARC 2002). The Bhopal gas tragedy, considered one of the largest industrial disasters causing the direct death of 3787 human beings (Government of Madhya Pradesh 2012) in addition to livestock, dogs and birds (Broughton 2005), involved the leakage of Methyl Isocyanide (MIC) (Varma and Varma 2005). In addition to MIC, the gas cloud emitted in the intervening night of December 2nd and 3rd of 1984, from the pesticide plant owned by Union Carbide India Limited (UCIL) contained gases like Chloroform, HCl,

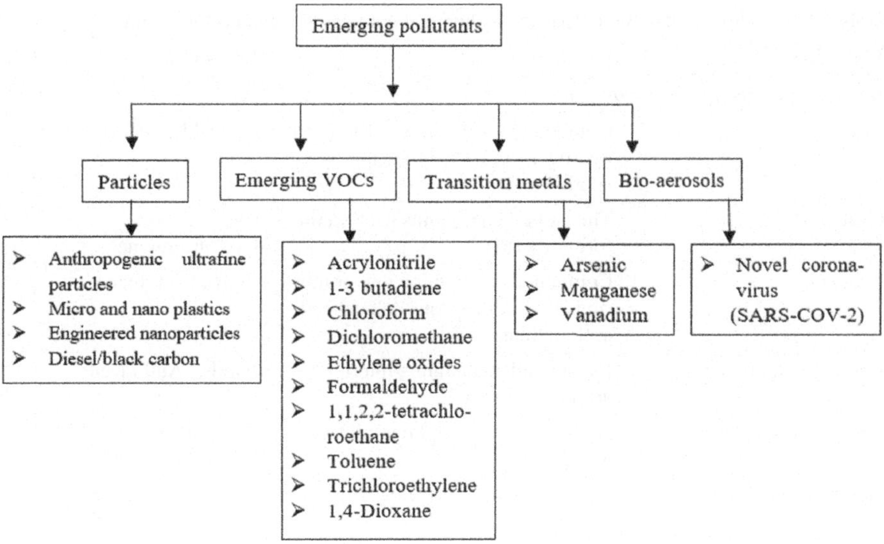

Fig. 6.1 Examples of emerging pollutants in the air (Ebere et al. 2020). Reproduced with permission. Copyright © 2021 Elsevier Ltd

dichloromethane, methylamine compounds and CO_2 (D'Silva 2006). In 2002, a fact-finding mission's investigation at Bhopal found that toxins like mercury, lead, 1,3,5 trichlorobenzene, dichloromethane and chloroform are present in the breast milk of nursing women (Sarangi 2002).

Chloroform is an emerging pollutant that causes serious health effects and is found abundantly in the environment, but not monitored frequently (Khalil and Rasmussen 1999). In atmospheric monitoring, chloroform comes under the broad category of Volatile Organic Compounds (VOCs). Classification of VOCs according to chemical composition and boiling point with examples are given in Table 6.1.

Montreal Protocol and its amendments regulate anthropogenic chlorine-containing compounds like chlorofluorocarbons, carbon tetrachloride and methyl chloroform. Due to this, there was a significant reduction in the emission of these compounds over the past twenty years. As a result, a sign of recovery was also observed in the Antarctic ozone hole (Solomon et al. 2016; Kuttippurath et al. 2018). However, the level of unregulated chlorocarbons like dichloromethane, perchloro ethane, methyl chloride and chloroform are increasing the stratospheric chlorine load. These compounds are not regulated because of their short atmospheric lifetime and substantial natural sources (Say et al. 2020). But in recent years, chlorocarbons have started to pose a threat to the healing of the ozone layer (Fang et al. 2019). The formation mechanism, properties and effects of chloroform are well characterized; however, a significant amount of uncertainty remains in understanding the source and impact on health and global climatic change. This may be due to the less availability of the monitoring data.

Table 6.1 Classification of VOCs based on chemical composition and boiling point

VOC class	Definition	Examples
Based on the chemical composition		
Aliphatic	Compounds containing carbon and hydrogen are linked together by single bonds	Formaldehyde, Acetone
Cyclic	The carbon chain joins itself in the form of a ring	Cyclohexane, Cyclohexanone
Aromatic	Compounds containing one or more rings with π electrons delocalized around them	Benzene, Toluene
Polycyclic aromatic	Compounds containing multiple aromatic rings	Pyrene, Anthracene
Nitrogen-containing	Hydrocarbons containing nitrogen group	Acetonitrile, Peroxy-acetyl-nitrate
Halogen containing	Hydrocarbons containing halogen group	Trichloroethylene, Chloroform
Based on boiling point °C		
Very volatile organic compounds	< 0 to 50–100	Methyl chloride, Chloroform
Volatile organic compounds	50–100 to 240–260	Ethanol, Toluene
Semi-volatile organic compounds	240–260 to 380–400	Phenol, Pesticides (DDT)
Non-volatile organic compounds	> 380	Organic particulate matter

This chapter summarizes the properties of chloroform and gives a wide spectrum of information on its sources, metabolic pathways, health effects and occurrence in the air medium.

6.1.2 Physio-Chemical and Toxicity Profile

Synonyms of Chloroform are Trichloroform, Trichloromethane, Methane trichloride, Methenyl trichloride and Methyl trichloride with the molecular formula $CHCl_3$. It is a volatile, heavy and colorless liquid at room temperature. It is nonflammable and possesses a sweet odor with an odor threshold of 85 ppm. The physio-chemical characteristics of $CHCl_3$ are provided in Table 6.2. It was first introduced clinically in the year 1847 as a general anesthetic (Estévez and Vilanova 2014). $CHCl_3$ gets evaporated rapidly when exposed to the atmosphere, having a vapor pressure of 21.3 kPa at 20 °C. When a concentrated emission occurs, the plume will tend to settle on the ground before dispersing because the density of $CHCl_3$ is four times that of air (Estévez and Vilanova 2014).

Table 6.2 Physio-chemical characteristics of $CHCl_3$

Property	Unit	Value
Boiling point @ 101.3 kPa	°C	61.3
Vapour pressure @ 20 °C	kPa	21.3
Water solubility @ 25 °C	g/L	7.2–9.3
Density @ 25 °C	g/cm^3	1.48
Henry's law constant @ 20 °C	$Pa.m^3/mol$	304
Log K_{ow}		1.97
Log K_{oc}		1.44–2.79

In general, the population is expected to get exposed to a low level of $CHCl_3$ via air, water and food. The expected total daily uptake is determined to be 2 μg/kg body weight. But high daily intake may be expected among the population that is occupationally exposed or those who live near to the potential sources of $CHCl_3$ (World Health Organization Geneva). According to USEPA guidelines, $CHCl_3$ was designated to the weight of evidence Group B2 i.e. probable human carcinogen. This categorization was made based on the increasing evidence of different types of tumors in rats and three strains of mice (US Environmental Protection Agency 1992). The reference concentration and carcinogenicity slope factor of $CHCl_3$ are listed in Table 6.3. Based on the development of hepatocellular carcinomas in mice in an oral gavage study, the estimated inhalation unit risk (IUR) is found to be 2.5 E-5 ($\mu g/m^3$)(US Environmental Protection Agency 1992).

Based on the number of halogen atoms, $CHCl_3$ is also classified under a chemical group called Trihalomethane (THM). These are organic chemical compounds in which three halogen atom replaces three among the four hydrogen atoms in methane (CH_4). There are four chemical compounds under this classification, it includes Chloroform ($CHCl_3$), bromoform ($CHBr_3$), bromodichloromethane ($CHBrCl_2$) and dibromochloromethane ($CHBr_2Cl$) (Estévez and Vilanova 2014). The International Agency for Research on Cancer (IARC) has formulated a system to determine the category of carcinogenicity based on scientific findings derived from human and

Table 6.3 Toxicity profile of $CHCl_3$ (RAIS 2021)

Toxicity parameter	Values	Unit	Reference
Inhalation unit risk	0.000023	$(\mu g/m^3)^{-1}$	IRIS
Inhalation RfC	0.488262 (Acute) 0.097652 (Chronic) 0.244131 (Sub-Chronic)	(mg/m^3)	ATSDR final
Oral RfC	0.3 (Acute) 0.01 (Chronic) 0.1 (Sub-Chronic)	(mg/kg-day)	ATSDR final and IRIS
Slope factor	6.10×10^{-3}(oral) 3.05×10^{-2} (dermal) 8.05×10^{-2} (inhalation)	$(mg/kg\text{-}day)^{-1}$	IRIS and RAIS

Table 6.4 Types of THMs

Chemical name	Common name	Chemical Formula	IARC Classification
Trichloro-methane	Chloroform	$CHCl_3$	2B
Dibromochloromethane		$CHClBr_2$	3
Bromodichloro-methane		$CHCl_2Br$	2B
Tribromo-methane	Bromoform	$CHBr_3$	3

2B—"Possibly carcinogenic to human"; 3—"Not classifiable as to its carcinogenicity to humans"

Table 6.5 Hazard classification of $CHCl_3$

	Hazard classification	Meaning
Toxicity category rating	Category-4	Harmful when swallowed
Carcinogenic classification	Category-2B	Probably carcinogen
Specific target organ toxicant classification	STOT RE 2	Likely to cause damage to organs by continuous or repeated exposure
Skin corrosion and irritation classification	Category 2	Causes skin irritation on contact
Eye irritation	Category 2	Causes serious eye irritation
Acute toxicity (inhalation)	Category 3	Toxic when inhaled
Reproductive toxicity	Category 2	Suspected to cause damage to the unborn child

animal experimental studies. The IARC classification of THMs is given in Table 6.4 (World Health Organization 2021).

Based on the criteria set by the "Classification, Labelling and Packaging" regulation from 2008 by the European Union (EU). $CHCl_3$ is designated under different categories as provided in Table 6.5 (Public Health England Chloroform Incident Management 2017).

6.2 Source

Earlier studies have reported that 90% of the chloroform flux in the atmosphere is due to natural sources (Mcculloch 2003) but recent studies reconstructed the earlier estimates and determined that 50% of the global total emission may be due to man-made sources (Trudinger et al. 2004; Wortan et al. 2006).

6.2.1 Natural Sources

One of the major sources is emission from the open ocean and other littoral and coastal sources due to seaweeds, micro and macroalgae (Khalil and Rasmussen 1999). Soil processes also contribute to the emission of $CHCl_3$ into the atmosphere due to the chlorination of soil acids, especially humic materials by hypochlorous acid (HOCl). The reaction happens due to the presence of hydrogen peroxide and chloride ions in the presence of chloroperoxidase enzyme as in Eqs. 6.1 and 6.2 (Mcculloch 2003). The highest emission is observed in woodchip-covered forest soil (Hoekstra et al. 2001). Human interventions in soil processes substantially increase the emission of $CHCl_3$. Studies have also shown the emission of chloroform from the peatland ecosystems (Dimmer et al. 2001). Accidental fire of vegetation also emits $CHCl_3$ to the atmosphere (Lobert et al. 1999). The level of $CHCl_3$ in termite mounds is observed to be a thousand times more than ambient concentrations. Hence, termites are one of the major sources contributing to $CHCl_3$ emission (Khalil et al. 1990). Volcanic activity is one of the significant contributors, as $CHCl_3$ is detected in volcanic gases (Isidorov et al. 1990). Emission of $CHCl_3$ was also observed in rice fields and biogas pits (Khalil and Rasmussen 1990). Laboratory simulations of tropospheric irradiation of trichloroethylene have shown the formation of $CHCl_3$ as one of the principal products and hence the photochemical transformation of trichloroethylene may also be one of the significant sources of chloroform in the atmosphere (US Environmental Protection Agency 1976).

$$H_2O_2 + H^+ + Cl^- \xrightarrow{\text{Chloroperoxidase}} HOCl + H_2O \tag{6.1}$$

$$\text{Humic material} + HOCl \rightarrow \text{Chlorinated humic material} + CHCl_3 + CH_3COOH \tag{6.2}$$

6.2.2 Anthropogenic Sources

$CHCl_3$ is one of the by-products obtained during the delignification of wood and cellulose pulps and also during bleaching of paper using chlorinated agents like chlorine, oxidants containing chlorine (Cl_2), chlorine dioxide, etc. during paper manufacturing (Juuti et al. 1996).

Disinfecting water, wastewater and swimming pools is highly essential to prevent waterborne diseases like typhoid and cholera (Powick 1989). The most commonly employed disinfectants include chlorine, bleaching powder, chlorine dioxide and others. Out of these, Cl_2 is the most common disinfecting agent because of its efficacy, cost and retentive power (Judd and Black 2000). The chlorinated disinfectants react with natural organic matter and lead to the formation of harmful byproducts, which are widely known as disinfection by-products (DBPs) (Pándics et al. 2018). There are more than 600 DBPs which have been detected in chlorinated waters and most of them are mutagens or probable carcinogens. Chloroform is one of the most commonly found DBPs (Anchal et al. 2020). Due to the volatile nature of THMs, these compounds are likely to get evaporated and remain in the air.

Hazardous waste sites and other sanitary landfills also emit chloroform to the atmosphere (US Department of Health and Human Services 1997). Natural and induced anaerobic fermentation in biogas generation also emits $CHCl_3$ to the atmospheric air (Khalil and Rasmussen 1990). In steam power generators, cooling water is used to absorb the heat generated. In this process, Cl_2 is often added to the cooling water to prevent fouling by killing the slime-forming organisms. During this process, $CHCl_3$ is formed in the aqueous reaction of Cl_2 with the organic matter present in the cooling water (Jolley et al. 1983). In chemical industries, it is used as an intermediate in the manufacturing of dyes, pesticides, insecticides and as a fumigant (US Environmental Protection Agency 2004). It is also widely used in pharmaceutical industries (eg. In the extraction process of antibiotics like penicillin and others), fertilizer industries, rubber industries, solvent manufacturing industries and others. Mainly it is used in the production of hydrochlorofluorocarbon-22 (Estévez and Vilanova 2014). $CHCl_3$ is a byproduct in the manufacturing polyvinyl chloride and other bulk chlorinated chemicals. Chloroform is also found as a cleansing agent in a wide range of products (For example- all-purpose cleaners, detergents, dryer sheets, etc. (Scranton 2011)). It is also an important building block in manufacturing fluorinated polymers and associated copolymers (European Union 1907). Though it is not possible to find chloroform in the ingredients list of certain household and personal care products it may also be present in the form of impurities. The presence of $CHCl_3$ was also detected in bleach, cleaning and fragrance products (Odabasi 2008; Kwon et al. 2008). $CHCl_3$ is very commonly used as a refrigerant (European Union 2007). Before banning chloroform in the usage of consumer products and drugs it was used in the products like toothpaste, linaments and cough syrup (US Environmental Protection Agency 1984).

6.3 Environmental Persistency

Chloroform readily dissolves in the water but does not adhere very well to the soil. Due to this reason, it travels down to the soil and reaches the groundwater system. It lasts for a longer duration in groundwater and air (US Department of Health and Human Services 1997). Because of its high volatility and low solubility, $CHCl_3$ will not remain in the water for a longer duration, but it may risk embryo-larval stages of several aquatic organisms when it is discharged into water bodies due to industrial discharges or spills (World Health Organization Geneva). In the air, the half-life of $CHCl_3$ ranges from 55–620 days (World Health Orgnization Geneva 2004). It decomposes slowly when exposed to sunlight with or without air and in the dark with air. When $CHCl_3$ is released into the atmosphere it comes in contact with photochemically generated hydroxyl radical (OH^-) in the troposphere, which leads to the formation of compounds like phosgene, dichloromethane, formyl chloride, CO, CO_2 and HCl (Park et al. 1997). It does not react with O_2 at a temperature up to 290 °C. But at 270 °C NO_2 oxidizes $CHCl_3$ to phosgene, HCl, H_2O and CO_2. At a temperature above 450 °C pyrolysis of $CHCl_3$ takes place to produce tetrachloroethylene, HCl and other chlorocarbons (International Programme on Chemical Safety 1994). On further chlorination of $CHCl_3$, carbon tetrachloride is formed by irradiation of elemental chlorine. In the temperature range of 225–275 °C $CHCl_3$ undergo bromination and yields bromochloromethane (International Programme on Chemical Safety 1994).

6.4 Metabolism of Chloroform

Chloroform enters the human body by three pathways- oral (food, beverages and water), inhalation (contaminated air) and dermal (swimming, cleaning and showering) as shown in Fig. 6.2. In the body, it is absorbed by the lungs, gastrointestinal tract and skin (US Department of Health and Human Services 1997; World Health Orgnization Geneva 1994). The metabolism of $CHCl_3$ happens in two different pathways i.e., via an oxidative or a reductive pathway as in Fig. 6.3 (Gemma et al. 2003). It is mainly excreted via lungs either unchanged or in the form of main metabolite CO_2 (US Environmental Protection Agency 2001). Other than lung exchange, a small amount of $CHCl_3$ is excreted via urine and feces. Around 60–80% of the inhaled $CHCl_3$ is absorbed in the human body. Studies have shown that high $CHCl_3$ concentrations are found in adipose tissue, brain, liver, kidney, adrenals and the blood of humans and animals after exposure via oral and inhalation routes.

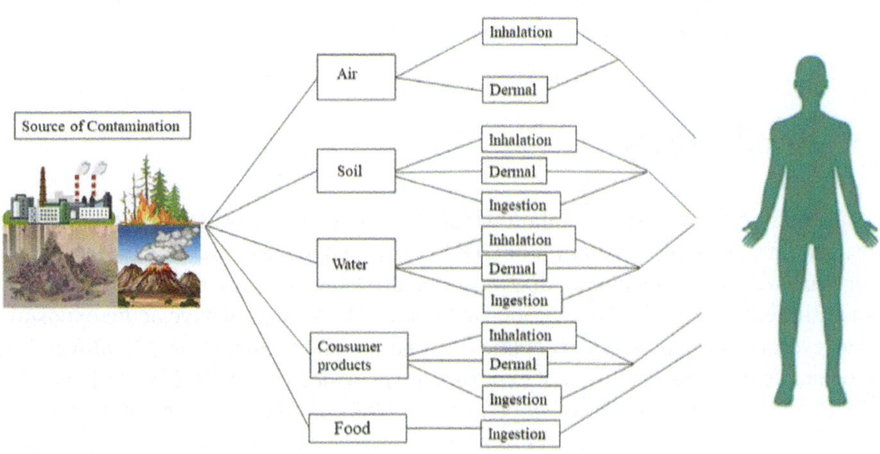

Fig. 6.2 Possible human exposure pathways of chloroform

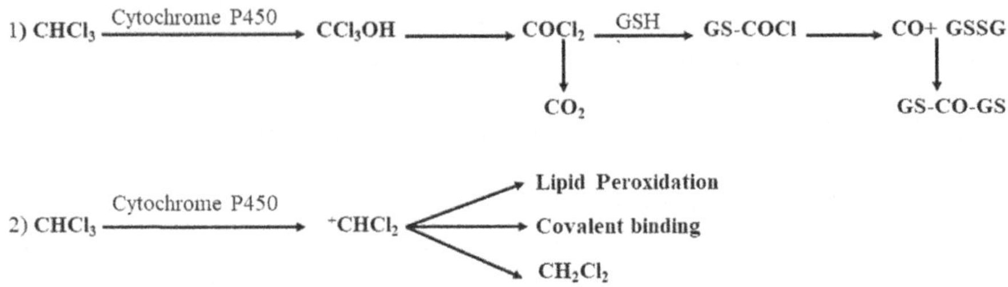

*GSH – reduced glutathione, GSSG – oxidative glutathione,

Fig. 6.3 Metabolic pathway of chloroform

In the oxidative pathway, $CHCl_3$ may be metabolized to trichloromethanol (CCl_3OH) which spontaneously decomposes to electrophilic phosgene ($COCl_2$). $COCl_2$ is highly reactive. Phosgene can react with cellular macromolecules such as enzymes, proteins or polar ends of phospholipids leading to the formation of covalent bonds. These supplements can interfere with molecular function and lead to loss of cell function and lead to cell death. Alternatively, $COCl_2$ may be hydrolyzed on reaction with water yielding CO_2 and HCl. In the reduction pathway, under anoxic or hypoxic conditions $CHCl_3$ is reduced to dichloromethyl free radical. It is highly reactive and forms covalent adducts with microsomal enzymes and fatty acid tails of phospholipids. This may lead to a loss in microsomal enzyme activity and may also lead to lipid peroxidation. Almost all the tissues are capable of metabolizing $CHCl_3$ and the highest metabolism occurs in the liver, kidney and nasal mucosa (US Environmental Protection Agency 1992; World Health Orgnization Geneva 1994; Health Protection Agency 2007). The relative ratio between oxidation and reduction pathway depends on O_2 partial pressure, $CHCl_3$ concentration, the rodent species, its gender and the organ exposed (Gemma et al. 1996).

6.5 Health Effects

Local effects observed after inhalation include shortness of breath and irritation of the nose and throat. Acute inhalation of $CHCl_3$ may lead to systemic effects like excitement, nausea, vomiting, ataxia, dizziness and drowsiness. Higher concentration may cause ailments like convulsion, coma and death. Those who survive acute exposure, after several days may develop hepatic dysfunction and renal damage. Health effects associated with different levels of exposure are provided in Table 6.6. Detailed health effects caused due to inhalation exposure of $CHCl_3$ are listed in Table 6.7.

Table 6.6 Health effects due to exposure by inhalation and ingestion (International Programme on Chemical Safety 1994, 2004)

Inhalation			Ingestion	
Level in ppm	Level in mg/m^3	Symptoms	Level in g	Symptoms
< 50	< 249	Discomfort	7.5	Serious illness
500	2490	Illness	45	Mean lethal oral dose (adult)
2000	9960	Several toxic effects on 60 min exposure		
4896–14,892	24,000–73,000	Induction of anesthesia		

6.6 Workplace Exposure Limits (WEL)

The workplace exposure limits set by different regulatory bodies are listed below.

- Occupational Safety and Health Administration (OSHA) Permissible Exposure Limits (PEL)—50 ppm (Should not exceed any time)
- National Institute for Occupational Safety and Health (NISOH) Recommended Airborne Exposure Limits (REL)—2 ppm (Should not exceed any time in a 60 min working hour)
- American Conference of Governmental Industrial Hygienists (ACGIH) Threshold limit value (TLV)—10 ppm (8 h TWA)
- Immediately dangerous to life or health (IDLH) Levels—500 ppm

Table 6.8 lists the Emergency response guideline value set by the American Industrial Hygiene Association (AIHA), Acute exposure guideline level (AEGL) set by USEPA and protective action criteria values.

6.7 Sampling and Analytical Techniques

6.7.1 Sampling Techniques

Sampling of any VOC includes either capturing of whole air or preconcentration of air sample on the adsorbent (Ras et al. 2009). All standard methods differ depending on the sampling method, the type and nature of the absorbent, the pollutant extraction method, and the determination technique. There are three major techniques employed in VOC sampling viz. whole air sampling, sorbent sampling and chemisorption (Kumar and Víden 2007). The general method of analysis includes direct column injection, sorption onto adsorbents followed by liquid or thermal desorption and subsequent analysis in Gas Chromatography (GC). In the case of $CHCl_3$, the air detection limit of 0.01 $\mu g/m^3$ can be achieved (World Health Organization Geneva).

Table 6.7 Health effects caused due to inhalation exposure

Type of effect	Explanation
Death	Death was reported among women exposed to $CHCl_3$ via anaesthesia. Hepatotoxicity was one of the major reasons for death. Hepatotoxicity is due to prolonged labour, starvation, dehydration and exhaustion (US Department of Health and Human Services 1997)
Respiratory effects	Alternation in the respiratory rate was found among the patients exposed to $CHCl_3$ via anaesthesia (Whitaker and Jones 1965)
Cardiovascular effects	$CHCl_3$ causes the cardiac effect. Other issues associated with $CHCl_3$ include bradycardia, cardiac arrhythmia, hypotension, first-degree Artio-ventricular block or complete block (Smith et al. 1973)
Gastrointestinal effects	On exposure to $CHCl_3$ between the range of 8000–22,500 ppm via anaesthesia causes side effects like nausea and vomiting. Other gastrointestinal symptoms observed include nausea, dry mouth and fullness of the stomach (Callen et al. 1980)
Hematological effects	Increased prothrombin was observed in the liver among individuals exposed to anaesthesia of 8000 ppm (Smith et al. 1973)
Musculoskeletal effects	Few musculoskeletal effects were observed in human and animal studies. $CHCl_3$ causes toxicity and regenerative cell proliferation among various tissues in mice (Larson et al. 1996)
Hepatic effects	Hepatotoxicity is one of the most important effects caused due to chloroform in humans and animals. When a patient was exposed to $CHCl_3$ of 8000–10,000 ppm via anaesthesia, Sulfobromophthalein retention was observed, indicating impairment in liver function. Hepatic necrosis was observed among women exposed to $CHCl_3$ in the form of anaesthesia during childbirth. Other effects found in women include jaundice, enlargement in the liver, tenderness, delirium, coma and even death. Hepatotoxicity was observed among swimmers in chlorinated swimming pools (Whitaker and Jones 1965; Smith et al. 1973)

(continued)

Table 6.7 (continued)

Type of effect	Explanation
Renal effects	Kidney toxicity was found in humans due to $CHCl_3$ inhalation. Renal toxicity was also observed among swimmers getting trained in chlorinated swimming pools. Elevation in β-2 microglobulin was observed among swimmers (Aiking et al. 1994; Royston 1924)
Body Weight effects	No studies were reported regarding the bodyweight effects caused due to $CHCl_3$. But in animal experimental studies, weight loss was found in the case of mice and rats (Larson et al. 1994)
Immunological and Lymphoreticular effects	In an occupationally exposed study conducted among workers in the pharmaceutical industry. Splenomegaly is the only immunological effect observed among the workers (Bornski et al. 1967)
Neurological effects	When exposed to $CHCl_3$ concentration of 920 ppm for 3 min, illnesses like dizziness and vertigo were found. With the increase in concentration, headache and slight intoxication were also found. Exhaustion was observed among women workers when exposed to a concentration greater than 22 ppm. Chronic $CHCl_3$ exposure of greater than 77 ppm causes exhaustion, concentration lack, depression or irritability among 9 out of 10 women. The other neurological symptoms include fatigue, insomnia, somnolence, increase in dreaming, hypomnesia, anorexia and palpitations (Whitaker and Jones 1965; Smith et al. 1973)
Reproductive effects	Exposures may also cause birth defect and miscarriage of pregnancy in humans (Williams et al. 2018). In the case of animals, rats exposed to 300 ppm of $CHCl_3$ during their gestation period have shown decreased conception rates (Schwetz et al. 1974). Exposures to even 30 ppm of $CHCl_3$ led to fetal resorptions (Land et al. 1979). Reduced ability to remain pregnant was observed when the exposure was 100 ppm (Murray et al. 1979)
Developmental effects	No studies were reported regarding the development's effects caused due to $CHCl_3$ in humans. But, $CHCl_3$ induced fetotoxicity and teratogenicity was found in the case of studies with animals (Schwetz et al. 1974)
Genotoxic effects	No studies were reported regarding the genotoxic effects caused due to $CHCl_3$ in humans but in the case of animals exposure to 400 ppm of $CHCl_3$ for a duration of 5 days showed an increase in abnormal sperm in mice (Land et al. 1979)

(continued)

Table 6.7 (continued)

Type of effect	Explanation
Cancer	IARC has reported that there are inadequate scientific evidences supporting carcinogenicity in humans. But sufficient evidence is available in studies with animals for carcinogenic property and hence it is classified as a possible carcinogen to humans (IARC 1993)

Table 6.8 Emergency response planning guideline value (ERPG) (American Industrial Hygiene Association (AIHA) 2016), Acute exposure guideline levels (AEGL) (National Academy and of Sciences 2012) in ppm and Protective action criteria

ERPG		AEGL						PAC	
			10 min	30 min	60 min	4 h	8 h		
ERPG-1	NA	AEGL-1	NA	NA	NA	NA	NA	PAC-1	2
ERPG-2	50	AEGL-2	120	80	64	40	29	PAC-2	64
ERPG-3	5000	AEGL-3	4000	4000	3200	2000	1600	PAC-3	3200

ERPG 1-mild effects; ERPG 2-serious effects; ERPG 3-life-threatening effects; AEGL 1-non-disabling; AEGL 2--disabling; AEGL 3-lethal; PAC 1-mild effects; PAC 2-Irreversible health effects; PAC 3-life-threatening effects

6.7.1.1 Whole Air Sampling

Whole air sampling is a method of collecting whole air in a pre-cleaned container of definite volume as shown in Fig. 6.4 (Berkowitz et al. 2005). This method is successfully used in the measurement of trace organic compounds. The most commonly used containers include stainless steel canisters, plastic bags and glass bulbs (Kumar and Víden 2007). Before sampling, the canister will be passed through a passivation process to minimize the internal surface reactivity. In addition to that, canisters must be cleaned and evacuated to minimize contamination (US Environmental Protection Agency 1999). The plastic bags (Tedlar, Teflon or Aluminized Tedlar) are very simple to use, economic and available in different sizes. Cleaning of plastic bags is done by filling pure nitrogen or ultrahigh purity zero air followed by evacuating it under pressure. Before sampling, bags will be filled with zero air overnight to check the leakage (Rudolph 1990). Bags are mostly used in measuring strong polluted air or gas which are relatively high in concentration with the levels ranging from parts per million (ppm) to parts per billion (ppb) (Zielinska and Faujita 1994).

6.7.1.2 Sorbent Sampling

A larger volume of gas can be collected in the sorbent tubes (Fig. 6.5) when compared with the canisters. In addition to that, sorbent tubes are easier to handle when

Fig. 6.4 Sketch of sampling canister

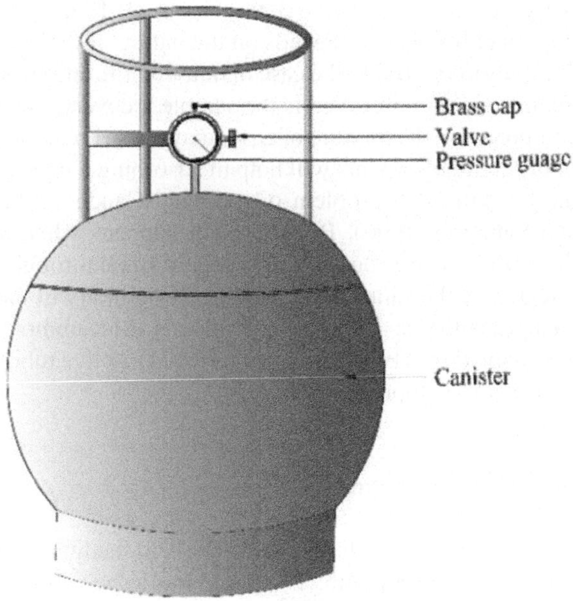

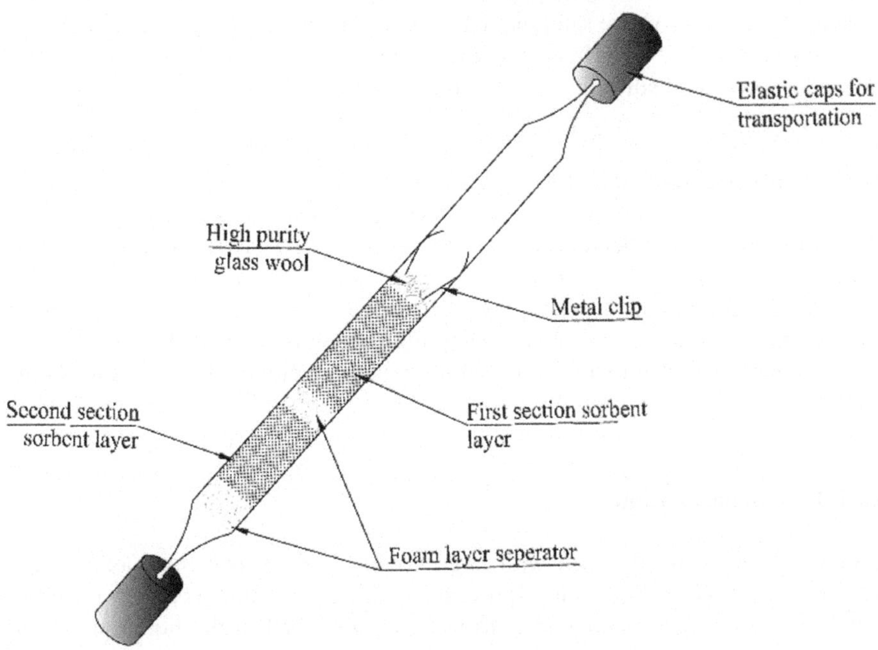

Fig. 6.5 Sketch of sorbent tube

compared with canisters (Batterman et al. 1998). The choice of adsorbent is very important because it depends on the nature of the matrix and the type of compound to be adsorbed. But in the case of non-volatile and strongly adsorbed compounds, the main limitation observed is the sample recovery. On the other hand, highly volatile component in the mixture of gases may cause breakthrough problems. A combination of different adsorbents will help in adsorbing a wide range of compounds to prevent the breakthrough problem (Camel and Caude 1995; Gawrys et al. 2001; Hansen and Palmgren 1996). Breakthrough happens when a contaminant passes through the adsorbent without getting trapped. Breakthrough capacity may be significantly affected by the sampling flow rate and humidity of the air. The collection efficiency changes with respect to the face velocity (Mccammon et al. 2016). Sorbent sampling is classified into two major categories (1) Active tube sampling and (2) Passive tube or diffusive sampling.

Active Tube Sampling

Active sampling on the packed adsorbent medium with the appropriate tube is done using a sampling pump. The pump has to be calibrated before use. This method is more convenient for determining workplace exposure because the sampler can be worn by the worker. Active sampling has a control over sampling rate and time. The most commonly used adsorbents include carbonaceous material, porous polymers, molecular sieves and silica gel (Kumar and Víden 2007).

Passive/Diffusive Tube Sampling

This sampling method depends on the free flow of the analyte from the sampling medium to the collection medium due to the difference in concentration between the two media. A continuous flow of analyte between the media occurs until equilibrium is reached in the system or when the sampling session is stopped. The advantage of passive sampling is that there is no need for expensive pumps (Gorecki and Namies 2002).

6.7.1.3 Chemisorption

Chemisorption is mostly used with large volumes of air. But this method needs solvent with a high boiling point to prevent any loss of the sample (Camel and Caude 1995). The identified compound is irreversibly absorbed in the liquid phase, and then it is further transformed by chemical reactions. In this process, an impactor and liquid-filled bubble are used. This method is mainly used to sample compounds with high boiling point, reactive or polar substances (Batterman et al. 1998). The major limitation of is technique is the contamination of absorbents during the sample preparation and storage (Dalene et al. 1992).

Table 6.9 Commonly used methods for determination of $CHCl_3$ in air

Sampling Process	Analysis techniques	USEPA Method	Reference
Adsorption onto Tenax sorbent tubes	GC/MS	USEPA TO 1	US Environmental Protection Agency (1984)
Adsorption onto carbon molecular sieve; thermal desorption	GC/MS	USEPA TO2	US Environmental Protection Agency (1984)
Cryogenic preconcentration; thermal transfer to GC	GC/ECD/FID	USEPA TO3	US Environmental Protection Agency (1984)
Cryogenic preconcentration	Direct FID	USEPA TO-12	US Environmental Protection Agency (1984)
Whole air sampling in polished SUMMA canisters	GC/FID/ECD or GC/MS	USEPA TO 14A	US Environmental Protection Agency (1999)
Collection in passive sampler; Thermal desorption	TD/GC	USEPA method 325 A & B	US Environmental Protection Agency (2017, 2019)
Collection in Tedlar or aluminized Mylar bags	GC/MS	USEPA Method 004 and 8260	US Environmental Protection Agency (1996, 2017)
Adsorption onto sorbent tubes	GC/MS	USEPA TO 17	US Environmental Protection Agecy (1999)
Adsorption onto bags/sorbent tubes; solvent extraction/thermal desorption	GC/MS	USEPA TO 18	US Environmental Protection Agency (2019)

6.7.2 Analytical Techniques

The most commonly used techniques for the determination of $CHCl_3$ in the air is Gas Chromatography with an electron capture detector (ECD), flame ionization detector (FID), photoionization detector (PID) or mass spectrometry (MS). Table 6.9 provides the list of standard methods by USEPA.

6.7.3 Sensor-Based Techniques

Development of low-cost sensors for monitoring of VOCs is emerging as a promising technology to monitor larger area with higher resolution at a lower cost when compared with the standard reference methods. The major disadvantage associated with sensor based technologies includes, too high limit of detection and poor sensitivity (Sekar et al. 2019; Spinelle et al. 2017).. The most commonly used sensors for the detection of VOC includes,

- Metal Oxide Semiconductor (Abdul Hamid et al. 2019)
- Photoionization Detector (Driscoll 2008)
- Nondispersive infrared sensors (Zhang et al. 2021)
- Electrochemical sensors (Abdul Hamid et al. 2019)
- Thermal sensors

Other than the above-mentioned sensors, Chemoresistors (rapheme, carbon nanotubes composites) (Tasaltin and Basarir 2014), Hybrid nanostructures (Athawale et al. 2002) and Optical sensors (Fong et al. 2015) are also used for detecting chloroform in air.

6.8 Levels of Chloroform in Different Environments

This section discusses the occurrence of $CHCl_3$ in different environments based on the data obtained from selected studies. A tabulation of the levels of chloroform observed in these studies are provided in Table 6.10.

6.8.1 Ambient Air

In a study conducted at Delhi to estimate exposure of non-methane hydrocarbons among the roadside shop vendors, it was found that between the years 2014–2015, the highest concentration was found at the ITO region with the value of 163 $\mu g/m^3$. It was also observed that 1,3-butadiene and $CHCl_3$ contributed higher cancer risk (Kumar et al. 2020). In several studies conducted in China, the atmospheric concentrations of $CHCl_3$ was at levels likely to cause cancer risk (Mozaffar and Zhang 2020; Li et al. 2020; Lyu et al. 2020). The mean concentration of $CHCl_3$ observed in deer park Texas during the year 2003 was 0.22 $\mu g/m^3$ and the level of $CHCl_3$ was not correlated with the concentration of BTEX compounds in the air (Smith et al. 2007). $CHCl_3$ was monitored in three different seasons (spring, summer and fall) at Minneapolis/ St. Paul metropolitan and the average concentrations found in outdoor, indoor and personal locations were 0.1, 1.5 and 1.5 $\mu g/m^3$, respectively (Sexton et al. 2004). A study was attempted in Mexico, to find the interrelationship between outdoor, indoor and personal exposures to $CHCl_3$. The geometric mean of indoor concentrations of chloroform was comparable to personal exposures but the geometric mean of personal exposure was much higher than the ambient air concentrations. The means values were found to be 0.1, 0.6 and 0.7 $\mu g/m^3$ in indoor, outdoor and personal exposure, respectively (Serrano-Trespalacios et al. 2004). The cancer risk associated

Table 6.10 Level of $CHCl_3$ observed in different air environments

Location type	Country/Location	Concentration ($\mu g/m^3$)	Reference
Ambient	India, Delhi	163	Kumar et al. (2020)
	US, Texas	0.22	Smith et al. (2007)
	Minneapolis/ St. Paul metropolitan	0.1	Sexton et al. (2004)
	Mexico	0.6	Serrano-Trespalacios et al. (2004)
Indoor	Residential	0.01–10.84	Wickliffe et al. (2020)
	Cars at Nanjing	72	Xing et al. (2018)
	Laundry rooms	4	Lin et al. (2021)
	Barbeque restaurant	3.5	Arı et al. (2019)
	School	2.1	Adgate et al. (2004)
Swimming pool	Jharkhand	0.033	Anchal et al. (2020)
	Beijing, China	269.60–516.22	Wang and Dong (2020)
	Portugal	13.0–369.3	Felgueiras et al. (2019)
	Northern Italy	46.1	Fantuzzi et al. (2001)
Industrial	Daegu dying industrial complex, South Korea	11.72	Shuai et al. (2018)
	Anthracite and bituminous coal unit	0.21 & 0.10 ppm	Pudasainee et al. (2010)
	Highly industrialized region in turkey	0.4	Dumanoglu et al. (2014)

with background concentrations was calculated in North American regions and the risk exceeded the benchmark value (one in million) in rural and remote areas (Lasa and Śliwka 2003).

6.8.2 *Indoor Air*

In a study conducted at residential indoor air in southeast Louisiana, an attempt was made to determine increased long-term health risks caused due to $CHCl_3$. The concentration was found to be in the range of 0.01–10.84 $\mu g/m^3$ and the cancer risk exceeded the guideline values. The possible sources of $CHCl_3$ in the indoor air were identified as chlorinated residential water, chlorine bleach and the use of products containing chlorinated VOCs (Wickliffe et al. 2020). In the random sampling conducted in five different cars of different brands at Nanjing, the mean value of $CHCl_3$ inside vehicles was found to be 72 $\mu g/m^3$ with the maximum level reaching up to 325 $\mu g/m^3$ (Xing et al. 2018). In a study reporting occupational personal exposure to chloroform among hotel housekeepers, the highest concentration was found in the laundry rooms with the value of 4 μg/m^3, a level sufficient to cause cancer risk among the workers (Lin et al. 2021). The mean concentration of $CHCl_3$ in a barbeque restaurant at Istanbul-Ankara Highway was 3.5 μg/m^3 and $CHCl_3$ had the highest cancer risk among the 60 selected VOCs (Arı et al. 2019). In a study conducted at microenvironments like residences, offices, restaurants and nightclubs, the highest concentration was observed in a sports complex with a swimming pool, where the level of $CHCl_3$ was found to be uncorrelated with the level of benzene (Hinwood et al. 2006). A similar study was conducted in Minneapolis, USA to determine indoor, outdoor and personal exposure to school children. The mean levels were observed to be 2.1, 0.8 and 2.5 μg/m^3 in indoor, outdoor and personal exposure, respectively (Adgate et al. 2004).

6.8.2.1 Swimming Pool

In the study conducted at five indoor swimming pools at Jharkhand, India it was observed that the amount of $CHCl_3$ in pool water and the air was 197.18 ± 16.31 μg/L and 0.033 μg/m^3, respectively. It was also observed that the risk caused due to inhalation exposure is higher and contributes 99% of total cancer risk (Anchal et al. 2020). In the study conducted at indoor swimming pools in Beijing, China, four species of THM were monitored (Chloroform, bromoform, dibromochloromethane and bromodichloromethane). Among the four compounds, $CHCl_3$ was found to be the most abundant. It contributed 97% of Trihalomethanes in swimming pool air and 93% in the air present in locker room, with concentration ranges of 269.60–516.22 μg/m^3 and 27.31–49.34 μg/m^3, respectively. It was also found that children between the age group of 6–17 were more at risk. Further, like other studies, inhalation exposure was found to contributes 98.7% of the total cancer risk caused due to THMs, of which $CHCl_3$ was responsible for 97.7% of total inhalation cancer risk (Wang and Dong 2020). In an Olympic-sized indoor pool in Portugal where disinfection is done by an automated chlorination system and provided with a new heating, ventilation and air conditioning (HVAC) system, the level of $CHCl_3$ was found to be in the range of 13.0–369.3 μg/m^3(Felgueiras et al. 2019). In five indoor swimming pools in the City of Modena, Northern Italy, the level of $CHCl_3$ was observed to be 46.1 ± 18.6 μg/m^3 in poolside air, 21.8 ± 19.3 μg/m^3 in the reception area, 21.7 ± 22.7 μg/m^3 in the indoor air and 18.7 ± 14.7 μg/m^3 in the alveolar air of the swimmers (Fantuzzi et al. 2001). In a study involving 17 swimmers at a pool where the concentration of $CHCl_3$ in pool water was 7.1–21.8 μg/L and pool air was 85–235 μg/m^3, the $CHCl_3$ concentration in the blood stream was 0.32 ± 0.26 μg/L for swimmers with scuba tanks, 0.99 ± 0.47 μg/L for swimmers without scuba tanks and 0.31 ± 0.25 μg/L for the persons walking around the pool. Correlations found using linear regression models showed a strong association between the level of chloroform in the air and blood stream, however, the correlation of blood concentration with water was less. The study concluded that the major exposure pathway is respiration (Erdinger et al. 2004).

Chloroform uptake during showers was estimated and it was observed that 70% of the inhaled $CHCl_3$ is absorbed and 2–9% of the absorbed chloroform is expired through lung excretion after 30 min. It was also observed that inhalation exposure during a 10–15 min shower is equivalent to ingesting 0.1–0.3 L of drinking water (Xu and Weisel 2005). In another study, exposure to $CHCl_3$ was estimated by measuring its alveolar concentrations in both swimmers and workers in the pool environment. The level of $CHCl_3$ found in the alveolar air of workers and swimmers was 2.4–5.6 $\mu g/m^3$. Among workers, after 2 h of exposure the mean concentration of $CHCl_3$ was 33 $\mu g/m^3$ and after 4 h exposure, the mean concentration was found to be 53 $\mu g/m^3$. In the case of swimmers, after 1 h of swimming the level was found to 82 $\mu g/m^3$ which was 3 times greater than in workers after 2 h. This indicates that dermal contact and accidental ingestion also contribute to total exposure (Caro and Gallego 2008). The increase in skin temperature and the higher respiration/breathing rate during swimming have been reported as contributing to the higher concentration of $CHCl_3$ in both the blood and breath of swimmers (Berkoff 1995). The multi-pathway exposure model was used to estimate incremental lifetime Cancer Risk caused due to chloroform and found that the values for male and female swimmers were 80×10^{-4} and 2.47×10^{-4}, respectively (Chen et al. 2011). The presence of chloroform was most predominant in the swimming pools which use chlorine as a disinfectant (Bessonneau et al. 2011).

6.8.3 *Industrial Air*

In a study conducted near the Daegu dying industrial complex, South Korea, there was no significant difference in the level of $CHCl_3$ during different seasons, with a mean value of 10.75 ± 4.13 and 11.72 ± 5.75 $\mu g/m^3$ in summer and autumn, respectively (Shuai et al. 2018). In the observations made in the ambient air near a paper and pulp industry in rural Idaho, USA, it was found that $CHCl_3$ is a strong signature compound to predict the emission from the paper and pulp industry. Temporal variation of $CHCl_3$ was prominent, the values were found to be higher in the morning hours, which may be due to the operation of paper mills and diurnal inversions. But similar trends in the level of $CHCl_3$ was observed during different seasons between the year 2017–2018 (Scott et al. 2020). Emission of $CHCl_3$ from coal and oil-fired power plants was determined and the level was observed to be 0.21 and 0.10 ppm in anthracite and bituminous coal unit and it was not detected in fuel oil-fired power plants (Pudasainee et al. 2010). In a study conducted in the highly industrialized region in turkey, the mean level was found to be 0.4 $\mu g/m^3$. The source profiling revealed that the emissions were from petrochemical industries and the vinyl- the process was responsible for it (Dumanoglu et al. 2014). Where $CHCl_3$ will be formed as a by-product during the production of ethylene dichloride (EDC) by direct dichlorination of ethylene & chlorine and by oxy-chlorination of ethylene & HCl. Finally, EDC will be subjected to thermal cracking to form vinyl chloride monomers (US Environmental Protection Agency 1984).

From Table 6.10, it is evident that there is huge variation in the concentrations observed across various environments and also across countries for similar environments.

6.9 Control/clean up of Chloroform in the Air

The control and clean-up methods adopted for VOCs in general can be adopted for Chloroform also. The control strategies for VOCs are classified into two broad categories (1) process and equipment modification, and (2) Add on control techniques. Control techniques can be further classified into destruction and recovery techniques (Khan and Ghoshal 2000). Further classification of destruction and recovery techniques is given in Fig. 6.6.

Regarding the clean-up of spills, the method suggested by NIOSH/OSHA includes, (1) ventilating the area of spill or leak, and (2) collecting the leaked pollutant for reclamation or absorption either in vermiculate (hydrous phyllosilicate mineral), dry sand, earth, or any other material of similar type (NIOSH 1981). Intermedia priority pollutant guidance documents suggest using water sprays to reduce the vapors (USEPA 1982). High-Efficiency Particulate Arrestor (HEPA) or charcoal filters can be used to minimize exposure. The collected filter should be stored in a plastic bag and should be sealed immediately and labeled properly (IARC 1975). The accidental release measure includes wearing respiratory protection aids and ensuring adequate ventilation (Sigma-Aldrich 2017).

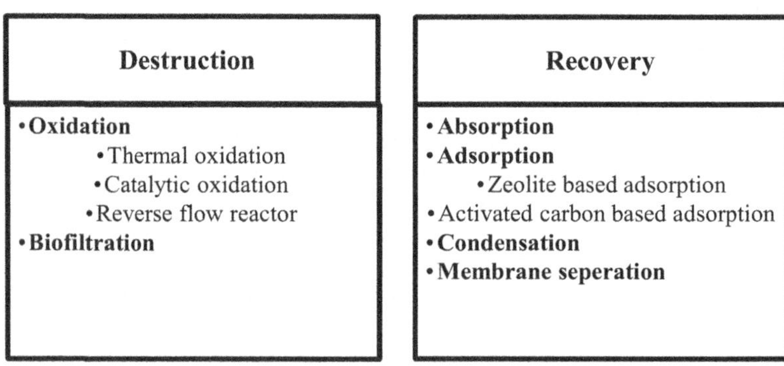

Fig. 6.6 Different types of destruction and recovery methods for removal of VOCs

6.10 Conclusion

Chloroform is categorized as a possible human carcinogen and studies have shown that it significantly affects the healing of the ozone hole. The percentages of natural and anthropogenic sources of $CHCl_3$ emissions are more or less the same. Photochemical reaction of $CHCl_3$ in the atmosphere and metabolism of $CHCl_3$ in the human body produces phosgene, which is also toxic to the environment and health of living organisms. Hepatotoxicity is one of the major health effects caused due to chronic exposure to $CHCl_3$. Several studies have shown that Incremental Lifetime Cancer Risk from exposures exceeded the permissible limits. In most of the studies, inhalation exposure was found to contribute 99% of the total cancer risk. Personal exposure to $CHCl_3$ is found to be higher when compared with levels in indoor and outdoor air. Competitive swimmers are highly exposed to $CHCl_3$ due to prolonged exposure and increased respiratory rate while swimming in pool water disinfected with chlorine. Due to its ubiquitous nature and harmful effects, $CHCl_3$ is one of the emerging pollutants in the air which needs higher attention and frequent monitoring. Consequently, there is an urgent requirement to include chloroforms in the ambient air monitoring programmes of countries across the globe.

References

Abdul Hamid H, Lockman Z, Hattori T, Abdul Razak K (2019) Sensitive and selective chloroform sensor using Fe_2O_3 nanoparticle-decorated ZnO nanorods in an aqueous solution. J Mater Sci Mater Electron 30(20):18990–19000. https://doi.org/10.1007/s10854-019-02256-2

Adgate JL, Eberly LE, Stroebel C, Pellizzari ED, Sexton K (2004) Personal, indoor, and outdoor VOC exposures in a probability sample of children. J Expo Anal Environ Epidemiol 14(SUPPL. 1). https://doi.org/10.1038/sj.jea.7500353

Aiking H, van Ackert MB, Schölten RJPM, Feenstra JF, Valkenburg HA (1994) Swimming pool chlorination: a health hazard? Toxicol Lett 72(1–3):375–380. https://doi.org/10.1016/0378-4274(94)90051-5

American Industrial Hygiene Association (AIHA) (2016) Emergency response planning guideline values

Anchal P, Kumari M, Gupta SK (2020) Human health risk estimation and predictive modeling of halogenated disinfection by-products (chloroform) in swimming pool waters: a case study of Dhanbad, Jharkhand, India. J Environ Health Sci Eng 18(2):1595–1605. https://doi.org/10.1007/s40201-020-00578-6

Arı A, Ertürk Arı P, Yenisoy-Karakaş S, Gaga EO (2020) Source characterization and risk assessment of occupational exposure to volatile organic compounds (VOCs) in a barbecue restaurant. Build Environ 174(December 2019). https://doi.org/10.1016/j.buildenv.2020.106791

Athawale AA, Pethkar S, Sharma S, Nirkhe C (2002) Chloroform vapour sensor based on copper/polyaniline nanocomposite. Sensors Actuators, B Chem. 85(1–2):131–136. https://doi.org/10.1016/S0925-4005(02)00064-3

Barcel6E D, Ljarrat E, Petrovic M (2005) Mass spectrometry I environmental applications. In: Worsfold P, Townshend A, Poole C (eds) Encyclopedia of analytical science

Batterman SA, Zhang G, Baumann M (1998) Analysis and stability of aldehydes and terpenes in electropolished canisters. Atmos Environ 32(10)

Berkoff DC (1995) Chloroform exposure and dose determination associated with competitive swimmers during a two-hour swim practice. The University of Montana Follow
Berkowitz CM, Spicer CW, Doskey PV (2005) Hydrocarbon observations and ozone production rates in Western Houston during the Texas 2000 air quality study. Atmos Environ 39:3383–3396. https://doi.org/10.1016/j.atmosenv.2004.12.007
Bessonneau V, Derbez M, Clément M, Thomas O (2011) Determinants of chlorination by-products in indoor swimming pools. Int J Hyg Environ Health 215(1):76–85. https://doi.org/10.1016/j.ijheh.2011.07.009
Bhattacharjee S (2020) Visakhapatnam gas leak | How negligence and violations led to a deadly disaster. The Hindu
Bornski H, Sobolewska A, Strakowski A (1967) Toxic damage of the liver by chloroform in chemical industry workers. Int Arch F Gewerbepathologie u, vol. Gewerbehyg, pp 127–134
Broughton E (2005) The Bhopal disaster and its aftermath: a review. Environ Health A Glob Access Sci Source 4:1–6. https://doi.org/10.1186/1476-069X-4-6
Callen D, Wolf C, Philpot R (1980) Cytochrome p-450 mediated genetic activity and cytotoxicity of seven halogenated aliphatic hydrocarbons in Saccharomyces cerevisiae. Mutat Res 77:55–63
Camel V, Caude M (1995) Trace enrichment methods for the determination of organic pollutants in ambient air. J Chromatogr A 710:3–19
Caro J, Gallego M (2008) Alveolar air and urine analyses as biomarkers of exposure to trihalomethanes in an indoor swimming pool. Environ Sci Technol 42(13):5002–5007. https://doi.org/10.1021/es800415p
Chen MJ, Lin CH, Duh JM, Chou WS, Hsu HT (2011) Development of a multi-pathway probabilistic health risk assessment model for swimmers exposed to chloroform in indoor swimming pools. J Hazard Mater 185(2–3):1037–1044. https://doi.org/10.1016/j.jhazmat.2010.10.011
CPCB (2020) National air monitoring programme. Ministry of Environment, Forest and Climate Change
D'Silva T (2006) The black box of Bhopal: a closer look at the world's deadliest industrial disaster. Trafford
Dalberg, CII, and CAF (2021) Air pollution and its impact on business—the silent pandemic
Dalene M, Persson P, Skarping G (1992) Determination of formaldehyde in air by chemisorption on glass filters impregnated with 2,4-dinitrophenylhydrazine using gas chromatography with thermionic specific detection. J Chromatogr A 626(2):284–288. https://doi.org/10.1016/0021-9673(92)85422-P
Dimmer CH, Simmonds PG, Nickless G, Bassford MR (2001) Biogenic fluxes of halomethanes from Irish peatland ecosystems. Atmos Environ 35:321–330
Driscoll J (2008) Measurement of chlorinated hydroarbons by PID with a long life 11.7 Lamp, no July. pp 1–4. [Online]. Available: https://www.hnu.com/papers/Measurement_of_chlorinated_HC_with_an_117_PID_608.pdf
Dumanoglu Y, Kara M, Altiok H, Odabasi M, Elbir T, Bayram A (2014) Spatial and seasonal variation and source apportionment of volatile organic compounds (VOCs) in a heavily industrialized region. Atmos Environ 98:168–178. https://doi.org/10.1016/j.atmosenv.2014.08.048
Ebere C et al (2020) An overview of emerging pollutants in air: method of analysis and potential public health concern from human environmental exposure. Trends Environ Anal Chem 28:e00107. https://doi.org/10.1016/j.teac.2020.e00107
Erdinger L et al (2004) Pathways of trihalomethane uptake in swimming pools. Int J Hyg Environ Health 207(6):571–575. https://doi.org/10.1078/1438-4639-00329
Estévez J, Vilanova E (2014) Chloroform. Encycl Toxicol Third Ed 1:885–890. https://doi.org/10.1016/B978-0-12-386454-3.00479-6
European Union (2006) Transitional measures regarding existing substances of REACH regulation (EC) 1907/2006, vol 136, no 3
European Union (2007) European Union risk assessment report-chloroform CAS 67-66-3
Fang X et al (2019) Rapid increase in ozone-depleting chloroform emissions from China. Nat Geosci 12(2):89–93. https://doi.org/10.1038/s41561-018-0278-2

Fantuzzi G, Righi E, Predieri G, Ceppelli G, Gobba F, Aggazzotti G (2001) Occupational exposure to trihalomethanes in indoor swimming pools. Sci Total Environ 264(3):257–265. https://doi.org/10.1016/S0048-9697(00)00722-1

Felgueiras F, Mourão Z, Morais C, Santos H, Fonseca M (2020) Comprehensive assessment of the indoor air quality in a chlorinated olympic-size swimming pool. Environ Int 136(July 2019), 105401. https://doi.org/10.1016/j.envint.2019.105401

Fong JK, Pena JK, Xue ZL, Alam MM, Sampathkumaran U, Goswami K (2015) Optical sensors for the detection of trace chloroform. Anal Chem 87(3):1569–1574. https://doi.org/10.1021/ac503920c

Gawrys M, Fastyn P, Gawlowski J, Gierczak T, Niedzielski J (2001) Prevention of water vapour adsorption by carbon molecular sieves in sampling humid gases. J Chromatogr A 933:107–116

Gemma S, Ade P, Sbraccia M, Testai E, Vittozzi L (1996) In vitro quantitative determination of phospholipid adducts of chloroform intermediates in hepatic and renal microsomes from different rodent strains. Environ Toxicol Pharmacol 2(2–3):233–242. https://doi.org/10.1016/S1382-6689(96)00060-9

Gemma S, Vittozzi L, Testai E (2003) Metabolism of chloroform in the human liver and identification of the competent P450s. Drug Metab Dispos 31(3):266–274. https://doi.org/10.1124/dmd.31.3.266

Gorecki T, Namies J (2002) Passive sampling. Trends Anal Chem 21(4):276–291

Government of Madhya Pradesh (2012) Bhopal gas tragedy relief and rehabilitation. https://web.archive.org/web/20120518020821/; http://www.mp.gov.in/bgtrrdmp/relief.htm. Accessed 26 Sep 2021

Hansen AB, Palmgren F (1996) VOC air pollutants in Copenhagen. Sci Total Environ 189(190):451–457

Health Protection Agency (2007) Chloroform toxicological overview, pp 1–12

Hinwood AL, Berko HN, Farrar D, Galbally IE, Weeks IA (2006) Volatile organic compounds in selected micro-environments. Chemosphere 63(3):421–429. https://doi.org/10.1016/j.chemosphere.2005.08.038

Hoekstra EJ, Duyzer JH, De Leer EWB, Brinkman UAT (2001) Chloroform—concentration gradients in soil air and atmospheric air, and emission fluxes from soil. Atmos Environ 35(1):61–70. https://doi.org/10.1016/S1352-2310(00)00285-5

IARC (1975) Handling chemical carcinogens in the laboratory: problems of safety

IARC (1993) IARC Monographs on the evaluation of carcinogenic risks to humans, 58th ed. Beryllium, Cadmium, Mercury, and exposures in the glass manufacturing industry

IARC (2002) International agency for research on cancer (IARC)—monograph 82-Styrene-7,8-oxide and Styrene, pp 72–78. [Online]. Available: https://monographs.iarc.who.int/wp-content/uploads/2018/06/TR42-9.pdf

International Programme on Chemical Safety (1994) Environmental health criteria 163: chloroform

International Programme on Chemical Safety (2004) Concise chemical international chemical assessment document 58. Chloroform

IQAir (2020) World air quality report-2020 (region & city PM2.5 ranking)

Isidorov V, Zenkevich I, Ioffe B (1990) Volatile organic compounds in solfataric gases. J Atmos Chem 329–340

Jolley RL, Brungs WA, Cotruvo JA, Cumming RB, Mattice VA, Jacobs JS (1983) Water chlorination: environmental impact and health effects. Book 1, Chemistry and water treatment, 4th ed. United States

Judd SJ, Black SH (2000) Disinfection by-product formation in swimming pool waters: a simple mass balance. Water Res 34(5):1611–1619. https://doi.org/10.1016/S0043-1354(99)00316-4

Juuti S, Vartiaine T, Joutsenoja P, Ruuskanen J (1996) Volatile organochlorine compounds formed in the bleaching of pulp with Clo2. Chemosphere 33(3):437–448

Khalil MAIC, Rasmussen RA (1990) Emissions of trace gases from Chinese rice fields and biogas generators: CH_4, N_2O, CO, CO_2, chlorocarbons, and hydrocarbons. Chemosphere 20(3):207–226

Khalil MAK, Rasmussen RA (1999) Atmospheric chloroform. Atmos Environ 33(7):1151–1158. https://doi.org/10.1016/S1352-2310(98)00233-7

Khalil MAK, Rasmussen RA, French J, Holt J (1990) The influence of termites on atmospheric trace gases: CH_4, CO_2, $CHCl_3$, N_2O, CO, H_2, and light hydrocarbon. J Geophys Res 95:3619–3634

Khan FI, Ghoshal AKr (2000) Removal of volatile organic compounds from polluted air. J Loss Prev Process Ind 13(6):527–545. https://doi.org/10.1016/S0950-4230(00)00007-3

Kumar A, Víden I (2007) Volatile organic compounds: sampling methods and their worldwide profile in ambient air. Environ Monit Assess 131(1–3):301–321. https://doi.org/10.1007/s10661-006-9477-1

Kumar RP, Kashyap P, Kumar R, Pandey AK, Kumar A, Kumar K (2020) Cancer and non-cancer health risk assessment associated with exposure to non-methane hydrocarbons among roadside vendors in Delhi, India. Hum Ecol Risk Assess 26(5):1285–1299. https://doi.org/10.1080/10807039.2019.1570077

Kuttippurath J, Kumar P, Nair PJ, Pandey PC Emergence of ozone recovery evidenced by reduction in the occurrence of Antarctic ozone loss saturation. NPJ Clim Atmos Sci 1(42). https://doi.org/10.1038/s41612-018-0052-6

Kwon KD, Jo WK, Lim HJ, Jeong WS (2008) Volatile pollutants emitted from selected liquid household products. Environ Sci Pollut Res 15(6):521–526. https://doi.org/10.1007/s11356-008-0028-x

Land P, Owen E, Linde H (1979) Metabolism and regulation. Anesthesiology 51(3)

Larson JL, Wolf DC, Morgan KT, Mery S, Butterworth BE (1994) The toxicity of 1-week exposures to inhaled chloroform in female B6C3F mice and male F-344 rats. Fundam Appl Toxicol 22:431–446

Larson J, Templin M, Wolf D (1996) A 90-day chloroform inhalation study in female and male B6C3Fl mice: implications for cancer risk assessment. Fundam Appl Toxicol 30:118–137

Lasa J, Śliwka I (2003) Long-term measurements of the concentrations of halocarbons in an urban area of Kraków, Poland. Appl Energy 75(3–4):155–163. https://doi.org/10.1016/S0306-2619(03)00028-X

Li X (2019) Air pollution: a global problem needs local fixes. Nature 570:437–439

Li C et al (2020) Environmental impact and health risk assessment of volatile organic compound emissions during different seasons in Beijing. J Environ Sci (china) 93:1–12. https://doi.org/10.1016/j.jes.2019.11.006

Lin N, Rosemberg M-A, Li W, Meza-Wilson E, Godwin C, Batterman S (2021) Occupational exposure and health risks of volatile organic compounds (VOCs) of hotel housekeepers. Indoor Air 31:26–39

Lobert JM, Keene WC, Logan JA, Yevich R (1999) Global chlorine emissions from biomass burning: reactive chlorine emissions inventory. J Geophys Res Atmos 104(D7):8373–8389. https://doi.org/10.1029/1998JD100077

Lyu X et al (2020) Hazardous volatile organic compounds in ambient air of China. Chemosphere 246:125731. https://doi.org/10.1016/j.chemosphere.2019.125731

Mccammon CS, Woebkenberg ML, General considerations for sampling airborne contaminants. NIOSH manual of analytical methods (NMAM), 5th edn sampling

Mcculloch A (2003) Chloroform in the environment: occurrence, sources, sinks and effects. Chemosphere 50:1291–1308

MoEF (2019) National clean air programme. [Online]. Available: https://moef.gov.in/wp-content/uploads/2019/05/NCAP_Report.pdf

Mozaffar A, Zhang YL (2020) Atmospheric volatile organic compounds (VOCs) in China: a review. Curr Pollut Reports 6(3):250–263. https://doi.org/10.1007/s40726-020-00149-1

Murray FJ, Schwetx BA, Mcbride JG, Staples RE (1979) Toxicity of inhaled chloroform in pregnant mice and their offspring. Toxicol Appl Pharmacol 50:515–522

National Academy of Sciences (2012) Chloroform: acute exposure guideline levels. In: Acute exposure guideline levels for selected airborne chemicals, vol 12

NIOSH (1981) Occupational Health guidelines for chemical hazards. In: Mackison FW, Stricoff RS, Partridge JLJ (eds) DHHS(NIOSH) Publication No. 81-123 (3 Vols). U.S. Government Printing Office, Washington, DC
Odabasi M (2008) Halogenated volatile organic compounds from the use of household products. Environ Sci Technol 42(5):1445–1451
Pándics T, Hofer Á, Dura G, Vargha M, Szigeti T, Tóth E (2018) Health risk of swimming pool disinfection by-products: a regulatory perspective. J Water Health 16(6):947–957. https://doi.org/10.2166/wh.2018.178
Park HR, Jeong YT, Ko SK (1997) Photochemical reaction of chloroform. J Photoscience 4(1):17–21
Powick PEJ (1989) Swimming pools-brief outline of water treatment and management. Water Sci Technol 21(2):151–160
Public Health England Chloroform Incident Management (2017) Compend Chem Hazards Chloroform no June, [Online]. Available: https://www.taylorfrancis.com/books/9781134476602/chapters/10.4324/9780203994559-14
Pudasainee D et al (2010) Hazardous air pollutants emission from coal and oil-fired power plants. Asia Pacific J Chem Eng 5:299–303. https://doi.org/10.1002/apj
RAIS (2021) The risk assessment information system-toxicity values. https://rais.ornl.gov/tutorials/toxvals.html
Ras MR, Borrull F, Marcé RM (2009) Sampling and preconcentration techniques for determination of volatile organic compounds in air samples. TrAC Trends Anal Chem 28(3):347–361. https://doi.org/10.1016/j.trac.2008.10.009
Ritchie H, Roser M (2017) Air pollution. OurWorldInData.org
Royston G (1924) Delayed chloroform poisoning following delivery. Am J Obs Gynecol 10:808–814
Rudolph J (1990) Sampling of organic volatiles in the atmosphere at moderate and low pollution levels. Anal Chim Acta 236:197–211
Sarangi S (2002) Crimes of Bhopal and the global campaign for justice. Soc Justice 29(3), 89
Say D et al (2020) Emissions and marine boundary layer concentrations of unregulated chlorocarbons measured at Cape Point, South Africa. Environ Sci Technol 54(17):10514–10523. https://doi.org/10.1021/acs.est.0c02057
Schwetz B, Leong BKJ, Gehring PJ (1974) Embryo and fetotoxicity of inhaled chloroform in rats. Toxicol Appl Pharmacol 28:442–451
Scott PS et al (2020) Observations of volatile organic and sulfur compounds in ambient air and health risk assessment near a paper mill in rural Idaho, U.S.A. Atmos Pollut Res 11(10):1870–1881. https://doi.org/10.1016/j.apr.2020.07.014
Scranton A (2011) Dirty secrets: what's hiding in your cleaning produtcs
Sekar A, Varghese GK, Ravi Varma MK (2019) Analysis of benzene air quality standards, monitoring methods and concentrations in indoor and outdoor environment. Heliyon 5(11), e02918. https://doi.org/10.1016/j.heliyon.2019.e02918
Serrano-Trespalacios PI, Ryan L, Spengler JD (2004) Ambient, indoor and personal exposure relationships of volatile organic compounds in Mexico City Metropolitan Area. J Expo Anal Environ Epidemiol 14(SUPPL 1). https://doi.org/10.1038/sj.jea.7500366
Sexton K et al (2004) Comparison of personal, indoor, and outdoor exposures to hazardous air pollutants in three urban communities. Environ Sci Technol 38(2):423–430. https://doi.org/10.1021/es030319u
Shuai J et al (2018) Health risk assessment of volatile organic compounds exposure near Daegu dyeing industrial complex in South Korea. BMC Public Health 18(1):1–13. https://doi.org/10.1186/s12889-018-5454-1
Sigma-Aldrich (2017) Safety data sheet for chloroform. Product Number: C2432, Version 3.17 (Revision Date 12 July 2017). http://www.sigmaaldrich.com/safety-center.html/. Accessed 24 Sept 2021
Smith LA et al (2007) Spatial analysis of volatile organic compounds from a community-based air toxics monitoring network in Deer Park, Texas, USA. Environ Monit Assess 128:369–379. https://doi.org/10.1007/s10661-006-9320-8

Smith AA, Volpitto PP, Gramling ZW, Devore MB, Glassman AB (1973) Chloroform, halothane, and regional anesthesia: a comparative study. Anesth Analg 52:1–11
Solomon S, Ivy DJ, Kinnison D, Mills MJ, Iii RRN, Schmidt A (2016) Emergence of healing in the Antarctic ozone layer. Science (80-) 353(6296):269–274
Spinelle L, Gerboles M, Kok G, Persijn S, Sauerwald T (2017) Review of portable and low-cost sensors for the ambient air monitoring of benzene and other volatile organic compounds. Sensors (Switzerland) 17(7). https://doi.org/10.3390/s17071520
Tasaltin C, Basarir F (2014) Preparation of flexible VOC sensor based on carbon nanotubes and gold nanoparticles. Sens Actuators, B Chem 194:173–179. https://doi.org/10.1016/j.snb.2013.12.063
Trudinger CM, Etheridge DM, Sturrock GA, Fraser PJ, Krummel PB, McCulloch A (2004) Atmospheric histories of halocarbons from analysis of Antarctic firn air: methyl bromide, methyl chloride, chloroform, and dichloromethane. J Geophys Res D Atmos 109(22):1–15. https://doi.org/10.1029/2004JD004932
US Department of Health and Human Services (1997) Toxicological profile for chloroform
U.S. Environmental Protection Agency (1976) Atmospheric freons and halogenated compounds, EPA-600/3-76-108. Environmental Sciences Research Laboratory, Research Triangle Park, NC
US Environmental Protection Agency (1984) Method TO-3: method for the determination of volatile organic compounds in ambient air using cryogenic preconcentration techniques and gas chromatography with flame ionization and electron capture detection
US Environmental Protection Agency (1984) Locating and estimating air emissions from sources of chloroform, no. EPA-450/4-84-007c. p 9
U.S. Environmental Protection Agency (1984) Method for the determination of non-methane organic compounds (NMOC) in ambient air using cryogenic preconcentration and direct flame ionization detection
US Environmental Protection Agency (1984) Method TO-2 method for the determinaion of volatile organic compounds in ambient air by carbon molecular sieve adsorption and gas chromatography/massspectrometry (GC/MS)
US Environmental Protection Agency (1984) Method TO-1, Method for the determination of volatile organic compounds in ambient air using TENAX adsorption and gas chromatography/mass spectrometry (GC/MS)
US Environmental Protection Agency (1992) Chloroform. Integrated Risk Information System (IRIS). Environmental Criteria and Assessment Office, Office of Health and Environmental Assessment, Cincinnati, OH
US Environmental Protection Agency (1992) Health effects assessment summary tables. Annual FY-92. Prepared by the Office of Health and Environmental Assessment, Environmental Criteria and Assessment Office, Cincinnati, OH, for the Office of Emergency and Remedial Response, Washington DC
US Environmental Protection Agency (1996) Method 0040-sampling of principal organic hazardous constituents from combustion sources using tedlar bags
U.S. Environmental Protection Agency (EPA) (1999) Compendium method TO-17: determination of volatile organic compounds in ambient air using active sampling onto sorbent tubes. Compend Methods Determ Toxic Org Compd Ambient Air (January):1–53
U.S. Environmental Protection Agency (EPA) (1999) Compendium method TO-14A determination of volatile organic compounds (VOCs) in ambient air using specially prepared canisters with subsequent analysis B. In: Compendium of methods for the determination of toxic organic compounds in ambient air, 2nd edn
US Environmental Protection Agency (1999) EPA compendium method TO-14A. Determination of volatile organic compounds (VOCs) in ambient air using specially prepared canisters with subsequent analysis by gas chromatography. EPA 625/R-96/010b. Office of Research and Development, Cincinnati, OH, 45268
US Environmental Protection Agency (2001) Toxicological review of chloroform
US Environmental Protection Agency (2004) Report on carcinogens, substance profile, chloroform CAS no. 67-66-3, 11th edn

US Environmental Protection Agency (2017) Method 325A-volatile organic compounds from fugitive and area sources

US Environmental Protection Agency (2017) Method 8260-volatile organic compounds by gas chromatography/mass spectrometry, Fourth revision

US Environmental Protection Agency (2019) Method 325B—volatile organic compounds from fugitive and area sources: sampler preparation and analysis. [Online]. Available: https://www.epa.gov/emc/method-325b-volatile-organic-compounds-fugitive-and-area-sources-sampler-preparation-and

US Environmental Protection Agency (2019) Method 18-measurement of gaseous organic compound emissions by gas chromatography

USEPA (1982) Intermedia priority pollutant guidance documents. U.S.Environmental Protection Agency, Washington, DC

Varma R, Varma DR (2005) The Bhopal disaster of 1984. Bull Sci Technol Soc 25(1):37–45. https://doi.org/10.1177/0270467604273822

Wang X, Dong S (2020) Assessment of exposure of children swimmers to trihalomethanes in an indoor swimming pool. J Water Heal 18(4):533–544. https://doi.org/10.2166/wh.2020.188

Whitaker A, Jones C (1965) Report of 1500 chloroform anesthetics administered with a precision vaporizer. Anesth Analg 44:60–65

Wickliffe JK et al (2020) Increased long-term health risks attributable to select volatile organic compounds in residential indoor air in southeast Louisiana. Sci Rep 10(1):1–12. https://doi.org/10.1038/s41598-020-78756-7

Williams AL, Bates CA, Pace ND, Leonhard MJ, Chang ET, DeSesso JM (2018) Impact of chloroform exposures on reproductive and developmental outcomes: a systematic review of the scientific literature. Birth Defects Res 110(17):1267–1313. https://doi.org/10.1002/bdr2.1382

World Health Organization (2021) IARC monographs on the identification of carcinogenic hazards to humans. Agents classified by the IARC monographs

World Health Organization Geneva. Chloroform: health and safety guide. (Health and safety guide; no. 87), Chloroform—adverse effects I. Series. [Online]. Available: http://www.inchem.org/documents/hsg/hsg/hsg87_e.htm

World Health Orgnization Geneva (1994) International programme on chemical safety (IPCS). Chloroform. Environmental Health Criteria 163

World Health Orgnization Geneva (2004) Concise International chemical assessment document (CICAD): Document 58

Wortan D, Sturges W, Schwander J, Mulvaney R, Barnola J, Chappellaz J (2006) 20th century trends and budget implications of chloroform and related tri-and dihalomethanes inferred from firn air. Atmos Chem Phys 6:2847–2863. https://doi.org/10.5194/acp-6-2847-2006

Xing L, Wang L, Zhang R (2018) Characteristics and health risk assessment of volatile organic compounds emitted from interior materials in vehicles: a case study from Nanjing, China. Environ Sci Pollut Res 25(15):14789–14798. https://doi.org/10.1007/s11356-018-1661-7

Xu X, Weisel CP (2005) Human respiratory uptake of chloroform and haloketones during showering. J Expo Anal Environ Epidemiol 15(1):6–16. https://doi.org/10.1038/sj.jea.7500374

Zhang T, Xing Y, Wang G, He S (2021) High sensitivity continuous monitoring of chloroform gas by using wavelength modulation photoacoustic spectroscopy in the near-infrared range. Appl Sci 11(15). https://doi.org/10.3390/app11156992

Zielinska B, Faujita E (1994) Environmental sampling for trace analysis, organic gas sampling. B. Mark, Federal Republic of Germany

Chapter 7
Phthalate Esters in the Environment: An Overview on the Occurrence, Toxicity, Detection, and Treatment Options

Ansaf V. Karim, Sukanya Krishnan, S. Sethulekshmi, and Amritanshu Shriwastav

Abstract Phthalate esters (PAEs) are industrial chemicals widely used in large quantities as additives, mainly in the plastic industry. They are colorless sticky liquids that are highly toxic and one of the most frequently detected persistent organic pollutants in the environment. They can leach out from plastic materials and cause severe health issues in living organisms. Their high occurrence in different environmental matrices results in widespread human exposure causing notorious endocrine-disrupting effects, subsequently producing adverse effects on reproduction, growth, and development in human beings as well as the marine ecosystem. Therefore, a better understanding of their occurrence and measurement is required for developing suitable treatment technologies. Recent research studies suggest that advanced treatment methods are sustainable approaches for removing these persistent compounds from the environmental system. This chapter provides an overview of the properties and different types of PAEs, their toxicity, exposure, detection, and measurement techniques in various environmental matrices. Further, the chapter also reviews the various treatment technologies for the removal of PAEs from the environment.

Keywords Phthalate esters · Occurrence · Toxicity · Quantification · Treatment

7.1 Introduction

Phthalate esters (PAE) are widely used industrial additives, primarily used as plasticizers for polyvinyl chloride, cellulosics, elastomers, adhesives etc. (Yuan et al. 2002). The demand for PAEs for industrial applications has grown rapidly in the last few decades owing to their ability to improve the flexibility and durability of polymers (Cheng et al. 2013). PAEs are colourless compounds with low volatility that are prepared by the reaction of phthalic acid with specific alcohol to form the desired ester (Zhang et al. 2015). The higher stability and fluidity of PAEs, which impart

A. V. Karim · S. Krishnan · S. Sethulekshmi · A. Shriwastav (✉)
Environmental Science and Engineering Department, Indian Institute of Technology Bombay, Mumbai 400076, India
e-mail: amritan@iitb.ac.in

S. P. Singh et al. (eds.), *New Trends in Emerging Environmental Contaminants*, Energy, Environment, and Sustainability,
https://doi.org/10.1007/978-981-16-8367-1_7

flexibility in polymers, makes them a highly suitable plasticizer. They are extensively used in plastics, building materials, personal care products etc. and culminate in the environment during their use or disposal. Since they are not chemically bound to the polymer matrix, they can easily leach out into the various environmental matrices (Wang et al. 2015; Kotowska et al. 2020). Currently, the pollution by PAEs is ubiquitous as they tend to get accumulated in the environmental matrices and can cause potential risk to organisms in numerous ways (Lu et al. 2009; Wei et al. 2020).

PAEs are one of the most frequently detected endocrine-disrupting compounds with teratogenicity, mutagenicity and carcinogenicity even at trace concentrations (Gao and Wen 2016; Wu et al. 2019). They are found in environmental matrices at measurable levels owing to their slow-release, refractory nature and tend to interfere with living organisms. Further, they get transmitted to different levels of organisms through the food chain and biomagnifies. Consumption of food containing PAEs can be a major source in the human body. They also reach the human body via inhalation, dermal absorption, and ingestion and causes health effects on children and older people (Wang et al. 2018a). Studies have also reported that exposure to PAEs can cause adverse effects on the reproductive system, neuroendocrine system, immune system, etc. (Boonnorat et al. 2014). Further, the toxicity of PAEs on organisms generally increases with the number of alkyl chains on the ester (Zhang et al. 2015). In this regard, understanding the fate and transport of phthalates esters is essential to developing remediation measures. Many treatment technologies, including physico-chemical, biological, chemical oxidation processes, effectively remove PAEs from water and other environmental matrices. The current chapter provides different classifications of PAEs, their fate and transport in the various environmental matrixes, toxicity and different quantification methods. Further, the chapter discusses the various approaches for the degradation of PAEs in aqueous solutions and soil medium, followed by the conclusion.

7.2 Classification of PAEs

PAEs are industrial chemicals that are dialkyl or alkyl aryl esters of 1,2-benzene dicarboxylic acid, also known as phthalic acid (Qureshi et al. 2016). They are esters of phthalic acid produced by the process of esterification of phthalic acid with different alcohols. The difference in physical–chemical properties among them is based on the type of alcohol employed (Ventrice et al. 2013). About 23 different types of PAEs are commercially available as plasticizers, solvents, and emulsifiers (Poopal et al. 2017). These are classified into two groups based on the molecular weight or carbon number of their alcohol as high molecular weight phthalates and low molecular weight phthalates. The molecular weight variation is attributed to the length of alkyl chains and the degree of branching (Nidens et al. 2021). High molecular weight phthalates are used as additives that impart flexibility to vinyl resins, whereas low molecular weight phthalates are used as plasticizers in non-vinyl resins. Some basic information about the structure, molecular weight, and application of commercially most important types of phthalate esters are given in Table 7.1.

Table 7.1 Details of commonly used Phthalate esters

PAEs	Abbreviation	IUPAC name	Molecular weight (g/mol)	Uses
Dimethyl phthalate	DMP	$C_{10}H_{10}O_4$	194	Insect repellent
Diethyl phthalate	DEP	$C_{12}H_{14}O_4$	222	Plasticizers, aerosol sprays, cosmetics
Di-iso-butyl phthalate	DiBP	$C_{16}H_{22}O_4$	278	Explosive materials, nail paints, lacquer
Di-n-butyl phthalate	DnBP	$C_{16}H_{22}O_4$	278	Cosmetics, home furnishings, pharmaceutical coatings, sealants, wrapping materials, etc.
Dutylbenzyl phthalate	BBP	$C_{19}H_{20}O_4$	312	Wrapping materials, conveyor belts, adhesives, perfumes, artificial leather
Di (2-ethylhexyl) phthalate	DEHP	$C_{24}H_{38}O_4$	390	Diapers, blood bags, tiles, shoes, children's dolls, air tubes, etc
Di-iso-nonyl phthalate	DiNP	$C_{26}H_{42}O_4$	419	Toys, spoons, shoes, teethers, drinking straws, etc.
Di-n-nonyl phthalate	DnOP	$C_{24}H_{38}O_4$	390	Conveyor belts, bottle cap liners, floorings, garden hoses, tarps etc
Di-iso-decyl phthalate	DiDP	$C_{28}H_{46}O_4$	447	Electrical cords, PVC floorings, leather for car interiors, etc

Source Heudorf et al. (2007), Cao (2010), Das et al. (2021)

High molecular weight PAEs can increase the flexibility and strength of products, whereas low molecular weight PAEs are used to maintain the color and fragrance of products or to give a glossy finish to the plastic products (Huang et al. 2021b). Di (2-ethylhexyl) phthalate (DEHP), di-iso-nonyl phthalate (DINP), di-isodecyl phthalate (DIDP), and butyl benzyl phthalate (BBP) are the most used phthalates in the world. Di-isononyl phthalate (DINP) and di-isodecyl phthalate (DIDP) are high molecular weight phthalates that constitute 80% of phthalates used in Europe and are not considered to be toxic to humans. According to REACH (Registration, Evaluation, Authorization, and Restriction of Chemical substances), a European Union regulation on chemicals, low molecular weight phthalates such as dibutyl phthalate (DBP), benzyl-butyl phthalate (BBP), di-2-ethyl-hexyl phthalate (DEHP) are classified as very dangerous compounds that are toxic to human health (Ventrice et al. 2013). The guidelines on the application of each type of phthalate compound are based on their nature of toxicity to organisms (Das et al. 2021). Based on the evaluation of possible toxic effects of these pollutants in the environment, United States Environmental Protection Agency (USEPA) declared six phthalate congeners as priority pollutants. They are di-methyl phthalate (DMP), di- ethyl phthalate (DEP), di-n-butyl phthalate (DnBP), di (2-ethylhexyl) phthalate (DEHP), di-n-octyl phthalate (DnOP) and butyl-benzyl phthalate (BBP) (Qureshi et al. 2016; Kashyap and Agarwal 2018).

7.3 Source of PAEs Pollution

Phthalates are commonly used in plasticizer industries to improve the plasticity of various food packaging, clothing and medical devices, building materials, etc. (Kotowska et al. 2020). Their presence has been found in different environmental matrices, including air, water, soil, and even in food, medicine, and personal care products (Huang et al. 2021a). DEHP possesses higher molecular weights, whereas DEP, DBP, and BBP of lower molecular weights among the various types of phthalates, commonly used as a plasticizer in polyvinyl products and personal care products (Cao 2010). The sources for the occurrence of phthalates in various environmental matrices, air, water, and soil, are listed in Fig. 7.1. The release of phthalates occurs

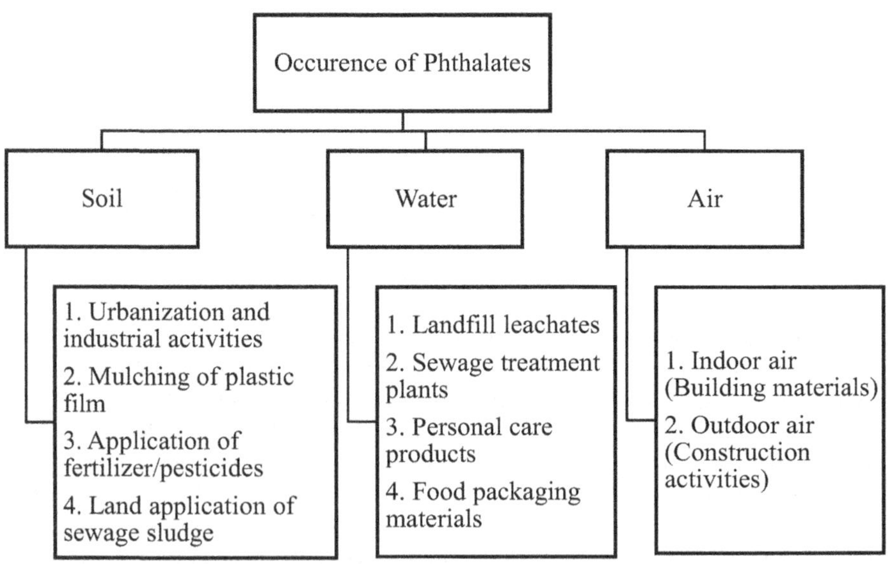

Fig. 7.1 Different sources of phthalates in environmental matrices (soil, water, and air)

in the entire life cycle from its production to disposal via leaching, discharge during industrial production and weathering, etc. (Das et al. 2021).

Moreover, different forms of phthalates have been used in the various daily used products, which will act as a source for the leaching into the environmental matrices. For example, building products (wallpaper, wire, and cable insulation), food packaging, children's products (toys, diapers), clothing, medical (blood bags and gloves), etc., are the potential source for DEHP (Heudorf et al. 2007). Similarly, DEP and DBP are used widely in personal care products and cosmetics. In addition, BBP is commonly used to prepare vinyl tiles, conveyor belts and artificial leather, etc. (Das et al. 2021).

7.4 Fate and Transport of PAEs

The major driving factors responsible for the fate and transport of phthalates in the various environmental matrices are meteorological and hydrological effects (He et al. 2013). Several studies particularly focus on the occurrence and fate and transport of phthalates in the river and lake water (Liu et al. 2014; Zheng et al. 2014), cultivation and urban soil (Sayyad et al. 2017; Wang et al. 2018), river sediments (Liu et al. 2014), indoor dust and air (Huang et al. 2021a; Liang et al. 2019; Bi et al. 2015) and municipal wastewater sludge (Salaudeen et al. 2018), etc. However, their sampling, sample pre-treatment, and detection and analysis in different environmental matrices (air, water, sludge, sediment, and soil) are different and tedious, thus become a challenging task (Net et al. 2015a).

7.4.1 Fate and Transport in the Aquatic Environment

The phthalates find their way to aquatic environments via several routes such as wastewater discharge from the sewage treatment plant, landfill leachates, and agricultural mulch, etc., as represented in Fig. 7.2. These plasticizers are disintegrated by the change in temperature, pH, irradiation, pressure, etc., into naked polymer and leachates. Once they enter the aquatic environment, they get transported through the food chain from high trophic levels to low trophic levels (Zhang et al. 2021). For instance, He et al. (2013) investigated the spatio-temporal distribution of phthalate esters in lake Chaohu, China. They found that the average concentration of 6 different PAEs to be in the range of 4.0 ± 3.9 $\mu g\ L^{-1}$ and the highest limit of 17.9 $\mu g\ L^{-1}$. In addition, DnBP was found to be the dominant form among the other types of phthalate esters with the highest concentration in September and lowest in June, which may be attributed to the variation in the rainfall-runoff during these times. Similarly, Liu et al. (2014) studied the distribution of phthalates in the Pearl river-sediments, China. They have identified 16 different phthalates esters with an average of 5.34 $\mu g\ g^{-1}$ DW (dry weight) with DiBP, DnBP and DEHP were found to be the

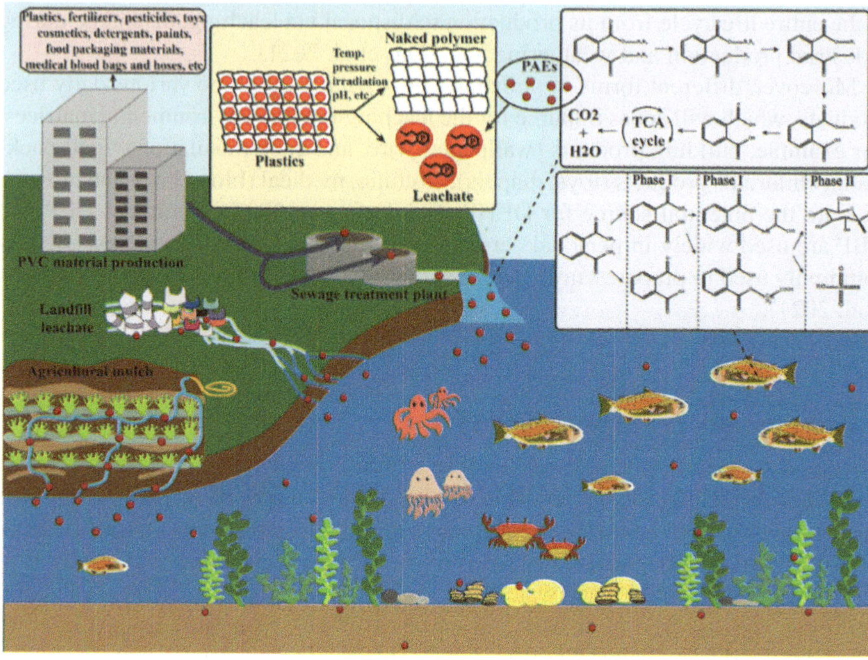

Fig. 7.2 Fate and transport of phthalates in the aquatic environment. Reprinted with permission from Zhang et al. (2021). Copyright © 2021 Elsevier Ltd.

most dominant type of PAEs. The higher concentration level is mainly due to the urbanized and industrialized zones in nearby areas (Liu et al. 2014).

The fate and transport model for phthalates in wastewater treatment plants by Fauser et al. (2003) provided a simple and robust tool for the removal of phthalates and the factors affecting the process, such as inlet concentration and flow conditions. They developed the model by considering the phthalates removal by biodegradation and sorption to sludge. It was observed that 70% of biodegradation and 28% of sorption to sludge and the remaining 2% are present in the treated water. Also, the predominant form in the wastewater was DEHP with a concentration of 35.4 $\mu g\ L^{-1}$. Similarly, Kotowska et al. (2020) observed the highest concentration of DEHP, around 143 mg L^{-1} from the effluent of municipal sewage wastewater and found the highest around 90% removal efficiency for DMP and DEP esters.

7.4.2 Fate and Transport in Air

The first kind of study by Bi et al. (2015) investigated the influence of temperature on the fate and transport of phthalates, particularly benzyl butyl phthalate (BBzP) and DEHP in the indoor environment. They found that there exists a significant effect in the indoor temperature on the airborne concentration of phthalates. As the temperature increases from 21 to 30 °C, the concentration becomes three times the initial concentration. Besides, there was enhanced sorption of phthalates on various indoor surfaces, including windows, fabric cloth, wood, dust and kitchen utensils, etc. Similarly, another study by Liang et al. (2019) comprehensively developed the mechanistic model for the fate and transport of phthalates esters in the indoor. Liang et al. (2019) developed a mechanistic model which predicted the average steady-state 2-ethyl hexyl phthalate (DEHP) concentrations of about 0.14 $\mu g\ m^{-3}$ in indoor air and ranged from 80 to 46,000 $\mu g\ g^{-1}$ in the settled dust on various indoor areas. Also, the major mechanism involved in the fate and transport model includes interaction between air, suspended, and settled particles with various interior surfaces.

Additionally, the studies reveal that indoor temperature, humidity, air exchange rate, building material, and indoor maintenance were found to be the major parameters influencing the indoor levels of phthalates in the air and dust (Kashyap and Agarwal 2018). Also, the study by Wu et al. (2018) revealed that particle/gas partition coefficient (K_p) is a significant parameter influencing the fate and transport of indoor semi-volatile organic compounds (SVOCs). They determined the K_p value of DEHP for ammonium sulphate (inorganic), oleic acid and squalene (organic) particles with the help of laminar flow tube chamber. The higher K_p value was observed for oleic acid (0.23 $m^3\ \mu g^{-1}$) and squalene (0.11 $m^3\ \mu g^{-1}$) as compared with ammonium sulphate (0.011 $m^3\ \mu g^{-1}$), clearly suggesting the mechanism by which particles accumulate in organic compounds will be different as compared with inorganic compounds.

7.4.3 Fate and Transport in Soil

Phthalates in urban, agricultural, coastal, and river soil sediments are causing greater risk to plants and other microorganisms. The study by Sayyad et al. (2017) developed the fate and transport model for the seven commonly detected phthalates in biosolid amended soil for corn cultivation. They simulated the discharge rate of water and phthalates mobility by HYDRUS-1D model and a two-site sorption model and found that with an increase in the carbon chain number of phthalates, the adsorption with soils also increases. Similarly, another study on the presence of phthalates in urban soil by Wang et al. (2018) found that soil texture and magnetic susceptibility were the major parameters influencing the phthalates concentrations in the urban soils. Besides, a higher level of phthalates in urban soil is mainly attributed to the application of plasticizers/additives, emission from construction activities and furnishing, industrial processes, etc. It was observed that DnBP and DEHP were the most predominant types in the urban soil analysis.

Moreover, these different classes of phthalates undergo degradation by biotic and abiotic pathways in the air, water, and soil/sediment environmental matrixes. Their global half-lives in air vary from few hours to days and few days to days months in soils. However, in water, it will undergo hydrolysis, photolysis, photooxidation, and biodegradation (Net et al. 2015b). Apart from urban soils/agricultural soils, the accumulation of phthalates in coastal sediments is of important concern. For instance, the study by Arfaeinia et al. (2019) estimated the distribution of different phthalates in the Persian Gulf region. They estimated the different phthalates concentrations present in the coastal sediments and found that highest of 28.15 $\mu g\ g^{-1}$ of DEHP content followed by 4.04 $\mu g\ g^{-1}$ of DnBP, 11.58 $\mu g\ g^{-1}$ BBP, and 1.78 $\mu g\ g^{-1}$ of DnOP, which is mainly attributed to the higher industrial and agricultural activities in that area (Arfaeinia et al. 2019). Recently, Li et al. (2021) modelled the fate of phthalates in the soil-air interface of the plastic film greenhouses using the Fugacity model. It was found that maximum deposition of DEHP from air to soil in autumn with mean deposition flux of 175.1 kg $(h\ km^2)^{-1}$ and 11.9 kg $(h\ km^2)^{-1}$ in spring. At the same time, diethyl phthalate was found to be escaped from the soil with a mean volatilization flux of 0.005 kg $(h\ km^2)^{-1}$ in autumn and 0.025 kg $(h\ km^2)^{-1}$ in spring. They concluded that the temperature differences majorly causes the fate and transport of phthalates in the soil-air interface during different seasons and their chemical properties.

7.5 Toxicity and Exposure of Phthalates

Human exposure to phthalates has recently become a significant health concern and the exposure route can be broadly classified as dermal absorption, oral/dietary ingestion, and inhalation (Wang et al. 2018; Huang et al. 2020). The analysis via these different routes can be integrated to get the total exposure value of phthalates to humans. However, determining the actual exposure dose of dermal, oral, and inhalation is very difficult. Hence, most researchers focused on developing analytical methods for identifying the internal exposure of phthalates metabolites in human samples such as urine, serum, breast milk, hair, and nails (Huang et al. 2021b). But these samples, especially urine, has the drawback of small sample size and limited representativeness. Therefore, the population exposure of phthalates is mainly determined by the study on wastewater-based epidemiology. For instance, the study by Du et al. (2018) investigated the population exposure of phthalates in major cities of China via wastewater-based epidemiology method for the first time. They have collected the wastewater samples containing urine from 54 wastewater treatment plants in 27 major cities of China for the study. An average exposure per person of 5 different phthalates was observed at around 2184 $\pm$ 1173 $\mu g\ day^{-1}$. Also, they identified that phthalates exposure imparts greater risk to children than to adults.

As the phthalates find their route to the human body via dermal, inhalation, and oral routes, they undergo different phases of metabolisms in the human body as represented in Fig. 7.3, such as de-esterification, glucuronide conjugation, and followed by excretion (via urine, sweat, and faeces) (Benjamin et al. 2017). Besides, the metabolites of phthalates formed after de-esterification can trigger various molecular signalling, which causes epigenetic modulation, insulin resistance, and type II diabetes, obesity, allergy and asthma, endocrine disruptions, etc. (Benjamin et al.

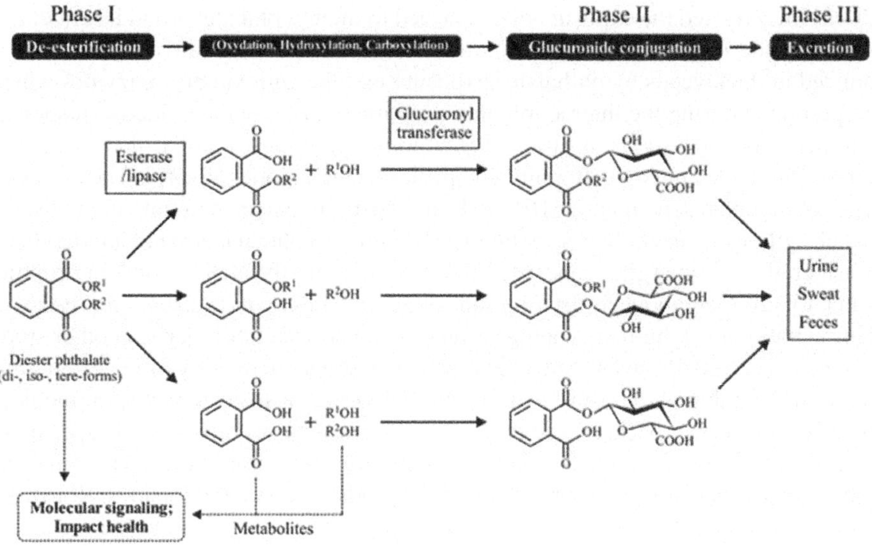

Fig. 7.3 Different phases of phthalates metabolism in humans. Reprinted with permission from Benjamin et al. (2017). Copyright © 2017 Elsevier Ltd.

2017). Therefore, once the phthalates enter the human body via various routes, it is distributed throughout the body tissues and then biotransformation occurs followed by excretion (Lyche et al. 2009). Park et al. (2017) studied the association of phthalates levels on thyroid hormone levels in the Korean adult population and found significant associations between urinary DEHP metabolite levels and thyroid hormones in adults. This further confirms the health risk associated with phthalates and their metabolites in humans. In addition, the studies revealed that the two priority phthalates BBP and DEHP possess potential cancer risk, coronary heart disease, and reproductive system toxicity in females compared with the other forms of phthalates (Wang et al. 2018; Lyche et al. 2009; Mariana et al. 2016). There are two important estimates such as estimated daily intake (EDI) and total estimated dose (TEDI), which have been preferred for evaluating the human health risk associated with the phthalates or any other contaminant. In addition, hazard quotient (HQ) for an individual compound and hazard index (HI) for cumulative health risks associated with the exposure of multiple compounds are other parameters intended for health risk assessment (Huang et al. 2021b).

Phthalates cause toxicity to aquatic organisms such as algae, microorganisms, fishes and other invertebrates (Staples et al. 1997). The recent studies reported that the total concentration of phthalates even exceeds 200 $\mu g\ L^{-1}$ in the aquatic environment (sea, river, and lake waters). The presence of phthalates has been detected in the tissues and organs of aquatic species and it causes adverse effects on their immune system, endocrine system, metabolism and developmental and behaviour changes, etc. (Zhang et al. 2021). For example, the study by Cheng and Cheng

(2012) investigated the exposure of duckweed to diethyl phthalates and its effects on oxidative and metabolic changes. It was observed that enhanced oxidative stress is induced in duckweeds by phthalates exposure and the antioxidative enzymes which are produced during the interaction play a significant role in the oxidative response. Another study by Adeogun et al. (2015) comprehensively studied the concentration of phthalates in two different fish species in the lagoon, Nigeria. The common types of phthalates, namely DEHP, DBP, and DEP, are detected mainly in the fishes' muscle, gill, liver, and kidney systems. The presence of phthalates in lagoons has been considered as an emerging concern as these lagoons are the major source of drinking water for the neighbouring Epe region, Nigeria. These can transport the PAEs or their metabolites to humans via ingestion of fish and water. Similarly, another study by Chen et al. (2020) investigated the reproductive system toxicity of male zebrafish due to the combined exposure of DBP and DiBP. Their major conclusion includes the combined effect of these two types of phthalates causes disruption in spermatogenesis rather than individual effects by inhibiting the steroid hormones signalling pathways that badly affect zebrafish's male reproductive system. Besides, Kotowska et al. (2020) identified that DMP, DPP, and DBP were more toxic to the algae and cyanobacteria, whereas DIBP showed an equal environmental risk to fish as well as to algae and cyanobacteria.

The presence of phthalates in soil was found to adversely impact the terrestrial invertebrates (Du et al. 2015). The effect of DBP on earthworms was studied by Du et al. (2015) and observed the development of genotoxicity of DBP on earthworms by the production of antioxidant enzymes which trigger DNA damage and oxidative stress etc. The max concentration of around 100 mg kg^{-1} of DBP causes the highest DNA damage to earthworm coelomocytes. Apart from microorganisms in the soil, the presence of phthalates significantly affects the plant species also. For instance, the study by Ma et al. (2015) investigated the phytotoxic effect of two major priority pollutants, namely DnBP and DEHP in the concentration of 0 to 500 mg kg^{-1} soil on seven plant species. They monitored the phytotoxicity via germination, root elongation, seedling growth, chlorophyll contents, and malondialdehyde (MDA) content variation in the plant species during exposure. It was observed that chlorophyll contents were greatly affected by both phthalates, which is an implication of potential damage to the photosynthetic system of plants. Whereas, the monitoring of increase in MDA content which is an indicator of free radical-mediated stress formed by the lipid peroxidation, it was found that DnBP is more toxic than DEHP, since it decreases root elongation, seedling growth, and biomass with an increase in DEHP exposure.

7.6 Method of Extraction and Detection of PAEs

The measurement of phthalates and their different metabolites in various environmental samples, including air, water, sludge, sediment, and soil, is challenging due to their trace concentrations and interference caused by the other components in the sample matrix. Therefore, the sample extraction or pre-treatment is an important step before the detection and quantification procedures (Net et al. 2015a). The extraction procedures for various environmental samples are different and must be simple and easy with a maximum yield of phthalates to minimize the interference caused by the sample matrix in the method of quantification (Otero et al. 2015; Russo et al. 2015). In general, phthalates in water and food samples can be extracted by liquid–liquid extraction (LLE), solid-phase extraction (SPE), solid-phase microextraction (SPME), and ultrasonication, whereas the Soxhlet extraction procedure was mostly adopted for particulates and soil samples (Kubwabo et al. 2013; Xu et al. 2014; Makkliang et al. 2015; He et al. 2019). However, a procedural blank for each sample is essential since phthalates present in the laboratory consumables can leach out into the real samples and affect the accurate measurement (Guo and Kannan 2012). Table 7.2 represents the method of extraction, quantification, and limit of quantification (LOQ) for phthalates samples from different environmental matrices.

Quantification of PAEs in the environmental matrices are carried out using sophisticated instrumentation facilities such as gas chromatography-mass spectrometry (GC–MS), high-performance liquid chromatography (HPLC), liquid chromatography-mass spectrometry (LC–MS) depending upon the level of concentrations of PAE. When the concentrations of PAEs are very low, high-resolution chromatographic analyses are usually conducted to quantify the compound. After the extraction step, the commonly adopted measurement techniques for phthalates include LC/MS and GC/MS. Since phthalates are semi-volatile and stable compounds, GC/MS is the most common and frequently used technique for separation with capillary columns coated with (5%-phenyl)-methyl polysiloxane (Cao 2010). Also, GC/MS provides high reproducibility and specificity for measuring semi-volatile compounds like phthalates compared with LC/MS methods (Otero et al. 2015). Indeed, the metabolites and degradation products of phthalates can be determined more accurately by LC/MS and ultra-high pressure liquid chromatography (UHPLC) (Net et al. 2015a). The most commonly used solvents in extraction methods are hexane, acetone, acetonitrile, dichloromethane, and methanol, etc. The solid-phase extraction (SPE) method requires preconditioned solid phase either cartridges or discs with sorbent phase and solvents (Zheng et al. 2019; Salaudeen et al. 2018; Calafat et al. 2004; Manzo et al. 2019).

The study by Salaudeen et al. (2018) utilized C18-U cartridges for extraction of phthalates from wastewater samples and elution was carried out with dichloromethane, hexane, and acetone solvents in gradient mode at a flow rate of 10 mL min^{-1} followed by concentrating the eluents in the rotary evaporator for GC/MS analysis. Similarly, the study by Manzo et al. (2019) extracted the phthalates present in the bottled water with rotating disc conditioned with Oasis® sorbent phase at 2000 rpm for 90 min. In this solid-phase extraction method, the adsorbed phthalates in the disc were extracted with methanol solvent followed by concentration in rotary evaporator and redissolution with dichloromethane solution before GC/MS analysis. Also, anhydrous sodium sulfate was commonly employed to concentrate the eluents by removing any traces of water in a rotary evaporation step-up (Zheng et al. 2019). In addition, fully automated solid-phase microextraction (SPME) with

Table 7.2 Method of extraction, detection techniques, and limit of quantification for phthalates from different environmental samples

Sample type	Extraction method	Detection method	Column used	Most abundant phthalates present	LOQ	References
Indoor dust	Ultrasonication	GC/MS	Zebron ZB-5HT	DEHP	9.8 $\mu g\ g^{-1}$	Kubwabo et al. (2013)
Packaged food	Magnetic micro-solid phase extraction	GC/MS	DB-5 fused silica capillary column	DBP and DEHP	121 and 88 ng mL^{-1}	Makkliang et al. (2015)
Food samples	LLE and SPE	LC/MS/MS	Poroshell 120 EC-C18 column	DIDP	100 $\mu g\ kg^{-1}$	Xu et al. (2014)
Bottled water	LLE	GC/MS	DB-5 ms Ultra Inert capillary column	DBP and DEHP	17.5 and 43.5 ng mL^{-1}	Otero et al. (2015)
Saliva of children (soft toys)	LLE	GC/MS	DB-17HT	DEHP	1 $\mu g\ L^{-1}$	Earls et al. (2003)
Urban soil	Soxhlet extraction	HPLC–DAD	Hypersil ODS2	DBP and DEHP	8.9 and 1.9 $\mu g\ kg^{-1}$	Wang et al. (2018)
Atmospheric particles	Soxhlet extraction	GC/MS	DB-5MS capillary column	DIBP and DBHP	356 and 1957 pg m^{-3}	He et al. (2019)
Coastal sediments	Solid phase extraction	GC/MS/MS	DB-5MS column	DEHP and DiNP	2700 and 553 $\mu g\ g^{-1}$	Lee et al. (2020)
Urine samples	LLE	UPLC-MS/MS	Betasil phenyl column	Metabolites of DEHP	0.28 $\mu g\ L^{-1}$	Park et al. (2017)

sorbents coated in the fused silica fibers was simpler and efficient technique for trace levels of phthalates extraction from the sample matrices (Yang et al. 2015). However, they are expensive and clogging of sorbent results in high backpressure. Recently, the sorbent of solid-phase extraction has been developed as magnetic solid-phase extraction (MSPE) by immobilising magnetic nanoparticles to enhance the yield of phthalates extraction with a minimum requirement of solvents (Makkliang et al. 2015).

The liquid–liquid extraction (LLE) method is a simple but time-consuming procedure and it requires a large amount of sample and organic solvents for the extraction protocols. For instance, LLE for wastewater treatment plant effluents samples were carried out by adding hexane (80%) dichloromethane (25%) into the 250 ml of unfiltered wastewater sample and kept for 15 min shaking followed by the collection of formed organic layer for further analysis (Gani and Kazmi 2016). The same procedure is usually repeated thrice in LLE to ensure the complete extraction of phthalates. Similarly, the phthalates detection in the bottled water samples was studied by Otero et al. (2015) and was first subjected to liquid–liquid extraction (LLE) with dichloromethane in the water-bath sonication for evaporation at 60 °C for 10 min. An improved LLE method by redissolving the residue collected with cyclohexane was employed before the GC/MS analysis with high reproducibility and specificity of measurements (Otero et al. 2015; Nantaba et al. 2021).

The sample extraction procedures followed for liquid and solid type food samples were different. For instance, the study by Xu et al. (2014) found that LLE for liquid food samples with acetonitrile with vortex mixing followed by centrifugation gave better results whereas solid-phase extraction (SPE) procedures were efficient for solid food samples (Xu et al. 2014). They determined about 23 phthalic esters in food samples (both liquid and solid) by liquid chromatography-tandem mass spectrometry (LC/MS/MS) which enables the direct measurement of isomeric mixtures of monoesters in food, including DINP, and DIDP. The selected mobile phase for LC analysis was methanol and 0.1% formic acid in deionized water at 0.7 mL min^{-1} in the gradient mode with Poroshell 120 EC-C18 column (Xu et al. 2014). Apart from food and water samples, human saliva and urine samples are found to be the most promising alternative matrices for identifying the human exposure and toxicity imposed by different phthalates but are challenging to quantify accurately (Earls et al. 2003; Park et al. 2017). For instance, the study by Silva et al. (2005) detected the different metabolites of phthalates in the saliva of adults using the LC/MS/MS method and extraction by LLE. The maximum concentration of 65.8 ng mL^{-1} was observed for the mono-n-butyl phthalates metabolite. For soil and sediment samples, the phthalates extraction is mainly by SPE and the Soxhlet extraction method. The study by Wang et al. (2018) utilized the Soxhlet extraction method which uses hexane and acetone (1:1) solution purified with a glass chromatography column of silica gel/neutral alumina followed by separation and detection with Hypersil ODS2 column in a high-performance liquid chromatograph equipped with a diode array detector (HPLC–DAD) for urban soil samples. Besides, Soxhlet extraction method is one of the traditional method for solid samples but are time-consuming (Net et al. 2015a).

Moreover, integration of biological sampling and environmental sampling, their efficient extraction procedures, and detection methods can provide an accurate estimation of human exposure and fate and transport scenario of phthalates in the environmental matrices. In another study conducted by Zeng et al. (2008), soil samples containing a mixture of 16 PAE compounds were extracted using DCM for 48 h at a rate of 4–6 cycles h^{-1} using Soxhlet extractor. Under a nitrogen atmosphere, the extracted samples were concentrated through anhydrous sodium sulphate and solvent exchanged to n-hexane and reduced to 1 mL concentrated extracts. These concentrated extracts were further cleaned using an alumina/silica gel glass column packed of 10 mm internal diameter and eluted with a mixed solvent of acetone/n-hexane.

7.7 Remediation Option for PAEs

7.7.1 Adsorption

Adsorption is a physicochemical treatment process that can be effectively used for the removal of phthalates from aqueous solutions. They are considered to be one of the efficient and low-cost treatment options for the removal of organic compounds. The hydrophobic distribution of PAEs, electron-accepting nature of ester groups, π-π interaction and electrostatic interaction are the underlying mechanisms for adsorptive removal of PAEs (Ye et al. 2021). A study conducted by Tümay Özer et al. (2016) confirmed that the pH of the solution could change the surface properties of the adsorbents by altering the functional groups and resulted in reducing the removal efficiency of PAEs. Activated carbon is the most commonly used adsorbent for PAEs. Adsorptive removal of DEP by activated carbon was studied by Mohan et al. (2007) and observed that sorption was highly dependent on the pH of the solution and initial DEP concentration. In another study, activated carbon derived from phoenix leaves was effectively used for the removal of DBP (Wang 2015). The higher specific surface area and the heterogeneous porous structure allowed a maximum DBP adsorption of 97.36% at pH 3. Though activated carbons are effective adsorbents, owing to their lower regeneration ability, several other adsorbents such as chitosan beads, clay minerals, zeolites etc., were used for PAE removal (Chen and Chung 2006; Wu et al. 2015; Mesdaghinia et al. 2017). Apart from the conventional adsorbents, several novel materials have received considerable attention for the removal of PAEs. In a study conducted by Yin et al. (2014), magnetic nanoparticles functionalized on graphene oxide (GO-MNPs) were used as an adsorbent to remove PAEs from different water sources. The studies concluded that π-π interactions between the adsorbents and the pollutants resulted in efficient adsorption, obtaining a maximum adsorption capacity of 8.71 mg g^{-1} within 30 min of contact time. Another adsorbent, phenyl-functionalized mesoporous silica material with meso-structure, higher surface charges and surface area effectively removed DBP from aqueous solution

(Fan et al. 2017). The catalyst showed excellent adsorption of DBP with a maximum adsorption capacity of 40 mg g^{-1} within 2 min of reaction.

Further, the catalyst with higher regeneration potential, stability, more surface functional groups, magnetic separability, etc., can effectively improve the adsorptive removal of PAEs from aqueous solutions (Mesdaghinia et al. 2017; Shaida et al. 2018). The presence of more amine and methyl functional groups on the surface of novel functional hyper-cross-linked polymer enhanced adsorption of PAEs through physical adsorption and ion exchange mechanism (Shi et al. 2017). In another study conducted by Wu et al. (2015), the adsorption capacity of kaolinite and montmorillonite clay minerals for the removal of DEP was investigated. The results from the studies concluded that montmorillonite clay saturated with K^+ ions resulted in higher adsorption of DEP owing to structure, hydrophobic surface and clay exchangeable cation. Another adsorbent, chitosan beads, was used to remove a mixture of PAEs and their degradation products (Salim et al. 2010). The results from the studies concluded that the hydrophobic interactions due to the presence of active polar groups enhanced the sorption between the adsorbent and PAEs. Among the different PAEs and their derivatives, the higher adsorption capacity was observed for MNP because of their higher hydrophobicity when compared to monoesters and phthalic acid. Further, an amphiphilic magnetic nanocomposite prepared by a modified sol–gel process using the hydrophilic amino groups and hydrophobic alkyl chains on Fe_3O_4 exhibited higher adsorption capacity for six PAEs (Gao et al. 2019). The adsorbent was stable enough for six reuse cycles with commendable regeneration properties and exhibited higher adsorption capacity and a faster desorption rate due to active surface functional groups.

7.7.2 *Biological Degradation*

The microorganisms present in wastewater treatment plants can enzymatically break down and subsequently degrade the recalcitrant PAEs by biodegradation, sorption or volatilization, etc. (Huang et al. 2008). Microbes under aerobic or anaerobic conditions have significant potential for PAE degradation in the aquatic and terrestrial systems owing to their lower cost and environmental friendliness (Gao and Wen 2016; Tang et al. 2016). The operational factors including, pH, nutrient conditions, substrate concentrations, temperature, etc., are significant factors during the biodegradation of PAEs. The most common microbial degradation mode of PAEs is by using bacterial strains isolated from different sources. The degradation pathway involves the action of esterase's converting the di-alkyl PAEs to mono-alkyl PAEs by the hydrolysis of ester bonds in PAEs (Huang et al. 2019; Xu et al. 2020). The ester bonds are hydrolysed stepwise and the common degradation product obtained is the phthalic acid which is finally converted to CO_2 and H_2O in the subsequent degradation process. Moreover, the hydrolytic enzymes synthesized by the microbes are specific for each ester molecule and doesn't interact with other complex bonding involved (Boll et al. 2020). Also, in most biodegradation studies, the bacterial strain used is substrate-specific and does not deal with a mixture of PAEs as in natural environments.

The biological processes can be aerobic or anaerobic, such as activated sludge process, anaerobic degradation etc., and several factors affecting the microbial properties such as hydraulic retention time, aeration rate, phthalate loading, etc., have significant effects on degradation potential. PAEs are degraded through the protocatechuate pathway followed by the ring cleavage to achieve complete mineralization in the aerobic degradation process (Pirsaheb et al. 2009). Activated sludge processes with an acclimatized mixed bacterial culture can effectively biodegrade phthalate compounds. The degradation of DEHP in an ASP was assessed and the effect of different operational parameters was studied by Gani et al. (2019). The overall performance of the ASP was depended on the mixed liquor suspended solids (MLSS) concentration and the sludge retention time (SRT). They have observed a maximum DEHP removal of 92% in the MLSS concentration of 3461–4972 mg L^{-1} at an SRT of 25 days. Further, the increase in the DEHP removal at reduced Food to Microorganism ratio suggested that starving conditions were more favourable for the degradation process. Another ASP study was effectively used for the removal of a mixture of four PAEs achieving more than 80% removal along with nitrogen and phosphorous removal (Roslev et al. 2007). Though significant PAEs removal was observed during the study, they concluded that the water solubility of PAEs significantly decreased the degradation ability of activated sludge.

Apart from the aerobic processes, anaerobic or anoxic systems are also capable of removing PAEs. An anaerobic fixed film fixed bed reactor (AnFFFBR) was successfully used by Ahmadi et al. (2017) for the removal of four short-chain PAEs, DAP, DMP, DEP and PA. Due to biofilm mass's higher bio-availability, all the four PAEs were degraded through de-esterification pathway, achieving significant biodegradability. Further, maximum mineralization and methane production was observed for DMP obtaining a maximum TOC removal of 80.39%. Another treatment system consisting of an anaerobic/anoxic/oxic (AAO) process was effectively used in laboratory scale for the degradation of DMP (Zhang et al. 2016). The microorganisms were able to degrade 73.8% DMP under an optimized condition of 18 h HRT and 15 d SRT and around 5.8% was released with the effluent and 19.3% accumulated in the system. Moving forward, a comparative study on AnFFFBR and up-flow AnFFFBR concluded that though both the treatment processes are capable of treating low to high strength wastewater containing DEP (Yousefzadeh et al. 2017). The former one was more efficient due to its more sludge yield and potential for bioenergy production. In a study conducted by Ahmadi et al. (2015), the effect of operational parameters affecting the degradation of DEP and DAP in a Moving Bed Biofilm Reactor (MBBR) was evaluated. Higher contact time of biofilms up to 12 h with the substrate and lower aeration rates of 60 L h^{-1} improved the performance of MBBR and resulted in more than 92% COD removal for 300 mg L^{-1} influent loading. In another study conducted by Cheng et al. (2010), the treatment efficiency of an attached growth system [Rotating Biological Contractor (RBC)] and a suspended biological system [Activated Sludge Process (ASP)] were studied for the degradation of DEHP. The studies concluded that the attached growth system was more tolerant to DEHP loading when compared to the suspended growth system and resulted in higher degradation owing to their partitioning behaviour and solubility of DEHP.

DBP was used as the carbon and energy source for a facultative bacterial strain, *Enterobacter sp.* isolated from a landfill bioreactor. The degradation studies were conducted for several environmental parameters and the optimum conditions for maximum biodegradation were pH 7 and 35 °C temperature, obtaining a half-life of 20.9 h when the concentrations of DBP were less than 1000 mg L^{-1}. In a similar kind of study, the bacterial strain *Agrobacterium sp.* derived from river sludge degraded DBP with a half-life of degradation of 10.4 h at an optimum condition of pH 8, temperature 30 °C and DMP concentration less than 200 mg L^{-1}. Apart from these studies, a similar trend in the biodegradation of diethyl phthalate was observed with a bacterial strain *Bacillus subtilis strain 3C3* isolated from an organic-solvent-tolerant bacterium from soil (Navacharoen and Vangnai 2011). The coexistence of PAEs in different environmental matrices may lead to the variation in biodegradation rate and extent of a single pollutant. However, a partial substrate inhibition trend was observed in all these studies. PAEs interact with each other under the concurrent presence of substrates and the microorganism played a key role in the metabolic breakdown of PAEs.

A better sense of the degradation characteristics and bacterial properties while dealing with a mixture of PAEs could help understand the potential of bioremediation for PAEs contaminated environmental matrices. *Mycobacterium sp.* YC-RL4 isolated from petroleum-contaminated soil was effectively used for the biodegradation of a mixture of PAEs comprising of DMP, DBP, DEP, DCHP and DEHP and achieved a degradation rate of 85% within five days (Ren et al. 2016). Owing to the higher cell surface hydrophobicity of the bacterial strain, they could easily adhere to the pollutant surface and accelerate the degradation process in different environmental conditions. Further degradation studies conducted on DEHP in three different environmental samples and achieving more than 65% degradation after ten days incubation revealed the potential of YC-RL4 strain in different PAE contaminated sites. In a similar study, another bacterial species *Bacillus mojavensis* B1811 was effectively used for the degradation of seven PAEs in a mineral salt medium (Zhang et al. 2018). Among the seven phthalates, only 5.9% and 42.9% degradation of DMP and DEP were respectively observed after four days, whereas significant degradation of DBP, BBP, DNOP, DEHP, and DPP was observed under the same condition. The decreasing trend in the degradation was attributed to the less affinity of bacterial strains on PAEs with shorter alkyl chains. However, the degradation pathway analysis confirmed that all the PAEs followed a similar degradation pathway.

Apart from the bacterial degradation of PAEs, other microorganisms can also degrade PAE under eco-friendly conditions. In a study conducted by (Pradeep et al. 2013), a mycelial fungal consortium consisting of *Penicillium lilacinum* strain BP13, *Aspergillus japonicus. japonicus* strain BP9 and *Purpureocillium brocae* strain BP6 were effectively used to degrade DEHP bound to PVC blood storage bags from heavily contaminated soil. Among the three fungal species, *P. lilacinum* independently showed significant degradation performance compared to that fungal consortium comprising all the species and completely utilized the DEHP bound to the blood storage bags.

Contamination of soil by the release of PAEs in the environment has led to considerable attention on their environmental fate and toxicity to different level of organisms. Microbial degradation of a mixture of four PAEs in soil was investigated and assessed the potential impacts on soil enzymatic activities (Xie et al. 2010). The degradation process followed the first-order kinetics with more persistence to microbial activity was observed for longer alkyl chains of PAE when compared to shorter chain PAEs. Further, they found that a higher concentration of PAEs lowered the β-glucosidase activity and inhibited the carbon substrate utilization. In another study, soil contaminated with a mixture of PAEs were degraded by microbes present in sewage sludge (Yuan et al. 2011). They concluded that smaller-sized sludge particles improved the biodegradation process, and the microbial strains F4 (*Rhodococcus sp.*) and F8 (*Microbacterium sp.*) dominated the degradation process. A novel 2D bacterial strains isolated from compost, *Providencia sp.* 2D utilized PA and DBP are carbon sources for the remediation of contaminated soil. Complete removal of DBP was observed and the degradation process continued irrespective of the aerobic or anaerobic conditions that prevailed during the treatment.

7.7.3 Advanced Oxidation Processes

Owing to the persistence of PAEs in wastewater, reliable and robust treatment options are required for their removal. Advanced Oxidation Processes (AOPs) are highly effective chemical treatment processes capable of mineralizing PAEs by producing highly reactive radicals. Among the different AOPs, photocatalysis, Fenton's, electrochemical, sulphate radical based etc., are extensively studied for the degradation of PAEs. Table 7.3 summarizes the different AOPs available for the removal of PAEs. The photocatalytic degradation of DEHP with an initial concentration of 75 $\mu g\ L^{-1}$ was carried out inside a UV reactor and complete removal of the compound was observed after 150 min irradiation (Chung and Chen 2009). The degradation followed the first-order kinetics and the dominant degradation mechanism was suggested as the cracking of the aliphatic part of DEHP while the aromatic part remained intact. In another UV/TiO_2 photocatalysis process, 78.6% degradation of DEP was observed within 3 h of irradiation under optimized conditions for initial DEP concentration of 150 mg L^{-1}, pH 7 and 1 g L^{-1} catalyst dosage (Mansouri and Bousselmi 2012). Improved photocatalytic performances were observed with surface modification of catalyst, including the visible light-mediated degradation of PAEs. In a study conducted by Ki et al. (2019), tungsten doped TiO_2 improved the degradation of DEP under blue light. Doped improved the overall photocatalytic performance and the degradation efficiency of the process was improved by 1.7–6.2 times under blue light irradiation.

Table 7.3 AOPs for PAEs removal

Pollutants	Treatment process	Highlights of the study	References
DBP	Photocatalysis	The attack of $^{\bullet}OH$ radicals on the aliphatic chain and aromatic ring of DBP resulted in faster mineralization under 30 min UV irradiation	Kaneco et al. (2006)
DMP	Photocatalysis	Under optimized conditions 8 g L^{-1} of TiO_2 supported on hollow glass microsphere and reaction time of 20 min, maximum $^{\bullet}OH$ are produced for the degradation of DMP under UV irradiation	Jiang et al. (2013)
MMP, DBP, DEP, DMP	Sonophotolysis	The more hydrophobic compound such as DBP got easily mineralized during the simultaneous sonophotolysis process	Xu et al. (2015)
DEP	Electro-peroxone	Highest mineralization of DEP was observed with carbon-PTFE cathode owing to its highest activity to produce H_2O_2	Hou et al. (2016)
DEP	Fenton's oxidation	Among the different clay minerals used for the degradation of DEP, nontronite clay with 19.2% Fe content accelerated the degradation process	Chen et al. (2016)
DEP	Photocatalysis	Ni-doped TiO_2 improved the photocatalytic activity compared to undoped TiO_2 particles due to the production of more electron–hole pairs	Singla et al. (2016)
Mixture of 6 PAEs	Fe-Ce oxide bimetallic/PS system	Fe^{3+}/Fe^{2+} and Ce^{4+}/Ce^{3+} redox couple enhanced the production of $SO_4^{-\bullet}$ with a maximum degradation of 86% observed at pH 2	Dong et al. (2019)

(continued)

Table 7.3 (continued)

Pollutants	Treatment process	Highlights of the study	References
DEP	Photo-Fenton	Among the different iron hydroxides used, maximum degradation of AEP was observed with 2-line ferrihydrite catalyst under UV irradiation owing to its higher O_2 yield and lower H_2O_2 consumption	Shuai et al. (2019)
DiBP	Electro-Fenton	93.07% degradation of DiBP was under optimum condition pH 5, current intensity 1A, Na_2SO_4 dosage of 2 g L^{-1} and H_2O_2 addition of 40 μL	Yang et al. (2020)
DMP, DEP, DBP	Electrochemical	Ce-doped Ti/PbO_2 electrode exhibited higher catalytic activity and stability and resulted in more than 80% degradation of PAEs at a current density of 25 mA cm^{-2}	Deng et al. (2020)

Fenton's reactions involving Fe^{2+} and H_2O_2 producing powerful radicals can easily decompose PAEs from polluted wastewater. Numerous degradation products were identified during Fenton's degradation of wastewater containing a mixture of DMP, DBP and DPP (Tay et al. 2011). The degradation process primarily occurs through the reaction of $^{\bullet}OH$ radicals and the hydroxylation reactions in the aromatic rings of PAEs resulted in the formation of transformation products. In another Fenton's oxidation study, an aqueous solution containing 100 mg L^{-1} of DAP was completely removed at an optimized condition of pH 3.2, H_2O_2 dosage 1000 mg L^{-1} and Fe^{2+} 50 mg L^{-1} with 120 min of reaction time. The TOC mineralization studies concluded that 95% mineralization was achieved with a major degradation pathway was through the opening of benzene ring followed by fragmentation of aliphatic compounds (Dbira et al. 2018). Besides the above studies, ultrasound irradiation that produces cavitation bubbles and reactive radicals such as $^{\bullet}OH$ and $HO_2^{\bullet-}$ can also oxidize and mineralize PAEs. A high-frequency ultrasound (US) treatment was able to degrade the hydrophobic compound, DMP through hydroxylation mechanism of the aromatic ring followed by the oxidation of the aliphatic chain (Xu et al. 2013a). The studies concluded that degradation efficiency decreased with an increase in the US frequency from 400 to 1200 kHz owing to the less energy release of the collapsed cavitational bubbles.

Electrochemical oxidation of DMP was carried out using a fluoride doped lead dioxide anode material in a filter-press reactor and the effect of operational parameters was studied (Souza et al. 2014). A higher degradation of DMP was observed in the presence of Na_2SO_4 electrolyte when compared to NaCl due to the poor reaction between Cl^- and $^{\bullet}OH$ species and the competition of Cl^- species for active electrodes of the anode. Further, owing to the higher stability of degradation intermediates, lower TOC was observed from the study. Using external energy sources such as US or light can improve the process efficiency of electrochemical processes by generating more reactive radical species. The photo-assisted electrochemical degradation of DMP using dimensionally stable anode (DSA) electrodes promoted the generation of a series of radical species including h^+, $Cl^{\bullet}$, $O_2^{\bullet-}$ and completed degraded the compound in 60 min of treatment (Souza et al. 2014).

Besides $^{\bullet}OH$ radical based AOPs, sulphate radicals are also powerful oxidizing agents for PAE. The sulphate radicals are usually generated with the help of some external energy sources or some catalyst activation of persulfate (PS) or peroxymonosulfate (PMS). In a study conducted by Wang et al. (2018b), 1 μM concentration of DBP was degraded by UV/PS system in an aqueous solution. They found that higher PS concentration and acidic pH conditions favoured the degradation process and both the $SO_4^{\bullet-}$ and $^{\bullet}OH$ radicals mediated the degradation process. In another study, A Zero valent iron (ZVI) /PS was effectively used for the degradation of DBP in aqueous solution and the role of solution pH and Fe corrosion products on the degradation efficiency was studied by (Li et al. 2014). The findings from the studies inferred that the corrosion products formed on the ZVI surface with a change in pH of solution significantly reduced the degradation efficiency by hindering the electron transfer from ZVI core to the surface-liquid interface. Moving forward, a similar observation was observed with Li et al. (2016) where free radicals generated in the ZVI/PS system varied with the initial pH of the solution and the maximum mineralization of DBP was observed at pH 3 where both $SO_4^{\bullet-}$ and $^{\bullet}OH$ were present. Apart from biodegradation processes, PAEs in sediment or soils were degraded using adsorption, oxidation process etc. (Di Gennaro et al. 2005; Bope et al. 2019; Yao et al. 2020). PS activation by Fe–Ce/biochar composites was used for the degradation of PAEs in marine sediments. The formation of $SO_4^{\bullet-}$ by Fe^{3+}/Fe^{2+} and Ce^{4+}/Ce^{3+} redox couple facilitated the degradation process. The underlying mechanism for the degradation process was the electrostatic and π–electron donor–acceptor interaction with the composite surface (Dong et al. 2020).

The presence of oxidants can improve the process efficiency of individual AOPs. The combined UV/H_2O_2 process effectively degraded DMP and the degradation was attributed to the enhanced production of OH radicals in the system (Xu et al. 2009). In another study, O_3/H_2O_2 improved the degradation efficiency of wastewater containing a mixture of four PAEs (Wen et al. 2011). The enhanced production of OH radicals in the system effectively mineralized the mixture of pollutants and reduced the overall toxicity of the wastewater. The combination of AOPs is also capable of removing PAEs from wastewater. In a catalysis free sonophotolytic process, the sequential and simultaneous application of US and UV irradiation for the removal of DMP had been evaluated (Xu et al. 2013b). The studies concluded higher UV irradiations improved the process efficiency and the simultaneous UV and US irradiation resulted in enhanced degradation with a synergy index of 2.6. In another study, an electro-Fenton process using an activated iron-doped carbon aerogel electrode resulted in 98% removal of 50 mg L^{-1} of DMP in 150 min reaction time (Zhao et al. 2017). The enhanced adsorption capacity and the self-cleaning ability of the electrode significantly improved the generation of OH radicals in the system and further oxidized the compound.

7.7.4 Other Treatment Options

Apart from the above treatment options, PAEs removal in aqueous solutions was also achieved through the membrane filtrations process. In a study conducted by Bhattacharya et al. (2021), a CuO/TiO_2 coated ceramic ultrafiltration membrane was used to remove a mixture of pollutants containing PAEs and parabens. Owing to the presence of functional groups that are capable of surface adsorption of pollutants, maximum removal of 99.7% was achieved after 120 min filtration. Further, the cytotoxicity evaluation on PA-1 cell lines after the treatment concluded that a lesser impact of toxicity was observed after the treatment. In another study, the thin-film composite nanofiltration membrane process removed trace concentrations of a mixture of five PAEs (Wei et al. 2016). The study concluded that hydrophobic adsorption of PAEs was significant and compounds with higher molecular weight had a higher rejection rate. Apart from these, combining electrocoagulation (EC) and electrofiltration (EF) with a graphene composite tubular membrane improved the removal of a mixture of micropollutants containing PAEs and pharmaceuticals. It was found that the higher removal of PAEs was observed with the conventional EC/EF crossflow filtration mode. Also, the removal efficiencies were primarily dependent on the basic properties of the pollutants including solubility and acid dissociation constant values (Wei et al. 2016).

7.8 Conclusion

PAEs are widely used industrial additives that are primarily used in the plastic industry. The extensive use of plastic materials containing PAEs has led to their release into different environmental matrices. Owing to their recalcitrant nature, they remain in the environment causing toxic effects to different levels of organisms. Continuous exposure to PAEs (potential carcinogens) can cause severe health effects on humans, viz. reproductive system toxicity in females and coronary heart diseases. Several studies were conducted to understand the fate and transport of PAEs in the environmental matrices. However, they were insufficient in explaining all the mechanisms and implications for the widespread occurrence of PAEs in the environment. Therefore, a better understanding of their occurrence and measurement is required for developing suitable treatment technologies. Quantification of PAEs and their metabolites from environmental samples require extraction before the chromatographic analysis. Highly sensitive and sophisticated analytical instruments such as HPLC, LC–MS, LC–MS/MS and GC–MS are used to quantify and identify degradation intermediates of PAEs. The removal of PAEs from different environmental matrices were achieved through biological processes, adsorption, advanced oxidation process etc. Further, the treatment options vary with the concentration of PAEs and its presence in different environmental matrices. Although individual treatment technologies are efficient in removing PAEs, the combination of treatment processes has shown significant improvement in degradation efficiency.

References

Adeogun AO, Ibor OR, Omogbemi ED et al (2015) Environmental occurrence and biota concentration of phthalate esters in Epe and Lagos Lagoons, Nigeria. Mar Environ Res 108:24–32. https://doi.org/10.1016/j.marenvres.2015.04.002

Ahmadi E, Gholami M, Farzadkia M et al (2015) Study of moving bed biofilm reactor in diethyl phthalate and diallyl phthalate removal from synthetic wastewater. Bioresour Technol 183:129–135. https://doi.org/10.1016/j.biortech.2015.01.122

Ahmadi E, Yousefzadeh S, Ansari M et al (2017) Performance, kinetic, and biodegradation pathway evaluation of anaerobic fixed film fixed bed reactor in removing phthalic acid esters from wastewater. Sci Rep 7:1–14. https://doi.org/10.1038/srep41020

Arfaeinia H, Fazlzadeh M, Taghizadeh F et al (2019) Phthalate acid esters (PAEs) accumulation in coastal sediments from regions with different land use configuration along the Persian Gulf. Ecotoxicol Environ Saf 169:496–506. https://doi.org/10.1016/j.ecoenv.2018.11.033

Benjamin S, Masai E, Kamimura N et al (2017) Phthalates impact human health: epidemiological evidences and plausible mechanism of action. J Hazard Mater 340:360–383. https://doi.org/10.1016/j.jhazmat.2017.06.036

Bhattacharya P, Mukherjee D, Deb N et al (2021) Indigenously developed CuO/TiO_2 coated ceramic ultrafiltration membrane for removal of emerging contaminants like phthalates and parabens: toxicity evaluation in PA-1 cell line. Mater Chem Phys 258:123920. https://doi.org/10.1016/j.matchemphys.2020.123920

Bi C, Liang Y, Xu Y (2015) Fate and transport of phthalates in indoor environments and the influence of temperature: a case study in a test house. Environ Sci Technol 49:9674–9681. https://doi.org/10.1021/acs.est.5b02787

Boll M, Geiger R, Junghare M, Schink B (2020) Microbial degradation of phthalates: biochemistry and environmental implications. Environ Microbiol Rep 12:3–15. https://doi.org/10.1111/1758-2229.12787

Boonnorat J, Chiemchaisri C, Chiemchaisri W, Yamamoto K (2014) Removals of phenolic compounds and phthalic acid esters in landfill leachate by microbial sludge of two-stage membrane bioreactor. J Hazard Mater 277:93–101. https://doi.org/10.1016/j.jhazmat.2014.02.044

Bope A, Haines SR, Hegarty B et al (2019) Degradation of phthalate esters in floor dust at elevated relative humidity. Environ Sci Process Impacts 21:1268–1279. https://doi.org/10.1039/c9em00050j

Calafat AM, Slakman AR, Silva MJ et al (2004) Automated solid phase extraction and quantitative analysis of human milk for 13 phthalate metabolites. J Chromatogr B Anal Technol Biomed Life Sci 805:49–56. https://doi.org/10.1016/j.jchromb.2004.02.006

Cao XL (2010) Phthalate Esters in Foods: Sources, Occurrence, and Analytical Methods. Compr Rev Food Sci Food Saf 9:21–43. https://doi.org/10.1111/j.1541-4337.2009.00093.x

Chen CY, Chung YC (2006) Removal of phthalate esters from aqueous solutions by chitosan bead. J Environ Sci Heal—Part A Toxic/hazard Subst Environ Eng 41:235–248. https://doi.org/10.1080/10934520500354573

Chen H, Chen K, Qiu X et al (2020) The reproductive toxicity and potential mechanisms of combined exposure to dibutyl phthalate and diisobutyl phthalate in male zebrafish (Danio rerio). Chemosphere 258:127238. https://doi.org/10.1016/j.chemosphere.2020.127238

Chen N, Fang G, Zhou D, Gao J (2016) Effects of clay minerals on diethyl phthalate degradation in Fenton reactions. Chemosphere 165:52–58. https://doi.org/10.1016/j.chemosphere.2016.09.016

Cheng HF, Kumar M, Lin JG (2010) Assessment of di-(2-ethylhexyl)phthalate (DEHP) removal in a rotating biological contactor and activated sludge process treating domestic wastewater. Sep Sci Technol 45:221–227. https://doi.org/10.1080/01496390903409385

Cheng LJ, Cheng TS (2012) Oxidative effects and metabolic changes following exposure of greater duckweed (Spirodela polyrhiza) to diethyl phthalate. Aquat Toxicol 109:166–175. https://doi.org/10.1016/j.aquatox.2011.10.003

Cheng Z, Nie XP, Wang HS, Wong MH (2013) Risk assessments of human exposure to bioaccessible phthalate esters through market fish consumption. Environ Int 57–58:75–80. https://doi.org/10.1016/j.envint.2013.04.005

Chung YC, Chen CY (2009) Degradation of Di-(2-ethylhexyl) phthalate (DEHP) by TiO_2 photocatalysis. Water Air Soil Pollut 200:191–198. https://doi.org/10.1007/s11270-008-9903-9

Das MT, Kumar SS, Ghosh P et al (2021) Remediation strategies for mitigation of phthalate pollution: Challenges and future perspectives. J Hazard Mater 409:124496. https://doi.org/10.1016/j.jhazmat.2020.124496

Dbira S, Bensalah N, Zagho MM, Bedoui A (2018) Degradation of diallyl phthalate (DAP) by fenton oxidation: mechanistic and kinetic studies. Appl Sci 9. https://doi.org/10.3390/app9010023

Deng D, Wu X, Li M et al (2020) Electrochemical degradation of three phthalate esters in synthetic wastewater by using a Ce-doped Ti/PbO2 electrode. Chemosphere 259:127488. https://doi.org/10.1016/j.chemosphere.2020.127488

Di Gennaro P, Collina E, Franzetti A et al (2005) Bioremediation of diethylhexyl phthalate contaminated soil: a feasibility study in slurry- and solid-phase reactors. Environ Sci Technol 39:325–330. https://doi.org/10.1021/es035420d

Dong DC, Chen CW, Nguyen TB et al (2020) Degradation of phthalate esters in marine sediments by persulfate over Fe–Ce/biochar composites. Chem Eng J 384:123301. https://doi.org/10.1016/j.cej.2019.123301

Dong DC, Huang CP, Nguyen TB et al (2019) The degradation of phthalate esters in marine sediments by persulfate over iron–cerium oxide catalyst. Sci Total Environ 696:133973. https://doi.org/10.1016/j.scitotenv.2019.133973

Du L, Li G, Liu M et al (2015) Evaluation of DNA damage and antioxidant system induced by di-n-butyl phthalates exposure in earthworms (Eisenia fetida). Ecotoxicol Environ Saf 115:75–82. https://doi.org/10.1016/j.ecoenv.2015.01.031

Du P, Zhou Z, Huang H et al (2018) Estimating population exposure to phthalate esters in major Chinese cities through wastewater-based epidemiology. Sci Total Environ 643:1602–1609. https://doi.org/10.1016/j.scitotenv.2018.06.325

Earls AO, Axford IP, Braybrook JH (2003) Gas chromatography-mass spectrometry determination of the migration of phthalate plasticisers from polyvinyl chloride toys and childcare articles. J Chromatogr A 983:237–246. https://doi.org/10.1016/S0021-9673(02)01736-3

Fan J, Wang X, Teng W et al (2017) Phenyl-functionalized mesoporous silica materials for the rapid and efficient removal of phthalate esters. J Colloid Interface Sci 487:354–359. https://doi.org/10.1016/j.jcis.2016.10.042

Fauser P, Vikelsøe J, Sørensen PB, Carlsen L (2003) Phthalates, nonylphenols and LAS in an alternately operated wastewater treatment plant—fate modelling based on measured concentrations

in wastewater and sludge. Water Res 37:1288–1295. https://doi.org/10.1016/S0043-1354(02)00482-7
Gani KM, Bux F, Kazmi AA (2019) Diethylhexyl phthalate removal in full scale activated sludge plants: effect of operational parameters. Chemosphere 234:885–892. https://doi.org/10.1016/j.chemosphere.2019.06.130
Gani KM, Kazmi AA (2016) Comparative assessment of phthalate removal and risk in biological wastewater treatment systems of developing countries and small communities. Sci Total Environ 569–570:661–671. https://doi.org/10.1016/j.scitotenv.2016.06.182
Gao DW, Wen ZD (2016) Phthalate esters in the environment: a critical review of their occurrence, biodegradation, and removal during wastewater treatment processes. Sci Total Environ 541:986–1001. https://doi.org/10.1016/j.scitotenv.2015.09.148
Gao L, Tang Y, Wang C et al (2019) Highly-efficient amphiphilic magnetic nanocomposites based on a simple sol-gel modification for adsorption of phthalate esters. J Colloid Interface Sci 552:142–152. https://doi.org/10.1016/j.jcis.2019.05.031
Guo Y, Kannan K (2012) Challenges encountered in the analysis of phthalate esters in foodstuffs and other biological matrices. Anal Bioanal Chem 404:2539–2554. https://doi.org/10.1007/s00216-012-5999-2
He W, Qin N, Kong X et al (2013) Spatio-temporal distributions and the ecological and health risks of phthalate esters (PAEs) in the surface water of a large, shallow Chinese lake. Sci Total Environ 461–462:672–680. https://doi.org/10.1016/j.scitotenv.2013.05.049
He Y, Wang Q, He W, Xu F (2019) Phthalate esters (PAEs) in atmospheric particles around a large shallow natural lake (Lake Chaohu, China). Sci Total Environ 687:297–308. https://doi.org/10.1016/j.scitotenv.2019.06.034
Heudorf U, Mersch-Sundermann V, Angerer J (2007) Phthalates: toxicology and exposure. Int J Hyg Environ Health 210:623–634. https://doi.org/10.1016/j.ijheh.2007.07.011
Hou M, Chu Y, Li X et al (2016) Electro-peroxone degradation of diethyl phthalate: cathode selection, operational parameters, and degradation mechanisms. J Hazard Mater 319:61–68. https://doi.org/10.1016/j.jhazmat.2015.12.054
Huang C, Zhang Liu Y-J, L-Y et al (2021a) Exposure to phthalates and correlations with phthalates in dust and air in South China homes Sci Total Environ 782:146806. https://doi.org/10.1016/j.scitotenv.2021.146806
Huang H, Zhang XY, Chen TL et al (2019) Biodegradation of structurally diverse phthalate esters by a newly identified esterase with catalytic activity toward Di(2-ethylhexyl) phthalate. J Agric Food Chem 67:8548–8558. https://doi.org/10.1021/acs.jafc.9b02655
Huang L, Qiao Y, Deng S et al (2020) Airborne phthalates in indoor environment: partition state and influential built environmental conditions. Chemosphere 254:126782. https://doi.org/10.1016/j.chemosphere.2020.126782
Huang M, Li Y, Gu G (2008) The effects of hydraulic retention time and sludge retention time on the fate of di-(2-ethylhexyl) phthalate in a laboratory-scale anaerobic-anoxic-aerobic activated sludge system. Bioresour Technol 99:8107–8111. https://doi.org/10.1016/j.biortech.2008.03.031
Huang S, Qi Z, Ma S et al (2021b) A critical review on human internal exposure of phthalate metabolites and the associated health risks. Environ Pollut 279:116941. https://doi.org/10.1016/j.envpol.2021.116941
Jiang W, Joens JA, Dionysiou DD, O'Shea KE (2013) Optimization of photocatalytic performance of TiO_2 coated glass microspheres using response surface methodology and the application for degradation of dimethyl phthalate. J Photochem Photobiol A Chem 262:7–13. https://doi.org/10.1016/j.jphotochem.2013.04.008
Kaneco S, Katsumata H, Suzuki T, Ohta K (2006) Titanium dioxide mediated photocatalytic degradation of dibutyl phthalate in aqueous solution-kinetics, mineralization and reaction mechanism. Chem Eng J 125:59–66. https://doi.org/10.1016/j.cej.2006.08.004
Kashyap D, Agarwal T (2018) Concentration and factors affecting the distribution of phthalates in the air and dust: a global scenario. Sci Total Environ 635:817–827. https://doi.org/10.1016/j.scitotenv.2018.04.158

SJ Ki YK Park JS Kim et al (2019) Facile preparation of tungsten oxide doped TiO_2 photocatalysts using liquid phase plasma process for enhanced degradation of diethyl phthalate Chem Eng J 377:120087 https://doi.org/10.1016/j.cej.2018.10.024

Kotowska U, Kapelewska J, Sawczuk R (2020) Occurrence, removal, and environmental risk of phthalates in wastewaters, landfill leachates, and groundwater in Poland. Environ Pollut 267. https://doi.org/10.1016/j.envpol.2020.115643

Kubwabo C, Rasmussen PE, Fan X et al (2013) Analysis of selected phthalates in Canadian indoor dust collected using household vacuum and standardized sampling techniques. Indoor Air 23:506–514. https://doi.org/10.1111/ina.12048

Lee YS, Lim JE, Lee S, Moon HB (2020) Phthalates and non-phthalate plasticizers in sediment from Korean coastal waters: occurrence, spatial distribution, and ecological risks. Mar Pollut Bull 154:111119. https://doi.org/10.1016/j.marpolbul.2020.111119

Li H, Wan J, Ma Y et al (2014) New insights into the role of zero-valent iron surface oxidation layers in persulfate oxidation of dibutyl phthalate solutions. Chem Eng J 250:137–147. https://doi.org/10.1016/j.cej.2014.03.092

Li H, Wan J, Ma Y, Wang Y (2016) Reaction pathway and oxidation mechanisms of dibutyl phthalate by persulfate activated with zero-valent iron. Sci Total Environ 562:889–897. https://doi.org/10.1016/j.scitotenv.2016.04.093

Li Y, Wang J, Yang S, Zhang S (2021) Occurrence, health risks and soil-air exchange of phthalate acid esters: a case study in plastic film greenhouses of Chongqing, China. Chemosphere 268:128821. https://doi.org/10.1016/j.chemosphere.2020.128821

Liang Y, Bi C, Wang X, Xu Y (2019) A general mechanistic model for predicting the fate and transport of phthalates in indoor environments. Indoor Air 29:55–69. https://doi.org/10.1111/ina.12514

Liu H, Cui K, Zeng F et al (2014) Occurrence and distribution of phthalate esters in riverine sediments from the Pearl River Delta region, South China. Mar Pollut Bull 83:358–365. https://doi.org/10.1016/j.marpolbul.2014.03.038

Lu Y, Tang F, Wang Y et al (2009) Biodegradation of dimethyl phthalate, diethyl phthalate and di-n-butyl phthalate by Rhodococcus sp. L4 isolated from activated sludge. J Hazard Mater 168:938–943. https://doi.org/10.1016/j.jhazmat.2009.02.126

Lyche JL, Gutleb AC, Bergman Å et al (2009) Reproductive and developmental toxicity of phthalates. J Toxicol Environ Heal—Part B Crit Rev 12:225–249. https://doi.org/10.1080/10937400903094091

Ma T, Teng Y, Christie P, Luo Y (2015) Phytotoxicity in seven higher plant species exposed to di-n-butyl phthalate or bis (2-ethylhexyl) phthalate. Front Environ Sci Eng 9:259–268. https://doi.org/10.1007/s11783-014-0652-2

Makkliang F, Kanatharana P, Thavarungkul P, Thammakhet C (2015) Development of magnetic micro-solid phase extraction for analysis of phthalate esters in packaged food. Food Chem 166:275–282. https://doi.org/10.1016/j.foodchem.2014.06.036

Mansouri L, Bousselmi L (2012) Degradation of diethyl phthalate (DEP) in aqueous solution using TiO_2/UV process. Desalin Water Treat 40:63–68. https://doi.org/10.1080/19443994.2012.671142

Manzo V, Becerra-Herrera M, Arismendi D et al (2019) Rotating-disk sorptive extraction coupled to gas chromatography mass spectrometry for the determination of phthalates in bottled water. Anal Methods 11:6111–6118. https://doi.org/10.1039/c9ay02076d

Mariana M, Feiteiro J, Verde I, Cairrao E (2016) The effects of phthalates in the cardiovascular and reproductive systems: a review. Environ Int 94:758–776. https://doi.org/10.1016/j.envint.2016.07.004

Mesdaghinia A, Azari A, Nodehi RN et al (2017) Removal of phthalate esters (PAEs) by zeolite/Fe_3O_4: investigation on the magnetic adsorption separation, catalytic degradation and toxicity bioassay. J Mol Liq 233:378–390. https://doi.org/10.1016/j.molliq.2017.02.094

Mohan SV, Shailaja S, Rama Krishna M, Sarma PN (2007) Adsorptive removal of phthalate ester (Di-ethyl phthalate) from aqueous phase by activated carbon: a kinetic study. J Hazard Mater 146:278–282. https://doi.org/10.1016/j.jhazmat.2006.12.020

Nantaba F, Palm WU, Wasswa J et al (2021) Temporal dynamics and ecotoxicological risk assessment of personal care products, phthalate ester plasticizers, and organophosphorus flame retardants in water from Lake Victoria, Uganda. Chemosphere 262:127716. https://doi.org/10.1016/j.chemosphere.2020.127716

Navacharoen A, Vangnai AS (2011) Biodegradation of diethyl phthalate by an organic-solvent-tolerant Bacillus subtilis strain 3C3 and effect of phthalate ester coexistence. Int Biodeterior Biodegrad 65:818–826. https://doi.org/10.1016/j.ibiod.2011.05.005

Net S, Delmont A, Sempéré R et al (2015a) Reliable quantification of phthalates in environmental matrices (air, water, sludge, sediment and soil): a review. Sci Total Environ 515–516:162–180. https://doi.org/10.1016/j.scitotenv.2015.02.013

Net S, Sempéré R, Delmont A et al (2015b) Occurrence, fate, behavior and ecotoxicological state of phthalates in different environmental matrices. Environ Sci Technol 49:4019–4035. https://doi.org/10.1021/es505233b

Nidens N, Vogel M, Körner A, Kiess W (2021) Prenatal exposure to phthalate esters and its impact on child development. Best Pract Res Clin Endocrinol Metab. https://doi.org/10.1016/j.beem.2020.101478

Otero P, Saha SK, Moane S et al (2015) Improved method for rapid detection of phthalates in bottled water by gas chromatography-mass spectrometry. J Chromatogr B Anal Technol Biomed Life Sci 997:229–235. https://doi.org/10.1016/j.jchromb.2015.05.036

Park C, Choi W, Hwang M et al (2017) Associations between urinary phthalate metabolites and bisphenol A levels, and serum thyroid hormones among the Korean adult population—Korean National Environmental Health Survey (KoNEHS) 2012–2014. Sci Total Environ 584–585:950–957. https://doi.org/10.1016/j.scitotenv.2017.01.144

Pirsaheb M, Mesdaghinia AR, Shahtaheri SJ, Zinatizadeh AA (2009) Kinetic evaluation and process performance of a fixed film bioreactor removing phthalic acid and dimethyl phthalate. J Hazard Mater 167:500–506. https://doi.org/10.1016/j.jhazmat.2009.01.003

Poopal RK, Ramesh M, Maruthappan V, Babu Rajendran R (2017) Potential effects of low molecular weight phthalate esters (C16H22O4 and C12H14O4) on the freshwater fish: Cyprinus carpio. Toxicol Res (camb) 6:505–520. https://doi.org/10.1039/c7tx00084g

Pradeep S, Faseela P, Josh MKS et al (2013) Fungal biodegradation of phthalate plasticizer in situ. Biodegradation 24:257–267. https://doi.org/10.1007/s10532-012-9584-3

Qureshi MS, Yusoff AR bin M, Wirzal MDH et al (2016) Methods for the determination of endocrine-disrupting phthalate esters. Crit Rev Anal Chem 46:146–159. https://doi.org/10.1080/10408347.2015.1004157

Ren L, Jia Y, Ruth N et al (2016) Biodegradation of phthalic acid esters by a newly isolated Mycobacterium sp. YC-RL4 and the bioprocess with environmental samples. Environ Sci Pollut Res 23:16609–16619. https://doi.org/10.1007/s11356-016-6829-4

Roslev P, Vorkamp K, Aarup J et al (2007) Degradation of phthalate esters in an activated sludge wastewater treatment plant. Water Res 41:969–976. https://doi.org/10.1016/j.watres.2006.11.049

Russo MV, Avino P, Perugini L, Notardonato I (2015) Extraction and GC-MS analysis of phthalate esters in food matrices: a review. RSC Adv 5:37023–37043. https://doi.org/10.1039/c5ra01916h

Salaudeen T, Okoh O, Agunbiade F, Okoh A (2018) Fate and impact of phthalates in activated sludge treated municipal wastewater on the water bodies in the Eastern Cape, South Africa. Chemosphere 203:336–344. https://doi.org/10.1016/j.chemosphere.2018.03.176

Salim CJ, Liu H, Kennedy JF (2010) Comparative study of the adsorption on chitosan beads of phthalate esters and their degradation products. Carbohydr Polym 81:640–644. https://doi.org/10.1016/j.carbpol.2010.03.024

Sayyad G, Price GW, Sharifi M, Khosravi K (2017) Fate and transport modeling of phthalate esters from biosolid amended soil under corn cultivation. J Hazard Mater 323:264–273. https://doi.org/10.1016/j.jhazmat.2016.07.032

Shaida MA, Dutta RK, Sen AK (2018) Removal of diethyl phthalate via adsorption on mineral rich waste coal modified with chitosan. J Mol Liq 261:271–282. https://doi.org/10.1016/j.molliq.2018.04.031

Shi J, Li F, Yin D et al (2017) Sorption and degradation of phthalate esters by a novel functional hyper-cross-linked polymer. Chemosphere 171:149–157. https://doi.org/10.1016/j.chemosphere.2016.12.048

Shuai W, Gu C, Fang G et al (2019) Effects of iron (hydr)oxides on the degradation of diethyl phthalate ester in heterogeneous (photo)-Fenton reactions. J Environ Sci (china) 80:5–13. https://doi.org/10.1016/j.jes.2018.06.015

Silva MJ, Reidy JA, Samandar E et al (2005) Detection of phthalate metabolites in human saliva. Arch Toxicol 79:647–652. https://doi.org/10.1007/s00204-005-0674-4

Singla P, Pandey OP, Singh K (2016) Study of photocatalytic degradation of environmentally harmful phthalate esters using Ni-doped TiO_2 nanoparticles. Int J Environ Sci Technol 13:849–856. https://doi.org/10.1007/s13762-015-0909-8

Souza FL, Aquino JM, Irikura K et al (2014) Electrochemical degradation of the dimethyl phthalate ester on a fluoride-doped Ti/β-PbO2 anode. Chemosphere 109:187–194. https://doi.org/10.1016/j.chemosphere.2014.02.018

Staples CA, Adams WJ, Parkerton TF et al (1997) Environmental toxicology review aquatic toxicity of eighteen phthalate esters. Environ Toxicol Chem 16:875–891

Tang WJ, Zhang LS, Fang Y et al (2016) Biodegradation of phthalate esters by newly isolated Rhizobium sp. LMB-1 and its biochemical pathway of di-n-butyl phthalate. J Appl Microbiol 121:177–186. https://doi.org/10.1111/jam.13123

Tay KS, Rahman NA, Bin AMR (2011) Fenton degradation of dialkylphthalates: products and mechanism. Environ Chem Lett 9:539–546. https://doi.org/10.1007/s10311-011-0317-3

Tümay Özer E, Göçenoğlu Sarıkaya A, Osman B (2016) Adsorption and removal of diethyl phthalate from aqueous media with poly(hydroxyethyl methacrylate) nanobeads. Desalin Water Treat 57:28864–28874. https://doi.org/10.1080/19443994.2016.1186568

Ventrice P, Ventrice D, Russo E, De Sarro G (2013) Phthalates: European regulation, chemistry, pharmacokinetic and related toxicity. Environ Toxicol Pharmacol 36:88–96. https://doi.org/10.1016/j.etap.2013.03.014

Wang L, Liu M, Tao W et al (2018a) Pollution characteristics and health risk assessment of phthalate esters in urban soil in the typical semi-arid city of Xi'an, Northwest China. Chemosphere 191:467–476. https://doi.org/10.1016/j.chemosphere.2017.10.066

Wang X, Song M, Guo M et al (2015) Pollution levels and characteristics of phthalate esters in indoor air in hospitals. J Environ Sci (china) 37:67–74. https://doi.org/10.1016/j.jes.2015.02.016

Wang Z (2015) Efficient adsorption of dibutyl phthalate from aqueous solution by activated carbon developed from phoenix leaves. Int J Environ Sci Technol 12:1923–1932. https://doi.org/10.1007/s13762-014-0554-7

Wang Z, Shao Y, Gao N, An N (2018b) Degradation kinetic of dibutyl phthalate (DBP) by sulfate radical- and hydroxyl radical-based advanced oxidation process in UV/persulfate system. Sep Purif Technol 195:92–100. https://doi.org/10.1016/j.seppur.2017.11.072

Wei L, Li Z, Sun J, Zhu L (2020) Pollution characteristics and health risk assessment of phthalate esters in agricultural soil and vegetables in the Yangtze River Delta of China. Sci Total Environ 726:137978. https://doi.org/10.1016/j.scitotenv.2020.137978

Wei X, Shi Y, Fei Y et al (2016) Removal of trace phthalate esters from water by thin-film composite nanofiltration hollow fiber membranes. Chem Eng J 292:382–388. https://doi.org/10.1016/j.cej.2016.02.037

Wen G, Ma J, Liu ZQ, Zhao L (2011) Ozonation kinetics for the degradation of phthalate esters in water and the reduction of toxicity in the process of O_3/H_2O_2. J Hazard Mater 195:371–377. https://doi.org/10.1016/j.jhazmat.2011.08.054

Wu J, Ma T, Zhou Z et al (2019) Occurrence and fate of phthalate esters in wastewater treatment plants in Qingdao, China. Hum Ecol Risk Assess 25:1547–1563. https://doi.org/10.1080/10807039.2018.1471341

Wu Y, Eichler CMA, Cao J et al (2018) Particle/gas partitioning of phthalates to organic and inorganic airborne particles in the indoor environment. Environ Sci Technol 52:3583–3590. https://doi.org/10.1021/acs.est.7b05982
Wu Y, Si Y, Zhou D, Gao J (2015) Adsorption of diethyl phthalate ester to clay minerals. Chemosphere 119:690–696. https://doi.org/10.1016/j.chemosphere.2014.07.063
Xie HJ, Shi YJ, Zhang J et al (2010) Degradation of phthalate esters (PAEs) in soil and the effects of PAEs on soilmicrocosm activity. J Chem Technol Biotechnol 85:1108–1116. https://doi.org/10.1002/jctb.2406
Xu B, Gao N, Cheng H et al (2009) Oxidative degradation of dimethyl phthalate (DMP) by UV/H_2O_2 process. J Hazard Mater 162:954–959. https://doi.org/10.1016/j.jhazmat.2008.05.122
Xu D, Deng X, Fang E et al (2014) Determination of 23 phthalic acid esters in food by liquid chromatography tandem mass spectrometry. J Chromatogr A 1324:49–56. https://doi.org/10.1016/j.chroma.2013.11.017
Xu LJ, Chu W, Graham N (2013a) A systematic study of the degradation of dimethyl phthalate using a high-frequency ultrasonic process. Ultrason Sonochem 20:892–899. https://doi.org/10.1016/j.ultsonch.2012.11.005
Xu LJ, Chu W, Graham N (2013b) Sonophotolytic degradation of dimethyl phthalate without catalyst: analysis of the synergistic effect and modeling. Water Res 47:1996–2004. https://doi.org/10.1016/j.watres.2013.01.015
Xu LJ, Chu W, Graham N (2015) Sonophotolytic degradation of phthalate acid esters in water and wastewater: influence of compound properties and degradation mechanisms. J Hazard Mater 288:43–50. https://doi.org/10.1016/j.jhazmat.2015.02.023
Xu Y, Minhazul KAHM, Wang X et al (2020) Biodegradation of phthalate esters by Paracoccus kondratievae BJQ0001 isolated from Jiuqu (Baijiu fermentation starter) and identification of the ester bond hydrolysis enzyme. Environ Pollut 263:114506. https://doi.org/10.1016/j.envpol.2020.114506
Yang J, Li Y, Wang Y et al (2015) Recent advances in analysis of phthalate esters in foods. TrAC—Trends Anal Chem 72:10–26. https://doi.org/10.1016/j.trac.2015.03.018
Yang Z, Chen H, Wang J et al (2020) Efficient degradation of diisobutyl phthalate in aqueous solution through electro-Fenton process with sacrificial anode. J Environ Chem Eng 8:104057. https://doi.org/10.1016/j.jece.2020.104057
Yao J, Yu Y, Qu R et al (2020) Fe-activated peroxymonosulfate enhances the degradation of dibutyl phthalate on ground quartz sand. Environ Sci Technol 54:9052–9061. https://doi.org/10.1021/acs.est.0c00793
Ye H, Zhao B, Zhou Y et al (2021) Recent advances in adsorbents for the removal of phthalate esters from water: Material, modification, and application. Chem Eng J 409:128127. https://doi.org/10.1016/j.cej.2020.128127
Yin L, Lin Y, Jia L (2014) Graphene oxide functionalized magnetic nanoparticles as adsorbents for removal of phthalate esters. Microchim Acta 181:957–965. https://doi.org/10.1007/s00604-014-1187-8
Yousefzadeh S, Ahmadi E, Gholami M et al (2017) A comparative study of anaerobic fixed film baffled reactor and up-flow anaerobic fixed film fixed bed reactor for biological removal of diethyl phthalate from wastewater: a performance, kinetic, biogas, and metabolic pathway study. Biotechnol Biofuels 10:1–15. https://doi.org/10.1186/s13068-017-0826-9
Yuan SY, Lin YY, Chang BV (2011) Biodegradation of phthalate esters in polluted soil by using organic amendment. J Environ Sci Heal—Part B Pestic Food Contam Agric Wastes 46:419–425. https://doi.org/10.1080/03601234.2011.572512
Yuan SY, Liu C, Liao CS, Chang BV (2002) Occurrence and microbial degradation of phthalate esters in Taiwan river sediments. Chemosphere 49:1295–1299. https://doi.org/10.1016/S0045-6535(02)00495-2
Zeng F, Cui K, Xie Z et al (2008) Phthalate esters (PAEs): Emerging organic contaminants in agricultural soils in peri-urban areas around Guangzhou, China. Environ Pollut 156:425–434. https://doi.org/10.1016/j.envpol.2008.01.045

Zhang J, Zhang C, Zhu Y et al (2018) Biodegradation of seven phthalate esters by Bacillus mojavensis B1811. Int Biodeterior Biodegrad 132:200–207. https://doi.org/10.1016/j.ibiod.2018.04.006
Zhang L, Liu J, Liu H et al (2015) The occurrence and ecological risk assessment of phthalate esters (PAEs) in urban aquatic environments of China. Ecotoxicology 24:967–984. https://doi.org/10.1007/s10646-015-1446-4
Zhang T, Huang Z, Chen X et al (2016) Degradation behavior of dimethyl phthalate in an anaerobic/anoxic/oxic system. J Environ Manage 184:281–288. https://doi.org/10.1016/j.jenvman.2016.10.008
Zhang Y, Jiao Y, Li Z et al (2021) Hazards of phthalates (PAEs) exposure: a review of aquatic animal toxicology studies. Sci Total Environ 771:145418. https://doi.org/10.1016/j.scitotenv.2021.145418
Zhao H, Wang Q, Chen Y et al (2017) Efficient removal of dimethyl phthalate with activated iron-doped carbon aerogel through an integrated adsorption and electro-Fenton oxidation process. Carbon N Y 124:111–122. https://doi.org/10.1016/j.carbon.2017.08.034
Zheng X, Yan Z, Liu P et al (2019) Derivation of aquatic life criteria for four phthalate esters and their ecological risk assessment in Liao River. Chemosphere 220:802–810. https://doi.org/10.1016/j.chemosphere.2018.12.047
Zheng X, Zhang BT, Teng Y (2014) Distribution of phthalate acid esters in lakes of Beijing and its relationship with anthropogenic activities. Sci Total Environ 476–477:107–113. https://doi.org/10.1016/j.scitotenv.2013.12.111

Chapter 8
Removal of Pharmaceutical Compounds: Overview of Treatment Methods

Chinthalapudi Naga Lakshmi and Narendra Singh

Abstract As human lifestyle is improved over the years due to technological, medical and many more developments but in the same time, it has also increased new environment and health challenges. The Emerging contaminants (ECs) are uncontrolled disposal of pollutants, they include group of synthetic chemicals in worldwide use such as water disinfection by-products, pharmaceuticals, chemical fertilizers, pesticides, gasoline additives, and man-made nanomaterials, etc. Pharmaceutical compounds (PCs) are the emerging pollutants, which need to be regulated. They have large potential of causing environment damage, ecosystems as well as human and animal health. Due to huge consumption of pharmaceuticals, effluents contains anti-diabetics, stimulant drugs, and antimicrobials compounds and many more. The decomposition of ECs is dependent on the chemical and biological persistence of ECs, their physicochemical characteristics, the technique applied, etc. The methods to trace and degrade/remove the PCs are very critical to address the issue. Several techniques have been adopted to minimize their hazards like adsorption, coagulation, constructed wetland, advanced oxidation process, membrane reactors (reverse osmosis, nanofiltration, microfiltration, etc.) and photocatalysis, etc. In this chapter, the different types of PCs and their affects will be summarized. The techniques involved to remove/degrade the compounds will also be discussed.

Keywords Emerging pollutants · Pharmaceutical compounds · Adsorption · Photocatalysis · Membrane bioreactor

8.1 Introduction

Emerging contaminants (ECs) are known as non-regulated trace micro-organic contaminants with a category of natural and synthetic chemicals (Grandclément

C. N. Lakshmi · N. Singh (✉)
Department of Chemical Engineering, Indian Institute of Technology Tirupati, Tirupati, Andhra Pradesh 517506, India
e-mail: narendra@iittp.ac.in

S. P. Singh et al. (eds.), *New Trends in Emerging Environmental Contaminants*, Energy, Environment, and Sustainability,
https://doi.org/10.1007/978-981-16-8367-1_8

2017). The resources and technologies have produced more compounds that accordingly grow the number of compounds recognized as causing potential environmental threats to the living organisms (Bolong et al. 2009). The conversion of these products occurring in water bodies throughout the earth. ECs include a wide range of various compounds and their transformation products such as pharmaceuticals, illicit drugs, personal care products, food additives, pesticides, veterinary products, and engineered nanomaterials (Lapworth et al. 2012; Wanda et al. 2017). These contaminants can quickly move from one place to another in air and water, as the life cycle of ECs is shown in Fig. 8.1 (Rehman et al. 2015).

In recent years, the main ECs reported in the literature include pharmaceuticals and personal care products (PPCPs), nanomaterials, flame retardants, UV filters, food additives, and siloxanes etc. (Richardson 2007). ECs represent a group of natural or synthetic compounds that are not generally observed within the environment but can enter the environment and cause effects for both environments and human health. Egbuna et al. identified the various ECs and mainly ECs are categorized into pharmaceutical compounds (PCs), personal care products, pesticides, gamma radiation, volatile organic compounds etc. (Egbuna 2021).

In the past two decades, pharmaceuticals as ECs have become greater attention to the researchers because of their high persistence, low biodegradable nature, bioaccumulation, potential toxicity, and relative stability. In the early 1970s, the PCs are in

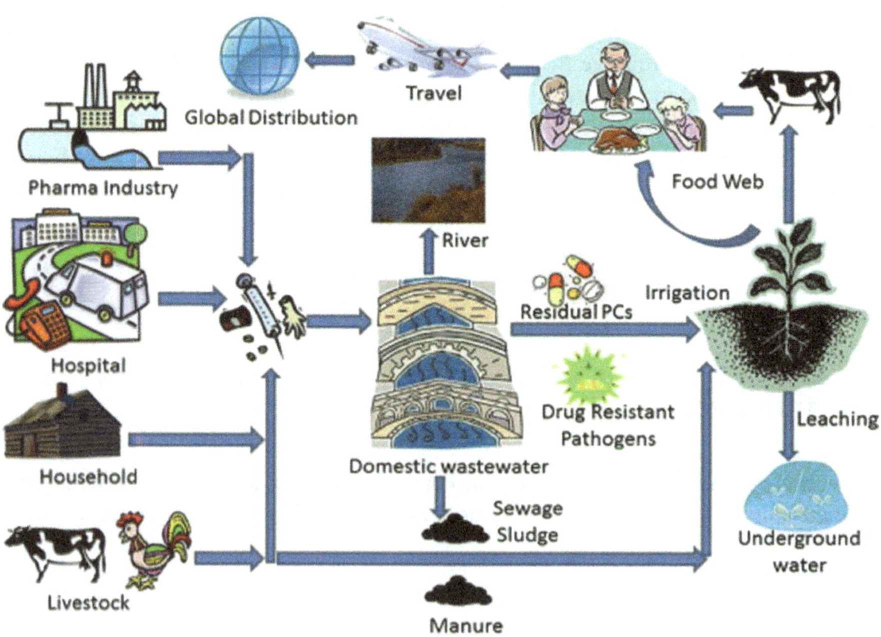

Fig. 8.1 Various sources and paths of PCs in different compartments of environment (Reproduced from Rehman et al. 2015)

wastewater, and surface water were not a big concern due to their lower concentration in the environment (Wilkinson et al. 2017). Nowadays, the concentration levels have been increased from ng/L to μg/L. Due to the huge consumption of pharmaceuticals or drugs, the contaminants of antibiotics, stimulant drugs, β-blockers, anti-inflammatory agents, and antimicrobial compounds and many more are occurring only a small portion in the environment, but causes severe problems to the ecosystem and human health. PCs are frequently entered into the environment at small concentrations range, but high impact on the ecosystem as well as human health (Yuan et al. 2009). There will be a significant health issue near the future due to various PCs in drinking water, groundwater, and surface water if proper treatment techniques are not developed (Dharupaneedi et al. 2019). The world is facing a drinking water crisis owing to several hazardous effluents in the environment. Waste Water Treatment plants (WWTPs) have been recognized as a major path for releasing PCs in an aqueous region. The available water resources are contaminated by industrialization and as well as human activities. The pharmaceuticals have accepted humankind with attributes consisting of curing diseases, extended life span, elaborating the quality of life are influenced a frequential threat to all the living organisms in the biosphere (Taoufik 2020). The global nature of antibiotics causes harmful effects on the rise of individuals. Although at low concentrations, the physicochemical nature of these antibiotics changes into microinvertebrates (Flaherty and Dodson 2005). The PCs of analgesics and antibiotics are mostly used drugs and are continuously growing in the environment. Among them, the occurrence of ibuprofen, acetaminophens are the primary non-steroidal anti-inflammatory drugs (NSAIDs) in the wastewater in India (Chandrashekar Kollarahithlu and Balakrishnan 2021).

Mezzelani and Regoli studied the main adverse effects of aquatic species revealed in field and laboratory conditions at various pharmaceuticals, including candiovascular, NSAIDs, hypocholesterolaemic drugs, and steroid hormones psychiatric, and antibiotics (Mezzelani et al. 2018). The PCs in the surface water and groundwater are recognized by many countries. Most pharmaceutical industries use active pharmaceutical ingredients to get pharmaceutical products that are pharmaceutically active, resistant to removal, potentially adverse events in microorganisms in the environment, and affect the environment and human health (Rivera-Utrilla et al. 2013). Most of the PCs have been shown to be lively past their expiry dates, and still, a few antibiotics occur in the environment, with as a minimum 90% of their compounds 28–40 years afterward their manufacturing dates. So, these PCs need to remove from water sources because of harmful effects on the ecosystem and human life (Velempini et al. 2021a). The techniques used to remove pharmaceutical ECs by primary/secondary treatment (flocculation, membrane separation process, constructed wetlands, etc.) Along with these, there are some advanced techniques such as adsorption, advanced oxidative process (AOP), photocatalysis was also included.

8.2 Sources and Occurrence of PCs

Among all ECs, PCs play a vital role in the environment due to their high stability, universal usage, and physicochemical properties. The biological effects of PCs can cause severe diseases to living organisms. Various sources like household activities, pharmaceuticals, food packaging, and hair products are released directly into aquatic media without pre-treatment. Moreover, the bioaccumulation and biomagnification affect the food chain and ecosystem. Along with these, climate change can also cause severe issues by concentrating them into rivers at the time of floods (Richardson 2007).

PPCPs, illegal drugs, anti-inflammatory drugs, antibiotics, etc. are the significant PCs due to their high existence in the aquatic media in the environment, and it causes various toxic effects on animals and humans. Figure 8.2 shows the basic contaminants routes of PCs. Pharmaceuticals are used by veterinary medicine, for the growth of promotors and other purposes. After that, these are excreted by the animals and reach the soil. These products may get into groundwater, and due to heavy rains, these may be transported to surface water from runoff. The flushed PCs and their metabolites enter into WWTP. The sewage treatment plants fail to complete remove PCs, and it enters into the fertile lands by the irrigation, and extent of dissolved components, and bio-adsorption. The chemical properties of these PCs are highly affecting the removal efficiency of the treatment plants. Acidic drugs and caffeine are significant effluents, which increase their solubility in WWTPs through sewage treatment plants (Comeau et al. 2008).

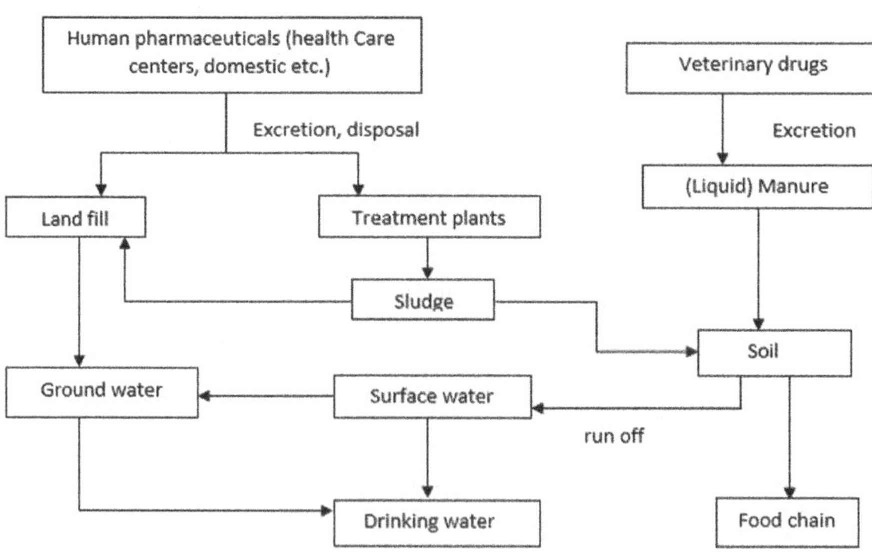

Fig. 8.2 Basic contaminants routes of PCs

Table 8.1 Concentration of commonly occurring PCs in wastewater in various regions over the world

Compound	Concentration (ng/L)	References
Carbamazepine	69	Chaves et al. (2020)
Sulfamethoxazole	7400	Blair et al. (2015)
Ibuprofen	450	Sun et al. (2019)
Naproxen	3000	Blair et al. (2015)
Diclofenac	105	Chaves et al. (2020)
Ketoprofen	220	Sun et al. (2019)
Atenolol	255	Roberts et al. (2016)
Florfenicol	396	Zhang (2020)
Propanolol	638	Fernández-López et al. (2016)
Erythromycen	320	Papageorgiou et al. (2016)
Gemfibrozil	15.5	Fram and Belitz (2011)

Focazio et al. instigated the visions about the environmental occurrence of a few PCs in the untreated source of drinking water in United States. This study was analyzed 63 of the 100 targeted analytes which included the sites 49 surface and 25 ground water sources of drinking water (Focazio et al. 2008). Another study revealed the occurrence of 32 drugs and 5 metabolites in German STP discharges into rivers. The acidic drugs like carbamazepine, antiphlogistics, and salicylic acid were mainly found in the rivers in the concentration range of ng/L (Ternes 1998). The presence of PCs in the influent and effluent of WWTPs in Spain, in the range of 0.34–26.5 μg/L (Fernández-López et al. 2016). The major pharmaceuticals such as atenolol, ibuprofen, carbamazepine, ciprofloxacin, gemfibrozil, and diclofenac drugs were reported in sewage treatment plant influents (Ebele et al. 2017). Table 8.1 showed the effluent concentrations of the most recently identified PCs in surface waters from the world.

The PCs can enter into the soil, air and water, which should be monitored regularly. There is a need to explore to research areas, which can reduce the risk of PCs in the water, air, and soil etc.

8.3 Effects of PCs

The PCs contaminate the water, air, and soil through various pathways, as discussed earlier, affecting the ecosystem and human health. There have been a lot of studies by the researchers, shown that >70% of PCs in the environment cause hazards and toxicity to human health (Dharupaneedi et al. 2019). The PCs are soluble in the water to consume by humans, and these compounds may not necessarily bioconcentrate in fish and that compounds that metabolized and excreted by the fish (Cunningham et al. 2009). Due to the beneficial health effects of human pharmaceuticals, the highly

Table 8.2 Effects of PCs on living organisms

Substance	Effect on living organisms	References
Ciprofloxacin	Vomiting, pale skin, headache, tiredness, abnormal liver	Villegas- Guzman et al. (2017)
Ibuprofen	Kidney issues in human	Khalil (2020)
Norfloxacin	Gastrointestinal upset, nausea, headache, rectal pain, muscle and joint pains, dizziness	Alkatheri et al. (2014)
Diclofenac	Tumors with chronic human exposure	Khalil (2020)
Doxycycline	Sensitivity of human intestinal microflora, skin allurgy	Korhonen et al. (2007)
Sulfathiazole	Changes in thyroid tissue, steroidogenesis in the human adrenocarcinoma (H295R) cell line	Ji et al. (2010a)
Amoxicillin	Diarrhoea, vomiting, and Nausea	Elizalde-Velázquez et al. (2016)
Sulphadiazine	Nausea, vomiting, diarrhoea, loss of appetite	López-Serna et al. (2013)

available evidence is used to evaluate any additional actions that might be required to decrease environmental impacts as a result of assessing the human health risks. The existence of the various antibiotics in the environment can cause toxic effects to the microbial bodies. Table 8.2 showed the effects of PCs on living organisms. Diclofenac is utilized for pain and joint stiffness relief. It was found that it may cause in some cases to tumors with chronic human exposure. Similarly, ibuprofen can cause various side effects like bleeding, kidney problems, vomiting, a cardiovascular risk (Khalil 2020).

The health effects caused by PCs are investigated by the light of its dose dependent resolution, bioaccumulation, and toxicity nature. The ratio of daily minimum therapeutic dose to daily input was considerable, so the PCs impacts of ingesting those components through drinking water could be negligible (Simazaki et al. 2015). It can be concluded that unregulated PCs consumption from various sources is dangerous for all living species and environments.

8.4 Detection and Analysis of PCs in the Environment

The detection and analysis of PCs are important in understanding the bulk materials, drugs, and their metabolites (Femina Carolin et al. 2021). The PCs detections can be explored through the different techniques such as the titrimetric, electrochemical, chromatography, spectroscopy, etc. (Femina Carolin et al. 2021). Sampling process, extraction procedure, concentration, cleaning, and chromatographic techniques are mainly used for the analysis of PCs. Few studies examine that the process of sample treatment gains up 80% of the analysis by using various approaches like liquid phase extraction, solid phase extraction, biosensor systems etc. (Narvaez and Jimenez 2012). Semi permeable membrane devices and organic chemical integrative sampling are the membranes used for sampler xenobiotics, metabolites, PCs, and antibiotics (Petrovic et al. 2008). The development of new techniques with a combination of gas chromatography and liquid chromatography along with mass spectroscopy has made to PPCPs at lower concentrations in the surface water and ground water (Roberts et al. 2016; Hernandez-Ruiz et al. 2012). Vieno et al. developed the analytical method to determine the PCs (e.g. acebutolol, carbamazepine, ciprofloxacin, ofloxacin, and norfloxacin). The method included the solid-phase extraction followed by huge-performance liquid chromatography, and detection were carried out through triple quadrupole mass spectrometry (Vieno et al. 2006). Similarly, the solid phase extraction-liquid chromatography-mass spectroscopy was utilized as a detection method by researchers (Park et al. 2021; Cimetiere et al. 2013). The detection of ciprofloxacin was carried out through the highly sensitive reversed-phase huge-performance liquid chromatography (Zotou and Miltiadou 2002). Researchers are still exploring the low concentration detection, multiple PCs detection, etc.

8.5 Treatment Methods of PCs

In the past few decades, the concentration of PCs has been increased and detected in most of the rivers, WWTPs, and STPs. Various review papers have been reported PCs in the environment, e.g. surface water and groundwater, and the treatment methods for their removal (Bolong et al. 2009; Lapworth et al. 2012; Kalaji and Rastogi 2017; Velempini et al. 2021b). Several removal methods are available for the removal of PCs as shown in Fig. 8.3. Due to their low biodegradable, and high stability, most of the PCs are unlikely to be efficiently removed by conventional treatment methods (primary and secondary). Later, various technologies were developed to obtain more PCs removal efficiency, such as adsorption, advanced oxidation process and photocatalysis.

8.5.1 Conventional Treatment Methods (Primary/Secondary Methods)

The primary/physical treatment for the PCs divided into flotation, flocculation, coagulation, and sedimentation. Flotation is a simple process, which is applicable to treat highly suspended solids in waste water with low energy consumption, inexpensive, and easy maintenance (Li and Li 2015). In addition, PPCPs can be removed from wastewater by adsorbing onto small aggregates or dissolving in the lipid fraction by flotation (Carballa et al. 2005). Bhagawan et al. investigated the coagulation

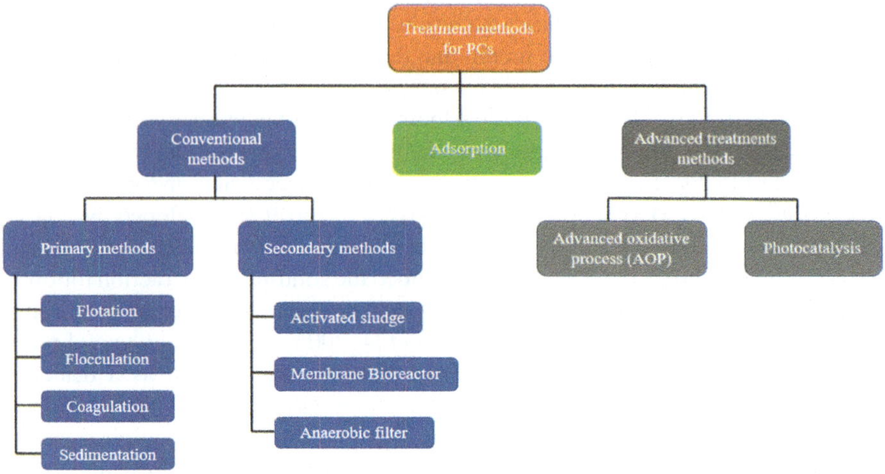

Fig. 8.3 Various treatment techniques for PCs removal

techniques for removing PCs in the environment using various coagulants like inorganic metal salts, alum, sulphate, and moringa oleifera. The application of inorganic compounds as a flocculant has been minimized due to its less efficiency in waste water with small dosage (Bhagawan et al. 2017). Among all these, coagulation with moringa oleifera is highly efficient for the treatment of PCs in wastewater. Polymeric flocculants are designed to improve the better flocculation process in wastewater (Lee et al. 2014). The efficiency of primary methods of PCs is very low, due to its large solid particles.

The secondary/biological treatment of PCs is categorized into activated sludge, membrane bioreactors, anaerobic filters, etc. In the activated sludge process, the PCs are treated by using bacteria and air. Oxygen is introduced into a mixture of primary treated wastewater and organisms to develop activated sludge. Biological processes reduce the concentration of biodegradable components in the aeration tank. The biological treatment of PPCPs has been attributed to the action of surface binding, volatilization, and microbial decomposition. The adsorption of PCs to the activated sludge is important for NSAIDs, indapamide and hydroxyzine (Salgado et al. 2012). Adsorption and volatilization of PPCPs is low in comparison to other treatment processes for other contaminates (Li et al. 2014). The advanced approach must be applied to increase the degradation efficiency of PPCPs. Plósz et al. designed a model activated sludge process for the removal of carbamazepine and diclofenac (Plósz et al. 2012).

Membrane technology is a type of phase-changing process with various applications in pharmaceutical ECs monitoring and removal. MBR has been increasing tremendously and employed for municipal and wastewater treatment. MBRs are also known as separation processes providing various applications in wastewater treatment (Dharupaneedi et al. 2019). Among all MBPs, only a few hybrid technologies have developed as possible water treatment technologies, which are highly successful for large-scale applications. MBR biological processes with some specifications include medium to high sludge retention times and aggregation of soluble microbial products. MBR allows for a higher biomass concentration to be maintained with the low small size of bioreactors. Schröder et al. investigated the treatment of NSAIDs and antibiotics in wastewater, performed by using two MBR reactors with sludge retention times over 28 days. It was shown that, compare to antibiotics, NSAIDs were removed with high efficiency in both of the MBRs (Schröder et al. 2012). Membrane filtration was used to remove the fine particles as well as dissolved substances from the PCs. Membrane filtration is mainly classified into reverse osmosis, ultrafiltration, nanofiltration, and microfiltration. The main advantages of the Membrane filtration are eco-friendly, less expensive, and low energy consumption over the other treatment techniques (Wenten 2003). Recently, Cuhorka et al. studied the removal of diclofenac and ibuprofen from waste water by using nanofiltration. The results showed that all membranes can be utilized for the removal of both drugs due to their higher fluxes and rejections (Cuhorka et al. 2020). Chon et al. evaluated a large-scale water reclamation unit as a combination of coagulation with ultrafiltration/reverse osmosis membranes to remove various PCs like atenolol, sulfamethoxazole, and carbamazepine with moderate efficiency. In this study, compared to the combined membranes process, reverse osmosis alone could give greater removal efficiency (Chon et al. 2012). A submerged MBR was used for the investigation of the removal of the PCs from the domestic wastewater. The model PCs were used acetaminophen, 17 β-estradiol, naproxen, diclofenac sodium, and carbamazepine and their removal efficiency was found to be 92.2%, 90.0%, 55.4%, 38.5% and 3.2%, respectively. In addition, sterilized sludge through the batch experiments showed the sludge adsorption efficiency of 7.9, 68.2, 60.1, 40.1, and 71.5%, for the five PCs, respectively (Fan et al. 2014). Park et al. carried out batch experiments to remove the 45 PPCPs using biomass from a MBR and studied the effect of ammonia oxidizing bacteria. The degradation of PPCPs was followed in the two stages i.e. cometabolic degradation and endogenous respiration (Park et al. 2017). Nanofiltration membrane was investigated to remove the cyclophosphamide and ciprofloxacine PCs and demonstrated the effective removal of chemical oxygen demand, total suspended solids and colour (Zaviska et al. 2013). The advanced integrated oxidation (UV/H_2O_2) and membrane filtration process was employed to remove the diclofenac and amoxicillin PCs (Żyłła et al. 2019). MBR and MBR associated with reverse osmosis/microfiltration/ultrafiltration can be classified as for removing PCs as shown in Fig. 8.4 (Goswami et al. 2018).

MBR removal efficacy influences by operating conditions such as biomass concentration, hydraulic retention time, chemical and structure properties of PCs, sludge retention time, temperature, and pH of the solution. So researchers have studied a different type of MBR concerning the effective removal of the ECs from the wastewater.

The treatment of pharmaceuticals by using constructed wetlands is highly potential, low cost, and less energy consumption in the developing countries (Li et al. 2014). Liu et al. reviewed the removal of 39 antibiotics through 106 several constructed wetland systems PCs (Liu 2019). The main components for the PCs removed in the wetlands are substrate, plant, and microbes. This method is environmentally friendly

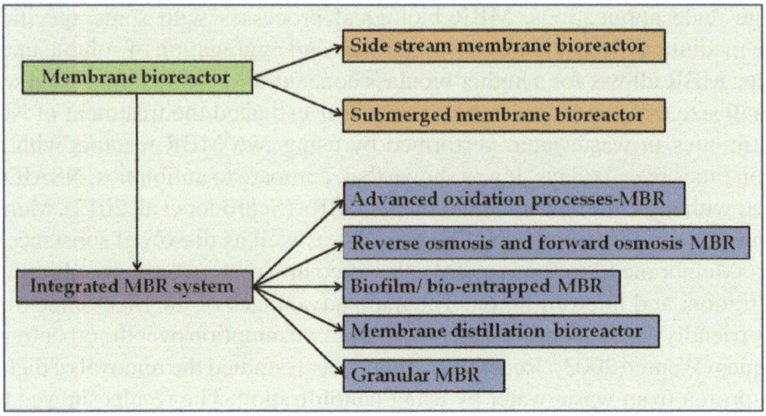

Fig. 8.4 An overview of MBR and integrated MBR systems (Reproduced from Goswami et al. 2018)

and cost-effective technology as mostly used biological process (Wang and Wang 2016). The conventional treatment methods are limited capacity to remove a large amount of PCs in wastewater. Among all conventional treatment methods, MBR with reverse osmosis and nanofiltration have showed appropriate for the removal of PCs.

8.5.2 Adsorption

In recent years, there have been many studies on the adsorption of PCs in the environment by various adsorbents. Figure 8.5 represents the sources of PCs and their treatment by various adsorbents like activated carbon, graphite, carbon nanotubes,

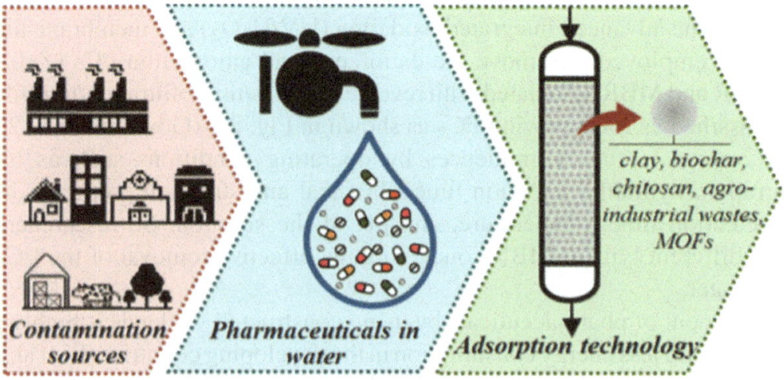

Fig. 8.5 Sources of PCs and their removal by adsorption technology (Reproduced from De Andrade et al. 2018)

biochar, and activated alumina etc. (De Andrade et al. 2018). Among all these adsorbents, activated carbon plays a vital role in PCs removal in the environment due to its high surface area, large porosity, and favorable pore size distribution. Carbamazepine and sildenafil citrate adsorption onto powdered activated carbon as a function of concentrations and physicochemical characteristics of water with the removal efficiencies greater than 85% was reported (Delgado et al. 2019).

In this study, coconut and bituminous activated carbon were employed for ibuprofen adsorption, which showed more than 50% reduction (Méndez-Arriaga et al. 2008). Activated carbon plays a vital role in the removal of PCs due to its high adsorptive ability. Many researchers studied various adsorption of nitroimidazoles on numerous types of activated carbon (Yoon et al. 2003). The low-cost activated carbon was obtained from the olive-waste cakes, and further their adsorption study was carried out to remove ibuprofen, ketoprofen, naproxen and diclofenac PCs. The adsorption kinetics of the activated carbon displayed a pseudo second-order kinetic model for the four PCs (Baccar et al. 2012). Ji et al. synthesized the KOH etched activated single-walled and multiwalled carbon nanotubes. The activated carbon nanotubes showed significant surface area enhancement from 410.7 m^2/g to 652.8 m^2/g and 157.3 m^2/g to 422.6 m^2/g in comparison to bare carbon nanotubes of single-walled and multiwalled, respectively. The activated single-walled and multiwalled carbon nanotubes showed the 2–3 and 3–8 times higher adsorption than the pristine carbon nanotubes towards phenol and nitrobenzene, tetracycline, sulfamethoxazole, and tylosin in aqueous solutions (Ji et al. 2010b). Pauletto et al. investigated the simultaneous adsorption of the paracetamol and nimesulide as a model PCs using the activate carbon. The simultaneous removal of the paracetamol and nimesulide was found to be 25.80 mg/g and 199.49 mg/g by activated carbon, while individual removal of paracetamol and nimesulide was 58.21 mg/g and 196.32 mg/g, respectively. The significant decrease in paracetamol removal in binary mixture was due to displacement of paracetamol active sites by nimesulide due to greater affinity between activated carbon and nimesulide (Pauletto et al. 2021). In another study, The graphene nanoplatelets were utilized for the adsorption of the aspirin, acetaminophen, and caffeine PCs (Al-Khateeb et al. 2014). Nowadays, graphene plays an ideal nano-adsorbent due to its high specific surface area along with two basal planes of single layer nature for the treatment of wastewater. Porous graphene with high surface area (679 m^2/g) was used as an adsorbent for the removal of atenolol, ibuprofen, carbamazepine, ciprofloxacin, gemfibrozil and diclofenac drugs from aqueous media with high efficiency (Khalil 2020).

The adsorption of Tramadol and Doxepin drugs in the wastewater by using Na-Mt smectite clay mineral as an adsorbent. Because of their high hydrophobic nature, more doxepin molecules probably adsorbed with weak molecular interaction to enhance the adsorption efficiency at low temperatures (Thiebault et al. 2015). By using bioadsorption process, the soluble molecules or ions of PPCPs are separated by the polymer substance of microalgae. The bioabsorption of PPCPs highly depends on structure and functional groups, hydrophobicity, and microalgae species (Hena et al. 2021). Choi and coworkers prepared the trimethylsilylated mesoporous SBA-15 to remove the mixture of 12 PCs and observed that it was influenced by the pH of the solution and PCs properties such as hydrophobicity and molecular charge (Bui et al. 2013). Zhang et al. prepared Zr-based MOFs of UiO-66-NH_2 and further investigated for the effective adsorptive removal of the salicylic acid and acetylsalicylic acid. The adsorption kinetics showed pseudo second order kinetics The main factors for the adsorption were hydrogen bonding, electrostatic interaction, and affinity between carboxyl groups and Zr-O clusters (Zhang et al. 2019).

The different materials such as activated carbon, mesoporous silica, bio adsorbent, MOFs are most studied materials to remove PCs. The combination of adsorption and other technique may address the PCs removal issues.

8.5.3 Advanced Treatment Techniques

8.5.3.1 Advanced Oxidative Processes (AOPs)

AOPs technologies are playing an significant role in the degradation of PCs by using the oxidation phenomena. AOPs are generally generate the free radicals such as HO·, O_2·, HO_2·. Hydroxyl radical is a highly active species that can react with organic molecules. AOPs are convert PCs organic compounds into carbon dioxide and water by the presence of ambient temperature and pressure with or without catalysts (Astas 2008). There are two stages mainly involved in the ozonation process the removal of PPCPs in water by using AOPs. Those are (1) accelerated ozone generates the OH radicals (2) the formed OH radicals contact and degrade the PPCPs (Li 2020). In the Fenton reaction, OH radicals are formed with the interaction of H_2O_2 and ferric ion. Figure 8.6. shows the different AOPs for the degradation of PCs.

The removal of Sulfamethazine by using Fenton like catalyst was designed (Oller and Malato 2021). Xiang et al. studied the UV/chlorine oxidation process to degrade the ibuprofen as a model water pollutant. The ibuprofen degradation followed the first-order kinetics using the UV/chlorine oxidation process and showed 3.3 higher activity than the UV/H_2O_2 oxidation process at pH 6 with a given chemical molar dosage. The degradation was found to be fast owing to hydroxyl radical and reactive chlorine species (Xiang et al. 2016). Ganzenko et al. studied the sequencing batch reactor followed by the electro-Fenton process towards removing the caffeine and 5-fluorouracil (Ganzenko et al. 2018). Electrochemical oxidation is an electrochemical technique that is useful for wastewater remediation. For treating non-chlorine solution, it is called anodic oxidation. The process-radiation has mainly formed the radicals, ions, and neutral molecules, highly reactive electrons via water exposure to high energy electromagnetic radiation (Burns and Sims 1981).

8.5.3.2 Photocatalysis

In the past few decades, photocatalytic technique utilizes solar energy to degrade and mineralize the various PCs (Li 2020). An ideal candidate for the photocatalyst

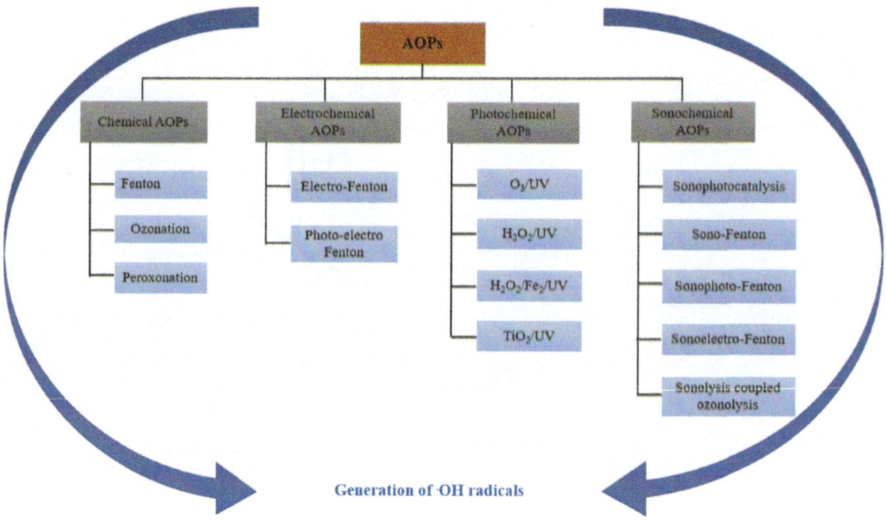

Fig. 8.6 Classification of AOPs for the removal of PCs

should be resistant to chemical, photoactive, photostable, and non-toxic (Kanakaraju et al. 2014). The photocatalytic degradation of PCs can be carried out by using various photocatalysts. Generally TiO_2, ZnO, CdS, WO_3, and Fe_2O_3 semiconductor nanomaterials are used as photocatalysts to remove PCs in the environment (Varma et al. 2020). TiO_2 is the most studied photocatalysis for removing various effluents in the environment among all photocatalysts. Figure 8.7 shows the photocatalytic mechanism for degradation of organic pollutants into CO_2 and H_2O (Paul et al. 2007). Metal and non-metal doping into TiO_2 is a highly effective method to enhance the photocatalytic activity via increasing the charge carriers generation under UV/visible light irradiation (Teoh et al. 2005).

Electrochromic TiO_2 nanotubes were synthesized through the electrochemical route followed by thermal treatment in N_2 atmosphere. The prepared TiO_2 nanotubes showed the high solar light absorbance, larger charge carrier generation, which is useful for the photocatalytic degradation of carbamazepine, ibuprofen, and caffeine organic compounds (Zheng et al. 2014). Badawy et al. studied the Ag doping in the TiO_2 and found that 0.1% Ag doped TiO_2 showed the higher photocatalytic activity towards the degradation of hospital wastewater (Badawy et al. 2014). In the other paper, B doped TiO_2 photocatalyst were synthesized through the solvothermal route. The bandgap of the materials was decreased, which result in the increased visible light absorption. The raw and B-doped TiO_2 were further utilized for the photocatalytic degradations of 2,4-dichlorophenol, bisphenol-A, ibuprofen, and flurbiprofen under exposure of visible light irradiation (Bilgin Simsek 2017). Ramasundaram et al. prepared the stainless steel filter integrated with TiO_2 nanofibers using the hot press technique, while polyvinylidene fluoride (PVDF) was used as a binder, as shown in Fig. 8.8. The photocatalytic oxidation of cimetidine was studied with the

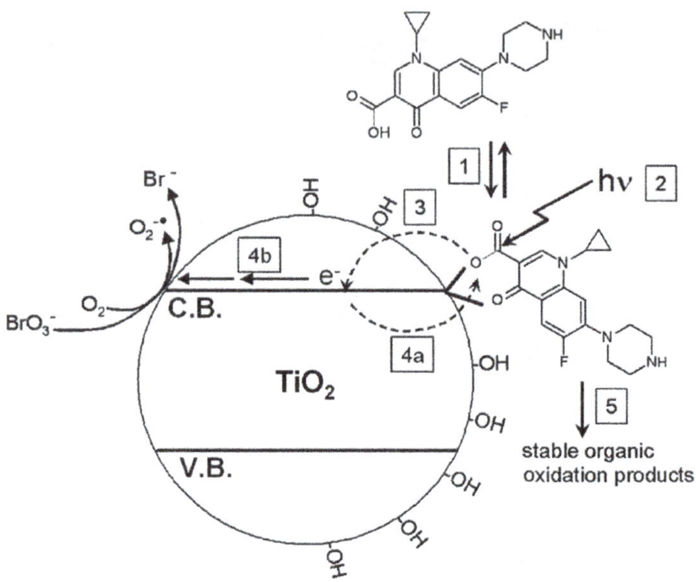

Fig. 8.7 Proposed mechanism for visible-light-mediated photocatalytic degradation of ciprofloxacin (Reproduced from Paul et al. 2007)

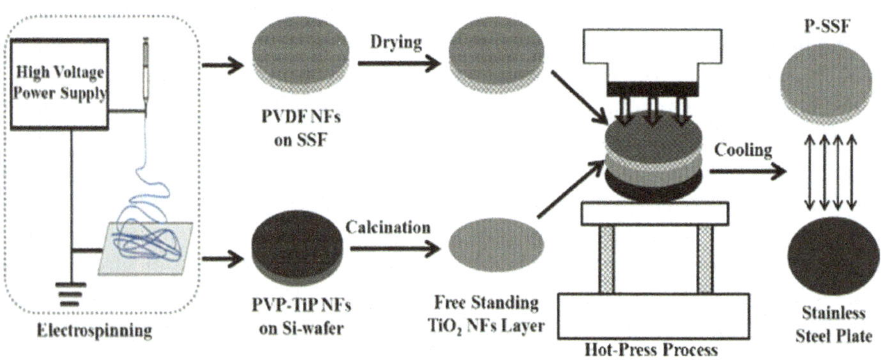

Fig. 8.8 Overall preparation process followed to integrate TiO_2 NFs on SSF using PVDF binder layer (Reproduced from Ramasundaram et al. 2013)

increase of the TiO_2 nanofibers thickness and found 29 μm as the optimum thickness. The photocatalytic activity was also decreased with an increase in water flux (Ramasundaram et al. 2013).

The TiO_2 nanoparticles were deposited over the commercial polymeric ultrafiltration membrane using the spray, vacuum, and sol–gel methods. They found the stable TiO_2 coated membrane in the sol–gel process, while other methods were not forming stable TiO_2 coating. TiO_2 coated polymeric hollow fiber membrane was further tested for the degradation of methylene blue and chlorhexidine digluconate

The degradation of compounds was found more than 30% and 40% respectively under the solar light irradiation along with filtrate conditions (Chakraborty et al. 2017). Méndez-Arriaga et al. studied the photolytic and photocatalytic degradation of Naproxen by using TiO_2 as photocatalyst. Photolysis process was highly efficient than photocatalysis for the Naproxen degradation (Méndez-Arriaga et al. 2008). The use of TiO_2-WO_3 photocatalysts for the degradation of ibuprofen in a municipal wastewater treatment plant with the degradation of organic compounds up to 64% under two hour solar light irradiation (Rey et al. 2014).

Hajiani & Co-workers investigated the effective photocatalytic degradation of diclofenac by novel magnetic TiO_2@$ZnFe_2O_4$/Pd nanocomposite, which is thermally stable, good stability, high photocatalytic activity, and low cost (Ahmadpour et al. 2020). Synthesis of multicomponent metal oxides to improve the photocatalytic efficiency was studied using Li-doped $Bi_{0.5}Na_{0.45}K_{0.5}TiO_3$-$BaTiO_3$. The multicomponent catalyst was used to obtain the high efficient photocatalytic degradation of ciprofloxacin (Velempini et al. 2021b). $LaFeO_3$ perovskite photocatalytic degradation of PPCP in wastewater under Visible light irradiation (Dhinesh Kumar and Jayavel 2014). $LaFeO_3$ (typical perovskite oxide) has attracted attention among researchers due to its high potential activity, such as catalytic oxidation, gas sensitive characters, unique optoelectronic properties, narrow band gap and surface electronic states (Tang et al. 2013; Yang et al. 2009). The Polyoxometalates/polymer composites were utilized for the 100% degradation of the ibuprofen under visible light irradiation, while only 20% degradation was observed in the case of visible light irradiation only (Brahmi 2021). The photocatalyst can be one of the best solutions to remove PCs as it utilizes the solar light for degradation of contaminants into non-toxic/less toxic compounds. Further, it can be integrated with the existing technologies to enhance the PCs removal efficiency.

8.6 Conclusions

Among all ECs, PCs plays a vital role in the environmental contamination. The occurrence of PCs as ECs in receiving environment should not be ignored and must be consistently observed and regulated. PPCPs, illegal drugs, anti-inflammatory drugs, antibiotics, etc. are the major PCs. Human pharmaceuticals (industries, households, hospitals, etc.) and veterinary medicine are the major sources for the PCs, which eventually into the soil, ground and surface water through the various pathways. The occurrence of PCs in the soil, and water were detected and analyzed by various techniques such as chromatography, spectroscopy, etc. The impact of PCs on the human health and ecosystem has urged the removal of these contaminants. The various methods such as conventional methods (primary and secondary treatments) and advanced treatment methods (AOP, photocatalysis, adsorption, etc.) utilized to remove these contaminants. Each removal technique has its own benefits and limitations as well.

References

Ahmadpour N, Sayadi MH, Sobhani S, Hajiani M (2020) Photocatalytic degradation of model pharmaceutical pollutant by novel magnetic TiO_2@$ZnFe_2O_4$/Pd nanocomposite with enhanced photocatalytic activity and stability under solar light irradiation. J Environ Manag 271:110964

Alkatheri AM, Alyousif SM, Alshabanah N, Abdulkareem M (2014) of Kidney diseases and transplantation original article medication adherence among adult patients on hemodialysis. Saudi J Kidney Dis Transpl 25:762–768

Al-Khateeb LA, Almotiry S, Salam MA (2014) Adsorption of pharmaceutical pollutants onto graphene nanoplatelets. Chem Eng J 248:191–199

Astas S (2008) Use of selected advanced oxidation processes (AOPs) for wastewater treatment—a mini review. Glob NEST J 10:376–385

Baccar R, Sarrà M, Bouzid J, Feki M, Blánquez P (2012) Removal of pharmaceutical compounds by activated carbon prepared from agricultural by-product. Chem Eng J 211–212:310–317

Badawy MI, Souaya EMR, Gad-Allah TA, Abdel-Wahed MS Ulbricht M (2014) Fabrication of Ag/TiO_2 photocatalyst for the treatment of simulated hospital wastewater under sunlight. Environ Prog Sustain Energy 33:886–894

Bhagawan D et al (2017) Industrial solid waste landfill leachate treatment using electrocoagulation and biological methods. Desalin Water Treat 68:137–142

Bilgin Simsek E (2017) Solvothermal synthesized boron doped TiO_2 catalysts: photocatalytic degradation of endocrine disrupting compounds and pharmaceuticals under visible light irradiation. Appl Catal B 200:309–322

Blair B, Nikolaus A, Hedman C, Klaper R, Grundl T (2015) Evaluating the degradation, sorption, and negative mass balances of pharmaceuticals and personal care products during wastewater treatment. Chemosphere 134:395–401

Bolong N, Ismail AF, Salim MR, Matsuura T (2009) A review of the effects of emerging contaminants in wastewater and options for their removal. Desalination 239:229–246

Brahmi C et al (2021) Performance improvement of the photocatalytic process for the degradation of pharmaceutical compounds using new POM/polymer photocatalysts. J Environ Chem Eng 9:106015

Bui TX, Pham VH, Le ST, Choi H (2013) Adsorption of pharmaceuticals onto trimethylsilylated mesoporous SBA-15. J Hazard Mater 254–255:345–353

Burns WG, Sims HE (1981) Effect of radiation type in water radiolysis. J Chem Soc, Faraday Trans 1: Phys Chem Condens Phases 77:2803–2813

Carballa M, Omil F, Lema JM (2005) Removal of cosmetic ingredients and pharmaceuticals in sewage primary treatment. Water Res 39:4790–4796

Chakraborty S et al (2017) Photocatalytic hollow fiber membranes for the degradation of pharmaceutical compounds in wastewater. J Environ Chem Eng 5:5014–5024

Chandrashekar Kollarahithlu S, Balakrishnan RM (2021) Adsorption of pharmaceuticals pollutants, ibuprofen, acetaminophen, and streptomycin from the aqueous phase using amine functionalized superparamagnetic silica nanocomposite. J Clean Prod 294:126155

Chaves M de JS et al. (2020) Pharmaceuticals and personal care products in a Brazilian wetland of international importance: occurrence and environmental risk assessment. Sci Total Environ 734:139374

Chon K, Kim SJ, Moon J, Cho J (2012) Combined coagulation-disk filtration process as a pretreatment of ultrafiltration and reverse osmosis membrane for wastewater reclamation: an autopsy study of a pilot plant. Water Res 46:1803–1816

Cimetiere N, Soutrel I, Lemasle M, Laplanche A, Crocq A (2013) Standard addition method for the determination of pharmaceutical residues in drinking water by SPE–LC–MS/MS. Environ Technol 34:3031–3041

Comeau F, Surette C, Brun GL, Losier R (2008) The occurrence of acidic drugs and caffeine in sewage effluents and receiving waters from three coastal watersheds in Atlantic Canada. Sci Total Environ 396:132–146

Cuhorka J, Wallace E, Mikulášek P (2020) Removal of micropollutants from water by commercially available nanofiltration membranes. Sci Total Environ 720:137474
Cunningham VL, Binks SP, Olson MJ (2009) Human health risk assessment from the presence of human pharmaceuticals in the aquatic environment. Regul Toxicol Pharmacol 53:39–45
De Andrade JR, Oliveira MF, Da Silva MGC, Vieira MGA (2018) Adsorption of pharmaceuticals from water and wastewater using nonconventional low-cost materials: a review. Ind Eng Chem Res 57:3103–3127
Delgado N, Capparelli A, Navarro A, Marino D (2019) Pharmaceutical emerging pollutants removal from water using powdered activated carbon: study of kinetics and adsorption equilibrium. J Environ Manage 236:301–308
Dharupaneedi SP et al (2019) Membrane-based separation of potential emerging pollutants. Sep Purif Technol 210:850–866
Dhinesh Kumar R, Jayavel R (2014) Facile hydrothermal synthesis and characterization of $LaFeO_3$ nanospheres for visible light photocatalytic applications. J Mater Sci: Mater Electron 25:3953–3961
Ebele AJ, Abou-Elwafa Abdallah M, Harrad S (2017) Pharmaceuticals and personal care products (PPCPs) in the freshwater aquatic environment. Emerg Contam 3:1–16
Egbuna C et al (2021) Emerging pollutants in Nigeria: a systematic review. Environ Toxicol Pharmacol 85:103638
Elizalde-Velázquez A et al (2016) Amoxicillin in the aquatic environment, its fate and environmental risk. Environ Health Risk—Hazard Factors Living Species. https://doi.org/10.5772/62049
Fan H, Li J, Zhang L, Feng L (2014) Contribution of sludge adsorption and biodegradation to the removal of five pharmaceuticals in a submerged membrane bioreactor. Biochem Eng J 88:101–107
Femina Carolin C, Senthil Kumar P, Janet Joshiba G, Vinoth Kumar V (2021) Analysis and removal of pharmaceutical residues from wastewater using membrane bioreactors: a review. Environ Chem Lett 19:329–343
Fernández-López C, Guillén-Navarro JM, Padilla JJ, Parsons JR (2016) Comparison of the removal efficiencies of selected pharmaceuticals in wastewater treatment plants in the region of Murcia, Spain. Ecol Eng 95:811–816
Flaherty CM, Dodson SI (2005) Effects of pharmaceuticals on Daphnia survival, growth, and reproduction. Chemosphere 61:200–207
Focazio MJ et al (2008) A national reconnaissance for pharmaceuticals and other organic wastewater contaminants in the United States—II) Untreated drinking water sources. Sci Total Environ 402:201–216
Fram MS, Belitz K (2011) Occurrence and concentrations of pharmaceutical compounds in groundwater used for public drinking-water supply in California. Sci Total Environ 409:3409–3417
Ganzenko O et al (2018) Bioelectro-Fenton: evaluation of a combined biological—advanced oxidation treatment for pharmaceutical wastewater. Environ Sci Pollut Res 25:20283–20292
Goswami L et al (2018) Membrane bioreactor and integrated membrane bioreactor systems for micropollutant removal from wastewater: a review. J Water Process Eng 26:314–328
Grandclément C et al (2017) From the conventional biological wastewater treatment to hybrid processes, the evaluation of organic micropollutant removal: a review. Water Res, 297–317
Hena S, Gutierrez L, Croué JP (2021) Removal of pharmaceutical and personal care products (PPCPs) from wastewater using microalgae: a review. J Hazard Mater 403
Hernandez-Ruiz S, Abrell L, Wickramasekara S, Chefetz B, Chorover J (2012) Quantifying PPCP interaction with dissolved organic matter in aqueous solution: combined use of fluorescence quenching and tandem mass spectrometry. Water Res 46:943–954
Ji K et al (2010a) Effects of sulfathiazole, oxytetracycline and chlortetracycline on steroidogenesis in the human adrenocarcinoma (H295R) cell line and freshwater fish Oryzias latipes. J Hazard Mater 182:494–502
Ji L, Shao Y, Xu Z, Zheng S, Zhu D (2010b) Adsorption of monoaromatic compounds and pharmaceutical antibiotics on carbon nanotubes activated by KOH etching. Environ Sci Technol 44:6429–6436

Kalaji HM, Rastogi A (2017) Pharmaceutical compounds: an emerging pollutant (A review on plant-pharmaceuticals interaction). Chiang Mai J Sci

Kanakaraju D, Glass BD, Oelgemöller M (2014) Titanium dioxide photocatalysis for pharmaceutical wastewater treatment. Environ Chem Lett 12:27–47

Khalil AME et al (2020) Nanostructured porous graphene for efficient removal of emerging contaminants (pharmaceuticals) from water. Chem Eng J 398:125440

Korhonen C, Peterson K, Bruder C, Jung P (2007) Self-reported adverse events associated with antimalarial chemoprophylaxis in peace corps volunteers. Am J Prev Med 33:194–199

Lapworth DJ, Baran N, Stuart ME, Ward RS (2012) Emerging organic contaminants in groundwater: a review of sources, fate and occurrence. Environ Pollut, 287–303

Lee CS, Robinson J, Chong MF (2014) A review on application of flocculants in wastewater treatment. Process Saf Environ Prot, 489–508

Li S et al (2020) Facile construction of novel Bi2WO6/Ta3N5 Z-scheme heterojunction nanofibers for efficient degradation of harmful pharmaceutical pollutants. Chem Eng J 402:126165

Li X, Li G (2015) A review: pharmaceutical wastewater treatment technology and research in China. In: Proceedings of the 2015 Asia-Pacific Energy Equipment Engineering Research Conference 9:345–348

Li Y, Zhu G, Ng WJ, Tan SK (2014) A review on removing pharmaceutical contaminants from wastewater by constructed wetlands: design, performance and mechanism. Sci Total Environ, 908–932

Liu X et al (2019) A review on removing antibiotics and antibiotic resistance genes from wastewater by constructed wetlands: performance and microbial response. Environ Pollut 254:112996

López-Serna R et al (2013) Occurrence of 95 pharmaceuticals and transformation products in urban groundwaters underlying the metropolis of Barcelona, Spain. Environ Pollut 174:305–315

Méndez-Arriaga F, Gimenez J, Esplugas S (2008) Photolysis and TiO_2 photocatalytic treatment of naproxen: degradation, mineralization, intermediates and toxicity. J Adv Oxid Technol 11:435–444

Mezzelani M, Gorbi S, Regoli F (2018) Pharmaceuticals in the aquatic environments: evidence of emerged threat and future challenges for marine organisms. Mar Environ Res 140:41–60

Narvaez JFV, Jimenez CC (2012) Pharmaceutical products in the environment: sources, effects and risks/Productos Farmacéuticos En El Ambiente: Fuentes, Efectos Y Riesgos. Vitae 19:93–108

Oller I, Malato S (2021) Photo-Fenton applied to the removal of pharmaceutical and other pollutants of emerging concern. Curr Opin Green Sustain Chem 29:100458

Papageorgiou M, Kosma C, Lambropoulou D (2016) Seasonal occurrence, removal, mass loading and environmental risk assessment of 55 pharmaceuticals and personal care products in a municipal wastewater treatment plant in Central Greece. Sci Total Environ 543:547–569

Park J, Yamashita N, Wu G, Tanaka H (2017) Removal of pharmaceuticals and personal care products by ammonia oxidizing bacteria acclimated in a membrane bioreactor: contributions of cometabolism and endogenous respiration. Sci Total Environ 605–606:18–25

Park J et al (2021) Determination of pharmaceuticals in solid samples in municipal wastewater treatment plants by online SPE LC–MS/MS using QuEChERS extraction. Environ Monit Assess 193:279

Paul T, Miller PL, Strathmann TJ (2007) Visible-light-mediated TiO_2 photocatalysis of fluoroquinolone antibacterial agents. Environ Sci Technol 41:4720–4727

Pauletto PS, Lütke SF, Dotto GL, Salau NPG (2021) Adsorption mechanisms of single and simultaneous removal of pharmaceutical compounds onto activated carbon: isotherm and thermodynamic modeling. J Mol Liq 336:116203

Petrovic M et al (2008) Emerging contaminants in waste waters: sources and occurrence. Handb Environ Chem, 5. Water Pollut 5 S1:1–35

Plósz BG, Langford KH, Thomas KV (2012) An activated sludge modeling framework for xenobiotic trace chemicals (ASM-X): assessment of diclofenac and carbamazepine. Biotechnol Bioeng 109:2757–2769

Ramasundaram S et al (2013) Titanium dioxide nanofibers integrated stainless steel filter for photocatalytic degradation of pharmaceutical compounds. J Hazard Mater 258–259:124–132
Rehman MSU et al (2015) Global risk of pharmaceutical contamination from highly populated developing countries. Chemosphere 138:1045–1055
Rey A et al (2014) WO3–TiO_2 based catalysts for the simulated solar radiation assisted photocatalytic ozonation of emerging contaminants in a municipal wastewater treatment plant effluent. Appl Catal B 154–155:274–284
Richardson SD (2007) Water analysis: emerging contaminants and current issues. Anal Chem, 4295–4323
Rivera-Utrilla J, Sánchez-Polo M, Ferro-García MÁ, Prados-Joya G, Ocampo-Pérez R (2013) Pharmaceuticals as emerging contaminants and their removal from water. A review. Chemosphere, 1268–1287
Roberts J et al (2016) Pharmaceuticals and personal care products (PPCPs) in Australia's largest inland sewage treatment plant, and its contribution to a major Australian river during high and low flow. Sci Total Environ 541:1625–1637
Salgado R et al (2012) Assessing the removal of pharmaceuticals and personal care products in a full-scale activated sludge plant. Environ Sci Pollut Res 19:1818–1827
Schröder HF et al (2012) The removal and degradation of pharmaceutical compounds during membrane bioreactor treatment. Water Sci Technol 65:833–839
Simazaki D et al (2015) Occurrence of selected pharmaceuticals at drinking water purification plants in Japan and implications for human health. Water Res 76:187–200
Sun H, Wang T, Yang Z, Yu C, Wu W (2019) Simultaneous removal of nitrogen and pharmaceutical and personal care products from the effluent of waste water treatment plants using aerated solid-phase denitrification system. Bioresour Technol 287:121389
Tang H et al (2013) Hydrothermal synthesis and visible-light photocatalytic activity of α-Fe_2O_3/TiO_2 composite hollow microspheres. Ceram Int 39:8633–8640
Taoufik N et al (2020) Removal of emerging pharmaceutical pollutants: a systematic mapping study review. J Environ Chem Eng 8
Teoh WY, Mädler L, Beydoun D, Pratsinis SE, Amal R (2005) Direct (one-step) synthesis of TiO_2 and Pt/TiO_2 nanoparticles for photocatalytic mineralisation of sucrose. Chem Eng Sci 60:5852–5861
Ternes TA (1998) Occurrence of drugs in German sewage treatment plants and rivers. Water Res 32:3245–3260
Thiebault T, Guégan R, Boussafir M (2015) Adsorption mechanisms of emerging micro-pollutants with a clay mineral: case of tramadol and doxepine pharmaceutical products. J Colloid Interface Sci 453:1–8
Varma KS et al (2020) Photocatalytic degradation of pharmaceutical and pesticide compounds (PPCs) using doped TiO_2 nanomaterials: a review. Water-Energy Nexus 3:46–61
Velempini T, Prabakaran E, Pillay K (2021a) Recent developments in the use of metal oxides for photocatalytic degradation of pharmaceutical pollutants in water—a review. Mater Today Chem
Velempini T, Prabakaran E, Pillay K (2021) Recent developments in the use of metal oxides for photocatalytic degradation of pharmaceutical pollutants in water—a review. Mater Today Chem 19:100380
Vieno NM, Tuhkanen T, Kronberg L (2006) Analysis of neutral and basic pharmaceuticals in sewage treatment plants and in recipient rivers using solid phase extraction and liquid chromatography–tandem mass spectrometry detection. J Chromatogr A 1134:101–111
Villegas- Guzman P et al (2017) A green solar photo-Fenton process for the elimination of bacteria and micropollutants in municipal wastewater treatment using mineral iron and natural organic acids. Appl Catal B: Environ 219:538–549
Wanda EMM, Nyoni H, Mamba BB, Msagati TAM (2017) Occurrence of emerging micropollutants in water systems in Gauteng, Mpumalanga, and North West provinces, South Africa. Int J Environ Res Public Health 14:8–20

Wang J, Wang S (2016) Removal of pharmaceuticals and personal care products (PPCPs) from wastewater: a review. J Environ Manage 182:620–640

Wenten IG (2003) Recent development in membrane science and its industrial applications. J Sci Technol 24:1009–1024

Wilkinson J, Hooda PS, Barker J, Barton S, Swinden J (2017) Occurrence, fate and transformation of emerging contaminants in water: an overarching review of the field. Environ Pollut 231:954–970

Xiang Y, Fang J, Shang C (2016) Kinetics and pathways of ibuprofen degradation by the UV/chlorine advanced oxidation process. Water Res 90:301–308

Yang J, Zhong H, Li M, Zhang L, Zhang Y (2009) Markedly enhancing the visible-light photocatalytic activity of $LaFeO_3$ by post-treatment in molten salt. React Kinet Catal Lett 97:269–274

Yoon Y, Westerhoff P, Snyder SA, Esparza M (2003) HPLC-fluorescence detection and adsorption of bisphenol A, 17β-estradiol, and 17α-ethynyl estradiol on powdered activated carbon. Water Res 37:3530–3537

Yuan F, Hu C, Hu X, Qu J, Yang M (2009) Degradation of selected pharmaceuticals in aqueous solution with UV and UV/H_2O_2. Water Res 43:1766–1774

Zaviska F, Drogui P, Grasmick A, Azais A, Héran M (2013) Nanofiltration membrane bioreactor for removing pharmaceutical compounds. J Membr Sci 429:121–129

Zhang R et al (2019) Highly effective removal of pharmaceutical compounds from Aqueous solution by magnetic Zr-based MOFs composites. Ind Eng Chem Res 58:3876–3884

Zhang L et al (2020) Occurrence, distribution, and ecological risk of pharmaceuticals in a seasonally ice-sealed river: From ice formation to melting. J Hazard Mater 389:122083

Zheng Q et al (2014) Electrochromic titania nanotube arrays for the enhanced photocatalytic degradation of phenol and pharmaceutical compounds. Chem Eng J 249:285–292

Zotou A, Miltiadou N (2002) Sensitive LC determination of ciprofloxacin in pharmaceutical preparations and biological fluids with fluorescence detection. J Pharm Biomed Anal 28:559–568

Żyłła R, Boruta T, Gmurek M, Milala R, Ledakowicz S (2019) Integration of advanced oxidation and membrane filtration for removal of micropollutants of emerging concern. Process Saf Environ Prot 130:67–76

Part II
Emerging Contaminants: Transport and Conventional and Advance Technologies for Treatment

Chapter 9
Microbial Degradation of Pharmaceuticals

Aishwarya Rastogi and Manoj Kumar Tiwari

Abstract In recent years, pharmaceuticals have surfaced as a novel class of pollutants due to their incomplete degradation in sewage treatment plants and their characteristic ability to promote physiological predicaments in humans even at low doses. Of the several physical, chemical and biological techniques studied for the degradation of drugs, microbial degradation is regarded preferential due to its energy intensive nature, lower ecological footprint as well as considerably lower production of toxic by-products. This chapter reviews the information on the microbial degradation of various classes of drugs including antibiotics, non-steroidal anti-inflammatory drugs, antihypertensives, antidepressants and anticancer drugs. Numerous bacterial, fungal and algal strains have shown substantial efficiency in degradation of various pharmaceuticals, mostly through co-metabolism using the drug as the secondary carbon source. A wide array of enzymes like dehydrogenase, hydrolase, oxidoreductase, oxidase, dioxygenase, monooxygenase, decarboxylase and many more are typically involved in the microbial degradation of pharmaceuticals. The ability of microbes to degrade pharmaceuticals is commonly attributed to their possession of functions analogous to mammalian CYP2C9 and CYP3A4. These isoforms of cytochrome P450 are majorly responsible for metabolism of drugs in humans. Hydroxylation by CYP450 is the most commonly reported initial step in the microbial degradation of drugs, and processes like decarboxylation, dehydrogenation, dechlorination, subsequent oxidation, demethylation, hydroxylation, cleavage of the ester group often constitute the degradation pathways. Although the efficiency and kinetics of degradation are known for most of the microbial strain, the degradation pathway is explained for very few because of the technical difficulties arising during the elucidation process. Though microbial degradation of pharmaceuticals is proving an attractive alternative in lab studies due to irrefutable advantages, the

A. Rastogi
School of Environmental Science and Engineering, Indian Institute of Technology, Kharagpur 721302, India

M. K. Tiwari (✉)
School of Water Resources, Indian Institute of Technology, Kharagpur 721302, India
e-mail: mktiwari@swr.iitkgp.ac.in

S. P. Singh et al. (eds.), *New Trends in Emerging Environmental Contaminants*, Energy, Environment, and Sustainability,
https://doi.org/10.1007/978-981-16-8367-1_9

scaling up process is tedious and challenging due to several limitations with respect to microbial growth conditions and subsequent handling of microbial contamination.

Keywords Biodegradation pathways · Metabolites · Antibiotics · NSAIDs · Cytochrome P450 · Degradation enzymes

9.1 Introduction

Pharmaceuticals are viewed as an important tool to facilitate the physical and mental well-being of the society and can be classified on the basis of usage and mechanisms of action. Drugs are metabolized in the liver and excreted from humans and animals body in the form of urine which is passed on to the environment directly or through sewage treatment plants in an intact or metabolized form. These pharmaceuticals reach the environment mainly via sewage or hospital wastewater treatment plants, leachate from solid waste landfills, solid waste management plants, or unintentional direct dumping by pharmaceutical industries.

Pharmaceuticals are found in the range of ng/L to μg/L in almost all aquatic environments including rivers, lakes, wells, groundwater as well as drinking waters all around the world (Kasprzyk-Hordern et al. 2008; Shanmugam et al. 2014; Brozinski et al. 2013; Vulliet and Cren-Olive 2011; Rabiet et al. 2006; Kondor et al. 2020). This is because of the absence of any strict monitoring and discharge levels for most pharmaceuticals. Groundwater contamination with drugs is a serious problem as it leads to increased persistence and difficulty in elimination because of somewhat reduced redox environments and a complete lack of photodegradation compared to other aquatic ecosystems (Peng et al. 2014). Typically, pharmaceuticals are present below 1 μg/L in aquatic environments, but some unusually high concentrations have also been reported. For example, a maximum concentration of 32 μg/L naproxen has been reported in the Malir and Lyari rivers of Pakistan (Scheurell et al. 2009). Similarly, an exceedingly high oxytetracycline concentration of 361.11 μg/L was found in the surface waters of China (Jiang et al. 2014). Although, the concentration of pharmaceuticals in groundwater is typically well below 500 ng/L, exceptions such as ciprofloxacin concentration of 14 μg/L in groundwater of India (Fick et al. 2009) are present. Meanwhile, the complexity of hospital and pharmaceutical industry effluents also make them a foremost source of drugs contamination and the drug concentration in these waters far exceeds that of Municipal Wastewater Treatment Plants (WWTPs) (Sim et al. 2011).

Due to a widespread occurrence of pharmaceuticals, the water quality and aquatic life in these ecosystems are highly affected, which ultimately leads to a reduction in the number of aquatic organisms and aquatic diversity. A grim possibility of bioaccumulation of drugs along the food chain is rapidly gaining ground because of the detection of several pharmaceuticals like diclofenac in liver or bile of rainbow trout (Schwaiger et al. 2004). Subtle impacts of pharmaceuticals have been suggested in terrestrial animals also. Several toxic effects of pharmaceuticals have been reported in model organisms which include physiological implications like genetic and cellular damage, modulation of proteins, enzymatic imbalance or difficulties in reproduction (Garcia-Medina et al. 2015; Gonzales-Rey and Bebianno 2012; Haap et al. 2008; Han et al. 2010).

The toxic effects and concentrations vary depending on the pharmaceutical as well as on the model organism used. For example, mortality was reported in *Ceriodaphnia dubia* for diclofenac with an EC50 of 22.7 mg/L (Ferrari et al. 2004). Also, 5-fluorouracil resulted in growth inhibition in *Pseudokirchneriella subcapitata* and *Synechococcus leopoliensis* at a concentration of 0.13 mg/L and 1.20 mg/L respectively (Brezovšek et al. 2014). The environmental risk due to these drugs is assessed by calculating the ratio of Predicted Environmental Concentration (PEC) to Predicted No-Effect Concentration (PNEC). This ratio is known as risk quotient and if its value is smaller than 1, there is no substantial risk associated with the environmental occurrence of the pharmaceutical. Though almost all studies invariably categorize the effects of pharmaceuticals on human health as very low risk (de Jesus Gaffney et al. 2015; Koopaei and Abdollahi 2017), this can lead to severe consequences in the future and therefore, is a topic of concern.

Due to the omnipresence of pharmaceuticals in the aquatic environment and long-term ecotoxic effects on individual components of the ecosystem, they are considered an immediate threat and an emerging contaminant of concern. There is a high focus on developing efficient drug removal techniques stressing on economic and environmental sustainability.

Highly selective and rapid removal methods, like advanced oxidation processes (AOPs), or adsorbents like activated carbon, are seen as preferred ways to remove pharmaceutical micropollutants. Other removal methods like Fenton oxidation, electrochemical oxidation, ion exchange, membrane filtration and soil aquifer treatment are also utilised in recent times. Membrane technologies are also used in a combination of physical, chemical, and biological processes to increase the removal efficiency. Many studies assessing and comparing the degradation efficiency for various pharmaceutical through these methods are also present (Tufail et al. 2020; Cherik et al. 2015).

Almost all of these removal methods are time efficient; however, they are costly and often lead to the generation of many potentially toxic by-products, and are essentially characterized by a large ecological footprint due to an elevated energy utilization (Schwarzenbach et al. 2006). Because of these reasons, microbial degradation can be viewed as a potential and efficient method to eradicate these micropollutants. Several pharmaceuticals have been successfully biodegraded and biotransformed using different microbes or microbial consortia in a laboratory setup. They demonstrate varying degrees of removal efficiencies with respect to time based on the pharmaceutical and the microbes used.

Microbial degradation of pharmaceuticals has attracted tremendous interests from the scientific community because it is regarded as a more sustainable approach for the remediation of several xenobiotics. Therefore, it is necessary to gain understanding about the microbial degradation potential, pathways and mechanisms of different classes of pharmaceuticals. This chapter aims to provide a comprehensive review of microbial degradation of pharmaceuticals, regarding their potential and the pathways employed. This chapter further delves into the application of microbial degradation of pharmaceuticals in WWTPs along with the possible limitations faced during its execution.

9.2 Degradation of Pharmaceuticals

9.2.1 Metabolism of Pharmaceuticals in Human Body

The liver is the chief site of drug metabolism in human body and contains concentrated enzymes involved in metabolism. Processes like hydrolysis, conjugation, oxidation, reduction, isomerization etc. are employed for drug metabolism to facilitate easier excretion of the metabolized drug. Metabolism occurs in two phases for all major drugs. Phase I reactions include oxidation, reduction, hydrolysis etc. and involve the production of a new or altered functional group along with cleavage. Phase II reactions involve conjugation with an endogenous substance like sulfate, acetate, glucuronic acid, glycine etc. which leads to the production of highly polar, readily excreted conjugates. These drug metabolism reactions yield lower lipophilic metabolites as compared to the original compound (Tillement and Tremblay 2007).

Cytochrome P450 (CYP450) is recognized as the principle enzyme system of phase I metabolism. The metabolism of xenobiotic compounds is carried out mainly by CYP1, CYP2, CYP3, and CYP4 families. Phase II metabolism is governed majorly by enzymes like Uridine 5′-diphospho(UDP)-glucuronosyltransferases, sulfotransferases, glutathione *S*-transferases, *N*-acetyltransferases, and epoxide hydrolases leading to the generation of easily excretable metabolites (Stanley 2017).

These metabolized drugs are finally excreted in the form of urine or faeces. In spite of drug metabolism in the liver, some portion of the drug is excreted in non-metabolized form (intact form) due to which a combination of parent drug and its metabolites are found simultaneously in the WWTPs. When the different classes of pharmaceuticals along with their metabolites present in WWTPs are taken into consideration, the resulting effect and toxicity on the microbes because of exposure to families of specific drugs becomes evident. The knowledge regarding this toxicity is important for implementation of microbial degradation of drugs in WWTPs.

9.2.2 Microbial Degradation of Pharmaceuticals

The microbial degradation of pharmaceuticals has acquired a tremendous interest from the research community because of their irrefutable advantages over other removal methods. Several microbial strains have shown the potential to degrade pharmaceuticals both as a primary and secondary substrate. Addition of a carbon substrate might enhance the removal efficiency due to co-metabolism. However, the addition of a carbon source might also result in a reduced removal efficiency because of competitive inhibition from secondary substrates.

Biodegradation of pharmaceuticals mainly depend on the specific microorganisms employed for degradation. Therefore, factors like optimum temperature, pH and substrate concentration needed for microbial growth impact the degradation efficiency. For example, microbial degradation of NSAIDs occur mainly at low pH (Vieno and Sillanpää 2014). Moreover, presence of biorefractory compounds such as heavy metals or selected aromatic compounds may reduce or completely inhibit the biodegradation. Inhibition of *Bacillus thuringiensis* B1 (2015b) was observed during ibuprofen biodegradation in the presence of Hg^{2+} (NOEC = 0.3 μM). Cd(II) and Cu(II) were also reported to be inhibitory to *B. thuringiensis* B1 (2015b) due to very low NOEC values (Marchlewicz et al. 2017). Slow degradation of ibuprofen and complete inhibition of ibuprofen degradation has been reported in presence of 2-chlorophenol and 4-chlorophenol respectively (Marchlewicz et al. 2017).

The major processes involved in the biotransformation of most pharmaceuticals was hydroxylation and oxidation catalyzed by cytochrome P450 (CYP). Hydroxylation can be defined as the introduction of hydroxyl group into an ion or radical usually by the replacement of hydrogen. This was supported by the decreases in drug concentrations on incubation of microbial cultures with 1-aminobenzotriazole (inhibitor of CYP) (Hata et al. 2010). This ability of microbes to degrade xenobiotic compounds is accredited to their possession of functions analogous to mammalian CYP2C9, CYP3A4 and many more. These isoforms of cytochrome P450 are majorly responsible for metabolism of drugs in humans. Purified laccase was also capable of catalyzing the biotransformation of some drugs; however, laccase is the enzymatic system responsible for only a few pharmaceuticals like naproxen and ciprofloxacin.

Apart from hydroxylation, other mechanisms like decarboxylation (elimination of a carboxyl group), dehydrogenation (removal of hydrogen), demethylation (removal of methyl group), dealkylation (removal of alkyl group), acetylation (introduction of an acetyl group), dechlorination (elimination of chlorine), detachment of acetate group, loss of hydroxyl and glucuronide group, and cleavage of the ester group are also employed in microbial pharmaceutical degradation. A wide arrays of enzymes are reported to be upregulated during the degradation process which indicate the possibility of their involvement in the microbial degradation of pharmaceuticals. Mass spectrometric and ^{1}H nuclear magnetic resonance analyses are an indispensable tool for the revelation and evaluation of the metabolites generated during biodegradation process.

Most of the microorganisms follow first-order or pseudo-first order kinetics for degradation of pharmaceuticals. Diclofenac and ciprofloxacin degradation by *Brevibacterium* sp. D4 and *Thermus* sp. respectively followed first order kinetics (Bessa et al. 2017; Pan et al. 2018). *Pseudoxanthomonas* DIN-3 also employed first-order degradation kinetics for removal of diclofenac, ibuprofen and naproxen (Lu et al. 2019). However, an enriched nitrifying culture was reported to degrade atenolol using pseudo-first order kinetics (Xu et al. 2017). Similarly, fluoxetine was also reported to be degraded by different microbial consortium following pseudo-first order kinetics (Velázquez and Nacheva 2017).

Even though the degradation pathways are assessed for a few pharmaceuticals, majority of microbial degradation pathways and mechanisms are still obscure. There is thought to be a relation between the degradation mechanism and the resistance mechanism of the microbe with respect to the drug which in turn depends on the mechanism of action for individual drugs. Therefore, the knowledge of mechanism of action is a prerequisite for understanding the microbial degradation mechanism of any drug. The primary mechanisms of antimicrobial resistance include enzymatic degradation of drugs, alteration in membrane permeability to drugs by inhibiting drug uptake or activating drug efflux pumps and modification of microbial drug targets. Drug resistance in bacteria can either be maintained on the bacterial chromosome or through plasmids.

Numerous studies evaluating the biodegradation potential of several algal, bacterial and fungal pure cultures were found. These microbes vary significantly with respect to the biodegradation potential, efficiency as well as time taken; however, a common result of decreased toxicity in the end-products was observed in all the studies. Studies examining the pharmaceutical degradation using activated sludge or a defined microbial culture are also present.

9.3 Biodegradation of Pharmaceuticals from Major Drug Classes

According to the Anatomical Therapeutic Chemical (ATC) Classification System, a drug can be classified based on the organ system the drug targets, therapeutic effect, mechanism of action, chemical properties and the chemical components of the drug. For convenience, the classification based on therapeutic effect is used for further discussion along with the description of their mechanism of action. Based on therapeutic effect, pharmaceuticals can be classified as:

1. Antibiotics
2. Non-Steroidal Anti-Inflammatory Drugs
3. Antidepressants
4. Antineoplastics
5. Beta blockers (antihypertensives).

Some other classes are also present, but our discussion will be focused on the aforementioned classes. The concentration of all the mentioned classes have been increasing consistently in our aquatic environments because of the increasing population as well as increasing cases of infections, mental disorders, cancer, cardiac problems and hypertension.

As can be concluded from Fig. 9.1, antibiotics had the highest number of articles regarding wastewater treatment, in spite of its highest representative (amoxicillin) dominating only 13% of the annual prescriptions. This was probably because of the threat of antibiotic resistance related to microbial degradation of antibiotics. Similarly, antidepressant drugs garnered only 6% of research even after its representative

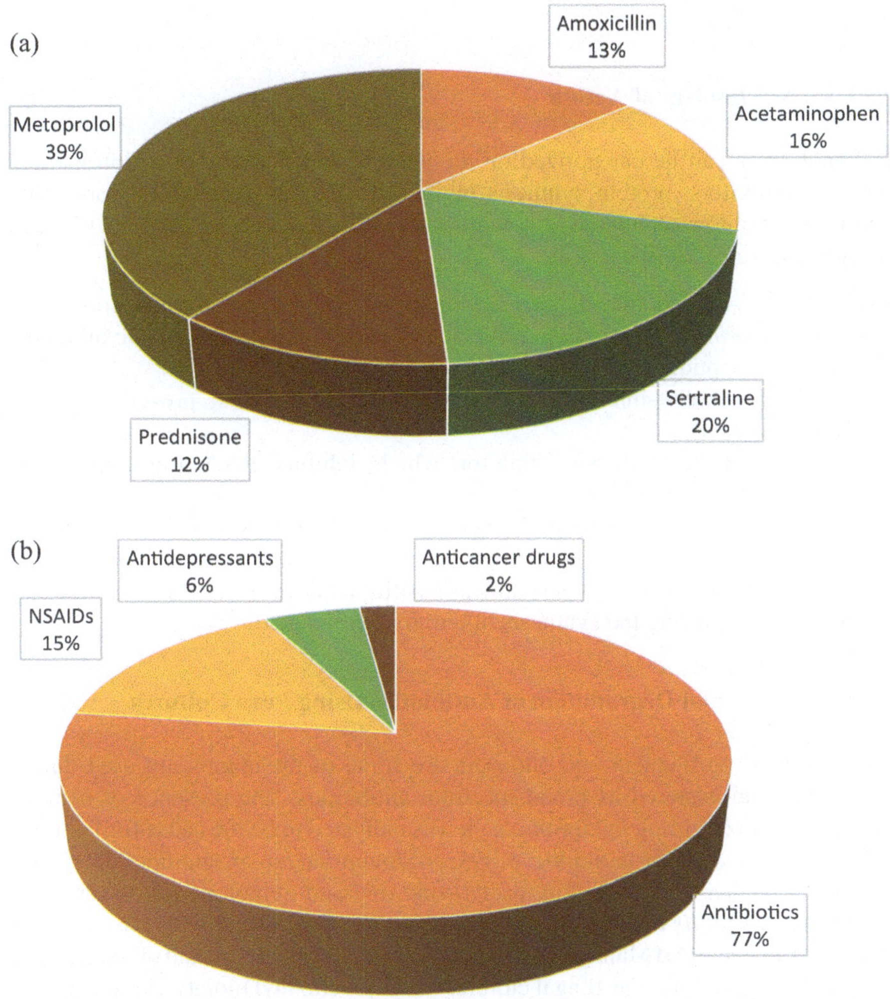

Fig. 9.1 **a** Annual prescriptions of representatives of major drug classes (NSAIDs-acetaminophen, Antidepressants-sertraline, Anticancer-prednisone, Antibiotic-amoxicillin, Beta-blockers-metoprolol) (Based on data from ClinCalc DrugStats Database-https://clincalc.com/DrugStats/Top300Drugs.aspx). **b** Number of research papers concerning wastewater treatment of each pharmaceutical group between 1998 and 2020 in Scopus (based on data from Peña et al. 2021)

(sertraline) having a 20% share in the annual prescriptions. Therefore, it is apparent that our research is not centred around the quantitative prescription of pharmaceuticals. However, the role of non-prescribed medications cannot be overlooked in this scenario. Further sections explore the microbial degradation of different classes of pharmaceuticals.

9.3.1 Antibiotics

9.3.1.1 Mechanism of Action

The antibiotics can be categorized on the basis of mechanisms of action namely, cell wall inhibitors, protein synthesis inhibitors, DNA synthesis inhibitors, RNA synthesis inhibitors and folic acid synthesis inhibitors (Kapoor et al. 2017). The different classes of antibiotics are:

- *Penicillins*—blocks cross linking leading to cell wall synthesis inhibition.
- *Cephalosporins*—hinders bacterial cell wall synthesis via competitive inhibition of the transpeptidase enzyme.
- *Fluoroquinolones*—inhibits DNA gyrase enzyme, thus preventing DNA synthesis.
- *Rifampin*—RNA synthesis inhibitor which inhibits RNA transcription by inhibiting RNA polymerase.
- *Sulfonamides*—mimic para-aminobenzoic acid (PABA) which further binds to pteredine synthetase, thus blocking folic acid and DNA synthesis.
- *Trimethoprims*—similar target pathway as sulfonamides and bind to dihydrofolate reductase, inhibiting the synthesis of tetrahydrofolic acid.

9.3.1.2 Microbial Degradation of Antibiotics Using Pure Cultures

Ciprofloxacin, norfloxacin and ofloxacin are some of the commonly used fluoroquinolones administered as broad-spectrum antibiotics. The presence of benzene ring and piperazinyl ring in ciprofloxacin was indicated to be the cause for failure of complete removal of the antibiotic. *L. gesseri*, *Enterobacter* sp. and *Bacillus sp.* was capable of changing the structure of pyrazine ring and quinolone ring whereas *M. luteus* and *B. subtilis* only reduced the quantity of ciprofloxacin with less structural changes (Liyanage and Manage 2018). Ciprofloxacin, norfloxacin and ofloxacin were preferably attacked by four fungal cultures at the piperazinyl moiety via substitution or/and decomposition. High residual activity was also reported following degradation which suggested that many metabolites had antibacterial activity (Čvančarová et al. 2015). Some important metabolites produced during ciprofloxacin biodegradation by fungal species are desethylene-N-ciprofloxacin, N-acetylciprofloxacin, desethylene N-acetylciprofloxacin, N-formylciprofloxacin, desethylene-N-formylciprofloxacin and hydroxymethyl-N-ciprofloxacin. The involvement of laccase in ciprofloxacin degradation has been indicated in this study (Čvančarová et al. 2015).

Labrys portucalensis F11 was capable of degrading 85% ciprofloxacin in 28 days along with a low dehalogenation with a 22% fluoride release. This slow removal of fluorine atom suggested that defluorination might not be the initial step in the biotransformation of ciprofloxacin. In a mixture, norfloxacin and ciprofloxacin were favourably utilized by strain F11 with a removal of 82 and 81%, respectively. The bacteria showed a higher degradation for ofloxacin, reaching 91% within 28 days (Amorim et al. 2014).

A study delving into the analysis of metabolites produced by degradation of norfloxacin by *Microbacterium* sp. reported four metabolites, which were identified as 8-hydroxynorfloxacin, 6-defluoro-6-hydroxynorfloxacin, desethylene norfloxacin, and *N*-acetylnorfloxacin. Defluorination and hydroxylation were reported to be the main mechanisms of biodegradation (Kim et al. 2011). Desethylene-N-acetylnorfloxacin and N-formylnorfloxacin are some of the newly reported metabolites generated during norfloxacin degradation by fungal strains (Čvančarová et al. 2015). Significant metabolites of ofloxacin include desmethyl-N-ofloxacin and desethylene-N-ofloxacin and the degradation process was similar to that of ciprofloxacin and norfloxacin examined in the same study (Čvančarová et al. 2015).

Amoxicillin is a penicillin derived broad-spectrum beta-lactam antibiotic which is favoured to treat streptococcal infections. Amoxicillin degradation products like benzothiazole, 2-amino-3-methoxyl benzoic acid, 4-hydroxy-2-methyl benzoic acid, 5-amino-2-methylphenol and 3,5-bis(tert-butyl)-2-hydroxybenzaldehyde have been reported after 14 days of incubation with *Alcaligenes* sp. MMA. A hydrogen bond was also reported between amoxicillin and porins when amoxicillin was present as a primary substrate (Sodhi et al. 2020). 4-aminobenzenesulfonic acid and 4-aminothiophenol are reported as the degradation metabolites for two sulfonamides namely, sulfapyridine and sulfamethoxazole (Mao et al. 2018) (Table 9.1).

9.3.1.3 Microbial Degradation of Antibiotics Using Mixed Cultures

A defined microbial consortium containing four microbial species including *Acinetobacter* sp. US1, *Bacillus halodurans, Pseudomonas putida, Acinetobacter* spp was reported to degrade >97% carbamazepine in 12 days. It was suggested that roughly 60% of this removal is accredited to microbial metabolism. On adding carbamazepine and diatrizoate simultaneously, the biotransformation rate of carbamazepine and diatrizoate was not found to be influenced by the presence of the other (Ha et al. 2016).

Another microbial consortium composed of two bacterial isolates: *Klebsiella pneumoniae* and *Achromobacter* sp. and two fungi *Candida manassasensis* and *Trichosporon asahii* was capable of degrading norfloxacin much faster than ofloxacin. Norfloxacin was degraded by 80% in 20 days whereas ofloxacin was 60% degraded after 28 days and had a relatively slower biodegradation rate (Jałowiecki et al. 2019).

Table 9.1 Microbial degradation of antibiotics using pure cultures

S. No.	Microorganism used	Removal %	Time taken	References
Ciprofloxacin				
1	*Irpex lacteus*	100	10 days	Čvančarová et al. (2015)
2	*Panus tigrinus*	60	14 days	
3	*Trametes versicolor*	100	14 days	
4	*Mucor ramannianus*	89	14 days	Parshikov et al. (1999)
5	*Labrys portucalensis F11*	85	28 days	Amorim et al. (2014)
6	*Lactobacillus gasseri*	100	14 days	Liyanage and Manage (2018)
7	*Micrococcus luteus*	56		
8	*Enterobacter* sp.	96		
9	*Bacillus* sp.,	74		
10	*Bacillus subtilis*	70		
11	*Gloeophyllum striatum*	>50	90 h	Wetzstein et al. (1999)
Norfloxacin				
1	*Irpex lacteus*	100	10 days	Čvančarová et al. (2015)
2	*Trametes versicolor*	85	14 days	
3	*Dichomitus squalens*	44	14 days	
4	*Labrys portucalensis F11*	85	28 days	Amorim et al. (2014)
Ofloxacin				
1	*Labrys portucalensis F11*	91	28 days	Amorim et al. (2014)
2	*Irpex lacteus*	100	10 days	Čvančarová et al. (2015)
3	*Trametes versicolor*	100	14 days	
Carbamazepine				
1	*Rhodococcus rhodochrous*	15	28 days	Gauthier et al. (2010)
2	*Pseudomonas* sp.CBZ-4	46.6	144 h	Li et al. (2013)
3	*Phanerochaete chrysosporium*	60–80	14 days	Li et al. (2015)
Amoxicillin				
1	*Bacillus cereus*	100	14 days	Liyanage and Manage (2016)

(continued)

Table 9.1 (continued)

S. No.	Microorganism used	Removal %	Time taken	References
2	*Enterobacter ludwigii*	75		
3	*Enterobacter* sp.	80		
4	*Alcaligenes* sp. MMA	84	14 days	Sodhi et al. (2020)
5	*Corynebacterium* sp.	43	30 days	Wasify and Hamed (2014)
6	*Flavobacterium* sp.	41		
7	*Pseudomonas aeruginosa*	77.8		
8	*Bacillus subtilis*	76.7		
9	*Acinetobacter lwoffi*	74.3		
Sulfonamide				
1	*Bacillus cereus*	80	14 days	Liyanage and Manage (2016)
2	*Enterobacter ludwigii*	60		
3	*Enterobacter* sp.	70		
4	*Rhodococcus rhodochrous*	14–20	12–36 days	Gauthier et al. (2010)
5	*Shewanella oneidensis* MR-1	~23–60	5 days	Mao et al. (2018)
6	*Shewanella* sp. strain MR-4	~23–64		

9.3.2 Non-Steroidal Anti-inflammatory Drugs

9.3.2.1 Mechanism of Action

NSAIDs have anti-inflammatory, antipyretic as well as analgesic properties. The antipyretic effect is exhibited due to the inhibition of generation of prostaglandins induced by interleukin-1 (IL-1) and interleukin-6 (IL-6) in the hypothalamus along with reorganizing of the thermoregulatory system resulting in the vasodilation and amplified heat loss (Osafo et al. 2017). The analgesic effect is associated with the peripheral inhibition of prostaglandin generation (Cashman 1996) whereas anti-inflammatory action of NSAIDs is because of the inhibition of cyclooxygenase (COX) enzyme, which is responsible for converting arachidonic acid to prostaglandins (Vane 1971).

9.3.2.2 Microbial Degradation of NSAIDs Using Pure Cultures

Diclofenac is a phenylacetic acid derivative NSAID used to treat inflammatory conditions such as rheumatoid arthritis and osteoarthritis along with injury-related inflammation due to surgery or physical trauma. Several enzymes were found to be possibly involved in degradation including dehydrogenase, hydrolase, oxidoreductase, oxidase, dioxygenase, monooxygenase, and decarboxylase with the probable involvement of hydroxylase and dehalogenase (Lu et al. 2019). Mono and di-hydroxylated metabolites along with fumarylacetoacetic acid, phenylacetic acid, acetoacetic acid and fumaric acid were reported to be the metabolites for microbial degradation of diclofenac by *Rhodococcus ruber* strain IEGM 346 (Ivshina et al. 2019). Benzoquinone imine was a significant metabolite generated by biodegradation by *Labrys portucalensis* F11 (Moreira et al. 2018).

Ibuprofen is a propionic acid derivative commonly sold over the counter medication for the management of mild to moderate pain in headache, spondylitis, migraine, osteoarthritis, postoperative dental pain, and rheumatoid arthritis. Murdoch and Hay (2005, 2015) explored the biodegradation pathways of ibuprofen in two bacterial species namely *Variovorax* Ibu-1 and *Sphingomonas* sp. Ibu-2. Catechols were the key metabolites in ibuprofen biotransformation (Murdoch and Hay 2005, 2015). The formation of 2-hydroxy ibuprofen through monooxygenase activity was found to be the rate-limiting step in *Bacillus thuringiensis* B1(2015b) (Bessa et al. 2017). Three hydroxylated metabolites namely 1-hydroxy ibuprofen; 2-hydroxy ibuprofen and 1,2-dihydroxy ibuprofen were reported by Marco-Urrea et al. (2009) on studying the degradation of ibuprofen by four fungal species (Marco-Urrea et al. 2009). High activity of phenol and hydroquinone monooxygenases and aliphatic monooxygenases was observed in Bacillus thuringiensis B1(2015b) (Marchlewicz et al. 2017).

Naproxen is used as a first-line therapy NSAID for dysmenorrhea, acute gouty arthritis, osteoarthritis, and musculoskeletal inflammation and pain. The activity of salicylate 1,2-dioxygease; phenol monooxygenase; tetrahydrofolate-dependent O-demethylase; catechol 1,2-dioxygenase; gentisate 1,2-dioxygenase; naphthalene dioxygenase, and hydroxyquinol 1,2-dioxygenase was related to the degradation of naproxen in different microbes. The intradiol cleavage of the ring catalyzed by hydroxyquinol 1,2-dioxygenase and fracture of naphthalene ring are reported to be the naproxen degradation processes in some microbial strains (Domaradzka et al. 2015; Wojcieszynska et al. 2014; Ding et al. 2017) along with the processes mentioned earlier. 6-O-desmethylnaproxen was a key metabolite reported in several microbial species such as *Bacillus thuringiensis* B1(2015b), *Aspergillus niger* and algal cultures of *Cymbella* sp. And *Scenedesmus Quadricauda* (Aracagök et al. 2017; Górny et al. 2019; Ding et al. 2017). A series of tyrosine conjugated metabolites were also observed in the algal cultures as a result of conjugation of 6-O-desmethylnaproxen with amino acid (Ding et al. 2017).

Ketoprofen is a propionic acid derivative NSAID used for the treatment of muscle pain along with rheumatoid arthritis, postoperative pain, osteoarthritis, dysmenorrhea and postpartum pain. Fungal degradation of ketoprofen has been studied in three white rot fungi namely, *Trametes versicolor*, *Pleurotus ostreatus,* and *Pleurotus djamor.* A major metabolite found in both *T. versicolor* and *P. ostreatus* was (2-[(3-hydroxy(phenyl)methyl)phenyl]propanoic acid). 2-[3-(4-hydroxybenzoyl)phenyl] propanoic acid was produced as a minor metabolite in *P. ostreatus* (Palli et al. 2017); whereas (2-(3-benzoyl-4-hydroxyphenyl)-propanoic acid) and 1(2-[3-(4-hydroxybenzoyl) phenyl]-propanoic acid) were generated as minor metabolites in *T. versicolor* (Marco-Urrea et al. 2010a; b, c). The biodegradation of ketoprofen was observed even in the presence of diclofenac and naproxen by *Pleurotus djamor*. 68% and 83% removal was obtained for ketoprofen after 6 and 48 h respectively, when it was added along with diclofenac and naproxen (Cruz-Ornelas et al. 2019) (Table 9.2).

9.3.2.3 Microbial Degradation of NSAIDs Using Mixed Cultures

Several studies investigated the use of mixed consortia for individual as well as a mixture of NSAIDs (Almeida et al. 2013b; Ismail et al. 2016; Aissaoui et al. 2017; Dawas-Massalha et al. 2014; Bragança et al. 2016). Several genera like *Nitratireductor*, *Pseudacidovorax*, *Asticcacaulis* and *Bacillus pseudomycoides, Rhodococcus ruber*, *Vibrio mediterranei* had a higher toxicity resistance towards diclofenac, ibuprofen, and ketoprofen or were observed to gain competitive advantages, which suggested that they might contribute to their respective biodegradation (Nguyen et al. 2019).

A mixed microbial consortia containing *Arthrobacter nicotianae. Pseudomonas* sp., *Enterobacter hormaechei and Citrobacter youngae* was capable of degrading 23.08% ibuprofen and 9.12% diclofenac in 48 h when both NSAIDs were utilized as primary substrates. In the presence of glucose, this degradation efficiency increased significantly resulting in complete degradation of ibuprofen and 56% diclofenac after 48 h (Aissaoui et al. 2017). Activated sludge, defined mixed microbial culture such as K_2 bacterial consortium or algal–bacterial consortium have also been used for the degradation of several NSAIDs (López-Serna et al. 2019; Ismail et al. 2016; Quintana et al. 2005). The metabolites generated and the degradation pathways were scarcely reported for mixed culture studies.

9.3.3 *Antidepressants*

9.3.3.1 Mechanism of Action

Antidepressants are the drugs which are used for reducing the symptoms of depressive disorders by modifying or rectifying the chemical discrepancies of neurotransmitters

Table 9.2 Microbial degradation of NSAIDs using pure cultures

S. No.	Microorganism used	Removal %	Time taken	References
Diclofenac				
1	*Trametes versicolor*	94	1 h	Marco-Urrea et al. (2010a)
2	*Phanerochaete sordida* YK-624	100	6 days	Hata et al. (2010)
3	*Rhodococcus ruber* strain IEGM 346	100	6 days	Ivshina et al. (2019)
4	*Labrys portucalensis* F11	70	30 days	Moreira et al. (2018)
5	*Brevibacterium sp.* D4	90	30 days	Bessa et al. (2017)
6	*Pseudoxanthomonas* sp. DIN-3	23	14 days	Lu et al. (2019)
7	*Klebsiella* sp. KSC	100	72 h	Stylianou et al. (2018)
8	*Ganoderma applanatum*	80	72 h	Bankole et al. (2020)
9	*Laetiporus sulphurous*	87		
10	*Enterobacter cloacae* D16	67.57	48 h	Aissaoui et al. (2017)
11	*Pseudomonas moorei* KB4	100	12 days	Żur et al. (2020)
Ibuprofen				
1	*Variovorax* Ibu-1	50	7 days	Murdoch and Hay (2015)
2	*Phanerochaete chrysosporium*	70–88	7 days	Marco-Urrea et al. (2009)
	Trametes versicolor	~100		
	Irpex lacteus			
	Ganoderma lucidum			
3	*Pseudoxanthomonas* sp. DIN-3	41	14 days	Lu et al. (2019)
4	*Patulibacter* sp. strain I11	62–92	90 h	Almeida et al. (2013a; b)
5	*Sphingobium yanoikuyae*	68	130 days	Balciunas et al. (2020)
6	*Bacillus thuringiensis* B1(2015b)	46.56	20 days	Marchlewicz et al. (2016)
7	*Ganoderma applanatum*	66	72 h	Bankole et al. (2020)
8	*Laetiporus sulphurous*	79		

(continued)

Table 9.2 (continued)

S. No.	Microorganism used	Removal %	Time taken	References
Naproxen				
1	*Stenotrophomonas maltophilia* KB2	78	35 days	Wojcieszynska et al. (2014)
2	*Pseudoxanthomonas* sp. DIN-3	39	14 days	Lu et al. (2019)
3	*Aspergillus niger*	98	48 h	Aracagök et al. (2017)
4	*Cymbella* sp.	97.1	30 days	Ding et al. (2017)
5	*Scenedesmus quadricauda*	58.8		
6	*Bacillus thuringiensis* B1(2015)	100	35 days	Marchlewicz et al. (2016)
7	*Phanerochaete chrysosporium*	86–90	7 days	Li et al. (2015)
8	*Planococcus* sp. S5	75–86	35 days	Domaradzka et al. (2015)
9	*Trametes versicolor*	100	6 h	Marco-Urrea et al. (2010c)
Ketoprofen				
1	*Pleurotus ostreatus*	90% in batch, while 70–85% in continuous stage	7–15 days	Palli et al. (2017)
2	*Trametes versicolor*	100%	5 h	Marco-Urrea et al. (2010b)
3	*Pleurotus djamor*	83%	72 h	Cruz-Ornelas et al. (2019)

in the brain. The antidepressants have been divided into four groups based on their mode of action (Sharma, 2017):

- *Tricyclic antidepressants (TCAs)*—inhibit the reuptake of both norepinephrine and serotonin.
- *Selective serotonin-reuptake inhibitors (SSRIs)*—inhibit the reuptake of serotonin and elevate the synaptic serotonin transmission but is not responsible for the reuptake of other neurotransmitters.
- *Monoamine oxidase inhibitors (MAOIs)*—inhibits the enzymatic transformation of serotonin and norepinephrine into their respective metabolites.
- *Serotonin-norepinephrine reuptake inhibitor (SNRI)*—has TCAs like activity with no side effects like sedation or hypotension.

9.3.3.2 Microbial Degradation of Antidepressants Using Pure Cultures

Citalopram, paroxetine, fluoxetine and sertraline are some examples of commonly used SSRIs whereas venlafaxine is an SNRI used for the treatment of major depression. Meanwhile, mianserin is an example of atypical tetracyclic antidepressant. Kózka et al. assessed the removal of several antidepressants using *Pleurotus ostreatus* (Kózka et al. 2020). High removal efficiencies were reported for sertraline (91.2%, 92.8%), paroxetine (86.1%, 93.7%), clomipramine (89.6%, 98.4%) and mianserin (63%, 94.01%) at 4 h and 96 h respectively. Fluoxetine had an almost stable concentration and showed very little change after the first hour to the end of 96 h (86.2%, 85.1%). Citalopram and venlafaxine had a comparatively lower removal efficiency of 50% and 22% respectively after 96 h. Enzymatic degradation and adsorption were suggested to be responsible for this removal. In the presence of drugs, the activity of laccase, lignin peroxidase and ligninolytic enzymes was three times higher, which indicated that the enzymes are responsible for the removal of the selected antidepressants. The degradation mechanisms for these antidepressants were also evaluated in this study (Kózka et al. 2020).

The microbial degradation of citalopram and fluoxetine was observed by *Bjerkandera* sp. R1, *Bjerkandera adusta* and *Phanerochaete chrysosporium*. *Bjerkandera* sp. R1 was able to degrade 58% citalopram in 4 days, whereas for *B. adusta* and *P. chrysosporium*, the removal efficiency was only 10% in 4 days. However, no trace of citalopram was present in any of the fungal strains at the end of 14 days. The degradation of fluoxetine was not substantial as only partial degradations of 23–46% were obtained. (Rodarte-Morales et al. 2011). This was mainly because of the highly recalcitrant nature of fluoxetine to photolysis, hydrolysis and microbial degradation (Kwon and Armbrust 2006).

Biodegradation of racemic fluoxetine and its enantiomers by *Labrys portucalensis* strain F11 was evaluated by Moreira et al. (2014). Complete removal of both enantiomers was achieved in 30 days on supplementing 2 μM of racemic fluoxetine as the primary substrate. This was coupled with the stoichiometric release of fluoride. On increasing the racemic fluoxetine concentration to 4 and 9 μM, partial biodegradation of the enantiomers was observed. On adding acetate as a supplementary carbon source, complete degradation was reported for lower concentrations.

9.3.3.3 Microbial Degradation of Antidepressants Using Mixed Cultures

Fluoxetine was degraded using different consortia by Velázquez and Nacheva (Velázquez and Nacheva, 2017). The ammonium-nitrite-oxidizing consortium was able to degrade 78.6% fluoxetine with a higher sorption in the biomass. This consortium displayed comparatively fast removal during the first 12 days which decreased with the decrease in the removal rate of NH_4-N. The heterotrophic consortium demonstrated the highest biodegradation of fluoxetine reaching 84.7%, with the sorption percentage of 8.8%. Very little fluoxetine biodegradation of only 22.6% was observed with the nitrite-oxidizing consortium, while the sorption percentage was 11.8%. It was suggested that the ammonium-nitrite-oxidizing and the heterotrophic consortia permitted a partial degradation of fluoxetine, whereas no significant degradation was anticipated using nitrite-oxidizing consortium.

The capability of sulphate-reducing bacteria to biodegrade fluoxetine as primary substrate in a microbial consortium has also been investigated (Palma and Costa, 2021). On using 20 mg/L of fluoxetine as a primary substrate, complete degradation under sulphate reducing conditions was observed, whereas on increasing the concentration to 50 mg/L, only 66% was degraded after 31 days. The bacterial community mainly comprised of *Desulfomicrobium* and *Desulfovibrio* whereas, while using fluoxetine as unique carbon source, an increase in *vadinBC27* wastewater-sludge group, *Macellibacteroidetes, Dethiosulfovibrio, Bacteroides, Tolumonas, Sulfuricurvum, f_Enterobacteriaceae*_OTU_18 was observed.

9.3.4 Antineoplastics

9.3.4.1 Mechanism of Action

Antineoplastic agents disrupt cellular replication by deterring the synthesis of DNA or by triggering irreversible damage to DNA. Although this affects both normal as well as cancerous cells, normal cells are generally able to repair minor damage and therefore continue living. Antineoplastic agents are divided by their mechanism of action (DeAngelis and Posner, 2009) and include:

- *Classic alkylating agents*—hinder DNA replication by DNA strand breaking, crosslinking DNA strands, and abnormal pairing of base pairs.
- *Antimetabolites*—inhibits DNA and RNA synthesis by acting as false metabolites, which are incorporated into the DNA strand or impede essential enzymes, which results in the prevention of DNA synthesis.
- *Antitumour antibiotics*—obstruct RNA and DNA synthesis.
- *Plant alkaloids*—bind to microtubule proteins during metaphase resulting in mitotic arrest which leads to death.
- *Hormonal agents*—alter the internal as well as extracellular environment.

9.3.4.2 Microbial Degradation of Antineoplastics Using Pure Cultures

Ferrando-Climent et al. studied the removal of anticancer drugs in hospital wastewater using a fluidized bed bioreactor inoculated with *Trametes versicolor*. High removals were observed for Azathioprine (100%), Ciprofloxacin (97%), Ifosfamide (61%), Etoposide (100%), and Tamoxifen (48%) after 8 days. Tamoxifen showed a higher degradation of 92% and 99% at the end of 1 h and 9 days respectively when synthetic wastewater was treated with *T. versicolor*. Two tamoxifen hydroxylated positional isomers were acknowledged as metabolites derived from biodegradation of tamoxifen (Ferrando-Climent et al. 2015).

High removal for antitumor antibiotics like sulfapyridine and sulfathiazole was also observed using *T. versicolor* in another study. Sulfathiazole removal was slightly slower than sulfapyridine. On one hand, complete removal of sulfapyridine was achieved after 48 h whereas on the other hand, more than 20% sulfathiazole was detected even after 72 h. Degradation intermediates of sulfapyridine were also observed where glycosylated products, desulfonated products, formyl metabolites and hydroxylated moiety of sulfapyridine were the commonly detected intermediates. For sulfathiazole, glycosylated and desulfonated products along with hydroxylated moiety of sulfathiazole and formyl sulfathiazole were the commonly observed metabolites (Rodríguez-Rodrígueza et al. 2012).

9.3.4.3 Microbial Degradation of Antineoplastics Using Mixed Cultures

The effect of different carbon utilizing cultures on the degradation of 5-fluorouracil, an antimetabolite antineoplastic drug, was reported by Rossmassler et al. (2019).The phenol utilizing culture degraded >90% 5-fluorouracil in 0.25 days, whereas same removal was obtained by the organic acids utilizing culture and the molasses utilizing culture within 0.75 days. However, the casamino acids utilizing culture and the high humic acid content-peptone mixture utilizing culture took 6 days to achieve similar biotransformation. Unclassified *Beijerinckiaceae, Beijerinckia, Sphingomonas,* unclassified *Sphingomonadaceae, Flavobacterium,* unclassified *Rhizobiales*, and *Nevskia* were also statistically linked with biotransformation of 5-fluorouracil in this study.

9.3.5 Beta Blockers

9.3.5.1 Mechanism of Action

Beta-blockers are used for the treatment of cardiovascular diseases, hypertension and other conditions. Three forms of beta receptors are prevalent in human body: β-1, β-2 and β-3. β-1 receptors are located in the heart and are responsible for mediation of cardiac activity. β-2 receptors control several features of metabolic activity and coordinate smooth muscle relaxation. β-3 receptors mainly prompt the breakdown of fat cells. Binding to β-1 receptors increases the cardiac automaticity and induces renin release leading to an elevated blood pressure. Binding to β-2 receptors causes relaxation of the smooth muscles with increased metabolic effects like glycogenolysis. These effects are inhibited by binding of beta blockers to β-1 and β-2 receptors. Therefore, the chronotropic and inotropic effects on the heart undergo inhibition leading to decrease in the heart rate. Beta-blockers reduce the blood pressure through several mechanisms which includes lowered renin and cardiac output (Farzam and Jan 2021).

9.3.5.2 Microbial Degradation of Beta-Blockers Using Pure Cultures

Carvedilol is an example of non-selective agent which binds to both β-1 and β-2 receptors. An 83% removal of carvedilol was reported after a period of 120 h using fungus *Cunninghamella echinulata*, where hydroxylated carvedilol metabolites were detected as the major biotransformation products. Two hydroxylated intermediates were identified as hydroxyphenyl-carvedilol and hydroxycarbazolyl-carvedilol whereas another metabolite was reported to be N-glucose-carbazolyl-carvedilol (Zawadzka et al. 2017).

9.3.5.3 Microbial Degradation of Beta-Blockers Using Mixed Cultures

Atenolol and metoprolol are β-1 receptor-selective blockers and are cardio-selective. Atenolol was approximately 50% biotransformed cometabolically using an enriched nitrifying culture in 240 h. The products formed during this biotransformation with ammonia oxidation included atenolol acid; 1-isopropylamino-2-propanol; 1-amino-3-phenoxy-2-propanol and an unknown product (Xu et al. 2017). The removal efficiencies of atenolol using conventional activated sludge (CAS) sludge and membrane bioreactor (MBR) sludge was also compared where MBR sludge showed better biodegradation with complete removal in 2 days. The biodegradation product atenololic acid was formed as a result of bacterial hydrolysis of amide bond which was also completely degraded under 20 days in case of MBR sludge (Radjenovic et al. 2008). On investigating the enantiomeric degradation of atenolol using activated sludge in presence and absence of acetate, 80% and 50% non-enantioselective biodegradation was reported respectively in 15 days (Ribeiro et al. 2013). Biodegradation of atenolol was also related to the activity of ammonia-oxidizing bacteria and heterotrophs. This was because of approximately 80% atenolol biodegradation in 25 h using nitrification enrichment culture under ammonia oxidation conditions (Sathyamoorthy et al. 2013).

Metoprolol was also degraded using different consortia by Velázquez and Nacheva (2017) as was seen for fluoxetine. The ammonium-nitrite-oxidizing consortium showed the highest biotransformation of metoprolol (64.5%) along with the lowest sorption in the biomass of metoprolol. The heterotrophic bacteria degraded 58.4% of metoprolol while the nitrite-oxidizing consortium biodegraded 42.7% metoprolol at a retention time of 24 days (Velázquez and Nacheva 2017). Both enantiomers of metoprolol were also degraded using activated sludge in presence and absence of acetate. The (S)-form of metoprolol was degraded at higher extents as compared to the (R)-form. The presence of acetate led to a slight elevation in the biodegradation

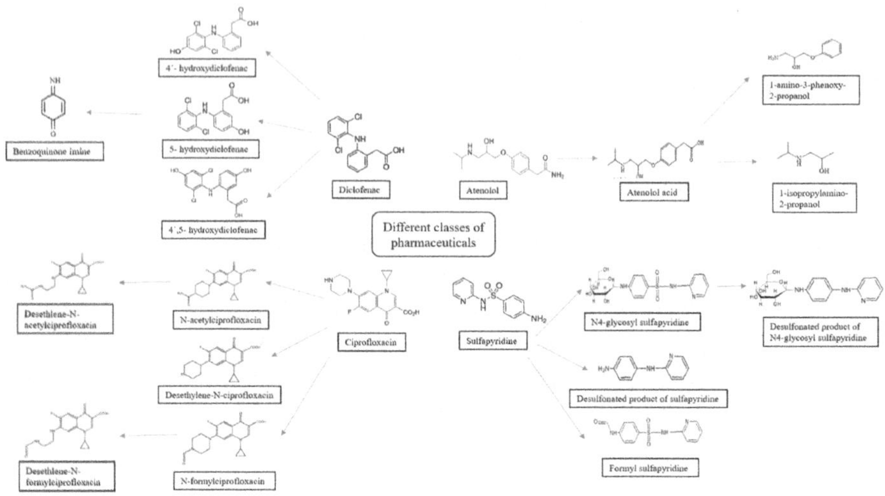

Fig. 9.2 Biodegradation pathways for representatives of major drug classes (NSAIDs-diclofenac, Beta-blockers-atenolol, antibiotics-ciprofloxacin, Antineoplastics-sulfapyridine)

rate of both the (S)-form and the (R)-form. 50–60% removal was observed for both forms in presence of acetate in 15 days (Ribeiro et al. 2013).

Figure 9.2 summarizes the degradation pathways of representatives of major drug classes.

9.3.6 Illicit Drugs

Phenylacetone is a precursor used in the manufacture of methamphetamine and amphetamine, where it is generally known as P2P (phenyl-2-propanone). Amphetamine is administered for the treatment of attention-deficit/hyperactivity disorders (ADHD) and central nervous system disorders like narcolepsy, but can also be used as an illicit drug. Amphetamine increases the extracellular dopamine and lengthens the dopamine receptor signalling in the striatum. P2P was known to be degraded to half of its initial concentration in about 4–8 days in four soil samples containing soil microbes, whereas in the same study, at least two-thirds of the methylamphetamine sulphate still persisted after 6 weeks. Six metabolites were generated during the degradation process, namely 1-phenyl-2-propanol; 1-phenyl-1,2-propanedione; 1-hydroxy-1-phenyl-2-propanone; 2-hydroxy-1-phenyl-1-propanone; syn- and anti-1- phenyl-1,2-propanediol. However, the degradation process was not explained in detail (Janusz et al. 2003).

Morphine is administered for managing chronic, moderate to severe pain by activating descending inhibitory pathways of the central nervous system (CNS) and inhibiting the nociceptive afferent neurons of the peripheral nervous system (PNS).

Pseudomonas putida M10 was capable of using morphine as a sole carbon and energy source. Morphinone was reported to be the key metabolite formed due to the oxidation of morphine catalysed by morphine dehydrogenase during the degradation of morphine by *P. putida* M10 (Bruce et al. 1990).

Chloroform earlier used as an anaesthetic, was almost completely degraded (96%) in 24 h using Methylosinus trichosporium OB3b which was grown under copper limitation (Oldenhuis et al. 1989). Greater than 93% of a 3 mg/L concentration of chloroform was also degraded using *Rhodococcus rhodochrous* ATCC21197. In the same study, 72% of an 8 mg/L concentration of chloroform was also degraded when present in a mixture of aromatic and chlorinated aliphatic compounds (Malachowsky et al. 1994). *Methylosinus trichosporium* OB3b, *Pseudomonas butanovora*, CF8 (a butane-grown enrichment culture), and *Mycobacterium vaccae* JOB5 readily degraded chloroform. *P. butanovora, M. vaccae* JOB5 and CF8 were capable of degrading chloroform in the range of 61–97% after 60 min which was similar to the degradation achieved by *M. trichosporium* OB3b (Oldenhuis et al. 1989; Hamamura et al. 1997).

9.4 Limitations in Execution to WWTPs

The applications of pharmaceutical microbial degradation at pilot scale or large wastewater treatment plants are limited. The execution and scaling up of this process in WWTPs might face several limitations. Several factors such as type of treated wastewater, type of microbes employed for degradation (pure or mixed culture), environmental conditions provided to the microbes during treatment, etc. might act as constraints for efficient removal of pharmaceuticals. Unfavourable conditions for the maintenance of pure culture conditions or even a defined consortium because of transient conditions in WWTPs is also an irredeemable factor during scaling up. For example, any significant variation in the temperature or pH of the incoming wastewater might prove detrimental to the desired microbial community during treatment.

In presence of antibiotics, the survival of pure or mixed cultures become even more challenging because of the bacteriostatic or bacteriocidal nature of antibiotics. Moreover, antimicrobial resistance developed from the antibiotic cocktail present in the wastewater is a huge concern. The spread of these developed resistant bacteria may lead to infections which are not curable by the available medical regime.

The location of the treatment reactor along with the hydraulic retention time (HRT) are important aspects for the treatment process. Significant sterility is necessary for the maintenance of microbial culture in the reactor which will be dependent on the location of the reactor. Moreover, the biodegradation pathways inside the reactor cannot be predicted easily since it might be a culmination of different factors which can be different at different times. The lab studies cannot be extrapolated to WWTPs successfully because the microbes responsible for degradation in lab studies when introduced in an operational WWTPs are less effective or completely useless because of the highly heterogeneous mixtures of micropollutants and varied microbial consortia. Another repercussion of this can be complete failure of the enzyme systems responsible for microbial degradation.

The need of a supplementary carbon source during cometabolism of drugs should also be considered, as its absence may render the process unfeasible. The limitation of studies regarding degradation of drugs in presence of metabolites is also a concerning issue because degradation metabolites and parent drug exists simultaneously in WWTPs. Therefore, the role of feedback inhibition cannot be ignored in these conditions. Therefore, further research regarding the in vivo conditions during treatment needs to be carried out for successful implementation of this drug removal process.

9.5 Conclusion

Pharmaceuticals are viewed as an emerging contaminant of concern because of their possible implications on the environment as well as human health. Microbial degradation of pharmaceuticals has become a centre of attraction for the scientific community because of its obvious advantages over traditional removal methods. This chapter explored the microbial degradation of different classes of pharmaceuticals using both pure and mixed microbial cultures. Hydroxylation by cytochrome P450 was observed to be the key step in degradation, though in case of naproxen and ciprofloxacin, purified laccase was also assumed to be the major mechanism based on the results obtained with the addition of 1-hydroxybenzotriazol to purified laccase. Other processes like decarboxylation, dechlorination, oxidation, demethylation, dehydrogenation and many more are also employed by these microbes for biodegradation of drugs. Different enzymes were upregulated during these degradation pathways. After a critical analysis of microbial degradation of drugs, several hindrances were thought to be prevalent during the execution of the process to field-scale WWTPs. Future areas of research for full-scale operation of the process commissioning microbial degradation were also recommended.

References

Aissaoui S, Ouled-Haddar H, Sifour M, Beggah C, Benhamada F (2017) Biological removal of the mixed pharmaceuticals: diclofenac, ibuprofen, and sulfamethoxazole using a bacterial consortium. Iranian J Biotech 15:e1530

Almeida B, Kjeldal H, Lolas I, Knudsen AD, Carvalho G, Nielsen KL, Barreto Crespo MT, Stensballe A, Nielsen JL (2013a) Quantitative proteomic analysis of ibuprofen degrading Patulibacter sp. strain I11. Biodegradation 24:615–630

Almeida B, Oehmen A, Marques R, Brito D, Carvalho G, Barreto Crespo MT (2013b) Modelling the biodegradation of non-steroidal anti-inflammatory drugs (NSAIDs) by activated sludge and a pure culture. Biores Technol 133:31–37

Amorim CL, Moreira IS, Maia AS, Tiritan ME, Castro PML (2014) Biodegradation of ofloxacin, norfloxacin, and ciprofloxacin as single and mixed substrates by Labrys portucalensis F11. Appl Microbiol Biotechnol 98:3181–3190. https://doi.org/10.1007/s00253-013-5333-8

Aracagök YD, Göker H, Cihangir NZ, Naturforsch (2017) Biodegradation of micropollutant naproxen with a selected fungal strain and identification of metabolites. Z. Naturforsch. 72:173–179

Balciunas ME, Kappelmeyer U, Harms H, Heipieper HJ (2020) Increasing ibuprofen degradation in constructed wetlands by bioaugmentation with gravel containing biofilms of an ibuprofen-degrading Sphingobium yanoikuyae. Eng Life Sci 20:160–167

Bankole PO, Adekunle AA, Jeon BH, Govindwar SP (2020) Novel cobiomass degradation of NSAIDs by two wood rot fungi, Ganoderma applanatum and Laetiporus sulphureus: ligninolytic enzymes induction, isotherm and kinetic studies. Ecotoxicol Environ Saf 203:110997

Bessa VS, Moreira IS, Tiritan ME, Castro PML (2017) Enrichment of bacterial strains for the biodegradation of diclofenac and carbamazepine from activated sludge. Int Biodeterior Biodegradation 120:135–142

Bragança I, Danko AS, Pacheco J, Frascari D, Delerue-Matos C, Domingues VF (2016) Cometabolic degradation of anti-inflammatory and analgesic pharmaceuticals by a pentane enrichment culture. Water Air Soil Pollut 227:227. https://doi.org/10.1007/s11270-016-2933-9

Brezovšek P, Eleršek T, Filipič M (2014) Toxicities of four anti-neoplastic drugs and their binary mixtures tested on the green alga Pseudokirchneriella subcapitata and the cyanobacterium Synechococcus leopoliensis. Water Res 52:168–177. https://doi.org/10.1016/j.watres.2014.01.007

Brozinski JM, Lahti M, Meierjohann A, Oikari A, Kronberg L (2013) The anti-inflammatory drugs diclofenac, naproxen and ibuprofen are found in the bile of wild fish caught downstream of a wastewater treatment plant. Environ Sci Technol 47:342–348

Bruce NC, Wilmot CJ, Jordan KN, Trebilcock AE, Stephens LDJ, Lowe CR (1990) Microbial degradation of the morphine alkaloids: identification of morphinone as an intermediate in the metabolism of morphine by *Pseudomonas putida* MI0. Arch Microbiol 154:465–470

Cashman JN (1996) The mechanism of action of NSAIDs in analgesia. Drugs 52:13–23

Cherik D, Benali M, Louhab K (2015) Occurrence, ecotoxicology, removal of diclofenac by adsorption on activated carbon and biodegradation and its effect on bacterial community: a review. World Sci News 10:116–144

Cruz-Ornelas R, Sanchez-Vazquez JE, Amaya-Delgado L, Guillen-Navarro K, Calixto-Romo A (2019) Biodegradation of NSAIDs and their effect on the activity of ligninolytic enzymes from Pleurotus djamor. Biotech 9:373. https://doi.org/10.1007/s13205-019-1904-4

Čvančarová M, Moeder M, Filipova A, Cajthaml T (2015) Biotransformation of fluoroquinolone antibiotics by ligninolytic fungi—metabolites, enzymes and residual antibacterial activity. Chemosphere https://doi.org/10.1016/j.chemosphere.2014.12.012

Dawas-Massalha A, Gur-Reznik S, Lerman S, Sabbah I, Dosoretz CG (2014) Cometabolic oxidation of pharmaceutical compounds by a nitrifying bacterial enrichment. Biores Technol 167:336–342

de Jesus Gaffney V, Almeida CM, Rodrigues A, Ferreira E, Benoliel MJ, Cardoso VV (2015) Occurrence of pharmaceuticals in a water supply system and related human health risk assessment. Water Res 72:199–208

DeAngelis LM, Posner JB (2009) Neurologic complications of cancer. Chapter 12-Side effects of chemotherapy

Ding T, Lin K, Yang B, Yang M, Li J, Li W, Gan J (2017) Biodegradation of Naproxen by freshwater algae Cymbella sp. and Scenedesmus quadricauda and the comparative toxicity. Biores Technol 238:164–173. https://doi.org/10.1016/j.biortech.2017.04.018

Domaradzka D, Guzik U, Hupert-Kocurek K, Wojcieszyńska D (2015) Cometabolic degradation of naproxen by Planococcus sp. strain S5. Water Air Soil Pollut 226:297

Farzam K, Jan A (2021). Beta blockers. StatPearls [Internet]. Treasure Island (FL): StatPearls Publishing. https://www.ncbi.nlm.nih.gov/books/NBK532906/

Ferrando-Climent L, Cruz-Morató C, Marco-Urrea E, Vicent T, Sarrà M, Rodriguez-Mozaz S, Barceló D (2015) Non conventional biological treatment based on Trametes versicolor for the elimination of recalcitrant anticancer drugs in hospital wastewater. Chemosphere 136:9–19

Ferrari B, Mons R, Vollat B, Fraysse B, Paxēaus N, Giudice RL, Pollio A, Garric J (2004) Environmental risk assessment of six human pharmaceuticals: are the current environmental risk assessment procedures sufficient for the protection of the aquatic environment? Environ Toxicol Chem 23:1344–1354

Fick J, Soderstrom H, Lindberg RH, Phan C, Tysklind M, Larsson DGJ (2009) Contamination of surface, ground, and drinking water from pharmaceutical production. Environ Toxicol Chem 28:2522–2527

Garcia-Medina AL, Galar-Martínez M, Garcia-Medina S, Gómez-Oliván LM, Razo-Estrada C (2015) Naproxen-enriched artificial sediment induces oxidative stress and genotoxicity in *Hyalella azteca*. Water Air Soil Pollut 226:195

Gauthier H, Yargeau V, Cooper DG (2010) Biodegradation of pharmaceuticals by *Rhodococcus rhodochrous* and *Aspergillus niger* by co-metabolism. Sci Total Environ 408:1701–1706

Gonzales-Rey M, Bebianno MJ (2012) Does non-steroidal anti-inflammatory (NSAID) ibuprofen induce antioxidant stress and endocrine disruption in mussel *Mytilus galloprovincialis*? Environ Toxicol Pharmacol 33:361–371

Górny D, Guzik U, Hupert-Kocurek K, Wojcieszyńska D (2019) A new pathway for naproxen utilisation by Bacillus thuringiensis B1(2015b) and its decomposition in the presence of organic and inorganic contaminants. J Environ Manage 239:1–7

Ha H, Mahanty B, Yoon S, Kim CG (2016) Degradation of the long-resistant pharmaceutical compounds carbamazepine and diatrizoate using mixed microbial culture. J Environ Sci Health, Part A 51:467–471. https://doi.org/10.1080/10934529.2015.1128712

Haap T, Triebskorn R, Kohler HR (2008) Acute effects of diclofenac and DMSO to Daphnia magna: immobilisation and hsp70-induction. Chemosphere 73:353–359

Hamamura N, Page C, Long T, Semprini L, Arp DJ (1997) Chloroform cometabolism by Butane-Grown CF8, *Pseudomonas butanovora*, and *Mycobacterium vaccae* JOB5 and Methane-Grown *Methylosinus trichosporium* OB3b. Appl Environ Microbiol 63:3607–3613

Han S, Choi K, Kim J, Ji K, Kim S, Ahn B, Yun J, Choi K, Khim JS, Zhang X, Giesy JP (2010) Endocrine disruption and consequences of chronic exposure to ibuprofen in Japanese medaka (*Oryzias latipes*) and freshwater cladocerans *Daphnia magna* and *Moina macrocopa*. Aquat Toxicol 98:256–264

Hata T, Kawai S, Okamura H, Nishida T (2010) Removal of diclofenac and mefenamic acid by the white rot fungus Phanerochaete sordida YK-624 and identification of their metabolites after fungal transformation. Biodegradation 21:681–689

Ismail MM, Essam TM, Ragab YM, Mourad FE (2016) Biodegradation of ketoprofen using a microalgal–bacterial consortium. Biotechnol Lett 38:1493–1502

Ivshina IB, Tyumina EA, Kuzmina MV, Vikhareva EV (2019) Features of diclofenac biodegradation by Rhodococcus ruber IEGM 346. Sci Rep 9:9159

Jałowiecki Ł, Płaza G, Ejhed H, Nawrotek M (2019) Aerobic biodegradation of norfloxacin and ofloxacin by a microbial consortium. Arch Environ Prot 45:40–47

Janusz A, Kirkbride KP, Scott TL, Naidu R, Perkins MV, Megharaj M (2003) Microbial degradation of illicit drugs, their precursors, and manufacturing by-products: implications for clandestine drug laboratory investigation and environmental assessment. Forensic Sci Int 134:62–71

Jiang YH, Li MX, Guo CS, An D, Xu J, Zhang Y, Xi BD (2014) Distribution and ecological risk of antibiotics in a typical effluent-receiving river (Wangyang River) in north China. Chemosphere 112:267–274

Kapoor G, Saigal S, Elongavan A (2017) Action and resistance mechanisms of antibiotics: a guide for clinicians. J Anaesthesiol Clin Pharmacol 33:300–305. https://doi.org/10.4103/joacp.JOACP_349_15

Kasprzyk-Hordern B, Dinsdale RM, Guwy AJ (2008) The occurrence of pharmaceuticals, personal care products, endocrine disruptors and illicit drugs in surface water in South Wales, UK. Water Res 42:3498–3518

Kim DW, Heinze TM, Kim BS, Schnackenberg LK, Woodling KA, Sutherland JB (2011) Modification of Norfloxacin by a *Microbacterium* sp. Strain Isolated from a wastewater treatment plant. Appl Environ Microbiol 77:6100–6108

Kondor AC, Jakab G, Vancsik A, Filep T, Szeberényi J, Szabó L, Maász G, Ferincz Á, Dobosy P, Szalai Z (2020) Occurrence of pharmaceuticals in the Danube and drinking water wells: efficiency of riverbank filtration. Environ Pollut 265:114893

Koopaei NN, Abdollahi M (2017) Health risks associated with the pharmaceuticals in wastewater. DARU J Pharma Sci 25:9

Kózka B, Nałęcz-Jawecki G, Turło J, Giebułtowicz J (2020) Application of *Pleurotus ostreatus* to efficient removal of selected antidepressants and immunosuppressant. J Environ Manag 273:111131

Kwon JW, Armbrust K (2006) Laboratory persistence and fate of fluoxetine in aquatic environments. Environ Toxicol Chem 25(10):2561–2568

Li A, Cai R, Cui D, Qiu T, Pang C, Yang J, Ma F, Ren N (2013) Characterization and biodegradation kinetics of a new cold-adapted carbamazepine-degrading bacterium, *Pseudomonas* sp. CBZ-4. J Environ Sci 25:2281–2290

Li X, de Toledo RA, Wang S, Shim H (2015) Removal of carbamazepine and naproxen by immobilized Phanerochaete chrysosporium under non-sterile condition. New Biotechnol 32:282–289. https://doi.org/10.1016/j.nbt.2015.01.003

Liyanage GY, Manage P (2016) Evaluation of amoxicillin and sulfonamide removal by Bacillus cereus, Enterobacter ludwigii and Enterobacter sp. Environ Nat Resour J 14:39–43. https://ph02.tci-thaijo.org/index.php/ennrj/article/view/68430

Liyanage GY, Manage PM (2018) Removal of Ciprofloxacin (CIP) by bacteria isolated from hospital effluent water and identification of degradation pathways. Int J Med, Pharm Drug Res 2:37–47. https://doi.org/10.22161/ijmpd.2.3.1

López-Serna R, Posadas E, García Encina PA, Muñoz R (2019) Removal of contaminants of emerging concern from urban wastewater in novel algal-bacterial photobioreactors. Sci Total Environ 662:32–40

Lu Z, Sun W, Li C, Ao X, Yang C, Li S (2019) Bioremoval of non-steroidal anti-inflammatory drugs by Pseudoxanthomonas sp. DIN-3 isolated from biological activated carbon process. Water Res 161:459–472

Malachowsky KJ, Phelps TJ, Teboli AB, Minnikin DE, White DC (1994) Aerobic mineralization of trichloroethylene, vinyl chloride, and aromatic compounds by Rhodococcus species. Appl Environ Microbiol 60:542–548

Mao F, Liu X, Wu K (2018) Biodegradation of sulfonamides by *Shewanella oneidensis* MR-1 and *Shewanella* sp. strain MR-4. Biodegradation 29:129–140. https://doi.org/10.1007/s10532-017-9818-5

Marchlewicz A, Domaradzka D, Guzik U, Wojcieszyńska D (2016) Bacillus thuringiensis b1(2015b) is a gram-positive bacteria able to degrade naproxen and ibuprofen. Water Air Soil Pollut 227:197

Marchlewicz A, Guzik U, Smułek W, Wojcieszyńska D (2017) Exploring the degradation of ibuprofen by Bacillus thuringiensis B1(2015b): the new pathway and factors affecting degradation. Molecules 22:1676

Marco-Urrea E, Pérez-Trujillo M, Vicent T, Caminal G (2009) Ability of white-rot fungi to remove selected pharmaceuticals and identification of degradation products of ibuprofen by Trametes versicolor. Chemosphere 74:765–772

Marco-Urrea E, Pérez-Trujillo M, Cruz-Morató C, Caminal G, Vicent T (2010a) Degradation of the drug sodium diclofenac by Trametes versicolor pellets and identification of some intermediates by NMR. J Hazard Mater 176:836–842

Marco-Urrea E, Pérez-Trujillo M, Cruz-Morató C, Caminal G, Vicent T (2010b) White rot fungus-mediated degradation of the analgesic ketoprofen and identification of intermediates by HPLC–DAD–MS and NMR. Chemosphere 78:474–481

Marco-Urrea E, Pérez-Trujillo M, Blánquez P, Vicent T, Caminal G (2010c) Biodegradation of the analgesic naproxen by Trametes versicolor and identification of intermediates using HPLC-DAD-MS and NMR. Biores Technol 101:2159–2166

Moreira IS, Bessa VS, Murgolo S, Piccirillo C, Mascolo G, Castro PML (2018) Biodegradation of Diclofenac by the bacterial strain Labrys portucalensis F11. Ecotoxicol Environ Saf 152:104–113

Moreira IS, Ribeiro AR, Afonso CM, Tiritan ME, Castro PML, (2014) Enantioselective biodegradation of fluoxetine by the bacterial strain Labrys portucalensis F11. Chemosphere 111:103–111

Murdoch RW, Hay AG (2005) Formation of Catechols via removal of acid side chains from ibuprofen and related aromatic acids. Appl Environ Microbiol, 6121–6125

Murdoch RW, Hay AG (2015) The biotransformation of ibuprofen to trihydroxyibuprofen in activated sludge and by Variovorax Ibu-1. Biodegradation 26:105–113

Nguyen LN, Nghiem LD, Pramanik BK, Oh S (2019) Cometabolic biotransformation and impacts of the anti-inflammatory drug diclofenac on activated sludge microbial communities. Sci Total Environ 657:739–745

Oldenhuis R, Vink RLJM, Janssen DB, Witholt B (1989) Degradation of chlorinated aliphatic hydrocarbons by *Methylosinus trichosporium* OB3b expressing soluble methane monooxygenase. Appl Environ Microbiol 55:2819–2826

Osafo N, Agyare C, Obiri DD, Antwi AO (2017) Mechanism of action of nonsteroidal anti-inflammatory drugs. Nonsteroidal anti-inflammatory drugs-Chapter 2, pp 13

Palli L, Castellet-Rovira F, Perez-Trujillo M, Caniani D, Sarra-Adroguer M, Gori R (2017) Preliminary evaluation of Pleurotus ostreatus for the removal of selected pharmaceuticals from hospital wastewater. Biotechnol Prog 33:1529–1537

Palma TL, Costa MC (2021) Anaerobic biodegradation of fluoxetine using a high-performance bacterial community. Anaerobe 68:102356

Pan L, Li J, Li C, Tang X, Yu G, Wang Y (2018) Study of ciprofloxacin biodegradation by a Thermus sp. isolated from pharmaceutical sludge. J Hazard Mater 343:59–67

Parshikov IA, Freeman JP, Lay JO Jr, Beger RD, Williams AJ, Sutherland JB (1999) Regioselective transformation of cipro£oxacin to N-acetylcipro£oxacin by the fungus Mucor ramannianus. FEMS Microbiol Lett 177:131–135

Peña OIG, Zavala MÁL, Ruelas HC (2021) Pharmaceuticals market, consumption trends and disease incidence are not driving the pharmaceutical research on water and wastewater. Int J Environ Res Public Health 18:2532. https://doi.org/10.3390/ijerph18052532

Peng X, Ou W, Wang C, Wang Z, Huang Q, Jin J, Tan J (2014) Occurrence and ecological potential of pharmaceuticals and personal care products in groundwater and reservoirs in the vicinity of municipal landfills in China. Sci Total Environ 490:889–898

Quintana JB, Weiss S, Reemtsma T (2005) Pathways and metabolites of microbial degradation of selected acidic pharmaceutical and their occurrence in municipal wastewater treated by a membrane bioreactor. Water Res 39:2654–2664

Rabiet M, Togola A, Brissaud F, Seidel JL, Budzinski H, Elbaz-Poulichet F (2006) Consequences of treated water recycling as regards pharmaceuticals and drugs in surface and ground waters of a medium-sized Mediterranean catchment. Environ Sci Technol 40:5282–5288

Radjenovic J, Pereza S, Petrović M, Barcelo D (2008) Identification and structural characterization of biodegradation products of atenolol and glibenclamide by liquid chromatography coupled to hybrid quadrupole time-of-flight and quadrupole ion trap mass spectrometry. J Chromatography A 1210:142–153

Ribeiro AR, Afonso CM, Castro PML, Tiritan ME (2013) Enantioselective biodegradation of pharmaceuticals, alprenolol and propranolol, by an activated sludge inoculum. Ecotoxicol Environ Saf 87:108–114

Rodarte-Morales AI, Feijoo G, Moreira MT, Lema JM (2011) Degradation of selected pharmaceutical and personal care products (PPCPs) by white-rot fungi. World J Microbiol Biotechnol 27:1839–1846. https://doi.org/10.1007/s11274-010-0642-x

Rodríguez-Rodrígueza CE, García-Galánb MJ, Blánquezd P, Díaz-Cruzb MS, Barcelób D, Caminala G, Vicent T (2012) Continuous degradation of a mixture of sulfonamides by Trametes versicolor and identification of metabolites from sulfapyridine and sulfathiazole. J Hazard Mater 213–214:347–354

Rossmassler K, Kim S, Broeckling CD, Galloway S, Prenni J, De Long SK (2019) Impact of primary carbon sources on microbiome shaping and biotransformation of pharmaceuticals and personal care products. Biodegradation 30:127–145. https://doi.org/10.1007/s10532-019-09871-0

Sathyamoorthy S, Chandran K, Ramsburg CA (2013) Biodegradation and cometabolic modeling of selected Beta blockers during ammonia oxidation. Environ Sci Technol 47:12835–12843

Scheurell M, Franke S, Shah RM, Huhnerfuss H (2009) Occurrence of diclofenac and its metabolites in surface water and effluent samples from Karachi, Pakistan. Chemosphere 77:870–876

Schwaiger J, Ferling H, Mallow U, Wintermayr H, Negele RD (2004) Toxic effects of the non-steroidal anti-inflammatory drug diclofenac. Part I: histopathological alterations and bioaccumulation in rainbow trout. Aquat Toxicol 68:141–150

Schwarzenbach RP, Escher BI, Fenner K (2006) The challenge of micropollutants in aquatic systems. Science 313:1072–1077

Shanmugam G, Sampath S, Selvaraj KK (2014) Non-steroidal anti-inflammatory drugs in Indian rivers. Environ Sci Pollut Res 21:921–931. https://doi.org/10.1007/s11356-013-1957-6

Sharma KB (2017) Antidepressants: mechanism of action, toxicity and possible amelioration. J Appl Biotechnol Bioeng. 3:437–448. https://doi.org/10.15406/jabb.2017.03.00082

Sim WJ, Lee JW, Lee ES, Shin SK, Hwang SR, Oh JE (2011) Occurrence and distribution of pharmaceuticals in wastewater from households, livestock farms, hospitals and pharmaceutical manufactures. Chemosphere 82:179–186

Sodhi KK, Kumar M, Singh DK (2020) Potential application in amoxicillin removal of Alcaligenes sp. MMA and enzymatic studies through molecular docking. Arch Microbiol 202:1489–1495. https://doi.org/10.1007/s00203-020-01868-1

Stanley LA (2017) Chapter 27-drug metabolism. *Pharmacognosy* Fundam, Appl Strat, 527–545. https://doi.org/10.1016/B978-0-12-802104-0.00027-5

Stylianou K, Hapeshi E, Vasquez M, Fatta-Kassinos D, Vyrides I (2018) Diclofenac biodegradation by newly isolated Klebsiella sp. KSC: microbial intermediates and ecotoxicological assessment. J Environ Chem Eng 6:3242–3248. https://doi.org/10.1016/j.jece.2018.04.052

Tillement JP, Tremblay D (2007) 5.02—Clinical pharmacokinetic criteria for drug research. Compr Med Chem II 5:11–30

Tufail A, Price W, Hai F (2020) A critical review on advanced oxidation processes for the removal of trace organic contaminants: a voyage from individual to integrated processes. Chemosphere 260:127460

Vane JR (1971) Inhibition of prostaglandin synthesis as a mechanism of action for aspirin-like drugs. Nat New Biol 231:232–235

Velázquez YF, Nacheva PM (2017) Biodegradability of fluoxetine, mefenamic acid, and metoprolol using different microbial consortiums. Environ Sci Pollut Res 24:6779–6793

Vieno N, Sillanpää M (2014) Fate of diclofenac in municipal wastewater treatment plant—a review. Environ Int 69:28–39

Vulliet E, Cren-Olive C (2011) Screening of pharmaceuticals and hormones at the regional scale, in surface and groundwaters intended to human consumption. Environ Pollut 159:292–293

Wasify R, Hamed S (2014) Bacterial biodegradation of crude oil using local isolates. Int J Bacteriol, 1–8

Wetzstein HG, Stadler M, Tichy HV, Dalhoff A, Karl W (1999) Degradation of ciprofloxacin by basidiomycetes and identification of metabolites generated by the brown rot fungus *Gloeophyllum striatum*. Appl Environ Microbiol 65:1556–1563

Wojcieszynska D, Domaradzka D, Hupert-Kocurek K, Guzik U (2014) Bacterial degradation of naproxen—undisclosed pollutant in the environment. J Environ Manage 145:157–161

Xu Y, Radjenovic J, Yuan Z, Ni BJ (2017) Biodegradation of atenolol by an enriched nitrifying sludge: products and pathways. https://doi.org/10.1016/j.cej.2016.11.153

Zawadzka K, Przemysł B, Felczak A, Lisowska K (2017) Microbial detoxification of carvedilol, a ß-adrenergic antagonist, by the filamentous fungus *Cunninghamella echinulate*. Chemosphere. https://doi.org/10.1016/j.chemosphere.2017.05.088

Żur J, Piński A, Wojcieszyńska D, Smułek W, Guzik U (2020) Diclofenac degradation—enzymes, genetic background and cellular alterations triggered in diclofenac-metabolizing strain Pseudomonas moorei KB4. Int J Mol Sci 21:6786

Chapter 10
Application of Biochar for Removal of Emerging Contaminants

Bikram Mishra, Mohit Aggarwal, and Neelancherry Remya

Abstract Biochar is a carbon-rich product obtained under reducing thermal conditions by the decomposition of biomass. The feedstock characteristics and production methods are the key factors in biochar production. In addition, the modification by different methods improved the morphological and physicochemical properties of biochar and offered diversified use in wastewater treatment. Biochar is extensively used in the removal of emerging contaminants (ECs) like personal care products (PCPs) and pharmaceutically active compounds (PhACs), which are not removed well during conventional wastewater treatment. The status quo of functionalization and characterization of the biochar and recent advancements in the use of as-derived or modified biochar in ECs removal from wastewater will be discussed in this chapter.

Keywords Biomass · Thermal decoposition · Biochar · Modification · Emerging contaminants · Adsorption · Removal

10.1 Introduction

In the last few decades, the consumption of essential commodities have increased due to increase in industrialization, population, and urbanization. Various human activities such as farming and manufacturing, generate many chemicals and their by-products. Advancements in new analytical methods have enabled us to detect these chemicals and their by-products in water bodies and the environment, even in trace levels (Dsikowitzky and Schwarzbauer 2014). Emerging contaminants (ECs) are a combination of synthetic, natural compounds and their by-products such as personal care products (PCPs), pharmaceutically active compounds (PhACs), and endocrine-disrupting chemicals (EDCs). They are found increasingly in water bodies in the last few decades (Saidulu et al. 2021). They can cause damage to the environment, ecosystems, by biological magnification and bioaccumulation (Zenker et al. 2014). It is challenging to control their spread, as they are an essential part of different

---

B. Mishra · M. Aggarwal · N. Remya (✉)
School of Infrastructure, Indian Institute of Technology Bhubaneswar, Argul, Odisha, India
e-mail: remya@iitbbs.ac.in

S. P. Singh et al. (eds.), *New Trends in Emerging Environmental Contaminants*,
Energy, Environment, and Sustainability,
https://doi.org/10.1007/978-981-16-8367-1_10

daily products. There is no toxicological data and suitable guidelines present for the management of ECs. The ECs in wastewater were further augmented in the covid pandemic due to increased usage of medicines, disinfectants, and cleaning agents.

The amount of ECs in the water bodies depends on various factors such as per capita water consumption, water usage pattern, sewer conditions, and environmental persistence. (Tran and Gin 2017; Patel et al. 2019). ECs concentrations in the water bodies are in the range of from ng/L to μg/L. Despite being quite low, if exposed for a long time, they can cause huge damage to the health and safety of the marine ecosystem and the living beings (Lutterbeck et al. 2020; Xiang et al. 2021).

Biochar is a carbon-rich material produced by pyrolysis wherein biomass is decomposed at a higher temperature (~700 °C) in the absence of oxygen. Biochar properties are determined by the feedstock characteristics, residence time, carrier gases, pyrolysis temperature, catalysts, and other pyrolysis conditions. The raw biochar is modified to improve its physicochemical properties and adsorption of pollutants on it. The mechanism of adsorption of pollutants on biochar depends on the type of pollutant and biochar (Cheng et al. 2021). ECs removal using biochar, an environmentally friendly adsorbent, has gained importance in recent years. A summary of sources, classification, effects of ECs, along with the treatment methods for ECs, biochar production, modification, and application for ECs removal is discussed in this chapter.

10.2 Sources, Classification, and Fate of Emerging Contaminants

ECs enter into the environment through different sources. The general classification of ECs and their sources are shown in Table 10.1. The sources are classified into five categories: effluent from industries, domestic wastewater, agricultural activities, animal husbandries, and hospitals. Effluent from chemical industries, pharmaceutical manufacturing units, pesticides, and fertilizers industries are significant contributors to ECs. Domestic wastewater is one of the major sources of ECs such as PCPs, PhACs, pesticides, and fertilizers. PCPs includes shampoos, body lotion, sunscreen lotion, cleansing solvents, and toothpaste (Yang et al. 2017). The medicines are excreted out of the human body in unchanged or metabolized forms into wastewater streams. Hospital effluents contain drugs, antibiotic-resistant genes, and pharmaceutical metabolites (Tiwari et al. 2017). Animal husbandries use medication to keep the animals healthy, which appears in the excreta and enters the environment (Barbosa et al. 2016). Agricultural runoff is the major carrier of pesticides and fertilizers. Some other minor sources of ECs include landfill leachate, reclaimed water for irrigation, and aquaculture discharge.

ECs discharged into the environment migrate to the aquatic environment through different pathways. The ECs concentration in the aquatic bodies varies primarily due to treatment efficiency, dilution, environmental persistency, etc. The ECs concentration in wastewater treatment plants depends on human usage patterns and seasonal

Table 10.1 Classification and sources of ECs (Rout et al. 2021; Shahid et al. 2021)

Class	Important sub-class	Representative ECs	Major sources
PhACs	Antidepressant, Beta-blockers, Hormones, Antibiotics, Nonsteroidal anti-inflammatory drugs (NSAID), Anticonvulsants	Diazepam, Metoprolol, Ciprofloxacin, Diclofenac, Carbamazepine, Clorfibric acid, Testosterone	Hospital effluent, Pharmaceutical industry effluent, Domestic wastewater, Effluent from aquaculture effluent
PCPs	Insect repellents, Sunscreen, Fragrances, UV filters	Diethyltoluamide, 4 benzophenone, Galaxolide, Tonalide	WWTP effluent, Surface water, Landfill leachate
Industrial chemicals	Plasticizers, Fire retardants	Dimethyl adipate (DMAD), Tris (1-chloro-2-propyl) phosphate (TCPP)	Domestic wastewater, Industrial effluent
EDCs	Bisphenol, Xenohormone, Phthalates	Dioctyl phthalate (DOP), Xenoestrogen, Bisphenol A (BPA)	Drinking water, Surface water, Sediments Secondary sludge, Soil
Biocides	Herbicide, Fungicide, Molluscicide	Metaldehyde, Epoxiconazole, Butachlor	Surface water, Aquaculture effluent, Agricultural runoff
Regulated compounds (RCs)	Pesticides, Polyaromatic hydrocarbons (PAHs)	Chlorpyrifos, Phenanthrene	Surface water, Sewage treatment plants, Soils, Agricultural runoffs, Sediments
Surfactants	Nonionic surfactants, Ionic surfactants	Polysorbates, Sodium lauryl sulfate (SLS)	Industrial effluents, Domestic wastewater

variability (Petrie, Barden, and Kasprzyk-Hordern 2015). For example, sunscreen agent's concentration is high during the summer, whereas during the winter, cough syrup liquids are in high concentration.

ECs in surface water bodies are mainly due to the discharge of wastewater treatment plant effluent into surface water. ECs concentrations in the surface water are low due to dilution. Rainfall helps dilute the surface water but may result in higher ECs due to the leaching of chemicals from building paints, pavement materials, increase runoff, and sewer overflow (K'oreje et al. 2020). The common ECs detected in surface water include carbamazepine, sulfamethoxazole, and triclosan. ECs pollution of groundwater is mainly due to landfill leachate percolation, groundwater recharge using reclaimed water, infiltration from agricultural lands, and seepage from sewers (Luo et al. 2014). The transportation of ECs through the soil into groundwater depends on the hydrophobic partitioning coefficient (octanol–water partitioning coefficient, log K_{ow}) of the ECs. LogK_{ow} <2.5 indicates high hydrophilic mobility, 2.5—4 specifies medium mobility; and log K_{ow} >4 specifies low mobility or high retention of ECs in the soil matrix (Rout et al. 2021).

From the aquatic environment, ECs enter into the food chain by different pathways, thereby causing the problem of biomagnification and bioaccumulation. Although the concentration of ECs is less, long-term exposure can have adverse effects on living beings and ecosystems. Significant damage happens to the aquatic ecosystem due to ECs (Mohapatra and Kirpalani 2019). Antibiotic resistance in microbial strains, alteration in microbial community structure, and low population of algae and bacteria are caused due to antibiotics exposure (Grenni et al. 2018). Fragrances (musk ketone, musk xylene) are toxic to aquatic organisms, carcinogenic to rodents, and may damage the human nervous system. Preservatives like methyl-paraben are responsible for weak estrogenic activity. Brominated fire retardants affect the brain, nervous system, hormone activity, and reproduction systems. Phthalates and bisphenol generally interfere with endocrine systems, causing birth defects (de Oliveira et al. 2020). Pesticide toxicity affects the endocrine systems too (Kim et al. 2017). Anti-inflammatory drugs like diclofenac and ibuprofen are responsible for increased risk of gastrointestinal ulcers, kidney diseases, and gill alterations in fishes. Bioluminescence inhibition and growth inhibition of microalgae are caused due to lipid regulators (gemfibrozil) exposure. Beta-blockers (atenolol, metoprolol) affect the reproduction and growth of fishes. Carbamazepine causes oxidative stress in rainbow trout and affects the central nervous system in humans. Hormones (estrone, testosterone) reduce fertility in fish, cause masculinization of females, and feminize males. Polyaromatic hydrocarbons are carcinogenic and responsible for cardiovascular disease. Perfluorinated alkylated substances can damage the liver, kidney and are responsible for thyroid. Nanoparticles and nanocomposites affect the respiratory system and are responsible for environmental toxicity.

10.3 Treatment Methods of ECs

ECs in surface/subsurface water undergo several natural transformation processes like biodegradation, transformation, photolysis, adsorption onto particles, etc., which reduces the concentration of ECs in surface water (Remya and Lin 2011; Pal et al. 2010). However, the efficacy of these processes is low, which necessitates the adoption of suitable treatment methods for their removal from water/wastewater. In conventional wastewater treatment plants, different physical, chemical, and biological techniques are used. These treatment techniques are seldom effective for the removal of ECs. The removal technique to be adapted for ECs removal depends upon the properties of ECs like charge, functional group, hydrophobicity, dissociation constant, morphology, and size. The physico-chemical methods generally adopted for ECs removal are membrane technology, advanced oxidation process, adsorption. Membrane technology uses certain membranes for the removal of ECs by sieving, adsorption, and electrostatic repulsion. Nanofiltration and reverse osmosis membranes are sustainable, feasible, and effective techniques compared to conventional methods in ECs removal. Membrane fouling and high energy consumption are the drawbacks of this technology (Shahid et al. 2021). The advanced oxidation process relies on the oxidation of pollutants through direct contact with oxidants or through the production of free radicals such as hydroxyl radicals (Prashanth et al. 2021). This technology includes processes like electrocatalytic oxidation, photocatalysis, and Fenton's reaction. Different ECs that have successfully been removed using advanced oxidation processes are methylparaben, caffeine, ketorolac, antipyrine, sulfamethoxazole, carbamazepine, diclofenac, etc. (Neelancherry Remya and Swain 2019; Rathi et al. 2021; Prashanth et al. 2021). The high cost of catalysts and reactor design for large-scale applications are some of the disadvantages of this technology (Rout et al. 2021).

Adsorption transfers the contaminants (adsorbate) from one phase to another, using either gas–solid, liquid–solid, or gas–liquid interface. Adsorbents are used to remove the contaminants via intermolecular forces such as electrostatic attraction, Vander Waals forces, hydrogen bonding, and π-π interactions. Adsorption is classified into physisorption and chemisorption. In physisorption, adsorbate and adsorbent are held by weak interactions and are reversible, whereas, in chemisorption, chemical interaction exists between adsorbate and adsorbent, irreversible in nature. Adsorption is highly effective for the removal of organic and inorganic pollutants from the water streams. Simple operational scheme, low cost, eco-friendly, and wide range of adsorbents are advantages of adsorption. Different adsorbents generally used for ECs removal are activated carbon, zeolite, biochar, clay, metal–organic framework, chitosan, and nano adsorbents (graphene, graphene oxide, carbon nanotubes, and nano-composites). Adsorbent and adsorbate characteristics, operational conditions, such as pH, ionic strength, concentration of pollutant, contact time, etc., affect the adsorption efficiency (Shahid et al. 2021). Activated carbon and biochar have been used to successfully remove ECs like caffeine, diclofenac, gallic acid, ibuprofen, salicylic acid, and ofloxacin (Rathi et al. 2021). Natural clay has been used to remove amoxicillin and trimethoprim (Rathi et al. 2021). Metal–organic frameworks have been applied to remove ECs such as bisphenol-A, naproxen, triclosan, benzoic, and ketoprofen (Shahid et al. 2021).

10.4 Biochar as Adsorbent for ECs Removal

10.4.1 Biochar Production

Biochar is a highly porous, lightweight, carbon-rich material obtained by thermally processing the waste biomass in the absence of oxygen (Dai et al. 2019; Wang et al. 2018). These waste biomass used for biochar production includes crops (willows, switchgrass, miscanthus), forestry residues (grain crops, sawdust, nutshells), organic waste (animal manure, yard waste), and sludge (Chao et al. 2018; Huang et al.

2017; Vause, Heaney, and Lin 2018). Biomass majorly contains lignin, cellulose, and hemicelluloses which can add up to more than 65% carbon on a dry weight basis. Biomass is converted to biochar using pyrolysis, in which biomass is burnt at high temperatures in the absence of oxygen (Sahoo and Remya 2020; Shukla et al. 2019). The first stage of pyrolysis is known as the drying stage, in which the moisture of the biomass evaporates. This stage only affects the physical properties of the biomass. When the temperature of the system is further increased, the unstable compounds of the biomass start transforming into carbon dioxide (CO_2), carbon monoxide (CO), acetic acid (CH_3COOH), and other stable compounds (Shaaban and Abid 2021). The next stage is the solid decomposition stage. The biomass is decomposed or degraded under anoxic conditions, generating products like methanol, wood tar, acetic acid and releasing certain gases like CO, CO_2, CH_4, H_2, etc. The third and final stage of pyrolysis is known as the carbonization stage, in which the non-volatile residues start to carbonize and become rich in carbon content and is called charcoal (Dai et al. 2019). Based on the biomass used in thermal processing, biochar can be divided into straw biochar, shell biochar, wood biochar, sludge biochar, animal fecal biochar, bamboo biochar, among others (Huang et al. 2018; Idrees et al. 2018; Chao et al. 2018; Vause et al. 2018).

10.4.2 Properties of Biochar

The nature of the biomass and the thermal degradation conditions (the heating rate, duration, temperature, and pressure) determines the physical and chemical properties of biochar (Bushra and Remya 2020; Kim et al. 2013; Shaaban et al. 2018). These properties include particle size, bulk and true density, proximate analysis (moisture and ash content), calorific value; percentage of lignin, cellulose, hemicellulose, fixed carbon, functional groups, inorganic substance, and volatile compounds (Angin 2013; Shivaram et al. 2013; Nartey and Zhao 2014). The carbon and oxygen content of the feedstock affects the functional groups present in the biochar. These groups act as additional active sites, hence, increasing the removal efficiency of biochar. The reaction temperature influences the biochar's physical and structural properties, such as ash content, porosity, nutrient content, electrical conductivity, pH, etc. Higher reaction temperatures and residence time generate biochar with higher ash content and nutrients such as P, K, Ca (Suliman et al. 2016).

Biochar is a heterogeneous organic product with stable carbon structures, ash, and labile components (Shaaban and Abid 2021). Biochar also contains aromatic groups with heteroatoms such as N, O, P, and S, making the biochar reactive (Tripathi et al. 2016). Compared to their feedstock, biochar has a lower H/C ratio, which indicates the occurrence of polymerization during the thermal degradation and recalcitrance capability of biochar (Cayuela et al. 2015).

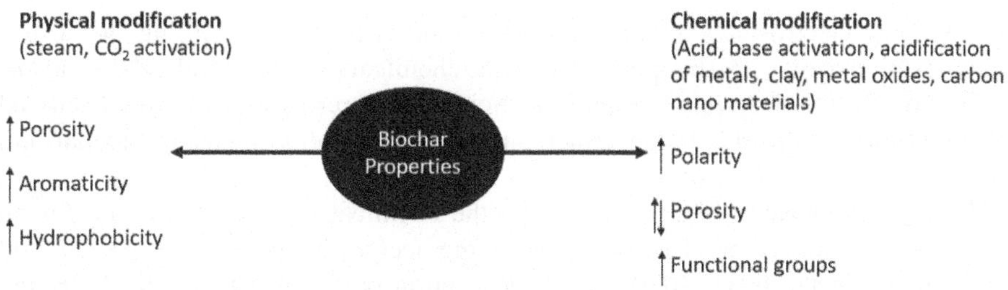

Fig. 10.1 Effect of physical and chemical modifications on biochar (Krasucka et al. 2021)

10.4.3 Emerging Contaminant Removal Using Biochar

The use of biochar for emerging contaminant removal has increased in recent years. As compared to activated carbon, biochar is much cheaper, which further enhances its potential for use. For example, biochar costs around \$350 to \$1,200 per ton, whereas activated carbon costs around \$1100–\$1700 per ton (Thompson et al. 2016). Properties of biochar like large surface area, porosity, structural integrity, and surface-active functional groups enable the adsorption of contaminants on the biochar surface (Tran et al. 2020). Surface functional groups like hydroxyl, carboxyl, carbonyl, amines, etc., influence the surface chemistry of biochar which includes surface charge and contaminant interaction (Nartey and Zhao 2014). The physicochemical properties of biochar can be improved to increase the adsorption by physical and chemical modifications (Bushra and Remya 2020; Lutterbeck et al. 2020; Xiang et al. 2021). Thus, biochar can be used to remove various types of contaminants like emerging contaminants, heavy metals, organic and inorganic pollutants, etc. (Zenker et al. 2014; Yang et al. 2017; Tiwari et al. 2017; Petrie, Barden, and Kasprzyk-Hordern 2015).

The biochar is usually modified to increase its removal capacity (Fig. 10.1). Physical modifications imcreases the porosity, aromaticity and hydrophobicity of the biochar.

The chemical modifications increases the polarity by changing the surface functional groups which also increases the biochar porosity. The addition of carbon nanomaterials and metals in the biochar leads to the formation of biochar-metal composites. Biochar-metal composites result in a change in surface charge and provides a functional group. Some metal compounds exhibit photocatalytic/catalytic activity, leading to catalytic degradation of ECs in the presence of electromagnetic radiation. In carbon nanomaterial and biochar combination, porosity increases, and functional groups are introduced. Adsorption on this compound is done by pore filling, interaction with the functional group, and π-π interaction.

Adsorption of contaminants on biochar occurs in four steps: (1) adsorbate molecule transport from bulk media (solution) to biochar; (2) adsorbate molecule diffusion through liquid film around biochar; (3) pore diffusion, i.e., transport of adsorbate from biochar surface to active sites along pore surface, and (4) adsorption. Various adsorbent-adsorbate interactions include dipole interactions, hydrophobic interactions, electrostatic interactions, and π interactions. Pore-filling or surface complexation or may also happen based on the chemical structure of adsorbent, adsorbate, and sorption/desorption conditions. Some recent research works are discussed below (Table 10.2) for a better understanding of ECs removal using biochar and modified biochar.

Magnetic biochar (MBC) prepared by the chemical precipitation of Fe_3O_4 on biochar was found to be effective in removing acetylsalicylic acid, ibuprofen, and caffeine (Liyanage et al. 2020). Their adsorption potential was 149.9, 39.9, and 75.1 mg/g, respectively, at 35 °C and pH 8. Adsorption occurred due to π-π electron interactions, dipole–dipole interactions, H-bonding, and electrostatic attraction. Fast adsorption kinetics and high adsorption capacity were due to large surface area, pore sizes, pore volumes provided by biochar, and additional adsorption sites due to iron oxides on MBC. Magnetic properties enabled the post-treatment magnetic separation of MBC.

Biochar as photocatalyst support enabled adsorption and photocatalytic degradation of ECs. TiO_2 supported on reed straw biochar (TiO_2-biochar composite) enhanced the degradation of sulfamethoxazole (SMX) by 3.5 times compared to reed straw biochar (Zhang et al. 2017). Despite reduced adsorption capacity, the removal of SMX by photocatalytic degradation was higher for TiO_2-biochar composite (91%) than pure TiO_2 (59%) due to a high degree of dispersion of TiO_2 in the biochar composite. The biochar intercepted the accumulation of TiO_2 particles and increased the photodegradation efficiency of SMX. After five cycles of regeneration, the SMX degradation was found to be 86%. Magnetic biochar prepared from chemically modified sludge biochar (SBC) modified by graphene oxide showed improved removal of insecticide imidacloprid (IMI) (Ma et al. 2021). Magnetic sludge biochar was prepared by adding SBC into cobalt nitrate and ferric chloride solution. The magnetic sludge biochar was then added to graphene oxide suspension to prepare GO/$CoFe_2O_4$-SBC novel composite. The hydroxyl group (-OH) of GO/$CoFe_2O_4$-SBC surface and the graphitic structure donated the π-electron, facilitating the π-π interaction between adsorbate and adsorbent. The larger surface area (184 m^2/g) and pore volume (0.346 cm^3/g) produced more binding sites on GO/$CoFe_2O_4$-SBC for IMI adsorption as compared to SBC with low surface area (53.1 m^2/g) and low pore volume (0.0816 cm^3/g). $CoFe_2O_4$ loading improved the magnetic sensitivity of GO/$CoFe_2O_4$-SBC, making the separation easier from the water sample.

Although both the physical and chemical modifications improvs ECs adsorption property of biochar. However, the chemical modification causes a higher increase in the specific surface are compared to the physical modification. After the chemical activation of malt bagasse biochar with $ZnCl_2$, the specific surface area increased to 545 m^2/g. However, the physical modification increased the specific surface area of CO_2-activated biochar to only 161 m^2/g (MacHado et al. 2020). Both the activated biochars showed an amorphous structure and hydroxyl functional groups. $ZnCl_2$-activated biochar removed 2-chlorophenol more efficiently than CO_2-activated biochar with a maximum adsorption capacity of 150 mg/g, with a removal efficiency of 98.21% from industrial effluents containing salts and phenolic

Table 10.2 ECs removal using biochar

Biochar feedstock	Pyrolysis conditions	Modifications	Surface area (m^2/g)	ECs	Adsorption condition	Maximum adsorption capacity (mg/g)	References
Waste wood	900–1000 °C	Iron Oxide nucleation	322	Caffeine Ibuprofen Acetylsalicylic acid	pH 8, 35 °C	75.1 39.9 149.9	Liyanage et al. (2020)
Reed Straw	500 °C, 6 h	TiO_2 Addition	102.16	Sulfamethoxazole	pH 4, 25 °C	6.592	Zhang et al. (2017)
Barley Malt Bagasse	900 °C, 50 min	$ZnCl_2$ activation	545	2-Chlorophenol	pH 6, 25 °C	150	MacHado et al. (2020)
Wastewater Sludge	500 °C, 2 h, N_2	Graphene oxide addition & $CoFe_2O_4$ nucleation	184	Imidacloprid	pH 2–12	–	Ma et al. (2021)
Sugarcane Bagasse	750 °C, O_2 Limited	-	540.36	Nanoplastics	pH 3–10, 25 °C	44.9	Ganie et al. (2021)
Soapnut seeds	450 °C, 2 h	Acid wash (Orthophosphoric acid)	2.21	Ciprofloxacin	pH 6	33.44	Velusamy et al. (2021)
Pecan nutshell	800 °C, 60 min	Zinc oxide addition	253.23	Reactive Red 97	pH 7	–	Leichtweis et al. (2020)
Wheat Straw	700 °C, 2 h	Alkali-activation (NaOH)	254.94	Bisphenol-A Tetracycline Ofloxacin	pH 9 pH 8 pH 8	71.43 101.01 54.05	Tang et al. (2021)
Anthriscus Sylvestris	300 °C, 15 min	Alkali-activation (NaOH)	–	Diclofenac Cephalexine	pH 2–10	392.94 724.54	Shirani et al. (2020)

compounds using a dosage of 10 g/L. Moreover, the $ZnCl_2$-activated biochar showed effective regeneration after multiple adsorption cycles.

Biochar also showed promising results for adsorptive removal of nano plastics (NPs). Sugar cane bagasse biochar, obtained by pyrolysis at 750 °C, with a surface area of 540.36 m^2/g showed 99% removal of NPs with a maximum adsorption capacity of 44.9 mg/g (Ganie et al. 2021). The high pyrolytic temperature decreased the negatively charged functional groups, thus decreasing the negative zeta potential.

Biochar derived from soap nut seeds was used for the removal of ciprofloxacin which is an antibiotic. Pyrolysis of soap nut seeds was carried out at 450 °C for 2 h in a tubular furnace under slow pyrolysis conditions. The biochar obtained was treated with orthophosphoric acid. The surface area of the biochar was found to be 2.21 m^2/g and showed a maximum adsorption capacity of 33.44 mg/g based on the Langmuir isotherm. The pseudo-second-order kinetics equation fitted well for removing ciprofloxacin by the soap nut seeds biochar. For the removal process, adsorption mechanisms such as π-π interactions, pore-filling effect, cation-exchange interactions were involved. The adsorption capacity of the biochar was enhanced by increasing the pH to 6 and later decreased with a further increase in pH. At pH 6, the maximum removal efficiency of ciprofloxacin of 95% was obtained (Velusamy et al. 2021).

Biochar derived from pecan nutshell was used as support material for Zinc oxide for degradation of reactive red 97 dye. Photocatalytic degradation of the dye occurred and 100% degradation was achieved at pH 7. As the pH further increased the degradation efficiency reduced. The removal rate of the pollutant was not affected even after 9 cycles of the run which shows the stability of zinc oxide on biochar (Leichtweis et al. 2020).

Biochar prepared from wheat straw and modified using sodium hydroxide was applied as an adsorbent to remove bisphenol-A, tetracycline and ofloxacin. The results suggested that the modified biochar exhibited 95% removal of the contaminants within 9 h due to a larger specific surface area and stronger hydrophobicity. Electrostatic attractions, π-π interactions as well as hydrophobic interactions, and pore-filling were the main adsorption mechanisms. The initial pH of water strongly affected the adsorption of ofloxacin and tetracycline. The maximum adsorption capacity at equilibrium was obtained when the pH was around 8. Whereas for bisphenol-A, there was no significant increase in removal efficiency up to pH 9. But on further increasing the pH, the removal efficiency decreased (Tang et al. 2021).

Alkali-activated (NaOH) biochar produced from Anthriscus sylvestris proved to be an effective adsorbent for removing diclofenac and cephalexine pharmaceuticals from synthetic wastewater. This might be due to a substantial increase in the porosity and functional groups of biochar as a result of activation. Based on the Langmuir model, the adsorption capacity of activated biochar for diclofenac and cephalexine was 392.94 and 724.54 mg/g, respectively. Electrostatic adsorption, hydrophobic interaction, and π-π bonding were the main mechanisms of adsorption of both diclofenac and cephalexine (Shirani et al. 2020).

10.5 Conclusion

With the extensive use of commodities such as pharmaceutical compounds, personal care products, pesticides, surfactants, etc., huge amounts ECs are released into the environment. Long-term exposure to ECs causes adverse effects on both humans and the environment. Thus, removal of ECs is necessary from wastewater before they are discharged into the water bodies. Conventional wastewater treatment techniques are not efficient in removing ECs from wastewater.

Adsorption is a popular process for the removal of ECs. Among the different adsorbents that can be used, biochar is one such potential adsorbent. Biochar is cheap and eco-friendly as it is produced from waste like municipal solid waste, agricultural waste, livestock waste, etc. Biochar production also reduces the volume of waste to be handled. The adsorption capacity of the biochar can be improved by modifying it. By combining photocatalysts with biochar, both adsorption and degradation of ECs can be achieved.

References

Angin D (2013) Effect of pyrolysis temperature and heating rate on biochar obtained from pyrolysis of safflower seed press Cake. Biores Technol 128(January):593–597. https://doi.org/10.1016/j.biortech.2012.10.150

Barbosa MO, Moreira NFF, Ribeiro AR, Pereira MFR, Silva AMT (2016) Occurrence and removal of organic micropollutants: an overview of the watch list of eu decision 2015/495. Water Res 94(May):257–279. https://doi.org/10.1016/J.WATRES.2016.02.047

Bushra B, Remya N (2020) Biochar from pyrolysis of rice husk biomass—characteristics, modification and environmental application. Biomass Convers Biorefinery. https://doi.org/10.1007/s13399-020-01092-3

Cayuela ML, Jeffery S, van Zwieten L (2015) the molar h: corg ratio of biochar is a key factor in mitigating N2O emissions from soil. Agr Ecosyst Environ 202(April):135–138. https://doi.org/10.1016/J.AGEE.2014.12.015

Chao X, Qian X, Han-hua Z, Shuai W, Qi-hong Z, Dao-you H, Yang-zhu Z (2018) Effect of biochar from peanut shell on speciation and availability of lead and zinc in an acidic paddy soil. Ecotoxicol Environ Saf 164(November):554–561. https://doi.org/10.1016/j.ecoenv.2018.08.057

Cheng N, Wang B, Pan W, Lee X, Xing Y, Chen M, Gao B (2021) Adsorption of emerging contaminants from water and wastewater by modified biochar: a review. Environ Pollut 273(March):116448. https://doi.org/10.1016/J.ENVPOL.2021.116448

Dai Y, Zhang N, Xing C, Cui Q, Sun Q (2019) "The adsorption, regeneration and engineering applications of biochar for removal organic pollutants: a review." Chemosphere. https://doi.org/10.1016/j.chemosphere.2019.01.161

Dsikowitzky L, Schwarzbauer J (2014) Industrial organic contaminants: identification, toxicity and fate in the environment. Environ Chem Lett 12(3):371–386. https://doi.org/10.1007/S10311-014-0467-1

Ganie ZA, Khandelwal N, Tiwari E, Singh N, Darbha GK (2021) Biochar-facilitated remediation of nanoplastic contaminated water: effect of pyrolysis temperature induced surface modifications. J Hazard Mater 417(September):126096. https://doi.org/10.1016/J.JHAZMAT.2021.126096

Grenni P, Ancona V, Caracciolo AB (2018) Ecological Effects of antibiotics on natural ecosystems: a review. Microchem J 136(January):25–39. https://doi.org/10.1016/J.MICROC.2017.02.006

Huang R, Tian D, Liu J, Lv S, He X, Gao M (2018) Responses of soil carbon pool and soil aggregates associated organic carbon to straw and straw-derived biochar addition in a dryland cropping mesocosm system. Agr Ecosyst Environ 265(October):576–586. https://doi.org/10.1016/j.agee.2018.07.013

Huang YF, Cheng PH, Chiueh PT, Lo SL (2017) “Leucaena biochar produced by microwave torrefaction: fuel properties and energy efficiency.” Appl Energy 204(Complete):1018–25. https://doi.org/10.1016/j.apenergy.2017.03.007

Idrees M, Batool S, Kalsoom T, Yasmeen S, Kalsoom A, Raina S, Zhuang Q, Kong J (2018) Animal manure-derived biochars produced via fast pyrolysis for the removal of divalent copper from aqueous media. J Environ Manage 213(May):109–118. https://doi.org/10.1016/J.JENVMAN.2018.02.003

K’oreje KO, Okoth M, Van Langenhove H, Demeestere K (2020) Occurrence and treatment of contaminants of emerging concern in the African aquatic environment: literature review and a look ahead. J Environ Manage 254(January):109752. https://doi.org/10.1016/J.JENVMAN.2019.109752

Kim KH, Kabir E, Jahan SA (2017) Exposure to pesticides and the associated human health effects. Sci Total Environ 575(January):525–535. https://doi.org/10.1016/J.SCITOTENV.2016.09.009

Kim KH, Kim TS, Lee SM, Choi D, Yeo H, Choi IG, Choi JW (2013) Comparison of physicochemical features of biooils and biochars produced from various woody biomasses by fast pyrolysis. Renew Energy 50(February):188–195. https://doi.org/10.1016/J.RENENE.2012.06.030

Krasucka P, Pan B, Ok YS, Mohan D, Sarkar B, Oleszczuk P (2021) Engineered biochar—a sustainable solution for the removal of antibiotics from water. Chem Eng J 405(February):126926. https://doi.org/10.1016/J.CEJ.2020.126926

Leichtweis J, Silvestri S, Carissimi E (2020) New composite of pecan nutshells biochar-ZnO for sequential removal of acid red 97 by adsorption and photocatalysis. Biomass Bioenerg 140(September):105648. https://doi.org/10.1016/J.BIOMBIOE.2020.105648

Liyanage AS, Canaday S, Pittman CU, Mlsna T (2020) Rapid remediation of pharmaceuticals from wastewater using magnetic Fe3O4/douglas fir biochar adsorbents. Chemosphere 258(November):127336. https://doi.org/10.1016/J.CHEMOSPHERE.2020.127336

Luo Y, Guo W, Ngo HH, Nghiem LD, Hai FI, Zhang J, Liang S, Wang XC (2014) A review on the occurrence of micropollutants in the aquatic environment and their fate and removal during wastewater treatment. Sci Total Environ 473–474(March):619–641. https://doi.org/10.1016/J.SCITOTENV.2013.12.065

Lutterbeck CA, Colares GS, Dell’Osbel N, da Silva FP, Kist LT, Machado ÊL (2020) “Hospital laundry wastewaters: a review on treatment alternatives, life cycle assessment and prognosis scenarios.” J Cleaner Prod 273(November). https://doi.org/10.1016/J.JCLEPRO.2020.122851

Ma Y, Li W, Li P, Yang L, He L, Chen S, Yang Y, Gao F, Qi X, Zhang Z (2021) A novel, efficient and sustainable magnetic sludge biochar modified by graphene oxide for environmental concentration imidacloprid removal. J Hazard Mater 407(April):124777. https://doi.org/10.1016/J.JHAZMAT.2020.124777

MacHado LMM, Lütke SF, Perondi D, Godinho M, Oliveira MLS, Collazzo GC, Dotto GL (2020) Treatment of effluents containing 2-chlorophenol by adsorption onto chemically and physically activated biochars. J Environ Chem Eng 8(6):104473. https://doi.org/10.1016/J.JECE.2020.104473

Mohapatra DP, Kirpalani DM (2019) “Advancement in treatment of wastewater: fate of emerging contaminants.” https://doi.org/10.1002/cjce.23533

Nartey OD, Zhao B (2014) “Biochar preparation, characterization, and adsorptive capacity and its effect on bioavailability of contaminants: an overview.” Adv Mater Sci Eng 2014. https://doi.org/10.1155/2014/715398

de Oliveira M, Farias BE Frihling, Velasques J, Corrêa Magalhães Filho FJ, Cavalheri PS, Migliolo L (2020) “Pharmaceuticals residues and xenobiotics contaminants: occurrence, analytical techniques and sustainable alternatives for wastewater treatment.” Sci Total Environ 705(February)

Pal A, Hoong Gin KY, Chen Lin AY, Reinhard M (2010) Impacts of Emerging Organic Contaminants on Freshwater Resources: Review of Recent Occurrences, Sources, Fate and Effects. Sci Total Environ 408(24):6062–6069. https://doi.org/10.1016/J.SCITOTENV.2010.09.026

Patel M, Kumar R, Kishor K, Mlsna T, Pittman CU, Mohan D (2019) Pharmaceuticals of emerging concern in aquatic systems: chemistry, occurrence, effects, and removal methods. Chem Rev 119(6):3510–3673. https://doi.org/10.1021/ACS.CHEMREV.8B00299

Petrie B, Barden R, Kasprzyk-Hordern B (2015) A review on emerging contaminants in wastewaters and the environment: current knowledge, understudied areas and recommendations for future monitoring. Water Res 72(April):3–27. https://doi.org/10.1016/J.WATRES.2014.08.053

Prashanth V, Priyanka K, Remya N (2021) Solar photocatalytic degradation of metformin by tio2synthesized using calotropis gigantea leaf extract. Water Sci Technol 83(5):1072–1084. https://doi.org/10.2166/wst.2021.040

Rathi BS, Kumar PS, Show PL (2021) A review on effective removal of emerging contaminants from aquatic systems: current trends and scope for further research. J Hazardous Mater 409(May):124413. https://doi.org/10.1016/J.JHAZMAT.2020.124413

Remya N, Lin JG (2011) "Carbofuran degradation by the application of mw-assisted H2O2 process." J Environ Sci Health—Part B Pesticides Food Contaminants Agricult Wastes 46(4). https://doi.org/10.1080/03601234.2011.559894

Remya N, Swain A. 2019. "Soft drink industry wastewater treatment in microwave photocatalytic system—exploration of removal efficiency and degradation mechanism." Separation Purif Tech. https://doi.org/10.1016/j.seppur.2018.08.051

Rout PR, Zhang TC, Bhunia P, Surampalli RY (2021) Treatment technologies for emerging contaminants in wastewater treatment plants: a review. Sci Total Environ 753(January):141990. https://doi.org/10.1016/J.SCITOTENV.2020.141990

Sahoo D, Remya N (2020) "Influence of operating parameters on the microwave pyrolysis of rice husk: biochar yield, energy yield, and property of biochar." Biomass Convers Biorefinery. https://doi.org/10.1007/s13399-020-00914-8

Saidulu D, Gupta B, Gupta AK, Ghosal PS (2021) "A review on occurrences, eco-toxic effects, and remediation of emerging contaminants from wastewater: special emphasis on biological treatment based hybrid systems." J Environ Chem Eng 9(4). https://doi.org/10.1016/J.JECE.2021.105282

Shaaban M, Abid M (2021) "Biochar as a sorbent for organic and inorganic pollutants." Sorbents Mater Control Environ Pollution, January:189–208. https://doi.org/10.1016/B978-0-12-820042-1.00001-8

Shaaban M, Van Zwieten L, Bashir S, Younas A, Núñez-Delgado A, Chhajro MA, Kubar KA et al (2018) A concise review of biochar application to agricultural soils to improve soil conditions and fight pollution. J Environ Manage 228(December):429–440. https://doi.org/10.1016/J.JENVMAN.2018.09.006

Shahid MK, Kashif A, Fuwad A, Choi Y (2021) Current advances in treatment technologies for removal of emerging contaminants from water—a critical review. Coord Chem Rev 442(September):213993. https://doi.org/10.1016/J.CCR.2021.213993

Shirani Z, Song H, Bhatnagar A (2020) Efficient removal of diclofenac and cephalexin from aqueous solution using anthriscus sylvestris-derived activated biochar. Sci Total Environ 745:140789. https://doi.org/10.1016/j.scitotenv.2020.140789

Shivaram P, Leong YK, Yang H, Zhang DK (2013) Flow and yield stress behaviour of ultrafine mallee biochar slurry fuels: the effect of particle size distribution and additives. Fuel 104(February):326–332. https://doi.org/10.1016/j.fuel.2012.09.015

Shukla N, Sahoo D, Remya N (2019) Biochar from microwave pyrolysis of rice husk for tertiary wastewater treatment and soil nourishment. J Clean Prod 235:1073–1079. https://doi.org/10.1016/j.jclepro.2019.07.042

Suliman W, Harsh JB, Abu-Lail NI, Fortuna AM, Dallmeyer I, Garcia-Perez M (2016) Influence of feedstock source and pyrolysis temperature on biochar bulk and surface properties. Biomass Bioenerg 84(January):37–48. https://doi.org/10.1016/J.BIOMBIOE.2015.11.010

Tang Y, Ye Li L, Zhan DW, Zhang S, Pang R, Xie B (2021) Removal of emerging contaminants (bisphenol a and antibiotics) from kitchen wastewater by alkali-modified biochar. Sci Total Environ 805:150158. https://doi.org/10.1016/j.scitotenv.2021.150158

Thompson KA, Shimabuku KK, Kearns JP, Knappe DRU, Scott Summers R, Cook SM (2016) Environmental comparison of biochar and activated carbon for tertiary wastewater treatment. Environ Sci Technol 50(20):11253–11262. https://doi.org/10.1021/ACS.EST.6B03239

Tiwari B, Sellamuthu B, Ouarda Y, Drogui P, Tyagi RD, Buelna G (2017) Review on fate and mechanism of removal of pharmaceutical pollutants from wastewater using biological approach. Biores Technol 224(January):1–12. https://doi.org/10.1016/J.BIORTECH.2016.11.042

Tran HN, Tomul F, Hoang Ha NT, Nguyen DT, Lima EC, Le GT, Chang CT, Masindi V, Woo SH (2020) "Innovative spherical biochar for pharmaceutical removal from water: insight into adsorption mechanism." J Hazardous Mater 394(July). https://doi.org/10.1016/J.JHAZMAT.2020.122255

Tran NH, Hoong Gin KY (2017) "Occurrence and removal of pharmaceuticals, hormones, personal care products, and endocrine disrupters in a full-scale water reclamation plant." Sci Total Environ 599–600 (December):1503–16. https://doi.org/10.1016/J.SCITOTENV.2017.05.097

Tripathi M, Sahu JN, Ganesan P (2016) Effect of process parameters on production of biochar from biomass waste through pyrolysis: a review. Renew Sustain Energy Rev 55(March):467–481. https://doi.org/10.1016/J.RSER.2015.10.122

Vause D, Heaney N, Lin C (2018) Differential release of sewage sludge biochar-borne elements by common low-molecular-weight organic acids. Ecotoxicol Environ Saf 165(December):219–223. https://doi.org/10.1016/j.ecoenv.2018.09.005

Velusamy K, Periyasamy S, Kumar PS, Jayaraj T, Krishnasamy R, Sindhu J, Sneka D, Subhashini B, Vo DVN (2021) Analysis on the removal of emerging contaminant from aqueous solution using biochar derived from soap nut seeds. Environ Pollut 287(October):117632. https://doi.org/10.1016/J.ENVPOL.2021.117632

Wang H, Wang X, Cui Y, Xue Z, Ba Y (2018) Slow Pyrolysis polygeneration of bamboo (phyllostachys pubescens): product yield prediction and biochar formation mechanism. Biores Technol 263(September):444–449. https://doi.org/10.1016/J.BIORTECH.2018.05.040

Xiang Y, Wu H, Li L, Ren M, Qie H, Lin A (2021) "A review of distribution and risk of pharmaceuticals and personal care products in the aquatic environment in China." Ecotoxicol Environ Safety 213(April). https://doi.org/10.1016/J.ECOENV.2021.112044

Yang Y, Ok YS, Kim KH, Kwon EE, Tsang YF (2017) Occurrences and removal of pharmaceuticals and personal care products (PPCPs) in drinking water and water/sewage treatment plants: a review. Sci Total Environ 596–597(October):303–320. https://doi.org/10.1016/J.SCITOTENV.2017.04.102

Zenker A, Cicero MR, Prestinaci F, Bottoni P, Carere M (2014) Bioaccumulation and biomagnification potential of pharmaceuticals with a focus to the aquatic environment. J Environ Manage 133(January):378–387. https://doi.org/10.1016/J.JENVMAN.2013.12.017

Zhang H, Wang Z, Li R, Guo J, Li Y, Zhu J, Xie X (2017) TiO2 supported on reed straw biochar as an adsorptive and photocatalytic composite for the efficient degradation of sulfamethoxazole in aqueous matrices. Chemosphere 185(October):351–360. https://doi.org/10.1016/J.CHEMOSPHERE.2017.07.025

Chapter 11
Remediation of Emerging Contaminants by Naturally Derived Adsorbents

Darakhshan Nayyar, Mohd Ahmed Naim Shaikh, and Tabish Nawaz

Abstract The topic of emerging contaminants has gathered attention all over the world. Emerging contaminants (EC) are anthropogenic compounds detected in the environmental matrices in minimal concentrations. The major classification of emerging contaminants are pharmaceuticals, endocrine disruptors, personal care products, per and polyfluoroalkyl substances, pesticides, nanomaterials and microplastics which are associated with diverse range of commercial and personal care products and throughout their life cycle are released into different environmental zones such as air, water, soil, biota and reach different receptors. The knowledge on remediation of these emerging contaminants from different environmental media is limited and still evolving. Among all the available treatment technologies, adsorption shows promising results for remediation due to low cost, ease of operation, and higher efficiencies. Moreover, adsorption is suited for trace contaminants as adsorbents can be functionalized with specific functional groups that show high affinity for the target contaminants. The choice of natural materials further enhances the sustainable aspect of the treatment process, as it would recycle and reuse them in the form of adsorbents. The natural material reviewed in the chapter includes, fruit waste, agricultural waste, Moringa oleifera, chitosan and rice husk for their potential for adsorption applications. The current review overviews adsorption of emerging contaminants, emphasizing its mechanisms and the factors affecting adsorption. The predominant adsorption mechanisms of emerging contaminants are diffusion, electrostatic interaction, hydrophobic interaction, ion exchange, hydrogen bonding and π-π bonding. It mainly describes remediation of significant classes of emerging contaminants through various natural adsorbents like activated carbon, biochar, food, and other agricultural waste.

Keywords Adsorption · Natural waste materials · Pharmaceuticals · Endocrine disruptors · Personal care products · Per and polyfluoroalkyl substances · Pesticides · Nanomaterials · Microplastics

---

D. Nayyar · M. A. N. Shaikh · T. Nawaz (✉)
Environmental Science and Engineering Department, Indian Institute of Technology Bombay, Powai, Mumbai, Maharashtra 400076, India
e-mail: tnawaz@iitb.ac.in

S. P. Singh et al. (eds.), *New Trends in Emerging Environmental Contaminants*, Energy, Environment, and Sustainability,
https://doi.org/10.1007/978-981-16-8367-1_11

11.1 Introduction

With an increase in economy, sophistication in living standard of human-kind and growing industries, aquatic media throughout the world have been adversely impacted, mainly because of the broad application of anthropogenic chemicals created by humans that were not considered as a threat in the past. These contaminants of emerging concern and their transformation products occurring in water bodies throughout the globe, are beginning to be tracked and investigated in the environment but have a high risk of causing environmental damage and have suspected detrimental effects on ecosystems and human health (Rosenfeld and Feng 2011). According to USEPA, "emerging contaminants (EC)" are substances that have the capability of causing harm or toxicity to human or other life forms or lack established health regulations (USEPA 2014). One of the primary criteria to consider a substance an EC, is that it lacks sufficient toxicological data on health threats to humans, together with the absence of regulation (Cordner and Brown 2013). Additionally, a substance can be considered EC because of newer modes of generation, utility, and disposal (Richter et al. 2018). These pollutants and their typical concentration range vary from ng/L to μg/L (Pal et al. 2014). Major classes of ECs as depicted in Fig. 11.1, are pharmaceuticals and personal care products (PPCPs), endocrine disruptors, pesticides, nanomaterials, microplastics and per- and polyfluoronated substances (PFAS) (Pereira et al. 2015). Day by day, newer chemicals are being added to the classes of ECs. ECs are found in all kinds of water systems such as surface, ground or wastewater. The concentration of EC in surface water bodies is of lesser value than sewage and wastewater sources due to natural processes like dilution, sorption, etc. (Pal et al.

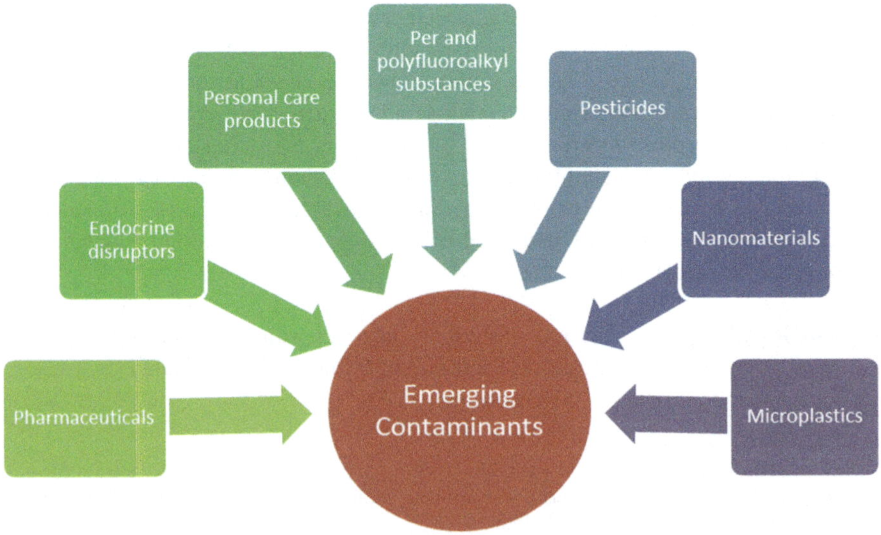

Fig. 11.1 Different classes of emerging contaminants

2010). Analgesics, anti-inflammatories, antimicrobials, β-blockers, and lipid regulators are the most commonly found pharmaceuticals. Other antiseptic/ antibacterial products like triclosan and synthetic fragrances like Nitromusk and Polycyclic musk are also toxic PPCPs. These chemicals generally enter the environment and drinking water supplies after being discharged from sewage treatment plants or leached from pesticides applied to farmland. Microplastics are used in toothpaste, cosmetics and are also produced by weathering of macroplastics, discarded plastics, fishing nets etc. and end up in municipal water treatment plant. PFAS are used in non-stick pans, paper, coatings on pan, stain proof carpets and textiles, waterproof packaging of food items and are used in cleaning agents, cosmetics, shampoos, paints, fabric softeners, adhesives, because of its surfactant property (Buck et al. 2011; Clara et al. 2008; Kissa 2001; Rahman et al. 2014). PFAS are introduced in the environment both by point and non-point sources. Some of the point sources of PFAS are manufacturing plants, application of aqueous film forming foams (AFFF) on areas, industrial and municipal sewage treatment plant, and some of the non-point sources are dry, wet atmospheric deposition and landfills (Hrens and Undschuh 2014).

EC possess different chemical properties, and their removal from environmental media largely depends on harnessing those properties. Thus, occurrence of the "unregulated" emerging contaminant requires advanced treatment technologies (Nawaz and Sengupta 2019). EC are found in different forms, and their removal mechanisms also differ from one to another, making their remediation from water-based media further difficult. Adsorption is a rapid, easy, and broadly practiced process as it can remove a wide range of pollutants with ease and high efficiency. It is one of the sustainable approaches for water purification due to its incredible effectiveness for contaminant removal without the generation of harmful by-products. This chapter summarises different modes of adsorption mechanisms for ECs during adsorption and the factors affecting them. It describes in detail various kinds of natural adsorbents used to remove the major EC classes, i.e. pharmaceuticals, personal care products, endocrine disruptors, pesticides, PFAS, nanomaterials and micro plastics, and the challenges faced for the same.

11.2 Adsorption

Adsorption is the partition of a compounds or particles from liquid or gases to a solid surface. Removal of EC by adsorption is an environmentally friendly approach that is popular due to its higher efficiency with lower sludge generation that can be achieved with ease of operation. The material being adsorbed is called adsorbate, and the adsorbing surface is termed as adsorbent. Based on the mechanism of interaction, there are two types of adsorption processes. One is called physisorption, in which physical interactions such as Van der Waal's forces (weak) are responsible for the interaction among the adsorbate and the adsorbent (Mariangela et al. 2012). Another one is known as chemisorption, in which chemical bonding takes place between the adsorbate and adsorbent. The complete process of adsorption involves three steps:

(i) transportation of contaminant from bulk phase to the surface/interface of the adsorbent, (ii) diffusion or transfer of adsorbate particles from the outer surface to its inner pores, and (iii) interaction of particles with the active sites of the adsorbent (Rashed 2012).

11.2.1 Mechanisms of Adsorption

This section discusses different mechanisms that are involved in emerging contaminant adsorption. As shown in Fig. 11.2, other adsorption mechanisms are diffusion, electrostatic interaction, hydrophobic interaction, ion exchange, hydrogen bonding, and π-π interactions. Diffusion occurs either by pore-filling or partition and was observed in the removal of emerging contaminants such as PFAS (Inyang and Dickenson 2017), PPCP (Zhu et al. 2014), and pesticides (Suo et al. 2019). One of the removal mechanisms of herbicide, Triazine on biochar made from corn waste, was through diffusion (Suo et al. 2019). Triclosan also interacted with magnetic hydrochars through pore filling in an investigation demonstrated by Zhu et al. (2014)

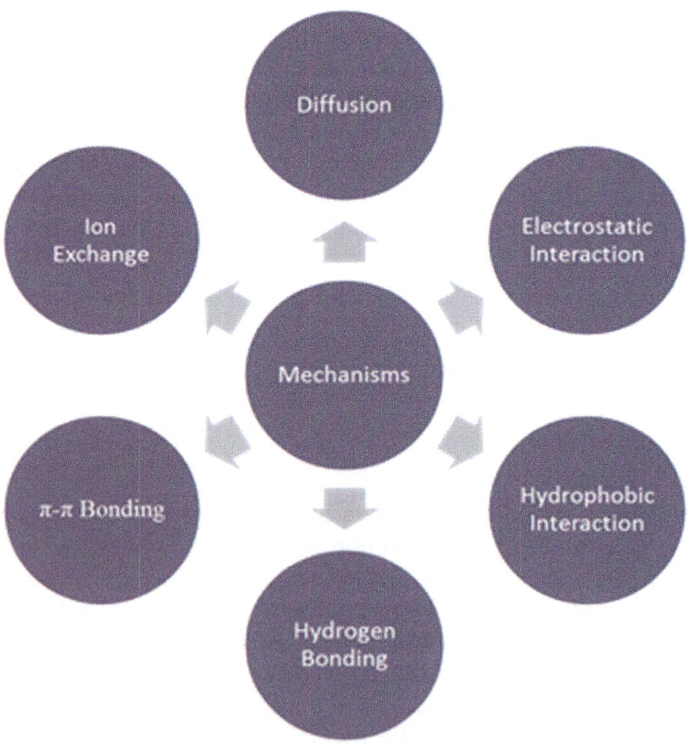

Fig. 11.2 Major adsorption mechanisms of emerging contaminants

due to small size of hydrochar. Electrostatic interaction plays a very important role in adsorption of most of EC such as microplastics, pesticides, PPCP, endocrine disruptors. It relies on the difference of charge between adsorbent and adsorbate. Li et al. (2018) found removal of ciprofloxacin removal using tea leaf biochar was due to electrostatic interaction, π-π interactions and hydrogen bonding. Electrostatic mechanism is responsible for Diclofenac removal by moringa pod adsorbent (Bagheri et al. 2020). Endocrine disruptors such as bisphenol A and 17-α-ethynyl estradiol are removed by natural adsorbents derived from tea leaves and montmorillonite clay through electrostatic interaction. Hydrophobic interaction is the phenomenon that causes the hydrophobic part to form a group or aggregation called micelles (Meyer et al. 2006). Hydrogen bonding is another plausible adsorption mechanism, for most of ECs. Various types of pharmaceuticals such as Acetaminophen (Lessa et al. 2018), ciprofloxacin (Li et al. 2018), Diclofenac (Bagheri et al. 2020), Carbamazepine (Torrellas et al. 2015), Triclosan bisphenol A (Weidemann et al. 2018) were removed by rice husk hydrochar through hydrogen bonding. In a study, it was found that oxygen contained in PFAS functional group reacts with hydrogen of water or other adsorbents (Cao et al. 2017; Xu et al. 2015). Chelation and ion exchange are one of the mechanisms in removal of tetracycline through waste sludge biochar modified by chitosan (Liu et al. 2019). Apart from that, several physical adsorption mechanisms like steric interaction, π–π interaction, Van der Waals forces, hydrogen bonds, hydrophobicity, and dipole induced dipole interactions also occur in ECs. It was found that π-π interaction was majorly responsible for interaction between aromatic part of paracetamol and the syringyl group of lignin (Sophia et al. 2018). Paracetamol was also removed by guaiacyl moieties of grape stalk through π-π interaction.

11.2.2 Factors Affecting Adsorption

Mechanism of adsorption predominantly relies on the characteristics of the solution (pH, presence of inorganic ions), properties of adsorbent and properties of adsorbate, properties of adsorbent such as pore shape and size, attached functional groups, etc. are important for adsorption. These parameters are discussed separately for different EC in later sections of the chapter. Solution pH, which plays a pivotal role in adsorption, also determines the surface charge of the adsorbent, and controls the strength of the adsorbent, ionization, and species formation in aqueous media (Gao et al. 2017; Higgins and Luthy 2006). Along with pH, presence of inorganic ions in solution also has pivotal role in adsorption of EC. Some of the studies showed that adsorption of PFAS increases in presence of inorganic ions in solution (Cao et al. 2017; Deng et al. 2010) whereas some investigation also found the opposite result (Du et al. 2015; Gong et al. 2016; Lu et al. 2016). Overall, the effect is the sum result of many smaller phenomena like surface-charge neutralization, compression of electric double layers, competitive adsorption, salting out effect, and bridging between PFASs (Gong et al. 2016; Lu et al. 2016; Wang et al. 2012). Divalent cations

(Ca^{2+}, Mg^{2+}) compress the electrical double layer better than monovalent cations and aid in adsorption (Fang et al. 2018; Zhang et al. 2019). Physical properties of an adsorbent such as specific surface area, pore distribution, particle size play a significant role in EC adsorption. Varying lengths of activated carbons showed different adsorption capacities. Apart from physical properties, chemical structure of the adsorbent also plays an important role in sorption. The high adsorption capacity of PFAS on amine modified materials such as cross-linked chitosan bead, quaternised cotton are related to basic functionality on adsorbent (Du et al. 2014). The presence of other co-contaminants, such as dissolved organic carbon, natural organic materials, humus, fulvic acid etc. can also affect the adsorption affinity of sorbents for EC. In the presence of these compounds with comparable molecular sizes with ECs, intense competition between co-contaminants and EC for sorbent surface will occur, and thus overall adsorption capacity decreases.

An ideal adsorbent should be low cost, sufficient material capability to endure water flow for a significant duration of time, high adsorption capacities and larger specific surface area, easily recyclable or reusable for water treatment's sustainable and economic development (Mahfoudhi and Boufi 2017). Adsorption has several advantages over other treatment methods in terms of simplicity, initial cost, ease of practice, etc., making it a feasible alternative for the treatment of emerging contaminants. It is considered to be insensitive against toxic substances (Kumar et al. 2019). One of the foremost benefits of adsorption is the use of cheap & environmentally-friendly resources with good adsorption properties (Ning et al. 2021). In current years, various studies investigated the elimination of emerging contaminants by adsorption method via commercially available adsorbents like activated carbons, metal oxides, synthetic resins, and polymers. AC is extensively employed to remove water repelling organic and microscopic pollutants due to its great efficiency. But, its application is inadequate due to higher costs. Therefore, the need of the hour is utilization of environment friendly, efficient and low-cost material, that offer sorption with better control on selectivity (Grégorio et al. 2019). To achieve these requirements, scientists are using environment-friendly adsorbents due to their ease in availability, and innocuous nature. Various types of economical adsorbents have been investigated to eliminate water and wastewater pollutants (Kaveeshwar et al. 2018). Low-cost adsorbents are generally of five types: (a) agricultural and domestic waste; (b) industrial by-products; (c) sea or aquatic species; (d) earth material or mineral and (e) novel low-cost adsorbents (De Gisi et al. 2016). Nowadays, to remove ECs from water, biochar, a cheap and more efficient adsorbent, is being used (Delgado-Moreno et al. 2021). An effective green technology was represented by a novel adsorbent of low cost (Saravanan et al. 2021). For an increment of their adsorption capacity, numerous chemical and/or physical modifications were employed to the fresh biomass adsorbents (Carolin et al. 2021). Although, in terms of sorption capacity, these green adsorbents were not much effective compared to commercially available adsorbents such as activated carbons, modified chitosan, inorganic substances, etc., but their accessibility and lower cost makes them superior (Carmalin et al. 2018).

11.3 Removal of EC by Naturally Derived Adsorbents

Generally, commercial adsorbents are preferred for pollutant removal, but extensive industrial application is limited due to their expensive nature. As an alternative, numerous natural waste material-based adsorbents are proposed. For example, farming wastes and various forest industry products including rice and coconut husk, tea waste, cellulosic waste, almond and olive shell, orange, banana, cucumber, pumpkin and moringa peels, coffee, peanut hull, lignocellulosic wastes, polysaccharides such as alginates and chitosan, and other various materials.

11.3.1 Pharmaceuticals

Generally, pharmaceuticals compounds are the materials that have medical purposes or healthcare for animals or humans. Those materials can be further divided into various groups based on their use and nature, like anti-inflammatory drugs, β-blockers, pain tablets, blood lipid regulatory agencies, antibiotics, fungicides, repellents and preservatives (Kinney et al. 2006). Table 11.1 shows summary of pharmaceuticals removal by natural adsorbents. For the remediation of ciprofloxacin, (Afzal et al. 2018) demonstrated that the adsorption process on composite of chitosan by using pomelo peels biochar. The results of experiments indicated that its sorption kinetics followed second-order model. Based on the Langmuir isotherm model, the highest removal efficiency was more than 76 mg g^{-1}, at 160 mg/L of initial concentration. The adsorbent showed a combined mechanism of ciprofloxacin removal through π- π electron donor–acceptor interaction, hydrogen bonding, and hydrophobic interaction. In another study, (Lessa et al. 2018) used a composite of Chitosan with waste coffee ground and poly-vinyl alcohol to remove four pharmaceuticals compounds through adsorption. Results of adsorption exhibit appreciable improvement (from 10 to 44%) for the pharmaceuticals compared to the original sample. Among them, the removal of acetylsalicylic acid was maximum. The adsorption data showed a better correlation by the Freundlich isotherm, and the kinetic data of the adsorption process follow the pseudo-second-order kinetics. The composite shows notable reusability up to five cycles by subsequent adsorbent regeneration with desorption solution. (Phasuphan et al. 2019) used chitosan to alter left-over tyre scrap rubber as a competent adsorbent for the treatment of naproxen, ibuprofen and diclofenac and achieved the highest uptake capacity of 2.3, 70 and 17.7 mg/g, respectively. The results of the study show that adsorption capacity increase on increasing chitosan content upto 23%. The main mechanism was electrostatic interaction among the amino groups of chitosan and carboxylic groups of pharmaceuticals molecules. A heterogeneous surface was suggested since the isotherm data fitted with the Freundlich model. (Teixeira et al. 2012) improved the adsorption capacity of the walnut shell by chemical treatment with HCl and NaOH and studied adsorption at varying pH values.

Table 11.1 Removal of pharmaceuticals, by naturally derived adsorbents

S.No	Target contaminant	Natural adsorbent	Uptake capacity	Conc. of contaminant	% removal	Experimental conditions	Mechanism of removal	References
1	Pharmaceuticals (metamizol (MET), acetylsalicylic acid (ASA), acetaminophen (ACE), and caffeine (CAF))	Chitosan/waste coffee-grounds composite	6.29 mg/g(MET), 9.92 mg/g(ASA),7.52 mg/g(ACE) and 8.21 mg/g(CAF)	2 mg/L		adsorbent dose 50 mg; volume 250 mL; temperature 25 °C; contact time 3 h; speed 100 rpm	H-bonding, van der Waals, and hydrophobic forces	Lessa et al. (2018)
2	Sulfamethoxazole	Walnut shells		0.5 mg/l		T = 20 °C; Adsorbent Dose = 10 g/L; pH = 2–8		Teixeira et al. (2012)
3	Dorzolamide, Pramipexole	Modified Potato peels Biochar	66 mg/g (pramipexole); 60 mg/g (dorzolamide)	50 mg L/l		V = 20 ml, T = 25 °C; adsorbent dose = 1 g/L; pH = 2; SS = 160 rpm	π-π bond, oxygenated functional groups (-OH, -C-O, SO3-), higher porosity of adsorbent	Lakkovikiotis et al. (2014)
4	Sulphamethoxazole	Rice straw biochar	1.963	0 – 200 mg/L		25 °C, pH = 6, 150 rpm, 24 h, adsorbent dose-2.5 g/L		Li et al. (2015)

(continued)

Table 11.1 (continued)

S.No	Target contaminant	Natural adsorbent	Uptake capacity	Conc. of contaminant	% removal	Experimental conditions	Mechanism of removal	References
5	ciprofloxacin (CIP)	used tea-leaf biochar	238.10 mg/g	150–250 mg/l		T = 20 °C; Adsorbent Dose = 10 g/L; pH = 2–8	π–π interactions, hydrogen bonding and electrostatic attraction are inferred as the main adsorption mechanisms	J. Li et al. (2018)
6	Diclofenac	Moringa oleifera pod	60.805 mg g1	50 mg/l		T -298 K, 6 h, pH-7; adsorbent dosage- 2 g/l	electrostatic attractions, p-p interactions and hydrogen bondings	Bagheri et al. (2020)
7	Flurbiprofen	Activated carbon from orange peel hydrochars	0.092 mmol/g	0.05 mM		V- 30 ml, T -25 C, 1–36 h, pH-2; adsorbent dosage- 0.015 g	electrostatic interaction due to abundance of oxygenated functional groups & higher development of porosity on surface of adsorbent	Reck et al. (2018)

(continued)

Table 11.1 (continued)

S.No	Target contaminant	Natural adsorbent	Uptake capacity	Conc. of contaminant	% removal	Experimental conditions	Mechanism of removal	References
8	Metoprolol (MTP)	Activated carbon of pine cones	62 mg/g	50 mg/l	89.20%	V- 50 ml, pH- 8.5, adsorbent dose- 1.5 g/L, 60 min	Due to high micropore surface area, micropore volume and electrostatic interaction	Naghipour et al. (2019)
9	Carbamazepine	Activated carbons from peach stones	335 mg/g	100 mg/l		V-50 ml, T -30 C,, adsorbent dosage- 0.12 g	hydrophobicity, p–p bond, electrostatic interaction and H-bonding interaction	Torrellas et al. (2015)
10	Triclosan bisphenol A	Rice husks Hydrochar	2 μg/g	10 μg/L	Almost complete removal	V-10 ml, T -20 C, 1–25 min., pH-6.5; adsorbent dosage- 50 mg	Polar rice husk promote the removal of more polar molecules via H-bonding	Weidemann et al. (2018)

The modified walnut shell exhibited higher adsorption capacity for sulfamethoxazole. The results of the adsorption process were extremely influenced by pH (2–8) and removal efficiency decreased for pH >2. The maximum removal capacities on raw HCl & NaOH modified walnut shells were found to be 91 and 78 and 97%, respectively. Similarly, (Lakkovikiotis et al. 2014) used modified AC from potato peels by activation with KOH to remediation of dorzolamide and pramipexole. The data of adsorption equilibrium was best fitted by Freundlich isotherm, and kinetic experiments follow pseudo-second-order equations. The main adsorption interaction among drug molecules and adsorbents was π-π assembling. The hydrothermally treated carbon exhibits excellent reusability up to 20 cycles by losing only 11% of its adsorption capacity. For sulphamethoxazole, (Li et al. 2015) demonstrated the sorption capacity of biochar prepared from rice straw and alligator flag. Author reported that data of adsorption equilibrium was best fitted by Langmuir equation (R^2 >0.94), and the highest sorption parameter (Q) was found to be 3650 mg kg^{-1} for rice straw. The removal efficiency of rice straw was more than that of alligator flag at pH <7, however, removal efficiency AF exceeded RS when pH $\geq$ 7. It was also found that due to the availability of Cu^{2+} and/or Cd^{2+} ions in fewer levels, sorption improved significantly on both adsorbents. In an almost identical investigation, (Li et al. 2018) evaluated the efficacy of used tea-leaf biochar for removing ciprofloxacin molecules from water-based media. After optimization, the authors obtained quite high ciprofloxacin sorption capacity, i.e., 238.10 mg/g. The main mechanisms of adsorption were π–π interactions, hydrogen bonding, and electrostatic attraction. Diclofenac adsorption was also studied on moringa seed biochar after chemical activation with phosphate (Bagheri et al. 2020). Isotherm data was best fitted using Sips model and the maximum sorption was found to be 100.876 mg/g by Sips model showing 82.8% removal efficiency at pH ~5. One more research was conducted, in which (Reck et al. 2018) used activated hydrochars of orange peels after several thermochemical processing for elimination of diclofenac, salicylic acid and flurbiprofen from water. The adsorbent exhibited the best adsorption properties with advanced development of porosity, showing surface areas ranging from 300 to 620 m^2/g. In another work, (Naghipour et al. 2019) investigated the elimination of metoprolol by AC prepared from pine cones using thermal activation process. The optimal abatement efficiency was found to be 89.2% at pH 8.5 in 60 min. with an initial concentration of 50 mg/l. Isotherm data were best fitted by Langmuir isotherms and kinetic data were better demonstrated by pseudo-second-order kinetic model. The effect of adsorbent dose and initial concentration was tested, adsorbent dose exhibits a positive effect on removal efficiency, and however, an increase of initial concentration reduces removal efficiency. (Torrellas et al. 2015) prepared AC from peach stones by chemical activation with H_3PO_4 for the removal of Carbamazepine from liquid media. The maximum adsorption capacity of this adsorbent towards carbamazepine was determined to be 335 mg/g. The adsorbent exhibits higher surface area values (1216 m^2/g) and pore volume (0.81 cm^3/g), favoring the adsorption process. The study reported that hydrophobicity and water-solubility were responsible for the high adsorption. (Yoong et al. 2020) developed composite of chitosan with activated carbon of magnetite rice husk. It was reported that the impregnation of chitosan on the AC improved the diclofenac adsorption capacity by 91%, but, when ratio of AC/chitosan increased from 1:2, the adsorption was reduced because of clogging of pores of AC. In acidic medium, amine group of chitosan attached on the surface of activated carbon enhance adsorption capacity to 270 mg/g which led to the attainment of 94% adsorption of diclofenac. In another study, (Yi et al. 2016), adsorption of levofloxacin was evaluated for rice husk and wood chip biochars. It was reported that elemental compositions of biochar change with pyrolysis temperature. On increasing temperature from 300 to 600 °C, carbon contents increase, however, oxygen and hydrogen contents decrease, resulting in biochar became less polar and more aromatic. Surface areas of biochar obtained at 600 °C were about 100 times more as compared to biochar at 300 °C. Their interaction between levofloxacin and biochar was more aromatic with less polar and hydrophobic interactions. Powder of moringa pods and husk was used by (Viotti et al. 2019) for the sorption of diclofenac. After chemical activation with HCl, thermal treatment was given in order to produce highly porous activated carbon. FTIR analysis of sorbent shows abundance of various functional groups that were responsible for diclofenac removal by hydrogen bonding, electrostatic interaction and π-π interactions. (Fernandez et al. 2015) used orange peels for synthesis of hydrochar by heating at 200 °C for 20 h and compared the results of adsorption of pharmaceuticals (flurbiprofen, diclofenac sodium and salicylic acid) after chemical activation with CO_2 & 50% phosphoric acid solution. Due to removal of volatile matter at 200 °C, carbon content and porosity of hydrochar significantly increased. Activation of hydrochar by CO_2 exhibited a highly basic character however activation by phosphoric acid developed an acidic character that increases oxygenated functional groups on the surface. At pH 2, the adsorption efficiency of the adsorbents significantly increased for the removal of flurbiprofen and salicylic acid.

11.3.2 Endocrine Disruptors

Several sorbents have been used for the elimination of these compounds from diverse water conditions. Table 11.2 gives removal summary of endocrine disruptors by naturally derived sorbents. Most of the works emphasize bisphenols, parabens, and triclosan. (Dehghani et al. 2016) investigated the adsorption of Bisphenol-A from liquid solutions using chitosan. The kinetic data revealed that synthesized chitosan as well as commercial chitosan, both follow the pseudo-second-order model. The experimental data of isotherm confirmed that both adsorbents follow the Langmuir model. This study found that the abatement efficacy of both chitosan decreases by increasing initial concentration and decreasing the dose of adsorbent. The maximum removal capacity of Bisphenol on synthesized chitosan and commercial chitosan were 27.02 and 34.48 mg/g respectively, at pH 5. In another study, (Weidemann et al. 2018) prepared hydrochars from agricultural waste for the adsorption mixture of ten

Table 11.2 Removal of endocrine disruptors, by naturally derived adsorbents

S.No	Target contaminant	Natural adsorbent	Uptake capacity	Conc. of contaminant	% removal	Experimental conditions	Mechanism of removal	References
1	Bisphenol-A (BPA)	Chitosan	34.489 mg/g at C-0.1 mg/l	0.25, 0.5, 1.0 & 2.0 mg/l	80%	adsorbent dose of 0.06 g/L, 75 min, pH-5		Dehghani et al., (2016)
2	Bisphenol A	Sulfonic acid functionalized carbonaceous adsorbent (TW-SO3H) from tea leaves	236.80 mg/g			V- 20 ml, adsorbent dose- 10 mg, 24 h,	electrostatic interaction and π–π stacking	Ahsan et al. (2018)
3	bisphenol A (BPA)	Argun Nut shells Hydrochar	1162.79 mg/g	60 mg/l	92%	V- 200 ml, T -293 K, 6 h, pH-6.5; adsorbent dosage- 0.01 g	π-π interactions, Hydrogen bonding and Pore filling	Zbair et al. (2020)
4	diuron	Argun Nut shells Hydrochar	833.33 mg/g	40 mg/l	95%	V-200 ml, T -293 K, 6 h, pH-6.5; adsorbent dosage- 0.01 g	π-π interactions, Hydrogen bonding and Pore filling	Zbair et al. (2018)
5	17-α-ethynyl estradiol	Montmorillonite rice husk hydrochar composites	69 mg/g	6 mg/l		V-100 ml, T- 25 C, 24 h, pH 2–8, adsorbent dosage- 5 mg	hydrophobicity, p–p bond, electrostatic interaction and H-bonding interaction	Tian et al. (2018)

(continued)

Table 11.2 (continued)

S.No	Target contaminant	Natural adsorbent	Uptake capacity	Conc. of contaminant	% removal	Experimental conditions	Mechanism of removal	References
6	Triclosan (TCS),	Hydrochar of olive oil wastes	13.7 mg/g	50 mg/l	98%	V-50 ml, T -20 C, 24 h, adsorbent dosage- 0.05 g	hydrogen bonding with the carboxyl and carbonyl groups of the hydrochars and through hydrophobic interaction with the aromatic surface	Delgado-Moreno et al. (2021)
7	p-methoxyphenol (PMP)	Activated carbon of olive husk	45.45 mg/g	50 mg/l	98%	V- 50 ml, adsorbent dose- 0.2 g, 24 h, pH-7, T- 25 C	hydrophobic effects, $\pi-\pi^*$ interaction and hydrogen-bonding	Hamadneh et al. (2020)
8	4-bromophenol (4-BrPhOH)	Activated carbon of shell of the cashew of Para (SCP)	488.2 mg/g	500 mg/l	98.64%	pH 7.0; adsorbent dose = 30 mg; temperature = 25 °C	microporous filling, hydrogen bonds, π-stacking interactions, and other Van der Waals interactions	Thue et al. (2020)

ECs (sulfamethoxazole, triclosan, fluconazole, octhilinone, paracetamol, trimethoprim, diphenhydramine, ciprofloxacin, bisphenol A, and diclofenac). The highest BET-determined specific surface areas (16.92 m^2/g) and maximum overall removal efficiency (66%) were exhibited by rice husk char. It demonstrated significant efficiency ($\geq$83%) of triclosan, octhilinone, and ciprofloxacin removal from the mixture. All constituents were removed rapidly and the removed amount saturated after 1–3 min. Similarly, (Ahsan et al. 2018) prepared sulfonic acid functionalized carbonaceous adsorbent from tea leaves for the sorption of bisphenol-A. The maximum sorption capacity of bisphenol was found to be 236.80 mg/g at 25 °C. The thermodynamic study indicated that the adsorption was exothermic and spontaneous. The $\pi - \pi$ interactions among the adsorbate and adsorbent molecules were responsible for the removal of a contaminant. The study also claimed a very nominal loss of adsorption capacity up to three cycles of adsorption/desorption. To remove bisphenol and diuron, (Zbair et al. 2020) synthesized hydrochar from argan nut shell. The hydrochar obtained at 200 °C showed more specific surface area (42 m^2/g) than hydrochar prepared at 180 °C (17 m^2/g). The hydrochars showed 92 and 95 adsorption efficiency for the abatement of bisphenol and diuron respectively. The maximum Langmuir adsorption capacity was 1162.79 mg/for Bisphenol A and 833.33 mg/g for diuron at room temperature (more than most investigated adsorbents). The thermodynamic study of adsorption process suggested that the process was spontaneous (ΔG° <0) and exothermic (ΔH° <0). Significant recyclability of adsorbent was claimed up to five cycles. (Tian et al. 2018) used hydrothermal carbonization process to synthesized montmorillonite/hydrochar (MMT/HC) composite for the sorption of 17b-estradiol (E2) and 17a-ethynylestradiol (EE2). After activation with 1% KOH, composite exhibited almost double removal capacity as compared to hydrochar, also the composites showed good adsorption capacity of E2 & EE2, for a pH range from 2 to 8. The results of study shown excellent removal capacity, approximately 80–82% of its initial capacity up to four cycles. For elimination of triclosan (TCS) from water, (Delgado-Moreno et al. 2021) used the pyrolysis and hydrothermal carbonization processes to obtained chars from olive oil production wastes. With the increase of pH of char suspensions, their adsorption capacities decreased. The highest adsorption rate for TCS was shown by a hydrochars was synthesised at the lowest temperatures ($\leq$240 °C), which was rich in oxygenated functional groups. For the adsorption of endocrine disrupting compounds (bisphenol A (BPA), atrazine (ATR), 17 α-ethinylestradiol (EE2)), Jung et al. (2013) characterized and used biochars of pine wood chips produced by chemical activation in the presence of oxygen (O-biochar) and in the absence of oxygen (N-biochar) and used After analysis, N-biochar showed higher polarity and more alkyl, O-alkyl, methoxyl, and carboxyl carbon content, while O-biochar exhibited more aromaticity as compared to N-biochar. The adsorption capacity of N-biochar was high toward contaminants because of greater pore volume and surface area. The H/C ratio indicated by biochars was 0.034 and 0.127 for O-biochar & N-biochar respectively, it showed that N-biochar was less carbonized (lesser hydrophobic) than O-biochar. (Hamadneh et al. 2020) reported that adsorption capacity of AC synthesised from olive husk for p-methoxyphenol (PMP) was increased from 45.455 to 98.039 mg/g after chemical activation by using $MgCl_2$. The adsorption mechanism study showed that PMP adsorption was primarily improved by introducing electron withdrawing and donating functional groups on the benzene ring. The result of equilibrium experiments exhibited that the maximum removal capacity (Q_{max}) reached at 45 °C was 488.2 mg g^{-1}.

11.3.3 Personal Care Products

In technical terms, Personal Care Products (PCPs) are complex combinations of chemicals in various forms like biocides, stabilizers, gels, creams, emulsifiers, lotions, semisolids, pH regulators, dyes, and many others. These compounds are generally used in lotions, toothpastes, fragrances, soaps, skincare products, conditioners, shampoos and insect repellents. Due to economical and suitability, biochar and activated carbon of various natural/waste materials have continuously been employed for the removal of these compounds (Ahmed et al. 2015). However, to improve the efficiency of removal process, various types of pre-treatments have been applied onto natural adsorbents. (Phasuphan et al. 2019) established adsorptive process of a new adsorbent synthesized by the waste tyre scrap rubber particles and having surfaces improved with adsorbed polymeric chitosan. The adsorption of naproxen was verified in this process with a Freundlich isotherm design of the maximum sorption capacity of 2.3 mg/g. The optimal adsorption was observed to be attained when the solution's pH was 6. Another contaminant that was removed from the system was the drug ibuprofen, which showed a maximum sorption capacity of 70 mg/g. To remove triclosan, (Zhu et al. 2014) proposed a new method for simultaneous magnetization and activation of hydrochar via pre-treatment with mixtures of ferric chloride ($FeCl_3$) and zinc chloride ($ZnCl_2$) followed by thermal pyrolysis. Activated hydrochar exhibited high values of specific surface area (1351 m^2/g) and pore volumes (0.549 cm^3/g), favoring triclosan adsorption. In an acidic medium (pH of 3.0 and 4.0), the novel adsorbent showed excellent removal of triclosan from aqueous solutions. Similarly, (Moreno-Marenco et al. 2019) presented the adsorption of parabens on activated carbons obtained from African palm shell after chemical modification with different metallic salts. The calorimetric results showed positive interaction enthalpy values (endothermic) for the adsorption of parabens. The investigation related adsorption process of the methylparaben with the occurrence of acid groups, especially lactonic groups by a complex "donor–acceptor" mechanism between the aromatic ring of methylparaben and carbonyl oxygen of lactone. It was also reported that both adsorbate–adsorbent interactions and adsorbent–solvent interactions were taking place. Another work (Torrellas et al. 2015) synthesized AC of peach stones by chemical modification with H_3PO_4(s), resulting in higher specific surface area and well-developed mesoporous materials. And adsorption of caffeine (stimulant) from the water was investigated. It was reported that there was a competition among the caffeine molecule and water molecules for the active sites, and acidic water pH, 4.8, favoured this competitive effect. Recently, (Czech et al. 2020) used a novel cotton-based adsorbent for sorption of naproxen (NAP). It was found that both adsorption capacity and mechanism of sorption were significantly affected by carbonization temperature and the presence of co-solute. The presence of co-solute caffeine increased NAP sorption and segmentation. The sorption of NAP onto adsorbent obtained at 1100°C was the highest without the co-solutes. (Li et al. 2017), studied a novel economical clay-biochar composite with potato stem and natural attapulgite for Norfloxacin removal from water. The result of batch experiment showed that maximum sorption capacity of 5.24 mg/g, which was almost 1.68 times more than the original biochar. The isotherm data shows better correlation by the Langmuir isotherm and kinetic data of the adsorption process followed closely the pseudo-second order model. The thermodynamic analysis indicated that the adsorption of the contaminant was spontaneous and exothermic. For the elimination of PCP from water, (Chauhan et al. 2020) prepared microporous aluminium-pillared clay (Al-PILC) by using Smectite clay (montmorillonite, MMT) from impregnation of Al-oxide pillars into the interlayer structure, which increased the surface area of natural clay to 258 m^2/g and its porosity to 0.16 cm^3/g. The abatement efficiency of novel sorbent was 681% more than natural clay, and shows a maximum adsorption capacity of 7.7 mg/g by following pseudo-second-order model. The Al-PILC showed effective regeneration up to three consecutive adsorption/desorption cycles. Table 11.3 shows removal of personal care products by natural adsorbents.

11.3.4 Per and Polyfluoroalkyl Substances

PFAS which stands of per and polyfluoroalkyl substances are predominantly removed by adsorption, some of the natural adsorbents for its removal are discussed in this section. Table 11.4 shows summary of different PFAS removal by natural adsorbents. Biochar which is formed from the combination of biomass and charcoal, is a porous high carbon rich solid derived from wood or manure through pyrolysis. It is a possible alternative to activated carbon AC and its effectiveness increases with the increase in surface area. Several mineral materials like activated alumina, silica, zeolite, and clay minerals like montmorillonite are being used for adsorption. Mineral has tuneable mesopores making them suitable for adsorption (Du et al. 2014). It is found that hydrotalcite is a better adsorbent for PFAS removal than other simple mineral adsorbents (Rattanaoudom et al. 2012). Rattanaoudom et al. (2012) also found that hydrotalcite had greater elimination efficiency for PFAS compared to normal mineral adsorbents because double layered hydroxide exhibits better affinity towards PFAS inked chitosan beads had greater uptake capacity for PFOS at pH 3 (Zhang et al. 2019). To increase adsorption performance and reduce cost of adsorbent, scientists have developed effective novel absorbents such as low-cost biomaterials. For example, a study conducted by Zhang et al. (2011) to remove PFOS by crossed linked chitosan beads, showed that sorption of PFOS was both dependent on pH and concentration of solution and the kinetic results showed that intraparticle diffusion was the rate limiting step of the reaction mechanism. The main mechanisms

Table 11.3 Removal of personal care products, by naturally derived adsorbents

S.No	Target contaminant	Natural adsorbent	Uptake capacity	Conc. of contaminant	Experimental conditions	Mechanism of removal	References
1	Naproxen	chitosan-modified waste tire crumb rubber adsorbent	2.3 mg/g	3 mg/l	pH 6, V-10 ml, T -25 C, 24 h, adsorbent dosage-3 g/l	interaction b/w adsorbate molecules and chitosan-modified surfaces, p–p bond	Phasuphan et al. 2019
2	Triclosan	Magnetic hydrochars from Salix psammophila	892.9 mg/g	50 mg/l	V-50 ml, T -30 C, 12 h, adsorbent dosage- 0.5 g/l	Pore filling, the high surface areas and pore volumes of carbonaceous material can promote organic matter adsorption, due to the pronounced microporefilling effect	Zhu et al. 2014
3	Methylparaben (MePB)	activated carbons obtained from African palm shell	167 mg/g	20–200 mg/l	V-50 ml, T -18 C, 3 week, adsorbent dosage- 0.025 g/l	hydrophobicity, p–p bond, electrostatic interaction and H-bonding interaction	Moreno-Marenco et al. 2019

(continued)

Table 11.3 (continued)

S.No	Target contaminant	Natural adsorbent	Uptake capacity	Conc. of contaminant	Experimental conditions	Mechanism of removal	References
4	Caffine	Activated carbons from peach stones	260 mg/g	100 mg/l	Voulume of 50 ml, 30 C,, adsorbent dosage of 0.12 g		Torrellas et al. 2015
5	naproxen (NAP)	cotton based adsorbent	19.73 mg g − 1	20 mg/l	Volume of 40 ml, T -23 + 2 C, 7 days, adsorbent dosage- 20 mg	heterogeneous adsorption, hydrophobic and electrostatic interactions, hydrogen bonds or π–π interactions	Czech et al. 2020
6	Amoxicillin (AMOX)	aluminum pillared clay	7.7 mg/g	100 mg/l	pH 2–12, V-50 ml, T -25 C, 0–180 min, adsorbent dosage- 0.1 g	The incorporated acidic sites in clay in form of Al_2O_3 pillars enhanced adsorption, sites interact with protonated amine and –OH groups of AMOX	Chauhan et al. 2020
7	Norfloxacin	Attapulgite modified biochar	5.24 mg/g	10 mg/l	T- 308 K, 24 h, adsorbent dose- 0.1 g, V- 25 ml, pH 2–11	electrostatic attraction/repulsion between NOR and biochar, hydrogen bonding formation, hydrophobic interaction, π-π electron–donor–acceptor	Li et al. 2017

Table 11.4 Removal of PFAS by naturally derived adsorbents

S.No	Target Contaminant	Natural Adsorbent (Identity & form)	Uptake Capacity	Conc. of EC	% removal	Experimental Conditions	Mechanism of removal	References
1	PFOS	Crossed linked chitosen beads	5.5 mmol/g	0.33 mmol/L		150 rpm,25 °C, pH 3	electrostatic and hydrophobic interactions; hemi-micelles and micelles formation	Zhang et al. 2011
2	PFOA	β-cyclodextrine based polymer network		1 μg/L	>99.99%			Xiao et al. 2017
3	PFOS & PFHxS	Cannabis Sativa L. (hemp)			>98%		adsorption with H bonding	Turner et al. 2019
4	PFOA	chitosan–ethylene glycol hydrogel	1275.9 mg/g,				Ionic hydrogen bonding	Long et al. 2019
5	PFOA	Fully stabilized magnetite (Fe3O4) nanoparticles	62.5 mg/g		62%			Gong et al. 2016
6	PFOA & PFOS	quaternized cotton	3.3 and 3.1 mmol/g for PFOS & PFOA			pH-5	anion exchange	Deng et al. 2012
7	PFOA, PFOS& PFBA	aminated rice husk	2.49, 1.70 and 2.65 mmol g1 for PFOA, PFBA and PFOS			pH5	electrostatic & hydrophobic interaction	Deng et al. 2013
8	PFCAs	hierarchically microporous biochar (HMB)	1269 mg/g (PFOA)		> 90%		electrostatic & hydrophobic interaction	Zhou et al. 2021

(continued)

Table 11.4 (continued)

S.No	Target Contaminant	Natural Adsorbent (Identity & form)	Uptake Capacity	Conc. of EC	% removal	Experimental Conditions	Mechanism of removal	References
9	PFOS	sawdust derived biochar (SDN600) & red mud modified biochar (RMSDN600)	RMSDN600-194.6 mg/g & SND600-178.6 mg/g	248.48 mg/L,		pH3.1, 4 h,	electrostatic & hydrophobic interaction	Hassan et al. 2020
10	PFOA & PFOS	Bamboo derived GAC	2.32 mmol/g for PFOS & 1.15 mmol/g for PFOA at pH 5.0					Deng et al. 2015
11	PFOA & PFOS	Vitis vinifera (grape) leaf litter	With AC-H_3PO_4 were 78.90 &75.13 mg/g for PFOA & PFOS		95 and 90% for PFOA using AC-H_3PO_4 & AC-KOH; & PFOS were 94 and 88%		physisorption	Fagbayigbo et al. 2017

of sorption for this experiment were electrostatic, hydrophobic interactions, hemi-micelles and micelles formation (Zhang et al. 2011). In another study, (Deng et al. 2013), successfully synthesized aminated rice husk adsorbent through radical polymerization, followed by amination reaction. After polymerization surface irregularity increases, some polymers with amine groups were attached on the biochar surface, which provides protonated amine sites for anionic perfluorinated compounds. The aminated rice husk showed fast and high sorption of perfluorinated compounds. The highest adsorption capacities of PFOS, PFOA, and PFBA were 2.65, 2.49, and 1.70 and mmol/g, respectively. Grape leaf litter powder was used to produce activated carbon that was further modified in both acidic (H_3PO_4) and basic media (KOH), removed PFOS and PFOA with greater than 88% efficiency (Fagbayigbo et al. 2017). Maximum adsorption capacities for AC-H_3PO_4 were 78.90 and 75.13 mg/g for PFOA and PFOS, respectively, achieved within 60 min (Fagbayigbo et al. 2017). Apart from that, its desorption was found to be 99% for the removal of PFOS and PFOA with 100% methanol, 96% using 100% acetonitrile and 85% with 0.1 M HCl. SDN600 (Thermally modified hydrophobic biochars) and RMSDN600 (metallic oxide modified biochars) removed PFOS in an aqueous medium with adsorption capacities of 178.1 mg/g and 194.6 mg/g at pH 3.1, respectively (Hassan et al. 2020). In another study, quaternised cotton which was prepared by atom transfer radicle polymerisation showed faster and higher removal capacities of PFOA and PFOS i.e. 3.1 and 3.3 mM/g respectively and the sorption equilibrium of PFOA and PFOS on quaternized cotton was achieved at 4 h and 12 h, respectively, much faster than that on the porous adsorbents reported (Deng et al. 2012). It was also observed that the solution pH has a negligible effect on adsorption (Deng et al. 2012). Crossed linked chitosan beads were able to remove PFOS concentration of 0.33 mmol/L with an uptake capacity of 5.5 mmol/g at room temperature and pH 3 through electrostatic, hydrophobic, micelles, and hemimicelles formation (Zhang et al. 2011). In a study demonstrated by Turner et al. (2019), hemp (Cannabis Sativa L.) protein was found to remove PFOS and PFHxS with a removal efficiency greater than 98% in contact time of just 1 h. Hierarchically microporous biochar (HMB), which were prepared via molten alkali (KOH) treatment on discarded coconut shells, provided uniform microporous structure and large surface area which in turn imparted a high adsorption capacity of 1269 mg/g for PFOA (Zhou et al. 2021).

11.3.5 Pesticides

Pesticide means any substance used to eliminate pest in any form pests or are intended for use as plant regulators (USEPA 2007). Table 11.5 shows a summary of different pesticides removal by naturally-derived adsorbent, found in literature. Different models such as Freundlich, Langmuir, BET, Temkin, Huttig and Jovanovic were investigated to study remediation of HCH pesticide from liquid media solutions (Ignatowicz et al. 2011). (Salman et al. 2011) prepared activated carbon of banana stem by chemical modification in the presence of CO_2 and nitrogen using

Table 11.5 Removal of pesticides by naturally derived adsorbents

S. No	Target Contaminant	Natural adsorbent	Uptake Capacity	Conc. of EC	% removal	Mechanism of removal	References
1	2,4-Dichlorophenoxyacetic acid	Mustard Plant Ash	0.76 mg/g		>90%		Trivedi et al. (2016)
2	Bentazon	Activated Carbon	169.4 mg/g			Chemisorption	Omri et al. 2016
3	Hexachlorocyclohexane	Activated carbon derived from coconut shell	650 mg/g				Ignatowicz (2011)
4	Atrazine	Biochar from Banana peel	14 mg/g				Chaparadza and Hossenlopp (2012)
5	2,4-Dichlorophenoxyacetic acid	Activated carbon from Langsat empty fruit bunch	261.2 mg/g				Njoku et al. (2015)
6	Ethylparaben	Activated biochar from oil palm fiber	349.65 mg/g				Zhou et al. (2019)

(continued)

Table 11.5 (continued)

S. No	Target Contaminant	Natural adsorbent	Uptake Capacity	Conc. of EC	% removal	Mechanism of removal	References
7	Triazine pesticides	biochar from corn straw	79.6 mg/g	2 mg/L	> 96%	Van der Waals' forces, hydrogen bonding, pore filling & electrostatic interactions	Suo et al. 2019
8	Diazinon	walnut shell modified activated carbon	34.98 mg/g	40 mg/L			Bayat et al. 2018
9	Atrazine and Imidacloprid	Biochars made from (eucalyptus bark, corn cob, rice husk, bamboo chips, rice straw,acid treated rice straw biochar (RSBC)			For RSBC, atrazine (37.5–70.7%) & Imidacloprid (39.9–77.8%)	entrapment of sorbed molecules inmeso-andmicroporous structures within mineral structures	Mandal, et al. 2017

(continued)

Table 11.5 (continued)

S. No	Target Contaminant	Natural adsorbent	Uptake Capacity	Conc. of EC	% removal	Mechanism of removal	References
11	Metalaxyl and Tricyclazole	Natural clays (Bentonite, Ghassoul clay, etc.)			Adsorption data- higher formetalaxyl than for tricyclazoleTricyclazole- π electrons of the aromatic rings, electron pairs of the S atom	Metalaxylelectrostatic interactions & hydrogen bonds	Azarkan et al. 2016
12	Metalaxyl and Fludioxonil	Natural clay	284 ± 4 μg g − 1) -adsorption of Metalax 176 ± 1 μg g − 1 -adsorption of Fludioxonil			Electrostatic interactions	Rodríguez-Liébana et al. 2016
13	Bentazon, Carbofuran and 2,4- Dichlorophenoxyacetic acid (2,4-D)	Activated carbon from palm oil fronds			8.2, 1.3 and 9.2% and yeild of palm oil frond was 5.6, Chemical impregnation ratio-2.38	π-π interactions Hydrophobic interactions	Salman 2014
14	Atrazine	Biochars from soybeans, corn stalks, rice stalks and manures of poultry, cattle, and pig	0.48, 0.96, 1.42, 2.32, & 3.05 mg/g soybean biochar	5,10,15,25,35 mg/L		H-bonding interactions π-π interactions, electron donor–acceptor interactions	Liu et al. 2015

KOH. The adsorption process was studied for the removal of bentazon and 2, 4-dichlorophenoxyacetic acid (pesticides) with a variation of temperature. It was reported that adsorption capacity of both pesticides decreased with the increase in temperature. This could be attributed to the enhancement of the desorption step of the adsorption mechanism due to the dislodgement of the adsorptive forces. The thermodynamic study presented that the adsorption of pesticides was exothermic, spontaneous and feasible. In a similar study, burned banana peel was used after phosphoric acid treatment for the adsorption of atrazine (Chaparadza and Hossenlopp 2012). The adsorption was observed to be pH and temperature dependent. More than 80% of atrazine removal was reported between 2 to 8 pH of solution. It was also reported that intraparticle diffusion and external mass transfer have significant effects on the adsorption process. With the increase of initial concentration of atrazine, adsorption capacity increased. The result showed that maximum adsorption capacity was 14 mg/g through monolayer adsorption. In another investigation by Suo et al. (2019), P doped biochar which was obtained from corn straw, efficiently removed triazine pesticides with removal greater than 96% and sorption capacity of 79.6 mg/g in just 20 min. Another advantage of this biochar is that it can be reused multiple times. The mechanisms responsible for its removal are Van der Waals' forces, hydrogen bonding, electrostatic interactions and pore filling (Suo et al. 2019). In a study demonstrated by Bayet et al. (2018), wherein walnut shell activated carbon were modified using H_3PO_4, resulting in increased surface area and resulted in removing diazinon pesticide from water media with significant efficiency under wide pH ranges. The highest removal abatement was observed to be 34.98 mg/g with an initial concentration of 40 mg/l (Bayet et al. 2018). Mandal et al. 2017 characterized rice husk and rice straw properties and studied for sorption of imidacloprid and atrazine. Rice husk and straw were pyrolyzed at 600 °C temperatures for 1 h and then and biochar of rice straw was chemically modified using H_3PO_4. Among these, the biochar of rice straw was most porous but after chemical activation surface area and void volume of biochar decreases, however, functional group was abundant. The biochar synthesised from the rice straw showed adsorption of atrazine and imidacloprid 37.5–70.7% and 39.9–77.7% respectively, which further enhanced after H_3PO_4 treatment. In the study conducted by Liu et al. (2015) properties of different biochars produced from bio-agricultural wastes such as soyabeans, corn stalks, rice stalks, poultry manure, cattle manure and pig manure were characterised for the removal of atrazine, out of which soyabean biochar showed maximum removal of 0.48, 0.96, 1.42, 2.32 and 3.05 mg/g when the initial concentration of atrazine were 5, 10, 15, 25, 35 g/L respectively. Also, it was observed that atrazine adsorption increased with increasing initial concentration and temperature, also pore volume and biochar pH play a significant role in its adsorption (Liu et al. 2015). A similar study was performed for the removal of pesticides, such as atrazine and imidacloprid removal using biochars made from different feed materials such as eucalyptus bark, corn cob, bamboo chips, rice husk and rice straw and acid treated rice husk (Mandal et al. 2017). He found that different feed had different physicochemical properties and acid modified rice straw showed maximum removal efficiency of atrazine and imidacloprid of 37.5–70.7% and (39.9–77.8%)

respectively, and it was also concluded that the factors that affect pesticide adsorption on biochar are aromaticity, polarity, pore-volume, pore diameter, pH and surface acidity. Activated carbon was prepared from palm oil fronds, using the physiochemical activation method, which consisted of potassium hydroxide (KOH) treatment and carbon dioxide (CO_2) gasification and removed pesticides such as bentazon, carbofuran and 2,4-Dichlorophenoxyacetic acid with efficacy of 8.2, 1.3 and 9.2%, respectively and for the yield of the palm oil frond activated carbon was 5.6 (Salman 2014).

11.3.6 Nanomaterials and Microplastics

Nanomaterials are particles whose size is one dimension lesser than 100 nm and are of various types (Ahmed et al. 2016). Silver nanoparticles (Ag NP) which are toxic to living organisms, were removed by sodium montmorillonite nanoclay in a study demonstrated by Zarei et al. (2015). They found maximum removal efficiency of AgNP at 7 pH, 10 min shaking time, initial concentration of 10 μg/m L Ag NP to be 71.4 mg/g (Zarei et al. 2015). Setyono and Valiyaveettil (2016) developed polyethyleneimine (PEI)-functionalized paper which showed significant removal efficiencies of Ag-PVP, Ag-citrate, Au-PVP, Ag-citrate nanoparticles to be 46, 79, 17 and 30 respectively. Table 11.6 shows summary of nanomaterials removal by natural adsorbents. Microplastics are particles of 5 mm or lesser diameter (Thompson et al. 2009), and get accumulated in the environment as they are resistant to degradation (Hossain et al. 2020). Various types of natural adsorbents used in literature that showed its significant removal of microplastics, are summarised in Table 11.7. Oat Protein Sponges (OPS), which was synthesised by chemical crosslinking method with 2% NaCl content, was able to remove microplastics with a removal efficiency of 81.20% at initial concentration of 100 mg/l, room temperature and 7 pH (Wang et al. 2021). In another study, iron modified biochar was pyrolysed at two different temperatures of 550 °C (FB 550) and 850 °C (FB 850). It was used to remove nanoplastics of various size and surface functionality i.e. NP1 (1000 nm, carboxylate-modified) and NP2 (1000 nm, amine modified) with adsorption capacities of 225.11 mg/g and 290.2 mg/g with FB 850 and NP 3 (30 nm, carboxylate-modified) with uptake capacity of 206.46 mg/g with FB 550 (Singh et al. 2021). In a study conducted by Siipola et al. (2020), low cost steam activated biochars could remove 2 g of microplastics altogether. Surface functionalised cellulose fibers removed polymer nanoparticles of 0.02 ml/L concentration with an efficiency of 98% with charge induced adsorption (Batool and Valiyaveettil et al. 2021). In the study demonstrated by Sun et al. (2020), which showed effective remediation of microplastics of varying functional groups i.e., polystyrene, carboxylate-modified polystyrene, and amine modified polystyrene, by using sponges made from chitin and graphene oxide, showed with a removal efficiency of 89.8%, 72.4%, and 88.9% respectively. Apart from high efficiency, these sponges showed good regenerability, better biodegradability, and eco-friendliness.

Table 11.6 Removal of nanomaterials by naturally derived adsorbents

S.No	Target Contaminant	Natural Adsorbent (Identity & form)	Conc. of EC	% removal	Experimental Conditions	Mechanism of removal	References
1	Microplastics (PNP-polymer nanoparticles)	Surface functionalized cellulose fibers	0.02 ml/L	(>98%)	aqueous solution of the scrub solution of nanoparticles (20 mL) containing the stained microplastic particles were mixed with PEI@CE (13.2 mg) and agitated at 250 rpm for 30 min on a mechanical shaker	charge induced adsorption	Batool and Valiyaveettil et al. 2021
2	Microplastics	steam-activated biochars	2 g	100%	25 C	unknown	Siipola 2020
3	Microplastics	Oat Sponges Protein(OSP)	100 mg/L	81.20%	OSP-2 (Oat Sponges Protein with 2% (w/v) NaCl final content calculated from Milli-Q water) at 25 ° C at 7 pH,	hydrophobic interactions and intra-particle diffusion	Wang et al. 2021
4	Microplastics	Ironmodified biochar- FB 850 (at 850 °C) & FB 550 (at 550 °C)	10 mg/L		200 RPM at 25 °C	surface complexation in sorption process	Singh et al. 2021

(continued)

Table 11.6 (continued)

S.No	Target Contaminant	Natural Adsorbent (Identity & form)	Conc. of EC	% removal	Experimental Conditions	Mechanism of removal	References
5	Microplastics-polystyrene, carboxylate-modified polystyrene, & amine modified polystyrene	chitin and graphene oxide based sponges	1 mg/L	89.8%, 72.4%, 88.9% for Microplastics polystyrene, carboxylate-modified polystyrene,and aminemodified polystyrene	24 h	electrostatic interactions, hydrogen bond interactions & $\pi - \pi$ interactions	Sun et al. 2020

Table 11.7 Removal of microplastics by naturally derived adsorbents

S. No	Target Contaminant	Natural Adsorbent (Identity and form)	Uptake Capacity mg/g	Conc. of EC	Experimental Conditions	References
1	Ag nanoparticles	sodium montmorillonite nanoclay	71.4	10 μg / mL	pH 7 at 25C in 10 min	Zarei and Barghak (2015)
2	Ag-PVP nanoparticles	polyethyleneimine (PEI)-functionalized paper	46	10 mL solutions of 2.5 × 10^{-4} M	24 h, neutral pH and under room temperature	Setyono and Valiyaveettil (2016)
3	Au-Citrate nanoparticles		30			
4	Ag-Citrate nanoparticles		79			
5	Au-PVP nanoparticles		17			

11.4 Conclusions and Scope

This chapter focuses on different natural waste-materials-based adsorbents, which can be applied to remove ECs. The cheap and efficient adsorbent will undoubtedly make an adsorption process a successful opportunity to remove ECs from water samples. Most of the EC, like PPCP, endocrine disruptors, PFAS, pesticides and microplastics, are suitable to be removed by natural adsorbents with higher efficiency. Various adsorbents such as chitosan, activated carbons, fruit and other agricultural wastes are most popularly used naturally derived adsorbents for remediation of ECs. Selection of suitable adsorbents for sorption of desired contaminant is a key step in the adsorption process. The adsorption capacity of adsorbents can further be improved by impregnating multi-functional material into them. Since, it is vital to prepare different modified adsorbents according to actual application needs; properties of adsorbents are altered by various thermal and chemical treatments. Regeneration and recyclability are some of the major issues that arise with the use of naturally derived sorbent. Electrostatic interaction, hydrophobic interaction, hydrogen bonding, and π-π interactions are the main mechanisms involved in the removal of EC through natural adsorbents. Many existing studies have explained this as a result of several interacting mechanisms. However, it is still unclear which mechanism plays a leading role and how much it contributes. The study of adsorption mechanism of modified adsorbent even requires more detailed information Clarification on these issues is of great significance for improving adsorption capacity and environmental applications of modified biochar. At the same time, the solution to these problems is also the key to upscale for industry applications. Very few studies have investigated the removal of microplastics and nanomaterials through natural adsorbents, which needs further investigation.

References

Afzal MZ, Sun XF, Liu J, Song C, Wang SG, Javed A (2018) Enhancement of ciprofloxacin sorption on chitosan/biochar hydrogel beads. Science of the Total Environment 639. https://doi.org/10.1016/j.scitotenv.2018.05.129

Ahmed MB, Zhou JL, Ngo HH, Guo W (2015) Adsorptive removal of antibiotics from water and wastewater: progress and challenges. Sci Total Environ 532. https://doi.org/10.1016/j.scitotenv.2015.05.130

Ahmed S, Ahmad M, Swami BL, Ikram S (2016) A review on plants extract mediated synthesis of silver nanoparticles for antimicrobial applications: a green expertise. J Adv Res 7:17–28

Ahsan MA, Islam MT, Hernandez C, Kim H, Lin Y, Curry ML, Gardea-Torresdey J, Noveron JC (2018) Adsorptive removal of sulfamethoxazole and bisphenol a from contaminated water using functionalized carbonaceous material derived from tea leaves. J Environ Chem Eng 6(4). https://doi.org/10.1016/j.jece.2018.06.022

Akhtar M, Moosa Hasany S, Bhanger MI, Iqbal S (2007) Sorption potential of Moringa oleifera pods for the removal of organic pollutants from aqueous solutions. J Hazardous Mater 141(3). https://doi.org/10.1016/j.jhazmat.2006.07.016

Azarkan S, Peña A, Draoui K, Sainz-Díaz CI (2016) Adsorption of two fungicides on natural clays of Morocco. Appl Clay Sci 123:37–46

Bagheri A, Abu-Danso E, Iqbal J, Bhatnagar A(2020) Modified biochar from Moringa seed powder for the removal of diclofenac from aqueous solution. Environ Sci Pollut Res 27(7). https://doi.org/10.1007/s11356-019-06844-x

Batool A, Valiyaveettil S (2021) Surface functionalized cellulose fibers–A renewable adsorbent for removal of plastic nanoparticles from water. J Hazardous Mater 413:125301

Bayat M, Alighardashi A, Sadeghasadi A (2018) Fixed-bed column and batch reactors performance in removal of diazinon pesticide from aqueous solutions by using walnut shell-modified activated carbon. Environ Technol Innov 12(2018):148–159

Buck RC, Franklin J, Berger U, Conder JM, Cousins IT, Voogt PD, Jensen AA, Kannan K, Mabury SA, van Leeuwen SPJ (2011) Perfluoroalkyl and polyfluoroalkyl substances in the environment: terminology, classification, and origins. Integr Environ Assess Manag 7(4):513–541. https://doi.org/10.1002/ieam.258

Cao FM, Wang L, Tian Y, Wu FC, Deng CB, Guo QW, Sun HW, Lu SY (2017) Synthesis and evaluation of molecularly imprinted polymers with binary functional monomers for the selective removal of perfluorooctanesulfonic acid and perfluorooctanoic acid. J Chromatogr 1516:42–53

Carolin CF, Kumar PS, Joshiba GJ (2021) Sustainable approach to decolourize methyl orange dye from aqueous solution using novel bacterial strain and its metabolites characterization. Clean Technol Environ Policy 23(1). https://doi.org/10.1007/s10098-020-01934-8

Carmalin, Limba EC (2018) Removal of emerging contaminants from the environment by adsorption. Ecotoxicol Environ Safety 150:1–17. https://doi.org/10.1016/j.ecoenv.2017.12.026

Chaparadza A, Hossenlopp JM (2012) Adsorption kinetics, isotherms and thermodynamics of atrazine removal using a banana peel based sorbent. Water Sci Technol 65(5). https://doi.org/10.2166/wst.2012.935

Chauhan M, Saini VK, Suthar S (2020) Removal of pharmaceuticals and personal care products (PPCPs) from water by adsorption on aluminum pillared clay. J Porous Mater 27(2). https://doi.org/10.1007/s10934-019-00817-8

Clara M, Scheffknecht C, Scharf S, Weiss S, Gans O (2008) Emissions of perfluorinated alkylated substances (PFAS) from point sources - Identification of relevant branches. Water Sci Technol 58(1):59–66. https://doi.org/10.2166/wst.2008.641

Cordner A, Brown P (2013) Moments of uncertainty: ethical considerations and emerging contaminants. Sociol Forum 28(3):469–494

Czech B, Shirvanimoghaddam K, Trojanowska E, Naebe M (2020) Sorption of pharmaceuticals and personal care products (PPCPs) onto a sustainable cotton based adsorbent. Sustain Chem Pharm 18. https://doi.org/10.1016/j.scp.2020.100324

Dehghani MH, Ghadermazi M, Bhatnagar A, Sadighara P, Jahed-Khaniki G, Heibati B, McKay G (2016) Adsorptive removal of endocrine disrupting bisphenol a from aqueous solution using chitosan. J Environ Chem Eng 4(3). https://doi.org/10.1016/j.jece.2016.05.011

Delgado-Moreno L, Bazhari S, Gasco G, Méndez A, El Azzouzi M, Romero E (2021) New insights into the efficient removal of emerging contaminants by biochars and hydrochars derived from olive oil wastes. Sci Total Environ 752. https://doi.org/10.1016/j.scitotenv.2020.141838

Deng S, Yu Q, Huang J, Yu G (2010) Removal of perfluorooctane sulfonate from wastewater by anion exchange resins: Effects of resin properties and solution chemistry. Water Res 44(18):5188–5195. https://doi.org/10.1016/j.watres.2010.06.038

Deng S, Zheng Y, Xu F, Wang B, Huang J, Yu G (2012) Highly efficient sorption of perfluorooctane sulfonate and perfluorooctanoate on a quaternized cotton prepared by atom transfer radical polymerization. Chem Eng J 193:154–160

Deng S, Niu L, Bei Y, Wang B, Huang J, Yu G (2013) Adsorption of perfluorinated compounds on aminated rice husk prepared by atom transfer radical polymerization. Chemosphere 91(2). https://doi.org/10.1016/j.chemosphere.2012.11.015

Deng S, Nie Y, Du Z, Huang Q, Meng P, Wang B, Yu G (2015) Enhanced adsorption of perfluorooctane sulfonate and perfluorooctanoate by bamboo-derived granular activated carbon. J Hazard Mater 282:150–157

Du ZW, Deng SB, Chen YG, Wang B, Huang J, Wang YJ, Yu G (2015) Removal of perfluorinated carboxylates from washing wastewater of perfluorooctanesulfonyl fluoride using activated carbons and resins. J Hazard Mater 286:136–143

Fagbayigbo BO, Opeolu BO, Fatoki OS, Akenga TA, Olatunji OS (2017) Removal of PFOA and PFOS from aqueous solutions using activated carbon produced from Vitis vinifera leaf litter. Environ Sci Pollut Res Int 24(14):13107–13120. https://doi.org/10.1007/s11356-017-8912-x

Fang C, Zhang X, Dong Z, Wang L, Megharaj M, Naidu R (2018) Smartphone app-based/portable sensor for the detection of fluoro-surfactant PFOA. Chemosphere 191:381–388. https://doi.org/10.1016/j.chemosphere.2017.10.057

Fernandez ME, Ledesma B, Román S, Bonelli PR, Cukierman AL (2015) Development and characterization of activated hydrochars from orange peels as potential adsorbents for emerging organic contaminants. Bioresource Technol 183. https://doi.org/10.1016/j.biortech.2015.02.035

Gao Y, Deng S, Du Z, Liu K, Yu G (2017) Adsorptive removal of emerging polyfluoroalky substances F-53B and PFOS by anion-exchange resin: a comparative study. J Hazardous Mater 323(April 2016):550–557. https://doi.org/10.1016/j.jhazmat.2016.04.069

Crini G, Lichtfouse E, Wilson LD, Morin-Crini N (2019) Conventional and non-conventional adsorbents for wastewater Treatment, Environ Chem Lett 195–213. https://doi.org/10.1007/s10311-018-0786-8

Gong YY, Wang L, Liu JC, Tang JC, Zhao DY (2016) Removal of aqueous perfluorooctanoic acid (PFOA) using starch-stabilized magnetite nanoparticles. Sci Total Environ 562:191–200

Hamadneh I, Abu-Zurayk RA, Al-Dujaili AH (2020) Removal of phenolic compounds from aqueous solution using MgCl2-impregnated activated carbons derived from olive husk: the effect of chemical structures. Water Sci Technol 81(11). https://doi.org/10.2166/wst.2020.297

Hassan M, Liu Y, Naidu R, Du J, Qi F (2020) Adsorption of Perfluorooctane sulfonate (PFOS) onto metal oxides modified biochar. Environ Technol Innovation 19:100816

Higgins CP, Luthy RG (2006) Sorption of perfluorinated surfactants on sediments. Environ Sci Technol 40(23):7251–7256

Hossain N, Bhuiyan MA, Pramanik BK, Nizamuddin S, Griffin G (2020) Waste materials for wastewater treatment and waste adsorbents for biofuel and cement supplement applications: a critical review. J Clean Prod 255:120261. https://doi.org/10.1016/j.jclepro.2020.120261

Hrens LUTZA, Undschuh MIB (2014) Critical Review fate and effects of poly- and perfluoroalkyl substances in the aquatic environment : a review 33(9):1921–1929. https://doi.org/10.1002/etc.2663

Ignatowicz K (2011) A mass transfer model for the adsorption of pesticide on coconut shell based activated carbon. Int J Heat Mass Transf 54:4931–4938

Inyang M, Dickenson ERV (2017) The use of carbon adsorbents for the removal of perfluoroalkyl acids from potable reuse systems. Chemosphere 184:168–175. https://doi.org/10.1016/j.chemosphere.2017.05.161ISSN0043-1354

Imran M, Islam AU, Tariq MA, Siddique MH, Shah NS, Khan ZUH, Amjad M, Din SU, Shah GM, Naeem MA, Nadeem M, Nawaz M, Rizwan M (2019) Synthesis of magnetite-based nanocomposites for effective removal of brilliant green dye from wastewater. Environ Sci Pollut Res 26(24). https://doi.org/10.1007/s11356-019-05706-w

Jung C, Park J, Lim KH, Park S, Heo J, Her N, Oh J, Yun S, Yoon Y (2013) Adsorption of selected endocrine disrupting compounds and pharmaceuticals on activated biochars. J Hazard Mater 263(Pt 2):702–710. https://doi.org/10.1016/j.jhazmat.2013.10.033

Kaveeshwar AR, Kumar PS, Revellame ED, Gang DD, Zappi ME, Subramaniam R (2018) Adsorption properties and mechanism of barium (II) and strontium (II) removal from fracking wastewater using pecan shell based activated carbon. J Clean Prod 193. https://doi.org/10.1016/j.jclepro.2018.05.041

Kinney CA, Furlong ET, Werner SL, Cahill JD (2006) Presence and distribution of wastewater-derived pharmaceuticals in soil irrigated with reclaimed water. Environ Toxicol Chem 25(2). https://doi.org/10.1897/05-187R.1

Kissa E (2001) Fluorinated surfactants and repellents. Marcel Dekker, New York

Kumar M, Chaminda T, Honda R, Furumai H (2019) Vulnerability of urban waters to emerging contaminants in India and Sri Lanka: resilience framework and strategy. APN Science Bulletin 9(1). https://doi.org/10.30852/sb.2019.799

Lakkovikiotis E, Kyzas G, Deliyanni E, Matis K (2014) Hyrdothermally prepared biochars from potato peels. Activation of biochars with phosphoric acid for use as sorbents for cobalt removal from wastewaters. EGU General Assembly Conference Abstracts, 16

Lessa EF, Nunes ML, Fajardo AR (2018) Chitosan/waste coffee-grounds composite: an efficient and eco-friendly adsorbent for removal of pharmaceutical contaminants from water. Carbohydrate Polym 189. https://doi.org/10.1016/j.carbpol.2018.02.018

Liu N, Charrua AB, Weng CH, Yuan X, Ding F (2015) Characterization of biochars derived from agriculture wastes and their adsorptive removal of atrazine from aqueous solution: A comparative study. Biores Technol 198:55–62. https://doi.org/10.1016/j.biortech.2015.08.129

Li T, Han X, Liang C, Shohag MJI, Yang X (2015) Sorption of sulphamethoxazole by the biochars derived from rice straw and alligator flag. Environmental Technology (United Kingdom) 36(2). https://doi.org/10.1080/09593330.2014.943299

Li Y, Wang Z, Xie X, Zhu J, Li R, Qin T (2017) Removal of Norfloxacin from aqueous solution by clay-biochar composite prepared from potato stem and natural attapulgite. Colloids Surf A: Physicochem Eng Aspects 514. https://doi.org/10.1016/j.colsurfa.2016.11.064

Li J, Yu G, Pan L, Li C, You F, Xie S, Wang Y, Ma J, Shang X (2018) Study of ciprofloxacin removal by biochar obtained from used tea leaves. J Environ Sci (China) 73. https://doi.org/10.1016/j.jes.2017.12.024

Liu J, Zhou B, Zhang H, Ma J, Mu B, Zhang W (2019) A novel Biochar modified by Chitosan-Fe/S for tetracycline adsorption and studies on site energy distribution. Bioresource Technol 294:122152. https://doi.org/10.1016/j.biortech.2019.122152

Long L, Hu X, Yan J, Zeng Y, Zhang J, Xue Y (2019) Novel chitosan-ethylene glycol hydrogel for the removal of aqueous perfluorooctanoic acid. J Environ Sci (china) 84:21–28. https://doi.org/10.1016/j.jes.2019.04.007

Lu XY, Deng SB, Wang B, Huang J, Wang YJ, Yu G (2016) Adsorption behavior and mechanism of perfluorooctane sulfonate on nanosized inorganic oxides. J Colloid Interface Sci 474:199–205

Mahfoudhi N, Boufi S (2017) Nanocellulose as a novel nanostructured adsorbent for environmental remediation: a review. Cellulose 24(3):1171–1197. https://doi.org/10.1007/s10570-017-1194-0

Mandal A, Singh N, Purakayastha TJ (2017) Characterization of pesticide sorption behaviour of slow pyrolysis biochars as low cost adsorbent for atrazine and imidacloprid removal. Sci Total Environ 577. https://doi.org/10.1016/j.scitotenv.2016.10.204

Grassi M, Kaykioglu G, Belgiorno V, Lofrano G (2012) Emerging compounds removal from wastewater. Springer Briefs in Green Chemistry for Sustainability. https://doi.org/10.1007/978-94-007-3916

Meyer EE, Rosenberg KJ, Israelachvili J (2006) Recent progress in understanding hydrophobic interactions. Proc Natl Acad Sci U S A 103(43):15739–15746

Moreno-Marenco AR, Giraldo L, Moreno-Piraján JC (2019) Parabens adsorption onto activated carbon: relation with chemical and structural properties. Molecules 24(23). https://doi.org/10.3390/molecules24234313

Naghipour D, Amouei A, Taher Ghasemi K, Taghavi K (2019) Removal of metoprolol from aqueous solutions by the activated carbon prepared from pine cones. Environ Health Eng Manage 6(2). https://doi.org/10.15171/ehem.2019.09

Nawaz T, Sengupta S (2019) Contaminants of emerging concern: occurrence, fate, and remediation. In Ahuja S (ed) Advances in water purification techniques (pp. 67–114), Advances in water purification techniques, Elsevier. https://doi.org/10.1016/B978-0-12-814790-0.00004-1

Cheng N, Wang B, Pan W, Lee X, Xing Y, Chen M (2021) Adsorption of emerging contaminants from water and wastewater by modified biochar: A review. Environ Pollut 273(2021):116448. https://doi.org/10.1016/j.envpol.2021.116448

Njoku VO, Islam MA, Asif M, Hameed BH (2015) Adsorption of 2, 4-dichlorophenoxyacetic acid by mesoporous activated carbon prepared from H3PO4-activated langsat empty fruit bunch. J Environ Manage 154:138–144

Omri A, Wali A, Benzina M (2016) Adsorption of bentazon on activated carbon prepared from Lawsonia inermis wood: Equilibrium, kinetic and thermodynamic studies. Arab J Chem 9:S1729–S1739

Pal A, Gin KY-H, Lin AY-C, Reinhard M (2010) Impacts of emerging organic contaminants on freshwater resources: Review of recent occurrences, sources, fate and effects. Sci Total Environ 408(24):6062–6069. https://doi.org/10.1016/j.scitotenv.2010.09.026 PMID:20934204

Pal A, He Y, Jekel M, Reinhard M, Gin KY (2014) Emerging contaminants of public health significance as water quality indicator compounds in the urban water cycle. Environ Int 71:46–62. https://doi.org/10.1016/j.envint.2014.05.025

Pereira LC, Oliveira de Souza A, Furio M, Bernardes F, Pazin M, Tasso MM et al (2015) A perspective on the potential risks of emerging contaminants to human and environmental health. Environ Sci Pollut Res 22:13800–13823

Phasuphan W, Praphairaksit N, Imyim A (2019) Removal of ibuprofen, diclofenac, and naproxen from water using chitosan-modified waste tire crumb rubber. J Mole Liquids 294. https://doi.org/10.1016/j.molliq.2019.111554

Rahman MF, Peldszus S, Anderson WB (2014) Behaviour and fate of perfluoroalkyl and polyfluoroalkyl substances (PFASs) in drinking water treatment: a review. Water Res 50:318–340. https://doi.org/10.1016/j.watres.2013.10.045

Rashed MN (2012) Adsorption technique for the removal of organic pollutants from water and wastewater. Organic Pollutants—Monitor Risk Treat 7. http://dx.doi.org/https://doi.org/10.5772/54048

Reck IM, Paixão RM, Bergamasco R, Vieira MF, Vieira AMS (2018) Removal of tartrazine from aqueous solutions using adsorbents based on activated carbon and Moringa oleifera seeds. J Clean Prod 171. https://doi.org/10.1016/j.jclepro.2017.09.237

Rattanaoudom R, Visvanathan C, Boontanon SK (2012) Removal of concentrated PFOS and PFOA in synthetic industrial wastewater by powder activated carbon and hydrotalcite. J Water Sustain 2(4):245–258

Richter L, Cordner A, Brown P (2018) Non-stick science: sixty years of research and (in)action on fluorinated compounds. Soc Stud Sci 48(5):691–714

Rodríguez-Liébana JA, López-Galindo A, Cisneros CJ, Gálvez A, Rozalén M, Sánchez-Espejo R, Caballero E, Peña A (2016) Adsorption/desorption of fungicides in natural clays from Southeastern Spain. Appl Clay Sci 132:402–411

Rosenfeld PE, Feng LGH (2011) Emerging contaminants. Risks Hazardous Wastes 215–222. https://doi.org/10.1016/b978-1-4377-7842-7.00016-7

Salman JM, Njoku VO, Hameed BH (2011) Adsorption of pesticides from aqueous solution onto banana stalk activated carbon. Chem Eng J 174(1). https://doi.org/10.1016/j.cej.2011.08.026

Salman J (2014) Optimization of preparation conditions for activated carbon from palm oil fronds using response surface methodology on removal of pesticides from aqueous solution. Arab J Chem 7:101–108

Saravanan A, Kumar PS, Govarthanan M, George CS, Vaishnavi S, Moulishwaran B, Kumar SP, Jeevanantham S, Yaashikaa PR (2021) Adsorption characteristics of magnetic nanoparticles coated mixed fungal biomass for toxic Cr(VI) ions in aquatic environment. Chemosphere 267. https://doi.org/10.1016/j.chemosphere.2020.129226

Shirani Z, Santhosh C, Iqbal J, Bhatnagar A (2018) Waste Moringa oleifera seed pods as green sorbent for efficient removal of toxic aquatic pollutants. J Environ Manage 227. https://doi.org/10.1016/j.jenvman.2018.08.077

Setyono D, Valiyaveettil S (2016) Functionalized paper–a readily accessible adsorbent for removal of dissolved heavy metal salts and nanoparticles from water. J Hazard Mater 302:120–128. https://doi.org/10.1016/j.jhazmat.2015.09.046

Siipola V, Pflugmacher S, Romar H, Wendling L, Koukkari P (2020) Low-cost biochar adsorbents for water purification including microplastics removal. Appl Sci 10(3):788

Singh N, Khandelwal N, Ganie ZA, Tiwari E, Darbha GK (2021) Eco-friendly magnetic biochar: An effective trap for nanoplastics of varying surface functionality and size in the aqueous environment. Chem Eng J 418:129405

Sophia AC, Lima EC (2018) Removal of emerging contaminants from the environment by adsorption. Ecotoxicol Environ Safety 150:1–17. https://doi.org/10.1016/j.ecoenv.2017.12.026

Sun C, Wang Z, Chen L, Li F (2020) Fabrication of robust and compressive chitin and graphene oxide sponges for removal of microplastics with different functional groups. Chem Eng J 393:124796

Suo F, You X, Ma Y, Li Y (2019) Rapid removal of triazine pesticides by P doped biochar and the adsorption mechanism. Chemosphere 235:918–925

Teixeira S, Delerue-Matos C, Santos L (2012) Removal of sulfamethoxazole from solution by raw and chemically treated walnut shells. Environ Sci Pollut Res 19(8). https://doi.org/10.1007/s11356-012-0853-9

Thompson RC, Moore CJ, Saal FSV, Swan SH (2009) Plastics, the environment and human health: current consensus and future trends. Phil Trans Biol Sci 364:2153e2166. https://doi.org/10.1098/rstb.2009.0053

Thue PS, Lima D, Naushad M, Lima EC, Albuquerque YR, Dias S, Cunha MR, Dotto G, Brum IA (2020) High removal of emerging contaminants from wastewater by activated carbons derived from the shell of cashew of Para. Carbon Letters 31:13–28

Tian SR, Liu YG, Liu SB, Zeng GM, Jiang LH, Tan XF, Huang XX, Yin ZH, Liu N, Li J (2018) Hydrothermal synthesis of montmorillonite/hydrochar nanocomposites and application for 17β-estradiol and 17α-ethynylestradiol removal. RSC Adv 8(8). https://doi.org/10.1039/c7ra12038a

Torrellas SÁ, García Lovera R, Escalona N, Sepúlveda C, Sotelo JL, García J (2015) Chemical-activated carbons from peach stones for the adsorption of emerging contaminants in aqueous solutions. Chem Eng J 279. https://doi.org/10.1016/j.cej.2015.05.104

Trivedi NS, Mandavgane SA, Kulkarni BD (2016) Mustard plant ash: a source of micronutrient and an adsorbent for removal of 2,4-dichlorophenoxyacetic acid. Environ Sci Pollut Res Int 23(20):20087–20099. https://doi.org/10.1007/s11356-016-6202-7

Turner BD, Sloan SW, Currell GR (2019) Novel remediation of per- and polyfluoroalkyl substances (PFASs) from contaminated groundwater using Cannabis Sativa L. (hemp)proteinpowder. Chemosphere 229:22–31. https://doi.org/10.1016/j.chemosphere.2019.04.139

United States Environmental Protection Agency (2014) Emerging contaminants—perfluorooctane sulfonate (PFOS) and perfluorooctanic acid (PFOA). United States Environmental Protection Agency, Washington, DC

USEPA.2007."Aboutpesticides."U.S.EPA, _http://www.epa.gov/pesticides/about/index.htm_ _Oct. 26, 2008_
Viotti PV, Moreira WM, Santos OAA. dos, Bergamasco R, Vieira AMS, Vieira MF (2019) Diclofenac removal from water by adsorption on Moringa oleifera pods and activated carbon: mechanism, kinetic and equilibrium study. J Clean Prod 219. https://doi.org/10.1016/j.jclepro.2019.02.129
Wang F, Liu CS, Shih KM (2012) Adsorption behavior of perfluorooctanesulfonate (PFOS) and perfluorooctanoate (PFOA) on boehmite. Chemosphere 89(8):1009–1014
Wang Z, Sun C, Li F, Chen L (2021) Fatigue resistance, re-usable and biodegradable sponge materials from plant protein with rapid water adsorption capacity for microplastics removal. Chem Eng J 415:129006
Weidemann E, Niinipuu M, Fick J, Jansson S (2018) Using carbonized low-cost materials for removal of chemicals of environmental concern from water. Environ Sci Pollut Res 25(16). https://doi.org/10.1007/s11356-018-1781-0
Xiao L, Ling Y, Alsbaiee A, Li C, Helbling DE, Dichtel WR (2017) β-cyclodextrin polymer network sequesters perfluorooctanoic acid at environmentally relevant concentrations. J Am Chem Soc 139(23):7689–7692. https://doi.org/10.1021/jacs.7b02381
Xu C, Chen H, Jiang F (2015) Adsorption of perflourooctane sulfonate (PFOS) and perfluorooctanoate (PFOA) on polyaniline nanotubes. Colloids Surf A 479:60–67
Yi S, Gao B, Sun Y, Wu J, Shi X, Wu B, Hu X (2016) Removal of levofloxacin from aqueous solution using rice-husk and wood-chip biochars. Chemosphere 150. https://doi.org/10.1016/j.chemosphere.2015.12.112
Yoong YH, Chong WC, Chung YT, Teoh HC (2020) Magnetite activated carbon/chitosan composite from biomass for removal of diclofenac in aqueous solution. IOP Conference Series: Earth Environ Sci 463(1). https://doi.org/10.1088/1755-1315/463/1/012183
Zarei AR, Barghak F (2015) Fast and efficient adsorption of ag nanoparticles by sodium montmorillonite nanoclay from aqueous systems. J Chem Res 39(9):542–545. https://doi.org/10.3184/174751915X14403380312326
Zbair M, Bottlinger M, Ainassaari K, Ojala S, Stein O, Keiski RL, Bensitel M, Brahmi R (2020) Hydrothermal carbonization of argan nut shell: functional mesoporous carbon with excellent performance in the adsorption of bisphenol a and diuron. Waste Biomass Valorization 11(4). https://doi.org/10.1007/s12649-018-00554-0
Zhang Q, Deng S, Yu G, Huang J (2011) Removal of perfluorooctane sulfonate from aqueous solution by crosslinked chitosan beads: sorption kinetics and uptake mechanism. Bioresourcetechnology 102(3):2265–2271. https://doi.org/10.1016/j.biortech.2010.10.040
Zhang DQ, Zhang WL, Liang YN (2019) Science of the total environment adsorption of perfluoroalkyl and poly fl uoroalkyl substances (PFASs) from aqueous solution—a review 694
Zhou X, Zhou J, Liu Y, He Y, Ren J, Guo J (2019) Adsorption of endocrine disrupting ethylparaben from aqueous solution by chemically activated biochar developed from oil palm fibre. Sep Sci Technol 54(5):683–695
Zhou Y, Xu M, Huang D, Xu L, Yu M, Zhu Y, Niu J (2021) Modulating hierarchically microporous biochar via molten alkali treatment for efficient adsorption removal of perfluorinated carboxylic acids from wastewater. Sci Total Environ 757:143719. https://doi.org/10.1016/j.scitotenv.2020.143719
Zhu X, Liu Y, Luo G, Qian F, Zhang S, Chen J (2014) Facile fabrication of magnetic carbon composites from hydrochar via simultaneous activation and magnetization for triclosan adsorption. Environ Sci Technol 48(10). https://doi.org/10.1021/es500531c

Chapter 12
Emerging Contaminants Removal from Wastewater by Nanotechnological Methods

Swathi Desireddy and P. C. Sabumon

Abstract The incidence of emerging contaminants in receiving water bodies is a threat to ecology and human health. Anthropogenic activities through industrialization and modernization contribute to the release of toxic pollutants such as pesticides, disinfectants, pharmaceuticals, personal care products, detergents, surfactants, wood preservatives, flame retardants, persistent organic pollutants, and their degradation products. These emerging contaminants do not have a defined set of regulations and are hazardous even if present in trace concentrations. Existing conventional physicochemical and biological wastewater treatment methods are not intended to remove emerging contaminants, owing to their trace occurrence (ng/L to μg/L), extremely recalcitrant nature, and associated bioaccumulation. Recently, nanotechnology has evolved as a fundamental significance to provide alternative and efficient wastewater treatment options. This chapter highlights the latest advances in Nanotechnological methods available to remove emerging contaminants from wastewaters. The potential use of nanoscale materials such as nanosorbents, nanofilters, and nanocatalysts in the degradation of emerging contaminants is discussed. In addition, the possible hazards related to the use of engineered nanomaterials, potential obstacles in the application, and the future perspectives are discussed.

Keywords Emerging contaminants · Wastewater treatment · Nanoscale materials · Nanofiltration · Nanoadsorption · Nanocatalysis

12.1 Introduction

Emerging Contaminants (ECs) have become a menace and present a universal water quality challenge. The shortage of clean water is detrimental to human welfare, industrialization, and viable development. It is very important to recycle and reuse

S. Desireddy · P. C. Sabumon (✉)
School of Civil Engineering, Vellore Institute of Technology (VIT), Chennai 600127, India
e-mail: pcsabumon@vit.ac.in

S. Desireddy
e-mail: desireddy.swathi@vit.ac.in

S. P. Singh et al. (eds.), *New Trends in Emerging Environmental Contaminants*, Energy, Environment, and Sustainability,
https://doi.org/10.1007/978-981-16-8367-1_12

wastewaters to meet the ever-increasing demand for clean water (Tzanakakis et al. 2020). Conventional treatment methods are effectively being employed to treat the wastewaters polluted with nutrients and organic contents to the acknowledged norms. Lately, the occurrence of ECs in wastewaters is posing several difficulties in reuse. ECs are defined as any synthetic or natural chemicals, and microbes that are not by and large examined in water but the reported prevalence has intensified recently and their impending dangerous effects on ecology and human health are uncertain. ECs are also termed as emerging pollutants, micropollutants, contaminants of emerging concern, and trace-organic compounds by diverse research groups (Rout et al. 2021). Different categories of ECs from several sectors as presented in Fig. 12.1 and are being detected in water supplies lately, varying from parts per trillion (ng/L) to parts per billion (μg/L), which is a major concern (Rizzo et al. 2019; Patel et al. 2020). To

Fig. 12.1 Sources and types of Emerging Contaminants

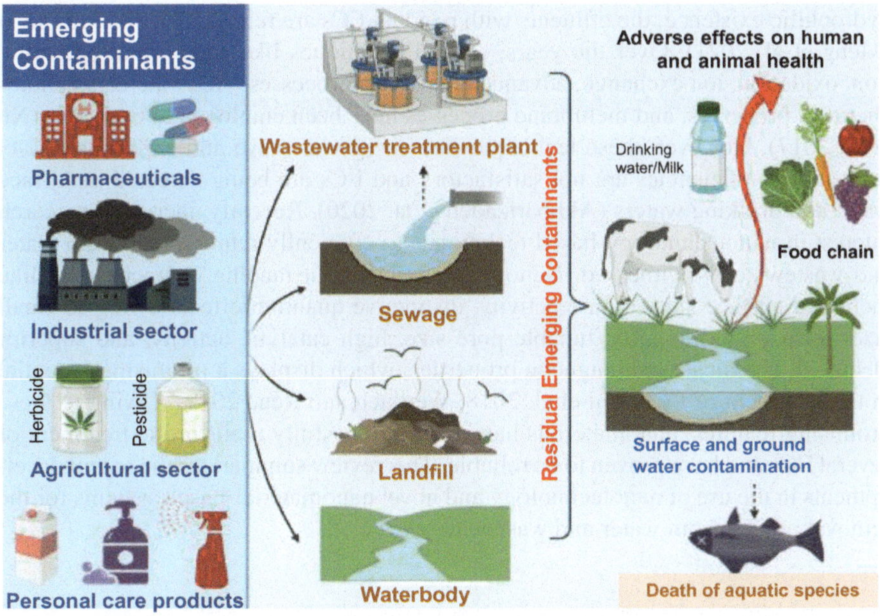

Fig. 12.2 Origin, transport pathways, and the fate of emerging contaminants

control the hazards of ECs, ample scientific and technical knowledge on the potential risks, metabolic pathways, transport pathways and fate, accumulating nature, ecotoxicology, quantification, evaluation, and treatment is required. Majority of ECs are not regulated in wastewater discharge standards and hence the suitable treatment methods have not been developed (Diaz-Sosa et al. 2020). ECs are persistent and accumulates in surface and drinking water as micro-pollutants. They have diverse physicochemical properties and it is hard to assess the fate of ECs in environment. The transport pathways and fate of ECs is depicted in Fig. 12.2. Though the concentration of ECs is very less in water sources, long-term exposure pose a severe risk to human, aquatic and animal life by causing chronic toxic effects such as kidney failure, reproductive impairment, different types of cancers, metabolic, and neural disorders (Wang et al. 2020a). Especially, ECs are reported to disrupt the endocrine system which is responsible for hormone production and regulation of vital body functions. ECs also increase microbial resistance to antibiotics, and World Health Organization (WHO) stated that resistance to antibiotics poses a severe threat to global health, food security, and development (Hu et al. 2020). Owing to these adverse effects of ECs, the utilization of advanced technologies to aid in their removal from water and wastewater is a prerequisite.

The main sources of ECs are wastewater treatment plants. Traditional wastewater treatment processes are intended to remove pathogens, biological oxygen demand (BOD), chemical oxygen demand (COD), nitrogen, and phosphorous compounds and are not competent to remove ECs from wastewater, and in fact, owing to its hydrophilic existence, the effluents with residual ECs are released into environment (Meng et al. 2021). Over the years, several techniques like adsorption, precipitation, oxidation, ion exchange, advanced oxidation processes (AOPs), electro/photochemical processes, and membrane processes have been employed to treat ECs (Xu et al. 2017). However, these techniques are energy-intensive and expensive. Also, the removal efficiencies are not satisfactory and ECs are being detected in surface water and drinking waters (Akhbarizadeh et al. 2020). Recently, increased research interest in nanotechnology-based techniques to efficiently remove ECs from water and wastewater is witnessed. Nanomaterials have outstanding characteristics like increased surface area, high reactivity, distinctive quantum effects owing to small size, surface modifiability, tunable pore size, high catalytic activity, and superior electrical, chemical, and magnetic properties, which displays a promising potential in the treatment of ECs (Cai et al. 2018; Werkneh and Rene 2019). Owing to these promising features, nanomaterials have been successfully used for the treatment of several ECs and have proven to be reliable. This review summarizes the recent developments in the use of nanotechnology and novel nanomaterial-based systems for the removal of ECs from water and wastewater.

12.2 Nanotechnological Methods for the Removal of ECs

Current wastewater treatment techniques are not efficient in removing ECs present in trace concentrations in waters and wastewaters. Details on the currently available conventional techniques that are used in the removal of ECs, their benefits, and shortcomings are given in Table 12.1. For efficient removal of ECs, several nanotechnology-based techniques have been developed in recent years. Nanoparticles have increased surface area because of the shrink in the diameter while bulk maintains the similar density. Small sizes provide high mobility enabling the spread of nanoparticles in the solution with relatively less material concentration. The enhanced aspect ratio also exposes multiple active sites at the surfaces thereby increasing the reactivity. These outstanding characteristics augment the ability of these nanomaterials to remove contaminants from wastewater. Major nanotechnology-based techniques employed in the treatment of ECs are detailed below.

12.2.1 Nano Adsorption

Adsorption is a surface process where contaminants in liquid or gas phases get adsorbed into a solid phase. Adsorption occurs due to diverse physical or weak chemical interactions such as hydrogen bonds, covalent bonds, electrostatic interactions, hydrophobic effects, and π,π-interactions among adsorbate and adsorbent (Zhao et al. 2018; Ersan et al. 2017). Adsorption is the widely accepted physicochemical method to eliminate pollutants from waters and wastewaters. Owing to their advantages like simple and flexible operation, low cost, diverse applicability, the possibility of recycling, and reuse of adsorbents, it is considered the most efficient and reliable process (Kumara et al. 2019; Rathi et al. 2021). However, the application of conventional synthetic adsorbents and other low-cost natural adsorbents on a large scale is limited due to various aspects like less surface area and active sites,

Table 12.1 Conventional treatment methods available for the removal of ECs

Process	Materials/Methods	Advantages	Disadvantages	References
Adsorption	• Zeolites • Organic soils • Organic-carbon framework • Metal–organic framework (MOF) • Activated carbon • Biochar • Activated hydrochars	• Easy operation • Can handle contaminants in micro levels • Environmental friendly • Recyclability	• Secondary sludge generation • High investment • Fast saturation of adsorbents	Sophia and Lima (2018)
Membrane filtration	• Microfiltration (MF) • Ultrafiltration (UF) • Nanofiltration (NF)	• Small footprint • Effective removal	• Membrane fouling • High energy requirement • High operational costs • Restricted flow rate	Gómez-Espinosa and Arizmendi-Cotero (2017)
Reverse osmosis	• RO membranes	• Can treat saline waters	• Energy-intensive • Membrane fouling • Corrosive nature of effluent	Egea-Corbacho Lopera et al. (2019)
Advanced oxidation processes	• O_3 • H_2O_2-O_3 • O_3-UV • H_2O_2-UV • Photo-catalysis • Fenton reaction • Sonolysis • Ozonation • UV irradiation • Microwave radiation • Electrolysis	• Effective degradation	• Cost and energy-intensive • Require pretreatment • Requires pH adjustment • Interference of radicals scavengers • Cannot handle large volumes of water • Separation and reuse	Fast et al. (2017)
Constructed wetlands	• Vertical flow • Lateral sub-surface flow	• Low cost	• Large foot-print • Higher retention times	Gorito et al. (2017)

(continued)

Table 12.1 (continued)

Process	Materials/Methods	Advantages	Disadvantages	References
Electrochemical methods	• Continuous Electrolytic Deionization • Capacitive Deionization	• Efficient	• Costly • Electrode inhibition due to sludge deposition • Pretreatment to remove certain components	Magro et al. (2020)
Biological methods	• Aerobic/Anaerobic bioreactors • Microbial treatment • Enzymatic treatment	• Cost-effective	• Slow start-up • Inefficient in ECs removal • Critical process control	Shah and Rather (2021)

poor selectivity, economic concerns as a large amount of adsorbents are required, low stability, and disposal problems (Chowdhury and Balasubramanian 2014).

On the contrary nanosorbents have proven to be superior in adsorbing target pollutants owed to their unique physicochemical properties including high specific surface area, small size, high reactivity, high catalytic activity, target specificity, tunable pore size, ease of separation, surface engineering to provide many functional groups, and active sites (Kunduru et al. 2017; Cincinelli et al. 2015). In the last decade, various nanostructured materials based on carbon, graphene, metals, MOFs, metal-oxides, magnetic and polymeric composites have gained significant acceptance as adsorbents to remove ECs from water and wastewater (Zhao et al. 2018; Chowdhury and Balasubramanian 2014; Kunduru et al. 2017; Yaashikaa et al. 2019; Song and Jhung 2017).

Carbon Based Nanosorbents: Carbon is a very flexible element, and materials with novel architecture can easily be fabricated. Narrow band gap and transfer of electrons among 2 s and 2p orbits of C allow it to be in diverse hybridizations. The malleable alignments of carbon result in stable carbonaceous assemblies even at the nano range (Karimi-Maleh et al. 2020). Research related to carbon nanostructures as adsorbents have gained increased attention in the last decade for their potential applications in the removal of ECs (Zhao et al. 2018). Ideal nanosorbent should have a high surface area and porosity. Low dimension carbon-based nanoparticles such as carbon nanotubes (CNTs), graphene, fullerenes, and their related structures display high stability, adsorption efficiency, and chemical inertness (Kumari et al. 2019).

CNTs are allotropes of carbon with cylindrical shapes and generally are conceptualized as rolled-graphite/graphene sheets. The diameter of CNTs is in nm range and lengths are in the μm range and hence CNTs are one-dimensional nanostructures (Cukierman et al. 2019). CNTs are synthesized via top-down or bottom-up approaches mostly with laser-ablation, arc-discharge, and chemical vapor deposition (Vaka et al. 2021). CNTs exist in several structures, differing in length, thickness, helicity, and number of layers. Based upon the direction of graphene sheets rolling, CNTs are classified into armchair, zigzag, and chiral varieties (He et al. 2013). The angle of rolling affects the physico-chemical properties of CNTs (Menezes et al. 2019). Based on the structure, CNTs are categorized as (a) single-walled CNTs (SWCNTs) which consists of single graphene sheets rolled into tubular structures; and (b) multi-walled CNTs (MWCNTs) which are formed by rolling up numerous graphene sheets to give tubular structures (Machado and Lima 2018). Literature shows that the CNTs are highly effective towards adsorption of ECs. CNTs are hydrophobic in nature and can readily adsorb non-polar components. Interestingly, the functionalization of CNTs improves the adsorption of polar molecules (Pan and Xing 2008). Engineering of CNTs gives diverse structured versatile nanosorbent materials with controlled alignments and shapes. The adsorption efficiency of CNTs is based on pore size, pore volume, and specific surface area. MWCNTs are mesoporous while SWCNTs are microporous with high surface area. CNTs get aggregated in aqueous conditions and present diverse pore sizes, with increased hydrophobic interior surface areas which lead to higher adsorption of polar molecules (Machado et al. 2016). SWCNTs are generally present as bundles and provide superior adsorption sites compared to individual SWCNTs as adsorption on SWCNT bundles occurs at four diverse locations; (i) exterior surface of bundles, (ii) channels at the contact sites of individual SWCNTs, (iii) nanotube pores, and (iv) the interstice passages amongst the bundle tubes (Ren et al. 2011). Conversely, MWCNTs are not present as bundles and the adsorption sites comprise pores aggregated in the interior surfaces of MWCNTs, at open ends, and on the exterior walls (Upadhyayula et al. 2009). Also, the occurrence of defects acts as reactive sites for adsorption (Machado and Lima 2018). Furthermore, CNTs are tunable which enhances the adsorption efficacy against particular contaminants via different physico-chemical, biological, and/or mechanical approaches (Kokkinos et al. 2020; Ahmad et al. 2019). Several such engineered CNTs are being employed as nanosorbents in the removal of ECs (Cukierman et al. 2019; Peng et al. 2012).

Graphene-based Nanosorbents: Graphene is a monolayer carbon nanomaterial made of sp2 hybridized C atoms organized in a hexagonal crystalline structure. Graphene is the elementary structural unit of other carbon allotropes like zero-dimensional fullerenes, one-dimensional CNTs, and 3D graphite (Chowdhury and Balasubramanian 2014). They make an ideal adsorbent due to high specific surface area (2600 m^2/g) (Jilani et al. 2018), ample functional groups, higher negative charge density, and a very hydrophilic nature (Khalil et al. 2020a; Qiu et al. 2018; Li et al. 2017).

Graphene-based nanosorbents are mainly classified as graphene, graphene oxide (GO), reduced graphene oxide (rGO), and graphene nanocomposites. GO is a strong oxidative form of graphene synthesized through the chemical exfoliation of graphite in the presence of a strong oxidant and a mineral acid. The oxygen functionalities (hydroxyl (OH), carboxylic (COOH), carbonyl (C = O), alkoxy groups (C–O–C), and epoxy (eCeOeCe) groups) increase the hydrophilic nature and adsorption ability to polar contaminants and helps in eliminating widespread classes of pollutants in wastewater (Molla et al. 2019; Perreault et al. 2015). GO is used as a substitute for

CNTs due to their low cost. Also, single layer of GO has two basal planes available for adsorption, while the interior walls of CNTs are not accessible for adsorption (Kumari et al. 2019; Zhao et al. 2014). These structural characteristics make GO an excellent adsorbent to remove ECs. Graphene and rGO are easily produced by the reduction of GO via a simple chemical exfoliation method. rGO can be produced by eliminating a portion of O_2 comprising functional groups from GO. rGO has improved physico-chemical properties and conjugate structure which leads to enhancement of specific surface area (530 m^2/g) and henceforth maximum adsorption capacity (Xiang et al. 2019; Wang et al. 2017). However, graphene in its bulk form gets agglomerated and restacked to give graphite in liquid processing. In contrast, GO has a meager binding affinity to anions because of strong electrostatic repulsions among them. In addition, the collection and separation of both graphene and GO from the treated waters are not easy which leads to secondary contamination (Chowdhury and Balasubramanian 2014).

Hence researchers have developed graphene-based nanocomposites with enhanced properties by combining with various other materials (Anirudhan and Deepa 2017). The functional groups that are generally attached include nanosized metal oxides, organic polymers, magnetic nanoparticles that help in easy separation, chitosan, SiO_2 etc. (Xiang et al. 2019; Wang et al. 2018; Huang et al. 2021). The combined materials also provide high selectivity and strong binding to target contaminants dependent on their structure, size, and crystallinity. GO-based nanocomposite materials are promising due to their versatile characteristics and several variables open for fine-tuning. The GO-based nanocomposite materials were broadly explored for the adsorptive removal of ECs from water and wastewater (Kar et al. 2016). These chemical functionalizations also help in the prevention of agglomeration by dispersing and stabilizing graphene, enhances the process ability and interaction with diverse organic and inorganic contaminants. Recently, to further enhance the adsorption capacity of ECs, multi-pore structures called graphene aerogels and GO hydrogels were developed by using cross-linking agents which showed high adsorption and extraction recovery (Rofouei et al. 2017; Yang et al. 2017).

Metal-based Nanosorbents: Recently, growing research attention is witnessed in engaging metal/metal-oxide/metal-oxyhydroxide nanoparticles (such as ZnO, TiO_2, CeO_2, MgO, and, so on) as nanosorbents in water and wastewater treatment (Biftu et al. 2020). All these find broad efficacy in various industries owing to their small size, high surface area, magnetic nature, high surface potential, surface porosity, superior physico-chemical properties like chemically inert, ease of synthesis, and ecofriendly (Lu et al. 2016). Among these metal-oxide nanosorbents such as oxides of Zn, Zr, Pb, Mn, Fe, Ti, Mg, Ce, and Al are the most prominent and promising ones. The surface of metals-oxides poses strong adsorption forces of varied natures such as electrostatic interactions, ion exchange, ion association affinities, complex formations, etc., and they play a vital part in increasing affinity among the contaminants and surface of the adsorbents. The physico-chemical adsorption is also caused by residual valences or defective sites in the metal-oxide matrices. However, the main downside of nano metal-oxides is that decrease in sizes from micro to nano range leads to increased surface energies thereby affecting the stability. Thus, at nano size, metal-oxides tend to form agglomerates and lead to decreased adsorption efficiency. Hence, to improve the mechanical strength and applicability, metal-oxides are impregnated into bigger sized porous supports such as zeolites, CNTs, Ag, CeO_2, γ-Mn_2O_3, etc. (Saravanan et al. 2014) to produce nano-composites with relatively superior properties and applications to form a composite adsorbent that helps in the efficient removal of ECs (Kumari et al. 2019).

Metal–Organic Frameworks (MOFs): OFs are an evolving group of porous coordination polymers made from inorganic metal nodes (atoms, clusters, chains, etc.) and organic linkers (carboxylates, phosphonates, azolates, etc.) which gets assembled into multi-dimensional periodic matrices (Yang and Gates 2019; Joseph et al. 2019). MOFs comprise active sites, where pollutants are chemisorbed and degraded. Apart from the properties of bulk MOF materials, nanoMOFs have exceptional benefits, like enhanced adsorption or desorption kinetics and accessible interior active sites (Rasheed et al. 2020). They are highly porous, well-ordered, crystalline, and characterized by enormous internal surface areas. Due to the small size and biocompatible nature, nanoMOFs are being broadly explored as adsorbents in the removal of ECs (Cai et al. 2020; Sarker et al. 2018; Seo et al. 2017; Moradi et al. 2016). Further, nanoMOFs can be tailor-made to perform shape-selective adsorption and can easily be manufactured in large-scale into numerous forms (monolith, pellet, or membranes) (Rojas and Horcajada 2020). The adsorption efficiency is considerably affected by the physico-chemical characteristics of MOF based nanosorbents, target contaminants (size, shape, hydrophobic nature, functional groups, and charge), and influent water (pH, temperature, concentration of solutes, organic compounds, and other anions or cations) (Joseph et al. 2019).

Another approach to increase the adsorption capacity of MOFs is to synthesize MOF-based nanocomposites which comprise MOF and a complex (with metal oxides, sulfides, noble metal nanoparticles, GO, etc.). They have superior characteristics compared to their discrete components such as target-specific, increased number of accessible active sites, extremely porous nature, and easy tunability of pores to attain required characteristics (Chen and Li 2016; Xiang et al. 2103). Several studies reported the use of nanoMOFs and MOF-based nanocomposites in the remediation (adsorption + degradation) of ECs (Rojas and Horcajada 2020). Comprehensive and critical information on recently employed nanoadsorption based techniques in the removal of ECs is given in Table 12.2.

12.2.2 Nanofiltration

Since majority of conventional physico-chemical techniques can remove only a narrow spectrum of ECs, membrane-based technologies including microfiltration (MF), ultrafiltration (UF), nanofiltration (NF), and reverse osmosis (RO) have gained wide attention owing to the production of high-quality water due to effective removal of dissolved solids and low molecular weight pollutants (Tang et al. 2018). Nanofiltration membranes have pore sizes from 1 to 10 nm, which falls between the pore sizes

Table 12.2 Comprehensive and critical information on recent nanoadsorption based techniques in the removal of ECs

Nanosorbent	Target ECs	Specific surface area (m^2/g)	Adsorbent capacity (mg/g)	Refs.
CNTs				
MWCNT	Tetracyclin	–	270	Yu et al. (2014)
MWCNT	Ciprofloxacin Hydrochloride	–	2	Avcı et al. (2020)
Magnetic MWCNT	Naproxen	120	21	İlbay et al. (2015)
N-doped CNTs encapsulated with Ni (Ni@NCNTs)	Sulfachloropyridazine	155	0.11 mol/g	Kang et al. (2018)
Nanodendrimer adsorbent on SiO_2/Al_2O_3 (SAPAMAA)	Salicylic acid	79	253	Arshadi et al. (2017)
CNTCOOH/MnO_2/Fe_3O_4 nanocomposite	Ibuprofen Paracetamol	114	103 81	Lung et al. (2021)
CNT/TiO_2/photocatalysts	Dibutylhydroxytoluene	156	100%	Flor et al. (2021)
Sludge biochar and MWCNT composite (CNT-SBC)	Sulfamethoxazole	119	2×10^4 μg/g	Ma et al. (2020)
CNTs-Fe-Cu catalyst combined with Al^0-CNTs	Sulfamerazine	111	85%	Chen et al. (2020)
Photo-regenerable CNT based membrane (Ag-CNT-FeOOH membrane)	Bisphenol A Amoxicillin p-Nitrophenol	–	13 8 7	Yang et al. (2021)
Al°-CNTs-Cu_2O composite	Sulfamerazine	19	74%	Liu et al. (2020)
Functionalized MWCNT	Tetracycline	–	253	Babaei et al. (2016)
MOF nanosorbents				
Isostructural functionalized MOF (UiO-66-R)	2,4-Dichlorophenoxy acetic acid Clofibric acid Diclofenac	1148	182 183 188	Zhao et al. (2019)
Hydrophobic perfluoroalkyl MOF sorbent (UiO-66-F)	Carbamazepine	–	34	Yuan et al. (2021)

(continued)

Table 12.2 (continued)

Nanosorbent	Target ECs	Specific surface area (m^2/g)	Adsorbent capacity (mg/g)	Refs.
C derived from Zn-based MOF	Phthalic acid Diethyl phthalate N,N-diethyl-3 methylbenzamide Chloroxylenol Oxybenzone	1395	450 540 340 560 620	Bhadra et al. (2020)
MOF loaded nano-zero-valent Fe (NZVI/MIL-101(Cr))	Tetracycline	786	625	Hou et al. (2020)
Graphene-based nanosorbents				
Novel porous-type graphene material	Atenolol Carbamazepine Ciprofloxacin Diclofenac Gemfibrozil Ibuprofen	679	36 35 40 30 29 29	Khalil et al. (2020b)
GO nanoparticles	Metformin	187	122	Balasubramani et al. (2020)
MnO_2/graphene	Tetracycline	106	198	Song et al. (2019)
Amino-functionalized alginate/graphene hydrogel beads (NH_2-DN)	Ciprofloxacin	0.6	301	Sun et al. (2020)
Magnetite/GO nanocomposite-rGO/Fe_3O_4 nanosorbent	Phenazopyridine	213	14	Karimi-Maleh et al. (2020)
Magnetic graphene/chitosan nanocomposite	2-naphthol	–	169	Rebekah et al. (2020)
Metal/metal-oxide based nanosorbents				
Zn–Al–O binary metal oxide	Benzotriazole	108	10	Xu et al. (2010)
Activated Carbon–MOF composite (Darco-KB-G-MIL-100Fe)	Caffeine Carbamazepine Clofibric acid 10,11epoxycarbam- -azepine Naproxen O-desmethylnaproxen Paraxanthine Salicylic acid	815	1426 1419 56 851 946 742 812 59	Muñoz-Senmache et al. (2020)

(continued)

Table 12.2 (continued)

Nanosorbent	Target ECs	Specific surface area (m^2/g)	Adsorbent capacity (mg/g)	Refs.
Superparamagnetic Fe_3O_4 nanoparticles with aminated β-cyclodextrin	Naproxen Carbamazepine Bisphenol A	–	0.4 0.5 0.3	Ghosh et al. (2013)
Magnetic Fe/silica-based nanoadsorbent	Ibuprofen Diclofenac Carbamazepine	90	33	Peralta et al. (2021)
TiO_2 nanoparticles	Penicillin	200	–	Peterson et al. (2012)
Zn–Fe bimetallic organic-framework (Zn-Fe-ZIF-NPC)	Bisphenol AF Bisphenol S Tetrabromobisphenol A Bisphenol A	578	439 423 330 252	Ma et al. (2019)

of membranes used in UF and RO. The NF membranes are classified as inorganic, organic polymer, and organic–inorganic membranes, each having typical factors in the removal of ECs (Aguilar-Pérez et al. 2020). NF membranes possess thin-film composite structures which include a support layer (offers mechanical strength with a low resistance to the flow of permeate), and a thin-film active layer (separates with higher selectivity). NF membranes employ high pressure and work on the principles of size exclusion, electrostatic repulsion, and adsorption (Huang et al. 2019). As a majority of ECs have moderately higher molecular weights than the limit of characteristic NF membranes, the elimination by size exclusion principle is effective. Filtration efficiency can further be increased through electrostatic repulsion by altering the NF membrane surfaces and pore charges (Wang et al. 2020b). Literature reports several studies with successful removal of various classes of ECs using NF. However, conventional NF membranes face certain hurdles like membrane fouling, low flux, concentrate disposal, and the requirement of costly materials (Imbrogno et al. 2019). Rejection of ECs by NF membranes is mainly reliant on the nanomembrane properties, while others being solutes physico-chemical properties, operational parameters, and influent wastewater characteristics (Cristóvão et al. 2019).

To overcome the barriers associated with conventional NF membranes, innovative nanomembranes that provide selective retention of ECs while permitting the passage of scale-forming components are imperative. Advancement in state-of-the-art synthesis techniques (layer-by-layer self-assembly, interfacial polymerization, coating, phase inversion technique, cross-linking, blending) of hybrid nanomaterials enabled the fabrication of several innovative and tailor-made NF membranes. This was done through molecular-level design developed by introducing selective ionizable functional groups to alter the surface charge, using novel nanoparticles in the synthesis of active layers, incorporation of carbon structures, such as graphene, GO, rGO, thin-film graphene membranes, or nano-composites in membranes (Tang et al. 2018; Liu et al. 2019; Kamrani et al. 2018; Bai et al. 2021). These innovative NF membranes are widely recognized in the removal of ECs owing to their benefits like (i) high mechanical strength, (ii) chemical stability, (iii) higher surface areas, low operational costs, (iv) recyclability, (v) superhydrophobic, (vi) ease of access, (vii) efficient uptake of contaminants, (viii) less energy requirement, (ix) biocompatible, (x) biologically persistent, (xi) resistance to fouling, (xii) ecofriendly fabrication methods, and (xiii) easy integration with other processes (Ouyang et al. 2019; Janssens et al. 2019).

Recently, magnetic ion-exchange resins (MIEX) are described as efficient antifouling strategies in NF membranes, as it removes foulants from the membrane surface (Imbrogno et al. 2018; López-Ortiz et al. 2018). Further, hi-tech NF membranes with dual charge were synthesized. Compared to commercial nanomembranes with a neutral and negative charge, membranes with dual charge have excellent properties like high sensitivity, controlled separation, environmental responsiveness, and self-regulation to parameters like pH, temperature, and electric-field. Such tunable and responsive hybrid membranes allow broad use in the remediation of wastewaters polluted with a broad range of ECs with diverse physico-chemical properties (Ouyang et al. 2019). A comprehensive review of recently employed nanofiltration based techniques in the removal of ECs is given in Table 12.3.

12.2.3 Nanocatalysis

The application of nanotechnology in AOPs is beneficial and an emerging trend for the removal of ECs. AOPs are emergent and proficient water remediation processes carried out at atmospheric temperature and pressure to generate various reactive oxygen species (HO•, O_2^{-}•, HO_2^{-}•, SO_4^{-}•, and ROO•) for degradation of pollutants (Amor et al. 2019). AOPs are widely employed as a post-treatment alternative for the oxidative degradation of ECs. AOPs can efficiently mineralize the contaminants, unlike the conventional strategies where toxic secondary metabolites are produced (Guo et al. 2021). AOPs are classified into various groups based on the oxidants used (O_3 and H_2O_2) and the usage of radiation, catalysts (Fe^{2+}, Fe^{3+}, or TiO_2), ultrasonic inserts, thermal inputs, and their combinations. AOPs are economical and can effectively handle higher or much lower concentrations of ECs. Among AOPs catalytic ozonation and photocatalysis have gained abundant attention in the treatment of ECs (Amor et al. 2019).

Catalytic ozonation is an AOP in which highly reactive radicals are produced by the oxidation of O_3 in the presence of a catalyst. The generated reactive oxygen species can effectively degrade and mineralize a diverse range of contaminants (Wang and Chen 2020). Heterogeneous photocatalytic oxidation is another AOP that is applied as a polishing step to treat a diverse range of ECs and recalcitrant organic compounds. In photocatalysis, solid semiconductors and radiation are used to generate stable colloidal suspensions which stimulate reactions. The type of catalysts

Table 12.3 Comprehensive review on recent nanofiltration based techniques in the removal of ECs

Nanofiltration membranes	Membrane synthesis technique	Target ECs	Surface area (cm^2)	Removal (%)	Refs
Dual charged polyelectrolyte membrane $(HTCC/PDA)_3$-NF	Layer-by-layer technique	Atenolol Carbamazepine Ibuprofen	–	81.67 92.5 89.85	Ouyang et al. (2019)
NF270-2540 membrane	Synthetic membrane	Caffeine Theobromine Theophylline Amoxicillin Penicillin	2.6 m^2	20 18 7 89 70	Egea-Corbacho et al. (2019)
Polyelectrolyte multilayer	Layer-by-layer assembly	Perfluoroalkyl substances	–	>85	Wang et al. (2020b)
Desal 5DK and NF270 membrane	GE Osmonics, and DOW FILMTEC, USA	Paclitaxel Etoposide Cyclophosphamide Ifosfamide	54	>89	Cristóvão et al. (2019)
Chitosan-modified acrylic membrane	Interfacial polymerization	Mebeverine	32	98	Kamrani et al. (2018)
Integrated adsorption-nanofiltration system	AC - F400 NF - 270	Caffeine Diclofenac Octylphenol	140	>95	Huang et al. (2019)
Fe^{2+}-activated peroxymonosulfate NF membrane	Cross-flow membrane module (CF042D, STERLITCH, USA)	Sulfamethoxazole	42	92	Bai et al. (2021)
Filtration-UV, UV-TiO_2 and UV-H_2O_2	Desal5 Dk membrane (MWCO 150 to 300 Da, GE Osmonics, USA)	Paclitaxel Etoposide Cyclophosphamide Ifosfamide	–	>85	Janssens et al. (2019)
MOF-UF (MIL-100(Fe) -MIL-101(Cr))	Material of Institute Lavoisier (MIL)	Ibuprofen 17α-ethinyl Estradiol	15	>40	Kim et al. (2020)

used plays a vital role in improving the efficacy of AOPs. Generally, semiconductor-catalysts (TiO_2, ZnO, Fe_2O_3, CdS, GaP, ZnS) are broadly used in the degradation of diverse pollutants into CO_2 and H_2O (Shah and Rather 2021). Although catalytic AOPs are advantageous compared to conventional counterparts, it has its limitations such as (i) meager control over the catalyst properties such as size, shape, composition, and surface exposition, (ii) limitations on stability, activity, and selectivity, (iii) leachate production, (iv) high costs associated with the use of certain chemicals, and (v) toxic intermediates.

To overcome the above problems nanocatalysts have emerged as promising alternatives. Exceptional qualities of nanocatalysts such as being highly stable, active, selective, and also effective in contaminant removal make them ideal candidates. The small size and high surface area of nanocatalysts provide an excellent interface for reactants even at mild concentrations due to maximized exposure of active sites. Nanocatalysts can also be easily separated, purified, and reused (Zhao et al. 2018; Corma 2016; Prinsen and Luque 2019; Wang et al. 2014). Nanocatalysts offer the benefits of both homogeneous (due to maximum exposure of active sites) and heterogeneous catalysis (recyclability) (Rodrigues et al. 2019). Further, tailor-made nanocatalysts can be synthesized by exploiting the physico-chemical properties like size, shape, composition, and structure, which gives rise to novel nanocatalysts with excellent electronic, physicochemical, and optical properties such as multiple active sites, high visible light absorption, the formation of efficient nano-junctions, suppression of the interfacial charge recombinations, etc. (Asgari et al. 2020). Recently, various such advanced nanomaterials including noble metals, zero-valent metals (Fe, Ag, Zn), metal-oxides (TiO_2, ZnO, MgO, MnO, WO_3), CNTs, GO, nanocomposites, and nano-oxide metals immobilized on membranes or combined with catalytic components ($ZnFe2O_4$, WO_3, $C3N_4$, Nb_2O_5, Co_3O_4, and Cu_2O) are being extensively used as catalysts in AOPs (Dang 2020). Also, combinations of these novel nanomaterials give rise to economic nano synergetic materials with enhanced removal efficiencies which have a great potential in the removal of ECs (Cheng et al. 2018). A comprehensive review of recently employed nanocatalysis-based techniques in the removal of ECs is given in Table 12.4.

12.3 Future Prospects

Nanoparticles are a boon to today's water remediation technologies and are employed in the treatment of various ECs. However, certain studies reported several nanoparticles themselves as ECs. Current knowledge on the fate of nanoscale particles in ecology and their toxicity is limited. Extensive usage of nanoparticles in day-to-day life is like a double-edged sword. The properties such as high chemical reactivity, small size, etc. which are generally considered to be beneficial in water remediation may be dangerous to human and animal life when entered into the food chain, absorbed through the skin, or inhaled. A thorough environmental risk assessment has to be carried out before scaling-up and safe disposal methods have to be followed. Also, the majority of the nanotechnology-based treatment methods discussed here gave efficient results at lab-scale studies with synthetic wastewaters, and hence extensive research is still needed before scaling up with real effluents, and long-term efficacy should be monitored to commercialize the developed techniques (Kunduru et al. 2017). On the other hand, lack of scientific knowledge on the exact pathways of ECs and unavailability of robust analytical techniques to quantify the ECs hinders the research and development of novel techniques. To date, most of the ECs remediation by nanoparticles was carried out as tertiary treatment techniques, and extending them

Table 12.4 Comprehensive review on recent nanocatalysis based techniques in the removal of ECs

Nanocatalysts	Synthesis methods	Target ECs	Removal (%)	Refs
TiO_2/UV/C doped with trivalent Fe (Fe^{3+}–TiO_2/UV/C)	Sol–gel	Amoxicillin	89	Olama et al. (2018)
Ce/TiO_2/O_3	Sol–gel	Caffeine	>88	Ndabankulu et al. (2019)
Mn activated with peroxymonosulfate	Thermal treatment	Sulfamethoxazole Sulfachloropyridazine Sulfamethazine Ciprofloxacin Azithromycin	99 100 88 100 25	Khan et al. (2021)
Fe-Carbon dots	Hydrothermal method	Tetracycline	95	Pirsaheb et al. (2020)
Fe-$BiVO_4$ photocatalyst + photo-fenton	Sonochemistry	Ciprofloxacin	55	Hernández-Uresti et al. (2019)
p-$ZnIn_2S_4$/rGO/ n-g-C_3N_4 Z	Hydrothermal method	Triclosan	83	Yu et al. (2020)
GO-based magnetic nanocatalyst (MFe_2O_4/xRGO)	Acile hydrothermal method	Ofloxacin	100	Kapoor et al. (2020)
Nano-MgO@CNT@Gr/O_3	Sol–gel	Diazinon	82	Asgari et al. (2020)
MOF-Engineered-FeS_2/C for heterogeneous electro-Fenton process	Carbonization and Sulfidation	Fluoxetine	100	Ye et al. (2020)
Ceramic membranes functionalized with $CoFe_2O_4$ nanocatalyst/ AOP (α-Al_2O_3/$CoFe_2O_4$)	Impregnation and low temperature calcination	Ibuprofen	99.5	Wang et al. (2020a)
F-doped ZnO (ZnO:F) nano and mesocrystals	Hydrothermal approach	Diclofenac	86	Vitiello et al. (2021)
Fe modified photocatalyst and ozonation (UV/Fe-TiO_2 + O_3)	Coating	Formaldehyde	92	Cheng et al. (2018)

as main stream treatment methods is a prerequisite (Rout et al. 2021). The future of nanotechnology in the removal of ECs seems fairly progressive if safe and standard protocols are followed which could be possible by collective efforts of academic, industrial, and government regulating bodies.

12.4 Conclusion

To combat the ever-increasing concentrations of ECs in water sources, a range of nanotechnology-based processes were developed and are successfully employed in their removal. Based on the literature, it is apparent that various nanomaterials such as CNTs, graphene, GO, rGO, fullerenes, nanofibers, nanodendrites, metal/metal-oxide/metal-oxyhydroxide nanoparticles, hybrid nanocomposites, synergetic nanosystems, and their combinations have tremendous potential to act as nanoadsorbents, nanocatalysts, and nanofiltration membranes. These novel nanomaterials offer several advantages such as high efficiency, high surface area, easy synthesis, separation and recyclability, large number of active sites, target specific designs, environmental responsive characteristics, biocompatibility, improved physicochemical, electrical, and optical properties, etc. Thus, nanotechnology has an immense role to play in the monitoring and removal of various classes of ECs. Choosing the most appropriate technique or combinations based on case-by-case situations (type of ECs, concentration, the volume of influent wastewaters, economic feasibility, etc.) is of primary importance to achieve maximum removal of ECs.

References

Aguilar-Pérez KM, Avilés-Castrillo JI, Ruiz-Pulido G (2020) Nano-sorbent materials for pharmaceutical-based wastewater effluents—an overview. Case Stud Chem Environ Eng 2:100028. https://doi.org/10.1016/j.cscee.2020.100028

Ahmad J, Naeem S, Ahmad M, Usman ARA, Al-Wabel MI (2019) A critical review on organic micropollutants contamination in wastewater and removal through carbon nanotubes. J Environ Manage 246:214–228. https://doi.org/10.1016/j.jenvman.2019.05.152

Akhbarizadeh R, Dobaradaran S, Schmidt TC, Nabipour I, Spitz J (2020) Worldwide bottled water occurrence of emerging contaminants: a review of the recent scientific literature. J Hazard Mater 392:122271. https://doi.org/10.1016/j.jhazmat.2020.122271

Amor C, Marchão L, Lucas MS, Peres JA (2019) Application of advanced oxidation processes for the treatment of recalcitrant agro-industrial wastewater: a review. Water 11:205. https://doi.org/10.3390/w11020205

Anirudhan TS, Deepa JR (2017) Nano-zinc oxide incorporated graphene oxide/nanocellulose composite for the adsorption and photo catalytic degradation of ciprofloxacin hydrochloride from aqueous solutions. J Colloid Interface Sci 490:343–356. https://doi.org/10.1016/j.jcis.2016.11.042

Arshadi M, Mousavinia F, Abdolmaleki MK, Amiri MJ, Khalafi-Nezhad A (2017) Removal of salicylic acid as an emerging contaminant by a polar nano-dendritic adsorbent from aqueous media. J Colloid Interface Sci 493:138–149. https://doi.org/10.1016/j.jcis.2017.01.017

Asgari G, Seidmohammadi A, Esrafili A, Faradmal J, Sepehr MN, Jafarinia M (2020) The catalytic ozonation of diazinon using nano-MgO@ CNT@ Gr as a new heterogenous catalyst: the optimization of effective factors by response surface methodology. RSC Adv 10:7718–7731. https://doi.org/10.1039/c9ra10095d

Avcı A, İnci İ, Baylan N (2020) Adsorption of ciprofloxacin hydrochloride on multiwall carbon nanotube. J Mol Struct 1206:127711. https://doi.org/10.1016/j.molstruc.2020.127711

Babaei AA, Lima EC, Takdastan A, Alavi N, Goudarzi G, Vosoughi M, Hassani G, Shirmardi M (2016) Removal of tetracycline antibiotic from contaminated water media by multi-walled carbon nanotubes: operational variables, kinetics, and equilibrium studies. Water Sci Technol 74:1202–1216. https://doi.org/10.2166/wst.2016.301

Bai L, Liu Z, Wang H, Li G, Liang H (2021) Fe(II)-activated peroxymonosulfate coupled with nanofiltration removes natural organic matter and sulfamethoxazole in natural surface water: performance and mechanisms. Sep Purif Technol 274:119088. https://doi.org/10.1016/j.seppur.2021.119088

Balasubramani K, Sivarajasekar N, Naushad M (2020) Effective adsorption of antidiabetic pharmaceutical (metformin) from aqueous medium using graphene oxide nanoparticles: Equilibrium and statistical modelling. J Mol Liq 301:112426. https://doi.org/10.1016/j.molliq.2019.112426

Bhadra BN, Yoo DK, Jhung SH (2020) Carbon-derived from metal-organic framework MOF-74: a remarkable adsorbent to remove a wide range of contaminants of emerging concern from water. Appl Surf Sci 504:144348. https://doi.org/10.1016/j.apsusc.2019.144348

Biftu WK, Ravindhranath K, Ramamoorty M (2020) New research trends in the processing and applications of iron-based nanoparticles as adsorbents in water remediation methods. Nanotechnol Environ Eng 5:1–12. https://doi.org/10.1007/s41204-020-00076-y

Cai Z, Dwivedi AD, Lee W-N, Zhao X, Liu W, Sillanpää M, Zhao D, Huang C-H, Fu J (2018) Application of nanotechnologies for removing pharmaceutically active compounds from water: development and future trends. Environ Sci Nano 5:27–47. https://doi.org/10.1039/C7EN00644F

Cai X, Xie Z, Li D, Kassymova M, Zang S-Q, Jiang H-L (2020) Nano-sized metal-organic frameworks: synthesis and applications. Coord Chem Rev 417:213366. https://doi.org/10.1016/j.ccr.2020.213366

Chen J, Li Y (2016) The road to mof-related functional materials and beyond: desire, design, decoration, and development. Chem Rec 16:1456–1476. https://doi.org/10.1002/tcr.201500304

Chen Y, Yang Z, Liu Y, Liu Y (2020) Fenton-like degradation of sulfamerazine at nearly neutral pH using Fe-Cu-CNTs and Al0-CNTs for in-situ generation of H2O2/OH/O2−. Chem Eng J 396:125329. https://doi.org/10.1016/j.cej.2020.125329

Cheng S-W, Li Y-H, Yuan C-S, Tsai P-Y, Shen H-Z, Hung C-H (2018) An innovative advanced oxidation technology for effective decomposition of formaldehyde by combining iron modified nano-TiO2 (Fe/TiO2) photocatalytic degradation with ozone oxidation. Aerosol Air Qual Res 18:3220–3233. https://doi.org/10.4209/aaqr.2018.05.0156

Chowdhury S, Balasubramanian R (2014) Recent advances in the use of graphene-family nanoadsorbents for removal of toxic pollutants from wastewater. Adv Coll Interface Sci 204:35–56. https://doi.org/10.1016/j.cis.2013.12.005

Cincinelli A, Martellini T, Coppini E, Fibbi D, Katsoyiannis A (2015) Nanotechnologies for removal of pharmaceuticals and personal care products from water and wastewater. A Review. J Nanosci Nanotechnol 15:3333–3347. https://doi.org/10.1166/jnn.2015.10036

Corma A (2016) Heterogeneous catalysis: understanding for designing, and designing for applications. https://doi.org/10.1002/anie.201601231

Cristóvão MB, Torrejais J, Janssens R, Luis P, Van der Bruggen B, Dubey KK, Mandal MK, Bronze MR, Crespo JG, Pereira VJ (2019) Treatment of anticancer drugs in hospital and wastewater effluents using nanofiltration. Sep Purif Technol 224:273–280. https://doi.org/10.1016/j.seppur.2019.05.016

Cukierman AL, Nunell GV, Bonelli PR (2019) Chapter 7—Removal of emerging pollutants from water through adsorption onto carbon-based materials. In: Mishra AK, Anawar HMD, Drouiche N (eds) Emerging and nanomaterial contaminants in wastewater. Elsevier, pp 159–213

Dang TT (2020) Nano-catalysts in ozone-based advanced oxidation processes for wastewater treatment. Curr Pollut Rep 6:217–229. https://doi.org/10.1007/s40726-020-00147-3

de la Flor MP, Camarillo R, Martínez F, Jiménez C, Quiles R, Rincón J (2021) Removal of emerging pollutant dibutylhydroxytoluene from water with CNT/TiO2 catalysts in a visible LED photoreactor. Environ Sci Pollut Res 28:23720–23730. https://doi.org/10.1007/s11356-020-11126-y

Diaz-Sosa VR, Tapia-Salazar M, Wanner J, Cardenas-Chavez DL (2020) Monitoring and ecotoxicity assessment of emerging contaminants in wastewater discharge in the City of Prague (Czech Republic). Water 12:1079. https://doi.org/10.3390/w12041079

Egea-Corbacho A, Gutiérrez Ruiz S, Quiroga Alonso JM (2019) Removal of emerging contaminants from wastewater using nanofiltration for its subsequent reuse: full–scale pilot plant. J CLean Prod 214:514–523. https://doi.org/10.1016/j.jclepro.2018.12.297

Egea-Corbacho Lopera A, Gutiérrez Ruiz S, Quiroga Alonso JM (2019) Removal of emerging contaminants from wastewater using reverse osmosis for its subsequent reuse: pilot plant. J Water Process Eng 29:100800. https://doi.org/10.1016/j.jwpe.2019.100800

Ersan G, Apul OG, Perreault F, Karanfil T (2017) Adsorption of organic contaminants by graphene nanosheets: a review. Water Res 126:385–398. https://doi.org/10.1016/j.watres.2017.08.010

Fast SA, Gude VG, Truax DD, Martin J, Magbanua BS (2017) A critical evaluation of advanced oxidation processes for emerging contaminants removal. Environ Process 4:283–302. https://doi.org/10.1007/s40710-017-0207-1

Ghosh S, Badruddoza AZM, Hidajat K, Uddin MS (2013) Adsorptive removal of emerging contaminants from water using superparamagnetic Fe3O4 nanoparticles bearing aminated β-cyclodextrin. J Environ Chem Eng 1:122–130. https://doi.org/10.1016/j.jece.2013.04.004

Gómez-Espinosa RM, Arizmendi-Cotero D (2017) Role of membrane on emerging contaminant removal. Ecopharmacovigilance, Springer, pp 157–174

Gorito AM, Ribeiro AR, Almeida CMR, Silva AMT (2017) A review on the application of constructed wetlands for the removal of priority substances and contaminants of emerging concern listed in recently launched EU legislation. Environ Pollut 227:428–443. https://doi.org/10.1016/j.envpol.2017.04.060

Guo D, You S, Li F, Liu Y (2021) Engineering carbon nanocatalysts towards efficient degradation of emerging organic contaminants via persulfate activation: a review. Chinese Chem Lett. https://doi.org/10.1016/j.cclet.2021.06.027

He H, Pham-Huy LA, Dramou P, Xiao D, Zuo P, Pham-Huy C (2013) Carbon nanotubes: applications in pharmacy and medicine. BioMed Res Int. https://doi.org/10.1155/2013/578290

Hernández-Uresti DB, Alanis-Moreno C, Sanchez-Martinez D (2019) Novel and stable Fe-BiVO4 nanocatalyst by efficient dual process in the ciprofloxacin degradation. Mater Sci Semicond Process 102:104585. https://doi.org/10.1016/j.mssp.2019.104585

Hou X, Shi J, Wang N, Wen Z, Sun M, Qu J, Hu Q (2020) Removal of antibiotic tetracycline by metal-organic framework MIL-101(Cr) loaded nano zero-valent iron. J Mol Liq 313:113512. https://doi.org/10.1016/j.molliq.2020.113512

Hu X-Y, Logue M, Robinson N (2020) Antimicrobial resistance is a global problem—a UK perspective. Euro J Integrative Med 36:101136. https://doi.org/10.1016/j.eujim.2020.101136

Huang Z, Gong B, Huang C-P, Pan S-Y, Wu P, Dang Z, Chiang P-C (2019) Performance evaluation of integrated adsorption-nanofiltration system for emerging compounds removal: exemplified by caffeine, diclofenac and octylphenol. J Environ Manage 231:121–128. https://doi.org/10.1016/j.jenvman.2018.09.092

Huang T, Tang X, Luo K, Wu Y, Hou X, Tang S (2021) An overview of graphene-based nanoadsorbent materials for environmental contaminants detection. TrAC Trends Anal Chem 139:116255. https://doi.org/10.1016/j.trac.2021.116255

İlbay Z, Şahin S, Kerkez Ö, Bayazit Ş (2015) Isolation of naproxen from wastewater using carbon-based magnetic adsorbents. Int J Environ Sci Technol 12:3541–3550. https://doi.org/10.1007/s13762-015-0775-4

Imbrogno A, Tiraferri A, Abbenante S, Weyand S, Schwaiger R, Luxbacher T, Schäfer AI (2018) Organic fouling control through magnetic ion exchange-nanofiltration (MIEX-NF) in water treatment. J Membr Sci 549:474–485. https://doi.org/10.1016/j.memsci.2017.12.041

Imbrogno A, Samanta P, Schäfer AI (2019) Fate of steroid hormone micropollutant estradiol in a hybrid magnetic ion exchange resin-nanofiltration process. Environ Chem 16:630–640. https://doi.org/10.1071/EN19126

Janssens R, Cristovao MB, Bronze MR, Crespo JG, Pereira VJ, Luis P (2019) Coupling of nanofiltration and UV, UV/TiO2 and UV/H2O2 processes for the removal of anti-cancer drugs from real secondary wastewater effluent. J Environ Chem Eng 7:103351. https://doi.org/10.1016/j.jece.2019.103351

Jilani A, Othman MHD, Ansari MO, Hussain SZ, Ismail AF, Khan IU (2018) Graphene and its derivatives: synthesis, modifications, and applications in wastewater treatment. Environ Chem Lett 16:1301–1323. https://doi.org/10.1007/s10311-018-0755-2

Joseph L, Jun B-M, Jang M, Park CM, Muñoz-Senmache JC, Hernández-Maldonado AJ, Heyden A, Yu M, Yoon Y (2019) Removal of contaminants of emerging concern by metal-organic framework nanoadsorbents: a review. Chem Eng J 369:928–946. https://doi.org/10.1016/j.cej.2019.03.173

Kamrani M, Akbari A, Yunessnia lehi (2018) A chitosan-modified acrylic nanofiltration membrane for efficient removal of pharmaceutical compounds. J Environ Chem Eng 6:583–587. https://doi.org/10.1016/j.jece.2017.12.044

Kang J, Duan X, Wang C, Sun H, Tan X, Tade MO, Wang S (2018) Nitrogen-doped bamboo-like carbon nanotubes with Ni encapsulation for persulfate activation to remove emerging contaminants with excellent catalytic stability. Chem Eng J 332:398–408. https://doi.org/10.1016/j.cej.2017.09.102

Kapoor S, Kumar V, Tikoo KB, Chudasama B, Goel N, Singhal S (2020) Strategically designed reduced graphene oxide based magnetic responsive nanocatalysts for the attenuation of recalcitrant pollutants. Ceram Int 46:2724–2742. https://doi.org/10.1016/j.ceramint.2019.09.262

Kar P, Sardar S, Liu B, Sreemany M, Lemmens P, Ghosh S, Pal SK (2016) Facile synthesis of reduced graphene oxide–gold nanohybrid for potential use in industrial waste-water treatment. Sci Technol Adv Mater 17:375–386. https://doi.org/10.1080/14686996.2016.1201413

Karimi-Maleh H, Shafieizadeh M, Taher MA, Opoku F, Kiarii EM, Govender PP, Ranjbari S, Rezapour M, Orooji Y (2020) The role of magnetite/graphene oxide nano-composite as a high-efficiency adsorbent for removal of phenazopyridine residues from water samples, an experimental/theoretical investigation. J Mol Liq 298:112040. https://doi.org/10.1016/j.molliq.2019.112040

Khalil U, Bilal Shakoor M, Ali S, Rizwan M, Nasser Alyemeni M, Wijaya L (2020a) Adsorption-reduction performance of tea waste and rice husk biochars for Cr(VI) elimination from wastewater. J Saudi Chem Soc 24:799–810. https://doi.org/10.1016/j.jscs.2020.07.001

Khalil AME, Memon FA, Tabish TA, Salmon D, Zhang S, Butler D (2020b) Nanostructured porous graphene for efficient removal f emerging contaminants (pharmaceuticals) from water. Chem Eng J 398:125440. https://doi.org/10.1016/j.cej.2020.125440

Khan A, Zhang K, Taraqqi-A-Kamal A, Wang X, Chen Y, Zhang Y (2021) Degradation of antibiotics in aqueous media using manganese nanocatalyst-activated peroxymonosulfate. J Colloid Interface Sci 599:805–818. https://doi.org/10.1016/j.jcis.2021.04.095

Kim S, Muñoz-Senmache JC, Jun B-M, Park CM, Jang A, Yu M, Hernández-Maldonado AJ, Yoon Y (2020) A metal organic framework-ultrafiltration hybrid system for removing selected pharmaceuticals and natural organic matter. Chem Eng J 382:122920. https://doi.org/10.1016/j.cej.2019.122920

Kokkinos P, Mantzavinos D, Venieri D (2020) Current trends in the application of nanomaterials for the removal of emerging micropollutants and pathogens from water. Molecules 25:2016. https://doi.org/10.3390/molecules25092016

Kumara PS, Joshibaa GJ, Feminaa CC, Varshinia P, Priyadharshinia S, Karthicka M, Jothiranib R (2019) A critical review on recent developments in the low-cost adsorption of dyes from wastewater. Desalin Water Treat 172:395–416. https://doi.org/10.1007/978-3-540-79210-9_3

Kumari P, Alam M, Siddiqi WA (2019) Usage of nanoparticles as adsorbents for waste water treatment: an emerging trend. Sustain Mater Technol 22:e00128. https://doi.org/10.1016/j.susmat.2019.e00128

Kunduru KR, Nazarkovsky M, Farah S, Pawar RP, Basu A, Domb AJ (2017) 2—Nanotechnology for water purification: applications of nanotechnology methods in wastewater treatment. In: Grumezescu AM (ed) Water purification. Academic Press, pp 33–74

Li T, Wang T, Qu G, Liang D, Hu S (2017) Synthesis and photocatalytic performance of reduced graphene oxide-TiO2 nanocomposites for orange II degradation under UV light irradiation. Environ Sci Pollut Res Int 24:12416. https://doi.org/10.1007/s11356-017-8927-3

Liu Y-L, Wang X-M, Yang H-W, Xie YF, Huang X (2019) Preparation of nanofiltration membranes for high rejection of organic micropollutants and low rejection of divalent cations. J Membr Sci 572:152–160. https://doi.org/10.1016/j.memsci.2018.11.013

Liu Y, Tan N, Guo J, Wang J (2020) Catalytic activation of O2 by Al0-CNTs-Cu2O composite for Fenton-like degradation of sulfamerazine antibiotic at wide pH range. J Hazard Mater 396:122751. https://doi.org/10.1016/j.jhazmat.2020.122751

López-Ortiz CM, Sentana-Gadea I, Varó-Galvañ P, Maestre-Pérez SE, Prats-Rico D (2018) The use of combined treatments for reducing parabens in surface waters: ion-exchange resin and nanofiltration. Sci Total Environ 639:228–236. https://doi.org/10.1016/j.scitotenv.2018.05.150

Lu H, Wang J, Stoller M, Wang T, Bao Y, Hao H (2016) An overview of nanomaterials for water and wastewater treatment. Adv Mater Sci Eng 2016. https://doi.org/10.1155/2016/4964828

Lung I, Soran M-L, Stegarescu A, Opris O, Gutoiu S, Leostean C, Lazar MD, Kacso I, Silipas T-D, Porav AS (2021) Evaluation of CNT-COOH/MnO2/Fe3O4 nanocomposite for ibuprofen and paracetamol removal from aqueous solutions. J Hazard Mater 403:123528. https://doi.org/10.1016/j.jhazmat.2020.123528

Ma S-Y, Wang X, Duan H-L, Wang J, Zhan H-Y, Zhang Z-Q (2019) A nanoporous carbon derived from bimetallic organic-framework for magnetic solid-phase extraction of bisphenol analogs. Talanta 202:479–485. https://doi.org/10.1016/j.talanta.2019.04.079

Ma Y, Yang L, Wu L, Li P, Qi X, He L, Cui S, Ding Y, Zhang Z (2020) Carbon nanotube supported sludge biochar as an efficient adsorbent for low concentrations of sulfamethoxazole removal. Sci Total Environ 718:137299. https://doi.org/10.1016/j.scitotenv.2020.137299

Machado FM, Carmalin SA, Lima EC, Dias SL, Prola LD, Saucier C, Jauris IM, Zanella I, Fagan SB (2016) Adsorption of alizarin red S dye by carbon nanotubes: an experimental and theoretical investigation. J Phys Chem C 120:18296–18306. https://doi.org/10.1021/acs.jpcc.6b03884

Machado FM, Lima ÉC (2018) Carbon nanoadsorbents for removal of organic contaminants from water. Bio Nanosorbents Natural Res. Springer, pp 21–53

Magro C, Mateus EP, Paz-Garcia JM, Ribeiro AB (2020) Emerging organic contaminants in wastewater: understanding electrochemical reactors for triclosan and its by-products degradation. Chemosphere 247:125758. https://doi.org/10.1016/j.chemosphere.2019.125758

de Menezes BRC, Rodrigues KF, da Silva Fonseca BC, Ribas RG, do Amaral Montanheiro TL, Thim GP (2019) Recent advances in the use of carbon nanotubes as smart biomaterials. J Mater Chem B 7:1343–1360. https://doi.org/10.1039/C8TB02419G

Meng Y, Liu W, Fiedler H, Zhang J, Wei X, Liu X, Peng M, Zhang T (2021) Fate and risk assessment of emerging contaminants in reclaimed water production processes. Front Environ Sci Eng 15:1–16. https://doi.org/10.1007/s11783-021-1392-8

Molla A, Li Y, Mandal B, Kang SG, Hur SH, Chung JS (2019) Selective adsorption of organic dyes on graphene oxide: theoretical and experimental analysis. Appl Surf Sci 464:170–177. https://doi.org/10.1016/j.apsusc.2018.09.056

Moradi SE, Shabani AMH, Dadfarnia S, Emami S (2016) Effective removal of ciprofloxacin from aqueous solutions using magnetic metal–organic framework sorbents: mechanisms, isotherms and kinetics. J Iran Chem Soc 13:1617–1627. https://doi.org/10.1007/s13738-016-0878-y

Muñoz-Senmache JC, Kim S, Arrieta-Pérez RR, Park CM, Yoon Y, Hernández-Maldonado AJ (2020) Activated carbon–metal organic framework composite for the adsorption of contaminants of emerging concern from water. ACS Appl Nano Mater 3:2928–2940. https://doi.org/10.1021/acsanm.0c00190

Ndabankulu VO, Maddila S, Jonnalagadda SB (2019) Ozone facilitated degradation of caffeine using Ce-TiO2 catalyst. J Environ Sci Health B 54:138–146. https://doi.org/10.1080/03601234.2018.1530549

Olama N, Dehghani M, Malakootian M (2018) The removal of amoxicillin from aquatic solutions using the TiO2/UV-C nanophotocatalytic method doped with trivalent iron. Appl Water Sci 8:97. https://doi.org/10.1007/s13201-018-0733-7

Ouyang Z, Huang Z, Tang X, Xiong C, Tang M, Lu Y (2019) A dually charged nanofiltration membrane by pH-responsive polydopamine for pharmaceuticals and personal care products removal. Sep Purif Technol 211:90–97. https://doi.org/10.1016/j.seppur.2018.09.059

Pan B, Xing B (2008) Adsorption mechanisms of organic chemicals on carbon nanotubes. Environ Sci Technol 42:9005–9013. https://doi.org/10.1021/es801777n

Patel N, Khan M, Shahane S, Rai D, Chauhan D, Kant C, Chaudhary V (2020) Emerging pollutants in aquatic environment: Source, effect, and challenges in biomonitoring and bioremediation-a review. Pollution 6:99–113. https://doi.org/10.22059/poll.2019.285116.646

Peng H, Pan B, Wu M, Liu Y, Zhang D, Xing B (2012) Adsorption of ofloxacin and norfloxacin on carbon nanotubes: Hydrophobicity- and structure-controlled process. J Hazard Mater 233–234:89–96. https://doi.org/10.1016/j.jhazmat.2012.06.058

Peralta ME, Mártire DO, Moreno MS, Parolo ME, Carlos L (2021) Versatile nanoadsorbents based on magnetic mesostructured silica nanoparticles with tailored surface properties for organic pollutants removal. J Environ Chem Eng 9:104841. https://doi.org/10.1016/j.jece.2020.104841

Perreault F, De Faria AF, Elimelech M (2015) Environmental applications of graphene-based nanomaterials. Chem Soc Rev 44:5861–5896. https://doi.org/10.1039/C5CS00021A

Peterson JW, Petrasky LJ, Seymour MD, Burkhart RS, Schuiling AB (2012) Adsorption and breakdown of penicillin antibiotic in the presence of titanium oxide nanoparticles in water. Chemosphere 87:911–917. https://doi.org/10.1016/j.chemosphere.2012.01.044

Pirsaheb M, Moradi S, Shahlaei M, Farhadian N (2020) Fenton-like removal of tetracycline from aqueous solution using iron-containing carbon dot nanocatalysts. New J Chem 44:17735–17743. https://doi.org/10.1039/D0NJ04014B

Prinsen P, Luque R (2019). Introduction to Nanocatalysts. https://doi.org/10.1039/9781788016292-00001

Qiu B, Xing M, Zhang J (2018) Recent advances in three-dimensional graphene based materials for catalysis applications. Chem Soc Rev 47:2165–2216. https://doi.org/10.1039/C7CS00904F

Rasheed T, Bilal M, Hassan AA, Nabeel F, Bharagava RN, Romanholo Ferreira LF, Tran HN, Iqbal HMN (2020) Environmental threatening concern and efficient removal of pharmaceutically active compounds using metal-organic frameworks as adsorbents. Environ Res 185:109436. https://doi.org/10.1016/j.envres.2020.109436

Rathi BS, Kumar PS, Show P-L (2021) A review on effective removal of emerging contaminants from aquatic systems: current trends and scope for further research. J Hazard Mater 409:124413. https://doi.org/10.1016/j.jhazmat.2020.124413

Rebekah A, Bharath G, Naushad M, Viswanathan C, Ponpandian N (2020) Magnetic graphene/chitosan nanocomposite: a promising nano-adsorbent for the removal of 2-naphthol from aqueous solution and their kinetic studies. Int J Biol Macromol 159:530–538. https://doi.org/10.1016/j.ijbiomac.2020.05.113

Ren X, Chen C, Nagatsu M, Wang X (2011) Carbon nanotubes as adsorbents in environmental pollution management: a review. Chem Eng J 170:395–410. https://doi.org/10.1016/j.cej.2010.08.045

Rizzo L, Malato S, Antakyali D, Beretsou VG, Đolić MB, Gernjak W, Heath E, Ivancev-Tumbas I, Karaolia P, Lado Ribeiro AR, Mascolo G, McArdell CS, Schaar H, Silva AMT, Fatta-Kassinos D (2019) Consolidated vs new advanced treatment methods for the removal of contaminants of emerging concern from urban wastewater. Sci Total Environ 655:986–1008. https://doi.org/10.1016/j.scitotenv.2018.11.265

Rodrigues TS, da Silva AG, Camargo PH (2019) Nanocatalysis by noble metal nanoparticles: controlled synthesis for the optimization and understanding of activities. J Mater Chem A 7:5857–5874. https://doi.org/10.1039/C9TA00074G

Rofouei MK, Jamshidi S, Seidi S, Saleh A (2017) A bucky gel consisting of Fe3O4 nanoparticles, graphene oxide and ionic liquid as an efficient sorbent for extraction of heavy metal ions from water prior to their determination by ICP-OES. Microchim Acta 184:3425–3432

Rojas S, Horcajada P (2020) Metal-organic frameworks for the removal of emerging organic contaminants in water. Chem Rev 120:8378–8415. https://doi.org/10.1021/acs.chemrev.9b00797

Rout PR, Zhang TC, Bhunia P, Surampalli RY (2021) Treatment technologies for emerging contaminants in wastewater treatment plants: a review. Sci Total Environ 753:141990. https://doi.org/10.1016/j.scitotenv.2020.141990

Saravanan R, Gupta VK, Narayanan V, Stephen A (2014) Visible light degradation of textile effluent using novel catalyst ZnO/γ-Mn2O3. J Taiwan Inst Chem Eng 45:1910–1917. https://doi.org/10.1016/j.jtice.2013.12.021

Sarker M, Song JY, Jhung SH (2018) Adsorptive removal of anti-inflammatory drugs from water using graphene oxide/metal-organic framework composites. Chem Eng J 335:74–81. https://doi.org/10.1016/j.cej.2017.10.138

Seo PW, Khan NA, Jhung SH (2017) Removal of nitroimidazole antibiotics from water by adsorption over metal–organic frameworks modified with urea or melamine. Chem Eng J 315:92–100. https://doi.org/10.1016/j.cej.2017.01.021

Shah AH, Rather MA (2021) Pharmaceutical residues: new emerging contaminants and their mitigation by nano-photocatalysis. Adv Nano Res 10:397–414

Song JY, Jhung SH (2017) Adsorption of pharmaceuticals and personal care products over metal-organic frameworks functionalized with hydroxyl groups: quantitative analyses of H-bonding in adsorption. Chem Eng J 322:366–374. https://doi.org/10.1016/j.cej.2017.04.036

Song Z, Ma Y-L, Li C-E (2019) The residual tetracycline in pharmaceutical wastewater was effectively removed by using MnO2/graphene nanocomposite. Sci Total Environ 651:580–590. https://doi.org/10.1016/j.scitotenv.2018.09.240

Sophia AC, Lima EC (2018) Removal of emerging contaminants from the environment by adsorption. Ecotoxicol Environ Safety 150:1–17. https://doi.org/10.1016/j.ecoenv.2017.12.026

Sun Y, Zhou T, Li W, Yu F, Ma J (2020) Amino-functionalized alginate/graphene double-network hydrogel beads for emerging contaminant removal from aqueous solution. Chemosphere 241:125110. https://doi.org/10.1016/j.chemosphere.2019.125110

Tang Y-J, Wang L-J, Xu Z-L, Zhang H-Z (2018) Novel chitosan-piperazine composite nanofiltration membranes for the desalination of brackish water and seawater. J Polym Res 25:1–12. https://doi.org/10.1007/s10965-018-1514-6

Tzanakakis VA, Paranychianakis NV, Angelakis AN (2020) Water Supply Water Scarcity. https://doi.org/10.3390/w12092347

Upadhyayula VKK, Deng S, Mitchell MC, Smith GB (2009) Application of carbon nanotube technology for removal of contaminants in drinking water: a review. Sci Total Environ 408:1–13. https://doi.org/10.1016/j.scitotenv.2009.09.027

Vaka M, Walvekar R, Yanamadala S (2021) Carbon nanotubes and their composites: from synthesis to applications. Contemporary Nanomater Mater Eng Appl. Springer, pp 37–67

Vitiello G, Iervolino G, Imparato C, Rea I, Borbone F, De Stefano L, Aronne A, Vaiano V (2021) F-doped ZnO nano- and meso-crystals with enhanced photocatalytic activity in diclofenac degradation. Sci Total Environ 762:143066. https://doi.org/10.1016/j.scitotenv.2020.143066

Wang J, Chen H (2020) Catalytic ozonation for water and wastewater treatment: recent advances and perspective. Sci Total Environ 704:135249. https://doi.org/10.1016/j.scitotenv.2019.135249

Wang J, Ando RA, Camargo PH (2014) Investigating the plasmon-mediated catalytic activity of AgAu nanoparticles as a function of composition: are two metals better than one? ACS Catal 4:3815–3819

Wang F, Ma S, Si Y, Dong L, Wang X, Yao J, Chen H, Yi Z, Yao W, Xing B (2017) Interaction mechanisms of antibiotic sulfamethoxazole with various graphene-based materials and multiwall carbon nanotubes and the effect of humic acid in water. Carbon 114:671–678. https://doi.org/10.1016/j.carbon.2016.12.080

Wang Z, Zhao P, Zhu B, Jiang Z, Guo X (2018) Magnetic solid-phase extraction based on Fe3O4/graphene nanocomposites for enantioselective determination of representative profens in the environmental water samples and molecular docking study on adsorption mechanism of graphene. J Pharm Biomed Anal 156:88–96. https://doi.org/10.1016/j.jpba.2018.04.023

Wang X, Li Y, Yu H, Yang F, Tang CY, Quan X, Dong Y (2020a) High-flux robust ceramic membranes functionally decorated with nano-catalyst for emerging micro-pollutant removal from water. J Membr Sci 611:118281. https://doi.org/10.1016/j.memsci.2020.118281

Wang Y, Zucker I, Boo C, Elimelech M (2020b) Removal of emerging wastewater organic contaminants by polyelectrolyte multilayer nanofiltration membranes with tailored selectivity. ACS ES&T Eng 1:404–414. https://doi.org/10.1021/acsestengg.0c00160

Werkneh AA, Rene ER (2019) Applications of nanotechnology and biotechnology for sustainable water and wastewater treatment. Water Wastewater Treatment Technol, Springer, pp 405–430

Xiang W, Zhang Y, Lin H, Liu CJ (2017) Nanoparticle/metal–organic framework composites for catalytic applications: current status and perspective. Molecules 22:2103. https://doi.org/10.3390/molecules22122103

Xiang Y, Xu Z, Wei Y, Zhou Y, Yang X, Yang Y, Yang J, Zhang J, Luo L, Zhou Z (2019) Carbon-based materials as adsorbent for antibiotics removal: mechanisms and influencing factors. J Environ Manage 237:128–138. https://doi.org/10.1016/j.jenvman.2019.02.068

Xu B, Wu F, Zhao X, Liao H (2010) Benzotriazole removal from water by Zn–Al–O binary metal oxide adsorbent: behavior, kinetics and mechanism. J Hazard Mater 184:147–155. https://doi.org/10.1016/j.jhazmat.2010.08.017

Xu Y, Liu T, Zhang Y, Ge F, Steel RM, Sun L (2017) Advances in technologies for pharmaceuticals and personal care products removal. J Mater Chem A 5:12001–12014. https://doi.org/10.1039/C7TA03698A

Yaashikaa PR, Senthil Kumar P, Varjani SJ, Saravanan A (2019) Advances in production and application of biochar from lignocellulosic feedstocks for remediation of environmental pollutants. Biores Technol 292:122030. https://doi.org/10.1016/j.biortech.2019.122030

Yang D, Gates BC (2019) Catalysis by metal organic frameworks: perspective and suggestions for future research. ACS Catal 9:1779–1798

Yang L, Han Q, Sun S, Ding M (2017) Evaluation of graphene aerogel monolith-based solid-phase extraction for the separation of pyrethroids from water samples. Chromatographia 80:1781–1787

Yang Y, Xiong Z, Wang Z, Liu Y, He Z, Cao A, Zhou L, Zhu L, Zhao S (2021) Super-adsorptive and photo-regenerable carbon nanotube based membrane for highly efficient water purification. J Membr Sci 621:119000. https://doi.org/10.1016/j.memsci.2020.119000

Ye Z, Padilla JA, Xuriguera E, Beltran JL, Alcaide F, Brillas E, Sirés I (2020) A highly stable metal-organic framework-engineered FeS2/C nanocatalyst for heterogeneous electro-fenton treatment: validation in wastewater at mild pH. Environ Sci Technol 54:4664–4674. https://doi.org/10.1021/acs.est.9b07604

Yu F, Ma J, Han S (2014) Adsorption of tetracycline from aqueous solutions onto multi-walled carbon nanotubes with different oxygen contents. Sci Rep 4:1–8

Yu T, Wu W, Liu L, Gao C, Yang T (2020) Novel ternary p-ZnIn2S4/rGO/n-g-C3N4 Z-scheme nanocatalyst with enhanced antibiotic degradation in a dark self-biased fuel cell. Ceram Int 46:9567–9574. https://doi.org/10.1016/j.ceramint.2019.12.221

Yuan N, Gong X-R, Han B-H (2021) Hydrophobic fluorous metal-organic framework nanoadsorbent for removal of hazardous wastes from water. ACS Applied Nano Materials 4:1576–1585. https://doi.org/10.1021/acsanm.0c03050

Zhao J, Wang Z, White JC, Xing B (2014) Graphene in the aquatic environment: adsorption, dispersion, toxicity and transformation. Environ Sci Technol 48:9995–10009. https://doi.org/10.1021/es5022679

Zhao L, Deng J, Sun P, Liu J, Ji Y, Nakada N, Qiao Z, Tanaka H, Yang Y (2018) Nanomaterials for treating emerging contaminants in water by adsorption and photocatalysis: systematic review and bibliometric analysis. Sci Total Environ 627:1253–1263. https://doi.org/10.1016/j.scitotenv.2018.02.006

Zhao P, Liu N, Jin C, Chen H, Zhang Z, Zhao L, Cheng P, Chen Y (2019) UiO-66: an advanced platform for investigating the influence of functionalization in the adsorption removal of pharmaceutical waste. Inorg Chem 58:8787–8792. https://doi.org/10.1021/acs.inorgchem.9b01172

Chapter 13
Magnetic Nanoparticles: Application in the Removal of Next-Generation Pollutants from Wastewater

Bhaskar Bhaduri, Ashutosh Kumar Dixit, and Kumud Malika Tripathi

Abstract Critical shortage of potable water is a grand challenge of current time due to our deliberate exploitation of one of our precious and limited resources. Water safety is threatened by a diverse range of pollutants including potentially toxic inorganic and organic substances, microalgae, bacteria and radioactive wastes because of the acute toxicities, mutagenic and carcinogenic nature of the pollutants. Moreover, pharmaceuticals and personal care products, antibiotics, microplastics and engineered nanomaterials are contaminants of emerging concern due to their endocrine disrupting nature and other acute effects. Magnetic nanoparticles (MNPs) are promising for wastewater treatment by virtue of their excellent catalytic and adsorption performance and ability to fast and facile separation by the application of external magnetic fields. Iron oxide NPs such as magnetite (Fe_3O_4), υ-Fe_2O_3, α-Fe_2O_3, and goethite (α-FeOOH) are the most commonly available MNPs till date. Application of MNPs and their hybrid composite offer unprecedented opportunities to improve water remediation technologies for the removal of multiple pollutants from traditional to emerging contaminants. In this chapter, recent progress in the development of MNPs and their hybrid materials-based techniques for the removal of emerging pollutants from wastewater have been surveyed. The potential development and execution barriers regarding both technical challenges and engineering concerns are surveyed along with future directions to overcome them and improve water security.

B. Bhaduri
Department of Chemical Engineering, Indian Institute of Technology, Kharagpur, West Bengal 721302, India
e-mail: bhaskar@che.iitkgp.ac.in

A. K. Dixit
Department of Electronics and Communication Engineering, VFSTR University, Vadlamudi, Guntur, Andhra Pradesh 522213, India
e-mail: ashutoshkrdikshit.rs.ece17@iitbhu.ac.in

K. M. Tripathi (✉)
Department of Chemistry, Indian Institute of Petroleum and Energy, Visakhapatnam, Andhra Pradesh 53003, India
e-mail: kumud@iipe.ac.in

S. P. Singh et al. (eds.), *New Trends in Emerging Environmental Contaminants*, Energy, Environment, and Sustainability,
https://doi.org/10.1007/978-981-16-8367-1_13

Keywords Magnetic nano-carbons · Nanocomposite · Wastewater remediation · Photocatalysis · Adsorption · Reactive oxygen species

13.1 Introduction

The accelerating economy, industrialization and worldwide burgeoning population have resulted in a host of numerous contaminants in aquatic environments with higher complexity and larger quantities than ever before (Qian et al. 2020). Water pollution by chemical threats, organic/inorganic contaminants altered the water cycle and challenged the efficiency of congenitally available water treatment technologies. The problem related to availability of drinking water is continuously rising, since number of contaminants including personal consumer products such as sunscreens, detergents, sanitizers or pharmaceuticals; microplastic, disinfection by-products, insecticides, engineered nanomaterials; and pesticides are continuously increasing in the water bodies due to unplanned human activities and industrial advancements (Zhang et al. 2020; Patel et al. 2019; Zhao et al. 2021). More than 800 pollutants of emerging contents are being researched globally (Gaston et al. 2019). Table 13.1 summarizes some commonly used pharmaceuticals with some physiochemical properties. The continuous increasing use of ionic-liquids as green solvent in numerous applications especially in energy storage devices and biofuel production are critical for the environment including groundwater because of their persistent nature and low sorption to soil particles (Oskarsson and Wright 2019). Another significant class of emerging pollutants are artificial sweeteners (Kokotou et al. 2012). This trace next-generation pollutants are considered as contamination of emerging concern due to their high structural diversity and complexity and their long-term exposure are having widespread and long-lasting threats on both human and the environment despite their low concentrations. The abatement of emerging pollutants from drinking and wastewater is of highest priority as they might produce genotoxicity development of pathogen resistance and endocrine disruption (Álvarez et al. 2010). Moreover, the occurrence, fate and ecotoxicological effects of many emerging pollutants are not fully known yet and a matter of further research (Magi et al. 2013). These pollutants may be introduced in wastewater by a number of routes as shown in Fig. 13.1 (Gaston et al. 2019). Many significant efforts have been made to alleviate the release of the contaminants into the environment. The wastewater treatment technologies generally use filtration, advanced oxidation, membrane filtration, adsorption and coagulation processes. However, many emerging pollutants are hard to remove by the aforementioned treatment technologies and may easily escape to water stream. The pollutants may further be released into ground water and finally accumulate in the soil.

Benefiting from the uniqueness and surface properties, nanomaterials have already shown extreme potential to deal with the wastewater hazards but concerns regarding their disposal, separation and environment exposure still exist. There may be probabilities that due to small sizes and low separation efficiency nanoparticles (NPs) can be involved in waste streams, have the risk of leaving cytotoxic residuals and pose

Table 13.1 Commonly used pharmaceuticals, their classes, therapeutic applications, physicochemical properties, and structures (Patel et al. 2019)

Pharmaceuticals (CAS number)	Class/therapeutic application	Chemical formula (Mol. Wt. in g/mol)	pKa	$logK_{ow}$	Predicted CEC[a] (ng/L)	Structure
Acebutolol (37517-30-9)	β-blocker/anti-hypertensive/ Human use	$C_{18}H_{28}N_2O_4$ (336.432)	13.8	1.53		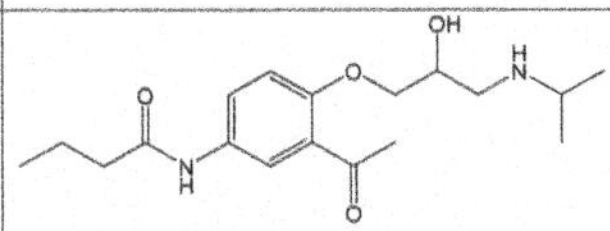
Acetaminophen (paracetamol) (103-90-2)	Analgesic/ anti-pyretic/ human and veterinary use	$C_8H_9NO_2$ (151.165)	9.38	0.46	2.4E+07	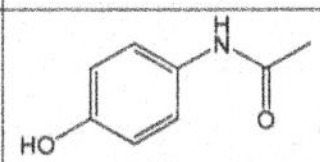
Albuterol (salbutamol) (18559-94-9)	Bronchodilator/ human use	$C_{13}H_{21}NO_3$ (239.315)	10.3	0.64	27669	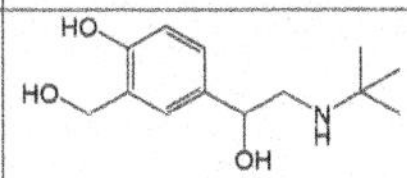
Alprazolam (28981-97-7)	Benzodiazepine/ anti-anxiety drug/Human use	$C_{17}H_{13}ClN_4$ (308.769)		2.12	57	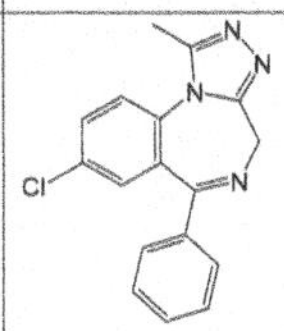
Amitriptyline (50-48-6)	Antidepressant/ human use	$C_{20}H_{23}N$ (277.411)	9.4	4.92	48	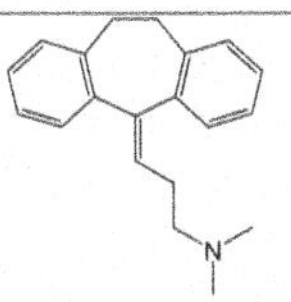
Amlodipine (88150-42-9)	Calcium channel agents/ antianginal agents/human use	$C_{20}H_{25}ClN_2O_5$ (408.879)	9.4	3.00	7003	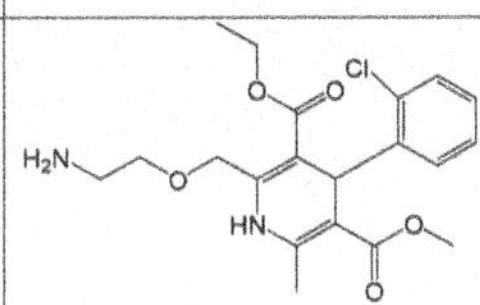
Amoxicillin (26787-78-0)	Antibiotic/ antibacterial/ β-lactums/ Human applications	$C_{16}H_{19}N_3O_5S$ (365.404)	3.2, 11.7	0.87	7.4E+06	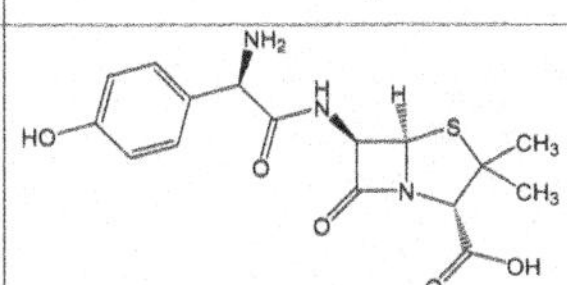
Ampicillin (69-53-4)	Antibiotic/ antibacterial/ β-lactums/Human applications	$C_{16}H_{19}N_3O_4S$ (349.405)	2.5, 7.3	1.35	12191	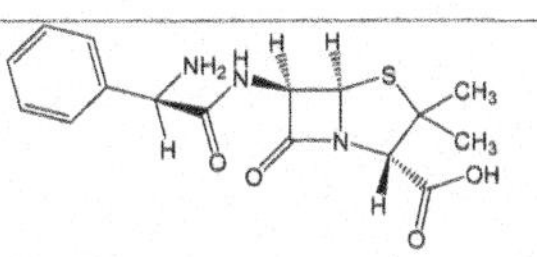
Aspirin (acetylsalicylic acid) (50-78-2)	Analgesic/ antipyritic/Human and veterinary use	$C_9H_8O_4$ (180.159)	3.49	1.19	3.4E+08	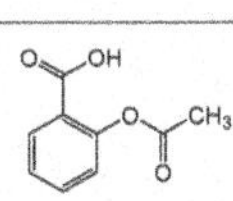
Atenolol (29122-68-7)	β-blocker/Anti-hypertensive/ human use	$C_{14}H_{22}N_2O_3$ (266.341)	9.6	0.16	792332	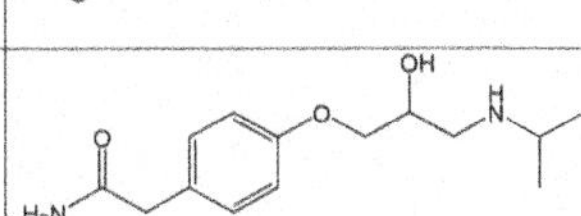

(continued)

Table 13.1 (continued)

Pharmaceuticals (CAS number)	Class/therapeutic application	Chemical formula (Mol. Wt. in g/mol)	pKa	$\log K_{ow}$	Predicted CEC[a] (ng/L)	Structure
Atorvastatin (134523-00-5)	Lipid lowering agent/ human applications	$C_{33}H_{35}FN_2O_5$ (558.64)	4.3, 14.9	6.36		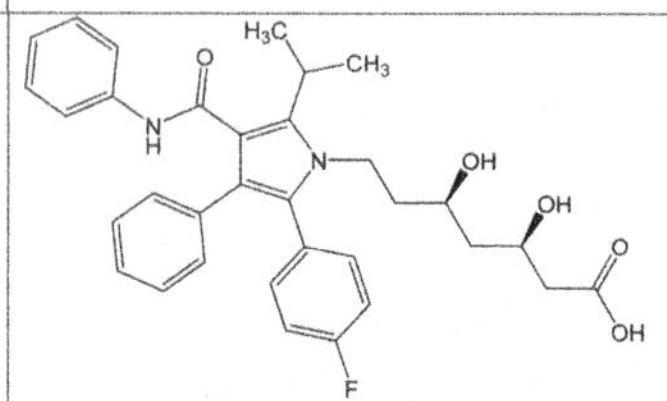
Azithromycin (83905-01-5)	Antibiotic/ antibacterial/ human use	$C_{38}H_{72}N_2O_{12}$ (748.996)	8.74	4.02	1302	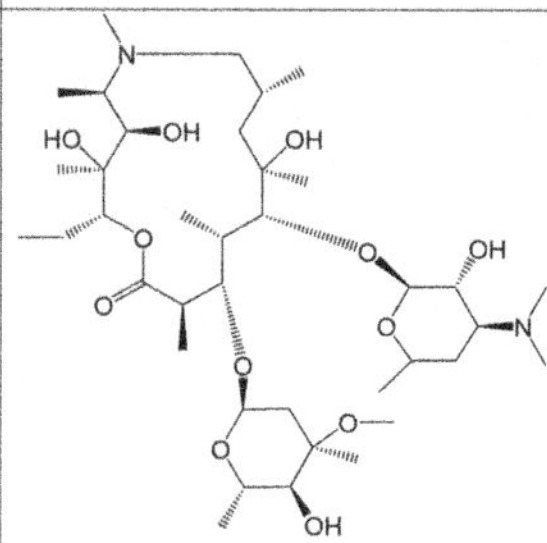
Budesonide (51333-22-3)	Corticosteroid drug/anti-allergic/Human use	$C_{25}H_{34}O_6$ (430.541)		3.20	86	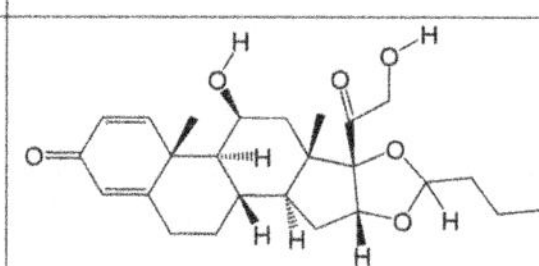
Bupropion (34911-55-2)	Antidepressant/ human use	$C_{13}H_{18}ClNO$ (239.743)	8.22	3.85	116	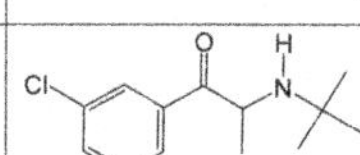
Carbamazepine (298-46-4)	Anti-epileptic/ Neuropathic pain reliever/Anti-psychotic	$C_{15}H_{12}N_2O$ (236.274)	13.9	2.45	346496	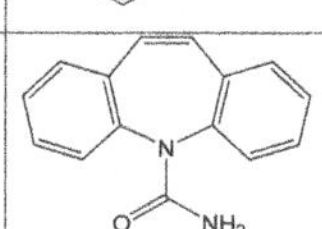
Cephalexin (15686-71-2)	Antibiotic/ antibacterial/ cephalosporins/ human applications	$C_{16}H_{17}N_3O_4S$ (347.389)	4.5	0.65	2.5E+08	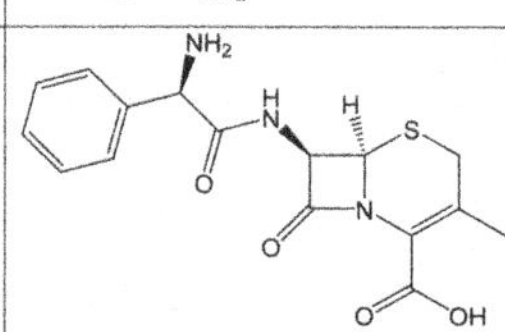
Cetirizine (83881-51-0)	Antihistamine/ antiallergic/ human Use	$C_{21}H_{25}ClN_2O_3$ (388.892)	2.70, 3.57, 7.56	1.70	423061	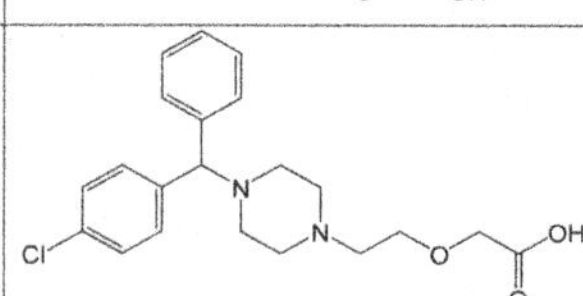
Chloramphenicol (56-75-7)	Antibiotic/ antibacterial/ broad spectrum antibiotic/ human applications	$C_{11}H_{12}Cl_2N_2O_5$ (323.126)		1.14	8.1E+06	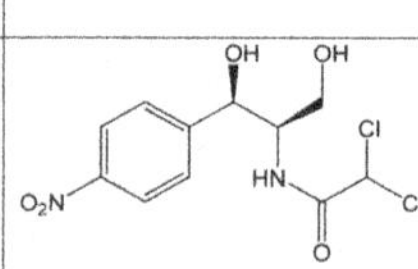

(continued)

Table 13.1 (continued)

Pharmaceuticals (CAS number)	Class/therapeutic application	Chemical formula (Mol. Wt. in g/mol)	pKa	$\log K_{ow}$	Predicted CEC[a] (ng/L)	Structure
Chlortetracycline (57-62-5)	Antibiotic/Human and veterinary applications/ growth promotor and aquaculture	$C_{22}H_{23}ClN_2O_8$ (478.882)			2.4E+07	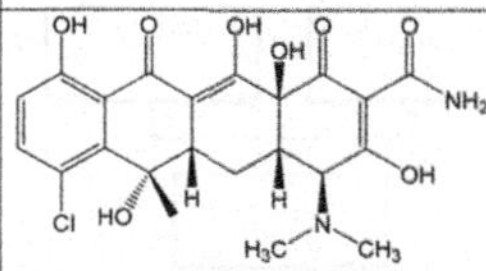
Cimetidine (51481-61-9)	Antihistamine/ antagonist/Human applications	$C_{10}H_{16}N_6S$ (252.340)	6.8	0.40	722637	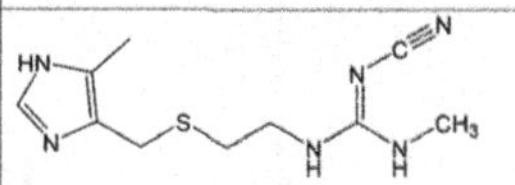
Ciprofloxacin (85721-33-1)	Antibiotic/ fluoroquinolone /human applications	$C_{17}H_{18}FN_3O_3$ (331.347)	6.09, 8.74	0.28	1.9E+07	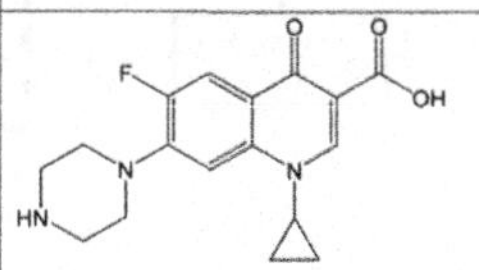
Citalopram (59729-33-8)	Antidepressant/ human use	$C_{20}H_{21}FN_2O$ (324.399)			141	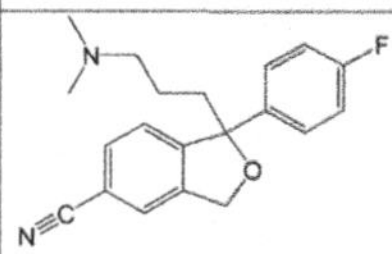
Clarithromycin (81103-11-9)	Antibiotic/ macrolide/Human applications	$C_{38}H_{69}NO_{13}$ (747.964)	8.99	3.16	7267	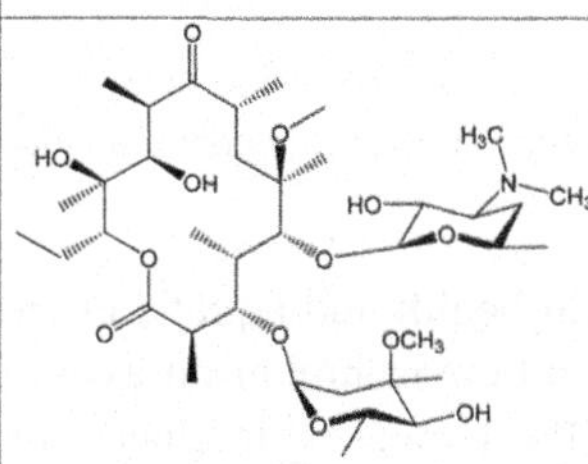
Clonazepam (1622-61-3)	Benzodiazepine/ tranquilizer/ human use	$C_{15}H_{10}ClN_3O_3$ (315.713)	1.5, 10.5	2.41	431	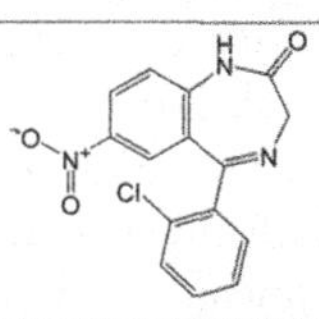
Clofibric acid (882-09-7)	Lipid lowering drugs metabolite/ human use and plant growth promotor	$C_{10}H_{11}ClO_3$ (214.645)	3.2	2.90		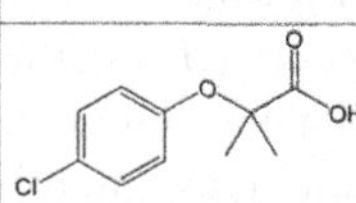
Clopidogrel (113665-84-2)	Antiplatelet drugs/ Human use	$C_{16}H_{16}ClNO_2S$ (321.819)	5.3	3.8		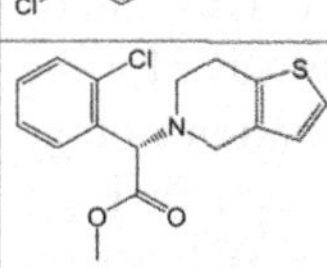
Diazepam (439-14-5)	Benzodiazepine/ antianxiety drug/Human use	$C_{16}H_{13}ClN_2O$ (284.743)	3.4	2.9	16219	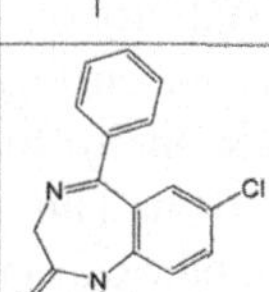

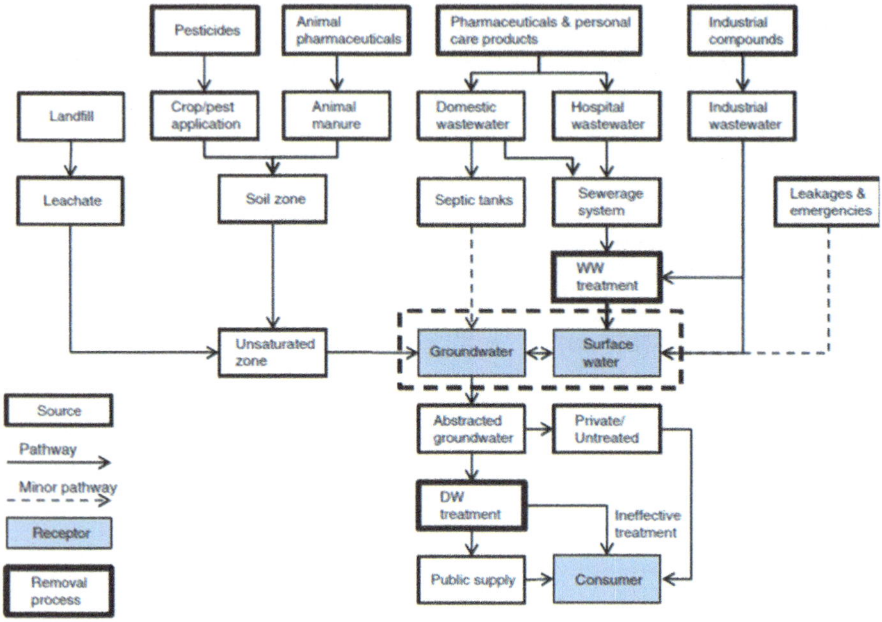

Fig. 13.1 Sources of emerging contaminants and pathways toward receptors (Gaston et al. 2019)

risk to human health and aquatic environment (Shi et al. 2013). Extensive regeneration steps and environment unfavourable disposal procedures are required to treat the spent NPs. However, magnetic nanoparticles (MNPs) can be fully recovered via magnetic separation than centrifugation and ultrafiltration and hence are quite safe and efficient (Shi et al. 2013). In this regard, MNPs have drawn considerable research interest for water remediation. Akin to separation ease, the most significant advantage of MNPs is uniform structure, versatility and low-cost. MNPs are particularly significant in wastewater remediation as they offer practical solutions to the enduring challenges associated with the recovery and recyclability of NPs. The feasibility of MNPs have been tested with diverse wastewater contaminants ranging from heavy metal ions, organic dyes, microbes to next generation pollutants. MNPs have been involved in diverse ways for the development of modern technologies, especially in catalysis, information storage, magnetic resonance imaging, drug delivery, sensors, energy storage, biomedicine, adsorption and environmental remediation (Lu et al. 2007). However, low stability over a longer period of time due to aggregations and consequently lessen active surface area are the major drawbacks. Moreover, MNPs tend to oxidize in air due to high chemical activity, which resulted in loss of dispersibility and magnetism. This issue can be overcome by surface modification, grafting and composite fabrication depending upon the targeted application (Lu et al. 2007). Composite or surface modified MNPs are very promising for application in adsorption and advanced oxidation processes. Toxicity of MNPs in the body is influenced by the surface chemistry and course of delivery of the MNPs (Shubayev et al. 2009). MNPs with sizes greater than 200 nm are known to be captured by the spleen via mechanical filtration. However, MNPs with sizes less than 10 nm can be removed through renal clearance. Thus, MNPs with diameter ranging between 10 and 100 nm is considered to be the optimal range of administration for specific applications (Gupta and Gupta 2005). The biodistribution patterns show that 80–90% of MNPs can be accumulated in liver. In comparison, 5–8% in the spleen and 1–2% in bone marrow (Duguet et al. 2006). The interaction and biodistribution of MNPs and the mechanism of internalization inside cells and organs, metabolism and possible toxicity all depend on any surface modifications (Nel et al. 2006).

Herein, in this chapter some of the most promising MNPs for the treatment of wastewater contaminants of emerging concern are discussed with the following goals: (1) the introduction of MNPs including metal based MNPs, magnetic carbon NPs and composite based MNPs; (2) fundamental aspects of pollutant removal process with MNPs; (3) prospects of their utilization to next generation pollutant removal by highlighting key points for further exploration.

13.2 Types of Magnetic Nanoparticles

13.2.1 Metal Based Magnetic Nanoparticles

Magnetic iron oxide (Fe_2O_3) offers significant potential over a wide range of applications owing to their amazing super-paramagnetism including magnetic storage, optoelectronics, medicine and biotechnologies due to its high stability towards oxidation (Lu et al. 2007). However, aggregation and aerial oxidation are still barriers for advanced applications (Baikousi et al. 2012). Titania (TiO_2) NPs were embedded in ordinary shock absorption foam to fabricate superhydrophobic foam with three-dimensional (3D) architecture for the selective adsorption of organic solvents and oil (Yu et al. 2016). Single domain MNPs have been widely investigated in diverse areas of material science and technology. Magnetic properties are related to chemical composition, morphology, shape and size of NPs. Core–shell morphology is particularly attractive due to synergistic effect ease of both components along with control in magnetic interaction between core and shell, which offer precise engineering of magnetic response (Oberdick et al. 2018). One such example of spin canting in Fe_3O_4/Mn-ferrite core/shell MNPs ($Fe_3O_4/Mn_xFe_{3-x}O_4$ MNP) is explored by Oberdick et al. via Dzyaloshinskii-Moriya interaction (DMI). The diameter of the Mn-ferrite shell was 0.5 nm having 3.2 nm of core as shown in Fig. 13.2a. The multiparticle correlations of these core–shell MNPs to the applied magnetic field in both parallel and perpendicular directions is depicted in Fig. 13.2b, which shows the aligned canting direction of particles (Oberdick et al. 2018). In an innovative approach Moon et al. reported that ultrathin shells can induce spin canting as an interface phenomenon and largely enhance magnetic coercivity due to the characteristic hard-soft exchange coupling and surface anisotropy (Moon et al. 2017).

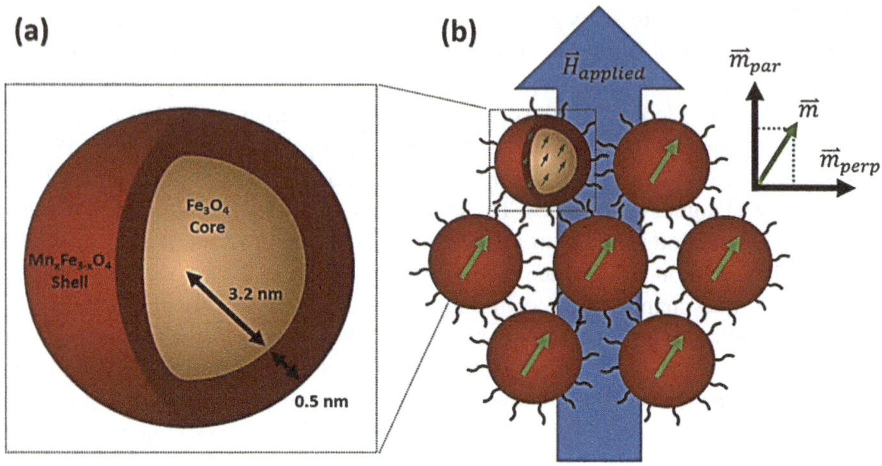

Fig. 13.2 **a** Schematic showing chemical and physical composition of core/shell nanoparticles. **b** Representative figure showing coherent spin canting across nanoparticles under application of applied magnetic field. Due to interparticle correlation of canted spins, the domain has a net canted magnetization that can be measured in terms of parallel and perpendicular components (Oberdick et al. 2018)

This approach offers the potential to synthesize of NPs with controlled magnetism and tuning of magnetic properties by modulating the core–shell parameters. The exchange coupling (EC) regime of core–shell NPs represent the regime of hard-soft exchange coupling, while enhanced spin canting (ESC) regime represent the population of canted spins as shown in Fig. 13.3 (Moon et al. 2017). Fluorescent core–shell Fe_3O_4@PDA@HL (polydopamine (PDA) and 3-aminopyridine-2-carboxaldehyde N(4)-methylthiosemicarbazone, HL) MNPs having multifunctional applications were synthesized via the chemical coprecipitation and conjugation method assisted by ultrasonication (Fang et al. 2019).

Zhang et al. used a co-assembly approach for the synthesis of magnetic silica microspheres (LPMMS) having core–shell morphology and mesoporous architecture using poly(ethylene oxide)-blockpoly(methyl methacrylate) (PEO-b-PMMA) soft template and cetyltrimethylammonium bromide (CTAB) as a cationic co-surfactant (Zhang et al. 2019). The multi-step synthetic route is shown in Fig. 13.4a. At first, nonporous dense silica ($nSiO_2$) were coted over solvothermally synthesized magnetic citrate-capped hydrophilic Fe_3O_4 NPs core followed by deposition of PEO-b-PMMA-SiO_2 layer via solvent evaporation process, which direct interphase co-assembly with CTAB micelles by hydrogen bonding and electrostatic interaction. After crosslinking with silica oligomer, organic surfactants were removed by pyrolysis in air to obtain large mesoporous LPMMS. These core–shell MNPs exhibit sphere like morphology as shown in Fig. 13.4b via transmission electron microscopic (TEM) analysis with rough surface (Fig. 13.4c). Besides confirming core–shell morphology TEM analysis also revealed porous architecture as shown in Fig. 13.4d

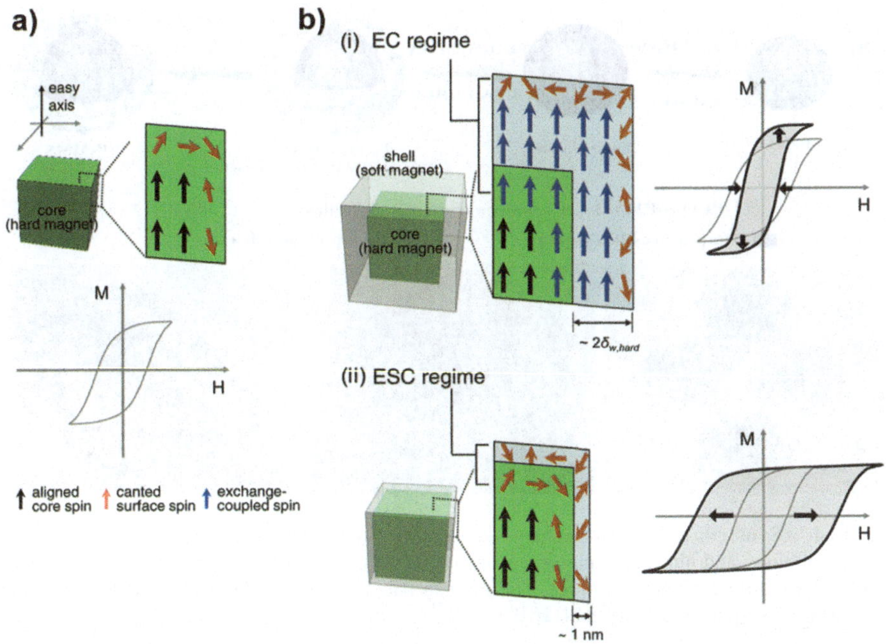

Fig. 13.3 **a** Magnetic spin state of a hard magnetic core nanoparticle. The core nanoparticle has aligned spins (black arrows) in its inner core and canted spins (orange arrows) on its surface. **(b)** Magnetic spin interactions in hard-soft bimagnetic core–shell nanoparticles. **(i)** In the EC regime, soft magnetic spins at the interface are pinned by hard magnetic spins. These spins are called exchange-coupled spins (blue arrows) and result in decreased magnetic coercivity and increased magnetism. **(ii)** In the ESC regime, where the shell thickness is similar to the size of a crystalline unit cell (~1 nm), the population of canted spins (orange arrows) increases (Moon et al. 2017)

(Zhang et al. 2019). Surface activation of MNPs is known to enhance the stability of NPs in solution and enhanced activity for a range of applications (Verma et al. 2019). For instance surface activation of hydrothermally synthesized Fe_3O_4 with imidazolium Gemini and tetraalkylammonium surfactant was found to significantly enhance the extraction of Ag and Au NPs via high hydrophobic and hydrophilic NPs-NPs interaction (Verma et al. 2019). Miao et al. reported the modification of Fe_3O_4@Au MNPs with DNA in a simpler approach of blending and aging to enhance selectivity and sensitivity for the detection of heavy metal ions in electrochemical sensing (Miao et al. 2017).

Hong et al. fabricated porous magnetic nanofibrous membranes (NiFe@SNM) composed of $NiFe_2O_4$@SiO_2 synthesized using zein dip-coating method to overcome the flexibility issue. The overall synthesis procedure is demonstrated in Fig. 13.5. Such fabricated membranes exhibit low density, high flexibility, larger surface area and excellent magnetic properties (Hong et al. 2015).

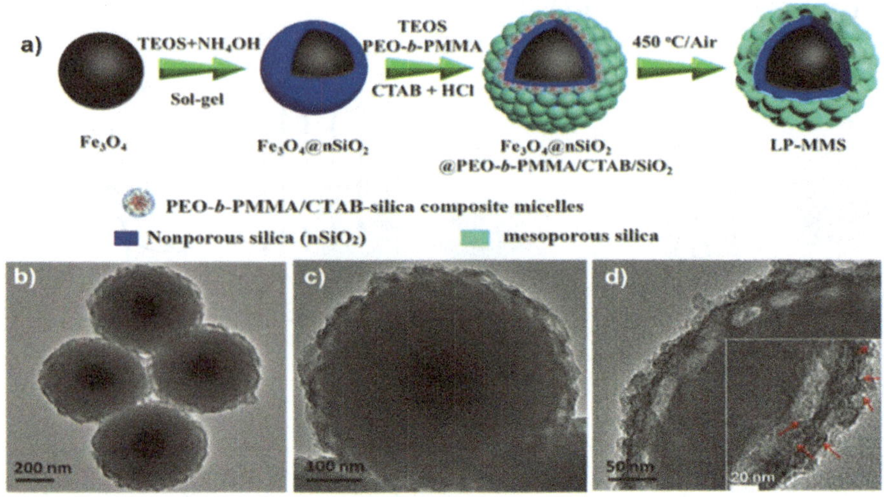

Fig. 13.4 **a** Synthetic route for magnetic mesoporous silica core–shell microspheres with rasberry-like morphology and ultra large mesopore size (LPMMS). **b–d** TEM images of LP-MMS microspheres with different magnifications. The inset in panel (F) is the enlarged TEM image of the LP-MMS microspheres (Zhang et al. 2019)

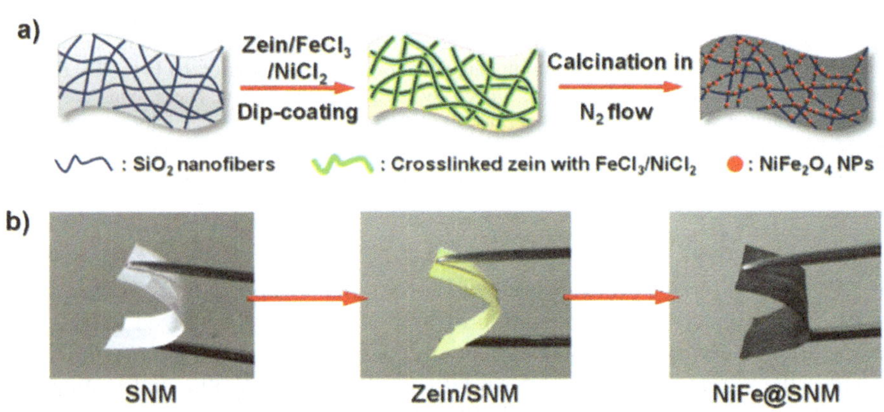

Fig. 13.5 **a** Schematic illustration of the fabriation steps of NiFe@SNM. **b** Optical images of the SNM, zein/SNM, and NiFe@SNM (Hong et al. 2015)

13.2.2 Nano-Carbons Based Magnetic Nanoparticles

Nano-carbon based magnetic NPs have received much interest due to their sustainable nature and the combined advantage of magnetic with inherent mechanical, optical, electrical and thermal properties (Siddiqui et al. 2019). Magnetic properties of nano-carbons offer an additional dimension of valuable merits that further widen the applications with enhanced performance (Siddiqui et al. 2019). A key feature of magnetic nano-carbons is their ease in separation from other media or matrices. In particular, the surface functionalities and magnetic characteristics are highly suitable for the removal of contamination from wastewater via adsorption. Unconventional ferromagnetism in nano-carbons is mainly due to the surface curvature, heteroatom (N, S, P, B) doping, mixed sp^2-/sp^3 hybridization and spin frustration. Shrivastava et al. synthesized ferromagnetic carbon nano-dots with blue PL emissions by the thermal decomposition of citric acid and l-lysine (Srivastava and Gajbhiye 2011). To align with environmental sustainability and reduce carbon foot print, utilization of pollutant soot or waste materials for their second life applications as value-added nano-carbons has remained active and appealing for the better utilization of waste. Pollutant soot derived nano-carbons have emerged as an appealing multifunctional material for energy storage and water remediation applications (Bhati et al. 2016; Jung et al. 2020). Petrol soot derived graphene nanosheets, after nitric acid treatment were found to have excellent photocatalytic activity not only for the single dye but also a mixture of dyes (Gunture et al. 2522). Even though soot derived nano-carbons have proven its good performance in water remediation towards dye degradation and metal ions sensing but their utilization for next-generation pollutant removal is still awaiting.

Water soluble version of CNTs (wsCNTs) with network structure (Fig. 13.6a) produced from discarded fullerene soot possess room temperature magnetic properties that are appealing for many applications including catalysis, adsorption and next-generation pollutant removal from wastewater (Dubey et al. 2013). Importantly, presence of various defects on graphitic tubular structure in the form of junctions, missing graphitic planes, bends and curvatures resulted in spin value of carbon free

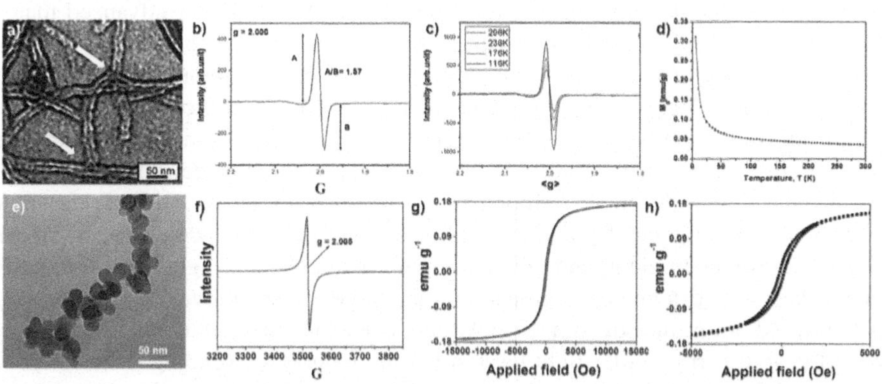

Fig. 13.6 **a** Low magnification TEM micrographs of wsCNTs (white arrow shows the presence of various types of junctions; **b** Solid-state ESR spectrum, and **c** temperature dependent ESR spectra of the wsCNTs network. **d** Temperature dependence of the mass magnetization (Mg), field cooled (FC) of the wsCNTs network measured at 5 kO_e in the temperature range $5 \leq T \leq 300$ K. (Dubey et al. 2013) **e** TEM image of OCNPs showing spherical morphology and continuous network. **f** ESR spectrum of the OCNPs due to the strong carbon-cantered radical; **g** magnetization (M-H) curve of the OCNPs at 300 K; **h** hysteresis loop of the OCNPs (Chowdhury et al. 2021)

radical in electron spin resonance (ESR) analysis as shown in Fig. 13.6b with asymmetric resonance (A/B = 1.57) line. Asymmetry was attributed to the contraction motion of the paramagnetic centre because of surface functionalization with the carboxylic acid group. Line shape i.e., asymmetric resonance was temperature dependent as shown in Fig. 13.6c, while spin value remains constant. Mass magnetization with field cooling (FC) shows positive magnetization and follows a non-Curie–Weiss system as shown in Fig. 13.6d (Dubey et al. 2013). In an interesting study by the same research group, anthropogenic multiwalled carbon nanotubes (MWCNTs) isolated from spider webs were observed to show ferromagnetic behaviour due to the presence of stable carbon centre radicals (Sonkar et al. 2014). Authors show that floating MWCNT were trapped on indoor spider webs due to the various routine burning activities and are able to generate reactive oxygen species (ROS) (Sonkar et al. 2014, 2009). In an attempt to reduce the environment pollution and fruitful utilization of black carbon (BC), onion-like carbon nanoparticles (OCNPs) with interconnected chain like conductive network Fig. 13.6e were isolated from diesel engine exhaust soot via solvent extraction with toluene and acetone (Chowdhury et al. 2021). Further annealing at 500 °C increases the graphitic carbon content. These OCNPs show room temperature magnetic properties with an average g value of carbon based radical (2.005) as shown in Fig. 13.6f. The magnetization-hysteresis (M–H) loop in Fig. 13.6g, h show saturation magnetization (M_s) of ~ 0.17 emu g^{-1}, attributed to spin frustrations due to high density of structural defects (Chowdhury et al. 2021).

13.2.3 Composite Based Magnetic Nanoparticles

To mitigate water scarcity, there is an emerging need to engineer advanced materials with multifunctional characteristics and applications (Wang et al. 2019). For instance, the inherent optoelectronic properties of single component materials like ZnO NPs, TiO_2 NPs, CNTs, graphene failed to provide sufficient use of solar light to degrade next-generation pollutants. Hence, composite fabrication through accumulation of guest material(s) to parent matrices to enhance inherent properties is attracting wide interest among researchers (Wang et al. 2019). Integration of delocalised conjugated polymers with semiconductor photocatalyst can create high extinction coefficient, and efficiently tailor band gap and light harvesting efficiency due to the extended π-conjugation for charge carriers movements (Pant et al. 2018). In addition, introduction of polymers can efficiently retard the recombination of charge-carriers after light harvesting via the hole scavenging. For instance, hybridization of $Ni_{0.5}Zn_{0.5}Fe_2O_4$ with polyaniline (PANI) forms an interface, which induces a synergistic role to enhance photocatalytic efficiency towards organic dye degradation. The inverse spinel structure weakened the interaction between Fe and Ni sublattices and consequently resulted in enhanced magnetic movement of Fe and catalytic separation (Pant et al. 2018). Typically, Li al. reported a facile method to encapsulate magnetic Fe_3O_4 core with amorphous carbon to synthesize bean-like Fe_3O_4@C@Cu_2O composites having core–shell morphology and

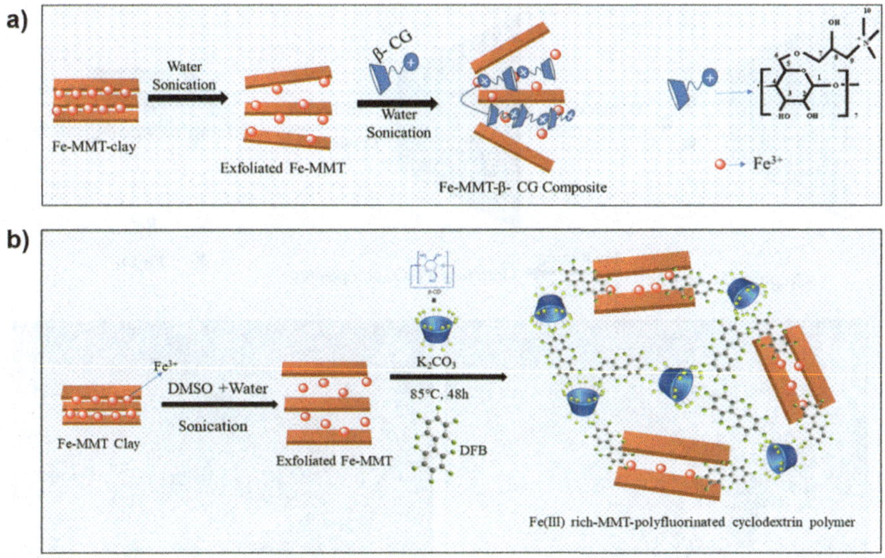

Fig. 13.7 Schematic illustration of synthesis of **a** Fe-MMT-βCG and **b** Fe-MMT-DFB-βCD composites (Kundu et al. 2020)

ferromagnetic nature by using a self-assembly approach. Such assembly protects the magnetic core from erosion and well screens the dipolar interaction (Li et al. 2011). Kundu et al. synthesized iron-clay-cyclodextrin (Fe-MMT-βCD) and iron-clay-cyclodextrin-polymers (Fe-MMT-βCD-DFB) composites via a simple process of sonication and electrostatic adsorption, respectively as shown in Fig. 13.7 (Kundu et al. 2020). Iron-oxide coated clay was first modified with CD monomers and polymers and purified via dialysis. Cd was chosen due to its structural advantage of both hydrophilic surface and hydrophobic cavity, which allows interaction with a range of organic molecules, while, polymers provide selective adsorption of pollutants (Kundu et al. 2020).

Hybridizing nano-carbons, polymer with magnetic metal NPs is a powerful strategy towards the fabrication of high-performance composite materials (Saini et al. 2020a). Composite NPs combine the fascinating merits of both of the constitutes with diverse properties by generating a strong synergism related to the intimately contacted interfaces (Saini et al. 2019). Numerous nano-carbon based nanocomposites with metal NPs, metal oxides/sulphide have been reported, which exhibit enhanced properties of the host materials (Park et al. 2019; Das et al. 2019a). An interesting review by Zhu et al. describes the different prospectus of nano-carbon based magnetic composite materials from synthesis to applications (Zhu and Diao 2011). Shao et al. reported the organic pollutant removal capacity of MWCNTs-polyaniline (PANI) composite. The magnetic nature of composite allows easy separation after adsorption, which can also be used in acidic medium (Shao et al. 2010). Su et al. designed graphene-Fe_3O_4 nanocomposites starting with GO, $FeCl_3$, ascorbic acid,

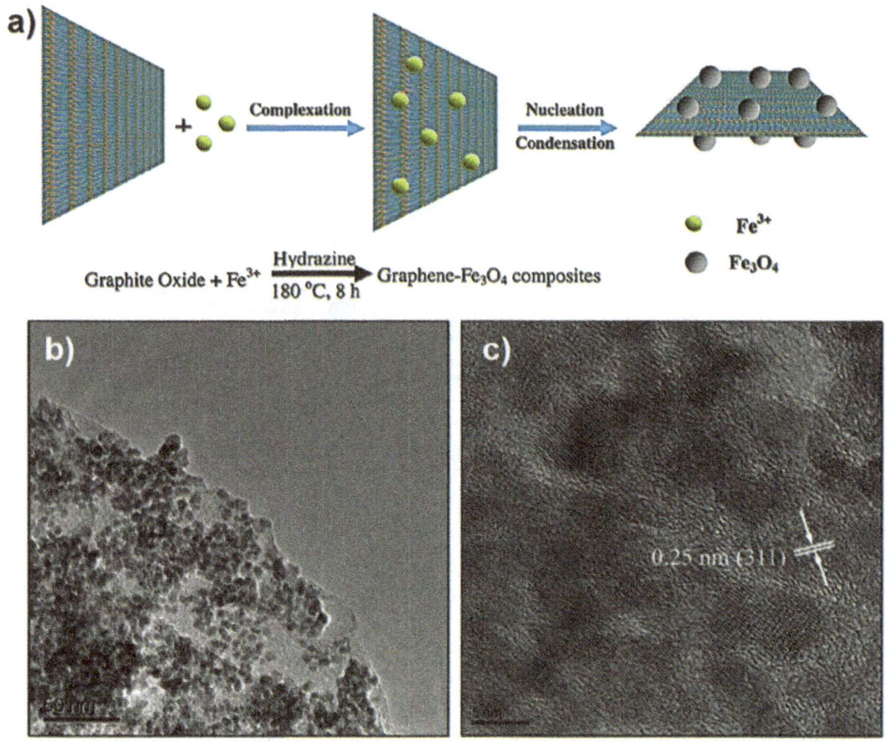

Fig. 13.8 **a** A schematic showing synthesis of graphene-Fe_3O_4 nanocomposites. **b** High magnification TEM image of graphene-Fe_3O_4 nanocomposites; **c** HRTEM image of graphene-Fe_3O_4 nanocomposites (Su et al. 2011)

and hydrazine hydrate via a one-pot hydrothermal technique (Fig. 13.8a).(Su et al. 2011) The reduction of GO to graphene, formation of Fe_3O_4 NPs and decoration of Fe_3O_4 NPs over graphene occurs at the same time. This composite shows high stability over a prolonged period of sonication and is highly electrically conductive. TEM image in Fig. 13.8b demonstrated composite fabrication and Fig. 13.8c shows clear lattice fringes of both Fe_3O_4 NPs and graphene (Su et al. 2011). Saini et al. used the waste printer ink as precursor for the synthesis of self-doped iron oxide nanocarbons in a cost effective approach (Saini et al. 2020a). These NPs were utilized for the degradation of Congo red and styrene oxidation under the influence of natural sunlight; and can be separated from the reaction medium by the use of external magnet (Saini et al. 2019, 2020a).

13.3 Magnetic Nanoparticles as Adsorbent for Next-Generation Pollutant Removal

Among many treatment methods, adsorption has been proved utmost cost-effective and attractive due to its high removal efficiency, diverse nature of adsorbents, which can be easily tailored for selectivity (Bajpai et al. 2019; Shukla et al. 2019). The effectiveness of adsorbent for the selective adsorption of next-generation pollutants depends upon the surface, chemical and textural properties of adsorbent and types, and the chemical nature of contaminants, so selection and development of an appropriate adsorbent is crucial. In an interesting approach, porous core–shell MNPs with Fe_3O_4 core and Covalent organic frameworks (COFs) shell were synthesized by using monomer-mediated in situ growth strategy for the selective adsorption of bisphenol A (BPA) and bisphenol AF (BPAF) (Li et al. 2017). 1,3,5-triformylphloroglucinol (Tp) functionalized spherical magnetic core act as epidemic centre for the controllable and uniform growth of COFs in all directions. Shell thickness were controlled by the variation in monomer concentration, while COF offered enhanced functionality to interact with endocrine-disrupting chemicals. These MNPs achieve adsorption equilibrium within 5 min for both BPA and BPAF (Li et al. 2017). Magnetic metal organic framework was explored for the selective removal of PPCPs. Selectivity was achieved by the use of seven deep eutectic solvents to regulate chelation (Wei et al. 2020).

Magnetic nano-carbon based nanocomposite combines the high adsorption capacity of nano-carbons due to high surface area and porosity and magnetic properties of MNPs. For example, magnetic mesoporous carbon composite (Fe_3O_4/C) was fabricated using a solvothermal method for the effective removal of CIP, one of the examples of third generation fluoroquinolones. Fe_3O_4/C with sphere like shape exhibit both n-π and π-π electron donor–acceptor interaction for the adsorption of CIP (Shi et al. 2013).

Ibuprofen (IB), a non-steroidal anti-inflammatories drug has become the pollutant of significant concern because of its wider use in society and consequent emergence in wastewater via urination (Vicente-Martínez et al. 2020). The complex chemical structure of IB is hard to degrade by nature and hence affecting the ability of aquatic species to move, feed and reproduce. IB has been adsorbed on Fe_3O_4@AgNPs surface from aqueous medium with high removal efficiency (93%) in 45 min only at neutral pH. The Fe_3O_4@AgNPs was recycled from the aqueous medium with magnet and reuse for two successive cycles after the desorption of IB with nitric acid. Ionic strength deteriorated the interaction between adsorbent and adsorbate and resulted in desorption of IB. This study proposed the potential of core–shell NPs for the removal of next-generation pollutants, which cross the barrier of current water treatment technologies because of their low concentrations (Vicente-Martínez et al. 2020). Another study by Guo et al. shows the removal of antibiotics oxytetracycline (OTC) from aqueous solution with Fe(0)-loaded magnetic mesoporous silica (Fe-MCM-41-A) NPs. Fast adsorption was attributed due to the presence of hexagonal channels and hence facilitated mass transfer, and enhance interaction between adsorbent and OTC (Guo et al. 2021).

Polycyclic aromatic hydrocarbons (PAHs) are ubiquitous in nature and have been released into environment as a result of incomplete combustion of fossil fuels, residential waste combustion, coal tar pitch and agricultural production (Zhu et al. 2004). PAHs are highly persistent due to their hydrophobic nature and complex organic core and are potentially toxic due to their carcinogenic and mutagenic nature (Zhu et al. 2004). Nisticó et al. reported the removal of PAHs by a simple adsorption process with hybrid composite of chitosan and magnetic Fe_2O_3/Fe_3O_4 NPs (Nisticò et al. 2017). The adsorbents were synthesized using a co-precipitation process followed by pyrolysis. To optimize process variables chitosan quantity was varied from 1% (MC1), 2% (MC2) and 3% (MC3) and pyrolysis temperature from 550 to 800 °C and referred as coded as MC1py550, MC1py800, MC2py550, MC2py800, MC3py550, or MC3py800. High-resolution transmission electron microscopy (HRTEM) as shown in Fig. 13.9a–c shows the amorphous parts together with crystal lattice fringes of crystalline part. These samples exhibited superparamagnetic behaviours revealed

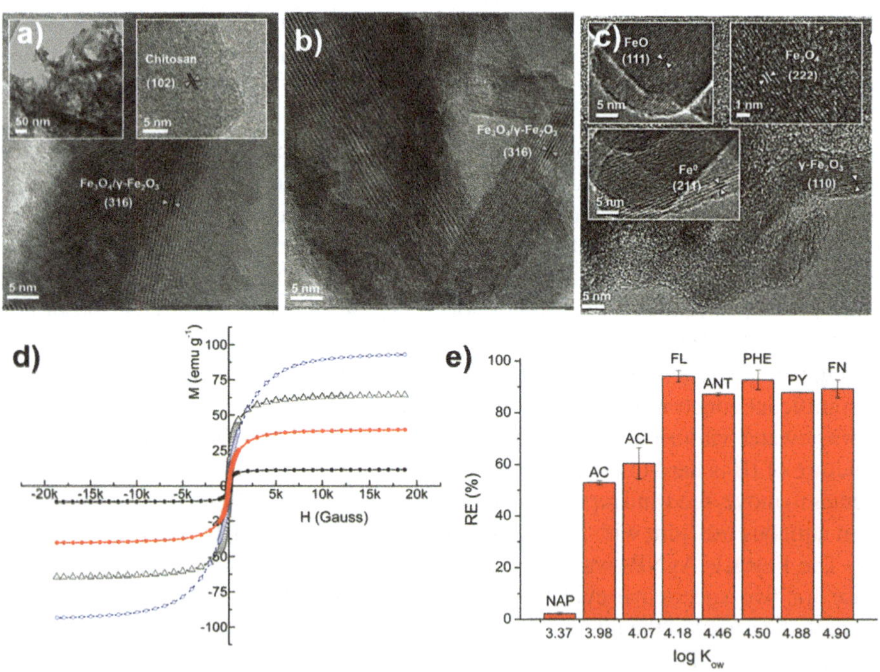

Fig. 13.9 HRTEM micrographs of MC3 (**a**), MC3py550 (**b**), and MC3py800 (**c**). Inset in panel A refers to MC3 collected at low magnification, whereas insets in panel C refer to the presence of different iron-containing phases. **d** Magnetization curves of MC3 (black circles, black solid curve), MC3py550 (red circles, red solid curve), and MC3py800 (white circles, blue dotted curve) as compared to neat magnetite/ maghemite M0 (white triangles, black dotted curve). **e** Adsorption experiments of a mixture of PAHs on MC3py550 (Nisticò et al. 2017)

from a very narrow hysteresis loop and low coercivity ($H_c \leq 10$ G) as shown in Fig. 13.9d. These hybrid composites were explored for the magnetic guided removal of 8 PAHs via adsorption. The adsorption efficiency was strongly dependent on hydrophobicity and extent of aromaticity in a linear fashion as shown in Fig. 13.9d due to enhanced π–π interaction (Nisticò et al. 2017).

13.4 Magnetic Nanoparticles as Oxidant for Next-Generation Pollutant Removal

Heterogeneous catalysis has been considered as a most promising and convenient approach for degrading refractory pollutants from wastewater due to the negligible generation of toxic by-products, high efficiency and mild reaction condition (Zhu et al. 2018; Das et al. 2019b). However, drawbacks such as requirement of particular pH, necessity of H_2O_2 or other radical generators and low reaction rate still exist (Kundu et al. 2020). While in photocatalysis, most of the semiconductor catalysts generally utilize only ~ 5% of the total available solar light and exhibit low quantum efficiency, which have to be overcome for practical applications (Zhu et al. 2018). Photocatalytic degradation has long been known as an environmentally friendly technology for wastewater remediation due to utilization of freely available Sunlight. Intrinsic characteristics of photocatalysts such as band gap, crystal phase, shape and charge transfer properties have considerable impact on photocatalytic efficiency (Abazari and Mahjoub 2017). In the photocatalytic process, described herein, core–shell architecture of TiO_2 was fabricated having magnetic Fe_3O_4 core for its easy separation from treated water. These TiO_2/Fe_3O_4 NPs showed a saturation magnetization ~ 40 emu g^{-1} but are less stable in solution under UV light due to photodissolution of iron. Which was overcome by the fabrication of TiO_2/ SiO_2/Fe_3O_4 with intermediate layer barrier of SiO_2 acting as charge carrier recombination centre. Both the MNPs were used for the photocatalytic degradation of PCPs including antipyrine, Acetaminophen, caffeine, bisphenol A and metoprolol (Álvarez et al. 2010).

Fluoroquinolones, a common antibiotic, when released in wastewater by human or industrial activities lead to the development of antimicrobial resistance and significant genotoxicity (Kaushik et al. 2019). Removal of fluoroquinolones from wastewater using conventional techniques is quite tough due to its high adsorption on soil and potential to migrate in the environment. Coupling adsorption and advanced oxidation can overcome the drawbacks of water purification technologies due to the simultaneous removal and degradation of pollutants, which allow the self-regeneration of adsorbent and boost the process efficiency (Kundu et al. 2020). In an interesting study by Kaushik et al. fluoroquinolones were removed by using both adsorption and degradation together under the irradiation of visible light in the presence of aluminium-doped TiO_2 (Kaushik et al. 2019). Besides ZnO and TiO_2, β-$AgVO_3$ is a promising photocatalyst due to narrow band gap, visible light harvesting efficiency and highly dispersed valence band. Abazar et al. combine β-$AgVO_3$ and $ZnFe_2O_4$ by using reverse micelle method in absence of any cosurfactants. The synergistic effect shows both catalytic and photocatalytic properties towards ammonium perchlorate and methylene blue degradation, respectively (Abazari and Mahjoub 2017). Iron-clay-cyclodextrin composites (Fe-MMT-DFB and Fe-MMT-βCD-DFB) were used to first adsorb and then degrade a range of pollutants including bisphenol A (BPA), perfluorooctanoic acid (PFOA) and carbamazepine (CBZ) assisted with H_2O_2. Cyclodextrin and polymers provide diverse adsorption sites, while H_2O_2 generates reactive oxygen species (ROS) for the degradation of pollutants. The catalytic efficiency was further validated by batch kinetic experiments as shown in Fig. 13.10. The composites were used for five successive cycles of pollutant removal (Kundu et al. 2020). Sulfamethoxazole (SMX), a model antibiotic was removed by the advanced oxidation process from aqueous solution using Fe-metal organic frameworks, which was synthesized by microwave assisted technique. An electron circulation theory-based mechanism was proposed for the degradation of SMX (Wan et al. 2020).

ROS include highly reactive oxygen free radicals, which can easily withdraw electrons from the aromatic or organic contaminants including various next-generation pollutants and result in their degradation into water and carbon dioxide (Bhati et al. 2016; Saini et al. 2020a; Tripathi et al. 2017). Hence, ROS are highly used for the removal of traditional, next-generation contaminants from wastewater as well as death of microorganisms (Bhati et al. 2016; Saini et al. 2020b; Tripathi et al. 2020; Das et al. 2020). ROS can be easily generated via appropriate photocatalysis and Fenton's reaction process. However, both the approaches have their own limitations. Photocatalysis efficiency is restricted by the light intensity and the nature of the

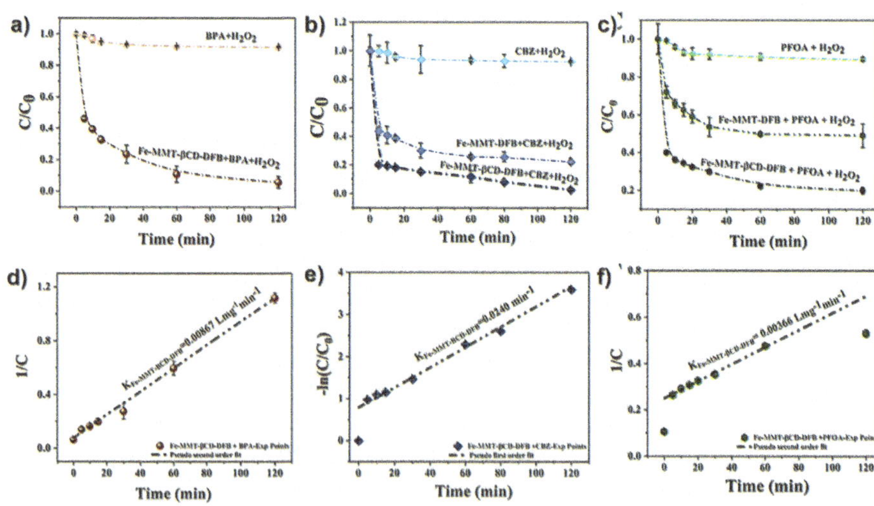

Fig. 13.10 a–c Kinetics of BPA, CBZ, and PFOA degradation by Fe-MMT-DFB and Fe-MMT-βCD-DFB in the presence of H_2O_2 (initial concentration: catalyst = 1 g L^{-1}; H_2O_2 = 20 mM). **d–f** Linear fitting of BPA, CBZ, and PFOA kinetic degradation data to rate equations (Kundu et al. 2020)

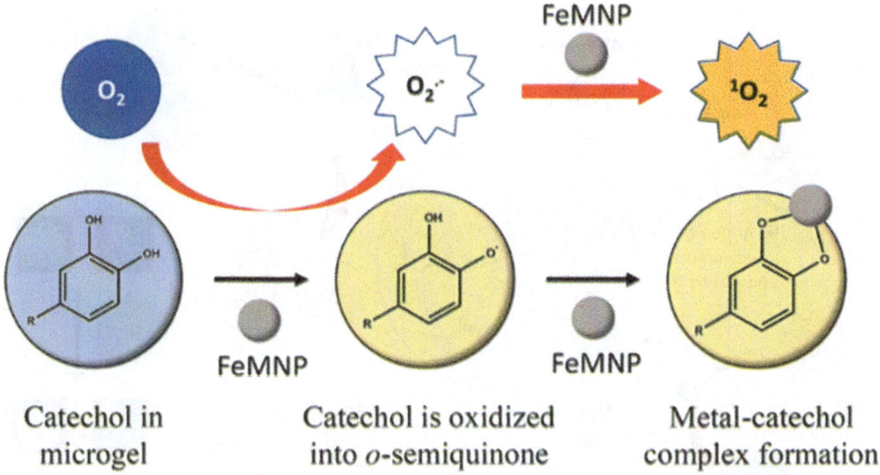

Fig. 13.11 Free radical generation mechanism of DMA microgel and FeMNPs (Zhang et al. 2020)

photocatalyst, while Fenton's reaction is limited by the need of acidic conditions for the generation of ROS (Zhang et al. 2020). In an innovative approach, Zhang et al. had synthesized magnetic composite of catechol-bearing dopamine methacrylamide (DMA) containing microgel and magnetic iron NPs (FeMNPs) with enhanced ROS generation activity over a wide pH range (Zhang et al. 2020). ROS were generated via the oxidation of catechol moiety in DMA into o-semiquinone radicals catalysed by Fe MNPs as shown in Fig. 13.11. These radicals were stabilized by the interfacial metal-catechol coordination bond and $^1O_2$ is generated by the oxidation of $O_2^{\bullet-}$ assisted by FeMNPs. This composite was utilized for the removal of a range of contaminants from wastewater including organic dyes (Malachite Green, Rhodamine B, Alizarin Red S, and Crystal Violet), antibiotic drug (ciprofloxacin (CIP)), metal ions and microbes (*Escherichia coli* and *Staphylococcus aureus*) (Fig. 13.12). CIP was removed via a cumulative approach of adsorption and photocatalysis; π–π interaction and H-bonding between microgel and CIP support adsorption, while ROS support photocatalysis (Zhang et al. 2020).

13.5 Conclusion

In summary, substantial progress in the MNPs have been made using various methods such as surface functionalization, doping, composite fabrication with nano-carbons, organic molecules and polymers. A comprehensive analysis of physio-chemical properties and underlying principles for their development as advanced functional materials is given. The fabrication procedure, functionalization and heteroatom doping have a pronounced effect on the properties of MNPs. Researchers are continuously

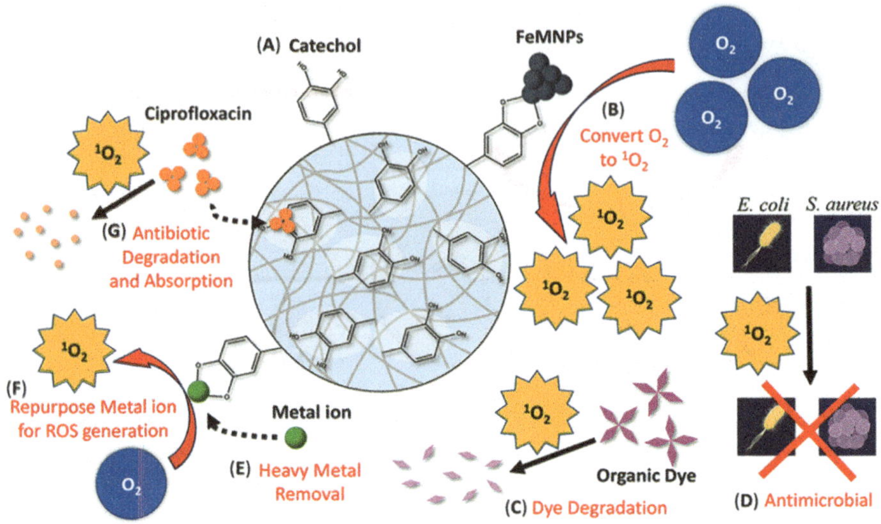

Fig. 13.12 Schematic of the Multifunctional DMA-Containing Microgel with the Ability to Generate ROS. **a** This microgel contains unoxidized catechol moiety. **b** In the presence of FeMNP, FeMNP-induced catechol oxidation converts molecular oxygen (O_2) into singlet oxygen ($^1O_2$), which can **c** degrade organic dyes and, **d** kill bacteria such as *E. coli* and *S. aureus*. **e** Catechol can chelate metal ions and remove heavy metal from the solution and **(F)** repurpose the metal ion for ROS generation. **g** The microgel can absorb or degrade antibiotic drugs such as ciprofloxacin (Zhang et al. 2020)

devoting themselves to simultaneous, fast, and effective removal of various inorganic/organic pollutants using MNPs. However, the factors that fully exploit the light absorption of these MNPs are not yet clear. A major challenge still is lack of selectivity and low process efficiency for the degradation of complex organic pollutants of next-generation such as microplastics and engineered NPs. Development of portable devices utilizing these MNPs; cost-effective and onsite water quality monitoring and remediation are required. Through the implementation of freely available solar energy, eco-friendly, cost effective materials and technology, MNPs offer considerable potential for the removal of a wide range of emerging pollutants. Looking forward, one can expect that research on these MNPs will persistently endorse the profound understanding of MNPs, along with their applied applications in widespread areas.

Acknowledgements KMT acknowledges financial assistance from the Department of Biotechnology (DBT), India, through the Ramalingaswami Faculty Award (BT/RLF/ Re-entry/45/2018).

References

Abazari R, Mahjoub AR (2017) Potential applications of magnetic β-$AgVO_3$/$ZnFe_2O_4$ nanocomposites in dyes, photocatalytic degradation, and catalytic thermal decomposition of ammonium perchlorate. Ind Eng Chem Res 56(3):623–634

Álvarez PM, Jaramillo J, López-Piñero F, Plucinski PK (2010) Preparation and characterization of magnetic TiO_2 nanoparticles and their utilization for the degradation of emerging pollutants in water. Appl Catal B Environ 100(1):338–345

Baikousi M, Bourlinos AB, Douvalis A, Bakas T, Anagnostopoulos DF, Tuček J, Šafářová K, Zboril R, Karakassides MA (2012) Synthesis and characterization of γ-Fe_2O_3/carbon hybrids and their application in removal of hexavalent chromium ions from aqueous solutions. Langmuir 28(8):3918–3930

Bajpai VK, Shukla S, Khan I, Kang S-M, Haldorai Y, Tripathi KM, Jung S, Chen L, Kim T, Huh YS, Han Y-K (2019) A sustainable graphene aerogel capable of the adsorptive elimination of biogenic amines and bacteria from soy sauce and highly efficient cell proliferation. ACS Appl Mater Interfaces 11(47):43949–43963

Bhati A, Singh A, Tripathi KM, Sonkar SK (2016) Sunlight-induced photochemical degradation of methylene blue by water-soluble carbon nanorods. Int J Photoenergy 2016:2583821

Chowdhury SN, Tung TT, Ta QTH, Gunture, Castro M, Feller JF, Sonkar SK, Tripathi KM (2021) Upgrading of diesel engine exhaust waste into onion-like carbon nanoparticles for integrated degradation sensing in nano-biocomposites. New J Chem 45(7):3675–3682

Das GS, Sarkar S, Aggarwal R, Sonkar SK, Park J-W, Tripathi KM, Kim T (2019a) Fluorescent microspheres of zinc 1,2-dicarbomethoxy-1,2-dithiolate complex decorated with carbon nanotubes. Carbon Lett. 29(6):595–603

Das GS, Tripathi KM, Kumar G, Paul S, Mehara S, Bhowmik S, Pakhira B, Sarkar S, Roy M, Kim T (2019b) Nitrogen-doped fluorescent graphene nanosheets as visible-light-driven photocatalysts for dye degradation and selective sensing of ascorbic acid. New J Chem 43(36):14575–14583

Das GS, Bhatnagar A, Yli-Pirilä P, Tripathi KM, Kim T (2020) Sustainable nitrogen-doped functionalized graphene nanosheets for visible-light-induced photocatalytic water splitting. Chem Commun 56(51):6953–6956

Dubey P, Sonkar SK, Majumder S, Tripathi KM, Sarkar S (2013) Isolation of water soluble carbon nanotubes with network structure possessing multipodal junctions and its magnetic property. RSC Adv 3(20):7306–7312

Duguet E, Vasseur S, Mornet S, Devoisselle J-M (2006) Magnetic nanoparticles and their applications in medicine. Nanomed 1(2):157–168

Fang Y, Xing C, Zhan S, Zhao M, Li M, Liu H, Wang C (2019) Multifunctional magnetic-fluorescent nanoparticle: fabrication, bioimaging, and potential antibacterial applications. ACS Biomater Sci Eng 5(12):6779–6793

Gaston L, Lapworth DJ, Stuart M, Arnscheidt J (2019) Prioritization approaches for substances of emerging concern in groundwater: a critical review. Environ Sci Technol 53(11):6107–6122

Guo Y, Chen B, Zhao Y, Yang T (2021) Fabrication of the magnetic mesoporous silica Fe-MCM-41-A as efficient adsorbent: performance, kinetics and mechanism. Sci Rep 11(1):2612

Gupta AK, Gupta M (2005) Synthesis and surface engineering of iron oxide nanoparticles for biomedical applications. Biomaterials 26(18):3995–4021

Hong F, Yan C, Si Y, He J, Yu J, Ding B (2015) Nickel Ferrite nanoparticles anchored onto silica nanofibers for designing magnetic and flexible nanofibrous membranes. ACS Appl Mater Interfaces 7(36):20200–20207

Jung S, Myung Y, Das GS, Bhatnagar A, Park J-W, Tripathi KM, Kim T (2020) Carbon nano-onions from waste oil for application in energy storage devices. New J Chem 44(18):7369–7375

Kaushik R, Samal PK, Halder A (2019) degradation of fluoroquinolone-based pollutants and bacterial inactivation by visible-light-active aluminum-doped TiO_2 Nanoflakes. ACS Appl Nano Mater 2(12):7898–7909

Kokotou MG, Asimakopoulos AG, Thomaidis NS (2012) Artificial sweeteners as emerging pollutants in the environment: analytical methodologies and environmental impact. Anal Methods 4(10):3057–3070

Kundu S, Korin Manor N, Radian A (2020) Iron–montmorillonite–cyclodextrin composites as recyclable sorbent catalysts for the adsorption and surface oxidation of organic pollutants. ACS Appl Mater Interfaces 12(47):52873–52887

Li S-K, Huang F-Z, Wang Y, Shen Y-H, Qiu L-G, Xie A-J, Xu S-J (2011) Magnetic Fe_3O_4@C@Cu_2O composites with bean-like core/shell nanostructures: Synthesis, properties and application in recyclable photocatalytic degradation of dye pollutants. J Mater Chem 21(20):7459–7466

Li Y, Yang C-X, Yan X-P (2017) Controllable preparation of core–shell magnetic covalent-organic framework nanospheres for efficient adsorption and removal of bisphenols in aqueous solution. Chem Commun 53(16):2511–2514

Lu A-H, Salabas EL, Schüth F (2007) Magnetic nanoparticles: synthesis, protection, functionalization, and application. Angew Chem Int Ed 46(8):1222–1244

Magi E, Scapolla C, Di Carro M, Rivaro P, Ngoc Nguyen KT (2013) Emerging pollutants in aquatic environments: monitoring of UV filters in urban wastewater treatment plants. Anal Methods 5(2):428–433

Miao P, Tang Y, Wang L (2017) DNA modified Fe_3O_4@Au magnetic nanoparticles as selective probes for simultaneous detection of heavy metal ions. ACS Appl Mater Interfaces 9(4):3940–3947

Moon SH, Noh S-H, Lee J-H, Shin T-H, Lim Y, Cheon J (2017) Ultrathin interface regime of core-shell magnetic nanoparticles for effective magnetism tailoring. Nano Lett 17(2):800–804

Nel A, Xia T, Mädler L, Li N (2006) Toxic potential of materials at the nanolevel. Science 311(5761):622–627

Nisticò R, Franzoso F, Cesano F, Scarano D, Magnacca G, Parolo ME, Carlos L (2017) Chitosan-derived iron oxide systems for magnetically guided and efficient water purification processes from polycyclic aromatic hydrocarbons. ACS Sustain Chem Eng 5(1):793–801

Oberdick SD, Abdelgawad A, Moya C, Mesbahi-Vasey S, Kepaptsoglou D, Lazarov VK, Evans RFL, Meilak D, Skoropata E, van Lierop J, Hunt-Isaak I, Pan H, Ijiri Y, Krycka KL, Borchers JA, Majetich SA (2018) Spin canting across core/shell Fe_3O_4/$MnxFe_{3-x}O_4$ nanoparticles. Sci Rep 8(1):3425

Oskarsson A, Wright MC (2019) Ionic liquids: New emerging pollutants, similarities with perfluorinated alkyl substances (PFASs). Environ Sci Technol 53(18):10539–10541

Pant A, Tanwar R, Kaur B, Mandal UK (2018) A magnetically recyclable photocatalyst with commendable dye degradation activity at ambient conditions. Sci Rep 8(1):14700

Park SJ, Das GS, Schütt F, Adelung R, Mishra YK, Tripathi KM, Kim T (2019) Visible-light photocatalysis by carbon-nano-onion-functionalized ZnO tetrapods: degradation of 2,4-dinitrophenol and a plant-model-based ecological assessment. NPG Asia Mater 11(1):8

Patel M, Kumar R, Kishor K, Mlsna T, Pittman CU, Mohan D (2019) Pharmaceuticals of emerging concern in aquatic systems: chemistry, occurrence, effects, and removal methods. Chem Rev 119(6):3510–3673

Qian J, Gao X, Pan B (2020) Nanoconfinement-mediated water treatment: from fundamental to application. Environ Sci Technol 54(14):8509–8526

Saini D, Aggarwal R, Anand SR, Sonkar SK (2019) Sunlight induced photodegradation of toxic azo dye by self-doped iron oxide nano-carbon from waste printer ink. Sol Energy 193:65–73

Saini D, Aggarwal R, Anand SR, Satrawala N, Joshi RK, Sonkar SK (2020) Sustainable feasibility of waste printer ink to magnetically separable iron oxide–doped nanocarbons for styrene oxidation. Mater Today Chem 16:100256

Saini D, Kaushik J, Garg AK, Dalal C, Sonkar SK (2020b) N, S-codoped carbon dots for nontoxic cell imaging and as a sunlight-active photocatalytic material for the removal of chromium. ACS Appl Bio Mater 3(6):3656–3663

Shao D, Hu J, Chen C, Sheng G, Ren X, Wang X (2010) Polyaniline multiwalled carbon nanotube magnetic composite prepared by plasma-induced graft technique and its application for removal of aniline and phenol. J Phys Chem C 114(49):21524–21530

Shi S, Fan Y, Huang Y (2013) Facile low temperature hydrothermal synthesis of magnetic mesoporous carbon nanocomposite for adsorption removal of ciprofloxacin antibiotics. Ind Eng Chem Res 52(7):2604–2612

Shubayev VI, Pisanic TR, Jin S (2009) Magnetic nanoparticles for theragnostics. Adv Drug Deliv Rev 61(6):467–477

Shukla S, Khan I, Bajpai VK, Lee H, Kim T, Upadhyay A, Huh YS, Han Y-K, Tripathi KM (2019) Sustainable graphene aerogel as an ecofriendly cell growth promoter and highly efficient adsorbent for histamine from red wine. ACS Appl Mater Interfaces 11(20):18165–18177

Siddiqui MTH, Nizamuddin S, Baloch HA, Mubarak NM, Al-Ali M, Mazari SA, Bhutto AW, Abro R, Srinivasan M, Griffin G (2019) Fabrication of advance magnetic carbon nano-materials and their potential applications: a review. J Environ Chem Eng **7**(1):102812

Singh GA, Bhati A, Khare P, Tripathi KM, Sonkar SK (2019) Soluble graphene nanosheets for the sunlight-induced photodegradation of the mixture of dyes and its environmental assessment. Sci Rep 9(1):2522

Sonkar SK, Tripathi S, Sarkar S (2009) Activation of aerial oxygen to superoxide radical by carbon nanotubes in indoor spider web trapped aerosol. Curr Sci 97(8):1227–1230

Sonkar SK, Tripathi KM, Sarkar S (2014) Ferromagnetic behaviour of anthropogenic multi-walled carbon nanotubes trapped in spider web indoor. J Nanosci Nanotechnol 14(3):2532–2538

Srivastava S, Gajbhiye NS (2011) Carbogenic nanodots: photoluminescence and room-temperature ferromagnetism. ChemPhysChem 12(14):2624–2632

Su J, Cao M, Ren L, Hu C (2011) Fe_3O_4–graphene nanocomposites with improved lithium storage and magnetism properties. J Phys Chem C 115(30):14469–14477

Tripathi KM, Tran TS, Kim YJ, Kim T (2017) Green fluorescent onion-like carbon nanoparticles from flaxseed oil for visible light induced photocatalytic applications and label-free detection of Al(III) Ions. ACS Sustain Chem Eng 5(5):3982–3992

Tripathi KM, Ahn HT, Chung M, Le XA, Saini D, Bhati A, Sonkar SK, Kim MI, Kim T (2020) N, S, and P-Co-doped carbon quantum dots: intrinsic peroxidase activity in a wide ph range and its antibacterial applications. ACS Biomater Sci Eng 6(10):5527–5537

Verma M, Singh K, Bakshi MS (2019) Surface active magnetic iron oxide nanoparticles for extracting metal nanoparticles across an aqueous–organic interface. J Mater Chem C 7(34):10623–10634

Vicente-Martínez Y, Caravaca M, Soto-Meca A, Solana-González R (2020) Magnetic core-modified silver nanoparticles for ibuprofen removal: an emerging pollutant in waters. Sci Rep 10(1):18288

Wan Y, Wan J, Zhao JR, Wang Y, Luo T, Yang S, Liu Y (2020) Facile preparation of iron oxide doped Fe-MOFs-MW as robust peroxydisulfate catalyst for emerging pollutants degradation. Chemosphere **254**:126798

Wang D, Saleh NB, Sun W, Park CM, Shen C, Aich N, Peijnenburg WJGM, Zhang W, Jin Y, Su C (2019) Next-generation multifunctional carbon-metal nanohybrids for energy and environmental applications. Environ Sci Technol 53(13):7265–7287

Wei X, Wang Y, Chen J, Xu F, Liu Z, He X, Li H, Zhou Y (2020) Adsorption of pharmaceuticals and personal care products by deep eutectic solvents-regulated magnetic metal-organic framework adsorbents: performance and mechanism. Chem Eng Sci 392:124808

Yu L, Zhou X, Jiang W (2016) Low-cost and superhydrophobic magnetic foam as an absorbent for oil and organic solvent removal. Ind Eng Chem Res 55(35):9498–9506

Zhang Y, Yue Q, Zagho MM, Zhang J, Elzatahry AA, Jiang Y, Deng Y (2019) Core-shell magnetic mesoporous silica microspheres with large mesopores for enzyme immobilization in biocatalysis. ACS Appl Mater Interfaces 11(10):10356–10363

Zhang Z, He X, Zhou C, Reaume M, Wu M, Liu B, Lee BP (2020) Iron magnetic nanoparticle-induced ROS generation from catechol-containing microgel for environmental and biomedical applications. ACS Appl Mater Interfaces 12(19):21210–21220

Zhao Y, Tong T, Wang X, Lin S, Reid EM, Chen Y (2021) Differentiating solutes with precise nanofiltration for next generation environmental separations: a review. Environ Sci Technol 55(3):1359–1376

Zhu M, Diao G (2011) Review on the progress in synthesis and application of magnetic carbon nanocomposites. Nanoscale 3(7):2748–2767

Zhu L, Chen B, Wang J, Shen H (2004) Pollution survey of polycyclic aromatic hydrocarbons in surface water of Hangzhou, China. Chemosphere 56(11):1085–1095

Zhu C, Chen X, Ma J, Gu C, Xian Q, Gong T, Sun C (2018) Carbon nitride-modified defective TiO_2–x@Carbon spheres for photocatalytic H2 evolution and pollutants removal: synergistic effect and mechanism insight. J Phys Chem C 122(35):20444–20458

Chapter 14
Clay Supported Zero Valent Iron Nanocomposites: Advancement in the Field of Green Catalyst for Abatement of Persistent Pollutant

Charulata Sivodia and Alok Sinha

Abstract The application of nanotechnology has revolutionised many industries to explore sustainable methods for the removal of persistent pollutants. Nano-zero valent iron (nZVI) is a significant nanomaterial having high surface area, reactivity, and adsorption capacity which is widely used for varying remediation techniques viz., nano-catalyst, nanofiltration, and adsorption. Moreover, nZVI particles tend to accumulate because of its magnetic interactions and oxidized when subjected to air. Hence, a capping agent or supporting material is required to stabilize the nZVI particles in aqueous solution. Materials like multiwalled carbon nanotubes, activated carbon, and polymers already been used for the stabilization of nZVI, however the cost incurred during synthesis and restoration restrict the use of such materials for environmental clean-up. Among different supporting materials clay minerals stand out as a substantial natural resource that could act as a potential candidate for nZVI stabilization. The chemical structure of clay is composed of different layers of **silicon**, calcium and aluminium ions that enhanced the ion exchange property and adsorption capacity of clay. Thus, allows clay to be an efficient supporting material for iron nanoparticles. This chapter depicts various aspects of surface modified clay and its utilization as a supporting material for iron nanoparticles. Later reaction mechanism of clay-nZVI on the removal of persistent pollutants also discussed.

Keywords Zero-valent iron · Nanocomposite · Surface modified clay · Water treatment

14.1 Introduction

The surge in industrial activities along with increased world population has radically augmented the water consumption. According to the World health organization (WHO), over 40% of the world's population are devoid of adequate supply of clean

C. Sivodia · A. Sinha (✉)
Indian Institute of Technology (Indian School of Mines), Dhanbad, India
e-mail: alok@iitism.ac.in

S. P. Singh et al. (eds.), *New Trends in Emerging Environmental Contaminants*, Energy, Environment, and Sustainability,
https://doi.org/10.1007/978-981-16-8367-1_14

water due to the lack of sufficient water resources (Fu et al. 2014; Adapa et al. 2016; Gude 2016; Ezzatahmadi et al. 2017). It is estimated that availability of fresh water will descend by 2050 and would create a major water stress around the world (CPCB 2014). Beside this, the problem of water pollution is pacing very fast owing to the release of different organic and inorganic contaminants from natural and anthropogenic sources. Of different water contaminants, the presence of persistent pollutants (PPs) such as biphenyls, dyes, pesticides, and pharmaceutical compounds are considered as more devastating to the environment. Once released in the environment, these compounds impose severe toxicity to the aquatic organisms since they are lipophilic in nature and tend to bioaccumulate in the food chain (Lallas 2001; Gaur et al. 2018; Balmer et al. 2019; Zhu et al. 2020). Dinç et al. (2021), reported the occurrence of different PPs including polycyclic aromatic hydrocarbons (PAHs), polychlorinated biphenyls and organochlorine pesticides in the surface sediments of Yeşilırmak River and Blak sea. Their study advocates that presence of organochlorine pesticides, and its derivative (4.4′-DDT) in the river samples were contributed by the industrial and port activities with the highest concentration of 0.05 $\mu g\ Kg^{-1}$. Similarly, Miraji et al. (2021), reviewed the distribution of PPs in western Indian ocean and stated that PPs have long range of mobility and accumulated in the oceanic ecosystem.

To address this issue efforts have been made towards the treatment of wastewater contaminated with PPs (Xie et al. 2016; Yadav et al. 2019). Technologies like advanced oxidation process, photodegradation, biological treatment, and membrane filtration were extensively used for the removal of persistent pollutant. Nevertheless, the set up and the cost applied for the establishment of such facilities is very high and required a lot of maintenance (Zekić et al. 2018; Tlili and Alkanhal 2019). Hence, development of a sustainable technology is the need of the hour to meet the demand of clean water supply. Recently, advancement has been made in the field of nanomaterials-based water remediation techniques to overcome the limitations of the existing technologies (Kamarudin et al. 2021).

Nanotechnology can be described as a field where matter is transformed to obtain unique function and specific characteristics by physical, chemical, and biological methods (Goswami et al. 2012; Taghizadeh et al. 2013). The materials which have size between 1–100 nm are defined as nanomaterials (Khan et al. 2019; Sheikholeslami et al. 2020a, b). The exceptional properties of nanomaterials such as high surface area to volume ratio has gained attraction of many researchers. Nanomaterials possess high adsorption capacity due to their specific surface area and number of available active sites for adsorption. Furthermore, they hold distinct catalytic, reactive, thermal, optical, and electrochemical properties which not only distinguished them as a priority adsorbent but also employed in the varying fields of materials science (Bulasara et al. 2011; Mokhtari et al. 2016; Asfaram et al. 2017; Mondal et al. 2020).

Among different nanomaterials zero-valent iron is of great interest for the treatment of wastewater and contaminated water bodies. Nano zero-valent iron (nZVI) particles can efficiently remove, transform, and degrade persistent pollutants such as dyes, heavy metals, pharmaceuticals, and other such hazardous pollutants. It can not only used for the surface water remediation but also applied in groundwater treatment in the form of reactive material in Permeable reactive barriers (PRBs) (Xi et al. 2011; Ruhl et al. 2012; Shimizu et al. 2012). Although, even after having a high surface area and catalytic activity nZVI particles get agglomerated in a chain like manner because of van der Waals interactions between the particles thereby reducing their reactivity (Sun et al. 2014; Guan et al. 2015). This limitation can be overcome by stabilizing the nZVI with a supporting material.

The use of activated carbon (Mortazavian et al. 2018), polymers (Ravikumar et al. 2018), and carbon nanotubes (Li et al. 2020) as matrix material for nZVI stabilization was well documented in the literature. Although, these materials cover the surface of iron nanoparticles in such a way that their magnetic property reduces, and separation becomes difficult. Recently, clay minerals were identified as an excellent supporting material for nZVI. Clay is found profusely in nature and being an effective adsorbent could work as an excellent supporting material (Ismadji et al. 2015; Rautureau et al. 2017). The presence of phyllosilicate minerals and 2D structure of clay provide some peculiar channels, edges, and interlayer spaces to tune the immobilization of nZVI and offered the formation of functionalised nanocomposite with distinct physico-chemical properties. Besides, the processing of clay minerals is easy as compared to other supporting materials like in case of activated carbon and biochar long hour of pyrolysis and activation is required to obtain the final product (Shichi and Takagi 2000; Li et al. 2016a). This chapter defines the role of clay as a supporting material for zero valent iron nanoparticles and discussed the interactions and reactions involved during the removal of contaminant.

14.2 Zero Valent Iron

Iron being the fourth abundant element in the earth is considered as a significant transition metal exclusively used in different environmental remediation process (Dale and Huber 2005). Zero-valent iron (ZVI) is reasonably a strong reducing material (**E^0 = − 0.44 V**) that can adsorb a wide array of contaminants viz., metalloids, heavy metals, and other persistent pollutants. Earlier studies reported use of ZVI in the form of nano iron fillings or powder on the reductive degradation of the halogenated aliphatic compounds (Gillham and O'Hannesin 1994). Later, Wang and Zhang (1997), proposed in situ remediation of aquifers and soil by incorporating two different sizes of ZVI. It was attributed that surface area to mass ratio of ZVI changes with particle size, where particles with < 10 μm have less mass **(<1 m^2g^{-1})** as compared to < 100 nm **(33.5 m^2g^{-1})**. This is because at microscale the number of atoms on the surface of the particle is less unlike at nanoscale where a great proportion of unsatisfied valence atoms are present (Nel et al. 2006). The nZVI is best defined as the core–shell structure where core comprised of ZVI (Fe^0) and shell consists of mixed valent oxide due to oxidation of the inner metallic iron in the core (Fig. 14.1) (Nurmi et al. 2005; Li et al. 2006). This structure is achieved by the virtue of air or water that reacts with nZVI during synthesis. The composition of the shell affects the size distribution, ageing, reactivity, and crystallinity of the nZVI particles

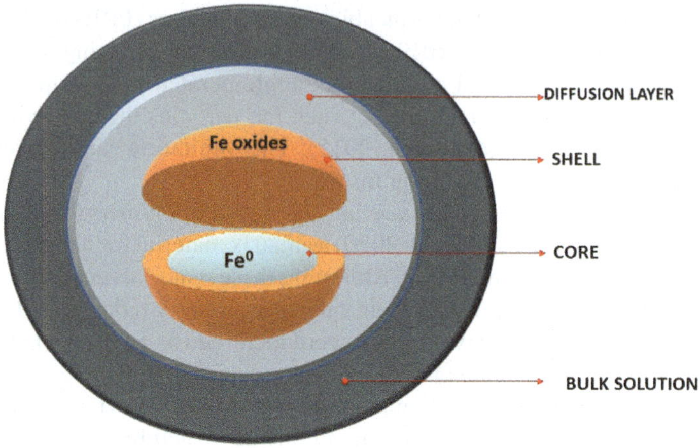

Fig. 14.1 Core–shell structure of nano-zero valent iron (Guan et al. 2015)

(Ramos et al. 2009). The adsorption phenomena are facilitated by the core–shell morphology attributable to the surface complexation and electrostatic interactions of different adsorbate. Contrary to this, the oxide layers of nZVI allows reduction phenomena due to increase in the electron transfer rate (Yen et al. 2011; O'Carroll et al. 2013).

14.2.1 Methods for Synthesis of nZVI Particles

The synthesis process for nZVI is differentiated into three types: physical, chemical, and biological by top-down or bottom-up approach. In this section the most common methods of nZVI synthesis by physical (milling), chemical (vapour reduction and chemical reduction), thermal, and biological (green and microorganism) process would be discussed.

14.2.2 Physical Methods

In physical method, the bulk iron material disintegrated into nanoscale through mechanical forces accompanied by high energy and pressure. Earlier techniques of nZVI synthesis includes severe plastic deformation (SPD). Valiev et al. (2000), processed nanostructured material by means of SPD method either by torsion straining or angular pressing under quasi-static pressure. However, it was hard to control such high pressure thereby limiting its application. Precision milling or ball milling is widely accepted physical method in which the samples were stirred by beads at a defined rotating speed to form nanoscale powder. In this system the iron particles were grinded in a high-speed rotating chamber known as planetary ball mill. The chamber consists of two compartments; first, a milling system and second is a cooling system. The milling system comprised of grinding chamber, an engine, an agitator, and steel balls. Whereas in cooling system, temperature of the suspended iron inside the mill and retention tank is controlled. The particles were continuously crushed by the steel spheres in the rotating chamber until the desired diameter is achieved. The cylinder set at the rotating chamber helps to separate the particle while retaining the steel balls (Li et al. 2009; Ioannou et al. 2012; Jung et al. 2015).

14.2.3 Chemical Methods

The prior chemical synthesis of nZVI includes decomposition and reduction method where iron pentacarbonyl ($Fe(CO)_5$) was decomposed into inert gas or in organic medium. Later involves reduction of iron with hydrogen. The chemical vapor deposition or evaporative condensation involves a heat source particularly an electron beam that sputtered and irradiate the raw material to gasify and condense them into nanotubes (Rong et al. 2006). This method is usually applied as films to coat products. Another way of nZVI synthesis is chemical reduction method either by a gas (hydrogen) or by a chemical (sodium borohydride). O'Carroll et al. 2013, used hydrogen gas to reduce hematite and geothite at high temperature to prepare nZVI (Eq. 14.1). Firstly, goethite and hematite particles were obtained by reaction of iron hydroxides or carbonates with compounds like alkaline carbonates and hydroxides in presence of oxygen gas. The magnetite ores thus obtained were transferred in a crucible and then placed in a thermogravimetric device. The temperature is increased gradually in an argon medium and when the reaction temperature is achieved hydrogen gas is injected in the system. When reduction is completed, the flow of argon gas again resumes, and later the resultant powder is removed from the system (Uegami et al. 2006). Nevertheless, sodium borohydride ($NaBH_4$) reduction method for iron salts (ferric (Fe^{+3}) or ferrous ions (Fe^{+2}) is usually applied for the synthesis of nZVI. In this method, a certain concentration of $NaBH_4$ is added dropwise to the iron solution in a controlled manner to reduce it to zero valence (Eq. 14.2) (Sun et al. 2006). The solution is continuously stirred for about an hour till the reaction is completed. The occurrence of black coloured solution indicates the formation of iron particles. Subsequently, the particles were separated through vacuum filtration followed by washing (ethanol) and drying in an inert media to check the oxidation of nZVI (Kanel et al. 2005; Barreto-Rodrigues et al. 2017; Martins et al. 2017).

$$Fe_3O_{4(s)} + 4H_{2(g)} \rightarrow 3Fe^0_{(s)} + 4H_2O_{(g)} \tag{14.1}$$

$$4Fe^{3+}_{(aq)} + 3BH^{-4} + 9H_2O_{(aq)} \rightarrow 12Fe^0_{(s)} + 3H_3BO_{3(aq)} + 12H^+_{(aq)} + 9H_{2(g)} \tag{14.2}$$

14.2.4 Carbothermal Reduction Method

Apart from gas and chemicals, carbothermal energy is also employed as reductant to obtain iron nanoparticles. In this process, the Fe^{+2} oxide particles (gaseous or salts) are subject to reduce at high temperature in the presence of carbon by using thermal energy (Eqs. 14.3–14.4) (Hoch et al. 2008; Crane and Scott 2012; Tan et al. 2021). Furthermore, the carbon not only reduce the iron salt but also act as a supporting material to stabilize and disperse the nZVI particles. Kong et al. 2016, carbonized the iron-rich sludge (at 600–800 °C) into iron nano-flakes and used as a catalyst to degrade Acid orange (II) dye in Fenton-like reaction. Their study revealed that the formation of nZVI mainly depends on the carbonization temperature, where the iron ions present in the sludge converted into Fe_3O_4 and Fe^0 at 500 °C and 700 °C respectively. The iron nano-flakes removed 98% of the dye at neutral pH. Apart from temperature, the carbon concentration can affect the yield of iron nanoparticles. Nisticò and Carlos (2019), reported synthesis of nZVI by carbothermal reduction method employing lignin-based biomass as carbon source. Their results attributed that by varying biomass concentration from 0.5 to 3 W% and obtained maximum yield of α-Fe particles.

$$Fe(C_2H_3O_2)_{2_{(aq)}} + C_{(s)} \rightarrow Fe^0_{(s)} + 2CH_2CO_{(g)} + CO_{(g)} + H_2O_{(g)} \quad (14.3)$$

$$Fe_3O_{4(aq)} + 2C_{(g)} \rightarrow 3Fe^0_{(s)} + 2CO_{2(g)} \quad (14.4)$$

14.2.5 Biological Methods

In biological method an eco-friendly approach is applied to produce nanoparticles by using plants and microorganisms, therefore also called as green synthesis process. It incorporates plants or microbial enzymes having high antioxidant and reducing properties viz., green tea, lemon, grape, sorgo, and coffee to reduce metallic compounds. The phytochemicals present in the plant are separated by heating the plant extract near to the boiling point. The residue collected is then mixed with the iron solution to reduce them to zero valent state (Machado et al. 2013; Stefaniuk et al. 2016; Yadav et al. 2017).

14.2.6 Electrochemical Mediated Synthesis of nZVI

In this method, electrolysis of iron ions (Fe^{+2}/Fe^{+3}) conducted at cathode and anode to obtain the nZVI particles. Unlike chemical reduction, this method is simple and inexpensive. However, iron atoms deposited at the cathode form clusters during the electrolysis process. Later, cationic surfactants along with ultrasonic waves was added to counteract the problem of agglomeration (Chen et al. 2004; Crane and Scott 2012). The surfactant stabilized the nZVI particles and the ultrasonic waves removes the iron nanoparticles from the cathode (Eq. 14.5) (Li et al. 2009).

$$Fe^{3+} + 3e^{+} + \text{stablizer} \rightarrow nFe^{0} \tag{14.5}$$

14.2.7 Ultrasound Mediated Synthesis of nZVI

In this process, the ultrasonic waves with frequency up to 20 kHz are employed along with the chemical reduction method to reduce the particle size. The acoustic energy generated by the ultrasonic waves create cavitation which enhanced the mixing and particle collision that leads to the disruption of iron particles into smaller fragments. The ultrasonic frequency was found to be a critical factor which governs the particle size (29–80 nm), shape (spherical, chain or needle) and surface area (42–145 $m^2\ g^{-1}$) of the iron nanoparticles (Jamei et al. 2014; Bagheri et al. 2021; Bounab et al. 2021).

14.3 Surface Modification of nZVI

Several studies reported the application of nZVI for the removal of persistent pollutant. Moreover, the magnetic property of nZVI leads to the clustering of particles thereby reduces its mobility and transportation. Additionally, the nZVI suffers from rapid oxidation due to its high reduction ability that in turn form layers of iron oxide. All these factors significantly affect the reactivity of the nZVI particles, therefore surface modification of bare nZVI is necessary to surmount these limitations. Process like coating, metal doping, and supporting matrix usually applied to inhibit the agglomeration and oxidation of nZVI surface (Stefaniuk et al. 2016).

14.3.1 Metal Doping

Metal doping is a distinguished technique which can alter the intrinsic properties (electrical and optical) of bare nZVI. This process constitutes the mixing of noble metals primarily palladium (Pd), platinum (Pt), silver (Ag), copper (Cu), and nickel (Ni) with the iron **at pre or post synthesis** steps to inhibit the passivation. Earlier studies on metal-doped nZVI includes removal of 4-cholorophenol, tetrachlorobisphenyl, and p-nitrophenol (Xu et al. 2012; Huang et al. 2013; Lai et al. 2014). Krasae and Wantala 2016, synthesized Cu-doped nZVI with TiO_2 as a photocatalyst to reduce nitrate and obtained 86% of removal. The crystallinity and oxidation rate of the nZVI particles mainly depends on the concentration of the doped metal.

14.3.2 Surface Coating

To enhance the longevity of nZVI, the addition of biopolymers, surfactants, and stabilizers like carboxymethyl cellulose (CMC), starch, rhamnolipid, polyacrylic acid, and guar gum are added to the surface of iron particles (Xiong et al. 2007; Tiraferri, et al. 2008; Qiu et al. 2012) The surface coating by such additive altered the surface of nZVI and brings down the electrostatic attraction between the particles. Also, the steric repulsion generated by the modifiers overpowered the van der Waals force and magnetic interaction of nZVI and stabilize them (Tesh and Scott 2014; Tosco et al. 2014). Basnet et al. 2015, studied the mobility of CMC-coated-Pd-nZVI particles in different sands (loamy, quartz, and clay) to model the groundwater environment. Eljamal et al. 2020, used four polymers viz., polyacrylamide (PAA), polyethylene sorbitan monolaurate, polyvinylpyrrolidone (PVP) and carboxymethyl cellulose to stabilize nZVI and studied its efficiency on the removal of nitrogen and phosphorous. The functional groups of the polymers viz., carbonyl and hydroxyl groups facilitated the binding and adsorption of the polymers on the nZVI surface. Among different polymers, CMC coated nZVI exhibit smallest particle size of 10 nm followed by PAA (106 nm) and PVP (108 nm). Zhou et al. 2020, extracted extracellular polymeric substance (EBS) from microbes to enhance the electron transfer rate of nZVI and utilized them to adsorb and reduce the antimony (Sb (V)). The EBS comprised of polysaccharides, proteins, and humic acid which improved the adsorption capacity (202 mg g^{-1}) and effectively reduce Sb (V) to Sb (III).

14.3.3 Supporting Matrix/material

Trapping of nZVI within the pores of a supporting material has made its separation easy from the matrix. Materials like zeolite, carbon, and membranes were mostly employed as a supporting matrix. Li et al. 2021, used zeolite supported nZVI to study the adsorption-reduction of uranium (U). The removal efficiency was found to be 96% with sorption capacity of 216 mg g^{-1}. In another study by Silva et al. (2021), polyvinylidene fluoride (PVDF) membranes coated with PAA were used to insert nZVI particles and later used for bisphenol- -A degradation. Moreover, microwaves were applied during the grafting of polymer to the membrane surface which enhanced the mechanical stability of the membrane. The nZVI-PAA-grafted membrane remove 52% of bisphenol-A with 50 L m^{-2} h^{-1} of permeate flux.

14.4 Clay-NZVI Hybrid Nanocomposite/Clay as Supporting Material for nZVI

Several adsorbents viz., biochar, resin, zeolite and activated carbon, have been used in water treatment process (Hanif et al. 2007; Özacar et al. 2008; Hernández-Montoya et al. 2013; Pirzadeh and Ghoreyshi 2014). Recently mineral like clay has emerged as an effective adsorbent by virtue of their remarkable chemical properties and ion exchange capacity (Frost et al. 2010). Clay is defined as phyllosilicate minerals comprised mainly of two layers of tetrahedral silicate ($Si_2O_6(OH)_4$) and one layer of octahedral aluminium ($Al_2(OH)_6$) (Fig. 14.2) (Guggenheim and Martin 1995). Based on their layer structure, clay is classified into two groups; 2:1, and 1:1 (Table 14.1). In 2:1 group, the clay has tetrahedral sheets of two silicon-oxygen sandwiched between octahedral sheets of aluminium-oxygenhydroxyl. While the 1:1 group have

Fig. 14.2 Structure of clay (Williams and Haydel 2010)

Table 14.1 Classification of clay

Silicate layer alignment	Morphology	Example	Chemical composition
1:1 (1 octahedral and 1 tetrahedral sheets)	Layer	Kaolinite	$(Al_2)(Si_2)O_5(OH_4)$
	Tube	Halloysite	$(Al_2)(Si_2)O_5(OH_4)0.2H_2O$
2:1 (2 tetrahedral and1 octahedral sheets)	Layer	Bentonite montmorillonite	$(Si_{12-x}M_x{}^{3+})O_{30}(OH)_4(OH_2)_4{\cdot}8H_2O$
	Chain	Palygorskite sepiolite	$(Si\ _{8-x}Mx^{3+})O_{20}(OH)_2(OH_2)_4{\cdot}4H_2O$

different layers of tetrahedral and octahedral sheets of silica and aluminium, respectively. These layers are further joined together by hydrogen bonds. The isomorphic substitution among the ions of tetrahedral (Al^{3+} and Si^{4+}) and octahedral sheets (Al^{3+} and Mg^{2+}) make the clay surface negatively charged which allows them to adsorb more cationic species (Shichi and Takagi 2000; Abu-Danso et al. 2020). The high ion-exchange capacity of clay minerals was endowed by the presence of natural cations (Ca^{2+}, Na^{2+} and Mg^{2+}) on the surface and interlayers of the clay that substantially increased its pollutant removal efficiency (Bergaya and Lagaly 2006). However, on hydration the interlayer space of clay decreases, that reduced the adsorption of large compounds. Additionally, clay has less affinity towards organic contaminants and have low surface area which limits its application during the remediation process (Borisover et al. 2008; Unuabonah and Taubert 2014).

On the other side, the iron nanoparticles owing to their strong van der Waals interaction and magnetic nature are prone to phenomena like oxidation and aggregation that eventually reduces its reactivity. Researchers stated that incorporation of an immobilizing agent or a supporting material to nZVI surface often increased the reactivity and bring down the effect of agglomeration (Bhosale et al. 2019; Zhao et al. 2019a). Hence, the limitations of both clay and iron nanoparticles can be overcome by synthesizing composite of clay-nZVI. The doping of metal nanoparticles with clay becomes more conducive by the presence of functional fragments of the clay. The metal particle intercalated with the clay will act as a "bridging agent" and thus increasing the adsorption capacity and overall physical and chemical strength of the hybrid composite (Zhang et al. 2020). The clay-nZVI nanocomposite is prepared by the liquid-phase reduction method, where iron ions ($FeSO_4$) were reduced by sodium borohydride ($NaBH_4$) in presence of clay mineral. In this method the clay-nZVI nanocomposite was prepared by mixing salt of $FeCl_3$ or $FeSO_4$ to the clay solution in a certain mass ratio (1:1, 4:1, and 2:1) and later dissolved in a 50 ml solution. Afterwards, $NaBH_4$ was added dropwise to the iron solution and stirred continuously. The resultant black particles were washed continuously with absolute ethyl alcohol to check further oxidation of nZVI (Weng et al. 2017).

14.5 Application of Clay-NZVI Nanocomposite

14.5.1 Adsorbent

Hybrid nanocomposite of clay-iron employed as an adsorbent for different organic and inorganic compounds (Table 14.2). Chen et al. (2011), synthesized bentonite-iron composite to degrade methyl orange dye from aqueous solution. They also compared the removal efficiency of the synthesized composite to the bare nZVI and with raw bentonite clay. It was found that clay-Fe^0 composite has significantly removed 79% of the dye, whereas bentonite and nZVI removes only 40% and 2% of the dye, respectively. The methyl orange degradation follows adsorption, iron (Fe^{+2})-dye

Table 14.2 Application of clay-nZVI nanocomposite as adsorbent on pollutant removal

Hybrid composite	Matrix	Target compound	Initial concentration	Dose	Removal efficiency (%); adsorption capacity	References
Bentonite-nZVI	Wastewater	Methyl orange (dye)	100 mg L^{-1}	1 g L^{-1}	79%	Chen et al. (2011)
Hangjin-2 clay-nZVI	Water	Methyl orange (dye)	50–200 mg L^{-1}	1 g L^{-1}	99%	Li et al. (2017)
Bentonite-nZVI	Water	Amoxicillin (antibiotic)	20 mg L^{-1}	4 g L	93%	Weng et al. (2017)
Attapulgite-nZVI	Water	Cr (VI) (Heavy metal)	40 mg L^{-1}	–	90%; 266 mg g^{-1}	Zhang et al. (2019)
Amino-functionalized Vermiculite	Chrome-electroplating wastewater	Cr (VI)	20 mg L^{-1}	^{1} 0.25 g L^{-1}	100%; 59 mg g^{-1}	Zhang et al. (2019)
Silty clay-nZVI	Water	Phenol	0.5 g L^{-1}	1 g L^{-1}	90%; 30 mg g^{-1}	Kadhum et al. (2021)

complexation, and later azo bond cleavage. Further, the potential of bentonite-Fe^0 composite was determined by using it on the azo dye wastewater. Their findings suggest that the presence of inorganic salts (Na_2SO_3) in the wastewater influenced the dye removal efficiency, since the sulphate ions compete with the dye molecules for the active sites at bentonite-Fe^0 composite. Li et al. (2017), employed Hangjin-2 clay to support nZVI particles for the degradation of methyl orange dye. It was attributed that the negative charge on clay surface attract the positively charged dye which concentrate the dye molecules from the aqueous solution to the composite surface. Further the absorbed dye molecules were immediately reduced by the nZVI particles. Apart from dyes, clay-iron nanocomposite was also applied for the removal of pharmaceuticals. Weng et al. (2017) described the adsorptive removal of amoxicillin by bentonite supported zero valent iron. Here also, the degradation mechanism followed adsorption and reduction of amoxicillin molecules, with more than 90% of removal efficiency. In their study, the reusability of the clay-iron adsorbent was also determined, and the synthesized adsorbent found to be efficient for up to five cycles of treatment. Kadhum et al. (2021), used silty clay to support Fe^0 nanoparticles for the adsorption of phenol. The highest removal efficiency was obtained as 90% in acidic medium with dosing of 1 g L^{-1} of clay-iron composite. The supported clay minerals not only reduce the agglomeration of iron nanoparticles but also increased the dispersibility of the nZVI on the clay surface.

The effect of loading mass ratio of clay to the iron was also counted as a key parameter that can affect the adsorption capacity. Zhang et al. (2019), studied the effect of mass loading of attapulgite supported nZVI on the removal of chromium from aqueous solution. Their study advocates that the lower attapulgite mass ratio (2:1) results in higher adsorption capacity (266 mg g^{-1}) as compared to the mass ratio of 4:1 (150 mg g^{-1}) and 8:1 (25 mg g^{-1}). The higher mass of attapulgite clay could block the surface-active sites for nZVI, thereby reducing the adsorption. Besides, the change in mass ratio of clay also affect the solution pH which indicate the occurrence of redox reaction. Moreover, the reaction mechanism for chromium follows an adsorption-reduction process, where the chromium ions (Cr^{+6}) first adsorbed on the clay surface and later reduced by Fe^0. Further, the reaction mechanism was also explained by them which suggests that AGT-nZVI synergistically removed chromium through adsorption-reduction process.

In order to enhance the adsorption capacity and reactivity, the clay surface has been modified via different physical and chemical methods. These methods include thermal modification, acid activation, organic modifiers, and pillaring (Table 14.3) (Sarkar et al. 2019). Wu et al. (2019), availed the properties of organo-montmorillonite-iron nanocomposite for persulfate activation to degrade sulfamethazine in the advanced oxidation process. Persulfate activation via nZVI results in the formation of SO_4^- ions which promote the degradation of sulfamethazine (antibiotic). The montmorillonite (MMT) clay was modified through a cationic surfactant (cetyltrimethylammonium bromide) to increase the lamellar space of the clay which eventually increased the adsorption capacity. The organo-modified clay-nZVI has shown higher kinetic rate (0.33 min^{-1}) than unmodified MMT-nZVI (0.22 min^{-1}). Zhao et al. (2019b), synthesized an amino functionalised vermiculite clay to support

Table 14.3 Modification method for clay

Method	Mechanism
Thermal	Clay is calcined at 400–500 °C that redistribute the interlayer cations modification and dehydroxylate the aluminium and silicon groups of the clay
Acid Treatment	Clay mineral is treated with acid (H_2SO_4 or HCl) to eliminate the interlayer cation which opened the octahedral and tetrahedral layers of the clay. This will increase the porosity and surface area of the clay
Organic modifiers	Clay is modified by grafting organic molecules viz., surfactants, or polymers to enhance the cation exchange capacity
Pillared clay	Pillaring of clay minerals is a two-step process where at first the clay is intercalated by a pillaring agent (Al, Fe, Co, Zr, Si and Mg) and then calcined to form oxide pillared clay. The pillaring agents provide high thermal stability, crystallinity, and adsorption ability to the clay

nZVI particles for chromium (Cr^{+6}) removal. It was found in their study that amino functionalised clay gets easily protonated ($-NH_3^+$) in acidic condition and results in higher removal efficiency (100%) for Cr^{+6} ions. The adsorbed Cr^{+6} reduced by Fe^0 to Cr^{+3} and subsequently the Fe^0 oxidized into Fe^{+2}. A part of Fe^{+2} released into the bulk solution also participate in the chromium reduction (Eq. 14.6–14.7) (Li et al. 2016b).

$$3Fe^0 + 2HCrO_4^- + NH_3^+ 14H^+ \rightarrow 2Cr^{3+} + 3Fe^{2+} + 8H_2O \tag{14.6}$$

$$3Fe^{2+} + 2HCrO_4^- + H^+ \rightarrow Cr^{3+} + 3Fe^{3+} + 4H_2O \tag{14.7}$$

14.5.2 Catalyst

The clay-nZVI nanocomposite utilized as a catalyst in various oxidation reaction to eliminate toxic compounds (Table 14.4). Lin et al. (2017), used kaolinite supported nZVI in Fenton-like reaction to degrade an azo dye Direct Black G and obtained 97% of removal efficiency in aqueous solution. The possible removal mechanism followed the surface adsorption of Direct black-G dye on the composite surface, then the nZVI breaks the azo bond ($-N=N-$) of the dye and formed the reduced by-products of the DB-G. Furthermore, the Fe^{+2} ions formed by the reduction of nZVI generated $^{\bullet}OH$ radicals which removed remaining by-products of the dye. Bao et al. (2020a), described the degradation mechanism of bisphenol by using bentonite-Fe^0 as catalyst in Fenton-like oxidation. The highest removal efficiency was achieved as 99–100% in acidic pH (<2) with H_2O_2 concentration of 0.7 ml. In acidic condition, the surface of nZVI corroded and converted into Fe^{+2} ions. The generated Fe^{+2} now react with the H_2O_2 and forms $^{\bullet}OH$ radicals. It was also found that the leaching of Fe

Table 14.4 Application of clay-nZVI nanocomposite as catalyst for pollutant removal

Clay	Matrix	Target compound	Mechanism	Initial pollutant concentration	Dose	Removal (%)	Reference
Bentonite, Kaolin, and native clay	Water	Rosso Zetanyl B-NG dye	Heterogenous Fenton oxidation	50–300 mg L^{-1}	112–224 mg L^{-1}	91	Kerkez et al. (2014)
Kaolinite	Water	Direct Black G (dye)	Fenton-like oxidation	100–500 mg L^{-1}	0.6 g L^{-1}	97	Lin et al. (2017)
Rectorite	Water	P-chlorophenol	Heterogenous Fenton oxidation	100 mg L^{-1}	0.005 g	99	Bao et al. (2019)
Bentonite	Water	4-nitrophenol	Catalytic reduction	0.2 mM	10 mg L^{-1}	96	Sarvanthi et al. (2019)
Organo-montmorillonite	Water	Sulfamethazine (sulfonamide antibiotic)		20 mg L^{-1}	1.5 mM	97	Wu et al. (2019)
Bentonite	Wastewater	Bisphenol	Fenton-like	50–100 mg L^{-1}	0.004–1 g L^{-1}	84–100	Bao et al. (2020a)
Muscovite oxidation	Water	P-chlorophenol	Fenton-like	100 mg L^{-1}	0.007 g	99	Bao et al. (2020b)
Montmorillonite	Soil	Polychlorinated biphenyls	Heterogeneous Fenton oxidation	1650 μg kg^{-1}	29.88 g kg^{-1}	76	Sun et al. (2021)

was lower in case of B-Fe^0 (4.38 mg L^{-1}) as compared to bare nZVI (8.34 mg L^{-1}). Sarvanthi et al. (2019), employed green synthesized bentonite clay-iron composite for the reduction of 4-nitrophenol by $NaBH_4$ from aqueous solution. The catalytic reduction of 4-nitrophenol follows diffusion and adsorption at the composite surface and further reduction through electron transfer by borohydride ions (BH_4) to 4-nitrophenol. Clay-nZVI was also found to be effective in soil matrix. Sun et al. (2021), used montmorillonite supported iron as a heterogenous catalyst in Fenton reaction to degrade PCB67 (2,3′,4,5-tetrachlorobiphenyl) from soil. The operating factors like pH, H_2O_2 dose, temperature and composite dosing affect the removal of PCB67. The reaction was diffusion controlled having activation energy calculated as 21 kJ mol^{-1}. The optimum removal was achieved as 78% at pH 3.5 within 80 min of reaction time. The acidic medium favours the dissolution of zero valent iron to soluble iron ions that substantially generate more $^{\bullet}OH$ radicals and increased the degradation rate. Additionally, their findings suggests that the buffering capacity of soil avert the formation of iron hydroxides which might reduce the scavenging of $^{\bullet}OH$ radicals. This allows the degradation of PCB67 in a wide pH range (at pH 7; removal efficiency = 66%) in the soil system.

14.6 Conclusions and Future Prospectus

The expeditious growth in the industrial sector has created many environmental issues including the problem of water pollution. The lack of proper drainage system, wastewater treatment and informal discharge of waste into water bodies has made the situation worse. Therefore, an efficient treatment method is essential to remediate contaminated water. This book chapter elaborates the various aspects of clay supported iron nanoparticles on the removal of persistent pollutants like dye, chlorinated compounds, phenol, and pharmaceuticals. The clay-iron composite widely applied as an adsorbent and catalyst in the treatment process. The excellent physico-chemical properties of clay (ion exchange and swelling capacity), ease of synthesis, economic feasibility and natural abundance makes them a reliable support material for iron nanoparticles by controlling the rate of dispersion and agglomeration of the nZVI. The adsorption-reduction mechanism prominently prevails in both adsorption and catalytic oxidation process. Besides, factors like pH, and clay-iron mixing ratio, significantly affect the rate of reaction for pollutant removal. Moreover, the surface modified clay has shown high porosity and increased surface area as compared to unmodified clay. However only few studies reported the working of clay-nZVI in real environment (groundwater and wastewater) and therefore, application of this hybrid nanocomposite should be explored at pilot scale level. Additionally, it is well known that bare nZVI imparts toxic effects on organisms when reached into different environmental matrix, so study on the clay-nZVI toxicity and iron leaching might give more insight on the feasibility of such nanocomposite in environment.

References

Abu-Danso E, Peräniemi S, Leiviskä T, Kim T, Tripathi KM, Bhatnagar A (2020) Synthesis of clay-cellulose biocomposite for the removal of toxic metal ions from aqueous medium. J Hazard Mater 381:120871

Adapa S, Bhullar N, de Souza SV (2016) A systematic review and agenda for using alternative water sources for consumer markets in Australia. J Clean Prod 124:14–20

Asfaram A, Ghaedi M, Purkait MK (2017) Novel synthesis of nanocomposite for the extraction of Sildenafil Citrate (Viagra) from water and urine samples: process screening and optimization. Ultrason Sonochem 38:463–472

Bagheri, M., Jafari, S. M., Eikani, M. H. (2021). Ultrasonic-assisted production of zero-valent iron-decorated graphene oxide/activated carbon nanocomposites: chemical transformation and structural evolution. Mater Sci Eng C 118:111362

Balmer B, Ylitalo G, Watwood S, Quigley B, Bolton J, Mullin K, Rosel P, Rowles T, Speakman T, Wilcox L, Schwacke L (2019) Comparison of persistent organic pollutants (POPs) between small cetaceans in coastal and estuarine waters of the northern Gulf of Mexico. Mar Pollut Bull 145:239–247

Bao T, Jin J, Damtie MM, Wu K, Yu ZM, Wang L, Chen J, Zhang Y, Frost RL (2019) Green synthesis and application of nanoscale zero-valent iron/rectorite composite material for P-chlorophenol degradation via heterogeneous Fenton reaction. J Saudi Chem Soc 23(7):864–878

Bao T, Damtie MM, Hosseinzadeh A, Frost RL, Yu ZM, Jin J, Wu K (2020a) Catalytic degradation of P-chlorophenol by muscovite-supported nano zero valent iron composite: synthesis, characterization, and mechanism studies. Appl Clay Sci 195:105735

Bao T, Damtie MM, Hosseinzadeh A, Wei W, Jin J, Vo HNP, Ye JS, Liu Y, Wang XF, Yu ZM, Ni BJ (2020b) Bentonite-supported nano zero-valent iron composite as a green catalyst for bisphenol A degradation: preparation, performance, and mechanism of action. J Environ Manag 260:110105

Barreto-Rodrigues M, Silveira J, Zazo JA, Rodriguez JJ (2017) Synthesis, characterization and application of nanoscale zero-valent iron in the degradation of the azo dye Disperse Red 1. J Environ Chem Eng 5(1):628–634

Basnet M, Tommaso CD, Ghoshal S, Tufenkji N (2015) Transport potential of a palladium-doped zero valent iron nanoparticle in a water saturated quartz, loamy and clay-amended sand. Aggregation Transp Nanoscale Zerovalent Iron Part Model Groundwater Syst Role Part Coat Biofilms 63

Bergaya F, Lagaly G (2006) General introduction: clays, clay minerals, and clay science. Dev Clay Sci 1:1–18

Bhosale SS, Rohiwal SS, Chaudhary LS, Pawar KD, Patil PS, Tiwari AP (2019) Photocatalytic decolorization of methyl violet dye using Rhamnolipid biosurfactant modified iron oxide nanoparticles for wastewater treatment. J Mater Sci Mater Electron 30(5):4590–4598

Borisover M, Gerstl Z, Burshtein F, Yariv S, Mingelgrin U (2008) Organic sorbate– organoclay interactions in aqueous and hydrophobic environments: sorbate– water competition. Environ Sci Technol 42(19):7201–7206

Bounab N, Duclaux L, Reinert L, Oumedjbeur A, Boukhalfa C, Penhoud P, Muller F (2021) Improvement of zero valent iron nanoparticles by ultrasound-assisted synthesis, study of Cr (VI) removal and application for the treatment of metal surface processing wastewater. J Environ Chem Eng 9(1):104773

Bulasara VK, Thakuria H, Uppaluri R, Purkait MK (2011) Effect of process parameters on electroless plating and nickel-ceramic composite membrane characteristics. Desalination 268(1–3):195–203

Chen SS, Hsu HD, Li CW (2004) A new method to produce nanoscale iron for nitrate removal. J Nanopart Res 6(6):639–647

Chen ZX, Jin XY, Chen Z, Megharaj M, Naidu R (2011) Removal of methyl orange from aqueous solution using bentonite-supported nanoscale zero-valent iron. J Colloid Interface Sci 363(2):601–607

Central pollution control board, status of water quality in India. monitoring of Indian National aquatic resources series (MINARS), vol 36, pp 1–237. Retrieved from http://www.cpcb.nic.in/WQ_Status_Report2012.pdf
Crane RA, Scott TB (2012) Nanoscale zero-valent iron: future prospects for an emerging water treatment technology. J Hazard Mater 211:112–125
Dale LH, Huber D (2005) Synthesis, properties, and applications of iron nanoparticles. Small 1(5):482–501
Dinç B, Çelebi A, Avaz G, Canlı O, Güzel B, Eren B, Yetis U (2021) Spatial distribution and source identification of persistent organic pollutants in the sediments of the Yeşilırmak River and coastal area in the Black Sea. Mar Pollut Bull 172:112884
Eljamal R, Eljamal O, Maamoun I, Yilmaz G, Sugihara Y (2020) Enhancing the characteristics and reactivity of nZVI: Polymers effect and mechanisms. J Mol Liq 315:113714
Ezzatahmadi N, Ayoko GA, Millar GJ, Speight R, Yan C, Li J, Zhu J, Xi Y (2017) Clay-supported nanoscale zero-valent iron composite materials for the remediation of contaminated aqueous solutions: a review. Chem Eng J 312:336–350
Frost RL, Xi Y, He H (2010) Synthesis, characterization of palygorskite supported zero-valent iron and its application for methylene blue adsorption. J Colloid Interface Sci 341(1):153–161
Fu F, Dionysiou DD, Liu H (2014) The use of zero-valent iron for groundwater remediation and wastewater treatment: a review. J Hazard Mater 267:194–205
Gaur N, Narasimhulu K, PydiSetty Y (2018) Recent advances in the bioremediation of persistent organic pollutants and its effect on environment. J Clean Prod 198:1602–1631
Gillham RW, O'Hannesin SF (1994) Enhanced degradation of halogenated aliphatics by zero-valent iron. Groundwater 32(6):958–967
Goswami A, Raul PK, Purkait MK (2012) Arsenic adsorption using copper (II) oxide nanoparticles. Chem Eng Res Des 90(9):1387–1396
Guan X, Sun Y, Qin H, Li J, Lo IM, He D, Dong H (2015) The limitations of applying zero-valent iron technology in contaminants sequestration and the corresponding countermeasures: the development in zero-valent iron technology in the last two decades (1994–2014). Water Res 75:224–248
Gude VG (2016) Desalination and sustainability—an appraisal and current perspective. Water Res 89:87–106
Guggenheim S, Martin RT (1995) Definition of clay and clay mineral: joint report of the AIPEA nomenclature and CMS nomenclature committees. Clays Clay Miner 43(2):255–256
Hanif MA, Nadeem R, Bhatti HN, Ahmad NR, Ansari TM (2007) Ni (II) biosorption by *Cassia fistula* (Golden Shower) biomass. J Hazard Mater 139(2):345–355
Hernández-Montoya V, Pérez-Cruz MA, Mendoza-Castillo DI, Moreno-Virgen MR, Bonilla-Petriciolet A (2013) Competitive adsorption of dyes and heavy metals on zeolitic structures. J Environ Manage 116:213–221
Hoch LB, Mack EJ, Hydutsky BW, Hershman JM, Skluzacek JM, Mallouk TE (2008) Carbothermal synthesis of carbon-supported nanoscale zero-valent iron particles for the remediation of hexavalent chromium. Environ Sci Technol 42(7):2600–2605
Huang Q, Liu W, Huang W (2013) Reductive dechlorination of tetrachlorobisphenol A by Pd/Fe bimetallic catalysts. J Hazard Mater 262:634–641
Ioannou M, Hatzikraniotis E, Lioutas C, Hassapis T, Altantzis T, Paraskevopoulos KM, Kyratsi T (2012) Fabrication of nanocrystalline Mg2Si via ball milling process: structural studies. Powder Technol 217:523–532
Ismadji, S., Soetaredjo, F. E., Ayucitra, A. (2015). Natural clay minerals as environmental cleaning agents. In: Clay materials for environmental remediation. Springer, Cham, pp 5–37
Jamei MR, Khosravi MR, Anvaripour B (2014) A novel ultrasound assisted method in synthesis of NZVI particles. Ultrason Sonochem 21(1):226–233
Jung HJ, Sohn Y, Sung HG, Hyun HS, Shin WG (2015) Physicochemical properties of ball milled boron particles: dry vs. wet ball milling process. Powder Technol 269:548–553

Kadhum ST, Alkindi GY, Albayati TM (2021) Eco friendly adsorbents for removal of phenol from aqueous solution employing nanoparticle zero-valent iron synthesized from modified green tea bio-waste and supported on silty clay. Chin J Chem Eng 36:19–28

Kamarudin NS, Jusoh R, Setiabudi HD, Sukor NF, Shariffuddin JH (2021) Potential nanomaterials application in wastewater treatment: physical, chemical and biological approaches. Mater Today Proc 42:107–114

Kanel SR, Manning B, Charlet L, Choi H (2005) Removal of arsenic (III) from groundwater by nanoscale zero-valent iron. Environ Sci Technol 39(5):1291–1298

Kerkez DV, Tomašević DD, Kozma G, Bečelić-Tomin MR, Prica MD, Rončević SD, Kónya Z (2014) Three different clay-supported nanoscale zero-valent iron materials for industrial azo dye degradation: a comparative study. J Taiwan Inst Chem Eng 45(5):2451–2461

Khan I, Saeed K, Khan I (2019) Nanoparticles: properties, applications and toxicities. Arab J Chem 12(7):908–931

Kong L, Zhu Y, Liu M, Chang X, Xiong Y, Chen D (2016) Conversion of Fe-rich waste sludge into nano-flake Fe-SC hybrid Fenton-like catalyst for degradation of AOII. Environ Pollut 216:568–574

Krasae N, Wantala K (2016) Enhanced nitrogen selectivity for nitrate reduction on Cu–nZVI by TiO_2 photocatalysts under UV irradiation. Appl Surf Sci 380:309–317

Lallas PL (2001) The Stockholm convention on persistent organic pollutants. Am J Int Law 95(3):692–708

Li XQ, Elliott DW, Zhang WX (2006) Zero-valent iron nanoparticles for abatement of environmental pollutants: materials and engineering aspects. Crit Rev Solid State Mater Sci 31(4):111–122

Li S, Yan W, Zhang WX (2009) Solvent-free production of nanoscale zero-valent iron (nZVI) with precision milling. Green Chem 11(10):1618–1626

Li X, Ai L, Jiang J (2016b) Nanoscale zerovalent iron decorated on graphene nanosheets for Cr (VI) removal from aqueous solution: surface corrosion retard induced the enhanced performance. Chem Eng J 288:789–797

Li X, Zhao Y, Xi B, Mao X, Gong B, Li R, Peng X, Liu H (2016a) Removal of nitrobenzene by immobilized nanoscale zero-valent iron: effect of clay support and efficiency optimization. Appl Surf Sci 370:260–269

Li X, Zhao Y, Xi B, Meng X, Gong B, Li R, Peng X, Liu H (2017) Decolorization of methyl orange by a new clay-supported nanoscale zero-valent iron: synergetic effect, efficiency optimization and mechanism. J Environ Sci 52:8–17

Li S, Tang J, Liu Q, Liu X, Gao B (2020) A novel stabilized carbon-coated nZVI as heterogeneous persulfate catalyst for enhanced degradation of 4-chlorophenol. Environment international 138:105639

Li Q, Wang H, Chen Z, He X, Liu Y, Qiu M, Wang X (2021). Adsorption-reduction strategy of U (VI) on NZVI-supported zeolite composites via batch, visual and XPS techniques. J Mol Liq 339:116719

Lin J, Sun M, Liu X, Chen Z (2017) Functional kaolin supported nanoscale zero-valent iron as a Fenton-like catalyst for the degradation of Direct Black G. Chemosphere 184:664–672

Machado S, Pinto SL, Grosso JP, Nouws HPA, Albergaria JT, Delerue-Matos C (2013) Green production of zero-valent iron nanoparticles using tree leaf extracts. Sci Total Environ 445:1–8

Martins F, Machado S, Albergaria T, Delerue-Matos C (2017) LCA applied to nano scale zero valent iron synthesis. Int J Life Cycle Assess 22(5):707–714

Miraji H, Ripanda A, Moto E (2021) A review on the occurrences of persistent organic pollutants in corals, sediments, fish, and waters of the Western Indian Ocean. Egyptian J Aquatic Res (in press)

Mokhtari P, Ghaedi M, Dashtian K, Rahimi MR, Purkait MK (2016) Removal of methyl orange by copper sulfide nanoparticles loaded activated carbon: kinetic and isotherm investigation. J Mol Liq 219:299–305

Mondal P, Anweshan A, Purkait MK (2020) Green synthesis and environmental application of iron-based nanomaterials and nanocomposite: a review. Chemosphere 127509

Mortazavian S, An H, Chun D, Moon J (2018) Activated carbon impregnated by zero-valent iron nanoparticles (AC/nZVI) optimized for simultaneous adsorption and reduction of aqueous hexavalent chromium: Material characterizations and kinetic studies. Chem Eng J 353:781–795

Nel A, Xia T, Mädler L, Li N (2006) Toxic potential of materials at the nanolevel. Science 311(5761):622–627

Nisticò R, Carlos L (2019) High yield of nano zero-valent iron (nZVI) from carbothermal synthesis using lignin-derived substances from municipal biowaste. J Anal Appl Pyrol 140:239–244

Nurmi JT, Tratnyek PG, Sarathy V, Baer DR, Amonette JE, Pecher K, Wang C, Linehan JC, Matson DW, Penn RL, Driessen MD (2005) Characterization and properties of metallic iron nanoparticles: spectroscopy, electrochemistry, and kinetics. Environ Sci Technol 39(5):1221–1230

O'Carroll D, Sleep B, Krol M, Boparai H, Kocur C (2013) Nanoscale zero valent iron and bimetallic particles for contaminated site remediation. Adv Water Resour 51:104–122

Özacar M, Şengil İA, Türkmenler H (2008) Equilibrium and kinetic data, and adsorption mechanism for adsorption of lead onto valonia tannin resin. Chem Eng J 143(1–3):32–42

Pirzadeh K, Ghoreyshi AA (2014) Phenol removal from aqueous phase by adsorption on activated carbon prepared from paper mill sludge. Desalin Water Treat 52(34–36):6505–6518

Qiu X, Fang Z, Yan X, Gu F, Jiang F (2012) Emergency remediation of simulated chromium (VI)-polluted river by nanoscale zero-valent iron: laboratory study and numerical simulation. Chem Eng J 193:358–365

Ramos MA, Yan W, Li XQ, Koel BE, Zhang WX (2009) Simultaneous oxidation and reduction of arsenic by zero-valent iron nanoparticles: understanding the significance of the core– shell structure. J Phys Chem C 113(33):14591–14594

Rautureau M, Gomes CDSF, Liewig N, Katouzian-Safadi M (2017) Clay and clay mineral definition. In: Clays and health. Springer, Cham, pp 5–31

Ravikumar KVG, Argulwar S, Sudakaran SV, Pulimi M, Chandrasekaran N, Mukherjee A (2018) Nano-bio sequential removal of hexavalent chromium using polymer-nZVI composite film and sulfate reducing bacteria under anaerobic condition. Environ Technol Innov 9:122–133

Rong C, Jianlong W, Weixian Z (2006) The research progress on degradation of halogenated organic compounds by nano iron. Prog Chem 18(01):93

Ruhl AS, Ünal N, Jekel M (2012) Evaluation of two-component Fe (0) fixed bed filters with porous materials for reductive dechlorination. Chem Eng J 209:401–406

Sarkar B, Rusmin R, Ugochukwu UC, Mukhopadhyay R, Manjaiah KM (2019) Modified clay minerals for environmental applications. In: Modified clay and zeolite nanocomposite materials. Elsevier, pp 113–127

Sheikholeslami M, Jafaryar M, Abohamzeh E, Shafee A, Babazadeh H (2020a) Energy and entropy evaluation and two-phase simulation of nanoparticles within a solar unit with impose of new turbulator. Sustain Energy Technol Assess 39:00727

Sheikholeslami M, Keshteli AN, Shafee A (2020b) Melting and solidification within an energy storage unit with triangular fin and CuO nano particles. J Energy Storage 32:101716

Shichi T, Takagi K (2000) Clay minerals as photochemical reaction fields. J Photochem Photobiol C 1(2):113–130

Shimizu A, Tokumura M, Nakajima K, Kawase Y (2012) Phenol removal using zero-valent iron powder in the presence of dissolved oxygen: roles of decomposition by the Fenton reaction and adsorption/precipitation. J Hazard Mater 201:60–67

Silva LL, Abdelraheem W, Nadagouda MN, Rocco AM, Dionysiou DD, Fonseca FV, Borges CP (2021) Novel microwave-driven synthesis of hydrophilic polyvinylidene fluoride/polyacrylic acid (PVDF/PAA) membranes and decoration with nano zero-valent-iron (nZVI) for water treatment applications. J Membr Sci 620:118817

Sravanthi K, Ayodhya D, Swamy PY (2019) Green synthesis, characterization and catalytic activity of 4-nitrophenol reduction and formation of benzimidazoles using bentonite supported zero valent iron nanoparticles. Mater Sci Energy Technol 2(2):298–307

Stefaniuk M, Oleszczuk P, Ok YS (2016) Review on nano zerovalent iron (nZVI): from synthesis to environmental applications. Chem Eng J 287:618–632

Sun YP, Li XQ, Cao J, Zhang WX, Wang HP (2006) Characterization of zero-valent iron nanoparticles. Adv Coll Interface Sci 120(1–3):47–56

Sun X, Wang X, Li J, Wang L (2014) Degradation of nitrobenzene in groundwater by nanoscale zero-valent iron particles incorporated inside the channels of SBA-15 rods. J Taiwan Inst Chem Eng 45(3):996–1000

Sun YM, Feng L, Yang L (2021) Degradation of PCB67 in soil using the heterogenous Fenton process induced by montmorillonite supported nanoscale zero-valent iron. J Hazard Mater 406:124305

Taghizadeh F, Ghaedi M, Kamali K, Sharifpour E, Sahraie R, Purkait MK (2013) Comparison of nickel and/or zinc selenide nanoparticle loaded on activated carbon as efficient adsorbents for kinetic and equilibrium study of removal of Arsenazo (III) dye. Powder Technol 245:217–226

Tesh SJ, Scott TB (2014) Nano-composites for water remediation: a review. Adv Mater 26(35):6056–6068

Tiraferri A, Chen KL, Sethi R, Elimelech M (2008) Reduced aggregation and sedimentation of zero-valent iron nanoparticles in the presence of guar gum. J Colloid Interface Sci 324(1–2):71–79

Tlili I, Alkanhal TA (2019) Nanotechnology for water purification: electrospun nanofibrous membrane in water and wastewater treatment. J Water Reuse Desalin 9(3):232–248

Tosco T, Papini MP, Viggi CC, Sethi R (2014) Nanoscale zerovalent iron particles for groundwater remediation: a review. J Clean Prod 77:10–21

Uegami M, Kawano J, Okita T, Fujii Y, Okinaka K, Kakuya K, Yatagi S (2006) U.S. Patent No. 7,022,256. Washington, DC: U.S. Patent and Trademark Office

Unuabonah EI, Taubert A (2014) Clay–polymer nanocomposites (CPNs): adsorbents of the future for water treatment. Appl Clay Sci 99:83–92

Valiev RZ, Islamgaliev RK, Alexandrov IV (2000) Bulk nanostructured materials from severe plastic deformation. Prog Mater Sci 45(2):103–189

Wang CB, Zhang WX (1997) Synthesizing nanoscale iron particles for rapid and complete dechlorination of TCE and PCBs. Environ Sci Technol 31(7):2154–2156

Weng X, Cai W, Lin S, Chen Z (2017) Degradation mechanism of amoxicillin using clay supported nanoscale zero-valent iron. Appl Clay Sci 147:137–142

Williams LB, Haydel SE (2010) Evaluation of the medicinal use of clay minerals as antibacterial agents. Int Geol Rev 52(7–8):745–770

Wu J, Wang B, Blaney L, Peng G, Chen P, Cui Y, Yu G (2019) Degradation of sulfamethazine by persulfate activated with organo-montmorillonite supported nano-zero valent iron. Chem Eng J 361:99–108

Xi Y, Megharaj M, Naidu R (2011) Dispersion of zerovalent iron nanoparticles onto bentonites and use of these catalysts for orange II decolourisation. Appl Clay Sci 53(4):716–722

Xie M, Shon HK, Gray SR, Elimelech M (2016) Membrane-based processes for wastewater nutrient recovery: technology, challenges, and future direction. Water Res 89:210–221

Xiong Z, Zhao D, Pan G (2007) Rapid and complete destruction of perchlorate in water and ion-exchange brine using stabilized zero-valent iron nanoparticles. Water Res 41(15):3497–3505

Xu F, Deng S, Xu J, Zhang W, Wu M, Wang B, Huang J, Yu G (2012) Highly active and stable Ni–Fe bimetal prepared by ball milling for catalytic hydrodechlorination of 4-chlorophenol. Environ Sci Technol 46(8):4576–4582

Yadav KK, Singh JK, Gupta N, Kumar VJJMES (2017) A review of nanobioremediation technologies for environmental cleanup: a novel biological approach. J Mater Environ Sci 8(2):740–757

Yadav VB, Gadi R, Kalra S (2019) Clay based nanocomposites for removal of heavy metals from water: a review. J Environ Manage 232:803–817

Yen CH, Chen KF, Kao CM, Liang SH, Chen TY (2011) Application of persulfate to remediate petroleum hydrocarbon-contaminated soil: feasibility and comparison with common oxidants. J Hazard Mater 186(2–3):2097–2102

Zekić E, Vuković Ž, Halkijević I (2018) Application of nanotechnology in wastewater treatment. Građevinar 70(4):315–323

Zhang W, Qian L, Ouyang D, Chen Y, Han L, Chen M (2019) Effective removal of Cr (VI) by attapulgite-supported nanoscale zero-valent iron from aqueous solution: enhanced adsorption and crystallization. Chemosphere 221:683–692

Zhang T, Wang W, Zhao Y, Bai H, Wen T, Kang S, Komarneni S (2020) Removal of heavy metals and dyes by clay-based adsorbents: from natural clays to 1D and 2D nano-composites. Chem Eng J 127574

Zhao N, Chang F, Hao B, Yu L, Morel JL, Zhang J (2019a) Removal of organic dye by biomass-based iron carbide composite with an improved stability and efficiency. J Hazard Mater 369:621–631

Zhao R, Zhou Z, Zhao X, Jing G (2019b) Enhanced Cr (VI) removal from simulated electroplating rinse wastewater by amino-functionalized vermiculite-supported nanoscale zero-valent iron. Chemosphere 218:458–467

Zhou L, Li A, Ma F, Zhao H, Deng F, Pi S, Tang A, Yang J (2020) Combining high electron transfer efficiency and oxidation resistance in nZVI with coatings of microbial extracellular polymeric substances to enhance Sb (V) reduction and adsorption. Chem Eng J 395:125168

Zhu Z, Ouyang S, Li P, Shan L, Ma R, Zhang P (2020) Persistent organic pollutants removal via hierarchical flower-like layered double hydroxide: adsorption behaviors and mechanism investigation. App Clay Sci 188:105500

Chapter 15
Application of Plasma-Assisted Advanced Oxidation Processes for Removal of Emerging Contaminants in Water

Shaik Mahamad Allabakshi, P. S. N. S. R. Srikar, Reetesh Kumar Gangwar, and Shihabudheen M. Maliyekkal

Abstract The increasing occurrence of contaminants such as pharmaceutically active compounds, endocrine-disrupting compounds, nanomaterials, surfactants, personal care products is an emerging concern in the water sector. Some of these emerging contaminants are toxic to all life forms, are bio-resistant, and can sustain after primary and secondary wastewater treatment. Conventional water treatment processes are also ineffective in removing these compounds. Advanced oxidation is a potential technique and can degrade and mineralize complex organic molecules. Advanced oxidation processes (AOPs) rely on the in-situ generation of reactive chemical species (RCS) such as hydroxyl radicals for degradation. Most of the RCS with higher oxidizing potential is short-lived, and hence the effective production of these compounds is crucial for this technology's success. The plasma-mediated AOP is an emerging technology superior to other conventional AOPs due to its ability to generate RCS at a controlled rate without using chemical agents. Moreover, the nonthermal plasma can also produce the RCS at controlled temperature and ambient pressure and, therefore, very suitable for commercial-scale processing. The plasma produces a cocktail of reactive species whose collective effect enhances the efficacy of the process. However, before translating this technology to the commercial scale, it is essential to make it affordable and energy-efficient. In this regard, significant studies are being carried focusing on reactor design and its optimization. The chapter reviews the recent developments in plasma-based reactors and their application in the

S. M. Allabakshi · S. M. Maliyekkal (✉)
Department of Civil and Environmental Engineering, Indian Institute of Technology Tirupati, Tirupati 517506, India
e-mail: shihab@iittp.ac.in

S. M. Allabakshi
e-mail: ce19d005@iittp.ac.in

P. S. N. S. R. Srikar · R. K. Gangwar
Department of Physics & CAMOST, Indian Institute of Technology Tirupati, Tirupati 517506, India
e-mail: ph19d503@iittp.ac.in

R. K. Gangwar
e-mail: reetesh@iittp.ac.in

S. P. Singh et al. (eds.), *New Trends in Emerging Environmental Contaminants*, Energy, Environment, and Sustainability,
https://doi.org/10.1007/978-981-16-8367-1_15

degradation of emerging contaminants. The chapter also highlights the current challenges and prospects of plasma-based technology in treating emerging contaminants and various operating parameters influencing the process.

Keywords Emerging contaminant · Plasma · Advanced oxidation · Reactors · Water treatment

15.1 Introduction

The decline in quality and quantity of fresh water is one of the most challenging problems faced by humanity in the twenty-first century. If adequate measures are not taken, these problems would aggravate in the coming decades and threaten the sustenance of human life and the ecosystem. According to a recent WHO report, at least 2 billion people lack access to clean water globally (WHO 2019). The United Nations World Water Development Report (UN WWDR) projected that by 2050, 6 billion people would face freshwater scarcity globally (Boretti and Rosa 2019). Further, the increasing use of chemicals in industrial processing, agriculture, and other daily needs has polluted many freshwater resources and are unavailable for direct use. The growing occurrence of new contaminants in the water is an emerging concern in the water sector. Some of these contaminants of emerging concern (ECs) are toxic to all life forms, are bio-resistant, flame retardants, and can sustain after primary and secondary wastewater treatment. In many cases, traces of these pollutants can pose serious health concerns (Rout et al. 2021). The widespread occurrence of ECs in water and their chemical toxicity demands the development of effective, efficient, and eco-friendly management and treatment strategies.

The NORMAN project, which enables the exchange of information and validation of ECs under the European Commission, has listed 1036 ECs by 2016 and is further labelled into 30 classes based on their origin and type (Dey et al. 2019). The ECs include pesticides, pharmaceutical active compounds (PhACs), disinfection by products, personal care products (PCPs), food additives, artificial sweeteners, flame retardants, endocrine-disrupting compounds (EDCs), nanomaterials, and surfactants (Lin et al. 2020). These contaminants exhibit very different physical and chemical properties, thus challenging to detect and remove from water. Many conventional treatment methods such as coagulation and flocculation, filtration, membrane treatment, and chlorination are ineffective in addressing ECs in water (Mondal et al. 2018). The high chemical usage and sludge production, biofouling, high energy requirements, and/or difficulty in operation and maintenance further limit their use. The frequent reports on the presence of ECs in treated water also support that the conventional treatment techniques are ineffective and cannot guarantee complete removal of ECs in water (Khetan and Collins 2007; Magureanu et al. 2015).

Advanced oxidation process (AOP) is a promising technique and can provide a potential route to degrade and mineralize complex organic molecules. The process relies on the in-situ generation of reactive chemical species (RCS) such as hydroxyl

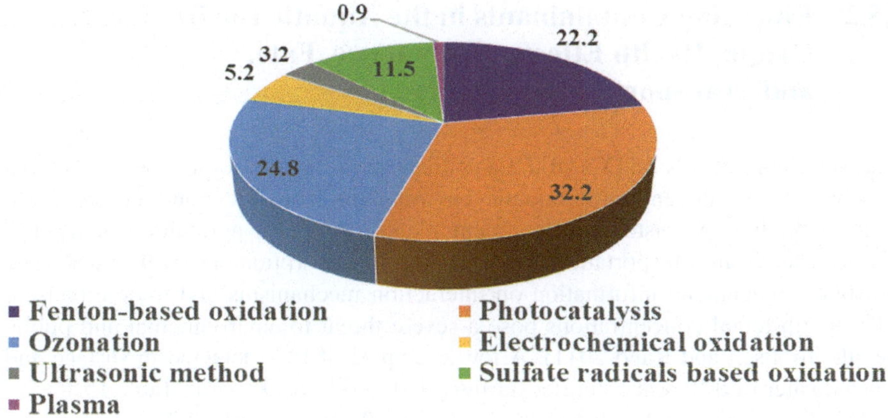

Fig. 15.1 The percentage number of articles published related to the application AOPs for the degradation of ECs in water and wastewater. This chart is made based on the number of articles published between 2000 and 2021 as per the scopus database

radicals, atomic oxygen, ozone, perhydroxyl radical, and hydrogen peroxide. These RCS can degrade or mineralize the ECs into less toxic, simple, and treatable compounds. Some existing AOPs include the Fenton and photo-Fenton processes, photocatalysis, sonolysis, catalytic wet air oxidation processes, and combinations thereof (Ribeiro et al. 2015). However, limitations like the addition of excess chemicals, catalyst regeneration, sludge generation, and low penetration efficiency of UV photons, formation of toxic byproducts, the negative influence of water quality parameters persist (Klamerth et al. 2010). Figure 15.1 summarizes the studies from 2000 to 2021 utilizing various AOPs to treat ECs in water.

Plasma-assisted AOP is an emerging technology that can simultaneously generate a wide variety of short and long-lived RCS at a controlled rate in an ambient environment. The technology also demonstrates good potential for commercial-scale processing (Pankaj et al. 2018). The significant attraction lies in the ability of plasma to produce a cocktail of reactive species without any additional chemical/catalyst (Gorbanev et al. 2018). The collective effect of plasma-generated RCS enhances the efficacy of the process. However, before translating this technology to the commercial scale, it is essential to make it affordable and energy-efficient. Significant efforts are being paid in this direction by developing novel reactors and hybridizing plasma technology with other approaches (Snoeckx and Bogaerts 2017). The following sections will outline the recent development in plasma-based reactors and their application in the degradation of ECs in water. The chapter will conclude by providing the current challenges and prospects of plasma-based technology in treating emerging contaminants.

15.2 Emerging Contaminants in the Aquatic Environment: Origin, Health Effects, Occurrence, Fate, and Transport

The significant levels of ECs in the aquatic system have become a serious global concern. They can enter the aquatic environment from the point or non-point sources, including wastewater treatment plants (WWTP), agricultural activities, surface runoff, air transportation and precipitation, adsorption onto soil, and sewage biosolids. Inadequate information on interaction mechanisms and toxic effects of ECs at sub-lethal concentrations pose a severe threat to environmental and public health (Brausch and Rand 2011). A few examples of ECs detected in surface and groundwater of different countries during 2010–2021 are shown in Table 15.1.

ECs like propranolol, ketoprofen, naproxen, E2, gemfibrozil, EE2, and ibuprofen are susceptible to photolysis (Khetan and Collins 2007). Recent studies report the occurrence of metabolites such as 3-hydroxycarbofuran, acephate, clethodim, acetochlor ethane sulfonic acid, and acetochlor oxanilic acid in water bodies. These metabolites are listed in contaminant candidate list 3 (Epa 2009). The human body cannot absorb drugs completely, and a good portion of the drug is often excreted with its metabolites. For instance, ibuprofen was found to be present as a parent chemical (1%) in surface water along with some of its metabolites like (+)-2–40-(2-carboxypropyl)-phenyl propionic acid (37%), (+)-2–4′-(2-hydroxy-2-methylpropyl)-phenyl propionic acid (25%) and conjugated ibuprofen (14%) (Kasprzyk-Hordern et al. 2008). Numerous metabolites are generated through various biodegradation pathways, and their toxicity is not often studied. Sorption of ECs onto dissolved organic matter is also responsible for the presence of ECs in an aqueous medium (Petrie et al. 2015).

Globally, only limited studies are available on the toxicity of ECs on humans and the environment. ECs like bisphenol A, organophosphate and brominated flame retardants, perchlorate, phthalate, polycyclic siloxanes, and triclosan can adversely affect human health (Covaci et al. 2012). Bisphenol A, a universal plasticizer, is responsible for severe endocrine disorders in salivary glands, thyroid, and men's genetic systems. Hormonal imbalance due to bisphenol A can lead to cancer (Bolong et al. 2009). Musk compound like xylol can indirectly affect the human tissues and leads to brain damage (Bolong et al. 2009; Lei et al. 2015). Prolonged exposure to herbicides like atrazine destroys adrenal glands and impairs steroid hormone metabolism (Debnath et al. 2019). About 70% increase in streptomycin, penicillin and tetracycline-resistant bacteria was identified in a dairy farm's soil manure (Qiao et al. 2011). Intake of phthalates leads to increased insulin resistance, abdominal obesity, and neurobehavioral disorder (Dewalque et al. 2014). EDCs like nonylphenol can cause feminization of aquatic organisms and hinder the hormonal system (Saidulu et al. 2021). Analysis of metabolites, in-depth survey of ECs existence, toxicology studies, risk assessment, mitigation measures, and awareness in public is required to plan appropriate mitigative measures and reduce the health impacts from ECs (Fig. 15.2).

Table 15.1 Emerging contaminants detected in surface and groundwater bodies

S. No.	Type of ECs	Name of ECs	Country	Type of water	Detection limit	References
1	Pesticides	Dichloro-diphenyl-trichloroethane	India	Surface water	13.5 μg/L	Nath et al. (2014)
		1*H*-1,2,4-triazole	USA	Ground water	20–430 ng/L	York (2014)
		Acetamiprid	Saudi Arabia	Surface water	1408 ng/L	Picó et al. (2021)
		Metalaxyl	China	Surface water	9.86 ng/L	Zhang et al. (2020)
		Atrazine	Argentina	Surface water	0.64–1.4 μg/L	Gerónimo et al. (2014)
		DEET	Zambian	Ground water	70–330 ng/L	Sorensen et al. (2015)
2	PhACs	Theophylline	Portugal	Surface water	184 ng/L	Gonzalez-Rey et al. (2015)
		Lincomycin	Canada	Surface water	143 ng/L	Kleywegt et al. (2011)
		Ibuprofen	India	Waste water	26.45 μg/L	Singh et al. (2014)
		Metoprolol	Spain	Surface water	380 ng/L	Huerta-Fontela et al. (2011)
		Norfloxacin	Saudi Arabia	Surface water	2763 ng/L	Picó et al. (2021)
		Carbamazepine	France	Drinking water	41.6 ng/L	Vulliet and Cren-Olivé 2009)
3	PCPs	Triclosan	India	Surface water	5160 ng/L	Gani and Kazmi (2016)
		Caffeine	Saudi Arabia	Surface water	12,374 ng/L	Picó et al. (2021)
		Benzophenone-3	Brazil	Drinking water	115 ng/L	Silva and Emídio (2015)
		Antipyrine	China	Drinking water	19.46 ng/L	Qiao et al. (2011)
		Naproxen	Spain	Surface water	278 ng/L	Carmona et al. (2014)
		4MBC	Germany	Surface water	235–2592 ng/L	Montes-Grajales et al. (2017)
4	EDC	Bisphenol F	Japan	Surface water	90–2850 ng/L	Yamazaki et al. (2015)
		Bisphenol S	India	Surface water	58–2100 ng/L	Yamazaki et al. (2015)
		Nonylphenol-tri-ethoxylate	Vietnam	Surface water	38–160 ng/L	Williams et al. (2019)

(continued)

Table 15.1 (continued)

S. No.	Type of ECs	Name of ECs	Country	Type of water	Detection limit	References
		Bisphenol A	Malaysia	Tap water	66.4 ng/L	Wee et al. (2020)
		Methyl paraben	Egypt	surface water	MDL-1780 ng/L	Radwan et al. (2019)
		4-nonylphenol	China	surface water	698–14,936 ng/L	Gong et al. (2016)
5	Flame retardants	Polybrominated diphenyl ether	Canada	Waste water	29–1000 ng/L	Kim and Kannan (2018)
		TCPP (tris (1-chloro-2-propyl) phosphate)	Saudi Arabia	Surface water	387 ng/L	Picó et al. (2021)
		TCPP (tris (1-chloro-2-propyl) phosphate)	China	Surface water	1.14–386 ng/L	Thi et al. (2016)
		Tris(2-butoxyethyl) phosphate (TBOEP)	USA	Surface water	2.53–366 ng/L	Kim and Kannan (2018)
		Tris(2-chloroethyl) phosphate (TCEP)	Korea	Drinking water	0.7–1400 ng/L	Lee et al. (2016)
		Polybrominated diphenyl ether	UK	Surface water	0.295 μg/L	Cristale et al. (2013)

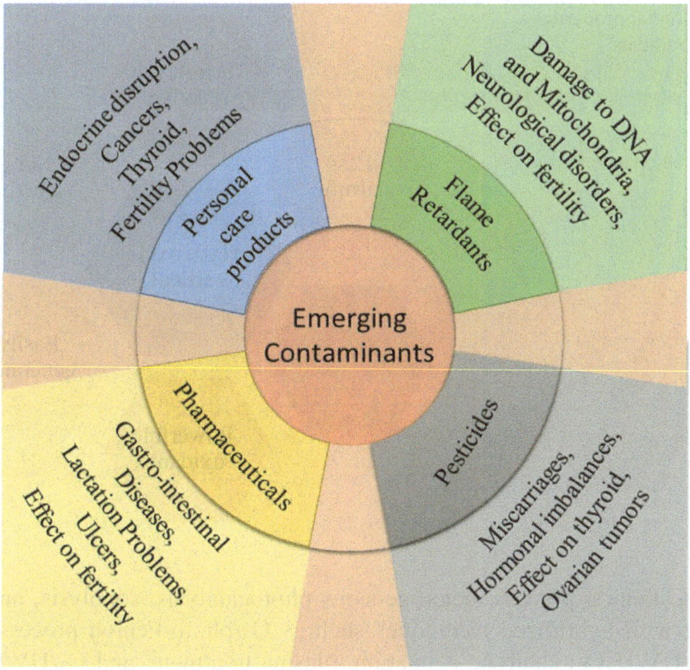

Fig. 15.2 Summary of effects of emerging contaminants on human health. The image is adapted with modifications from Rathi et al. (2021) with the permission from Elsevier

15.3 Advanced Oxidation Process for Removal of ECs in Water

Owing to the widespread occurrence and possible toxic effects on humans and other living organisms, removing ECs from water is imperative. However, most conventional treatment technologies are ineffective in removing ECs due to their complex nature and various physicochemical properties. Even popular technologies such as adsorption and membrane filtration are not attractive due to ECs complex nature and generation of contaminated sludge. In this context, developing more reliable and effective technologies is the need of the hour. The recent advancement in AOP promises a potential treatment option for removing ECs in water. The AOP can degrade and mineralize the ECs to simple carbon dioxide, water, and inorganic acids or ions. The technique utilizes ROS like hydroxyl radicals (•OH), hydrogen peroxide (H_2O_2), ozone (O_3), nascent oxygen (•O), and hydroperoxyl (•HO_2) for degradation. Among these, the •OH is the most reactive radical with an oxidation potential of 2.8 V. The highlights of its properties are shown in Fig. 15.3 (Brillas et al. 2009).

Most of the higher oxidizing potential radicals are short-lived; therefore, in-situ production of RCS is a prime requirement for efficient processing. Many AOPs

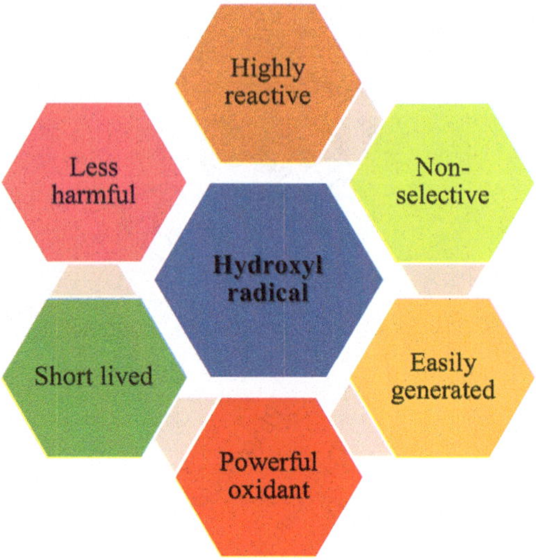

Fig. 15.3 Crucial properties of hydroxyl radicals

such as the Fenton process, heterogeneous photocatalysis, sonolysis, and ozonation along with hybridized techniques such as O_3/photo-Fenton process, O_3/UV, O_3/H_2O_2, UV/H_2O_2, catalytic ozonation, plasma treatment, and O_3/UV/H_2O_2 are developed. A description of pathways of radical production in these processes is shown in Table 15.2.

Though AOPs are promising, the radical generation pathway, water matrices, and experimental conditions significantly influence the degradation/mineralization of ECs. Despite higher oxidation potential, low mineralization values are reported depending on the persistence of the pollutant. For example, Zhao et al. achieved complete degradation of 2.5 μM indomethacin with ozone dosage of 2–35 mg/L in 7 min (Zhao et al. 2017). However, only 50% mineralization was achieved even after extending the treatment to 30 min. Similarly, 0.38 mM of propranolol was degraded entirely in 8 min using the ozone dosage of 0.47 mM. Still, the total organic carbon (TOC) value remained close to 5% even after 60 min exposure with ozone dosages of 3.54 mM (Dantas et al. 2011). Even though ozonation is a proven technique, the limitations like inefficient mass transfer and the formation of secondary intermediates restrict operational capability (Shin et al. 1999). The presence of organic matter in WWTP effluent or surface waters can significantly demand higher ozone dosage and reduce the process efficacy (Almomani et al. 2016).

Fenton and photo-Fenton are extensively studied AOPs. Though the process is promising, large chemical requirements, the generation of chemical sludge, and the high sensitivity of the process are major concerns (Diya'uddeen and Daud 2012). Photocatalysis is also a promising AOPs suitable for removing complex organic molecules in water. However, catalyst fouling, slow rate of degradation, and low penetration of UV photons are some of the disadvantages of photocatalysis despite the less operational cost. The plasma-assisted AOP is an emerging approach that can

Table 15.2 Chemical pathways showing the radical production in various AOPs. This is adapted with modification from Salimi et al. (2017)

Process	Reaction	References
H_2O_2/O_3	$H_2O_2 \leftrightarrows H^+ + HO_2^-$	Gligorovski et al. (2018)
	$HO_2^- + O_3 \rightarrow \bullet OH + O_2 + O_2^{\bullet -}$	
H_2O_2/UV	$H_2O_2 + h\,\nu \rightarrow 2\bullet OH$	Cruz et al. (2013)
$UV/H_2O_2/O_3$	$O_3 + OH^- \rightarrow \bullet OH$	Joseph et al. (2009)
	$3O_3 + h\,\nu \rightarrow 2\bullet OH$	
	$H_2O_2 + h\,\nu \rightarrow 2\bullet OH$	
	$H_2O_2 + O_3 \rightarrow 2\bullet OH$	
	$H_2O_2 + O_3 + h\,\nu \rightarrow \bullet OH$	
Fenton	$Fe^{2+} + H_2O_2 \rightarrow Fe^{3+} + HO^- + \bullet HO$	Asghar et al. (2015)
Ultra sound (US)	$H_2O + US \rightarrow \bullet OH + OH^-$	Sathishkumar et al. (2016)
	$H_2O + US \rightarrow (1/2)H_2 + (1/2)H_2O_2$	
Ultra sound/Fenton	$Fe^{2+} + H_2O_2 \rightarrow Fe^{3+} + HO^- + \bullet HO$	Bagal and Gogate (2014)
	$H_2O_2 + Fe^{3+} \rightarrow Fe^-OOH^{2+} + H^+$	
	$Fe^-OOH^{2+} + US \rightarrow Fe^{2+} + \bullet HOO$	
	$Fe^{2+} + \bullet HOO \rightarrow Fe^{2+} + H^+ + O_2$	
	$Fe^{2+} + H_2O_2 \rightarrow Fe^{3+} + HO^- + \bullet HO$	
US/advanced Fenton	$Fe^0 + 2H^+ \rightarrow Fe^{2+} + H_2$	Bagal and Gogate (2014)
	$Fe^{2+} + H_2O_2 \rightarrow Fe^{3+} + HO^- + \bullet HO$	
	$H_2O_2 + US \rightarrow 2\bullet HO$	
	$H_2O_2 + \bullet HO \rightarrow H_2O + \bullet HO_2$	
	$Fe^{3+} + H_2O_2 \rightarrow Fe(OOH)^{2+} + H^+$	
	$Fe(OOH)^{2+} + US \rightarrow Fe^{2+} + \bullet HO_2$	
	$Fe^0 + 2Fe^{3+} \rightarrow 3Fe^{2+}$	
	$Fe^{3+} + HO_2^{\bullet} \rightarrow Fe^{2+} + H^+ + O_2$	
Sono-Fenton	$Fe^{3+} + H_2O_2 \rightarrow Fe^{2+} + \bullet OOH + H^+$	Dindarsafa et al. (2017)
	$Fe^{2+} + H_2O_2 \rightarrow Fe^{3+} + \bullet OH + H^-$	

(continued)

Table 15.2 (continued)

Process	Reaction	References
	$H_2O + US \rightarrow {\bullet}OH + {\bullet}H$	
	${\bullet}OH + {\bullet}OH \rightarrow H_2O_2$	
Photocatalysis	$TiO_2 \rightarrow {}^{h\nu}e^- + H^+$	Mahamuni and Adewuyi (2010)
	$TiO_2(H^+) + H_2O_{ad} \rightarrow TiO_2 + {\bullet}HO_{ad} + H^+$	
	$TiO_2(H^+) + H\ O_{ad} \rightarrow TiO_2 + {\bullet}OH_{ad}$	
	$TiO_2(H^+) + RX_{ad} \rightarrow TiO_2 + RX^{\bullet +}{}_{ad}$	
Plasma	$H_2O + e^- \rightarrow {\bullet}OH + H{\bullet} + e^-$	Itikawa and Mason (2005)
	$H_2O + e^- \rightarrow 2e^- + H_2O^+$	
	$H_2O^+ + H_2O \rightarrow {\bullet}OH + H_3O^+$	
	${\bullet}OH + {\bullet}OH \rightarrow H_2O_2$	
	${\bullet}OH + H_2O_2 \rightarrow H_2O + {\bullet}HO_2$	

offer an efficient and eco-friendly way to degrade and mineralize ECs in a relatively lower treatment time. The plasma-assisted AOP can produce short and long-lived RCS simultaneously, and its combined effect can significantly enhance the efficacy of the process (Nippatla and Philip 2019). The plasma can also be nonthermal and can produce RCS at ambient pressure, which is very suitable for commercial-scale processing. It can also eliminate the use of any additional precursor for the generation of RCS. The following section discusses the plasma-assisted AOP technique in detail and the recent studies on the degradation of emerging contaminants. The current challenges and prospects of plasma-based technology in treating emerging contaminants and various operating parameters influencing the process are also discussed.

15.4 Introduction to Plasma

As shown in Fig. 15.4, the phase transition from the solid-state to the gaseous state of matter can occur by increasing the system's internal energy. The addition of energy in the gaseous system enhances inelastic collisions that can prompt ionization and dissociation processes. Under appropriate conditions, this shall lead to the transition from the gaseous phase to the plasma, often referred to as the fourth state of matter. For the first time (1928), the "Plasma" ("mouldable substance" or "jelly" in Greek) (Fridman 2008) term was coined by well-known chemist Irving Langmuir to describe the ionized gas system, which is macroscopically neutral and exhibits collective behaviour. All ionized systems cannot be classified as plasma, and the plasma state must fulfil a set of criteria to ensure bulk quasi properties. One of the most exciting properties of plasma is bulk quasi neutrality. In simple terms, quasi neutrality means

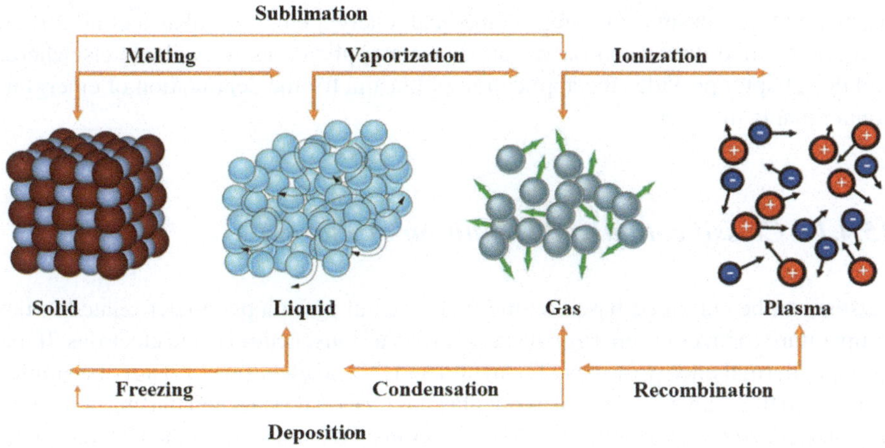

Fig. 15.4 Phase transition diagram highlighting the various states of the matter and underlying processes

that the density of negatively charged species is equal to the density of positively charged species. To achieve quasi neutrality, the smallest dimension (L) defining the volume occupied by the plasma should be much larger than the Debye length (λ_d), i.e. ($L \gg \lambda_d$). The plasma is a collection of charged particles. Therefore, the effective potential due to charge inside plasma at a point is screened Coulomb potential. This screening can be parameterized through the Debye length. The self-screening effect of plasma limits the influence of any given charge particle within the Debye sphere (a sphere with a radius λ_d) and allows quasi neutrality in bulk plasma. The λ_d depends on species temperature and density and can be expressed in Eq. 15.1 (Mozetiˇc et al. 2019).

$$\lambda_d = \sqrt{\frac{\varepsilon_0 k_B T_e}{e^2 n_e}} \tag{15.1}$$

For further details of basic properties, the readers are requested to refer the standard textbooks on introduction to plasma (Fridman 2008; Chen 2003). The plasma state can be characterized through some basic plasma parameters associated with the temperature and density of plasma constituents. Since the electron-induced kinetics is dominant in low-temperature plasma (LTP), the electron temperature (T_e) and electron density (n_e) are crucial plasma parameters. Another key parameter is the gas temperature to ensure the non-thermal nature of the plasma during processing. However, the most exciting feature of the plasma state is that it provides an environment where electrons, positive ions, photons, reactive species (excited molecules and atoms), and radicals can coexist simultaneously. This makes plasma a very suitable candidate for a wide range of applications, including material deposition, etching, surface modification, waste management, material processing, plasma cutting, plasma spraying, biomedical, energy, etc. (Samukawa et al. 2012). A more detailed discussion on various plasma applications is available elsewhere, and this chapter provides the application of plasma for the degradation of emerging contaminants in water.

15.4.1 Classification and Definition of Plasma

Plasma can be classified based on the fundamental plasma parameter related to the temperature (or average energy) of species such as ions, neutrals, and electrons. If the average thermal energy of ions (T_i), neutrals (T_g), and electrons (T_e) are of a similar order, then they can be classified as hot plasmas (HP). The stars and plasma generated in fusion reactors come under HP. Most laboratory plasmas fall in the LTP category, where electron temperature is higher than other heavy mass species' temperatures. The LTPs significantly deviate from equilibrium nature and can also be sustained at ambient pressure and temperature. In few cases where the electron density is very high such as welding or laser-induced plasmas, Local Thermal Equilibrium (LTE) may exist (Mal 2019; Rezaei et al. 2020). Further, in some plasmas, such as surface wave discharges or rotating gliding arcs, the partial local thermal equilibrium may also exist for higher-lying levels due to lower energy spacing (Ananthanarasimhan et al. 2021). Table 15.3 presents the classification of the plasmas commonly described in the literature (Fridman 2008; Yuri et al. 1997).

In recent years, the application domain of non-thermal plasma (NTP) assisted processing is significantly increasing. The major attraction lies in its ability to generate reactive species (ions, metastable, radicals, neutrals) at a relatively lower temperature and often close to room temperature (Shashurin et al. 2008). The generation of such plasmas is challenging as the operating condition are relatively very limited. However, over the years, various schemes to sustain the NTP are proposed.

Table 15.3 The classification of plasma-based on the temperature of its constituents. The information is adapted with modification from Lee et al. (2017)

	Low-temperature		High-temperature
	Non-thermal plasmas	Thermal plasmas	
Species temperature (T_x)	T_i, $T_g \ll T_e$, $T_g \ll 500$ K	$T_i \approx T_g \approx T_e$ $T_g < 10^5$ K	$T_i \approx T_g \approx T_e$ $T_g \gg 10^5$ K
Species densities (n_e)	$\approx 10^{20}/m^3$	$>10^{20}/m^3$	$>10^{20}/m^3$
Classification	Non- thermal plasma (NTP)	Local thermal equilibrium (LTE)	Thermal equilibrium
Examples	DBD's, Corona plasma, glow discharge	Arc torch, laser-induced plasma	Stars, fusion reactors

The following section will overview this NTP generation scheme for water treatment application.

15.4.2 Brief Note of Nonthermal Plasma/Discharges

The critical factor in sustaining the NTP is to limit the collisions between low mass (electron) and heavy mass (ion, neutral) species (Foster 2017). The mass difference leads to different mobilities, and plasma can simultaneously attain different equilibrium temperatures for other mass species. It is worth mentioning that this non-equilibrium regime is susceptible to plasma environment and operating conditions. Therefore, for water treatment application one needs to design the plasma generation scheme carefully so that plasma water interaction (direct or indirect) does not disturb the nonthermal characteristics of the plasma. Typically, a high voltage signal is applied between the two electrodes. When the voltage reaches above the inter-medium breakdown potential, an avalanche of electrons is being produced in the form of streamers. Once the streamers are generated, there are high chances of the thermal runaway due to the electric field application for long durations; this leads to the transition of non-thermal to the thermal regime. Here, the collision frequency of electron-neutral/ion increases and causes gas heating phenomena (Foster 2017; Wang et al. 2013). Thus, through such collisions, a gradual merge can be observed in the electron and gas/ion temperatures. To avoid the transition from non-thermal to thermal plasma or from streamer to arc discharge transition, one can employ a variety of techniques: (1) use of dielectric barriers, (2) shot pulse voltage signal, (3) gas composition, (4) flow rate, (5) variation in frequency, etc. (Bruggeman et al. 2016; Vanraes and Bogaerts 2018).

Over the years, numerous plasma reactors for water treatment applications were reported. Therefore, the classification of these reactors is challenging (Vanraes and Bogaerts 2018). Various criteria can be used to classify these plasmas. For example, based on typical current–voltage characteristics, these plasmas can be classified into three discharge regimes viz. Townsend, glow, and arc discharges (Yuri et al. 1997). Similarly, one can also classify these plasmas based on the applied signal such as DC, low frequency (kHz range), radiofrequency (MHz range), microwave (GHz range), or pulsed wave discharges. In context to plasma water treatment application, the most popular way is to classify the plasma based on the plasma interaction with the water surfaces. In this approach, the plasma can be divide into three categories: (1) Direct (discharge inside the water), (2) Indirect (discharge at the surface of the water without contact), and (3) Hybrid (multi-phase) discharges (Bruggeman et al. 2016). Most of the plasmas employed for the AOP process in water treatment are non-equilibrium discharges that are sustained at high-pressure conditions. Therefore, the present discussion will be limited to such discharge.

In the direct discharge or liquid-phase discharge, the discharge is sustained inside the water. To maintain such discharge, a very high electric field of the order of ~1000 kV/cm is applied between the electrodes. The liquid-phase discharges are

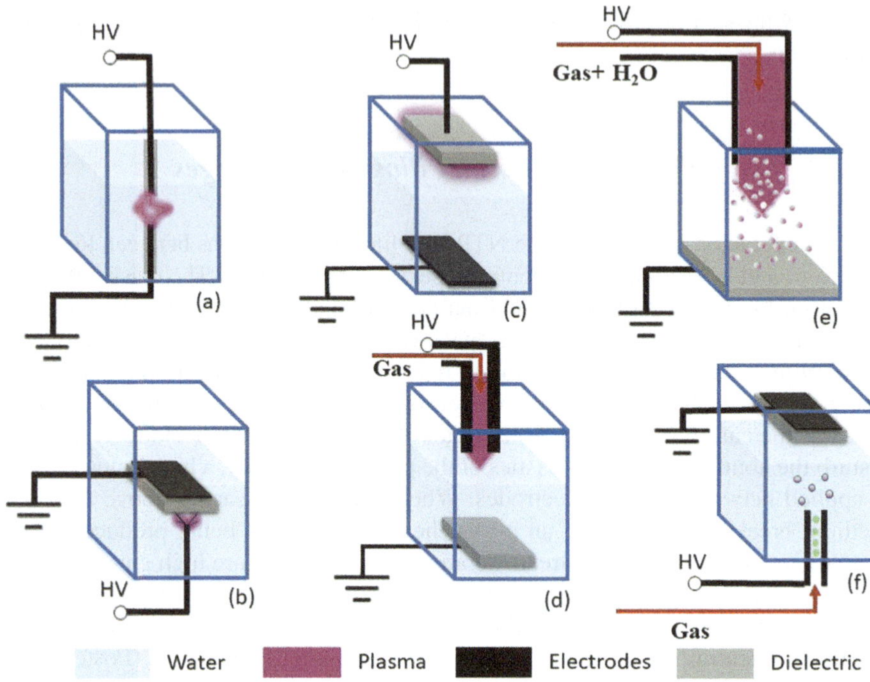

Fig. 15.5 Illustration of different reactors based on plasma water interactions: **a, b** direct discharge (liquid phase discharge); **c, d** indirect discharge (gas phase discharge); **e, f** hybrid discharge (multi-phase discharge). The image is adapted with modifications from Vanraes et al. (2016) (open source subjected to creative common attribution 3.0 license)

energy-intensive, and to limit the heating and electrode erosion, the short pulses of micro-second order are applied. As shown schematically through Fig. 15.5 (a), (b), the pin-to-pin and pin-to-plate are the most common schemes to generate liquid phase discharges. The discharge is highly transient and often comes under the steamer or corona discharge category. Compare to other classes, the electron density in liquid-phase discharge is at least 5 to 6 orders (10^{24}–10^{26}/m^3) higher and also produces 3 to 4 times higher radical density (~10^{24} m^{-3} s for OH radical) (Bruggeman et al. 2016; Locke and Thagard 2012).

In the indirect or gas-phase discharge, the discharge is sustained in contact with water. As shown schematically in Fig. 15.5 (c), (d), the discharge is often created between the electrode plate and water surface or remotely generated above the water surface without direct contact. Normally for water treatment applications, the employed gas-phase discharges are based on dielectric barrier discharge concepts. However, spark or corona discharges can also be operated using pin-water electrode configuration. In recent years, remote discharges such as the atmospheric pressure plasma jets (APPJs) and surface dielectric barrier discharge are becoming popular choices because the discharge characteristics are independent of the properties of the treated water sample. The gas-phase discharge can be sustained by a wide range of high voltage signals ranging from DC, kHz, MHz to GHz frequencies (Foest et al. 2005). Compare to liquid-phase discharges, the gas phase discharges consume low power. Consequently, the electron density is of the order of 10^{18}–10^{21} m^{-3} and the radical density of the order of 10^{19}–10^{23} m^{-3} s (Bobkova et al. 2014; Wang et al. 2021a).

In the hybrid or multi-phase discharge category, the widely used approach is to sustain the discharge in the dispersed gas (Bubbles) inside the liquid. The water treatment can also be achieved by dispersing the water droplets (aerosols) in the gas phase plasmas. The schematic representation of both the approaches is shown in Fig. 15.5 (e), (f). The discharge is formed inside the bubbles by creating a sufficiently high electric field gradient around the bubbles. The studies reported that the discharge could sustain for ~100 ps before the bubble breaks off. The discharge is of steamer type and initiated at the bubble water interface. The electron density in hybrid discharges is reported as high as 10^{22} cm^{-3} (Bruggeman et al. 2010). The reactive species are mainly being produced in gas-phase in both hybrid as well as indirect discharges. One of the major advantages of the hybrid discharge approach is that reactive species are immediately dispersed in the water. The highest oxidizing potential species such as OH radicals are short-lived, and therefore the immediate dispersion can significantly enhance the efficacy of water treatment process.

It is worth mentioning that only brief details of the reactors and underlying discharge mechanism are given here and for further information, the readers are direct to published articles (Bruggeman et al. 2016; Vanraes and Bogaerts 2018). In fact, over the years, hundreds of geometries under the categories mentioned above are reported to address the engineering aspect as well as to improve performance (Samukawa et al. 2012; Vanraes et al. 2016).

15.4.3 Plasma Induced Reactive Species

The advanced oxidation process is achieved by the plasma water interaction, which is the source of the generation of various reactive chemical species (RCS). One of the significant advantages is that the RCS generation does not require any additional consumables. Figure 15.6 schematically shows the generation of RCS through plasma-water interactions (Samukawa et al. 2012). The RCS generated immediately after collision with electrons from the plasma is called primary species. Electrons (e^-), excited neutrals (M^*), ion (M^+), neutrals (M), radicals (•OH, N, O), UV photon, etc. are primary species produced in plasma-gas phase interaction. These species usually have less lifetime. For example, the lifetimes of OH radicals, NO, and O_2^* are 2.7 μs ~1.2 μs, 1.4 μs, and ~1.3 μs respectively (Attri et al. 2015). However, due to higher reactivity rates, these species get quickly consumed, and secondary species are formed such as H_2O_2, NO_2, NO_3, and O_3. The secondary species have relatively long lifetimes and are dissolved in water media. The secondary species radicals further oxides to form tertiary species such as O_3, H_2O_2, NO_3 that can last

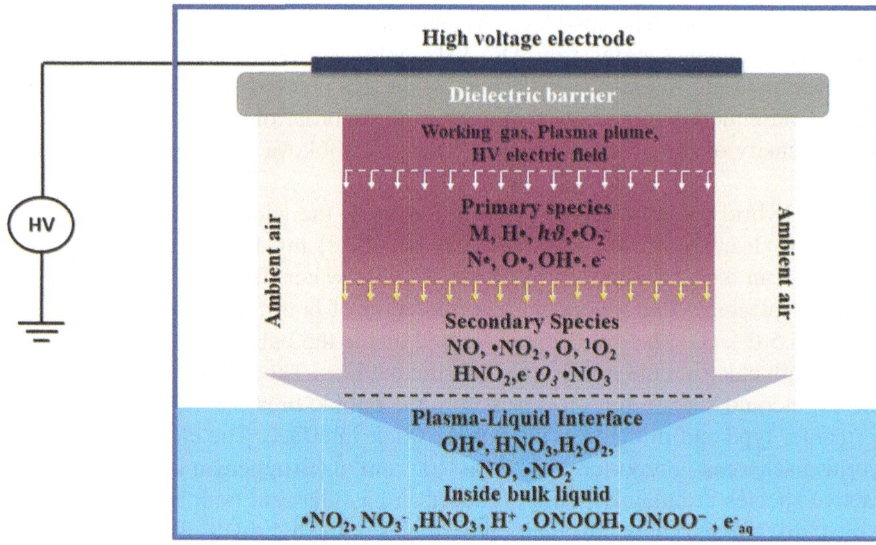

Fig. 15.6 Schematic representation of the formation of reactive chemical species through plasma-water interactions. The image is adapted with modifications from Barjasteh et al. (2021)

up to several days (Kondeti et al. 2018). The RCS generated through plasma water interaction either in gas-phase or inside liquid can broadly be classified into reactive oxygen species (ROS) and reactive nitrogen species (RNS). Sometimes both classes can simultaneously be referred to as reactive nitrogen and oxygen species (RONS).

15.4.3.1 Reactive Oxygen Species

Diatomic oxygen (O_2) contains two unpaired electrons in the valance shell, limiting its scope to react and degrade ECs in water. An electron or energy transfer process can convert O_2 into ROS, atoms or molecules with at least one unpaired electron. These include both radicals and non-radical such as hydroxyl radical (•OH), hydrogen peroxide (H_2O_2), singlet oxygen ($^1O_2$), superoxide anion ($•O_2^-$) and ozone (O_3). (Nosaka and Nosaka 2017). Figure 15.7 represents an overview of various possible reaction routes for ROS production during the plasma-assisted advanced oxidation process. A few of the significant ROS species are discussed below.

Hydroxyl radical (•OH): Hydroxyl radicals are the most potent ROS with the oxidation potential value of 2.8 V, which is only next to the oxidation potential of fluorine (3.03 V). The high oxidising potential and short lifetime (~1 ns) allow OH radicals to react with most organic molecules, including ECs in water and degrade them. In the plasma-mediated treatment of water, the OH radicals are produced by the interaction of high energetic plasma electrons with water molecules by dissociating them into •OH and H• and relatively less energetic secondary electrons. Ozone

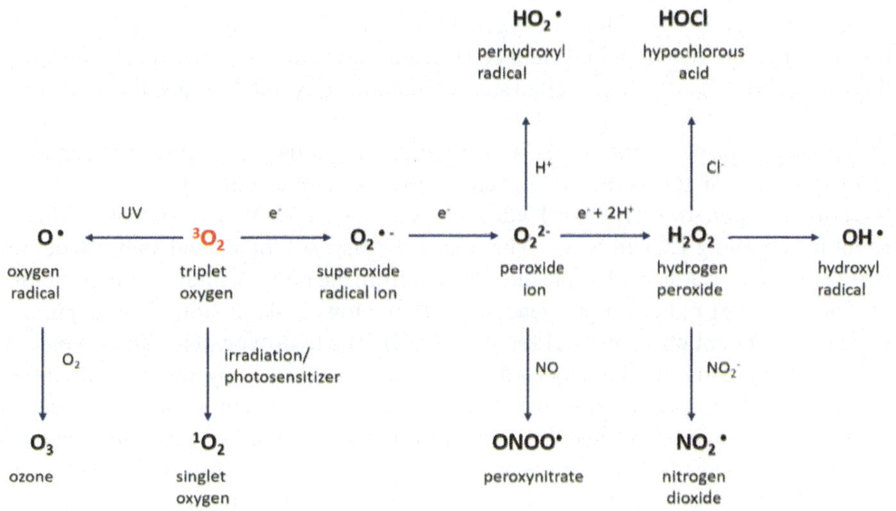

Fig. 15.7 Overview of the mechanisms of RCS production during plasma assisted advanced oxidation process. Reprinted from Mitra et al. (2019)

generated during the plasma process can react with water molecules and produces diatomic oxygen and OH radicals. Plasma also produces UV light along with a variety of RCS. This UV light lies in VUV to UV region and break H_2O_2 into two OH radicals (Mitra et al. 2019).

$$H_2O + e^- \rightarrow \bullet OH + \bullet H + e^- \quad (15.2)$$

$$3O_3 + H_2O \rightarrow 2 \bullet OH + 4O_{2^-} \quad (15.3)$$

$$H_2O_2 + h\nu \rightarrow 2 \bullet OH \quad (15.4)$$

Superoxide anion ($\bullet O_2^-$): The monovalent reduction of diatomic oxygen (O_2) by electron impact ionisation produces a superoxide anion ($\bullet O_2^-$). It is one of the few molecules among the ROS that are considered radical and anion with low reactivity. Due to its unstable nature and low reactivity, it quickly disintegrates into two weak potential reactive species, namely oxygen and hydroperoxide anion (HO_2^-), as shown below (Hayyan et al. 2016).

$$O_2 + e^- \rightarrow \bullet O_{2^-} \quad (15.5)$$

$$2 \bullet O_2 + H_2O_2 \rightarrow O_2 + HO_2^- + OH^- \quad (15.6)$$

Singlet oxygen ($^1O_2$): Singlet oxygen is not radical but highly reactive. It is usually formed when the two unpaired valance electrons attain anti-parallel spin and occupy the same orbital. It is the first excited state of diatomic oxygen (O_2), and the removal of spin restriction makes it a powerful enough oxidant. Singlet oxygen is thus capable of reacting with electrons of all spin orientations, unlike molecular oxygen. The primary source of $^1O_2$ is the photoexcitation process (Wu et al. 2011).

Hydrogen peroxide (H_2O_2): Hydrogen peroxide (1.75 V) is a potent oxidizing agent in removing ECs in water. The relatively high solubility and stability help it participate in reactions with molecules distant from the point of their origin. Its ability to reproduce OH radicals upon reacting with photons make it suitable to degrade a wide range of contaminants (Kehrer et al. 2010). The hydrogen peroxide is typically formed in aqueous media subjected to plasma treatment is by the recombination of OH radicals as per the reaction. They can also form when plasma is produced from gases containing O and H_2, but the route is perilous as both the reactants are highly combustible. When plasma is in direct contact with water, it can entail hydrogen peroxide by the relatively less probable reaction mechanism, as shown below (Nosaka and Nosaka 2017).

$$\bullet OH + \bullet OH \rightarrow H_2O_2 \tag{15.7}$$

$$H_2 + O_2 \rightarrow H_2O_2 \tag{15.8}$$

$$\bullet OH + H_2O^* \rightarrow H_2O_2 + H \tag{15.9}$$

$$H_2O_2 + h\nu \rightarrow \bullet OH + \bullet OH \tag{15.10}$$

Ozone (O_3)*:* Ozone (2.07 V) is a powerful oxidising agent extensively employed in water and wastewater treatment. Ozone is produced through a three-body recombination mechanism where the atomic oxygen is reacted with the diatomic oxygen in the presence of a third colliding body produces ozone, where the electron-O_2 collisions produce the atomic oxygen (Zangouei and Haynes 2019).

$$O + O_2 + M \rightarrow O_3 \tag{15.11}$$

15.4.3.2 Types of Reactive Nitrogen Species

Nitrogen molecules have numerous excited states and metastable states, which often leads to many interactions with the liquid media. As a result, it is not easy to understand the mechanism of nitrogen plasmas or plasmas produced with ambient air. Nitrogen subjected to excitation has high vibrational energies ranging from a few

100–1000 K and less rotational energies lying around room temperature. This distinctive characteristic results from the low probability for collisions between neutral-vibrationally excited N_2 molecules. As the dissociation energies lie well below the excitation energy of N_2, the collision between these exited atoms leads to ionization, resulting in the formation of positive ions and electrons inside the plasma. Due to the richness in exited states and metastable stable states, nitrogen species are formed with oxygen available in ambient air or liquid media. One explanation for this phenomenon is that the dissociation energy of water molecules and oxygen lies par below some nitrogen species' excitation energy (Bradu et al. 2020).

Nitric oxide (•NO)*:* During the gas phase discharge of N_2 in the presence of O_2 or ambient air, the RNS reacts with ROS and leads to the formation of nitric oxide, which is one of the prominent secondary species of nitrogen (Tian and Kushner 2014).The key mechanisms through which •NO produced are given in Eqs. (15.12)–(15.16).

$$N_2 + \bullet O \rightarrow \bullet NO + \bullet N \quad (15.12)$$

$$\bullet N + \bullet O + M \rightarrow \bullet NO + M \quad (15.13)$$

$$\bullet N + O_2 \rightarrow \bullet NO + \bullet O \quad (15.14)$$

$$\bullet N + \bullet OH \rightarrow \bullet NO + H \quad (15.15)$$

$$\bullet N + O_3 \rightarrow \bullet NO + O_2 \quad (15.16)$$

Nitrogen dioxide (•NO_2): In oxygen-rich plasmas, nitric oxide produced can interact with species that can donate oxygen or O-donors such as •OH, •O and O_3 to produce nitrogen radicals. In dry gases, the ozone plays a significant role in nitrogen dioxide formation, whereas in humid gases, it is substantially less (Chen and Davidson 2002; Chen and Wang 2005). The possible chemical pathways of •NO_2 formation are:

$$\bullet NO + \bullet O + M \rightarrow \bullet NO_2^- + M \quad (15.17)$$

$$\bullet NO + O_3 \rightarrow \bullet NO_2 + O_2 \quad (15.18)$$

$$\bullet NO + \bullet HO_2 \rightarrow \bullet NO_2 + \bullet OH \quad (15.19)$$

$$\bullet NO + O_2 \rightarrow \bullet NO_2 + O \quad (15.20)$$

Nitrous and nitric acids (HNO_2, HNO_3): Discharges happening during the humid air or N_2/O_2 mixtures form nitric and nitrous acids through the following reactions (Tas et al. 1997; Chase and Hunt 2002)

$$\bullet NO + \bullet OH + M \rightarrow HNO_2 + M \tag{15.21}$$

$$\bullet NO + \bullet HO_2 + M \rightarrow HNO_3 + M \tag{15.22}$$

$$\bullet NO_2 + \bullet OH + M \rightarrow HNO_3 + M \tag{15.23}$$

$$N_2O_5 + H_2O \rightarrow 2HNO \tag{15.24}$$

Nitrogen species in aqueous media: Alongside the humid air, the interaction of plasmas formed in nitrogen-rich environments can also produce RNS in the liquid–gas phase and inside the liquid phase. The most dominating ones are the nitrogen oxy, and per-oxy molecules, nitric acids, nitrites and nitrates are formed from the interaction of water molecules and oxygen radicals such as •OH, •O with the dissolved nitrogen oxides produced in gas phase interactions (Bruggeman et al. 2008; Parvulescu et al. 2012).The interaction of nitrogen dioxide (•NO_2) species with oxygen radicals such as •OH forms the nitric acid in aqueous media similar to the reaction in the gas phase. Peroxy-nitrous acid can be formed by the interaction of •NO_2 with H_2O. Hydrogen peroxide and several other ROS can also produce peroxy-nitrous acid by reacting with nitrites, nitrogen dioxide and nitric oxide, which are listed below (Tian and Kushner 2014).

$$\bullet NO_{2\,(aq)} + H_2O \rightarrow NO_{2^-(aq)} + NO_{3^-(aq)} + 2H^+ \tag{15.25}$$

$$\bullet NO_{2(aq)} + \bullet NO_{(aq)} + H_2O \rightarrow 2NO_{2^-(aq)} + 2H^+_{(aq)} \tag{15.26}$$

$$\bullet NO_{3(aq)} + \bullet HO_{2(aq)} \rightarrow NO_{3^-(aq)} + O_{2(aq)} + H^+_{(aq)} \tag{15.27}$$

$$\bullet NO_{2(aq)} + \bullet OH_{(aq)} \rightarrow HNO_{3(aq)} \tag{15.28}$$

$$\bullet NO_{2(aq)} + \bullet OH_{(aq)} \rightarrow ONOOH_{(aq)} \tag{15.29}$$

$$\bullet NO_{(aq)} + \bullet HO_{2\,(aq)} \rightarrow ONOOH_{(aq)} \tag{15.30}$$

$$\bullet NO_{(aq)} + \bullet O_{2^-(aq)} \rightarrow ONOO^-_{(aq)} \tag{15.31}$$

15.4.4 Quantification of Reactive Chemical Species

The quantification of whole range of RCS is a challenging task due to uncertainty in formation pathways of these chemical species. So, the quantification of the radicals like •OH, $^1O_2$, H_2O_2, and O_3 through well-known characterisation methods is discussed briefly in the following section.

15.4.4.1 Quantification of OH Radicals

Direct measurement of OH radicals in an aqueous solution is difficult due to its low concentrations and short lifetime. One must rely on relatively more straightforward methods, like the chemical probe method, to quantify the OH radicals. Among all the available techniques, sodium terephthalate as OH radical probe has many advantages. Its symmetric structure forms only a single ring compound, 2- hydroxy terephthalic acid, upon interaction with OH radical, as given in Fig. 15.8. In addition to this, it has low limits of detection, typically 2 nM and strongly fluorescent (Gonzalez et al. 2018; Tampieri et al. 2021).

Another popular technique used to detect the presence of OH radicals is Electron Spin Resonance (ESR). It is a conventional spin trapping method for which DMPO (5,5-dimethyl-1-pyrroline N-oxide) is used as a spin trapping compound. During the plasma treatment of water, the OH radicals react with DMPO to form stable DMPO-OH radicals and can be detected by ESR spectroscopy (Dvoranová et al. 2014).

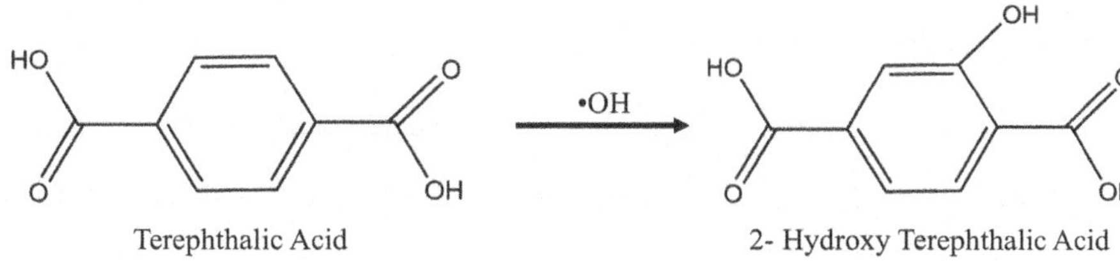

Fig. 15.8 Formation of 2-Hydroxy terephthalic acid from terephthalic acid

15.4.4.2 Quantification of H_2O_2

Hydrogen peroxide is typically quantified by a colorimetric method using an aqueous titanium reagent solution as an indicator. Metal complexes are formed in acidic solution, resulting in the formation of pale-yellow solution whose absorbance can be measured using a UV–Visible spectrophotometer at 410 nm (Eisenberg 1943).

$$Ti^{4+} + H_2O_2 + 2H_2O \rightarrow H_2TiO_4(\text{perititanic acid}) + 4H^+ \qquad (15.32)$$

With titanium-oxy-sulphate solution as a colorimetric probe, other probes such as N, N-diethyl-p-phenylenediamine (DPD) method are based on the oxidation of DPD on reacting with H_2O_2 catalysed using horseradish peroxidase. The reactant DPD generated during the reaction can form long-lasting colour and exhibit an absorption peak at 551 nm (Wu et al. 1999).

15.4.4.3 Quantification of ($^1O_2$)

Sterically hindered cyclic amines such as 4-hydroxy-2,2,6,6- tetramethylpiperidine (HTMP) generate 1-oxyl radical by reacting with singlet oxygen. The reaction produces stable compounds such as 4-hydroxy-2,2,6,6-tetramethylpiperidine 1-oxyl, also called nitroxide radical. The chemical reaction is shown below. This can be detected using ESR spectroscopy (Dimitrijevic et al. 2009; Brezová et al. 2005) (Fig. 15.9).

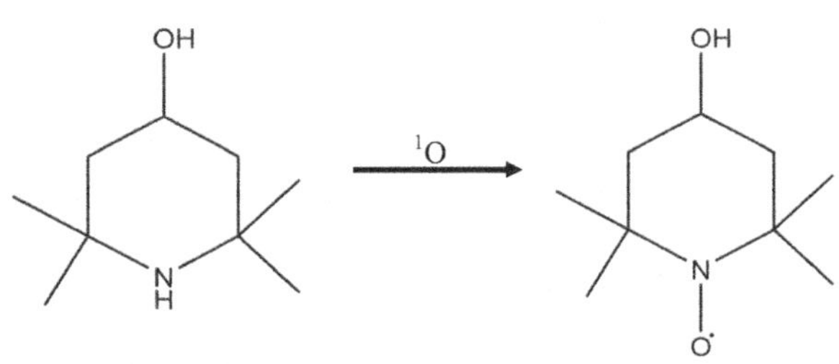

Fig. 15.9 Formation of TEMPOL from HTMP

15.4.4.4 Quantification of O_3

In recent years a rapid detection method for the ozone based on N-diethy1-1,4-phenylen-ediammonium sulphate coupled with spectrophotometry is proposed. When ozone reacts with potassium iodide in a buffered solution (pH 3.35) iodine is formed. This iodine would undergo a reaction with DPD reagent and produces a characteristic pink colour which gives an absorption peak at 510 nm (Song et al. 2000).

$$O_3 + 2I^- + 2H^+ \rightarrow O_2 + I_2 + H_2O \tag{15.33}$$

15.5 Plasma Assisted Degradation Process for ECs

The past decade has witnessed several works from the scientific community to enhance plasmas efficiency, especially NTP, for removing ECs in water. Phenols, azo dyes, pharmaceutical compounds, and pesticides have dominated this spectrum primarily due to their extensive usage in household and commercial purposes. Knowledge of plasma operating parameters like input voltage, electrode distance, flow rate, gas composition, reactor geometry, and type of discharge is essential. The water quality parameters such as pH, conductivity, concentration of organic contaminants, and background ions can also affect the process. The effect of these parameters on the plasma process is discussed in the following subsections.

15.5.1 Effect of Initial Pollutant Concentration

The influence of initial pollutant concentration on degradation kinetics is systematically studied. It is found that the rate of reaction increases with an increase in initial pollutant concentration but fails to keep up when the concentration further increased. The observed decrease in the rate constant at higher pollutant concentrations is attributed to the limited generation of RCS. Wang et al. studied the effect of triclocarban (TCC) concentration on the degradation efficiency in a DBD reactor. They observed a significant improvement in reaction rate when the pollutant concentration is varied from 2 to 7 mg/L. However, the increase in TCC above 7 mg/L did not keep up the trend (Wang et al. 2016). A similar study was performed on Rhodamine B (RhB) solution by Panomsuwan et al. by changing the initial concentration from 1 to 15 mg/L. The first-order kinetic model fits well for the lower concentrations than the higher concentrations (Panomsuwan et al. 2016). The data reveals that there is much competition between dye molecules and their byproducts at higher pollutant concentrations to react with available RCS, which leads to the deviation of the first-order kinetics model.

15.5.2 *Effect of pH*

pH is one of the critical factors that decide the efficiency of the plasma degradation process. Previous studies report a proportional relationship between the pH of the sample and the decomposition rate (Reddy et al. 2013). We can explain this phenomenon based on the change in OH radicals' production when medium pH changes from alkaline to acidic. In an alkaline medium, ozone decomposes to form more powerful OH radicals because of the favorable presence of OH^- ions, as shown in the equations below (Peleg 1976).

$$O_3 + H_2O \rightarrow HO_3^+ + OH^- \tag{15.34}$$

$$HO_3^+ + OH^- \rightarrow 2H_2O \tag{15.35}$$

$$3O_3 + H_2O \rightarrow 2\bullet OH + 4O_2 \tag{15.36}$$

In alkaline conditions, the ozone can be further reacted with water to form a superoxide ion and the recombination of OH radicals to form hydrogen peroxide. In acidic media ozone participates in degradation process more actively compared to basic medium because of relatively higher oxidation potential ranging from 1.4 to 2.8 V. Chen et al. have studied the influence of pH on methylene blue (MB) degradation (Chen et al. 2019). They reported that the shorted lived species have limited penetration inside the water; the long-lived species performed better under acidic conditions. OH radicals dominate the higher pH, but the degradation efficiency is suffered from the recombination of OH radicals to form peroxides. Although the initial pH is high while the experiment is initiated, due to the presence of N_2, O_2 in feeder gas produces secondary and tertiary species like HNO_2, HNO_3 along with peroxy-nitrous acid, which results in the reduction of pH values as low as 2. Peroxy-nitrous acid is a powerful oxidant and very unstable in acidic conditions, thus enhances the degradation of MB.

Ghodbane et al. reported that in acidic conditions, increased efficiency of AB25 dye is due to the protonation of $SO_3{}^-$ results in increased hydrophobic character of the dye molecules, which leads to the greater reactivity with reactive species generating higher degradation rates (Ghodbane et al. 2015; Ghodbane and Hamdaoui 2009). Zhao et al. performed the degradation of p-nitrophenol (PNP) using the microwave ambient air discharge observed that in the first minute of treatment the pH of the solution dropped significantly from 5.5 to 4 (Zhao et al. 2021). It slowed down during the next 11 min of treatment time. Further analysis with IC confirmed that the drop in the pH is due to the formation of tertiary compounds like acetic acid and formic acid when PNP is reacted with secondary species formed from the N_2 discharge interacting with the liquid phase.

Wu et al. have also studied the effect of pH on the degradation of MB using a dielectric barrier discharge and observed that pH of all the samples reached a specific value, which is in the range of pH 2–4 (Wu et al. 2019). It is inferred that the fall in pH is due to the solubility of nitrogen species in water and forming nitrous and nitric acid, as shown in Fig. 15.6.

15.5.3 Effect of Water Conductivity

The effect of the initial conductivity of the water sample significantly affects the discharge propagation, initiation in underwater discharges, and the number density of species. Hamdan et al. studied the effect of water conductivity using microwave underwater bubble discharge using argon as a trace gas (Hamdan et al. 2020). They have reported that the plasma electron density has increased with an increase in the conductivity of water from 10 to 100 μS/cm. No significant change in degradation was observed with a further increase in conductivity of sample from 100 to 1000 μS/cm. However, the further increase in the conductivity did not show any significant impact on the degradation due to the interaction of electrons with ions available in liquid media, leading to the formation of lower oxidative species.

The conductivity of the solution can also be increased by plasma exposure. Rancev et al. investigated the degradation of reactive orange 16 in a nonthermal multipin electrode atmospheric plasma reactor as a function of plasma exposure time and initial dye concentrations (Rancev et al. 2018). The conductivity of the sample was increased from 11.1 to 77.4 μS/cm within a minute of plasma exposure time. In some cases, the sample having high initial concentrations of the pollutant upon subjected to plasma treatment accumulates reactive nitrogen species like nitrite, nitrate ions and peroxy-nitric acid as discussed in the above Sect. 4.3 will increases the system conductivity and shifts the pH towards acidic.

15.5.4 Effect of Feed Gas Composition and Flow Rate

The choice of the feed gas is crucial in the plasma process as it influences the choice of input voltage, reactor configuration, electrode gap, input waveform etc. It can also affect the density of RCS formed during the plasma operation. It is observed that the use of oxygen as feeder gas results better degradation efficiencies followed by noble gas mixtures and ambient air. The flow rate of the feed gas is another important parameter that decides the total number of reactive chemical species being generated at a particular instant of time. The higher the flow rate, the more species will be generated as long as the applied electric field has enough strength to ionize more atoms. This increased species density results in a higher removal rate. Yamatake et al. studied the influence of gas flow at different velocities with inlet diameter changing from 200 to 300 μm (Yamatake et al. 2006). The degradation rate increases with an increase in a gas velocity proportional to the flow rate. The system's efficiency is higher at inlet diameters 200 μm compared to 300 μm because of a substantial increase in gas velocity due to smaller diameters. A possible conclusion from this study is that the generation and lifetime of ROS like •OH and •O are in the range of milliseconds, so they need to be transported fast to achieve faster degradation rates.

15.5.5 Effect of Reactor Configuration and Plasma Discharge

The last decade has witnessed several reactor configurations deployed to study degradation of ECs in terms of power consumption, formation of byproducts, toxicity of plasma-treated water and complete mineralization. It can be observed that DBD reactors have been most successful and leading the battle against the removal of ECs from the water with low power consumption and breaking of ECs into less nontoxic byproducts and time required to mineralize the target contaminant completely.

In summary, numerous reactors under various plasma conditions were employed for water treatment; commonly used configurations are summarized in Fig. 15.5. Further, the reactors mentioned in Table 15.4 are being often utilized for the removal of ECs. The DBD discharges, also called silent discharges, are quite energy-efficient. DBDs can be produced by covering either or both the electrodes using a dielectric material. The role of dielectric is to prevent the transition from glow/Townsend mode to arc mode. A schematic configuration is illustrated in Fig. 15.5 (c) (Magureanu et al. 2011). In DBDs the electron density is of the order of 10^{18}–10^{21} m^{-3} (Bruggeman et al. 2016). Micro-arc discharges both in-liquid and above the liquid surface are energy-intensive. However, they produce a high amount of RCS of the order of ~10^{24} m^{-3} s (Chen et al. 2019). Nevertheless, electrode erosion during this type of direct plasma treatment is a significant concern. A schematic discharge of this category can be seen in Fig. 15.5 (a). Bubble discharges are hybrid discharges and similar to those mentioned in Fig. 15.5 (e). The hybrid discharges are not pure in-liquid or in-gas phase discharges but discharges induced in liquid media inside the gas bubble (Shimizu et al. 2010). In hybrid discharge, the energy consumption is relatively lower than the direct discharges inside the liquid. The electron density in hybrid discharge is in the range of 10^{21}–10^{23} m^{-3} (Bruggeman et al. 2010). Gliding arc discharges are formed between two electrodes exposed to open air, one of which is ground and another connected to a high voltage source. At atmospheric pressure, this discharge can produce high electron densities (10^{19}–10^{21} m^{-3}) (Bruggeman et al. 2016; Du et al. 2007).

Further, some innovative reactor designs were also reported to enhance the degradation efficiency, including plasma-catalyst hybrid processes. Shang et al. has studied the DBD reactor with the addition of persulfate (PS) (Shang et al. 2017). The PS greatly enhanced the degradation efficiency of acid orange 7 (AO7) by almost 60%. The interaction of plasma with PS enhanced the production of •OH and $•SO_4^-$ radicals. It also improved the TOC removal efficiency. Ajo et al. (2017) have proposed

Table 15.4 An overview of various plasma reactors employed in EC removal applications

S. No.	Reactor type	Chemical compound	Concentration (mg/L)	Time (min)	Degradation (%)	Power/voltage	References
1	DBD	Amoxicillin	100	10	100	2 W	Magureanu et al. (2011)
		Dichlorvos	2	8	78.9	80 kV	Sarangapani et al. (2016)
		Crystal violet	100	25	94.2	18 kV	Reddy and Subrahmanyam (2012)
		Carbamazepine	23.6	30	38	250 W	Krause et al. (2009)
2	Micro needle	Atrazine	5	10	100	60 W	Mededovic and Locke (2007)
		Carbofuran	30	10	100	101.5 W	Singh et al. (2016)
		Chlorobenzene	200	12	100	17 kV	Jose and Philip (2019)
		Methyl orange	10	15	84.3	22 kV	Sun et al. (2012)
3	Gliding arc	Paracetamol	200	60	98	10 kV	Slamani et al. (2018)
		Bisphenol	28	30	100	10 kV	Abdelmalek et al. (2006)
		Perfluoroalkyl carboxylates	100	60	98	150 W	Lewis et al. (2020)
		Acid orange	100	20	98	10 kV	Yan et al. (2008)
4	Bubble discharge	Ibuprofen	60	25	82	11.6 kV	Wang et al. (2021b)
		Phenol	100	120	100	21 kV	Wang et al. (2008)
		Aniline	10	60	82.7	38.2 W	Liu et al. (2018)
		Alizarin yellow	30.9	10	100	14.4 kV	Zhou et al. (2021)

a pulsed corona discharge (PCD) reactor consisting of a vertical wire plate configuration where the water is sprayed using five atomizers from the top and plasma directly contact the water. It is reported that the gas–liquid interface is a barrier for the effective decomposition of ECs in aqueous media by OH radical. The generation of OH radicals is possible through the enhanced interaction of plasma-liquid interface through which high energetic plasma electrons directly collide with water molecules to generate •OH. Raji et al. added ZnO nanoparticles in a non-thermal atmospheric pressure plasma jet and compared the efficiency under various feed gases like oxygen, argon, and ambient air. The degradation of their pharmaceutical compound valsartan has decomposed more effectively by the addition of ZnO nanoparticles (Raji et al. 2020).

15.5.6 Effect of the Electrode Gap

An optimal electrode gap is crucial to achieving the maximum degradation capacity of the given configuration. The higher or lower gap can have adverse effects on degradation efficiency. Sugiarto and Sato have investigated the impact of electrode gap in removing phenols from water in the needle to plate configuration (Sugiarto and Sato 2001). Three configurations with 45, 15 and, 6 mm electrode gaps were investigated, and the 6 mm electrode gap yielded the highest removal rate. The formation of multiple small steamers to single spark transition creates shock waves, intense UV radiation, and high-energy electrons produce copious amounts of radicals. At 45 mm electrode gap, it is observed that there are multiple streamers, yielding less amount of RCS and lower degradation efficiency. Shang et al. studied the effect of different air discharge gaps in the degradation of acid orange 7 (Shang et al. 2017). The study reported that the degradation efficiency increases as the electrode gap increases from 2 to 4 mm. The further increase in the gap from 4 to 6 mm showed no significant improvement in the degradation efficiency. It is worth mentioning that at a low electrode gap, the discharge can significantly disturb by the water surface turbulence. Chen et al. has pointed out cathode fall as an important factor in producing reactive species (Chen et al. 2019).

15.5.7 Effect of Input Power

Several studies have been conducted to explore the input power or voltage influence on the degradation efficiency of ECs in water. As the input voltage increases, it is reported that the overall degradation efficiency increases. Energy yield increases proportion to the applied voltage up to a certain point and decreases, because production of reactive species reaches to saturation stage which cannot be further enhanced unless external parameters like feeder gas composition and flow rate are varied. The increase in removal rate can be attributed to high electric field strength and associated increase in electron temperature. High collision rates with the gas molecules lead to the rise in the concentration of reactive species, high-intensity UV-light and induced ionization of oxygen and water molecules.

Wu et al. studied the removal of tetracycline antibiotics (TCs) from water using a DBD reactor (Wu et al. 2021). The reactor was operated at different voltages (12, 15 and 18 kV) to investigate the influence of voltage on degradation kinetics of TC. The degradation efficiency increases proportionally to the rise in voltage. Reddy et al. also studied methylene blue (MB) removal using a non-thermal DBD discharge at different operating voltages (14, 16 and 18 kV). It is observed that 90% decolorization is achieved with an initial concentration of MB 100 mg/L in the first 25 min of plasma exposure at an applied voltage of 14 kV. The efficiency slightly increased to 94% as the voltage increased to 18 kV. The TOC removal % also slightly improved from 18 to 20% (Reddy et al. 2013). Pankaj et al. also reported the degradation kinetics of various dyes like methyl red, crystal violet and fast green FCF by deploying a high voltage atmospheric air and modified air cold plasma (Pankaj et al. 2017). It was reported that all the dyes showed higher removal rates at a maximum applied voltage of 80 kV.

15.6 Conclusion and Future Prospects

The widespread occurrence of emerging contaminants in water bodies is a serious concern in the water sector worldwide. Unlike conventional water pollutants, these contaminants are complex, diverse in their physicochemical properties. As a result, traditional water and wastewater treatment processes are ineffective in removing emerging contaminants from water. However, many of these compounds are potentially toxic to all life forms, and their removal from water is imperative to protect public health and the ecosystem. Considering the various physicochemical properties and complex fate and transport, a thorough understanding of various aspects of emerging contaminants, including their source, occurrence, classification, fate and transport, detection method, and treatment techniques, are essential in developing an effective management strategy.

The advanced oxidation process is a promising treatment technology for the removal of ECs in water. Among advanced oxidation processes, NTP looks attractive due to its ability to generate RCS at a controlled rate and ambient conditions. The process is also attractive due to its ability to produce reactive species like ROS and RONS, non-selective behaviour towards target pollutants, and sludge-free operation. However, the application of NTP in removing organic contaminants is mainly limited to small-scale laboratory experiments. The biggest challenge in scale-up is high energy consumption and lack of information on the process that dictates liquid plasma interaction. To overcome these issues, new models that include all the effects ranging from transport phenomenon, fluid dynamics effects, complex plasma chemistry, and fundamental collision process need to be set up, explaining the plasma liquid interaction phenomenon alongside the experimental observations.

However, the plasma modelling studies with experimental validation are limited. The plasma diagnostics in the liquid phase are also challenging due to the interference of many species; most of the available methods are not selective. Moreover, physical techniques like optical emission spectroscopy (OES) and optical absorption spectroscopy (OAS) are limited to estimating the species densities in the gas phase. The liquid surrounding the plasma or species poses a severe challenge in quantification using spectroscopic techniques. Besides, radical transport in liquid and liquid surfaces is still unknown. The effect of UV radiation and shockwaves produced by plasma on intermediate compounds and the plasma decomposition of water is clueless. For plasma technology to become dominant, these complex problems have to be addressed.

Acknowledgements Authors gratefully acknowledge the Government of India, Ministry of Science and Technology, Department of Science and Technology (DST), Technology Mission Division [Grant no: DST/TM/WTI/WIC/2K17/82(C)] and SERB-DST New Delhi, Government of India for financial support through Grant Nos. CRG/2018/000419 and CVD/2020/000458 for supporting the work. All the authors also thank IIT Tirupati for the support.

References

Abdelmalek F, Ghezzar MR, Belhadj M, Addou A, Brisset JL (2006) Bleaching and degradation of textile dyes by nonthermal plasma process at atmospheric pressure. Ind Eng Chem Res 45:23–29. https://doi.org/10.1021/ie050058s

Ajo P, Kornev I, Preis S (2017) Pulsed corona discharge induced hydroxyl radical transfer through the gas-liquid interface. Sci Rep 7:1–6. https://doi.org/10.1038/s41598-017-16333-1

Almomani FA, Shawaqfah M, Bhosale RR, Kumar A (2016) Removal of emerging pharmaceuticals from wastewater by ozone-based advanced oxidation processes. Environ Prog Sustain Energy 35:982–995. https://doi.org/10.1002/EP.12306

Ananthanarasimhan J, Gangwar RK, Leelesh P, Srikar PSNSR, Shivapuji AM, Rao L (2021) Estimation of electron density and temperature in an argon rotating gliding arc using optical and electrical measurements. J Appl Phys 129:223301. https://doi.org/10.1063/5.0044014

Asghar A, Raman AAA, Daud WMAW (2015) Advanced oxidation processes for in-situ production of hydrogen peroxide/hydroxyl radical for textile wastewater treatment: a review. J Clean Prod 87:826–838. https://doi.org/10.1016/J.JCLEPRO.2014.09.010

Attri P, Kim YH, Park DH, Park JH, Hong YJ, Uhm HS, Kim KN, Fridman A, Choi EH (2015) Generation mechanism of hydroxyl radical species and its lifetime prediction during the plasma-initiated ultraviolet (UV) photolysis. Sci Rep 5:1–8. https://doi.org/10.1038/srep09332

Bagal MV, Gogate PR (2014) Wastewater treatment using hybrid treatment schemes based on cavitation and fenton chemistry: a review. Ultrason Sonochem 21:1–14. https://doi.org/10.1016/J.Ultsonch.2013.07.009

Barjasteh A, Dehghani Z, Lamichhane P, Kaushik N, Choi EH, Kaushik NK (2021) Recent progress in applications of non-thermal plasma for water purification, bio-sterilization, and decontamination. Appl Sci 11:3372 11:3372 . https://doi.org/10.3390/APP11083372

Bobkova ES, Smirnov SA, Zalipaeva Y V., Rybkin V V. (2014) Modeling chemical composition for an atmospheric pressure dc discharge in air with water cathode by 0-D model. Plasma Chem Plasma Process 34:721–743. https://doi.org/10.1007/s11090-014-9539-z

Bolong N, Ismail AF, Salim MR, Matsuura T (2009) A review of the effects of emerging contaminants in wastewater and options for their removal. Desalination 239:229–246. https://doi.org/10.1016/j.desal.2008.03.020

Boretti A, Rosa L (2019) Reassessing the projections of the world water development report. npj Clean Water 21(2):1–6 . https://doi.org/10.1038/s41545-019-0039-9

Bradu C, Kutasi K, Magureanu M, Puač N, Živković S (2020) Reactive nitrogen species in plasma-activated water: generation, chemistry and application in agriculture. J Phys D Appl Phys 53. https://doi.org/10.1088/1361-6463/ab795a

Brausch JM, Rand GM (2011) A review of personal care products in the aquatic environment: environmental concentrations and toxicity. Chemosphere 82:1518–1532. https://doi.org/10.1016/j.chemosphere.2010.11.018

Brezová V, Gabčová S, Dvoranová D, Staško A (2005) Reactive oxygen species produced upon photoexcitation of sunscreens containing titanium dioxide (an EPR study). J Photochem Photobiol B Biol 79:121–134. https://doi.org/10.1016/j.jphotobiol.2004.12.006

Brillas E, Sirés I, Oturan MA (2009) Electro-fenton process and related electrochemical technologies based on Fenton's reaction chemistry. Chem Rev 109:6570–6631. https://doi.org/10.1021/cr900136g

Bruggeman P, Degroote J, Vierendeels J, Leys C (2008) Dc electrical breakdown between a metal electrode and a water surface. In: GD 2008—17th Int Conf Gas Discharges Their Appl 36:321–324

Bruggeman P, Verreycken T, González MÁ, Walsh JL, Kong MG, Leys C, Schram DC (2010) Optical emission spectroscopy as a diagnostic for plasmas in liquids: opportunities and pitfalls. J Phys D Appl Phys 43:124005. https://doi.org/10.1088/0022-3727/43/12/124005

Bruggeman PJ, Kushner MJ, Locke BR, Gardeniers JGE, Graham WG, Graves DB, Hofman-Caris RCHM, Maric D, Reid JP, Ceriani E, Fernandez Rivas D, Foster JE, Garrick SC, Gorbanev Y, Hamaguchi S, Iza F, Jablonowski H, Klimova E, Kolb J, Krcma F, Lukes P, MacHala Z, Marinov I, Mariotti D, Mededovic Thagard S, Minakata D, Neyts EC, Pawlat J, Petrovic ZL, Pflieger R, Reuter S, Schram DC, Schröter S, Shiraiwa M, Tarabová B, Tsai PA, Verlet JRR, Von Woedtke T, Wilson KR, Yasui K, Zvereva G (2016) Plasma-liquid interactions: a review and roadmap. Plasma Sourc Sci Technol 25:053002. https://doi.org/10.1088/0963-0252/25/5/053002

Carmona E, Andreu V, Picó Y (2014) Occurrence of acidic pharmaceuticals and personal care products in Turia river Asin: from waste to drinking water. Sci Total Environ 484:53–63. https://doi.org/10.1016/J.SCITOTENV.2014.02.085

Chase WJ, Hunt JW (2002) Solvation time of the electron in polar liquids. Water and alcohols. J Phys Chem 79:2835–2845 . https://doi.org/10.1021/J100593A007

Chen, Francis F. JPC (2003) Principles of plasma processing

Chen J, Davidson JH (2002) Ozone production in the positive DC corona discharge: model and comparison to experiments. Plasma Chem Plasma Process 22:495–522. https://doi.org/10.1023/A:1021315412208

Chen J, Wang P (2005) Effect of relative humidity on electron distribution and ozone production by DC coronas in air. IEEE Trans Plasma Sci 33:808–812. https://doi.org/10.1109/TPS.2005.844530

Chen Q, He B, Ma Y, Wang X, Xiong Q, Li J, Liu QH (2019) Influence of the pH value on the degradation of an azo dye of methyl orange by air discharge plasma. Plasma Process Polym 16:1800152. https://doi.org/10.1002/ppap.201800152

Covaci A, Geens T, Roosens L, Ali N, Eede Van den N, Ionas AC, Malarvannan G, Dirtu AC (2012) Human exposure and health risks to emerging organic contaminants, 20:243–305. https://doi.org/10.1007/978-3-642-28132-7

Cristale J, García Vázquez A, Barata C, Lacorte S (2013) Priority and emerging flame retardants in rivers: Occurrence in water and sediment, daphnia magna toxicity and risk assessment. Environ Int 59:232–243. https://doi.org/10.1016/J.ENVINT.2013.06.011

da Silva CP, Emídio ES (2015) Marchi MRR de (2015) The occurrence of UV filters in natural and drinking water in São Paulo State (Brazil). Environ Sci Pollut Res 2224(22):19706–19715. https://doi.org/10.1007/S11356-015-5174-3

Dantas RF, Sans C, Esplugas S (2011) Ozonation of propranolol: transformation, biodegradability, and toxicity assessment. J Environ Eng 137:754–759. https://doi.org/10.1061/(ASCE)EE.1943-7870.0000377

De Gerónimo E, Aparicio VC, Bárbaro S, Portocarrero R, Jaime S, Costa JL (2014) Presence of pesticides in surface water from four sub-basins in Argentina. Chemosphere 107:423–431. https://doi.org/10.1016/J.CHEMOSPHERE.2014.01.039

De la Cruz N, Esquius L, Grandjean D, Magnet A, Tungler A, de Alencastro LF, Pulgarín C (2013) Degradation of emergent contaminants by UV, UV/H_2O_2 and neutral photo-Fenton at pilot scale in a domestic wastewater treatment plant. Water Res 47:5836–5845. https://doi.org/10.1016/J.WATRES.2013.07.005

Debnath D, Gupta AK, Ghosal PS (2019) Recent advances in the development of tailored functional materials for the treatment of pesticides in aqueous media: a review. J Ind Eng Chem 70:51–69. https://doi.org/10.1016/j.jiec.2018.10.014

Dewalque L, Charlier C, Pirard C (2014) Estimated daily intake and cumulative risk assessment of phthalate diesters in a Belgian general population. Toxicol Lett 231:161–168. https://doi.org/10.1016/j.toxlet.2014.06.028

Dey S, Bano F, Malik A (2019) Pharmaceuticals and personal care product (PPCP) contamination-a global discharge inventory. Elsevier Inc.

Dimitrijevic NM, Rozhkova E, Rajh T (2009) Dynamics of localized charges in dopamine-modified TiO and their effect on the formation of reactive oxygen species. J Am Chem Soc 131:2893–2899. https://doi.org/10.1021/ja807654k

Dindarsafa M, Khataee A, Kaymak B, Vahid B, Karimi A, Rahmani A (2017) Heterogeneous sono-Fenton-like process using martite nanocatalyst prepared by high energy planetary ball milling for treatment of a textile dye. Ultrason Sonochem 34:389–399. https://doi.org/10.1016/J.ULTSONCH.2016.06.016

Diya'uddeen BH, A.R. AA, Daud WMAW (2012) On the limitation of Fenton oxidation operational parameters: a review. Int J Chem React Eng 10. https://doi.org/10.1515/1542-6580.2913

Du CM, Yan JH, Cheron BG (2007) Degradation of 4-chlorophenol using a gas-liquid gliding arc discharge plasma reactor. Plasma Chem Plasma Process 27:635–646. https://doi.org/10.1007/s11090-007-9092-0

Dvoranová D, Barbieriková Z, Brezová V (2014) Radical intermediates in photoinduced reactions on TiO_2 (An EPR spin trapping study). Molecules 19:17279–17304. https://doi.org/10.3390/molecules191117279

Eisenberg GM (1943) Colorimetric determination of hydrogen peroxide. Ind Eng Chem Anal Ed 15:327–328. https://doi.org/10.1021/i560117a011

Epa U, of Water O, of Groundwater O, Water D (2009) Office of water (4607M) EPA EPA 815F09001 fact sheet: final third drinking water contaminant candidate list (CCL 3)

Foest R, Kindel E, Ohl A, Stieber M, Weltmann KD (2005) Non-thermal atmospheric pressure discharges for surface modification. Plasma Phys Control Fusion 47. https://doi.org/10.1088/0741-3335/47/12B/S38

Foster JE (2017) Plasma-based water purification: Challenges and prospects for the future. Phys Plasmas 24:055501. https://doi.org/10.1063/1.4977921

Fridman A (2008) Plasma Chemistry (Cambridge 2008)-Cambridge University Press (2008), 1st ed. Cambridge University Press

Gani KM, Kazmi AA (2016) Contamination of emerging contaminants in Indian aquatic sources: first overview of the situation. J Hazard Toxic Radioact Waste 21:04016026. https://doi.org/10.1061/(ASCE)HZ.2153-5515.0000348

Ghodbane H, Hamdaoui O (2009) Degradation of acid blue 25 in aqueous media using 1700 kHz ultrasonic irradiation: ultrasound/Fe(II) and ultrasound/H_2O_2 combinations. Ultrason Sonochem 16:593–598. https://doi.org/10.1016/j.ultsonch.2008.11.006

Ghodbane H, Hamdaoui O, Vandamme J, Van Durme J, Vanraes P, Leys C, Nikiforov AY (2015) Degradation of AB25 dye in liquid medium by atmospheric pressure non-thermal plasma and plasma combination with photocatalyst TiO_2. Open Chem 13:325–331. https://doi.org/10.1515/chem-2015-0040

Gligorovski S, Strekowski R, Barbati S, Vione D (2018) Addition and correction to environmental implications of hydroxyl radicals (•OH). Chem Rev 118:2296. https://doi.org/10.1021/ACS.CHEMREV.7B00742

Gong J, Duan D, Yang Y, Ran Y, Chen D (2016) Seasonal variation and partitioning of endocrine disrupting chemicals in waters and sediments of the pearl river system, south china. Environ Pollut 219:735–741. https://doi.org/10.1016/J.ENVPOL.2016.07.015

Gonzalez DH, Kuang XM, Scott JA, Rocha GO, Paulson SE (2018) Terephthalate probe for hydroxyl radicals: Yield of 2-hydroxyterephthalic acid and transition metal interference. Anal Lett 51:2488–2497. https://doi.org/10.1080/00032719.2018.1431246

Gonzalez-Rey M, Tapie N, Le Menach K, Dévier MH, Budzinski H, Bebianno MJ (2015) Occurrence of pharmaceutical compounds and pesticides in aquatic systems. Mar Pollut Bull 96:384–400. https://doi.org/10.1016/J.MARPOLBUL.2015.04.029

Gorbanev Y, Privat-Maldonado A, Bogaerts A (2018) Analysis of short-lived reactive species in plasma-air-water systems: the dos and the do nots. Anal Chem 90:13151–13158. https://doi.org/10.1021/acs.analchem.8b03336

Hamdan A, Profili J, Cha MS (2020) Microwave plasma jet in water: effect of water electrical conductivity on plasma characteristics. Plasma Chem Plasma Process 40:169–185. https://doi.org/10.1007/s11090-019-10034-5

Hayyan M, Hashim MA, Alnashef IM (2016) Superoxide ion: generation and chemical implications. Chem Rev 116:3029–3085. https://doi.org/10.1021/acs.chemrev.5b00407

Huerta-Fontela M, Galceran MT, Ventura F (2011) Occurrence and removal of pharmaceuticals and hormones through drinking water treatment. Water Res 45:1432–1442. https://doi.org/10.1016/J.WATRES.2010.10.036

Itikawa Y, Mason N (2005) Cross sections for electron collisions with water molecules. J Phys Chem Ref Data 34:1–22. https://doi.org/10.1063/1.1799251

Lewis J, Thomas Joyce A, Hadaya M, Ebrahimi F, Dragiev I, Giardetti N, Yang J, Fridman G, Rabinovich A, Fridman A, McKenzie R, EM Sales C (2020) Rapid degradation of PFAS in aqueous solutions by reverse vortex flow gliding arc plasma. Environ Sci Water Res Technol 6:1044–1057. https://doi.org/10.1039/C9EW01050E

Jose J, Philip L (2019) Degradation of chlorobenzene in aqueous solution by pulsed power plasma: mechanism and effect of operational parameters. J Environ Chem Eng 7:103476. https://doi.org/10.1016/j.jece.2019.103476

Joseph CG, Li Puma G, Bono A, Krishnaiah D (2009) Sonophotocatalysis in advanced oxidation process: a short review. Ultrason Sonochem 16:583–589. https://doi.org/10.1016/J.ULTSONCH.2009.02.002

Kasprzyk-Hordern B, Dinsdale RM, Guwy AJ (2008) The occurrence of pharmaceuticals, personal care products, endocrine disruptors and illicit drugs in surface water in South Wales, UK. Water Res 42:3498–3518. https://doi.org/10.1016/j.watres.2008.04.026

Kehrer JP, Robertson JD, Smith CV (2010) Free radicals and reactive oxygen species. Compr Toxicol Second Ed 1–14:277–307. https://doi.org/10.1016/B978-0-08-046884-6.00114-7

Khetan SK, Collins TJ (2007) Human pharmaceuticals in the aquatic environment: a challenge to green chemisty. Chem Rev 107:2319–2364. https://doi.org/10.1021/cr020441w

Kim U-J, Kannan K (2018) Occurrence and distribution of organophosphate flame retardants/plasticizers in surface waters, tap water, and rainwater: Implications for human exposure. Environ Sci Technol 52:5625–5633. https://doi.org/10.1021/ACS.EST.8B00727

Klamerth N, Malato S, Maldonado MI, Agüera A, Fernández-Alba AR (2010) Application of photo-fenton as a tertiary treatment of emerging contaminants in municipal wastewater. Environ Sci Technol 44:1792–1798. https://doi.org/10.1021/es903455p

Kleywegt S, Pileggi V, Yang P, Hao C, Zhao X, Rocks C, Thach S, Cheung P, Whitehead B (2011) Pharmaceuticals, hormones and bisphenol A in untreated source and finished drinking water in

Ontario, Canada—occurrence and treatment efficiency. Sci Total Environ 409:1481–1488. https://doi.org/10.1016/j.scitotenv.2011.01.010

Kondeti VSSK, Phan CQ, Wende K, Jablonowski H, Gangal U, Granick JL, Hunter RC, Bruggeman PJ (2018) Long-lived and short-lived reactive species produced by a cold atmospheric pressure plasma jet for the inactivation of *Pseudomonas aeruginosa* and *Staphylococcus aureus*. Free Radic Biol Med 124:275–287. https://doi.org/10.1016/j.freeradbiomed.2018.05.083

Krause H, Schweiger B, Schuhmacher J, Scholl S, Steinfeld U (2009) Degradation of the endocrine disrupting chemicals (EDCs) carbamazepine, clofibric acid, and iopromide by corona discharge over water. Chemosphere 75:163–168. https://doi.org/10.1016/J.CHEMOSPHERE.2008.12.020

Le Thi MT, Nguyen Phuoc D, Dinh Quoc T, Ngo HH, Do Hong Lan C (2016) Presence of e-EDCs in surface water and effluents of pollution sources in Sai Gon and Dong Nai river basin. Sustain Environ Res 26:20–27. https://doi.org/10.1016/J.SERJ.2015.09.001

Lee S, Jeong W, Kannan K, Moon HB (2016) Occurrence and exposure assessment of organophosphate flame retardants (OPFRs) through the consumption of drinking water in Korea. Water Res 103:182–188. https://doi.org/10.1016/J.WATRES.2016.07.034

Lee J, Lee CW, Yong HI, Lee HJ, Jo C, Jung S (2017) Use of atmospheric pressure cold plasma for meat industry. Korean J Food Sci Anim Resour 37:477–485. https://doi.org/10.5851/kosfa.2017.37.4.477

Lei M, Zhang L, Lei J, Zong L, Li J, Wu Z, Wang Z (2015) Overview of emerging contaminants and associated human health effects. Biomed Res Int 2015:1–12. https://doi.org/10.1155/2015/404796

Lin X, Xu J, Keller AA, He L, Gu Y, Zheng W, Sun D, Lu Z, Huang J, Huang X, Li G (2020) Occurrence and risk assessment of emerging contaminants in a water reclamation and ecological reuse project. Sci Total Environ 744:140977. https://doi.org/10.1016/j.scitotenv.2020.140977

Liu Y, Zhang H, Sun J, Liu J, Shen X, Zhan J, Zhang A, Ognier S, Cavadias S, Li P (2018) Degradation of aniline in aqueous solution using non-thermal plasma generated in microbubbles. Chem Eng J 345:679–687. https://doi.org/10.1016/J.CEJ.2018.01.057

Locke BR, Thagard SM (2012) Analysis and review of chemical reactions and transport processes in pulsed electrical discharge plasma formed directly in liquid water. Plasma Chem Plasma Process 32:875–917. https://doi.org/10.1007/s11090-012-9403-y

Magureanu M, Mandache NB, Parvulescu VI (2015) Degradation of pharmaceutical compounds in water by non-thermal plasma treatment. Water Res 81:124–136. https://doi.org/10.1016/j.watres.2015.05.037

Magureanu M, Piroi D, Mandache NB, David V, Medvedovici A, Bradu C, Parvulescu VI (2011) Degradation of antibiotics in water by non-thermal plasma treatment. Water Res 45:3407–3416. https://doi.org/10.1016/j.watres.2011.03.057

Mahamuni NN, Adewuyi YG (2010) Advanced oxidation processes (AOPs) involving ultrasound for waste water treatment: a review with emphasis on cost estimation. Ultrason Sonochem 17:990–1003. https://doi.org/10.1016/J.ULTSONCH.2009.09.005

Mal E (2019) Laser induced plasma studies on W, Mo and alloys of Cu in air via time and space resolved LIBS and application in single line calibration free -LIBS. Doctoral dissertation, Indian Institute of Technology Guwahati

Mededovic S, Locke BR (2007) Side-chain degradation of atrazine by pulsed electrical discharge in water. Ind Eng Chem Res 46:2702–2709. https://doi.org/10.1021/IE070020A

Mitra S, Nguyen LN, Akter M, Park G, Choi EH, Kaushik NK (2019) Impact of ROS generated by chemical, physical, and plasma techniques on cancer attenuation. Cancers 11: 1030 11:1030 . https://doi.org/10.3390/CANCERS11071030

Mondal S, Purkait MK, De S (2018) Advances in dye removal technologies. Springer Singapore, Singapore

Montes-Grajales D, Fennix-Agudelo M, Miranda-Castro W (2017) Occurrence of personal care products as emerging chemicals of concern in water resources: a review. Sci Total Environ 595:601–614. https://doi.org/10.1016/J.SCITOTENV.2017.03.286

Mozetiˇc M, Vesel A, Gregor P, Zaplotnik R (2019) Introduction to plasma and plasma diagnostics. In: Non-thermal plasma technology for polymeric materials, 1st ed. Elsevier, p 459
Nath A, Vendan SE, Priyanka JKS, Singh CK, Kumar S (2014) Carcinogenic pesticides residue detection in cow milk and water samples from Patna, India. Curr Trends Biotechnol Chem Res 3:1–7
Nippatla N, Philip L (2019) Electrocoagulation-floatation assisted pulsed power plasma technology for the complete mineralization of potentially toxic dyes and real textile wastewater. Process Saf Environ Prot 125:143–156. https://doi.org/10.1016/J.PSEP.2019.03.012
Nosaka Y, Nosaka AY (2017) Generation and detection of reactive oxygen species in photocatalysis. Chem Rev 117:11302–11336. https://doi.org/10.1021/acs.chemrev.7b00161
Pankaj SK, Wan Z, Keener KM (2018) Effects of cold plasma on food quality: a review. Foods 7:1–21. https://doi.org/10.3390/foods7010004
Pankaj SK, Wan Z, Colonna W, Keener KM (2017) Degradation kinetics of organic dyes in water by high voltage atmospheric air and modified air cold plasma. Water Sci Technol 76:567–574. https://doi.org/10.2166/wst.2017.169
Panomsuwan G, Morishita T, Kang J, Rujiravanit R, Ueno T, Saito N (2016) Degradation of synthetic dye in water by solution plasma process. J Korean Soc Mar Eng 40:888–893. https://doi.org/10.5916/jkosme.2016.40.10.888
Parvulescu VI, Magureanu M, Lukes P (2012) Plasma chemistry and catalysis in gases and liquids. Plasma Chem Catal Gases Liq. https://doi.org/10.1002/9783527649525
Peleg M (1976) The chemistry of ozone in the treatment of water. Water Res 10:361–365. https://doi.org/10.1016/0043-1354(76)90052-X
Petrie B, Barden R, Kasprzyk-Hordern B (2015) A review on emerging contaminants in wastewaters and the environment: current knowledge, understudied areas and recommendations for future monitoring. Water Res 72:3–27. https://doi.org/10.1016/j.watres.2014.08.053
Picó Y, Campo J, Alfarhan AH, El-Sheikh MA, Barceló D (2021) A reconnaissance study of pharmaceuticals, pesticides, perfluoroalkyl substances and organophosphorus flame retardants in the aquatic environment, wild plants and vegetables of two Saudi Arabia urban areas: environmental and human health risk assessment. Sci Total Environ 776:145843. https://doi.org/10.1016/J.SCITOTENV.2021.145843
Qiao T, Zhengrong Yu, Xihui Zhang TAuDW (2011) Occurrence and fate of pharmaceuticals and personal care products in drinking water in southern China. J Environ Monit 13:3097–3103. https://doi.org/10.1039/C1EM10318K
Radwan EK, Ibrahim MBM, Adel A, Farouk M (2019) The occurrence and risk assessment of phenolic endocrine-disrupting chemicals in Egypt's drinking and source water. Environ Sci Pollut Res 272(27):1776–1788. https://doi.org/10.1007/S11356-019-06887-0
Raji A, Pandiyaraj KN, Vasu D, Ramkumar MC, Deshmukh RR, Kandavelu V (2020) Non-equilibrium atmospheric pressure plasma assisted degradation of the pharmaceutical drug valsartan: Influence of catalyst and degradation environment. RSC Adv 10:35709–35717. https://doi.org/10.1039/d0ra05608a
Rancev S, Petrovic M, Bojic A, Radivojevic D, Maluckov C, Radovic M (2018) Degradation of reactive orange 16 using a prototype atmospheric-pressure non-thermal plasma reactor. Facta Univ Ser Phys Chem Technol 16:285–295. https://doi.org/10.2298/fupct1803285r
Rathi BS, Kumar PS, Show PL (2021) A review on effective removal of emerging contaminants from aquatic systems: current trends and scope for further research. J Hazard Mater 409:124413. https://doi.org/10.1016/j.jhazmat.2020.124413
Reddy PMK, Subrahmanyam C (2012) Green approach for wastewater treatment—degradation and mineralization of aqueous organic pollutants by discharge plasma. Ind Eng Chem Res 51:11097–11103. https://doi.org/10.1021/IE301122P
Reddy P, Rama Raju B, Karuppiah J, Linga Reddy E, Subrahmanyam C (2013) Degradation and mineralization of methylene blue by dielectric barrier discharge non-thermal plasma reactor. Chem Eng J 217:41–47. https://doi.org/10.1016/j.cej.2012.11.116

Rezaei F, Cristoforetti G, Tognoni E, Legnaioli S, Palleschi V, Safi A (2020) A review of the current analytical approaches for evaluating, compensating and exploiting self-absorption in laser induced breakdown spectroscopy. Spectrochim Acta Part B At Spectrosc 169:105878. https://doi.org/10.1016/j.sab.2020.105878

Ribeiro AR, Nunes OC, Pereira MFR, Silva AMT (2015) An overview on the advanced oxidation processes applied for the treatment of water pollutants defined in the recently launched directive 2013/39/EU. Environ Int 75:33–51. https://doi.org/10.1016/j.envint.2014.10.027

Rout PR, Zhang TC, Bhunia P, Surampalli RY (2021) Treatment technologies for emerging contaminants in wastewater treatment plants: a review. Sci Total Environ 753:141990. https://doi.org/10.1016/j.scitotenv.2020.141990

Saidulu D, Gupta B, Gupta AK, Ghosal PS (2021) A review on occurrences, eco-toxic effects, and remediation of emerging contaminants from wastewater: special emphasis on biological treatment based hybrid systems. J Environ Chem Eng 9:105282. https://doi.org/10.1016/j.jece.2021.105282

Salimi M, Esrafili A, Gholami M, Jafari AJ, Kalantary RR, Farzadkia M, Kermani M, Sobhi HR (2017) Contaminants of emerging concern: a review of new approach in AOP technologies. Environ Monit Assess 1898(189):1–22 . https://doi.org/10.1007/S10661-017-6097-X

Samukawa S, Hori M, Rauf S, Tachibana K, Bruggeman P, Kroesen G, Whitehead JC, Murphy AB, Gutsol AF, Starikovskaia S, Kortshagen U, Boeuf JP, Sommerer TJ, Kushner MJ, Czarnetzki U, Mason N (2012) The 2012 plasma roadmap. J Phys D Appl Phys 45. https://doi.org/10.1088/0022-3727/45/25/253001

Sarangapani C, Misra NN, Milosavljevic V, Bourke P, O'Regan F, Cullen PJ (2016) Pesticide degradation in water using atmospheric air cold plasma. J Water Process Eng 9:225–232. https://doi.org/10.1016/J.JWPE.2016.01.003

Sathishkumar P, Mangalaraja RV, Anandan S (2016) Review on the recent improvements in sonochemical and combined sonochemical oxidation processes—a powerful tool for destruction of environmental contaminants. Renew Sustain Energy Rev 55:426–454. https://doi.org/10.1016/j.rser.2015.10.139

Shang K, Wang X, Li J, Wang H, Lu N, Jiang N, Wu Y (2017) Synergetic degradation of acid orange 7 (AO7) dye by DBD plasma and persulfate. Chem Eng J 311:378–384. https://doi.org/10.1016/j.cej.2016.11.103

Shashurin A, Keidar M, Bronnikov S, Jurjus RA, Stepp MA (2008) Living tissue under treatment of cold plasma atmospheric jet. Appl Phys Lett 93:181501. https://doi.org/10.1063/1.3020223

Shimizu K, Muramatsu S, Sonoda T, Blajan M, Supply APP (2010) Water treatment by low voltage discharge in water. Int J Plasma Environ Sci Technol 4:58–64

Shin WT, Mirmiran A, Yiacoumi S, Tsouris C (1999) Ozonation using microbubbles formed by electric fields. Sep Purif Technol 15:271–282. https://doi.org/10.1016/S1383-5866(98)00107-5

Singh KP, Rai P, Singh AK, Verma P (2014) Gupta S (2014) Occurrence of pharmaceuticals in urban wastewater of north Indian cities and risk assessment. Environ Monit Assess 18610(186):6663–6682. https://doi.org/10.1007/S10661-014-3881-8

Singh RK, Philip L, Ramanujam S (2016) Rapid removal of arbofuran from aqueous solution by pulsed corona discharge treatment: kinetic study, oxidative, reductive degradation pathway, and toxicity assay. Ind Eng Chem Res 55:7201–7209. https://doi.org/10.1021/ACS.IECR.6B01191

Slamani S, Abdelmalek F, Ghezzar MR, Addou A (2018) Initiation of Fenton process by plasma gliding arc discharge for the degradation of paracetamol in water. J Photochem Photobiol A Chem 359:1–10. https://doi.org/10.1016/J.JPHOTOCHEM.2018.03.032

Snoeckx R, Bogaerts A (2017) Plasma technology-a novel solution for CO_2 conversion? Chem Soc Rev 46:5805–5863. https://doi.org/10.1039/c6cs00066e

Song Y, Cai S, Zhang W (2000) Rapid determination of the ozone in water. J Hyg Res 29:151–153

Sorensen JPR, Lapworth DJ, Nkhuwa DCW, Stuart ME, Gooddy DC, Bell RA, Chirwa M, Kabika J, Liemisa M, Chibesa M, Pedley S (2015) Emerging contaminants in urban groundwater sources in Africa. Water Res 72:51–63. https://doi.org/10.1016/J.WATRES.2014.08.002

Sugiarto AT, Sato M (2001) Pulsed plasma processing of organic compounds in aqueous solution. Thin Solid Films 386:295–299. https://doi.org/10.1016/S0040-6090(00)01669-2

Sun B, Aye NN, Gao Z, Lv D, Zhu X, Sato M (2012) Characteristics of gas-liquid pulsed discharge plasma reactor and dye decoloration efficiency. J Environ Sci 24:840–845. https://doi.org/10.1016/S1001-0742(11)60837-1

Tampieri F, Ginebra MP, Canal C (2021) Quantification of plasma-produced hydroxyl radicals in solution and their dependence on the pH. Anal Chem 93:3666-3670. https://doi.org/10.1021/acs.analchem.0c04906

Tas MA, Van Hardeveld R, Van Veldhuizen EM (1997) Reactions of NO in a positive streamer corona plasma. Plasma Chem Plasma Process 17:371–391. https://doi.org/10.1023/A:1021818313047

Tian W, Kushner MJ (2014) Atmospheric pressure dielectric barrier discharges interacting with liquid covered tissue. J Phys D Appl Phys 47. https://doi.org/10.1088/0022-3727/47/16/165201

Vanraes P, Bogaerts A (2018) Plasma physics of liquids—a focused review. Appl Phys Rev 5:031103. https://doi.org/10.1063/1.5020511

Vanraes P, Nikiforov, Anton Y, Christophe L (2016) Electrical discharge in water treatment technology for micropollutant decomposition. In: Mieno T (ed) Plasma Science and Technology: Progress in Physical States and Chemical Reactions. InTech, pp 457–506

Vulliet E, Cren-Olivé C, Grenier-Loustalot M-F (2009) Occurrence of pharmaceuticals and hormones in drinking water treated from surface waters. Environ Chem Lett 91(9):103–114 .https://doi.org/10.1007/S10311-009-0253-7

Wang H, Li J, Quan X, Wu Y (2008) Enhanced generation of oxidative species and phenol degradation in a discharge plasma system coupled with TiO_2 photocatalysis. Appl Catal B Environ 83:72–77. https://doi.org/10.1016/J.APCATB.2008.02.004

Wang Y, Sun J, Wang D (2013) Modelling of atmospheric pressure plasmas. Low Temp Plasma Technol 61–96. https://doi.org/10.1201/b15153

Wang J, Sun Y, Feng J, Xin L, Ma J (2016) Degradation of triclocarban in water by dielectric barrier discharge plasma combined with TiO_2/activated carbon fibers: effect of operating parameters and byproducts identification. Chem Eng J 300:36–46. https://doi.org/10.1016/J.CEJ.2016.04.041

Wang J, Simeni Simeni M, Rong M, Bruggeman P (2021) Absolute OH density and gas temperature measurements by laser induced fluorescence in a microsecond pulsed discharge generated in a conductive NaCl solution. Plasma Sourc Sci Technol 30:075016. https://doi.org/10.1088/1361-6595/abf71c

Wang X, Luo J, Huang Y, Mei J, Chen Y (2021) Degradation of pharmaceutical contaminants by bubbling gas phase surface discharge plasma combined with g-C 3 N 4 photocatalysis. Environ Sci Water Res Technol 7:610–621. https://doi.org/10.1039/D0EW00985G

Wee SY, Aris AZ, Yusoff FM, Praveena SM (2020) Occurrence of multiclass endocrine disrupting compounds in a drinking water supply system and associated risks. Sci Rep 101 10:1–12. https://doi.org/10.1038/s41598-020-74061-5

WHO (2019) Drinking-water. https://www.who.int/en/news-room/fact-sheets/detail/drinking-water. Accessed 1 Aug 2021

Williams M, Kookana RS, Mehta A, Yadav SK, Tailor BL, Maheshwari B (2019) Emerging contaminants in a river receiving untreated wastewater from an Indian urban centre. Sci Total Environ 647:1256–1265. https://doi.org/10.1016/J.SCITOTENV.2018.08.084

Wu T, Liu G, Zhao J, Hidaka H, Serpone N (1999) Evidence for H2O2 generation during the TiO_2-assisted photodegradation of dyes in aqueous dispersions under visible light illumination. J Phys Chem B 103:4862–4867. https://doi.org/10.1021/jp9846678

Wu H, Song Q, Ran G, Lu X, Xu B (2011) Recent developments in the detection of singlet oxygen with molecular spectroscopic methods. TrAC Trends Anal Chem 30:133–141. https://doi.org/10.1016/j.trac.2010.08.009

Wu L, Xie Q, Lv Y, Wu Z, Liang X, Lu M, Nie Y (2019) Degradation of methylene blue via dielectric barrier discharge plasma treatment. Water (Switzerland) 11. https://doi.org/10.3390/w11091818

Wu H, Fan J, Liu F, Shu L, Yin B (2021) Degradation of tetracycline in aqueous solution by persulphate assisted gas-liquid dielectric barrier discharge. Water Environ J 35:902–912. https://doi.org/10.1111/wej.12678

Yamatake A, Fletcher J, Yasuoka K, Ishii S (2006) Water treatment by fast oxygen radical flow with dc-driven microhollow cathode discharge. IEEE Trans Plasma Sci 34:1375–1381. https://doi.org/10.1109/TPS.2006.877249

Yamazaki E, Yamashita N, Taniyasu S, Lam J, Lam PKS, Moon HB, Jeong Y, Kannan P, Achyuthan H, Munuswamy N, Kannan K (2015) Bisphenol A and other bisphenol analogues including BPS and BPF in surface water samples from Japan, China, Korea and India. Ecotoxicol Environ Saf 122:565–572. https://doi.org/10.1016/J.ECOENV.2015.09.029

Yan JH, Liu YN, Bo Z, Li XD, Cen KF (2008) Degradation of gas–liquid gliding arc discharge on Acid Orange II. J Hazard Mater 157:441–447. https://doi.org/10.1016/J.JHAZMAT.2008.01.007

York N (2014) Water quality monitoring data for pesticides on long island, NY, 1–172

Yuri, P R, John, E A, V, I K (1997) Gas discharge Physics, 1st ed. Springer-verlag

Zangouei M, Haynes BS (2019) The role of atomic oxygen and ozone in the plasma and post-plasma catalytic removal of N_2O. Plasma Chem Plasma Process 39:89–108. https://doi.org/10.1007/s11090-018-9926-y

Zhang Y, Qin P, Lu S, Liu X, Zhai J, Xu J, Wang Y, Zhang G, Liu X (2020) Wan Z (2020) Occurrence and risk evaluation of organophosphorus pesticides in typical water bodies of Beijing. China. Environ Sci Pollut Res 282(28):1454–1463. https://doi.org/10.1007/S11356-020-10288-Z

Zhao Y, Kuang J, Zhang S, Li X, Wang B, Huang J, Deng S, Wang Y, Yu G (2017) Ozonation of indomethacin: Kinetics, mechanisms and toxicity. J Hazard Mater 323:460–470. https://doi.org/10.1016/J.JHAZMAT.2016.05.023

Zhao C, Xue L, Zhou Y, Zhang Y, Huang K (2021) A microwave atmospheric plasma strategy for fast and efficient degradation of aqueous p-nitrophenol. J Hazard Mater 409:124473 . https://doi.org/10.1016/j.jhazmat.2020.124473

Zhou R, Zhang T, Zhou R, Mai-Prochnow A, Ponraj SB, Fang Z, Masood H, Kananagh J, McClure D, Alam D, Ostrikov K (Ken), Cullen PJ (2021) Underwater microplasma bubbles for efficient and simultaneous degradation of mixed dye pollutants. Sci Total Environ 750:142295. https://doi.org/10.1016/J.SCITOTENV.2020.142295

Chapter 16
Graphene Modified Photocatalysts for the Abatement of Emerging Contaminants in Water

Afrah Harafan, Shibil Abdul Gafoor, Tadimeti Divya Kusuma, and Shihabudheen M. Maliyekkal

Abstract The increasing occurrence of emerging contaminants in the water bodies is a significant concern worldwide due to their potential health effects and recalcitrant nature. The conventional treatment methods are not efficient in removing these complex organic molecules to a safe level. In this context, access to state-of-the-art techniques holds the key to removing these contaminants and protect public health and other forms of life. The advanced oxidation processes, which rely on highly reactive chemical species, are suitable for treating these pollutants. The semiconductor-based photocatalytic process is economical, relatively greener, and well documented among the advanced oxidation processes reported. The materials, including titanium dioxide, zinc oxide, cadmium sulfide, molybdenum oxide, and other oxide forms of transition metals, and their derivatives are investigated as photocatalysts. However, the technology's success depends mainly on the sunlight utilization capability of the photocatalyst, low recombination rate, and photocatalyst stability. These limitations can be overcome by doping/co-doping/supporting these catalysts with carbon and its allotropes, metal and metal oxide, non-metal, or rare-earth metals. However, doping with graphene is gaining interest due to its excellent surface-to-volume ratio, charge carrier capability, thermal and mechanical stability, electron conductive and storage properties, and can form visible light-activated semiconductor nanocomposite. This chapter reviews the recent development of graphene-modified semiconductor photocatalyst employed for the photocatalytic degradation of emerging contaminants. The synthesis protocol, mechanism of degradation, and factors influencing the efficiency of the degradation are discussed.

Keywords Advanced oxidation · Photocatalyst · Graphene · Titanium dioxide · Water treatment

---

A. Harafan · S. A. Gafoor · T. D. Kusuma · S. M. Maliyekkal (✉)
Department of Civil and Environmental Engineering, Indian Institute of Technology Tirupati, Tirupati 517506, India
e-mail: shihab@iittp.ac.in

S. P. Singh et al. (eds.), *New Trends in Emerging Environmental Contaminants*, Energy, Environment, and Sustainability,
https://doi.org/10.1007/978-981-16-8367-1_16

16.1 Introduction

Emerging contaminants (ECs) or contaminants of emerging concern are discrete chemicals or a mixture of chemical compounds, natural or synthetic, existing in water bodies; perceived as a threat to the ecosystem due to its limited toxicological information, detection limit or unidentified source and route of human exposure (Dey et al. 2019; Geissen et al. 2015). Since these chemicals exist at nanogram and picogram levels in the environment, they are also called micropollutants or trace organic compounds. Examples of ECs include pharmaceuticals, personal care products (PCPs), pesticides, disinfection by-products, algal toxins, biocides and metabolites, and bioterrorism devices (Abdulrazaq et al. 2020; Kümmerer 2011). Due to its obsolete nature, around 38 ECs have been categorized as persistent organic pollutants (POPs) by the Stockholm convention (United Nations Environment Program Stockholm Convention 2021).

The first report mentioning the presence of ECs in water dates back to 1962 by Rachel Carson (Carson 2002). Currently, more than 110,000 ECs have been identified in the water bodies by the European Commission (NORMAN Substance Database 2021). The removal of ECs and their transformed products by the conventional water treatment plant is challenging owing to its variation in physicochemical properties, negligible concentration, and unique characteristics (Ahmed 2021). This has led to the ubiquitous presence of ECs in various environmental matrices. The translocation of ECs into environmental compartments is dependent on their physical and chemical properties (Feiyan et al. 2021). Figure 16.1 shows the possible sources and pathways of EC contamination in water, sediment and soil matrices. Based on hydrogeological parameters, economic growth, and access to treatment technologies, the magnitude of the ECs varies between regions (Barron et al. 2010). Figure 16.2 represents the number of ECs detected in surface water (SW), groundwater (GW) and drinking water (DW) worldwide. Considering the geographical area, the number of ECs identified in Russia is comparatively less. This can be probably due to a lower number of publications, resulting in a lack of sufficient data (Howard et al. 2020). Another possible reason can be linked to the initiatives taken by the Russian government to control POPs released into the environment (Kovner 2009). Over the last two decades, the US, UK, Canada, Brazil, China and India have been considered as the most prolific countries in terms of publications related to ECs (Howard et al. 2020). Figure 16.2 also presents the distribution of some of the commonly recognized ECs in these areas. Concentrations of triclosan, a key ingredient in PCPs, was reported high in UK and US waters compared to other countries. This increase can be credited to the large-scale utilization of PCPs in the US and UK (~350,000–450,000 kg) to satisfy the modern lifestyle of the population (Lalonde et al. 2019; Perez et al. 2013). Meanwhile, high levels of bisphenol A (BPA) were reported in Asian countries, which can result from the mass production of plastic commodities in these regions. Besides, the mismanagement of plastic wastes, especially in India and China, might also be a possible reason for these statistics. Among the ECs, pharmaceuticals are of significant concern due to their increased consumption and detection in water

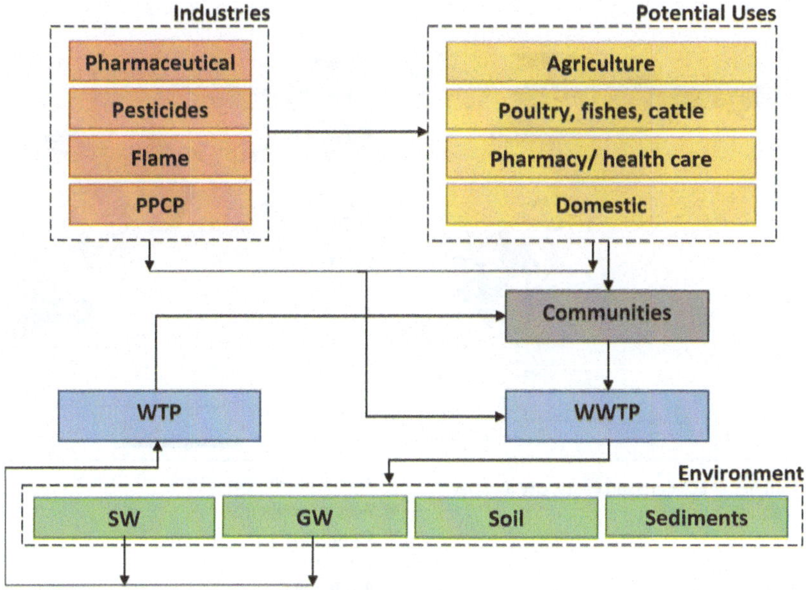

WTP: Water treatment plant; WWTP: wastewater treatment plant; SW: Surface water; GW: Ground water; DW: Distilled water

Fig. 16.1 Sources and pathways of EC contamination in environmental matrices (Data obtained from Ahmed (2021); Mateo-Sagasta et al. 2017))

bodies. Global trends in antibiotic consumption have revealed a 36% increase in Brazil, Russia, China, India and South Africa between 2000 and 2010 (Danner et al. 2019). In 2019, the US and China emerged as the largest pharmaceutical market accounting for 45% of revenue worldwide. Consequently, this has led to the elevated presence of pharmaceuticals in the US and China, as seen in Fig. 16.2. Several studies in India have also revealed the presence of pharmaceuticals and other emerging contaminants in the river streams (Wei et al. 2012; Hughes et al. 2013). However, the average concentration of acetaminophen in India reported is relatively low due to their decreased consumption. Other commonly recorded ECs also include diclofenac, sulfamethoxazole (SMX), chlorpyrifos and 17-β estradiol.

Most of these ECs possess high molecular weight and are resistant to biodegradation, thereby resulting in bioaccumulation. Also, the majority of these organics produce toxic products on solar exposure due to the presence of π electron donors (Pharma Manual 2021). Studies revealed that prolonged exposures to various ECs could cause mutagenicity, teratogenicity, and carcinogenicity in humans, animals and microorganisms (Feiyan et al. 2021). Globally, around 214,000 neonatal sepsis death annually is reportedly linked to the presence of antibiotic-resistant genes (ARGs) in water bodies (Pant et al. 2021). Due to their recalcitrant and synergistic effects, developing appropriate technologies is substantial to eradicate these ECs from the water bodies (Richardson and Kimura 2017).

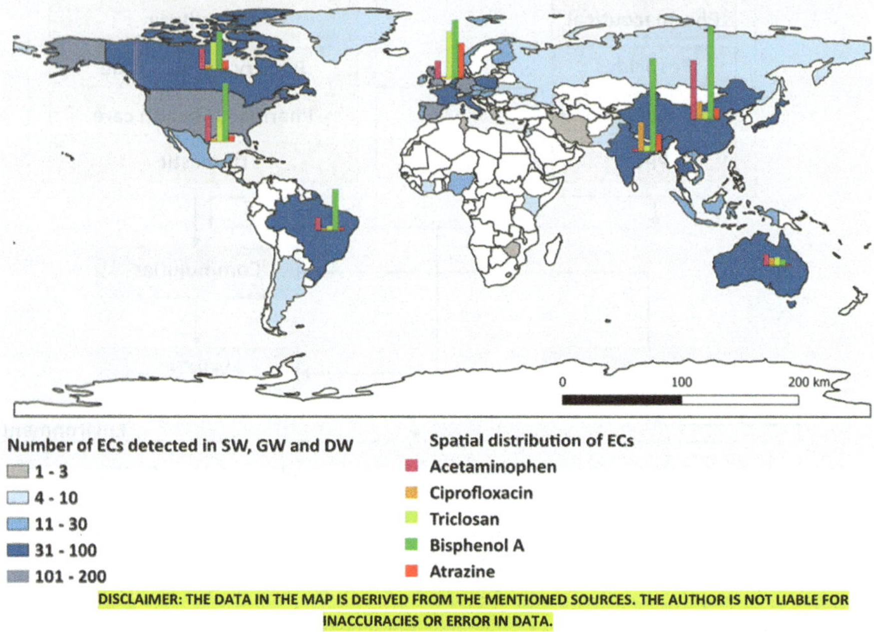

Fig. 16.2 Distribution of ECs in SW, GW and DW worldwide (Data obtained from Lalonde et al. 2019; Wei et al. 2012; Hughes et al. 2013; Pulicharla et al. 2021; Canada and CC 2020; Cuprys et al. 2019; Balakrishna et al. 2017; Balakrishna et al. 2017; Manamsa et al. 2016; Jiang et al. 2011; Zhao et al. 2010)

The advancements in chromatographic and spectroscopic techniques have increased our ability to detect and analyze pollutants in trace quantities existing in various phases. This, in turn, has assisted in choosing suitable treatment alternatives based on the available concentrations. Various treatment methods such as advanced oxidation process (AOPs), adsorption, membrane filtration, membrane bioreactors have been developed to supply EC free water (Kanakaraju et al. 2018a). The advanced oxidation processes such as Fenton oxidation, hydrogen peroxide oxidation, ozonation, photocatalysis, and cold plasma processes and their combinations have gained much interest as it complies with the EC removal goals (Kanakaraju et al. 2018a). The ability of these methods to either degrade or mineralize the organic compounds by a redox reaction with the generation of reactive oxygen species (ROS) have made it emerge as the most preferred treatment technique in the removal of ECs from water matrices. Among the AOPs, photocatalysis has gained widespread interest in the degradation of ECs present in the gaseous or aqueous phase. This effectiveness can be credited to process stability, easy operation, high photoactivity, minimal secondary pollution and recoverability of photocatalyst. However, its high cost in scaling up and energy consumption restrict its usage on an industrial scale (Adesina 2004). Despite its drawback, the technique of photocatalysis has been extended to various fields like energy, medicine, agriculture, transportation, and water and wastewater treatment.

Over the years, numerous semiconductor oxides such as Bi_2O_3, MnO_2, TiO_2, α-Fe_2O_3, ZnO, WO_3, CdSe, MoS_2, SnO_2 and graphitic-C_3N_4 have been explored as photocatalysts (Khan et al. 2015). Nevertheless, TiO_2 was widely used due to its high stability, photoactivity and low cost (Fujishima et al. 2000). However, owing to its quick recombination of electron–hole pairs, resulting in poor quantum efficiency, the photoactivity exhibited by the semiconductors was not efficient for practical applications (Belver et al. 2020). Recently, graphene nanomaterials (GNM) have paved the way as efficient photocatalysts due to their improved reactivity under light irradiation. Various graphene derivatives have been developed to alter their characteristics, such as hydrophilicity, dispersibility and active sites (Pastrana et al. 2018). These are usually coupled or modified with transition/noble metals to function as heterogeneous photocatalysts (Tang et al. 2018). This chapter reviews the application of various graphene-modified photocatalysts in the treatment of emerging contaminants in water. The working principle, factors influencing the removal, a brief history, challenges and prospects in the development of photocatalysis are discussed. A brief outlook concerning the source and occurrence of ECs are also summarized.

16.2 Milestones in ECs Detection and Removal in Water

With the advancements in analytical chemistry, the ability to detect and analyze pollutants in trace quantities have increased significantly. The evolution of sensitive analytical techniques has helped us to detect the presence of ECs, measure their trace level concentration, and identify their source and transport mechanisms. Figure 16.3 shows the timeline of the development of various detection techniques. UV–vis absorbance spectroscopy, colourimetry and electrospray ionization technique (ESI) are some of the earliest methods used to detect ECs. Currently, chromatographic methods are extensively used to identify and detect various emerging compounds in environmental samples. Prior to detection, sampling of contaminants has to be carried out either from solid, liquid or gaseous phase. This depends on the polarizability, volatility, and thermal properties of the target organics (Ismail and Mokhtar 2020). Hence, based on the nature of contaminants, a well-established sampling protocol is necessary to ensure that the extracts are compatible with the instrument's limit of detection (LOD) and to minimize the interferences (Petrović et al. 2003). Usually, nonpolar, thermostable, and volatile ECs, and some pesticides, are determined by gas chromatography (GC), while non-volatile, polar, and thermolabile ECs are analyzed by liquid chromatography (LC). The coupling of chromatography with mass spectroscopic (MS) techniques such as quadrupole (Q), ion traps, time of flight (TOF), and the magnetic sensor has proven revolutionary in the detection of ECs (Ryu et al. 2021). The ability of chromatography to separate the components from chemical mixture and MS to identify the extracted components is desirable due to

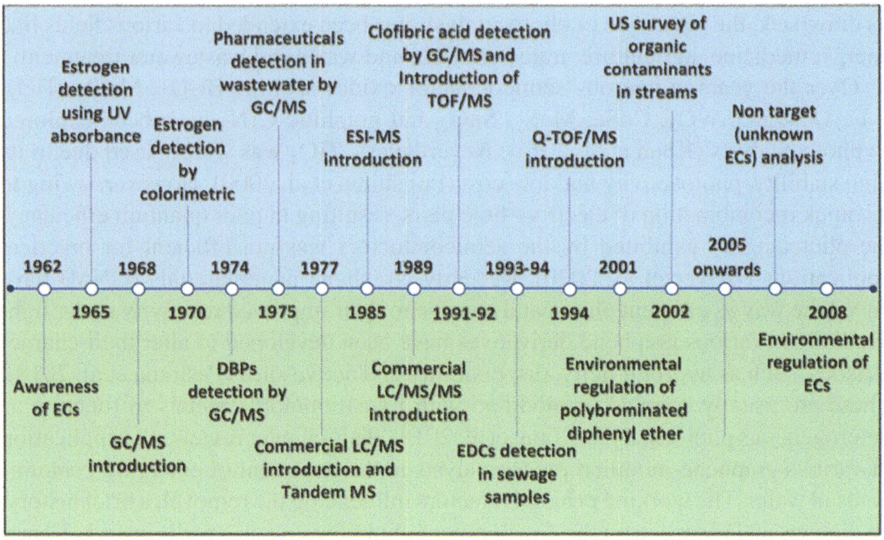

Fig. 16.3 Timeline of the development of various detection methods (Adapted from Long et al. 2020 with the permission of Elsevier)

its sensitive and highly specific nature (James et al. 2018). Although GC/MS is the convention technique, most ECs are responsive to LC/MS. In recent years, significant advancements in LC/MS have made the technique vital in tracking down thousands of organic contaminants present in various matrices.

Generally, the monitoring of ECs is associated with the implementation of regulatory measures as there is limited information regarding their harmful effects (Ismail and Mokhtar 2020). Hence, this has enabled the progression of various treatment technologies to either eliminate ECs or achieve maximum contamination levels. Although many chemicals categorized under a particular class of EC have similar physical, chemical, and biological parameters, there is considerable disparity in their removal efficiency (Stefanakis and Becker 2020). Figure 16.4 shows the timeline of various treatment techniques used in EC removal. The adsorption method is capable of removing ECs by attracting them onto the surface of the adsorbent from aqueous media. Over the years, various materials have been probed and tailored as adsorbents for the effective removal of pharmaceuticals, endocrine disrupting compounds (EDCs), and PCPs (Rossner et al. 2009; Ghosh et al. 2013). A recent study by (Dongli et al. 2020) investigated the adsorptive removal of an EDC, BPA using magnetic covalent organic frameworks. The adsorbent was able to remove BPA with 98.8% removal efficiency through hydrogen bonding interaction. However, the practical application of the technology is limited due to its decreased efficiency in recycling, increased cost of adsorbent regeneration, and non-selectivity. Moreover, as the technique follows the phase transfer mechanism, it poses a constraint for the sustainable disposal of spent adsorbents (Aris et al. 2020). Membrane-based technologies have proven their effectiveness in removing organics through a semipermeable membrane.

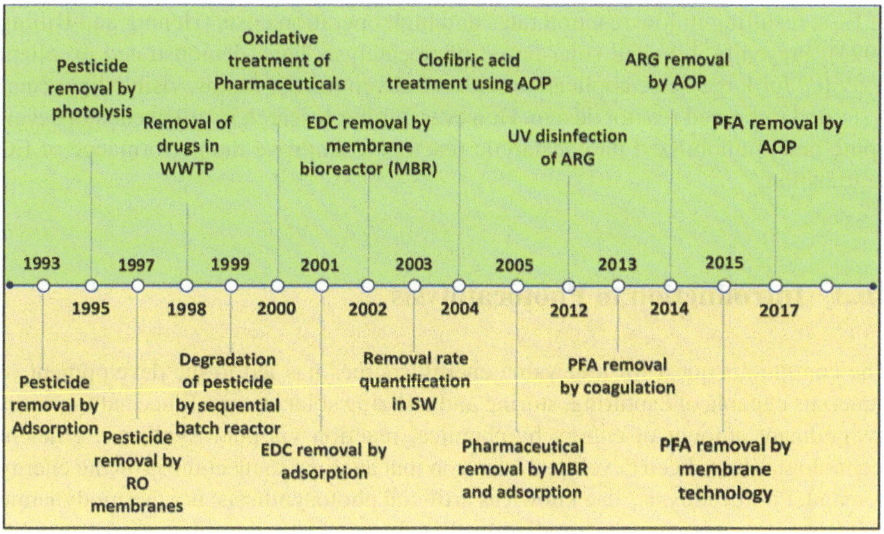

Fig. 16.4 Major highlights in the EC removal technologies (Data obtained from Chua and Chen 1995; Gander et al. 2000; Luprano et al. 2016; Zwiener et al. 1995; Mangat and Elefsiniotis 1999; Andreozzi et al. 2002; Wintgens et al. 2002; El-Gohary et al. 1995; Ternes 1998; Zwiener and Frimmel 2000; Nakada et al. 2006; Aranami and James 2007; Najm et al. 1993; Zaggia et al. 2016)

Various membranes such as reverse osmosis (RO), nanofiltration, microfiltration, and ultrafiltration are used to separate perfluoroalkyl compounds (PFA) and other contaminants in water. In a study, (Dong et al. 2022) investigated the performance of functionalized polyamide (PA) forward osmosis membranes via the chelation of Fe^{3+} and carboxylate to remove trace organics. Unlike conventional PA, the functionalized membranes were able to achieve 92% rejection of organics and exhibited favourable resistance to fouling (Dong et al. 2022). The membrane characteristics such as porosity, surface charge and hydrophobicity play a significant role in determining the efficacy of the process. However, its high operation cost and susceptibility to fouling limit its usage.

Numerous biological treatment systems have also been used to treat ECs. This includes activated sludge, constructed wetland, aerobic and anaerobic bioreactor, trickling filter, rotating biological reactor and biosorption (Ahmed et al. 2017). Nonetheless, the low concentration and biodegradability of ECs and the risk of formation of ARGs pose a drawback in utilizing this method. AOPs based treatment processes are now gaining interest as an alternative to conventional treatment technologies (Kurt et al. 2017). Among the AOPs, ozonation was reported most effective in removing concentrations of hormones and pharmaceuticals from water. A study investigated the removal of 38 pharmaceuticals and PCPs detected in the secondary effluent by ozonation. Amidst them, 31 pharmaceuticals and PCPs degraded to or below their LOD at the ozone dosage of 6 mg/L in 10 min (Paucar et al. 2019). Nevertheless, their utilization on an industrial scale is constrained by the low concentration of ECs, resulting in low reaction rates and high operation costs (Hennig and Billing 1993). Presently, UV and solar-based photocatalysis have demonstrated excellent potential for large-scale application with the advent of UV lamps, visible light catalysts, and improved reactor design. However, further research is still needed in developing new immobilized photocatalytic reactors to improve the performance of EC degradation.

16.3 Introduction to Photocatalysis

The continuous quest for renewable energy sources has led to the development of materials capable of capturing, storing and utilizing solar energy. Since nature stores a significant amount of energy by chemical reaction via photosynthesis, efficient technologies have been developed based on that analogy to meet the growing energy demand. Photocatalysis, also known as artificial photosynthesis, is a thermodynamic uphill reaction powered by irradiation. Briefly, it can be described as the process where photons are absorbed directly or indirectly to activate the semiconductor known as a photocatalyst and modify the rate of a chemical reaction (Linsebigler et al. 2002; Radhika et al. 2019). The earliest use of the name photocatalysis dates back to 1911, when Prussian blue was decolourized by illuminating ZnO (Ohtani 2013). This laid the basis of using ZnO for other reactions such as photoinduced reduction of Ag^+ to Ag. Similarly, the potential of TiO_2 to perform light-induced catalytic reduction was also investigated. Nevertheless, it was not until 1969 that the practical applicability of photocatalysis was demonstrated. The theory of photo-electrolysis by Honda and Fujishima marked an important event in the history of photocatalysis (Honda Fujishima Effect 2021). This increased the popularity of TiO_2 as a 'UV excited catalyst' for a wide range of applications (Landmann et al. 2012; Tang et al. 1994). Amidst them, the ability of TiO_2 to degrade organic contaminants made it emerge as a valuable material for water purification. Consequently, studies were performed to understand the mechanism of degradation in photocatalysis. During 1960–93, hydroxyl and superoxide radicals were proposed to be the active species during the degradation reaction (Long et al. 2020). However, holes were later proved to be the primary oxidizing species and could react directly with the contaminants or water molecules. Table 16.1 highlights the significant milestones in the field.

Generally, photocatalysts are differentiated from conventional catalysts based on the thermodynamic state of a reaction. Conventional catalyst performance is limited to thermodynamically feasible reactions, thereby decreasing the activation energy within the system (Ohtani 2013). Conversely, a photocatalyst proceeds a non-spontaneous reaction with the production of electrons by a light source. Also, photocatalysts do not undergo change during the reaction as the redox process occurs at the interfacial region between the excited solid and the solution (Ohtani 2013). Figure 16.5 shows an illustration between a photocatalyst and a conventional catalyst. A semiconductor is usually chosen as a photocatalyst because they possess a void energy region, unlike metals with a continuum electronic state (Swift 2019).

Table 16.1 Milestones in the field of photocatalysis

Year	Milestones
1911	Coined the term photocatalysis: bleaching of Prussian blue by ZnO illumination
1924	Luminescent reduction of Ag^+ to Ag
1932	TiO_2 and Nb_2O_5 as photocatalyst for reduction of $AgNO_3$ and $AuCl_3$ to Ag and Au, respectively
1938	TiO_2 as a photosensitizer to bleach dyes
Around 1960	Usage of ZnO, NiO and TiO_2 as photocatalyst
1972	H_2 production using TiO_2 electrode and UV irradiation
	Degradation of isopropanol
1977	Photocatalytic splitting of water in the presence of UV
1979	Photocatalytic CO_2 reduction using inorganic semiconductors as catalysts
1980	Studies on TiO_2 as an effective photocatalyst
Around 1993	Usage of WO_3, CdS, ZnS and Fe_2O_3 as photocatalyst
1997	Photon induced super hydrophilicity effect of TiO_2
2001–10	Visible light-absorbing photocatalyst; commercialization
2011-present	Design of high-efficiency photocatalytic systems; coupling with other systems; usage of solar energy

The energy of this void region is known as the bandgap or forbidden gap energy. For semiconductors, this bandgap energy extends from the top of the filled valence band to the bottom of the vacant conduction band. Usually, materials with a suitable bandgap, morphology, high surface area, stability and reusability are desired as photocatalysts. A series of oxides of metal such as Zn, Ni, CdS, W has been explored to address a wide range of pollutants. Since most of these materials utilize a small percentage of solar energy for activation, numerous studies have been carried out to improve photo-responsiveness to visible light by modifying photocatalysts for better energy conversion and reduction in the recombination rate of electrons and holes.

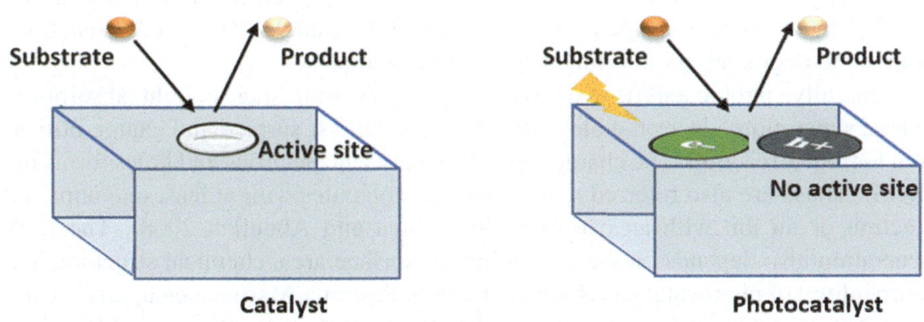

Fig. 16.5 Illustration of the difference between a photocatalyst and a conventional catalyst

During 2001–10, nanotechnology marked a new era in the development of photocatalysts. It gained widespread attention due to its exceptional properties resulting in improved mineralization. Their exceptional ability to fine-tune semiconductor properties and inhibit charge carriers' recombination has gained attention in the photocatalytic degradation of ECs. Also, these materials provide more active sites for chemical reactions to occur (You et al. 2019). Furthermore, they are classified as surface plasmon resonance-mediated, metal–organic charge transfer-based, and semiconductor-based nano photocatalysts. In the recent years, photocatalytic recycling, regeneration and reactor configuration has been extensively studied to facilitate the scaling-up of photocatalysis technique in industries. Currently, photocatalysts are used for a multitude of applications such as hydrogen production, CO_2 remediation, and air purification.

16.3.1 Principle of Photocatalysis

The photocatalysis reaction is initiated by illuminating photons of a suitable wavelength on the surface of semiconductors to promote excitation, thus generating electrons and holes (Mecha and Chollom 2020). Figure 16.6a shows an overview of the mechanism of photocatalysis. The feasibility of reaction depends on the ability of the semiconductors to transfer an electron from the highest occupied molecular orbital to the lowest molecular orbital (Karaush-Karmazin et al. 2020). This stimulation of semiconductors depends on the bandgap of the material used. Table 16.2 contains the bandgap energy of some of the conventionally used photocatalysts (Hennig 2015). Typically, the energy of incident light should be of higher energy than the bandgap for excitation (Minella et al. 2016). When excited from the valence band to the conduction band, the electron leaves a hole of positive charge in the valence band. This hole is generally assumed as a pseudo particle because of electron deficiency. Generally, for every 250 femtoseconds, a few electrons try to recombine within the body or on the surface, and the remaining electrons get trapped on the surface of the photocatalyst (Salasoo et al. 2002; Jones 2016). Thus, trapped electron–hole pairs act as reducing and oxidizing agents on the surface of the catalyst. These redox reactions help in the degradation of pollutants in aquatic media.

Generally, photocatalytic reactions progress as four stages: light absorption, charge separation, electron–hole transfer to the catalyst surface and charge utilization for redox reaction. The charge separation between electrons and holes forms the (ROS). These are also referred to as atoms or molecules with at least one unpaired electron or an ion with an extra electron (Gaya and Abdullah 2008). The ROS generation also depends on the availability of surface area, chemical structure, and morphology of photocatalyst (Hong et al. 2006; Pastrana-Martínez et al. 2013; Choi et al. 2007; Su et al. 2011; Awfa et al. 2018; Zhong et al. 2010; Ren et al. 2017). The ROS initiates the photocatalytic process by simultaneous reduction and oxidation on the photocatalyst surface at the conduction band and valency band, respectively. Hydroxyl radical ($HO\cdot$), superoxide anion radical ($O_2\cdot^-$), and hydroperoxyl radical

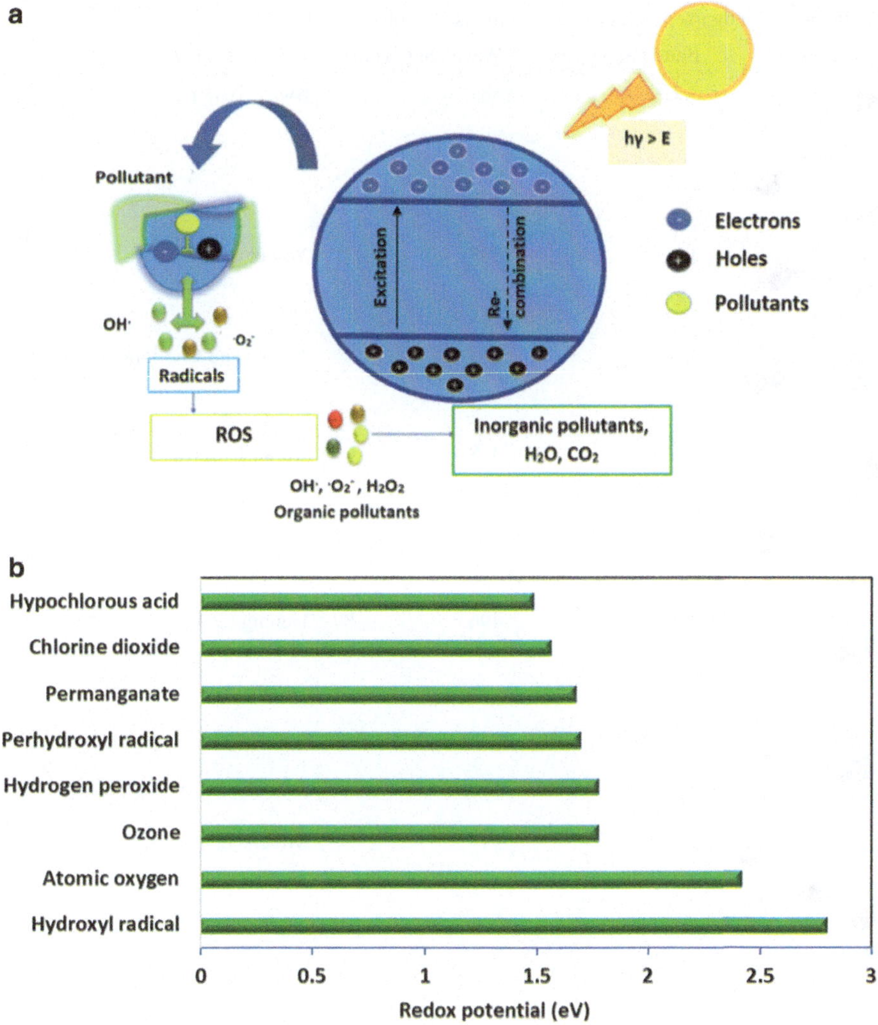

Fig. 16.6 **a** Mechanism of photocatalysis, **b** Redox potential of various oxidizing agents (Adapted from Phaniendra et al. 2015)

($HO_2\cdot$) are some of the examples of ROS. The electronic configuration of the radicals, empty or half-filled σ and π-orbitals, makes them highly reactive towards organics compounds (Phaniendra et al. 2015). Based on the nature of ROS generation, photocatalysts are further classified as homogeneous and heterogeneous catalysts. In the case of homogeneous photocatalysts (UV/TiO_2, VL/ZnO), the substrate itself generates ROS, whereas in heterogeneous, a semiconductor photocatalyst is required to generate ROS (H_2O_2/UV, Fe^{3+}/UV) (Rivera-Utrilla et al. 2013).

Table 16.2 Bandgap energies of commonly used photocatalysts

Photocatalyst	Band energy (eV)	Wavelength (nm)	Relevant study
 TiO_2	3.2	387	Swift (2019)
 ZnO	3.2	387	You et al. (2019)
 ZnS	3.6	335	Mecha and Chollom (2020)
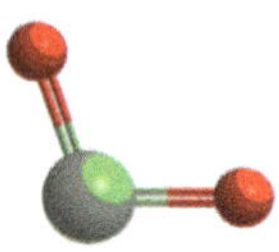 SnO_2	3.5	318	Karaush-Karmazin et al. (2020)
 CdS	2.4	496	Hennig (2015)
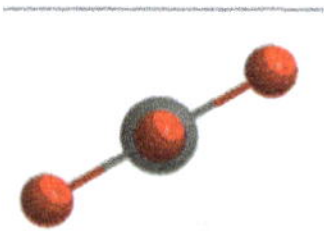 WO_3	2.4–2.8	443	Minella et al. (2016)

Primarily, radicals are unselective, reactive, and have powerful oxidizing capacities. Figure 16.6b represent the redox potential of various oxidizing agents. Despite these, radicals are very short-lived species (Guo et al. 2020). The superoxide radical has a residence time in the range of microseconds followed by peroxide radicals. However, the superoxide radical is the precursor of most radical species. The hydrogen peroxide is formed from the reaction between two superoxide radicals, as in Eq. 16.2. Hydrogen peroxide reduces to hydroxyl radical, and complete reduction forms water and oxygen as Eqs. 16.3 and 16.4. The most reactive, the hydroxyl radical, lives only in the nanoseconds range (Jaishankar et al. 2014). Despite its life span, hydroxyl radical has more oxidation capacity compared to the remaining ROS. As the target analytes are aromatic in structure, the hydroxyl radical aids in dissociating the benzene ring resulting in a linear compound, which makes further degradation easier, as in Eq. 16.5 (Rasheed et al. 2019).

$$O_2 + e^- \rightarrow O_2^{\bullet} + e^- \tag{16.1}$$

$$O_2^{\bullet-} + O_2^{\bullet-} + 2H^+ \rightarrow H_2O_2 + O_2 \tag{16.2}$$

$$H_2O_2 \rightarrow 2OH^{\bullet} \tag{16.3}$$

$$2H_2O_2 \rightarrow 2H_2O + O_2 \tag{16.4}$$

$$RH + OH^{\bullet} \rightarrow R^{\bullet} + H_2O;\quad R^{\bullet} \xrightarrow{O_2} ROO^{\bullet} \tag{16.5}$$

16.3.2 Factors Affecting Photocatalysis

As photocatalysts get illuminated, photogenerated charges transfer across the solid or the electrolyte to aid chemical reactions. There is also a considerable chance for the electron–hole pairs to recombine within the catalyst during this process. Hence, there is a need to improve the photoactivity of catalysts to enhance their remediation capacity. Several factors such as redox levels of substrate, conduction band, and valency band influence the rate of a reaction and the degradation efficiency of pollutants (Dhangar and Kumar 2020). Essentially, the excitation of semiconductors with defined photonic energy depends on the bandgap or forbidden band of the semiconductor. As per the Plancks energy equation, the energy of incident light should be higher than the bandgap energy of the semiconductor for the electron to jump to the conduction band (Kumar and Pandey 2017). Since the solar energy is within the 280–4000 nm wavelength range, both UV and visible spectrum absorbing catalysts are desired for sustainable purposes. However, a reduction in redox capacity was observed at increased wavelength due to limited energy (Ohtani 2013). Additionally, to avoid the recombination of electron–hole pairs, the conduction and valence bands are expected to be more cathodic and anodic than the standard potential of reductant and oxidant, respectively. Figure 16.7 shows the orientation of conduction and valence band and their redox potential. Nowadays, the substrate has been modified to decrease the recombination rate for the feasibility of both reduction and oxidation reactions.

The size of the photocatalyst also plays a role in determining the number of photons absorbed by the material (Zhu and Wang 2017). This controls the rate of photocatalytic reaction, which implies that reaction occurs only at the absorbed

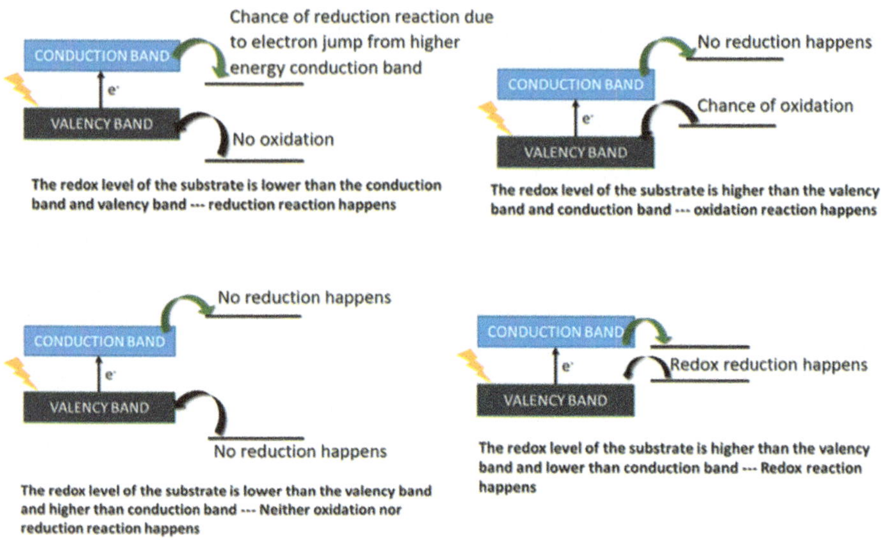

Fig. 16.7 Orientation of conduction and valence band and their redox potential. Modified from Sohrabi et al. 2019

phase of the catalyst. With the increase in size, the number of photons striking the photocatalyst per unit time increases. Consequently, the time interval between photon strikes is reduced, limiting surface chemical reactions. In addition, a small catalyst implies fewer grain boundaries within the bulk of the particle, thereby reducing the chance for recombination (Qian et al. 2019; Saravanan et al. 2017). Increased surface areas of small-sized particles also improve the adsorption capacity of the analyte. However, small-sized materials tend to aggregate over time due to their high reactivity.

Another factor influencing the rate of photolysis is the formation of ROS. The potential of ROS to degrade EC depend on both surface area and excitation wavelength. Increased charge transfer was observed when the pollutants were pre-adsorbed onto the surface of the photocatalyst. Hence, the adsorption capacity of the material plays a significant role in the removal process. Another parameter, light intensity, also assists in determining the rate of photolysis initiation (Sohrabi et al. 2019). It is necessary to obtain a lamp of quantified critical wavelength to emit photons for augmenting photocatalytic activity. Additionally, the radicals generated at the conduction band are more accessible than at the valency band for many of the primary photocatalysts because of their higher oxidation potential (Kanakaraju et al. 2018b). The continuous supply of reactants to the photocatalysts, the nature of diffusion of the products away from the reaction sites, and the regeneration capacity of the photocatalysts are some of the factors that control the system's functionality.

Besides the above factors, the photocatalytic system is also governed by a number of operational parameters such as analyte concentration, solution pH, the temperature of the reaction, amount of catalyst, nature of the analyte, nature of photocatalyst and oxygen concentration. A brief overview of factors affecting photocatalysis along with their significance is given in Table 16.3. Further analyses on these factors are discussed in the upcoming section.

16.3.3 Application of Graphene Modified Photocatalysts in Removing ECs in Water

The photocatalytic units can be molecules dissolved (homogeneous photocatalyst) or solids dispersed/ suspended (heterogeneous photocatalyst) in an aqueous medium. In recent years, heterogeneous photocatalysis has emerged as the most appealing technique for sustainably harnessing solar energy. Generally, semiconductors such as metal oxides are preferred as heterogeneous catalysts due to their ability to generate charge carriers resulting in photocatalytic activity. Among all metal oxides, TiO_2 is widely used due to its excellent photoactive properties. In a study, Cai and Hu demonstrated the removal of antibiotics SMX and trimethoprim using TiO_2, photo-assisted by a UVA light source. The study reported a removal efficiency of >90% for an analyte concentration of 400 mg/L and a TiO_2 dosage of 0.05 g/L (Cai and Hu 2018). ZnO is also another commonly used photocatalyst because of its low cost and reduced toxic effects. However, it displays less photocatalytic activity when compared to TiO_2. Most metal oxides' drawback lies in their large bandgap, stability, cost, and efficiency (Khan et al. 2015). Thenceforth, several strategies have been devised to synthesize new materials or modify existing materials responsive to visible light in the past few years. This includes optimization of crystal structure, surface area, morphology or incorporation of defects (Djurišić et al. 2020).

Due to the limitations posed by the conventional photocatalysts, there has been increased attention towards the integrated photocatalytic system (IPS) or multi-component heterogeneous photocatalyst. IPS include combining a heterogeneous photocatalyst with narrow bandwidth semiconductors or a co-catalyst such as metal, noble metal or carbon. Recently, IPS with graphene modifications have received increased attention due to its intrinsic properties, such as increased surface area and enhanced charge mobility. Moreover, GNMs aids the photocatalytic process by acting as a substrate for fabricating various photo materials with controllable size, composition or morphology. Sordello and his coworkers demonstrated the ability of functionalized GNM to serve as a support to provide nucleation sites for the growth of TiO_2 nanoparticles (NP). These in-situ synthesized TiO_2 NPs exhibited higher photocatalytic efficiency in degrading organic contaminants (Minella et al. 2017). So far, several kinds of research have been carried out to modify GNM to improve degradation efficiency. Graphene susceptibility to recombination owing to its bandgap (0 eV) has been improved by oxidizing it to graphene oxide (GO) and reduced graphene oxide (rGO). This increase in bandgap is because of the broken sp^2 conjugated network in GO and the restriction of π electrons in the sp^2 domain (Li et al. 2016). Moreover, the oxygen-containing functional groups also improve their dispersibility in solvents and serve as nucleation sites for the semiconductors to bind evenly. Another possible way to improve the catalytical efficiency is to

Table 16.3 Factors effecting photocatalysis along with their significance

Factors influencing photocatalysis		Significance	References
Photocatalyst			
Catalyst amount		The photodegradation increases proportionally with the catalyst amount due to increases the number of active sites on the photocatalyst surface. However, beyond a certain limit of catalyst, irradiation gets blocked and decreases degradation	Kanakaraju et al. (2018b)
Morphology	Surface area	Increased surface area can increase the adsorption of ECs, thereby facilitating degradation; Finer particles are dispersed more homogeneously and can absorb more photons	Cai and Hu (2018)
	Crystallinity	Increase in crystallinity increases density of defects, thereby decreasing the recombination rates of photogenerated holes and electrons	
Light source		Depending on the bandgap energy, UV or visible light source can be preferred for photodegradation	Djurišić et al. (2020)
Reaction medium			
Initial concentration		The degradation efficacy decreases with increasing amount of target analyte, while keeping a fixed amount of catalyst due to reduced number of available photons to the catalyst surface	Sakar et al. (2019)
pH		Variation in solution pH changes the surface charge of the catalyst and shifts the potentials of photocatalytic reactions	Minella et al. (2017)
Temperature		Photocatalytic activity increases with reaction temperature. However, beyond a critical temperature the reaction promotes the recombination of charge carriers and disfavour the adsorption of ECs	
Presence of electrolytes		Based on the concentration, the presence of cations or anions exhibited minimal to significant effects on degradation efficacy. Increased presence of electrolytes diminishes the colloidal stability, increases mass transfer and reduces the surface contact between the analyte and the photocatalyst. Moreover, it can also scavenge both the hole and the hydroxyl radicals	Li et al. (2016)

(continued)

Table 16.3 (continued)

Factors influencing photocatalysis	Significance	References
Dissolved oxygen	Dissolved oxygen involves in the stabilization of radical intermediates, mineralization and direct photocatalytic reactions. Its presence is also known to induce the cleavage mechanism for aromatic rings in organic pollutants that are present in the water matrices	

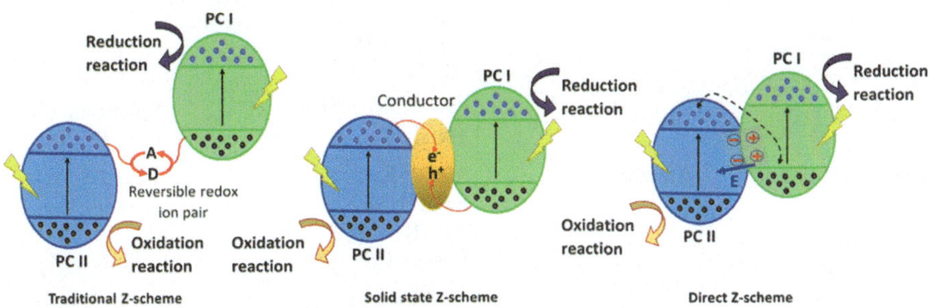

Fig. 16.8 Schematic representation of heterojunction in an IPS (Adapted from Sakar et al. 2019) (Open source subjected to creative common attribution 4.0 license)

improve interfacial contact/heterojunction between graphene derivatives and semiconductors via ex-situ or in-situ processes. This includes mechanical mixing, sol–gel process, solvothermal treatment, microwave or sonication assisted chemical deposition. Based on the methodology, various graphene-based heterojunction such as Schottky, Type-II, Z-scheme, in-plane junctions has been devised. Figure 16.8 shows a schematic representation of some commonly reported heterojunctions. Depending upon the nature of synthesis, various precursors react at graphene interfaces, transforming them to expected semiconductors with specific nanostructures to form a graphene-semiconductor composite (Hu et al. 2013).

The ability of graphene to interact with the organic compounds by π-π bonding renders it a suitable material for removing ECs. Table 16.4 presents an overview of graphene-modified photocatalyst used in the treatment of various ECs in water. Liu et al. (2019) investigated the degradation efficiency of sulfamethazine (SMT) using graphene grafted anatase titania nanosheets (G/A/TNS) via the in-situ hydrothermal method. During the process, graphene interacted with titania through C–O–Ti linkages, which were critical for the electron transfer chain during the reaction process. The photocatalytic experiments were carried out in a sealed quartz reactor using a 450 W xenon lamp for illumination at an intensity of 90 mW/cm^2. The initial concentration of the pollutant, photocatalyst dosage, pH and temperature was maintained as

Table 16.4 Overview of graphene-modified photocatalyst used in the treatment of various ECs in water

Photocatalyst	Pollutant	Experiment condition	Degradation (%)	References
Pesticides				
Ag/GO	Lindane	Solar light, pH: 7–9, catalyst: 0.01 mM	99.9%	Gupta et al. (2015a)
$CoFe_2O_4$ @TiO_2/GO	Chlorpyrifos	400 W UV lamp, Co: 5 mg/L, catalyst: 0.4 g/L, pH 5.8	89.9% in 60 min	Gupta et al. (2015b)
TiO_2/rGO	Pyrimethanil	Co: 200 μg/L	100% in 25 min	Luna-Sanguino et al. (2020)
CdS/C_3N_4/rGO	Atrazine	350 W Xe lamp (UV and IR filters), Co: 10 mg/l	NIL	Jo and Selvam (2017)
Pharmaceuticals				
Magnetic GO-Ce-TiO_2	Tetracycline	Visible light, catalyst: 0.5 g/L	83% in 60 min	Cao et al. (2016)
TiO_2-rGO coated SOFs[1]	Sulfamethoxazole	High pressure UV light	92% in 180 min	Lin et al. (2017)
TiO_2-NS[2]/Pt/GO	Chlortetracycline	Visible light, catalyst: 0.2 g/L	45% in 90 min	Liang et al. (2017)
Fe_3O_4@TiO_2-GO	Enrofloxacin	Visible light, catalyst: 0.4 g/L	96% in 240 min	Yu et al. (2017)
ZnO/rGO	Metronidazole	Visible light, catalyst: 1 g/L	49.3% in 160 min	Dong et al. (2014a)
$ZnSnO_3$/rGO	Metronidazole	Visible light, catalyst: 1 g/L	72.5% in 180 min	Dong et al. (2014b)

(continued)

Table 16.4 (continued)

Photocatalyst	Pollutant	Experiment condition	Degradation (%)	References
nano-ZnO/CdS/GO	Ciprofloxacin	Visible light, catalyst: 0.5 g/L	86% in 60 min	Huo et al. (2016)
rGO-$BiVO_4$	Ciprofloxacin	Visible light, catalyst: 0.2 g/L	68.2% in 60 min	Yan et al. (2013)
Cu/$Bi_2Ti_2O_7$/rGO	Tetracycline	Visible light, catalyst: 5 g/L	92% in 90 min	Scopus preview - Scopus - Welcome to Scopus. (2021)
CQDs-CdSe-rGO	Tetracycline	Visible light, catalyst: 0.8 g/L	90% in 60 min	Huo et al. (2017)
Au/C_3N_4/graphene	Ciprofloxacin	Visible light, catalyst: 1 g/L	96.7% in 240 min	Xue et al. (2015)
GO/Ag_2CrO_4/g-C_3N_4	Oxytetracycline	Visible light, catalyst: 0.2 g/L	81.34% in 90 min	Chen et al. (2017)
rGO-WO_3	Sulfamethoxazole	Visible light, catalyst: 1 g/L	98% in 180 min	Zhu et al. (2017)
MIL-68(In)-NH_2/GO	Amoxicillin	Visible light, catalyst: 0.6 g/L	93% in 120 min	Yang et al. (2017)
Graphene/TNT[1]	Amoxicillin	UVA light	96.94% in 175 min	Song et al. (2016)
nano-GO/magnetite	Ciprofloxacin	Catalyst: 0.5 g/L	93% in 30 min	DESWATER (2021).
ARGs				
TiO_2-rGO-Ag	sul1, ampC, ermB, mecA[1]	Solar treatment	Survival rate-12.2% after 10 min	O'Dowd et al. (2021)
rGO	pUC 18[1] and pBR 322[1]	Visible light	NIL	Saha et al. (2019)

(continued)

Table 16.4 (continued)

Photocatalyst	Pollutant	Experiment condition	Degradation (%)	References
PCPs				
nZnO/graphene	Triclosan	Visible light, catalyst: 1 g/L, Co: 8 mg/L	45% in 1 h	Rincón et al. (2018)
$Ag/BiVO_4/rGO$	Triclosan	Visible light, catalyst: 1 g/L, Co: 1 μg/L	100% in 100 min	Li et al. (2019)
rGO/CdS	Methyl paraben	Visible light, catalyst: 0.75 g/L, Co: 30 mg/L, pH 3	100% in 90 min	Mohan et al. (2021)
EDCs				
rGO/TiO_2	Decabromodephenyl ether	UV, catalyst: 0.5 g/L	Degradation of 72% and debromination of 59.4% in 12 h	Lei et al. (2014)
$Graphene/TiO_2$ nanotube	Dibutyl phthalate	UV lamp (150 W), Co: 10 mg/L	87.9% in 90 min	Wang et al. (2019)

1. *SOF* Side glowing optical fibre; *NS* Nanosheets; *TNT* Titanium nanotubes, *mec* methicillin-resistant gene, *pBR 322* Plasmids with ARGs of ampicillin and tetracycline, *pUC 18* Plasmids with ARG of ampicillin and its growth promoter

5 mg/L, 0.8 g/L, 6 and 25 °C, respectively. The studies revealed that the graphene-modified catalyst achieved a degradation efficiency of 96.1% within a period of 4 h at a graphene loading of 0.5 wt%. This can be attributed to the increased surface area of graphene, which enabled SMT preconcentration onto the composite surface via π-π and hydrophobic interactions. Besides, the potential of graphene to function as an electron mediator and acceptor also inhibited the recombination of electron–hole pairs, thereby improving the degradation rate. Furthermore, the discrete performance of G/A/TNS in the visible light region was also probed by the authors. It was reported that visible light contributed to higher degradation efficiency (70.9%) of SMT than UV, validating the synergistic potential of graphene in the efficient removal of SMT. In addition, the influence of pH on catalytical efficiency was also evaluated. It was reported that at pH < 6, titania particles tend to leach out of the

composite. Contrastingly, the preconcentration of SMT decreased due to 'negative charge' buildup within the adsorbent at pH > 6. Hence, pH 6 was optimized for the studies. At the optimized pH, the reusability of the photocatalyst showed no loss in removal efficiency and Ti leaching for five consecutive cycles (Liu et al. 2019). Similarly, studies on 4-chlorophenol (5 mg/L) using G/A/TNS reported removal efficiency of 99.2% in 120 min for an optimized graphene content of 2 wt%, pH 7, and dosage of 0.2 g/L. The reaction was driven by •OH, causing the cleavage of bonds that are more prone to electrophile attack (Li et al. 2018). In another study, Ag_3PO_4-graphene synthesized via chemical precipitation exhibited enhanced photocatalytic activity in removing SMX. The decomposition of SMX (1 mg/L) was observed within 30 min. Additionally, the behaviour of the catalyst with increasing silver NPs was also investigated. It was observed that increased loading hindered the light penetration efficiency resulting in a deteriorated performance of the catalyst (Xiang et al. 2015).

Anirudhan and Deepa synthesized a novel photocatalyst, zinc oxide /graphene oxide/ nanocellulose (ZnO-GO/NC), for the photodegradation of ciprofloxacin (CIP). The composite enabled the tuning of the GO bandgap from 1.7 to 2.8 eV, thereby promoting excitation within the visible light spectrum. The adsorption kinetics followed pseudo-first-order with Sips isotherm serving the best fit. Compared to the composite, GO exhibited slower kinetics due to surface recombination of GO with CIP molecules. The presence of COOH and OH groups in GO enhanced the adsorption of CIP molecules, favouring rapid photodegradation within a time interval of 60 min. Consequently, the photocatalytic degradation efficiency of 98% was achieved due to electrostatic interaction between the functional groups on the catalyst and CIP molecules. The study also reported the photosensitizer ability of CIP molecules, thereby resulting in excited CIP to eject electrons onto the conduction band of ZnO-GO/NC. During the process, the oxidized form of CIP molecules underwent decomposition to yield degraded products. The reusability of the composite after five consecutive cycles portrays it as a potential material for removing CIP from the aquatic environment (Anirudhan and Deepa 2017). Another study by the same authors reported a similar trend in the photodegradation of aureomycin hydrochloride (AMH) by chemically modified GO/ nanohydroxyapatite (CMGO/nHA) out using a solar source of intensity 100 mW/cm^2. Based on the experimental study, the initial concentration of the analyte played a significant role in the removal efficiency of the analyte. Since concentration gradient is the driving force, the adsorption capacity correlated positively with initial concentration, enhancing photoactivity at high concentrations. The composite exhibited a degradation ability of 97.0% towards AMH within 42 min at an optimum pH of 5.5 due to COOH, NH_2 and OH groups, which easily interacted with AMH molecules through hydrogen bonding (Anirudhan et al. 2016).

Photostable GO/Fe based Material of Institute Lavoisier (MIL) was chosen as effective candidates for the remediation of atrazine (ATZ) and diazinon (DIZ). The tests were conducted under visible light using a 400 W mercury lamp for an initial concentration of 30 mg/L and dosage of 0.4 g/L. The concentrations of the analytes after oxidation were detected using high-performance liquid chromatography. Based on adsorption experiments, the removal efficiency of ATZ was higher than DIZ due to π-π interaction between triazine ring and unlocalized electrons on GO, van der Waals interactions and hydrogen bonding. The photoactivity of GO/MIL-Fe achieved 100% and 81% degradation efficiencies for DIZ and ATZ, respectively, with pH playing an insignificant role in the degradation reaction. The performance may be attributed to the elevated production of •OH by GO by reducing Fe^{3+} to Fe^{2+} Fakhri et al. (2020).

EDCs and PFAs such as perfluorooctanoic acid (PFOA) and perfluoro-octane sulfonate are refractory pollutants known to cause cancer, obesity, respiratory issues, congenital disabilities and developmental disorders in humans. Hence, the elucidation of the photocatalysis mechanism is necessary to identify their roles in the degradation activity. A study evaluated the degradation of 4-nitrophenol, an EDC by graphene/TiO_2 over a high-pressure Hg lamp source. Batch studies were carried out in a 100 mL flask with 10 mg/L of 4-nitrophenol and graphene/TiO_2 dosage of 1 g/L for 30 min to reach adsorption equilibrium. The effect of GO concentration on the degradation of EDC was evaluated. It was observed that when the concentration was increased from 0.125–0.5 g/L, there was an enhancement in electron transfer. However, beyond 0.5 g/L, the catalytic efficiency decreased due to the shielding effect. The degradation reaction proceeds via electrophilic •OH attacking dense carbon atoms forming 4-nitrocatechol. Due to the presence of OH groups, the electron density is high at ortho and para positions of 4-nitrophenol. Subsequently, the electrophilic substitution of •OH occur at the $–NO_2$ group position yielding 1,2,4-benzenetriol as metabolites (Liu et al. 2017).

In another study, PFOA degradation was examined using GNS/In_2O_3 for a PFOA concentration of 30 mg/L and photocatalyst dosage of 0.5 g/L, using a 15 W UV light source at neutral pH. On incorporating GNS to In_2O_3, it enwrapped on top of In_2O_3 NPs in Li et al. (2013). Li and his coworkers concluded that when GNS covered the NPs completely, it reduced the efficiency due to a reduction in exposed surface area. However, on partial coverage, the GNS functioned as an electron carrier from the semiconductor to the electron acceptor, contributing to the efficient separation of the photogenerated hole–electron pairs. The composite In_2O_3-graphene achieved 87% and 60.9% of degradation and defluorination, respectively (Li et al. 2013). In another study, Huang et al. (2016) investigated PFOA decomposition by Si/graphene in a cylindrical reactor with a UV lamp source (5 W). Preliminary experiments were conducted in the dark for 60 min until no further adsorption occurred. Decomposition experiments were carried out by suspending 40 mg of the composite to 0.12 mmol/L of PFOA solution for 8 h. The author was the first to report the photoinduced hydro-defluorination mechanism by the catalyst. It was proposed that F atoms in perfluoroalkyl groups could be replaced by H atoms owing to the nucleophilic substitution reaction, as shown in Fig. 16.9c. The mechanism followed the development of Si–H bonds generated under UV light excitation and the replacement of F atoms by H atoms of the Si–H bonds to form $C_nF_{2n}HCOOH$. Moreover, the photogenerated electrons from Si/graphene catalyst also reacted with PFOA to generate C_7F_{15}• radicals (Huang et al. 2016).

Subsequently, the presence of excessive oxygen-containing groups in GO can result in a decrease in intrinsic features such as conductivity. With properties similar

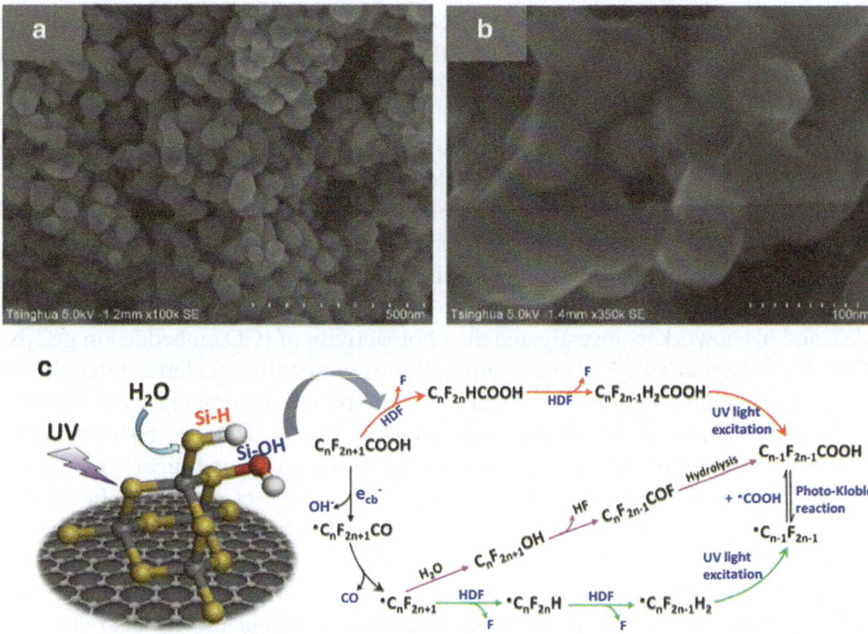

Fig. 16.9 **a** SEM image of pristine In_2O_3, **b** SEM image of GNS coated In_2O_3, **c** Nucleophilic substitution reaction between perfluoroalkyl groups and Si/graphene. Reprinted with the permission of Elsevier and ACS publications (Li et al. 2013; Huang et al. 2016)

to graphene, rGO improves the light efficiency for photocatalysts by accelerating charge transfer from excited semiconductors to electron acceptors. Tang et al. (2015) investigated the photoactivity of rGO-CdS/ZnS in degrading tetracycline (TC) under visible light. It was reported that excited electron from the conduction band of CdS is driven to ZnS by the band potential. The photogenerated free electrons from ZnS are then captured by rGO, which aids in the formation of singlet oxygen and H_2O_2 radicals. Simultaneously, the holes in ZnS migrate to the valence band in CdS, thereby oxidizing ~85% of TC in 60 min into less hazardous products. However, OH radicals' formation was not significant, thus indicating holes as the main reactive species in the system. The photostable rGO-CdS/ZnS exhibited a light scattering effect at high rGO loading, thereby decreasing the catalyst performance (Pastrana-Martínez et al. 2013). In another study, norfloxacin (NOR) was mineralized using rGO/Bi_2WO_6 under visible light irradiation by the oxidation of piperazine ring and hydroxy group attack on quinolone ring. Based on the results analyzed by reactive-species-trapping experiments, OH and electrons played a significant role during the process. The incorporation of rGO facilitated the transfer of electrons, thereby reducing NOR to CO_2, H_2O and small molecules (Zhao et al. 2021). Chen et al.(2017) reported the photoactivity of Z-scheme catalyst, Ag_3PO_4/Ag/$BiVO_4$/rGO showed the optimal degradation performance, 96.2% in 10 min, with catalysis dosage of 0.5 g/L, pH 6.75 and 0.5 M of Na_2SO_4 as the supporting electrolyte. The addition of Ag improved the electron transfer capability from Ag_3PO_4 to graphene and thus hindered the recombination of electron–hole pairs. He further explored the influence of coexisting ions and water sources on the degradation activity of TC. In the presence of electrolytes (0.05M), TC removal was observed to decrease in the order $Na_2SO_4 > NaCl > Na_2CO_3$. This can be attributed to the competition between Cl^-, CO_3^{2-}, SO_4^{2-} and TC on the catalyst surface. To identify the composite's field compatibility, experiments were carried out on various water sources. The deionized, medical, municipal, river and tap wastewater followed a removal efficiency of 94.96%, 84.21%, 83.43%, 80.21% and 87.23%, respectively, indicating the long term potential of the composite in water purification (Chen et al. 2017).

Yu and his coworkers investigated the photoactivity of rGO embedded in g-C_3N_4 (rGO/gC_3N_4) against BPA. This incorporation significantly accelerated the photolysis rate of BPA under visible light than its counterparts at neutral pH. Consequently, a photo mineralization rate of 80% was achieved in 4 h. This can result from the formation of singlet oxygen through electron reduction of O_2 and superoxide radical formed by the oxidation of photogenerated holes (Awfa et al. 2018). In another study, experimental studies were conducted using rGO-WO_3 (RW) heterojunction to remove SMX effectively. About 98% of SMX was achieved under 3 h of visible light irradiation using RW composite synthesized by varying amounts of tungsten trioxide. However, with the increase in WO_3, the degradation efficiency was also observed to decrease, as in Fig. 16.10. Meanwhile, as displayed in the inset, RW composites showed higher TOC removal efficiency than that WO_3. The improved photocatalytic performance was explained as the result of increased surface area, which improves adsorption. The study concluded that the p–n junction formed enhanced the electron–hole pair generation and suppressed recombination. To study the role of various radicals during the photocatalytic process, sodium oxalate (SO), isopropanol (IPA) and benzoquinone (BQ) acting as holes, OH^- radical and O_2^-radical scavengers were added as in Fig. 16.10b. The presence of BQ and IPA remarkably inhibited

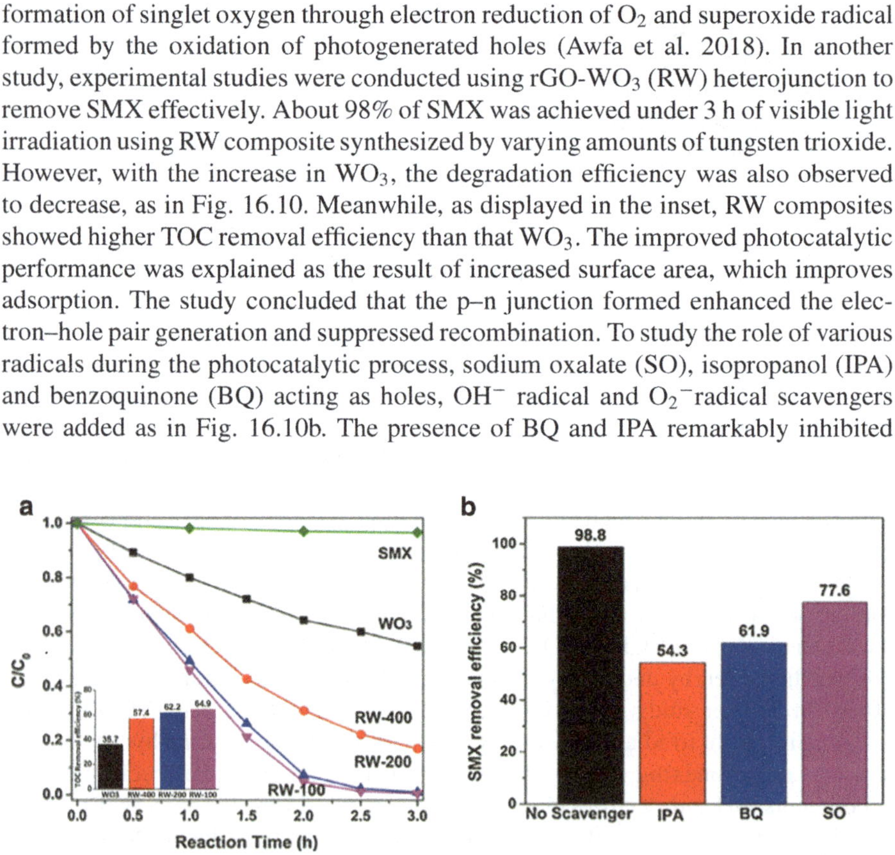

Fig. 16.10 Degradation of SMX under visible light over WO_3, and RW. Inset in (**a**) shows TOC removal efficiency of SMX with WO_3, and RW, **b** degradation performance of RW in SMX removal with the presence of various scavengers. Reprinted with the permission of Elsevier (Zhu et al. 2017)

the degradation of SMX under visible light, which demonstrates that h^+ played a minor role in the degradation process (Zhu et al. 2017). In another study, Cu_2O/rGO photocatalyst was prepared by a simple wet-chemical route using $CuSO_4 \cdot 5H_2O$ and GO as precursors and tested for photodegradation SMX under visible light (Liu et al. 2016). The increased photocatalytic activity was ascribed to enhanced dispersion of Cu_2O NPs, a p-type semiconductor with small bandgap energy of 2.0 eV, on the surface of graphene sheets. In addition, rGO also facilitated the removal by serving as an effective adsorbent of the pollutant species and as a charge acceptor to promote the separation and transfer of photogenerated carriers.

The elevated presence of ECs in water bodies has led to the formation of ARGs within organisms. This resistance can result in the decreased performance of antibiotics, thereby increasing the illness, cost and length of treatment. Karaolia et al. (2018) set up a laboratory-scale solar simulator (1 kW Xenon lamp) to carry out photocatalytic experiments on ARG using graphene-based TiO_2 composites. The experiments were done at the inherent pH of MBR effluent, which was slightly acidic under air-equilibrated conditions. A catalyst concentration of 100 mg/L was suspended in the solution using sonication (10 min) followed by magnetic stirring at 500 rpm at 60 min for ARG inactivation in the dark to ensure complete equilibration of adsorption/desorption onto the catalyst surface. Later, the solar lamp was turned on for a photocatalytic treatment time of 180 min for ARG inactivation. The analyses reported minor visible changes for sulfonamide and macrolide-resistant genes: *sul*1, *ermB*. However, abundant reduction of the ampicillin-resistant gene, ampC, was observed due to intracellular DNA damages of bacteria produced by photo-irradiated TiO_2-rGO (Karaolia et al. 2018).

16.4 Photocatalytic Reactor Design

The large-scale setting up of photocatalytic reactors pose constraints due to significant energy consumption, uneven photon distribution, inappropriate flow rate, or catalyst deactivation, directly affecting the photocatalytic efficiency. Hence, optimizing a reactor is vital to balance the treated solution's quantity and the degraded ECs. Generally, the photoreactors are categorized based on geometrical shape, photocatalyst size, morphology, fluid dynamics or applications. Figure 16.11 represent a brief classification of photocatalytic reactors reported in various literatures.

The simplest type of photoreactor design, slurry photoreactor, have a three-phase interface: the photocatalyst as the solid, water as the liquid, and vapours as the gaseous phase. The system is more dependent on mass transfer limitation due to negligible exposure of the catalyst to the light source and can be operated as continuous, batch or plug flow. Amidst this, studies on continuous and plug flow reactors are vital for scaling up the reactor for industrial-scale applications. However, the use of fixed beds in continuous flow reactors reduces photocatalytic efficiency. Generally, this limitation is overcome by the use of micro-sized powders as catalysts dispersed in the system. Nevertheless, the susceptibility of the system catalyst fouling has a

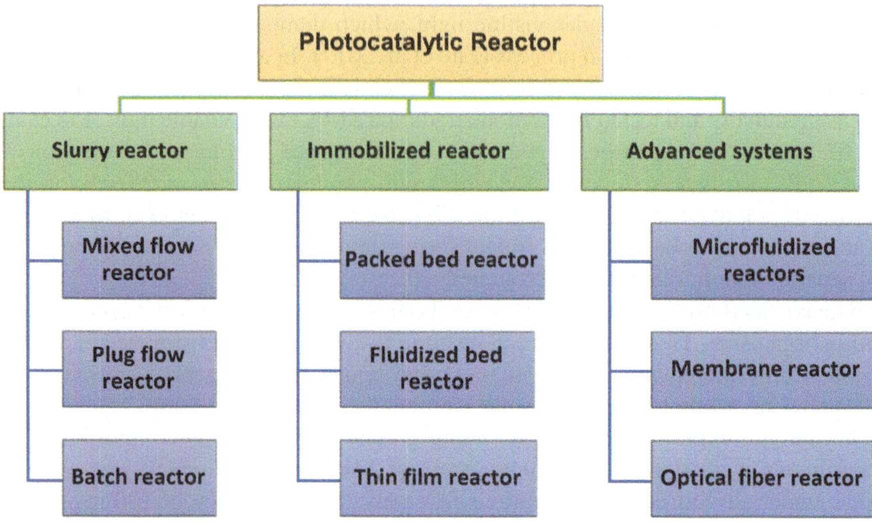

Fig. 16.11 Classification of photocatalytic reactor models

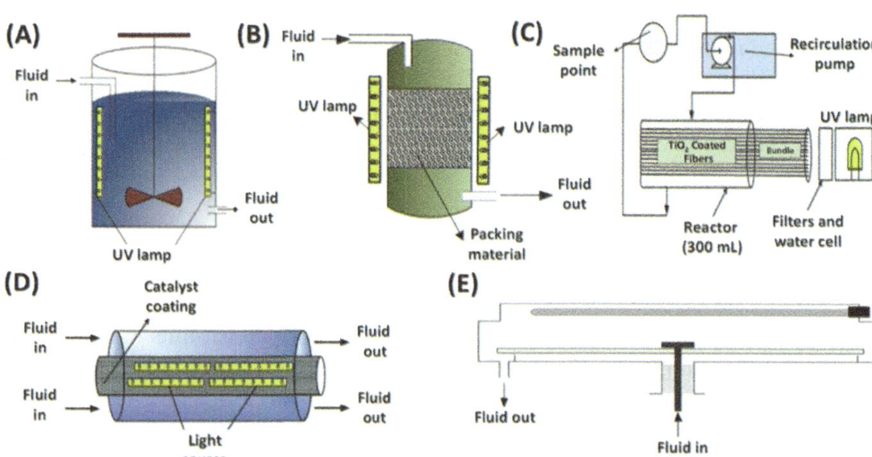

Fig. 16.12 Optimized configurations of various photocatalytic reactors. **a** Mixed flow reactor. **b** Packed bed reactor. **c** Optical fiber membrane reactor. **d** Annular reactor. **e** Thin film reactor. (Modified from Sargolzaei and Hedayati Moghaddam 2015; Azevedo et al. 2009; Gan et al. 2007)

tendency to hinder the continuous flow of the reactor (Yu and Yu 2008; Das and Mahalingam 2020).

Figure 16.12a, b show an illustration of slurry-based reactors. The immobilized reactor system is more advanced than the slurry systems in terms of catalyst arrangement, thereby cutting down downstream processing costs. In this system, the photocatalyst is supported on packing materials or coated on the transparent sheets during the synthesis phase, as in Fig. 16.12d, e. This enhances the surface area, which consequently augments the mass transfer rate. The advanced reactors are a more customized form of photoreactors to address the conventional system's drawbacks and increase efficiency and quantum yield. Unlike the usage of lamps in conventional systems, the advanced system utilizes optical fibre (OFC) and light-emitting diodes (LED), as shown in Fig. 16.12c (Claes et al. 2019; Tugaoen et al. 2018; Levchuk and Sillanpää 2020). These light sources have an excellent ability to reduce variations in wavelength, compatibility in immersed systems, and thereby improve the quantum yield. Since the catalyst is coated on support material like OFC or LED, the type of support material also influences the reaction process, reactor efficiency, and cost of fabrication/operation (Dijkstra et al. 2003).

16.5 Conclusion

Adequate access to fresh water is critical for the survival of humankind. The mismanagement of water resources, water pollution, and the increasing demand for freshwater are causing significant stress on the existing water resources. Along with conventional organic, inorganic, and biological pollutants, the rise of emerging contaminants in water has been a massive area of concern for the past few years. These contaminants coexist in various matrices based on their different physicochemical properties. Prolonged exposure to emerging contaminants has been linked to multiple health effects, including cancerous and non-cancerous diseases. Several treatment technologies are developed to remove them in water, given their toxicity and widespread occurrence. Photocatalysis is an advanced oxidation process extensively studied for the removal of emerging contaminants in water. However, the large-scale deployment of the process is hindered due to catalytic fouling, interferences from other ions, poor photostability, high energy consumption, and recombination of charges.

The recent advancement in nanotechnology has facilitated the synthesis of catalysts at the nanoscale with enhanced properties, including visible-range illumination, enhanced separation of photogenerated charges, improved photostability, reduced recombination, and photo corrosion. Semiconductors modified with graphene-based derivatives are promising catalysts for the purification of water contaminated with emerging contaminants. They are preferred due to their high surface area, reduced recombination, enhanced photoinduced separation, and increased mobility of charge carriers. However, the mechanistic details of the functioning of graphene-modified catalysts are not fully understood. The substantial advantages of graphene over its allotropic family are not fully exploited due to the lack of rational design and engineering of graphene surfaces. Moreover, the conventional methodology of dispersing semiconducting materials over graphene surfaces is ineffective in utilizing its structural advantage. Consequently, poor interfacial contact between the semiconductor and graphene is obtained, thereby affecting photoactivity. Hence there is a need to develop more efficient synthesis methods or strategies to make full use of the structural and electronic merits of graphene. Though a few reports confirm the formation of a Schottky junction between semiconductor and graphene, it is still under debate and not understood fully. In addition, interfacial mediators such as transition metal ions could optimize the charge transfer in graphene-modified photocatalysts' interface. In this regard, a combined effort between experiment and theory would be desirable to have an in-depth insight into the microscopic charge transfer pathway. This can act as a guide towards the design target of smarter graphene-based photocatalysts for potential applications.

Another critical aspect that requires attention is concerning the approaches for producing graphene and its derivatives with high quality, such as high purity, few defects, and controlled size. So far, strong oxidative exfoliation of graphite followed by chemical reduction of GO is used to prepare graphene. Although this approach is effective for the mass production of graphene, it inevitably produces many defects that could affect the electrical conductivity and mechanical strength of graphene, thereby affecting the photodegradation efficacy. Hence, the large-scale production of graphene composite at an affordable cost is with minimal defects or oxidation sites still poses a challenge for field applications. Hence, strategic planning and design of photocatalysis setup are essential for fabricating energy-efficient and cost-effective reactors. Until now, no precise regulation has been developed on which photoreactor model is suitable for all cases. Hence, optimization of a photoreactor design required continuous effort to be applicable in large-scale treatment systems. Coupling with other treatment methods can also boost the performance and represent a possible alternative to the future.

References

Alexandros IS , Julie AB (2016) A review of emerging contaminants in water: Classification, sources, and potential risks. In Impact of Water Pollution on Human Health and Environmental Sustainability, pp 55–80

Abdulrazaq Y, Abdulsalam A, Rotimi AL et al (2020) Classification, potential routes and risk of emerging pollutants/contaminant. In: Emerging contaminants. https://doi.org/10.5772/intechopen.94447

Adesina AA (2004) Industrial exploitation of photocatalysis: progress, perspectives and prospects. Catal Surv Asia 8:265–273. https://doi.org/10.1007/s10563-004-9117-0

Amit P, Mackraj I, Govender T (2021) Advances in sepsis diagnosis and management: A paradigm shift towards nanotechnology. In Journal of Biomedical Science:1–30. https://dx.doi.org/10.1186%2Fs12929-020-00702-6

Ahmed MB, Zhou JL, Ngo HH et al (2017) Progress in the biological and chemical treatment technologies for emerging contaminant removal from wastewater: a critical review. J Hazard Mater 323:274–298. https://doi.org/10.1016/j.jhazmat.2016.04.045

Ahmed, Forruque S (2021) Progress and challenges of contaminate removal from wastewater using microalgae biomass. In: Chemosphere. https://www.academia.edu/50345530/Progress_and_challenges_of_contaminate_removal_from_wastewater_using_microalgae_biomass. Last accessed on 8 Aug 2021

Andreozzi R, Raffaele M, Gabriele P, Antonino P (2002) Carbamazepine in water: persistence in the environment, ozonation treatment and preliminary assessment on algal toxicity. Water Res 36:2869–2877. https://doi.org/10.1016/S0043-1354(01)00500-0

Andreozzi R, Vincenzo C, Raffaele M, Anita R (2003) Ozonation and H_2O_2/UV treatment of clofibric acid in water: a kinetic investigation. J Hazard Mater 103:233–246. https://doi.org/10.1016/j.jhazmat.2003.07.001

Anirudhan TS, Deepa JR (2017) Nano-zinc oxide incorporated graphene oxide/nanocellulose composite for the adsorption and photo catalytic degradation of ciprofloxacin hydrochloride from aqueous solutions. J Colloid Interface Sci 490:343–356. https://doi.org/10.1016/j.jcis.2016.11.042

Anirudhan TS, Deepa JR, Nair AS (2016) Fabrication of chemically modified graphene oxide/nano hydroxyapatite composite for adsorption and subsequent photocatalytic degradation of aureomycine hydrochloride. J Ind Eng Chem 47. https://doi.org/10.1016/j.jiec.2016.12.014

Aranami K, James WR (2007) Photolytic degradation of triclosan in freshwater and seawater. Chemosphere 66:1052–1056. https://doi.org/10.1016/j.chemosphere.2006.07.010

Aris AZ, Mohd Hir ZA, Razak MR (2020) Metal-organic frameworks (MOFs) for the adsorptive removal of selected endocrine disrupting compounds (EDCs) from aqueous solution: a review. Appl Mater Today 21:100796. https://doi.org/10.1016/j.apmt.2020.100796

Awfa D, Ateia M, Fujii M et al (2018) Photodegradation of pharmaceuticals and personal care products in water treatment using carbonaceous-TiO_2 composites: a critical review of recent literature. Water Res 142:26–45. https://doi.org/10.1016/j.watres.2018.05.036

Azevedo EB, Tôrres AR, Neto A et al (2009) TiO_2-Photocatalyzed degradation of phenol in saline media in an annular reactor: hydrodynamics, lumped kinetics, intermediates, and acute toxicity. Braz J Chem Eng 26(1):75–87. https://doi.org/10.1590/S0104-66322009000100008

Balakrishna K, Rath A, Praveenkumarreddy Y et al (2017) A review of the occurrence of pharmaceuticals and personal care products in Indian water bodies. Ecotoxicol Environ Saf 137:113–120. https://doi.org/10.1016/j.ecoenv.2016.11.014

Barron L, Nesterenko E, Hart K, Power E, Quinn B, Kelleher B, Paull B (2010) Holistic visualization of the multimodal transport and fate of twelve pharmaceuticals in biosolid enriched topsoils. In Analytical and Bioanalytical Chemistry 287–296: https://doi.org/10.1007/s00216-010-3494-1

Belver C, Bedia J, Peñas-Garzón M et al (2020) Structured photocatalysts for the removal of emerging contaminants under visible or solar light. In: Visible light active structured photocatalysts for the removal of emerging contaminants, 41–98

Cai Q, Hu J (2018) Effect of UVA/LED/TiO_2 photocatalysis treated sulfamethoxazole and trimethoprim containing wastewater on antibiotic resistance development in sequencing batch reactors. Water Res 140:251–260. https://doi.org/10.1016/j.watres.2018.04.053

Canada E, CC (2020) Bisphenol A in water and sediment. https://www.canada.ca/en/environment-climate-change/services/environmental-indicators/bisphenol-a-water-sediment.html. Last accessed on 10 Aug 2021

Cao M, Wang P, Ao Y et al (2016) Visible light activated photocatalytic degradation of tetracycline by a magnetically separable composite photocatalyst: Graphene oxide/magnetite/cerium-doped Titania. J Colloid Interface Sci 467:129–139. https://doi.org/10.1016/j.jcis.2016.01.005

Carson R (2002) Silent spring https://rachelcarson.org/SilentSpring.aspx. Last accessed on 30 July 2021

Chen F, Yang Q, Wang S et al (2017) Graphene oxide and carbon nitride nanosheets co-modified silver chromate nanoparticles with enhanced visible-light photoactivity and anti-photo corrosion properties towards multiple refractory pollutants degradation. Appl Catal b Environ 209:493–505. https://doi.org/10.1016/j.apcatb.2017.03.026

Chen F, Yang Q, Li X, Zeng G, Wang D, Niu C, Deng Y (2017) Hierarchical assembly of graphene-bridged Ag_3PO_4/Ag/$BiVO_4$ Z-scheme photocatalyst: An efficient, sustainable and heterogeneous catalyst with enhanced visible-light photoactivity towards tetracycline degradation under visible light irradiation. In Appl Catal B:330–342. https://dx.doi.org/10.1016/j.apcatb.2016.07.021

Choi H, Stathatos E, Dionysiou DD (2007) Photocatalytic TiO_2 films and membranes for the development of efficient wastewater treatment and reuse systems. Desalination 202:199–206. https://doi.org/10.1016/j.desal.2005.12.055

Chua H, Chen YF (1995) A novel adsorption-anaerobiosis column for the removal of persistent organics from contaminated water. Mar Pollution Bull 31:313–316. https://doi.org/10.1016/0025-326X(95)00173-K

Claes T, Dilissen A, Leblebici ME et al (2019) Translucent packed bed structures for high throughput photocatalytic reactors. Chem Eng J 361:725–735. https://doi.org/10.1016/j.cej.2018.12.107

Cuprys A, Lecka J, Proulx F et al (2019) Appearance of ciprofloxacin/chlortetracycline-resistant bacteria in waters of Québec City in Canada. J Infect Public Health 12:897–899. https://doi.org/10.1016/j.jiph.2019.04.012

Danner MC, Robertson A, Behrends V et al (2019) Antibiotic pollution in surface fresh waters: occurrence and effects. Sci Total Environ 664:793–804. https://doi.org/10.1016/j.scitotenv.2019.01.406

Das S, Mahalingam H (2020) Novel immobilized ternary photocatalytic polymer film-based airlift reactor for efficient degradation of complex phthalocyanine dye wastewater. J Hazard Mater 383:121219. https://doi.org/10.1016/j.jhazmat.2019.121219

DESWATER (2021) https://www.deswater.com/home.php. Last accessed on 30 July 2021

Dey S, Bano F, Malik A (2019). Pharmaceuticals and personal care product (PPCP) contamination—a global discharge inventory. In: Pharmaceuticals and personal care products: waste management and treatment technology, 1–26. https://doi.org/10.1016/B978-0-12-816189-0.00001-9

Dhangar K, Kumar M (2020) Tricks and tracks in removal of emerging contaminants from the wastewater through hybrid treatment systems: a review. Sci Total Environ 140320. https://doi.org/10.1016/j.scitotenv.2020.140320

Dijkstra MFJ, Koerts ECB, Beenackers AACM et al (2003) Performance of immobilized photocatalytic reactors in continuous mode. In AIChE Journal 49:734–744. https://doi.org/10.1002/aic.690490317

Djurišić AB, He Y, Ng AMC (2020) Visible-light photocatalysts: prospects and challenges. APL Mater 8:030903. https://doi.org/10.1063/1.5140497

Dongli W, Li J, Chen Z, Liang L, Ma J, Wei M, Ai Y, Wang X (2020) Understanding bisphenol-A adsorption in magnetic modified covalent organic frameworks: Experiments coupled with DFT calculations. In Journal of Molecular Liquids. https://doi.org/10.1016/j.scitotenv.2020.142177

Dong S, Li Y, Sun J et al (2014a) Facile synthesis of novel ZnO/rGO hybrid nanocomposites with enhanced catalytic performance for visible-light-driven photodegradation of metronidazole. Mater Chem Phys 145:357–365. https://doi.org/10.1016/j.matchemphys.2014.02.024

Dong S, Sun J, Li Y et al (2014b) $ZnSnO_3$ hollow nanospheres/reduced graphene oxide nanocomposites as high-performance photocatalysts for degradation of metronidazole. Appl Catal B Environ 144:386–393. https://doi.org/10.1016/j.apcatb.2013.07.043

Dong X, Meng QW, Hu W, Chen R, Ge Q (2022) Forward osmosis membrane developed from the chelation of Fe3+ and carboxylatefor trace organic contaminants removal. In Chemical Engineering Journal. https://doi.org/10.1016/j.cej.2021.131091

El-Gohary FA, Abou-Elela SI, Aly HI (1995) Evaluation of biological technologies for wastewater treatment in the pharmaceutical industry. Water Sci Technol 32:13–20. https://doi.org/10.1016/0273-1223(96)00112-6

Fakhri H, Farzadkia M, Boukherroub R, Srivastava V, Sillanpaa M (2020) Design and preparation of core-shell structured magnetic graphene oxide@ MIL-101 (Fe): Photocatalysis under shell to remove diazinon and atrazine pesticides. In Solar Energy 990–1000: https://doi.org/10.1016/j.solener.2020.08.050

Feiyan W, Xin L, Sumei Y et al (2021) Chemical factors affecting uptake and translocation of six pesticides in soil by maize (Zea mays L.). J Hazardous Mater. https://doi.org/10.1016/j.jhazmat.2020.124269.

Fujishima A, Rao TN, Tryk DA (2000) Titanium dioxide photocatalysis. J Photochem Photobiol C 1:1–21. https://doi.org/10.1016/S1389-5567(00)00002-2

Gan WY, Lam SW, Chiang K et al (2007) Novel TiO_2 thin film with non-UV activated superwetting and antifogging behaviours. J Mater Chem 17(10):952–954. https://doi.org/10.1039/B618280A

Gander M, Jefferson B, Judd S (2000) Aerobic MBRs for domestic wastewater treatment: a review with cost considerations. Sep Purif Technol 18:119–130. https://doi.org/10.1016/S1383-5866(99)00056-8

Gaya UI, Abdullah AH (2008) Heterogeneous photocatalytic degradation of organic contaminants over titanium dioxide: a review of fundamentals, progress and problems. J Photochem Photobiol c Photochem Rev 9:1–12. https://doi.org/10.1016/j.jphotochemrev.2007.12.003

Geissen V, Mol H, Klumpp E et al (2015) Emerging pollutants in the environment: a challenge for water resource management. Int Soil Water Conserv Res 3:57–65. https://doi.org/10.1016/j.iswcr.2015.03.002

Ghosh S, Badruddoza AZMd, Hidajat K, Uddin MS (2013) Adsorptive removal of emerging contaminants from water using superparamagnetic Fe_3O_4 nanoparticles bearing aminated β-cyclodextrin. J Chem Eng 1:122–130. https://doi.org/10.1016/j.jece.2013.04.004

Guo K, Wu Z, Fang J (2020) Chapter 10—UV-based advanced oxidation process for the treatment of pharmaceuticals and personal care products. In: Contaminants of emerging concern in water and wastewater: 367–408. https://doi.org/10.1016/B978-0-12-813561-7.00010-9

Gupta SS, Chakraborty I, Maliyekkal SM et al (2015a) Simultaneous dehalogenation and removal of persistent halocarbon pesticides from water using graphene nanocomposites: a case study of lindane. ACS Sustain Chem Eng 3:1155–1163. https://doi.org/10.1021/acssuschemeng.5b00080

Gupta VK, Eren T, Atar N et al (2015) $CoFe_2O_4$@TiO_2 decorated reduced graphene oxide nanocomposite for photocatalytic degradation of chlorpyrifos. J Mol Liquids 208: 122–129. https://doi.org/10.1016/j.molliq.2015.04.032

Hennig H (2015) Semiconductor photocatalysis: principles and applications. Angew Chem Int Edn 54: 4429–4429. Wiley. https://doi.org/10.1002/anie.201501876

Hennig H, Billing R (1993) Advantages and disadvantages of photocatalysis induced by light-sensitive coordination compounds. Coord Chem Rev 125:89–100. https://doi.org/10.1016/0010-8545(93)85010-2

Hong R, Pan T, Qian J et al (2006) Synthesis and surface modification of ZnO nanoparticles. Chem Eng J 119:71–81. https://doi.org/10.1016/j.cej.2006.03.003

Howard R, Diego H, Quiñones M, Diego M (2020) Emerging contaminants as global environmental hazards: a bibliometric analysis. In: Emerging contaminants, 179–193. https://doi.org/10.1016/j.emcon.2020.05.001

Hu C, Lu T, Chen F et al (2013) A brief review of graphene–metal oxide composites synthesis and applications in photocatalysis. J Chin Adv Mater Soc 1:21–39. https://doi.org/10.1080/22243682.2013.771917

Huang D, Yin L, Niu J (2016) Photoinduced Hydrodefluorination mechanisms of perfluorooctanoic acid by the SiC/graphene catalyst. Environ Sci Technol 50:5857–5863. https://doi.org/10.1021/acs.est.6b0065251

Hughes SR, Kay P, Brown LE (2013) Global synthesis and critical evaluation of pharmaceutical data sets collected from river systems. Environ Sci Technol 47:661–677. https://doi.org/10.1021/es3030148

Huo P, Zhou M, Tang Y et al (2016) Incorporation of N-ZnO/CdS/Graphene oxide composite photocatalyst for enhanced photocatalytic activity under visible light. J Alloys Compd 670:198–209. https://doi.org/10.1016/j.jallcom.2016.01.247

Huo P, Guan J, Zhou M et al (2017) Carbon quantum dots modified CdSe loaded reduced graphene oxide for enhancing photocatalytic activity. J Ind Eng Chem 50:147–154. https://doi.org/10.1016/j.jiec.2017.02.008

Ismail WNW, Mokhtar SU (2020) various methods for removal, treatment, and detection of emerging water contaminants. In: Emerging contaminants. IntechOpen. https://doi.org/10.5772/intechopen.93375

Jaishankar M, Tseten T, Anbalagan N et al (2014) Toxicity, mechanism and health effects of some heavy metals. Interdisc Toxicol 7:60–72. https://doi.org/10.2478/intox-2014-0009

James G, Jyoti B, Viki RC, Kathiresan P, Rathore Anurag S (2018) Monitoring and control of bioethanol production from lignocellulosic biomass. In Waste Biorefinery 727–749: https://doi.org/10.1016/B978-0-444-63992-9.00025-2

Jiang L, Hu X, Yin D et al (2011) Occurrence, distribution and seasonal variation of antibiotics in the Huangpu River, Shanghai, China. Chemosphere 82:822–828. https://doi.org/10.1016/j.chemosphere.2010.11.028

Jo W, Selvam NCS (2017) Z-scheme CdS/g-C3N4 composites with rGO as an electron mediator for efficient photocatalytic H_2 production and pollutant degradation. Chem Eng J 317:913–924. https://doi.org/10.1016/j.cej.2017.02.129

Jones K (2016) Solar energy conversion and storage. Photochemical modes. Chromatographia 79: 1209–1209. https://doi.org/10.1007/s10337-016-3112-2

Kanakaraju D, Glass BD, Oelgemöller M (2018b) Advanced oxidation process-mediated removal of pharmaceuticals from water: a review. J Environ Manag 219:189–207. https://doi.org/10.1016/j.jenvman.2018.04.103

Kanakaraju D, Glass BD, Oelgemöller M (2018) Advanced oxidation process-mediated removal of pharmaceuticals from water: a review. J Environ Manage, 189–207. https://doi.org/10.1016/j.jenvman.2018.04.103

Karaolia P, Michael-Kordatou I, Hapeshi E et al (2018) Removal of antibiotics, antibiotic-resistant bacteria and their associated genes by graphene-based TiO_2 composite photocatalysts under solar radiation in urban wastewaters. Appl Catal b Environ 224:810–824. https://doi.org/10.1016/j.apcatb.2017.11.020

Karaush-Karmazin NN, Baryshnikov GV, Agren H et al (2020). Furans and their benzo derivatives: structure☆. In: Reference module in chemistry, molecular sciences and chemical engineering. https://doi.org/10.1016/B978-0-12-409547-2.14768-7

Khan MM, Adil SF, Al-Mayouf A (2015) Metal oxides as photocatalysts. J Saudi Chem Soc 19:462–464. https://doi.org/10.1016/j.jscs.2015.04.003

Kovner K (2009) Persistent organic pollutants: a global issue, a global response. https://www.epa.gov/international-cooperation/persistent-organic-pollutants-global-issue-global-response. Last accessed on 10 Aug 2021

Kumar A, Pandey G (2017) A review on the factors affecting the photocatalytic degradation of hazardous materials. Mater Sci Eng Int J 1:106–114. https://doi.org/10.15406/mseij.2017.01.00018

Kümmerer K (2011) Emerging contaminants. In: Treatise on water science, 69–87. https://doi.org/10.1016/B978-0-444-53199-5.00052-X

Kurt A, Mert BK, Özengin N, Sivrioğlu Ö, Yonar T (2017) Treatment of antibiotics in wastewater using advanced oxidation processes. In Physico-Chemical Wastewater Treatment and Resource Recovery. https://doi.org/10.5772/67538

Lalonde B, Garron C, Dove A et al (2019) Investigation of spatial distributions and temporal trends of triclosan in Canadian surface waters. Arch Environ Contam Toxicol 76:231–245. https://doi.org/10.1007/s00244-018-0576-0

Landmann M, Rauls E, Schmidt WG (2012). The electronic structure and optical response of rutile, anatase and brookite TiO_2. J Phys Condensed Matter 24: 195503. IOPScience. https://doi.org/10.1088/0953-8984/24/19/195503

Lei M, Wang N, Zhu L et al (2014) A peculiar mechanism for the photocatalytic reduction of decabromodiphenyl ether over reduced graphene oxide–TiO_2 photocatalyst. Chem Eng J 241:207–215. https://doi.org/10.1016/j.cej.2013.12.032

Levchuk I, Sillanpää M (2020) Chapter 1—Titanium dioxide–based nanomaterials for photocatalytic water treatment. In: Advanced water treatment, 1–56. https://doi.org/10.1016/b978-0-12-819225-2.00001-6

Li Z, Zhang P, Li J, Shao T, Jin L (2013) Synthesis of In_2O_3-graphene composites and their photocatalytic performance towards perfluorooctanoic acid decomposition. In J Photochem Photobiol A 271:111–116. https://doi.org/10.1016/j.jphotochem.2013.08.012

Li X, Yu J, Wageh S et al (2016) Graphene in photocatalysis: a review. Small 12:6640–6696. https://doi.org/10.1002/smll.201600382

Li F, Du P, Liu W et al (2018) Hydrothermal synthesis of graphene grafted titania/titanate nanosheets for photocatalytic degradation of 4-chlorophenol: solar-light-driven photocatalytic activity and computational chemistry analysis. Chem Eng J 331:685–694. https://doi.org/10.1016/j.cej.2017.09.036

Li M, Xu G, Guan Z et al (2019) Synthesis of Ag/$BiVO_4$/rGO composite with enhanced photocatalytic degradation of triclosan. Sci Total Environ 664. https://doi.org/10.1016/j.scitotenv.2019.02.027

Liang S, Zhou Y, Kang K et al (2017) Synthesis and characterization of porous TiO_2-NS/Pt/GO aerogel: a novel three-dimensional composite with enhanced visible-light photoactivity in degradation of chlortetracycline. J Photochem Photobiol Chem 346:1–9. https://doi.org/10.1016/j.jphotochem.2017.05.036

Lin L, Wang H, Xu P (2017) Immobilized TiO_2-reduced graphene oxide nanocomposites on optical fibers as high performance photocatalysts for degradation of pharmaceuticals. Chem Eng J 310:389–398. https://doi.org/10.1016/j.cej.2016.04.024

Linsebigler AL, Lu G, Jr JTY (2002) Photocatalysis on TiO_2 surfaces: principles, mechanisms, and selected results. In: American Chemical Society. https://doi.org/10.1021/cr00035a013

Liu SH, Wei YS, Lu JS (2016) Visible-light-driven photodegradation of sulfamethoxazole and methylene blue by Cu_2O/rGO photocatalysts. In Chemosphere:118–123. https://doi.org/10.1016/j.chemosphere.2016.03.107

Liu X, Ji H, Li S et al (2019) Graphene modified anatase/titanate nanosheets with enhanced photocatalytic activity for efficient degradation of sulfamethazine under simulated solar light. Chemosphere 233:198–206. https://doi.org/10.1016/j.chemosphere.2019.05.229

Liu X, Zhao L, Lai H, Li S, Yi Z (2017) Efficient photocatalytic degradation of 4-nitrophenol over graphene modified TiO_2. In Journal of Chemical Technology and Biotechnology 2417–2424: https://dx.doi.org/10.1002/jctb.5251

Long Z, Li Q, Wei T et al (2020) Historical development and prospects of photocatalysts for pollutant removal in water. J Hazard Mater 395:122599. https://doi.org/10.1016/j.jhazmat.2020.122599

Luna-Sanguino G, Ruíz-Delgado A, Tolosana-Moranchel A et al (2020) Solar photocatalytic degradation of pesticides over TiO_2-rGO nanocomposites at pilot plant scale. Sci Total Environ 737: 140286. https://doi.org/10.1016/j.scitotenv.2020.140286

Luprano ML, Marco DS, Guido DM et al (2016) Antibiotic resistance genes fate and removal by a technological treatment solution for water reuse in agriculture. Sci Total Environ 571:809–818. https://doi.org/10.1016/j.scitotenv.2016.07.055

Manamsa K, Crane E, Stuart M et al (2016) A national-scale assessment of micro-organic contaminants in groundwater of England and Wales. Sci Total Environ 568:712–726. https://doi.org/10.1016/j.scitotenv.2016.03.017

Mangat SS, Elefsiniotis P (1999) Biodegradation of the herbicide 2,4-dichlorophenoxyacetic acid (2,4-D) in sequencing batch reactors. Water Res 33:861–867. https://doi.org/10.1016/S0043-1354(98)00259-0

Mateo-Sagasta J, Zadeh SM, Turral H, Burke J (2017) Water pollution from agriculture: a global review. https://www.researchgate.net/publication/345153510_Water_pollution_from_Agriculture_a_global_review_Executive_summary. Last accessed on 30 July 2021

Mecha AC, Chollom MN (2020) Photocatalytic ozonation of wastewater: a review. Environ Chem Lett 18: 1491–1507. SpringerLink. https://doi.org/10.1007/s10311-020-01020-x

Minella M, Sordello F, Minero C (2017) Photocatalytic process in TiO_2/graphene hybrid materials. Evidence of charge separation by electron transfer from reduced graphene oxide to TiO_2. Catal Today 281:29–37. https://doi.org/10.1016/j.cattod.2016.03.040

Minella M, Maurino V, Minero C et al (2016) A model assessment of the ability of lake water in Terra Nova Bay, Antarctica, to induce the photochemical degradation of emerging contaminants. Chemosphere 162: 91–98. https://doi.org/10.1016/j.chemosphere.2016.07.049

Mohan H, Ramalingam V, Karthi N et al (2021) Enhanced visible light-driven photocatalytic activity of reduced graphene oxide/cadmium sulfide composite: methylparaben degradation mechanism and toxicity. Chemosphere 264:128481. https://doi.org/10.1016/j.chemosphere.2020.128481

Najm IN, Vernon LS, Yves R (1993) Removal of 2,4,6-trichlorophenol and natural organic matter from water supplies using PAC in floc-blanket reactors. Water Res 27:551–560. https://doi.org/10.1016/0043-1354(93)90164-D

Nakada N, Toshikatsu T, Hiroyuki S et al (2006) Pharmaceutical chemicals and endocrine disrupters in municipal wastewater in Tokyo and their removal during activated sludge treatment. Water Res 40:3297–3303. https://doi.org/10.1016/j.watres.2006.06.039

NORMAN Substance Database (2021) https://www.norman-network.com/nds/susdat/. Last accessed on 9 Aug 2021

O'Dowd K, Nair KM, Pillai SC (2021) Photocatalytic degradation of antibiotic-resistant genes and bacteria using 2D nanomaterials: what is known and what are the challenges? Curr Opin Green Sustain Chem 30:100471. https://doi.org/10.1016/j.cogsc.2021.100471

Ohtani B (2013) Principle of photocatalysis and design of active photocatalysts. In: New and future developments in catalysis. https://doi.org/10.1016/B978-0-444-53872-7.00006-6

Pastrana LM, Morales-Torres S, Figueiredo JL et al (2018) Graphene photocatalysts. In: Multifunctional photocatalytic materials for energy, 79–101. https://doi.org/10.1016/B978-0-08-101977-1.00006-5

Pastrana-Martínez LM, Morales-Torres S, Papageorgiou SK et al (2013) Photocatalytic behaviour of nanocarbon-TiO_2 composites and immobilization into hollow fibres. Appl Catal B: Environ 142–143:101–111. https://doi.org/10.1016/j.apcatb.2013.04.074

Paucar NE, Kim I, Tanaka H, Sato C (2019) Effect of O_3 dose on the O_3/UV treatment process for the removal of pharmaceuticals and personal care products in secondary effluent. ChemEngineering. https://doi.org/10.3390/chemengineering3020053

Perez AL, De Sylor MA, Slocombe AJ et al (2013) Triclosan occurrence in freshwater systems in the United States (1999–2012): a meta-analysis. Environ Toxicol Chem 32:1479–1487. https://doi.org/10.1002/etc.2217

Petrović M, Gonzalez S, Barceló D (2003) Analysis and removal of emerging contaminants in wastewater and drinking water. Trends Anal Chem 22:685–696. https://doi.org/10.1016/S0165-9936(03)01105-1

Phaniendra A, Jestadi DB, Periyasamy L (2015) Free radicals: properties, sources, targets, and their implication in various diseases. Indian J Clin Biochem 30:11–26. https://doi.org/10.1007/s12291-014-0446-0

Protection of light sensitive drugs: pharma manual (2021) http://pharmamanual.com/photostability-of-drugs/. Last accessed on 9 Aug 2021

Pulicharla, Rama, François Proulx, Sonja Behmel, Jean-B. Sérodes, and Manuel J. Rodriguez (2021) Occurrence and seasonality of raw and drinking water contaminants of emerging interest in five water facilities. In: Science of the total environment 751: 141748. https://doi.org/10.1016/j.scitotenv.2020.141748

Qian R, Zong H, Schneider J et al (2019) Charge carrier trapping, recombination and transfer during TiO_2 photocatalysis: an overview. In: Advances in photo (Electro)catalysis for environmental applications and chemical synthesis: 78–90. https://doi.org/10.1016/j.cattod.2018.10.053

Radhika NP, Selvin R, Kakkar R, Umar A (2019) Recent advances in nano-photocatalysts for organic synthesis. Arab J Chem 12:4550–4578. https://doi.org/10.1016/j.arabjc.2016.07.007

Rasheed T, Bilal M, Nabeel F et al (2019) Environmentally-related contaminants of high concern: potential sources and analytical modalities for detection, quantification, and treatment. Environ Int 122:52–66. https://doi.org/10.1016/j.envint.2018.11.038

Ren H, Koshy P, Chen W et al (2017) Photocatalytic materials and technologies for air purification. J Hazard Mater 325:340–366. https://doi.org/10.1016/j.jhazmat.2016.08.072

Richardson SD, Kimura SY (2017) Emerging environmental contaminants: challenges facing our next generation and potential engineering solutions. Environ Technol Innov 8:40–56. https://doi.org/10.1016/j.eti.2017.04.002

Rincón NC, Hammouda SB, Sillanpaa M et al (2018) Enhanced photocatalytic performance of zinc oxide nanostructures via photoirradiation hybridization with graphene oxide for the degradation of triclosan under visible light: Synthesis, characterization and mechanistic study. J Environ Chem Eng 6:6554–6567. https://doi.org/10.1016/j.jece.2018.09.064

Rivera-Utrilla J, Sánchez-Polo M, Ferro-García MA et al (2013) Pharmaceuticals as emerging contaminants and their removal from water. A Rev Chemosphere 93:1268–1287. https://doi.org/10.1016/j.chemosphere.2013.07.059

Rossner A, Snyder SA, Knappe DRU (2009) Removal of emerging contaminants of concern by alternative adsorbents. Water Res 43:3787–3796. https://doi.org/10.1016/j.watres.2009.06.009
Ryu H, Li B, De Guise S et al (2021) Recent progress in the detection of emerging contaminants PFASs. J Hazard Mater 408:124437. https://doi.org/10.1016/j.jhazmat.2020.124437
Saha D, Visconti MC, Desipio MM et al (2019) Inactivation of antibiotic resistance gene by ternary nanocomposites of carbon nitride, reduced graphene oxide and iron oxide under visible light. Chem Eng J 382:122857. https://doi.org/10.1016/j.cej.2019.122857
Sakar M, Prakash RM, Do T (2019) Insights into the TiO_2-based photocatalytic systems and their mechanisms. Catalysts 9:680. https://doi.org/10.3390/catal9080680
Salasoo L, Inzinna LP, Linsebigler AL et al (2002) Electron backscatter in X-ray tubes: experiment and analysis. In: Third IEEE international vacuum electronics conference: 255–256. https://doi.org/10.1109/IVELEC.2002.999366
Saravanan R, Gracia F, Stephen A (2017) Basic principles, mechanism, and challenges of photocatalysis. In: Khan MM, Pradhan D, Sohn Y (eds) Nanocomposites for visible light-induced photocatalysis. Springer series on polymer and composite materials, pp 19–40. https://doi.org/10.1007/978-3-319-62446-4_2
Sargolzaei J, Hedayati Moghaddam A et al (2015) Modeling the removal of phenol dyes using a photocatalytic reactor with SnO_2/Fe_3O_4 nanoparticles by intelligent system. J Dispersion Sci Technol 36(4):540–548. https://doi.org/10.1080/01932691.2014.916222
Scopus preview—scopus—welcome to scopus. 2021. https://www.scopus.com/home.uri. Last accessed on 30 July 2021
Sohrabi S, Moraveji MK, Iranshahi D (2019) A review on the design and development of photocatalyst synthesis and application in microfluidic reactors: challenges and opportunities. Rev Chem Eng, 1–36. https://doi.org/10.1515/revce-2018-0013
Song J, Xu Z, Liu W et al (2016) $KBrO_3$ and graphene as double and enhanced collaborative catalysts for the photocatalytic degradation of amoxicillin by UVA/Ti_O2 nanotube processes. Mater Sci Semicond Process 52:32–37. https://doi.org/10.1016/j.mssp.2016.04.011
Su R, Bechstein R, Sø L et al (2011) How the anatase-to-rutile ratio influences the photoreactivity of TiO_2. J Phys Chem C 115:24287–24292. https://doi.org/10.1021/jp2086768
Swift E (2019) A durable semiconductor photocatalyst. Science 365:320–321. https://doi.org/10.1126/science.aax8940
Tang H, Berger H, Schmid PE et al (1994) Optical properties of anatase (TiO_2). Solid State Commun 92: 267–271. Elsevier. https://doi.org/10.1016/0038-1098(94)90889-3
Tang B, Chen H, Peng H et al (2018) Graphene modified TiO_2 composite photocatalysts: mechanism, progress and perspective. In: Nanomaterials: 105. https://doi.org/10.3390/nano8020105
Ternes TA (1998) Occurrence of drugs in German sewage treatment plants and rivers. Water Res 32:3245–3260. https://doi.org/10.1016/S0043-1354(98)00099-2
The Honda-Fujishima effect: using titanium dioxide as a photocatalyst (2021) https://born2invest.com/articles/the-honda-fujishima-effect-using-titanium-dioxide-as-a-photocatalyst/. Last accessed on 30 July 2021
Tugaoen HO, Garcia-Segura S, Hristovski K et al (2018) Compact light-emitting diode optical fiber immobilized TiO_2 reactor for photocatalytic water treatment. Sci Total Environ 613–614:1331–1338. https://doi.org/10.1016/j.scitotenv.2017.09.242
United Nations Environment Program Stockholm Convention (2021) http://chm.pops.int/TheConvention/ThePOPs/AllPOPs/tabid/2509/Default.aspx. Last accessed on 30 July 2021
Wang G, Zhang Q, Chen Q et al (2019) Photocatalytic degradation performance and mechanism of dibutyl phthalate by graphene/TiO_2 nanotube array photoelectrodes. Chem Eng J 358:1083–1090. https://doi.org/10.1016/j.cej.2018.10.039
Wei R, Ge F, Chen M, Wang R (2012) Occurrence of ciprofloxacin, enrofloxacin, and florfenicol in animal wastewater and water resources. J Environ Qual 41:1481–1486. https://doi.org/10.2134/jeq2012.0014
Wintgens T, Martin G, Melin T (2002) Endocrine disrupter removal from wastewater using membrane bioreactor and nanofiltration technology. Desalination 146:387–391. https://doi.org/10.1016/S0011-9164(02)00519-2

Xiang Q, Lang D, Shen T et al (2015) Graphene-modified nanosized Ag_3PO_4 photocatalysts for enhanced visible-light photocatalytic activity and stability. Appl Catal b Environ 162:196–203. https://doi.org/10.1016/j.apcatb.2014.06.051

Xue J, Ma S, Zhou Y et al (2015) Au-loaded porous graphitic C_3N_4/graphene layered composite as a ternary plasmonic photocatalyst and its visible-light photocatalytic performance. RSC Adv 5:88249–88257. https://doi.org/10.1039/C5RA17719G

Yan Y, Sun S, Song Y et al (2013) Microwave-assisted in situ synthesis of reduced graphene oxide-BiVO4 composite photocatalysts and their enhanced photocatalytic performance for the degradation of ciprofloxacin. J Hazard Mater 250–251:106–114. https://doi.org/10.1016/j.jhazmat.2013.01.051

Yang C, You X, Cheng J et al (2017) A novel visible-light-driven In-based MOF/graphene oxide composite photocatalyst with enhanced photocatalytic activity toward the degradation of amoxicillin. Appl Catal b Environ 200:673–680. https://doi.org/10.1016/j.apcatb.2016.07.057

You J, Guo Y, Guo R et al (2019) A review of visible light-active photocatalysts for water disinfection: features and prospects. Chem Eng J 373: 624–641. https://doi.org/10.1016/j.cej.2019.05.071.

Yu J, Yu X (2008) Hydrothermal synthesis and photocatalytic activity of zinc oxide hollow spheres. Environ Sci Technol 42:4902–4907. https://doi.org/10.1021/es800036n

Yu Y, Yan L, Cheng J et al (2017) Mechanistic insights into TiO_2 thickness in Fe_3O_4@TiO_2-GO composites for enrofloxacin photodegradation. Chem Eng J 325:647–654. https://doi.org/10.1016/j.cej.2017.05.092

Zaggia A, Lino C, Luigi F (2016) Use of strong anion exchange resins for the removal of perfluoroalkylated substances from contaminated drinking water in batch and continuous pilot plants. Water Res 91:137–146. https://doi.org/10.1016/j.watres.2015.12.039

Zhao JL, Ying GG, Liu YS et al (2010) Occurrence and risks of triclosan and triclocarban in the pearl river system, South China: from source to the receiving environment. J Hazard Mater: 215–222. https://doi.org/10.1016/j.jhazmat.2010.02.082

Zhao Y, Liang X, Hu X, Fan J (2021) rGO/Bi_2WO_6 composite as a highly efficient and stable visible-light photocatalyst for norfloxacin degradation in aqueous environment. In Journal of Colloid and Interface Science 336–346: https://doi.org/10.1016/j.chemosphere.2019.03.117

Zhong L, Haghighat F, Blondeau P et al (2010) Modeling and physical interpretation of photocatalytic oxidation efficiency in indoor air applications. Build Environ 45:2689–2697. https://doi.org/10.1016/j.buildenv.2010.05.029

Zhu S, Wang D (2017) Photocatalysis: basic principles, diverse forms of implementations and emerging scientific opportunities. Adv Energy Mater 7:1700841. https://doi.org/10.1002/aenm.201700841

Zhu W, Sun F, Goei R et al (2017) Facile fabrication of RGO-WO_3 composites for effective visible light photocatalytic degradation of sulfamethoxazole. Appl Catal b Environ 207:93–102. https://doi.org/10.1016/j.apcatb.2017.02.012

Zwiener C, Weil L, Niessner R (1995) Atrazine and parathion-methyl removal by UV and UV/O_3 in drinking water treatment. Int J Environ Anal Chem. Taylor & Francis: 247–264. https://doi.org/10.1080/03067319508033128

Zwiener C, Frimmel FH (2000) Oxidative treatment of pharmaceuticals in water. Water Res 34:1881–1885. https://doi.org/10.1016/S0043-1354(99)00338-3

Chapter 17
Reverse Osmosis (RO) and Nanofiltration (NF) Membranes for Emerging Contaminants (ECs) Removal

Monisha Monachan, Nandini Dixit, Shihabudheen M. Maliyekkal, and Swatantra P. Singh

Abstract In recent years, the effect of emerging contaminants (ECs) has been a threat to humans and wildlife. The advancement in water and wastewater treatment technologies is not very efficient enough for the removal of ECs. ECs have very high toxic levels, and even their traces can cause a dangerous effect on the ecological system and are rarely traceable in wastewater effluent. Due to difficulties in conventional processes for removing ECs, new methods like membrane technologies are being used. Therefore, membrane processes like Reverse Osmosis (RO) and Nanofiltration (NF) have broad applicability. Even though these treatment processes are energy-intensive, these technology's removal efficiencies are better than conventional methods. This chapter consists of recent studies on the removal of ECs using reverse osmosis and nanofiltration. Different pharmaceuticals, antibiotics, Endocrine Disruptor (EDCs), and pesticides are mentioned, and their removal efficiencies by RO and NF methods are reviewed from recent works of literature. The chapter also highlights the characteristics and effects of emerging pollutants on water bodies. The newer technologies like Membrane Bio-reactor (MBR) are also reviewed along

Monisha Monachan and Nandini Dixit with equal contribution

M. Monachan · N. Dixit · S. P. Singh (✉)
Environmental Science and Engineering Department (ESED), Indian Institute of Technology Bombay, Mumbai 400076, India
e-mail: swatantra@iitb.ac.in

S. M. Maliyekkal
Department of Civil and Environmental Engineering, Indian Institute of Technology Tirupati, Tirupati 517506, India

Department of Physics & CAMOST, Indian Institute of Technology Tirupati, Tirupati 517506, India

S. P. Singh
Centre for Research in Nanotechnology & Science (CRNTS), Indian Institute of Technology Bombay, Mumbai 400076, India

Interdisciplinary Program in Climate Studies, Indian Institute of Technology Bombay, Mumbai 400076, India

S. P. Singh et al. (eds.), *New Trends in Emerging Environmental Contaminants*, Energy, Environment, and Sustainability,
https://doi.org/10.1007/978-981-16-8367-1_17

with various technologies like Advanced Oxidation Processes (AOPs) combined with RO/NF process for ECs removal. Finally, an outline of current knowledge gaps and future research scope related to the application of RO and NF for wastewater treatment is recommended.

Keywords Emerging contaminants · Reverse osmosis · Nanofiltration · Water treatment · Wastewater treatment · Membrane processes

17.1 Introduction

In recent years, water recycling has become very important to overcome the increasing demand for water because of the rise in population, climate change, and overuse (Al-Rifai et al. 2011). Nearly 71% of the Earth is covered with salt water, and only 29% of freshwater sustains life. According to World Health Organisation (WHO), about 2.6 billion people have no access to clean drinking water (Pandey et al. 2014). As water pollution is increasing, ~4000 children die every single day from waterborne diseases. Water sources are getting polluted daily because of fast industrialization, leading to environmental crises (Deblonde et al. 2011). Various research studies have been conducted to develop energy-efficient and economic processes such as membrane-based technologies to treat contaminated wastewater. There are also hybrid membrane-based technologies combined with other methods to purify and recycle wastewater (Ortego et al. 1995; Sairam et al. 2006).

Studies suggest that if ECs exceed 70%, it is environmentally harmful to human health (Eurostat 2014). As per the United States Environmental Protection Agency (USEPA), ECs are categorized as hazardous contaminants with no regulatory standards. Polyaromatic hydrocarbons (PAHs) are solely listed EC among hazardous ECs by USEPA (Barbosa et al. 2016). There are various kinds of ECs present in wastewater, but the most common and widely found ECs are nonylphenol (NP) and bisphenol (BPA) (Rasheed et al. 2019) (Corcoll et al. 2014). The presence of ECs in water can cause neurological, reproductive, developmental, and immune effects in humans as well as wildlife (Fisher and Eugster 2014; Gomes et al. 2018). Figure 17.1 typically shows how ECs in the wastewater travel different routes and finally reach the water source that people and animals consume.

Conventional treatment has proven inefficient in removing ECs from wastewater as it is unlikely to remove steroids from wastewater even after complete treatment (Tijani et al. 2013). Various conventional water treatment techniques such as adsorption, ion exchange, coagulation, bio-oxidation, sedimentation, filtration, UV radiation, and chlorination have been broadly investigated in the literature; however, a greater part of these methodologies are deficient for a compelling treatment of wastewater (Jones et al. 2005). Apart from these, other water treatment technologies include membrane-based techniques, which produce high-quality potable water with low maintenance costs (Mohammad et al. 2015). The Membrane-based technologies are classified as Ultrafiltration (UF), Microfiltration (MF), Forward osmosis

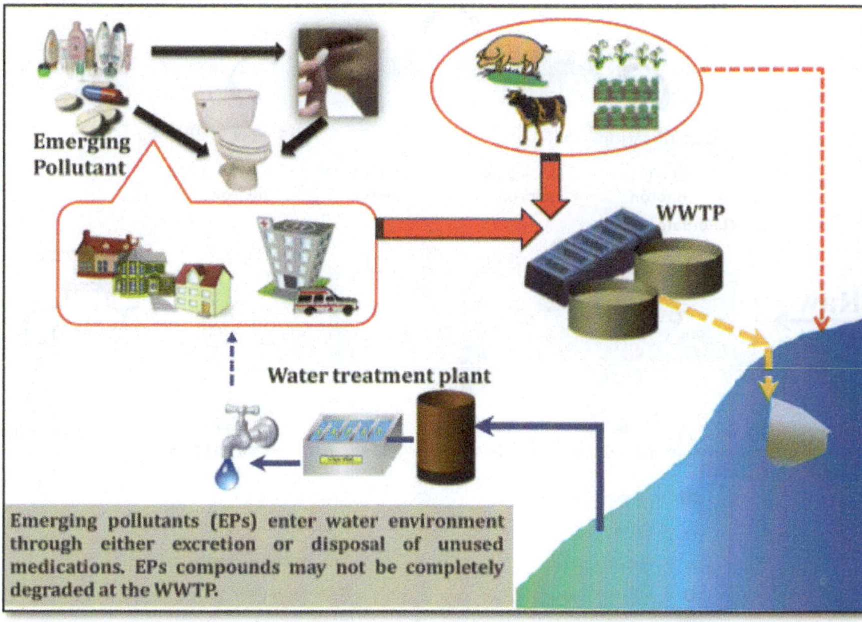

Fig. 17.1 A schematic showing the prevalent sources of ECs. Reprinted from Reference (Gogoi et al. 2018), Copyright (2018) Elsevier

(FO), Reverse osmosis (RO), and Nanofiltration (NF), which are emerged for reclamation of wastewater (Rodriguez-Narvaez et al. 2017). Microfiltration is a widely used technique as it can be taken under atmospheric pressure but cannot remove pollutants <1 μm and so is unable to remove emerging contaminants, whose size ranges from 1 to 100 nm (Zhou et al. 2010; Lu and Liu 2010). Ultrafiltration has the potential to remove ECs as it has a pore size ranging from 0.001 to 0.1 μm size (Acero et al. 2010; Lidén and Persson 2016). Although FO, RO, and NF are the most commonly used techniques for ECs rejection, as they have a pore size ranging from 10 to 100 Å respectively (Lee et al. 2011; Malaeb and Ayoub 2011). The pore size and the morphologies of membranes are shown in Fig. 17.2. The main objective of this study focuses on an in-depth review of the reverse osmosis (RO) process and nanofiltration (NF) process for the removal of emerging contaminants (ECs) in water and wastewater. The significant advancements in the area of membrane technology with combined processes like MBR are also being covered. Along with this, various types of emerging contaminants and their removal efficiencies have also been discussed. This study also concludes the future research directions in this area and the effectiveness of these technologies as they can be an environmentally benign and alternative water treatment source.

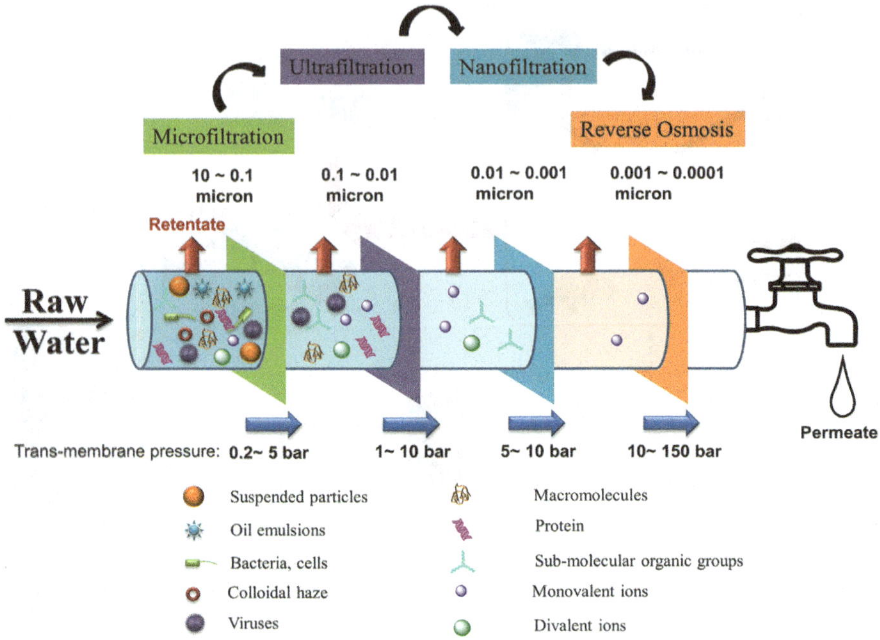

Fig. 17.2 Schematic representation membrane-based separation process. Reprinted with permission from Reference (Liao et al. 2018), Copyright (2018) Elsevier

17.2 Emerging Contaminants and Their Effects

17.2.1 Pharmaceuticals (PhACs)

PhACs are a type of emerging contaminants generated because of their utilization in human and veterinary medication. Antibiotics, biologics, biotics, fragrances, nutritionally functional foods, diagnostic products, etc., have a considerable contribution to form complex compounds of ECs when mixed in the aquatic ecology (Fatta-Kassinos et al. 2011).

The concentration of ECs in wastewater ranges from 2.2 to 34,228 ng/L, and their removal rate varies from 0 to 97%. Aspirin is one of the most commonly found ECs in wastewater which can be easily reduced to salicylic acid, a more active chemical form (Al-Odaini et al. 2013). Apart from this, the removal efficiency of phthalates is around > 90%, but for antibiotics, the removal rate varies from 50 to 70%. Anti-inflammatory, β blockers, and analgesics have less degradation efficiency of 30–40%. Compared to this, ECs like tetracycline and codeine have meager removal rates, which is around 10% (Deblonde et al. 2011). The compounds like polychlorinated biphenyls (PCBs) are found in fatty tissues of humans resulting due to unwanted exposure in the environment via industries. It also has a harmful effect on the environment as it gets deposited in the sediments because of lower solubility in water (Tavakoly Sany et al.

2015). Ibuprofen is another drug that is commonly found in surface water. These types of ECs are found in minimal concentrations in the environment and are also recognized as persistent pseudo-pollutants (Gogoi et al. 2018). Marine life is one of the significantly affected species to these ECs as they are exposed to wastewater in their entire lifespan. However, the level of toxicity of pharmaceuticals on aquatic life is not entirely explored, and the knowledge of its effects is inadequate.

17.2.2 Endocrine Disruptors (EDCs)

Endocrine disruptors are synthetic chemicals that can imitate hormones of the human body when consumed. It has been found that EDCs can copy the functioning of natural estrogens, thyroid, and testosterone hormones in the human body which can lead to behavioural, hormonal, reproductive, and developmental changes (Snyder et al. 2003). Bisphenol A is a priority EDCs and is studied widely to under its fate and occurrences in wastewater for its effective removal process. The concentration of nonylphenol (NP) is up to 1.1×10^3 ng/L and is observed in the effluent of household and industrial wastewater (Fan et al. 2013). Because of their long-term exposure to the water, it can have an adverse effect and which is a major concern even if the concentration present in the wastewater is low.

17.2.3 Personal Care Products

Personal care products (PCPs) are another type of emerging contaminant that is used for individual care purposes. PCPs present in personal care products include steroids, fragrances, shampoos, hormones, conditioners, etc. (Pal et al. 2014). Sunscreen containing UV filters is proved to have estrogenic activities, according to studies. These UV filters are widely present in groundwater and other surface water. At lower concentrations, these personal care products are biologically active, increasing the risk of aquatic organisms (Sharma et al. 2019). Dioctyl phthalate (DOP) commonly used in food packaging, personal care product, blood containers, and tubings have very high toxicity levels which are difficult to eliminate by water treatment processes (Blanchard et al. 2004; Qureshi et al. 2016). These PCPs get converted into carbon dioxide and water in their mineralized or natural form in wastewater treatment plants.

17.2.4 Pesticides and Biocides

Pesticides are used for agricultural purposes for the production of quality food. The excessive usage of pesticides has caused its remnant to pollute the environmental surrounding, which has dangerous and toxic outcomes. The consumption of pesticides can cause cancer, neurological disorders, and deformity. The drinking water can have a permissible limit of up to 0.5 μg/L of combined active ingredients of pesticides (Karabelas et al. 2009). Biocides are widely used in the food industry as food preservatives. It is used to destroy fungicides and herbicides that lead to microbial growth in food containers (Margot et al. 2015). The chemicals including spiroxamine, carbendazime, terbutylazine, acetochlor, etc. are a range of fungicide and herbicides that get alloyed with the groundwater and runoff water, causing harmful consequences humans as well as aquatic organisms (Comber et al. 2015; Ender et al. 2017; Gao et al. 2015; Weston et al. 2015).

17.3 Quantification of ECs

In recent years it has been seen that various ECs have been detected in the surface water, groundwater, drinking water, and wastewater in their derivative forms with original compounds (Nabeel et al. 2020). Due to the chemical composition of ECs, they are easily dissolved in water which is risky for aquatic life and human life by the water cycle. The detection of these ECs is seen at nano- gram per liter (ng/l) scale (Thomaidis et al. 2012). Such small detection requires susceptible analytical techniques. Thus, more efficient monitoring techniques for the identification of ECs are becoming important day by day. Different analytical methods have been developed to determine the number of ECs present in groundwater, wastewater, drinking water, and surface water. One of the majorly used techniques by regulatory agencies involves gas or liquid chromatography combined with mass spectroscopy (Rasheed et al. 2019). This method is effective in identifying ECs in complex wastewater because of its sensitivity, selectivity, and specificity (Thomaidis et al. 2012). Identification of ECs can lead to a more accurate, cost-effective, and fugacious approach towards the treatment of water compared to conventional techniques such as High-Resolution Accurate Mass Spectroscopy (HRAM), which are expensive and challenging. For analytical methods, pre-treatment for the sample is considered a crucial step. This step helps in reducing the lower detection limits (LOD) and low quantification limits.

Advanced techniques such as ion trap and triple quadrupole have also been explored when assessing the nanograms (ng) ranges (Agüera et al. 2013). To interpret the structural form of transformed products, more developed methods such as quadrupole, linear ion traps quadrupole, quadrupole-time of flight, triple quadrupole, and quadrupole-linear ion trap are used (Nikolaou 2013). Other analytical methods include immunoanalytical techniques, microbiological assays, or capillary electrophoresis. Compared to the liquid chromatography technique, capillary electrophoresis is less sensitive, although immunoanalytical technique depends more upon antibody used, and microbiological assays are widely sampled nature dependent (Buchberger 2011). The use of solid-phase extraction has been increased in recent studies compared to liquid–liquid extraction because of its simple, relevant, and efficient applications. Nanomaterial with a combined effect of magnetism, including iron nanoparticles, has also got attention as a substrate in solid-phase extraction (Moliner-Martínez et al. 2011). Despite showing promising results in ECs identification, a more sensitive determination technique needs to be developed.

17.4 Reverse Osmosis

17.4.1 Removal of ECs by RO Process

Reverse osmosis (RO) is one of the most important and energy-efficient technologies currently available for water and wastewater treatment. In RO, the monovalent ions like Na^+ and Cl^- are separated from water through a semi-permeable membrane. Due to higher external pressure than the osmotic pressure, the water molecules are forced to flow in the opposite direction than that of the osmosis phenomenon. Reverse osmosis works on high pressure and the pressure vary from 20 to 70 bar and the permeate flow ratio ranges from 10 to 100L/h/m^2 (Gómez-Espinosa and Arizmendi-Cotero 2019). Because of the hydrophilic nature of the membrane, water readily diffuses in and out of the polymer structure. The mechanism of reverse osmosis is shown in Fig. 17.3. Even though RO membranes are widely used for wastewater treatment, steric exclusion, electrostatic interaction, and hydrophobic/absorptive

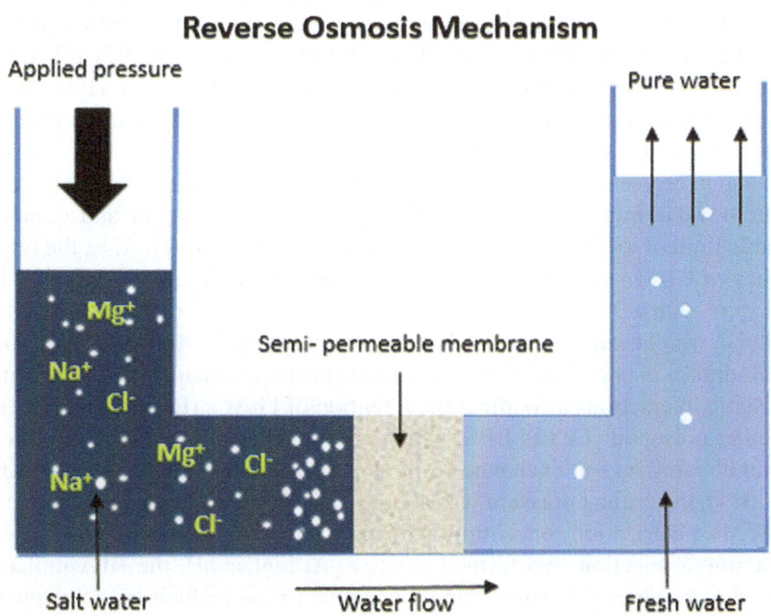

Fig. 17.3 Mechanism of reverse osmosis. Adapted and modified from Reference (Quatrini et al. 2021)

interaction needs to be investigated for the removal of different types of ECs based on charge, size, and hydrophobicity, etc. The aromatic polyamide present in the reverse osmosis membrane is responsible for its high retention efficiency. Along with this, size exclusion, electrostatic repulsion, and hydrophobic/supermolecular interactions are also responsible. Hybrid techniques, including Reverse osmosis and carbon filters equipped with drinking water treatment plants, are also found to be effective for removing several ECs. Even compounds having smaller molecular weights like nitrosodimethyl amine, dioxane, and 2-methylsoborneol have a greater removal rate via carbon filters. Also, 98% of antibodies are removed by RO and NF process, where retention was ineffective for low MWCO (Molecular Weight cut-off) NF membrane (Wang et al. 2018). Additionally, in another study conducted, it was seen that RO has better retention efficiency than FO membranes. The average retention was as follows: ATZ (93.7%) > CBM (84.3%) > SMX (75.2%) > 4 CP (60.9%) > PHN (47.3%) for RO membrane (Boleda et al. 2010; Heo et al. 2013). The principal mechanism for removal of these compounds was steric hindrance. Molecular dynamics simulations also revealed the high adsorption of phenolic compounds with the RO membranes. Also the small size and hydrophilic nature of compounds leads to the enhanced transportation in the DS side of FO system, which helped in better removal efficiency (Heo et al. 2013).

A potent ECs, namely N-nitrosamine, is generally found in drinking water as a disinfection by-product. The highest N-nitrosamine is NDMA (N-nitrosodimethylamine), having a critical level concentration of about 10 ng/ L in potable water. In this study, it was explored that, through the integration of reverse osmosis and UV, the NDMA rejection was ranged from 24 to 56% and 43% to 66%, respectively, which was observed by solid-phase extraction (SPE). The overall removal of NDMA by advanced treatment process was 69 $\pm$ 7% (Plumlee et al. 2008). RO process combined with membrane bioreactor (MBR) has also been used to treat raw sewage and secondary effluent. The retention rate of RO – MBR has been found to be greater than 99% for 20 PhACs, while retention of RO also showed robust removal of micropollutants like ETM, MTP, clarithromycin, and atenolol to a detectable limit of less than or equal to 10 ng/L (Joss et al. 2011). Also, the retention efficiency of CBM was found to be greater than 99%, along with SMX, MTP, and sotalol were with a 98% removal (Radjenović et al. 2008). Due to the high MWCO (Molecular weight cut off) of the RO membranes (TR70-4021-HF) i.e., 100 Da, steric hindrance is one of the potential size exclusion mechanisms. Apart from this, electrostatic interactions also affect the retention of PhACs (Dolar et al. 2012).

Another poisonous ECs is BPA, which is produced by condensation of acetone and phenol using an ion exchange resin as a catalyst. In a study done by Khazali et al. (2014), the critical pressure was about (408–476 kPa). Due to this high critical pressure, the polarization concentration forms, which reduces the rejection efficiency and maximum rejection was formed at pH 8. At higher pH, the BPA molecule is converted to bisphenolate with a single negative charge, which led to electrostatic interactions between the O- and OH- groups respectively, thereby decreasing the rejection rate. Furthermore, the effect of feed pH on the UF/RO was studied by Qin et al. (2003), wherein they found that rejection was increased with an increase in pH but the membrane flux was decreased. Moreover, with an increase in pH of the feed concentration of TOC was decreased. It is because of the enhanced electrostatic repulsion between charged organic solute and the surface of the membrane. Overall, the optimum pH of the feed water for the integrated UF/RO process was found to be around 3.7–5.6 for the treatment (Qin et al. 2003).

The most common drug used for chemotherapy is cyclophosphamide (CP) which can have a toxic effect on the living organism present in water bodies during discharge. CP rejection in feed water was investigated by NF, RO, and MBR processes. CP rejection was most efficiently observed with the RO process with almost 90% rejection. In the NF process, the CP rejection was observed to be 20–40%. The hybrid process of MBR-RO and MBR-NF was suggested to successfully eliminate CP in real wastewater treatment (Wang et al. 2009). A range of micro ECs were investigated at various conditions in wastewater recycling plants. Eleven PhACs, two EDCs, and BPA (6–23 μg/L) removal were observed using MF and RO processes, indicating the higher concentration of salicylic acid in wastewater treatment plant effluent. ECs concentration decreases majorly from primary to secondary treatment. The retention efficiency was greater than 97% in treated water concentrations which was <0.1 μg/L for most ECs, and the concentration of BPA was <0.5 μg/L. Although the presence of BPA in the treated water is 0.5 μg/L is still a major concern (Abdelmelek et al. 2011). To remove PhACs, organic matter, and other inorganic contaminants, a combined advanced oxidation process with reverse osmosis can also be utilized. The OH radicals in combination with proteins in RO retentate proved to be effective for removing PhACs.

Beta-blockers such as metoprolol and propranolol are classified as a potential risk to aquatic as well as human life if present in water. A study was conducted for the removal of β blockers using ozonation. Tests were carried out on chlorinated and non-chlorinated effluent of wastewater treatment plants, and a decrease of OH radical exposure shown effective removal of β blockers in the RO process (Benner et al. 2008). A comparison study between Forward Osmosis and Reverse Osmosis was conducted to examine the rejection of Bisphenol A, triclosan, and diclofenac. The pore size of the commercial cellulose-based FO membrane was 0.74 nm (Benner et al. 2008). BPA and triclosan rejection were seen higher in the FO membrane. RO and FO membrane successfully rejected the triclosan because the molecular width of triclosan is larger than the membrane pore size. Furthermore, as NaCl was used as feed solution, there was a major change in filtration behaviour of BPA on adsorption surface of FO and RO. In FO, the concentration of BPA was reduced from 0.5 to 0.47 ng/l in the first 100 min. But in RO stable value of 2.4 was reached in 200 min. Hence, the adsorption of BPA on the RO surface was seen more consistent than on the surface of the FO membrane (Xie et al. 2012).

According to a study, NF 90 membrane was compared with the pilot and full-scale experiments with two RO membranes (BW30LE and ESPA2). The results show that ACT removal was 72% in RO and NF 90 removal efficiency was 67% which is a minor difference. Investigation using NDMA revealed a 45% removal by NF 90 and 35% with ESPA2. NF 90, BW30LE, and ESPA2 were capable of removing pollutants with MW > 200 with rejections over 90%. Contaminants like salicylic acid (SAC) which has a smaller molecular weight, can be removed by NF and RO. The removal efficiency of NF was 82% for neutral contaminants and 97% for ionic pollutants. Removal efficiencies of RO membrane for natural and ionic contaminants are 85% and 99%, respectively. Hence, RO has better removal efficiency compared to NF membrane (Yangali-Quintanilla et al. 2011). Oxidation processes have the potential to degrade organics from contaminated water streams. A pilot-scale study was introduced to treat sewage wastewater in which a combined AOP-RO system was developed to eliminate ECs from secondary municipal wastewaters. By using this process, more than 99% of ECs were removed, but the effective removal of N-nitrosodimethylamine (NDMA) and H_2O_2 was difficult. Pilot-scale experimentation was also carried out to study ciprofloxacin's removal by using a spiral wound membrane in which more than 90% removal of ciprofloxacin was observed (James et al. 2014). The recent development and ongoing studies prove that RO membrane is efficient for desalination and effectively removes various contaminants from wastewater streams. The removal efficiency of the RO membrane is summarized in Table 17.1.

Table 17.1 Summary of removal efficiencies of ECs by RO filtration process

	S. No.	Emerging contaminants (ECs)	Removal efficiencies (%)	References
RO	1	Atrazine (ATZ)	93.7	Heo et al. (2013)
	2	4-Chlorophenol (4CP)	60.9	
	3	Phenol (PHN)	47.3	
	4	Carbamazepine (CBM)	84.3	
	5	Sulfamethoxazole (SMX)	75.25	
	6	N-nitrosodimethylamine (NDMA)	50–65	Plumlee et al. (2008)
	7	Bisphenol A (BPA)	87	Khazaali et al. (2014)
	8	Salicylic acid	85	Yangali-Quintanilla et al. (2011)
RO-MBR	1	ETM	99	Dolar et al. (2012)
	2	Atenolol		
	3	Clarithromycin		
	4	Sotalon	98	
	5			
	6	Cyclophosphamide (CP)	90	Lv et al. (2009) and Wang et al. (2009)
RO-AOP	1	NDMA	99	James et al. (2014)
	2	Ciprofloxacin	90	

17.5 Nanofiltration

17.5.1 Removal of ECs by Nanofiltration Process

A study showed BPA rejection by NF membrane (NF- 4040-70; Saehan MWCO ~ 200 Da) with less than 74.1%, along with the rejection of IBP or salicylic acid, which were found to be 98.1% and 97%, respectively. The pH of BPA remained constant throughout the study; however, IBP and salicylic acid were noticed a negative charge. Due to this, low BPA rejection has been shown due to the absence of repulsion between BPA and membrane surface. The molecular weight of IBP and salicylic acid was seen much higher than BPA, and because of that, IBP and salicylic acid had high size exclusion and electrostatic repulsion. BPA showing no charge difference at pH seven and being hydrophobic must have absorbed the hydrophobic surface of membrane till its saturate which led to less rejection of BPA with operational time (Kim et al. 2008).

NF modified with polymer brushes has improved the retention of BPA with 96.9% removal efficiency. The steric hindrance coupled with polymer chains causes enhanced retention of BPA. Also, the hydrophilic polymerized surface of the membrane increases the adsorption of BPA. Steric hindrance increased by 60 min due to the production of longer polymer chains. The polymer chains produced at 15 min were not very effective in increasing the steric hindrance. IBP and salicylic acid removal was also noted to be 98.1% and 99.7%, respectively. It was due to more steric hindrance and increased negative charge surface that made the removal much more effective (Kim et al. 2008). A study was carried out on agricultural wastewater to eliminate BPA, triclosan, benzyl phthalate, carbamazepine, and 4- nonylphenol with NF-270 at 800 kPa pressure. An observation was made that the retention of BPA was reduced because of water solubility. The retention of hydrophobic compounds increased because of the adsorption surface of the membrane. The flux of NF- 270 was 87-30 L/m^2/hr. The removal of ECs on NF was around 90%. The rejection of two major pesticides, dimethoate, and atrazine, was explored. The NF membrane used for demonstration was NF-90, NF-200, NF-270, and DK of ~200 MWCO. NF-90 showed the highest removal of acetaminophens (López-Muñoz et al. 2012). A new form of ECs known as Perfluoro octane sulfonates (PFOS) is commonly used in surfactants, fire retardants, lubricants, coating material, and metal plating solutions (Deng et al. 2010). PFOS in a very minimum quantity is also considered very toxic in the water. NF-270 was used to study the removal of PFOSA. The water used for the investigation was stimulated surface water with calcium ions. The concentration of calcium chloride in the feed water was 10 ppb. The results showed that the rejection of PFOS increased from 94 to 99% because of calcium chloride concentration (Zhao et al. 2013).

Polyamide membranes have low separation performance due to their adsorption property. A study was conducted using surface adsorption as a parameter. Polysulfone, polyester, and polyamide membrane were employed for rejection of estrone and estradiol. NF polyamide membranes have shown the highest adsorption among other membranes. Surface morphology and pore size played an important role in surface adsorption. The pH of the medium played another crucial role in the removal of ECs. As at neutral pH, electrostatic repulsion between the surface of membrane and solute was observed more compared to that of pH 11 (Nghiem et al. 2005). In a different study, tert-butyl phenol retentions were examined by using commercial NF-270 membrane, and it was seen that 90% of ECs were retained, giving an effective removal (Khouni et al. 2011). A recent study has observed that steric exclusion is also crucial for the separation of hormone-like compounds. pH, ionic strength, and organic matter are also responsible for the separation of ECs (Darko-Kagya et al. 2010).

Commercial NF-90 and NF-270 membranes were used for removal of ECs like sulfamethoxazole, diclofenac sodium, hydrochlorothiazide, 4- acetamidoantipyrine, nicotine, and ranitidine hydrochloride. At low pressure, the desired retention of ECs was observed because of steric hindrance and dynamic interaction between membrane surface and solutes. The retention of solutes was >95% in NF-90 while NF-270 showed less retention ranging from 75 to 95%. It was concluded that this is because of permeate flux and physiochemical properties of ECs (López-Muñoz et al. 2012). Removal of cyclophosphamide was demonstrated with NF hollow fibre membrane having charged surface. The membrane was designed using hyperbranched polyethyleneimine (PEI) as crosslinkers to the polyamide-imide backbone. The flux of the membrane was very high because of the sponge-like structure porous membrane. Membrane pore size decreased immensely after crosslinking with PEI. This made the surface of the membrane hydrophobic with a positive charge. The molecular weight of PEI is 60000, which contributed to the maximum rejection of ciprofloxacin (Bedor et al. 2010; Nghiem et al. 2006). Multi-walled carbon nanotubes were used to remove ECs from the primary effluent. In this process, as a pre-treatment, coagulation was used before the nanofiltration process to reduce the interference of humic acid and biopolymers. This method was not very efficient in the removal of ECs, and the removal concentrations were deficient (Boleda et al. 2010).

The advanced oxidation processes and adsorption processes have also been widely explored to remove ECs from wastewater. An experiment was performed to investigate the retention percentages of antibiotics like norfloxacin (NOR), ofloxacin (OFL), roxithromycin (ROX), and azithromycin from effluent streams. UV/O_3 process has reduced the concentration of ECs in the wastewater by 87%. An increase of BOD/COD ratio of 4.6 times, dissolved organic carbon (DOC), and reduction of acute toxicity has shown the removal efficiencies of 40% and 58%. NF resulted in the highest removal efficiency of 98%. Thus, the overall study confirms NF is efficient in removing antibiotics from wastewater streams, but the UV/ O_3 process was able to remove antibiotics from NF concentrate (Liu et al. 2014). PES- NF membranes were used to recover 1-(5-bromo-fur-2-il)-2-bromo-2-nitroethane in crossflow NF model having >80% retention. Membrane prepared with 25% and 75% PES on NF-270 has given the maximum results (Martínez et al. 2012). Chlorination is widely used for the disinfection of wastewater. The organic matter of chlorine can cause the generation of toxic by-products like trihalomethane (THM). NF-200 membranes have shown the efficiency of removal of trihalomethanes from the wastewater (Uyak et al. 2008).

The wastewater from the textile mills will combine with municipal wastewater, sewage effluent, and surface water contaminating the groundwater, surface water, and soil. Metals and organic dye used in the preparation of the dye can also contaminate the food and increase human health risks. Dopamine (DA)—modified halloysite nanotubes (HNT) was combined and blended with PVDF to investigate the removal of direct red-28 (DR-28), direct blue-14 (DB-14), and direct yellow-4 (DY-4). The removal efficiencies for DR-28 were 86%, DB-14 was 94%, and DY-4 was 85% (Zeng et al. 2017). The water flux of the membrane was improved by about 80%. The removal efficiency of the NF membrane is summarized in Table 17.2.

Table 17.2 Summary of removal efficiencies of ECs by NF process

		S. No.	Emerging Contaminants (ECs)	Removal efficiencies (%)	References
NF	NF 90	1	BPA	74.1	Kim et al. (2008)
		2	IBP	98	
		3	Salicylic acid	97	
		4	Acetaminophen	95	López-Muñoz et al. (2012)
		5	Sulfamethoxazole		
		6	Diclofenac sodium		
		7	Hydrochlorothiazide		
		8	4-Acetamidoantipyrine		
		9	Nicotine		
		10	Ranitidine		
	NF 270	1	Triclosan	90	López-Muñoz et al. (2012)
		2	Benzyl phthalate		
		3	Carbamazepine		
		4	4- nonylphenol		
		5	Tert-butyl phenol		
		6	Perfluoro octane sulfonates (PFOS)	99	Zhao et al. (2013)
		7	Sulfamethoxazole	75–95	Nghiem et al. (2006)
		8	Diclofenac sodium		
		9	Hydrochlorothiazide		
		10	4- Acetamidoantipyrine		
		11	Nicotine		
		12	Ranitidine		
NF-AOP		1	Norfloxacin (NOR)	98	Liu et al. (2014)
		2	Ofloxacin (OFL)		
		3	Roxithromycin (ROX)		
		4	Azithromycin		

17.6 Conclusion

In recent years many kinds of literature have mentioned the presence of ECs present in wastewater that is not easily removable by conventional waste treatment processes. The presence of ECs in wastewater harms the environment, and researchers are constantly developing newer methods and strategies to treat the effluent that can be recycled and reused to reduce water scarcity. These ECs are originated from different sources in the form of synthetic chemicals, pharmaceutical compounds, naturally occurring compounds, and microorganisms. Therefore, different technologies need to be adopted for different wastewater nature. Membrane-based technologies have proved to eliminate some of the toxic ECs presents in nature; however, the viability in terms of energy and economic becomes the need of the hour. Among different membrane processes, the RO and NF process has proven suitable for eliminating ECs from wastewater. The hybrid technologies combined with RO and NF have given excellent results as per the literature. The recently developed technologies, like the MBR process, have also shown prominent removal of ECs when integrated with RO/NF processes. Current developments in the field of RO and NF processes are highlighted with recommendations for future improvements. Efforts are needed to identify the ECs in wastewater because only <5% of the pharmaceuticals, personal care products, antibiotics, pesticides, etc., are detected. The toxicity level of each emerging contaminant should be evaluated for developing treatment methods that can efficiently separate and degrade these ECs. The efficiency of the hybrid process like MBR and AOPs combined with the RO and NF process should focus on improving the efficiency of removal of ECs to obtain good quality water after treatment. Parameter optimization of the integrated process like AOPs combined with RO/NF process is essential with a proper flow sheet. This approach can achieve notable results over the currently used treatment method, i.e., activated carbon. Despite its excellent results over a wide range of contaminants, membrane-based processes still suffer some limitations due to the different physicochemical properties of the materials used in the fabrication processes. In addition to this, a change in separation mechanism could occur due to the change in membrane properties due to adsorption, which can lead to membrane breakdown (Kárászová et al. 2020). Therefore, a trade-off between selectivity and permeability needs to be achieved for optimal operations. Apart from this, a major challenge occurs in Reverse Osmosis membranes processes, i.e., the generation of a concentrated stream of contaminants which requires additional treatment steps with the expense of high energy. These limitations could be addressed with new generation membrane materials and hybrid processes such as forward osmosis. In addition to this, social and legislative actions should improvise the research and development for implementing new generation technologies for treating the wastewater.

Acknowledgements The corresponding author acknowledges the funding received from Water Technology Initiative Programme (WTI), Department of Science and Technology, Government of India for carrying out this work.

References

Abdelmelek SB, Greaves J, Ishida KP, Cooper WJ, Song W (2011) Removal of pharmaceutical and personal care products from reverse osmosis retentate using advanced oxidation processes. Environ Sci Technol 45(8):3665–3671. https://doi.org/10.1021/es104287n

Acero JL, Benitez FJ, Teva F, Leal AI (2010) Retention of emerging micropollutants from UP water and a municipal secondary effluent by ultrafiltration and nanofiltration. Chem Eng J 163(3):264–272. https://doi.org/10.1016/j.cej.2010.07.060

Agüera A, Martínez Bueno MJ, Fernández-Alba AR (2013) New trends in the analytical determination of emerging contaminants and their transformation products in environmental waters. Environ Sci Pollut Res 20(6):3496–3515. https://doi.org/10.1007/s11356-013-1586-0

Al-Odaini NA, Zakaria MP, Yaziz MI, Surif S, Abdulghani M (2013) The occurrence of human pharmaceuticals in wastewater effluents and surface water of Langat River and its tributaries, Malaysia. Int J Environ Anal Chem 93(3):245–264. https://doi.org/10.1080/03067319.2011.592949

Al-Rifai JH, Khabbaz H, Schäfer AI (2011) Removal of pharmaceuticals and endocrine disrupting compounds in a water recycling process using reverse osmosis systems. Sep Purif Technol 77(1):60–67. https://doi.org/10.1016/j.seppur.2010.11.020

Barbosa MO, Moreira NFF, Ribeiro AR, Pereira MFR, Silva AMT (2016) Occurrence and removal of organic micropollutants: An overview of the watch list of EU Decision 2015/495. Water Res 94:257–279. https://doi.org/10.1016/j.watres.2016.02.047

Bedor CNG, Morais RJL, Cavalcanti LS, Ferreira JV, Pavão AC (2010) Carcinogenic potential of endosulfan and its metabolites based on a quantum chemical model. Sci Total Environ 408(24):6281–6284. https://doi.org/10.1016/j.scitotenv.2010.09.014

Benner J, Salhi E, Ternes T, von Gunten U (2008) Ozonation of reverse osmosis concentrate: Kinetics and efficiency of beta blocker oxidation. Water Res 42(12):3003–3012. https://doi.org/10.1016/j.watres.2008.04.002

Blanchard M, Teil MJ, Ollivon D, Legenti L, Chevreuil M (2004) Polycyclic aromatic hydrocarbons and polychlorobiphenyls in wastewaters and sewage sludges from the Paris area (France). Environ Res 95(2):184–197. https://doi.org/10.1016/j.envres.2003.07.003

Boleda R, Majamaa K, Aerts P, Gómez V, Galceran T, Ventura F (2010) Removal of drugs of abuse from municipal wastewater using reverse osmosis membranes. Desalin Water Treat 21(1–3):122–130. https://doi.org/10.5004/dwt.2010.1294

Buchberger W (2011) Current approaches to trace analysis of pharmaceuticals and personal care products in the environment. J Chromatogr A 1218(4):603–618. https://doi.org/10.1016/j.chroma.2010.10.040

Comber S, Gardner M, Jones V, Ellor B (2015) Source apportionment of trace contaminants in urban sewer catchments. Environ Technol (united Kingdom) 36(5):573–587. https://doi.org/10.1080/09593330.2014.953599

Corcoll N, Acuña V, Barceló D, Casellas M, Guasch H, Huerta B, Petrovic M, Ponsatí L, Rodríguez-Mozaz S, Sabater S (2014) Pollution-induced community tolerance to non-steroidal anti-inflammatory drugs (NSAIDs) in fluvial biofilm communities affected by WWTP effluents. Chemosphere 112:185–193. https://doi.org/10.1016/j.chemosphere.2014.03.128

Darko-Kagya K, Khodadoust AP, Reddy KR (2010) Reactivity of lactate-modified nanoscale iron particles with 2,4-dinitrotoluene in soils. J Hazard Mater 182(1–3):177–183. https://doi.org/10.1016/j.jhazmat.2010.06.012

Deblonde T, Cossu-Leguille C, Hartemann P (2011) Emerging pollutants in wastewater: a review of the literature. Int J Hyg Environ Health 214(6):442–448. https://doi.org/10.1016/j.ijheh.2011.08.002

Deng S, Yu Q, Huang J, Yu G (2010) Removal of perfluorooctane sulfonate from wastewater by anion exchange resins: Effects of resin properties and solution chemistry. Water Res 44(18):5188–5195. https://doi.org/10.1016/j.watres.2010.06.038

Dolar D, Gros M, Rodriguez-Mozaz S, Moreno J, Comas J, Rodriguez-Roda I, Barceló D (2012) Removal of emerging contaminants from municipal wastewater with an integrated membrane system, MBR-RO. J Hazard Mater 239–240:64–69. https://doi.org/10.1016/j.jhazmat.2012.03.029

Ender A, Goeppert N, Grimmeisen F, Goldscheider N (2017) Evaluation of β-D-glucuronidase and particle-size distribution for microbiological water quality monitoring in Northern Vietnam. Sci Total Environ 580:996–1006. https://doi.org/10.1016/j.scitotenv.2016.12.054

Eurostat. (2014). Chemicals production statistics—Statistics explained

Fan Z, Hu J, An W, Yang M (2013) Detection and occurrence of chlorinated byproducts of bisphenol a, nonylphenol, and estrogens in drinking water of China: comparison to the parent compounds. Environ Sci Technol 47(19):10841–10850. https://doi.org/10.1021/es401504a

Fatta-Kassinos D, Meric S, Nikolaou A (2011) Pharmaceutical residues in environmental waters and wastewater: current state of knowledge and future research. Anal Bioanal Chem 399(1):251–275. https://doi.org/10.1007/s00216-010-4300-9

Fisher MM, Eugster EA (2014) What is in our environment that effects puberty? Reprod Toxicol 44:7–14. https://doi.org/10.1016/j.reprotox.2013.03.012

Gao P, He S, Huang S, Li K, Liu Z, Xue G, Sun W (2015) Impacts of coexisting antibiotics, antibacterial residues, and heavy metals on the occurrence of erythromycin resistance genes in urban wastewater. Appl Microbiol Biotechnol 99(9):3971–3980. https://doi.org/10.1007/s00253-015-6404-9

Gogoi A, Mazumder P, Tyagi VK (2018) Tushara Chaminda GG, An AK, Kumar M (2018) Occurrence and fate of emerging contaminants in water environment: a review. Groundwater Sustain Dev 6(September 2017):169–180. https://doi.org/10.1016/j.gsd.2017.12.009

Gomes IB, Simões LC, Simões M (2018) The effects of emerging environmental contaminants on Stenotrophomonas maltophilia isolated from drinking water in planktonic and sessile states. Sci Total Environ 643:1348–1356. https://doi.org/10.1016/j.scitotenv.2018.06.263

Gómez-Espinosa RM, Arizmendi-Cotero D (2019) Role of membrane on emerging contaminant removal. In: Handbook of environmental chemistry, vol 66, pp 157–174. https://doi.org/10.1007/698_2017_149

Heo J, Boateng LK, Flora JRV, Lee H, Her N, Park YG, Yoon Y (2013) Comparison of flux behavior and synthetic organic compound removal by forward osmosis and reverse osmosis membranes. J Membr Sci 443:69–82. https://doi.org/10.1016/j.memsci.2013.04.063

James CP, Germain E, Judd S (2014) Micropollutant removal by advanced oxidation of microfiltered secondary effluent for water reuse. Sep Purif Technol 127:77–83. https://doi.org/10.1016/j.seppur.2014.02.016

Jones OAH, Voulvoulis N, Lester JN (2005) Human pharmaceuticals in wastewater treatment processes. Crit Rev Environ Sci Technol 35(4):401–427. https://doi.org/10.1080/10643380590956966

Joss A, Baenninger C, Foa P, Koepke S, Krauss M, McArdell CS, Rottermann K, Wei Y, Zapata A, Siegrist H (2011) Water reuse: >90% water yield in MBR/RO through concentrate recycling and CO2 addition as scaling control. Water Res 45(18):6141–6151. https://doi.org/10.1016/j.watres.2011.09.011

Karabelas AJ, Plakas KV, Solomou ES, Drossou V, Sarigiannis DA (2009) Impact of European legislation on marketed pesticides—A view from the standpoint of health impact assessment studies. Environ Int 35(7):1096–1107. https://doi.org/10.1016/j.envint.2009.06.011

Kárászová M, Bourassi M, Gaálová J (2020) Membrane removal of emerging contaminants from water: which kind of membranes should we use? Membranes 10(11):1–23. https://doi.org/10.3390/membranes10110305

Khazaali F, Kargari A, Rokhsaran M (2014) Application of low-pressure reverse osmosis for effective recovery of bisphenol A from aqueous wastes. Desalin Water Treat 52(40–42):7543–7551. https://doi.org/10.1080/19443994.2013.831795

Khouni I, Marrot B, Moulin P, Ben Amar R (2011) Decolourization of the reconstituted textile effluent by different process treatments: Enzymatic catalysis, coagulation/flocculation and nanofiltration processes. Desalination 268(1–3):27–37. https://doi.org/10.1016/j.desal.2010.09.046

Kim JH, Park PK, Lee CH, Kwon HH (2008) Surface modification of nanofiltration membranes to improve the removal of organic micro-pollutants (EDCs and PhACs) in drinking water treatment: Graft polymerization and cross-linking followed by functional group substitution. J Membr Sci 321(2):190–198. https://doi.org/10.1016/j.memsci.2008.04.055

Lee KP, Arnot TC, Mattia D (2011) A review of reverse osmosis membrane materials for desalination—Development to date and future potential. J Membr Sci 370(1–2):1–22. https://doi.org/10.1016/j.memsci.2010.12.036

Liao Y, Loh CH, Tian M, Wang R, Fane AG (2018) Progress in electrospun polymeric nanofibrous membranes for water treatment: fabrication, modification and applications. Prog Polym Sci 77:69–94. https://doi.org/10.1016/j.progpolymsci.2017.10.003

Lidén A, Persson KM (2016) Comparison between ultrafiltration and nanofiltration hollow-fiber membranes for removal of natural organic matter: a pilot study. J Water Supply: Res Technol AQUA 65(1):43–53. https://doi.org/10.2166/aqua.2015.065

Liu P, Zhang H, Feng Y, Yang F, Zhang J (2014) Removal of trace antibiotics from wastewater: a systematic study of nanofiltration combined with ozone-based advanced oxidation processes. Chem Eng J 240:211–220. https://doi.org/10.1016/j.cej.2013.11.057

López-Muñoz MJ, Sotto A, Arsuaga JM (2012) Nanofiltration removal of pharmaceutically active compounds. Desalin Water Treat 42(1–3):138–143. https://doi.org/10.1080/19443994.2012.683099

Lu NC, Liu JC (2010) Removal of phosphate and fluoride from wastewater by a hybrid precipitation-microfiltration process. Sep Purif Technol 74(3):329–335. https://doi.org/10.1016/j.seppur.2010.06.023

Lv Y, Liu H, Wang Z, Liu S, Hao L, Sang Y, Liu D, Wang J, Boughton RI (2009) Silver nanoparticle-decorated porous ceramic composite for water treatment. J Membr Sci 331(1–2):50–56. https://doi.org/10.1016/j.memsci.2009.01.007

Malaeb L, Ayoub GM (2011) Reverse osmosis technology for water treatment: state of the art review. Desalination 267(1):1–8. https://doi.org/10.1016/j.desal.2010.09.001

Margot J, Rossi L, Barry DA, Holliger C (2015) A review of the fate of micropollutants in wastewater treatment plants. Wires Water 2(5):457–487. https://doi.org/10.1002/wat2.1090

Martínez MB, Van der Bruggen B, Negrin ZR, Luis Alconero P (2012) Separation of a high-value pharmaceutical compound from waste ethanol by nanofiltration. J Ind Eng Chem 18(5):1635–1641. https://doi.org/10.1016/j.jiec.2012.02.024

Mohammad AW, Teow YH, Ang WL, Chung YT, Oatley-Radcliffe DL, Hilal N (2015) Nanofiltration membranes review: recent advances and future prospects. Desalination 356:226–254. https://doi.org/10.1016/j.desal.2014.10.043

Moliner-Martínez Y, Ribera A, Coronado E, Campíns-Falcó P (2011) Preconcentration of emerging contaminants in environmental water samples by using silica supported Fe_3O_4 magnetic nanoparticles for improving mass detection in capillary liquid chromatography. J Chromatogr A 1218(16):2276–2283. https://doi.org/10.1016/j.chroma.2011.02.036

Nabeel F, Rasheed T, Bilal M, Li C, Yu C, Iqbal HMN (2020) Bio-inspired supramolecular membranes: a pathway to separation and purification of emerging pollutants. Sep Purif Rev 49(1):20–36. https://doi.org/10.1080/15422119.2018.1500919

Nghiem LD, Schäfer AI, Elimelech M (2005) Nanofiltration of hormone mimicking trace organic contaminants. Sep Sci Technol 40(13):2633–2649. https://doi.org/10.1080/01496390500283340

Nghiem LD, Schäfer AI, Elimelech M (2006) Role of electrostatic interactions in the retention of pharmaceutically active contaminants by a loose nanofiltration membrane. J Membr Sci 286(1–2):52–59. https://doi.org/10.1016/j.memsci.2006.09.011

Nikolaou A (2013) Pharmaceuticals and related compounds as emerging pollutants in water: analytical aspects. Glob Nest J 15(1):1–12. https://doi.org/10.30955/gnj.000969

Ortego JD, Aminabhavi TM, Harlapur SF, Balundgi RH (1995) A review of polymeric geosynthetics used in hazardous waste facilities. J Hazard Mater 42(2):115–156. https://doi.org/10.1016/0304-3894(95)00008-I

Pal A, He Y, Jekel M, Reinhard M, Gin KYH (2014) Emerging contaminants of public health significance as water quality indicator compounds in the urban water cycle. Environ Int 71:46–62. https://doi.org/10.1016/j.envint.2014.05.025

Pandey PK, Kass PH, Soupir ML, Biswas S, Singh VP (2014) Contamination of water resources by pathogenic bacteria. AMB Express 4(1):1–16. https://doi.org/10.1186/s13568-014-0051-x

Plumlee MH, López-Mesas M, Heidlberger A, Ishida KP, Reinhard M (2008) N-nitrosodimethylamine (NDMA) removal by reverse osmosis and UV treatment and analysis via LC-MS/MS. Water Res 42(1–2):347–355. https://doi.org/10.1016/j.watres.2007.07.022

Progress on sanitation and drinking water: 2015 update and MDG assessment. https://www.wssinfo.org/fileadmin/user_upload/resources/JMP-Update-report-2015_English.pdf. Accessed on 05 Aug 2021. ISBN 978 92 4 150914 5, UNICEF and World Health Organization 2015EC

Qin JJ, Oo MH, Wai MN, Wong FS (2003) Effect of feed pH on an integrated membrane process for the reclamation of a combined rinse water from electroless nickel plating. J Membr Sci 217(1–2):261–268. https://doi.org/10.1016/S0376-7388(03)00141-8

Quatrini E, Costantino F, Mba D, Li X, Gan TH (2021) Monitoring a reverse osmosis process with kernel principal component analysis: a preliminary approach. Appl Sci (Switzerland) 11(14). https://doi.org/10.3390/app11146370

Qureshi MS, Yusoff AR, Bin M, Wirzal MD H, Sirajuddin Barek J, Afridi HI, Üstündag Z (2016) Methods for the determination of endocrine-disrupting phthalate esters. Crit Rev Anal Chem 46(2):146–159. https://doi.org/10.1080/10408347.2015.1004157

Radjenović J, Petrović M, Ventura F, Barceló D (2008) Rejection of pharmaceuticals in nanofiltration and reverse osmosis membrane drinking water treatment. Water Res 42(14):3601–3610. https://doi.org/10.1016/j.watres.2008.05.020

Rasheed T, Bilal M, Nabeel F, Adeel M, Iqbal HMN (2019) Environmentally-related contaminants of high concern: Potential sources and analytical modalities for detection, quantification, and treatment. Environ Int 122(September 2018):52–66. https://doi.org/10.1016/j.envint.2018.11.038

Rodriguez-Narvaez OM, Peralta-Hernandez JM, Goonetilleke A, Bandala ER (2017) Treatment technologies for emerging contaminants in water: a review. Chem Eng J 323:361–380. https://doi.org/10.1016/j.cej.2017.04.106

Sairam M, Nataraj SK, Aminabhavi TM, Roy S, Madhusoodana CD (2006) Polyaniline membranes for separation and purification of gases, liquids, and electrolyte solutions. Sep Purif Rev 35(4):249–283. https://doi.org/10.1080/15422110600859727

Sharma BM, Bečanová J, Scheringer M, Sharma A, Bharat GK, Whitehead PG, Klánová J, Nizzetto L (2019) Health and ecological risk assessment of emerging contaminants (pharmaceuticals, personal care products, and artificial sweeteners) in surface and groundwater (drinking water) in the Ganges River Basin, India. Sci Total Environ 646:1459–1467. https://doi.org/10.1016/j.scitotenv.2018.07.235

Snyder S, Westerhoff P, Yoon Y, Sedlak D (2003) Disruptors in water: implications for the water industry. Environ Eng Sci 20(5):449–469

Tavakoly Sany SB, Hashim R, Salleh A, Rezayi M, Karlen DJ, Razavizadeh BBM, Abouzari-lotf E (2015) Dioxin risk assessment: mechanisms of action and possible toxicity in human health. Environ Sci Pollut Res 22(24):19434–19450. https://doi.org/10.1007/s11356-015-5597-x

Thomaidis NS, Asimakopoulos AG, Bletsou AA (2012) Emerging contaminants: a tutorial mini-review. Glob Nest J 14(1):72–79. https://doi.org/10.30955/gnj.000823

Tijani JO, Fatoba OO, Petrik LF (2013) A review of pharmaceuticals and endocrine-disrupting compounds: sources, effects, removal, and detections. Water Air Soil Pollut 224(11). https://doi.org/10.1007/s11270-013-1770-3

Uyak V, Koyuncu I, Oktem I, Cakmakci M, Toroz I (2008) Removal of trihalomethanes from drinking water by nanofiltration membranes. J Hazard Mater 152(2):789–794. https://doi.org/10.1016/j.jhazmat.2007.07.082

Wang L, Albasi C, Faucet-Marquis V, Pfohl-Leszkowicz A, Dorandeu C, Marion B, Causserand C (2009) Cyclophosphamide removal from water by nanofiltration and reverse osmosis membrane. Water Res 43(17):4115–4122. https://doi.org/10.1016/j.watres.2009.06.007

Wang Y, Huang H, Wei X (2018) Influence of wastewater precoagulation on adsorptive filtration of pharmaceutical and personal care products by carbon nanotube membranes. Chem Eng J 333(19):66–75. https://doi.org/10.1016/j.cej.2017.09.149

Weston DP, Schlenk D, Riar N, Lydy MJ, Brooks ML (2015) Effects of pyrethroid insecticides in urban runoff on Chinook salmon, steelhead trout, and their invertebrate prey. Environ Toxicol Chem 34(3):649–657. https://doi.org/10.1002/etc.2850

Xie M, Nghiem LD, Price WE, Elimelech M (2012) Comparison of the removal of hydrophobic trace organic contaminants by forward osmosis and reverse osmosis. Water Res 46(8):2683–2692. https://doi.org/10.1016/j.watres.2012.02.023

Yangali-Quintanilla V, Maeng SK, Fujioka T, Kennedy M, Li Z, Amy G (2011). Nanofiltration vs. reverse osmosis for the removal of emerging organic contaminants in water reuse. Desalinat Water Treatment 34(1–3):50–56. https://doi.org/10.5004/dwt.2011.2860

Zeng G, Ye Z, He Y, Yang X, Ma J, Shi H, Feng Z (2017) Application of dopamine-modified halloysite nanotubes/PVDF blend membranes for direct dyes removal from wastewater. Chem Eng J 323:572–583. https://doi.org/10.1016/j.cej.2017.04.131

Zhao C, Zhang J, He G, Wang T, Hou D, Luan Z (2013) Perfluorooctane sulfonate removal by nanofiltration membrane the role of calcium ions. Chem Eng J 233:224–232. https://doi.org/10.1016/j.cej.2013.08.027

Zhou JE, Chang Q, Wang Y, Wang J, Meng G (2010) Separation of stable oil-water emulsion by the hydrophilic nano-sized ZrO2 modified Al_2O_3 microfiltration membrane. Sep Purif Technol 75(3):243–248. https://doi.org/10.1016/j.seppur.2010.08.008

Chapter 18
Introduction to Membrane Distillation and Its Application in Emerging Contaminants Removal

Utkarsh Misra, Vartika Nishad, and Swatantra P. Singh

Abstract Emerging contaminants (ECs) have huge impacts on all living beings, and conventional treatment processes like coagulation, precipitation, and chlorination have limited capability for removal. So, a tertiary and combined treatment process is required. Alternative treatment technologies include adsorption, chemical treatment, and membrane filtration. However, the associated operating cost, ECs rejection, fouling propensity, and by-product formation are some of the drawbacks. Membrane distillation (MD) is one of the promising membrane technologies for emerging contaminants removal. In MD, The vapor pressure difference between the hot feed and cold permeate is a driving force. MD technology has some added advantages like low-pressure requirements, less fouling susceptibility, low-temperature requirements, and only vapor mass transfer, i.e., 100% non-volatile compounds retention. MD employs a low temperature and pressure so fouling is less compact and is easily cleanable. MD technology has been studied for desalination, hypersaline brine treatment, chemical separation and can potentially remove emerging contaminants. The MD technology does not require very high-quality heat; solar heat, waste heat, or cogeneration-based heat utilization is possible. This way, MD can be operated on renewable energy and becomes sustainable and carbon neutral. MD technology has also been integrated with other efficient treatment technologies like Forward Osmosis (FO), Reverse osmosis (RO), and Nanofiltration (NF), providing a leading edge compared to other treatment methods. This chapter elaborates on various available MD technologies, possible materials, configurations, operating parameters, and

U. Misra · S. P. Singh (✉)
Centre for Research in Nanotechnology & Science (CRNTS), Indian Institute of Technology Bombay, Mumbai 400076, India
e-mail: swatantra@iitb.ac.in

V. Nishad · S. P. Singh
Environmental Science and Engineering Department (ESED), Indian Institute of Technology Bombay, Mumbai 400076, India

S. P. Singh
Interdisciplinary Program in Climate Studies, Indian Institute of Technology Bombay, Mumbai 400076, India

S. P. Singh et al. (eds.), *New Trends in Emerging Environmental Contaminants*,
Energy, Environment, and Sustainability,
https://doi.org/10.1007/978-981-16-8367-1_18

energy requirements. We have also highlighted future research trends and challenges for MD treatment technology's sustainable and commercial application.

Keywords Membrane distillation · Hybrid technologies · Renewable energy · Micropollutants · Sustainability · Carbon neutral · Energy efficiency

Abbreviations

MD	Membrane distillation
EP	Emerging pollutants
EC	Emerging contaminants
USEPA	United States Environmental Protection Agency
EDHs	Endocrine-disrupting hormones
PFOS	Perfluorooctanesulfonic acid
PFOA	Perfluorooctanoic acid
AO	Advanced oxidation
ED	Electrodialysis
DCMD	Direct Contact MD
AGMD	Air Gap MD
SGMD	Sweeping Gas MD
VMD	Vacuum MD
LEP	Liquid Entry Pressure
NOM	Natural Organic Matter
BR	Biological reactor
FO	Forward Osmosis
EMBR	Enzymatic Membrane Bioreactor
ROS	Reactive Oxygen Species
PPCP	Pharmaceuticals and Personal Care Products
SV	Semi-Volatile
NV	Non-volatile
ED	Electrodialysis
TrOCs	Trace Organic Compounds
DBPs	Disinfection by-products
PFPeA	Perfluoropentanoic acid

18.1 Introduction

Water is an integral part of our life, and with improving living standards and high consumer care specific demands, pollution in every vertical has increased. With the advent of modern science, detection and discovery of reasonably low concentrations of chemicals and their derivatives have also become possible. With the detection of the number of known pollutants existing in society vicinity, the related concerns have increased. These compounds being detected recently as a pollutant in water bodies are also known as Emerging pollutants (Vasilachi et al. 2021). Emerging pollutants (EP) or Emerging contaminants (EC) consist of many anthropogenic chemicals used for household and industrial applications, which are vital for modern society. NORMAN database has 1036 compounds identified as emerging (NORMAN 2021). Norman database classified the emerging contaminants into seven broad categories of Pharmaceuticals, Personal care products, Industrial chemicals (food additives, pesticides, polychlorinated biphenyl, flame retardant, antimicrobial substances), Disinfection by-products, Algal toxins (toxic released from some algae: Cyanotoxins, microcystin), Biocides and their metabolites (plants and agricultural preventive agents [pesticides]), Bioterrorism and disruption devices (Biological and Chemical weapons). Other classifications depend on their suspected health effects, sources, uses, and chemistries (Hernández-Maldonado and Blaney 2019; Richardson and Ternes 2011; Thomaidis et al. 2012; Fawell and Ong 2012; Yang et al. 2014). United States Environmental Protection Agency (USEPA) lists the contaminants candidate chemicals and has published four such lists by 2015, and a new list is still in draft condition since 2018 (US Enviromental Protection Agency 2009). Many of these ECs have no standard for regulation, and thus at times, chronic exposures are also observed but remain non-conclusive in the absence of any concrete facts.

ECs are considered highly toxic, as ng/L exposure can severely affect both aquatic and human life. These effects include genotoxicity, endocrine disruption, and immune toxicity. The long-term impact is not yet evaluated and is difficult to be estimated (Vasilachi et al. 2021). Quantitative and qualitative number of Pharmaceuticals and endocrine-disrupting hormones (EDHs) observed in the aquatic environment has been reported in various reports. They can cause antibiotic resistance, increasing the cancers rate, and cause organ damage (Bound and Voulvoulis 2005; Milić et al. 2013). EDHs may affect sexual and reproductive behaviour (Bradley and Journey 2014). Sunscreens, UV filters, and other personal care products like siloxane, benzophenone, and aminobenzoic acid get into the aquatic environment through washing, bathing, swimming, and domestic wastes (Lei et al. 2015). Many studies suggest some Endocrine-disrupting effects of these compounds (Krause et al. 2012). Disinfection by-products (DBPs) get into the human body by direct ingestion, showering, dermal absorption. Long-term exposure with DBPs results in genotoxicity, carcinogenic effects (Li and Mitch 2018). PFOS and PFOA are persistent and toxic compounds, and they also bioaccumulate in biological systems. Fire retardants are less soluble in water, less volatile, and they are proposed to be causing neurotoxicity, hormonal disruption, and cancers. Nutrient-rich water systems get algal blooms, and blue-green algae produce toxins affecting the toddlers and elders alot (Miraji et al. 2016). Perfluorinated compounds are demonstrated to induce benign dysfunction in rodent models (Klaunig et al. 2012). Figure 18.1 summarizes how the ECs end up affecting human life and give rise to various complications.

New-generation contaminants arise from industries, agriculture, mining, and domestic residential sources (Ali et al. 2019). Biological treatment methods like

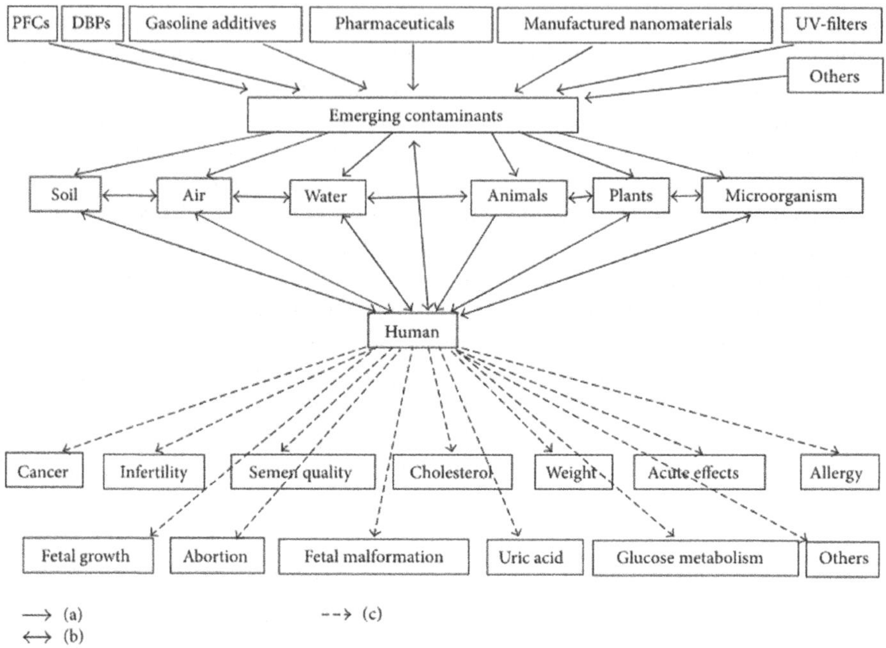

Fig. 18.1 Emerging contaminants with their associated human exposure and related problems. Reprinted from reference (Lei et al. 2015) with permission from Hindawi

filtration and activated sludge process are not efficient for removing many contaminants such as drugs, heavy metals, and other emerging contaminants (Rajapaksha et al. 2018). Other physical methods like coagulation and flocculation also have limited capabilities for the removal of emerging contaminants. At the same time, chlorination may result in undesirable taste and odor, producing some carcinogenic by-products also (Rajapaksha et al. 2018; Varma et al. 2013; Li et al. 2019; Junussova and Chicherin 2020). Membrane separation through size exclusion, steric effect, or electrostatic interactions is an alternative and effective method for removing these contaminants. Table 18.1 lists some studies for the removal of EC using membranes. Advanced oxidation (AO) methods generate reactive oxygen species to mineralize the recalcitrant pollutant, but it requires substantial energy. As ECs exist in very low concentrations affecting the efficiency of the AO.

Membrane bioreactor (MBR) is another efficient treatment method capable of producing effluents of reuse standards (Melin et al. 2006). MBRs also degrade TrOCs that are biodegradable and hydrophobic but have challenges in removing recalcitrant compounds through biological processes only (Clara et al. 2005; Radjenović et al. 2009; Tadkaew et al. 2011; Wijekoon et al. 2013). So MBR systems integrated with NF (Choi et al. 2002), FO (Achilli et al. 2009; Alturki et al. 2012), and MD (Goh et al. 2013; Khaing et al. 2010), also known as high retention MBR, have been proposed. An MD-MBR process consists of an MD system to remove the low molecular weight

Table 18.1 Recent studies for membrane-based water treatment

Type	Membrane material	Target compound	References
Microfiltration	Industrial membrane—hydrophobic and negatively charged PVDF	Ciprofloxacin removal	Guo et al. (2018)
Ultrafiltration	Polyether sulfone (PES)	Caffeine removal	Acero et al. (2017) and Guo et al. (2018)
Ultrafiltration	Guanidyl-functionalized graphene/poly-sulfone mixed matrix	Antibacterial and antimicrobial	Zhang et al. (2019)
Nanofiltration	Graphene oxide/attapulgite composite membranes	Dyes removal	Wang et al. (2019a)
Nanofiltration	Polyacrylonitrile and Fe_2O_3	Diazinon removal	Pordel et al. (2019)
Nanofiltration	Chitosan and graphene oxide	Diazinon removal	Chen et al. (2019)
Nanofiltration/Reverse osmosis	Polyamide thin-film composite (TFC) membranes	Pesticides	Nikbakht Fini et al. (2019)
Filtration adsorption	β-cyclodextrin polymers	Bisphenol A removal	Wang et al. (2019b)
Solid-phase extraction	Porous carbon derived from amine-functionalized MIL-125 metal–organic framework	Bisphenol A and 4-tertbutylphenol	Sánchez et al. (2019)

calcitrant compound from the feed of MBR. This MBR operates in thermophilic conditions to facilitate MD operation. Thermophilic conditions also favor enhanced degradation and lesser sludge production (Lapara and Alleman 1999).

Various Biological reactor (BR)-MD configurations have been suggested in literature like submerged (Goh et al. 2013; Tomaszewska and Białończyk 2016; Zhang et al. 2015), continuous feed recirculation (Phattaranawik et al. 2008; Jacob et al. 2015; Asif et al. 2018; Song et al. 2018; Pal et al. 2018). Forward osmosis (FO), being less energy-intensive with low fouling propensity, has applications in water, effluent treatment, food processing, and desalination (Cath et al. 2006). FO-BR-MD system where MD regenerates the draw solution (Siddique et al. 2018; Siddique et al. 2018). Electrodialysis (ED) uses the applied electric potential for the transport of ions across the ion-exchange membrane. ED has been used in industries before RO with a limitation that it can remove only ionic species (Warsinger et al. 2018). So, ED has been used for desalination only and has the potential limited to remove charged emerging contaminants only (Metcalf et al. 2007). Roman et al. (2019) studied transport and adsorption for positively and negatively charged organic micropollutants in ED and found the highest adsorption up to 100% for positively charged species.

Fonseca Couto et al. (2018) have reviewed the various membrane-based treatment methods for removing the pharmaceutically active compounds and suggest that MD is a promising technology for preventing health and environmental problems caused by PhAC in the environment (Fonseca Couto et al. 2018).

Membrane distillation (MD) is a microfiltration technique that uses heat energy for desalting highly saline waters. This procedure allows only vapor molecules to pass over a hydrophobic microporous sheet (Alkhudhiri et al. 2012). In contrast with traditional procedures, low operating temperature is an attractive feature of the MD process. In reverse osmosis (RO), hydrostatic pressure is more than in MD so, membranes are less fouled and thus get a longer lifetime (Alkhudhiri et al. 2012). MD has been explored to recover whey protein (Hausmann et al. 2013), olive oil processing (El-Abbassi et al. 2012), orange juice cconcentration (Alves and Coelhoso 2006), textile and petroleum industry wastewater (Khaing et al. 2010; Jacob et al. 2015; Wu et al. 2018; Leaper et al. 2019; Sivakumar et al. 2013); radioactivity contaminated water (Jia et al. 2018), municipal reuse of water (Mericq et al. 2011; Cath et al. 2005), resources recovery (Zarebska et al. 2014; Quist-Jensen et al. 2016), and fermentation broth separation (Gryta et al. 2013).

This chapter consists of a perspective of MD for Emerging contaminants removal. We have started with a basic introduction to emerging contaminants with issues related to ECs, and their conventional removal techniques. Next, we have discussed the MD as a technique for EC remediation with its mechanism, configuration, and material requirements. Further, in the next section, we have discussed the EC removal by MD with literature and its perspective. At last, we have given a conclusive summary of the economics and energy consumption of MD with a future trend in the research of MD.

18.2 Membrane Distillation

Membrane-based treatment systems are efficient and have a low footprint, whereas thermal systems can use various alternative thermal sources. A Membrane Distillation system was proposed to take advantage of both systems. To perform separation by MD system, we need a hydrophobic and porous membrane. A membrane is a thin barrier that allows one or more constituents to move across it when placed between two phases or mediums in the presence of an appropriate driving force. MD process has three phenomena, namely evaporation followed by vapor molecule migration and subsequent condensation. At high temperatures, evaporation of feed fluid takes place at the fluid/vapor interface. The vapor produced during the process crosses the membrane pores and condenses on the other side of the membrane with a colder draw solution. Typically feed temperature ranges from 3060 °C, allowing the use of low-quality thermal energy.

MD is driven by the vapor pressure difference across the hydrophobic membrane, which is sandwiched between a hot feed and a compartment. This compartment may sweep gas, air, or vacuum depending upon the type of technology, making Direct

Contact MD (DCMD), Air Gap MD (AGMD), Sweeping Gas MD (SGMD), and Vacuum MD (VMD) four common types of MD systems (Alkhudhiri and Hilal 2018). MD is supposed to have a capacity of working efficiently at higher flux and high salinity (Drioli et al. 2015; Ashoor et al. 2016). In an MD system, the temperature involved is low, and the hydrophobicity of the membrane tends to make the process less susceptible to fouling. Theoretically, 100% rejection can be obtained for non-volatile solutes (Moradi et al. 2016; Wijekoon et al. 2014a), and solute concentrations do not impact the process efficiency for MD process (Silva et al. 2018). Due to the low temperature and pressure, the fouling, if developed in MD, is less compacted and quickly cleaned (Alkhudhiri et al. 2013). Silva et al. (2018) showed no change in performance on increasing the solute concentration by 1000 times (Silva et al. 2018). Organic and inorganic membranes can be used for water and wastewater treatment. Highly selective, chemically stable, resistant to fouling, and suitable mechanical strength membranes are preferred.

In an MD system, the membranes are made of hydrophobic polymeric materials like PTFE, PVDF, or PP and have pores at some standard distances at about 0.1–0.5 μm. A strong dipole in water prevents the material from being wetted by the water (Warsinger et al. 2017). However, the pores in the membrane are bigger than the particles in the feed; the high Liquid entry pressure of the surface hinders the entry of the liquid phase in the pores. A convex meniscus forms into the pore due to a phenomenon known as capillary action. MD process relies upon the following conditions (Franken et al. 1987).

- The porosity of the membrane
- Non-wettability due to fluids
- The membrane at one side should be in direct contact with the solution
- No capillary action.

The rejection mechanism of organic compounds is DCMD depends on their volatility, surface charge, and hydrophobicity (Naidu et al. 2017). In MD procedures, mass and heat transfer takes place from the hot side to the cold side. The feed temperature T_f drops while moving from the bulk feed side boundary to T_1 at the membrane surface. Evaporation of water occurs during the process and gets transported across the membrane. At the same time, the conduction of heat to the cold (permeate) side occurs. As the water vapor condenses on the permeate side, temperature increases from T_p to T_2. The driving force for this phenomenon is the vapor-pressure gradient caused due to two temperatures T_1 and T_2, which is less than the vapor pressure change between T_f and T_p. This phenomenon is also called temperature polarization. Schofield et al. have defined this temperature polarization coefficient (TP) by following expression (Camacho et al. 2013):

$$TP = \frac{(T_1 - T_2)}{(T_f - T_p)} \tag{18.1}$$

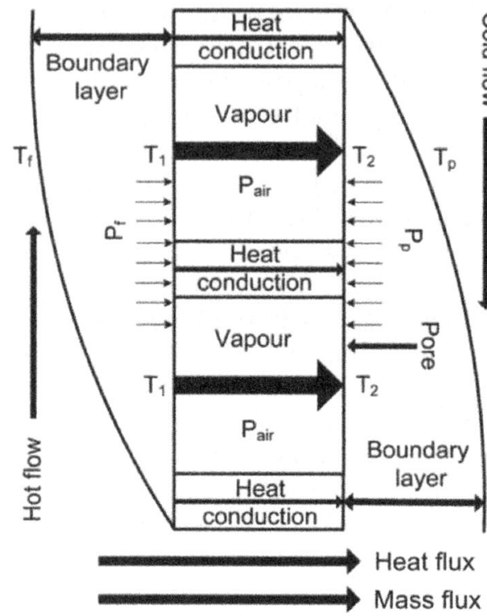

Fig. 18.2 Mass and Heat transfer in DCMD. Reprinted from reference (Camacho et al. 2013) with permission from MDPI

where T_1 = Feed side boundary temperature, T_2 = Permeate side boundary temperature, T_f = Bulk feed side temperature, T_p = Bulk permeate side temperature (Fig. 18.2).

Mass exchange in the MD system incorporates three stages: (1) vaporization of hot feed at the gas/liquid interface, (2) vapor transport through hydrophobic pores due to vapor pressure difference from hot to cold side, and (3) condensation of vapor into the cold streamside. Therefore, mass is controlled by two major factors: the difference in vapor pressure and membrane permeability. Considering fluid dynamics conditions on both sides of the membrane, the limiting step for mass transfer in MD is the mass transfer across the membrane. The following properties influence membrane permeability (Eykens et al. 2017):

- Since the membrane is not 100% porous, the effective area for mass transfer is less than the total membrane area.
- Membrane pores are not straight throughout the membrane; instead, they have some tortuosity induced due to their fabrication methods. Thus, the vapor transport path is greater than the membrane thickness.
- The inside walls of the pores increase the resistance to diffusion by decreasing the momentum of the vapor molecules.

Poiseuille-flow (P), Knudsen-diffusion (K), and Molecular-diffusion (M), or a combination of these, govern mass transport and are called the transition mechanism. The dominant mass transfer mechanism in the pores is represented using the Knudsen number (Kn). Other popular models for mass transfer in MD are Scofield model (Schofield et al. 1990) and Dusty-gas model (Mason et al. 1983).

Two steps are involved in heat transfer from the feed side to the permeate side. First is the heat transfer for sensible heating and latent heating on the cold side. Sensible heating is caused through the membranes, and the latent heat transfers through the water vapors. This transfer forms a temperature difference in the bulk and boundary layer; this temperature difference drives a flow from bulk to the boundary through convection (Eykens et al. 2017). In the case of an AGMD, the sensible heating is less, and thus the system performs better than DCMD. If we use a flow of air in this gap using Stripping gas, mass transfer is boosted, and more sensible heating resistance is obtained. Since gas flowing and subsequent condensation are energy-intensive, VMD is developed where sensible heating losses are neglected at very low vacuum levels.

18.2.1 Different Membrane Distillation Configuration

Membrane Distillation systems have four configurations, classified according to the conditions near the membrane on the draw side. As the technology has advanced, the use of vacuum, sweep gas, and air gap has emerged to improve the distillation system's performance. The most popular system used is direct contact membrane distillation. Here we have briefly described different configurations of Membrane distillation.

18.2.1.1 Direct Contact Membrane Distillation (DCMD)

In the DCMD process, hot feed solution is directly contacted on one side of the membrane. Over the permeate side, the cold draw solution is passed. The membranes used for the DCMD process consist of a thin selective, hydrophobic, and microporous layer on the feed side with a support layer that is thicker, less hydrophobic, and porous at the distillate side. The steps involved in the process are (Hsu et al. 2002):

- Vaporization of solvent at the hot side
- Dispersion through the apertures
- Condensation at the cold draw solution side.

More porosity, thickness, and less hydrophobicity lead to polarization effect on the distillate side and, consequently, distillate flux declines. DCMD process operates on low hydrostatic pressure. The concentration of microbes, chemicals, and ions on the feed side due to leakage, evaporation, or wind action leads to corrosion. Qu et al. (2013) performed a study on removing ammonia from industrial and municipal wastewater (Qu et al. 2013*).* A capillary PVDF membrane with 80% porosity, an average pore size of 0.22 μm, LEP of 250 kPa, and a surface contact angle of 87° were used for the experiments. Feed samples were prepared by dissolving ammonia chloride into distilled water, and the pH values were adjusted by adding HCl and NaOH to the feed solution. It was shown that there was 52%, 88%, and 99.5% ammonia

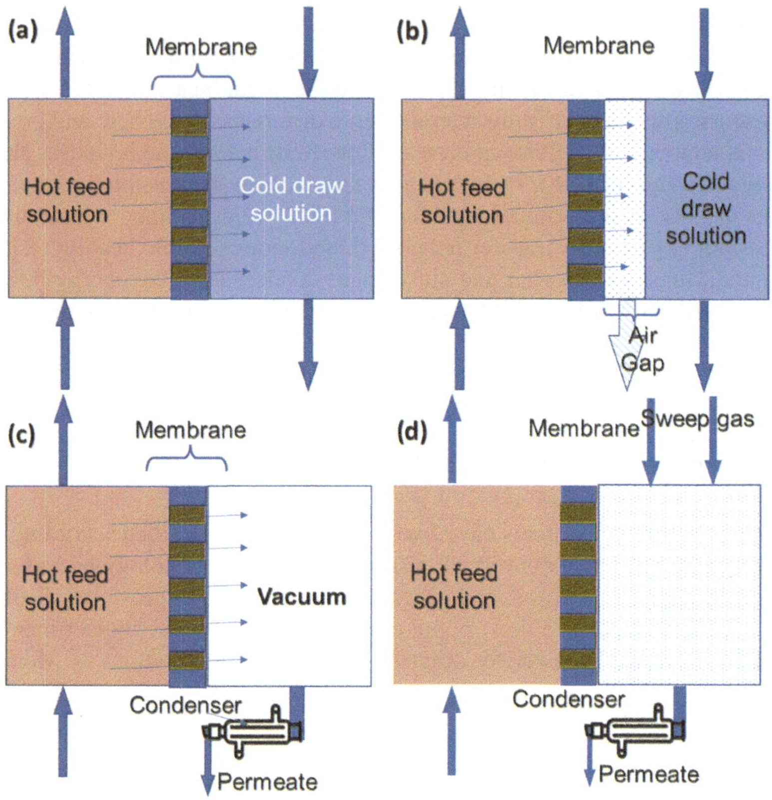

Fig. 18.3 Different membrane distillation configurations. **a** Direct contact MD, **b** Air gap MD, **c** Vacuum MD, **d** Sweep gas MD

removal efficiency, respectively, within 105 min, when the feed solution was passed through three DCMD configurations. These configurations were (1) conventional DCMD, (2) hollow fiber membrane contactor, and (3) a modified DCMD apparatus. This study concluded that the DCMD process could be used for ammonia removal from wastewater (Qu et al. 2013). Heat loss by conduction is one of the drawbacks of Direct contact configuration design (Hsu et al. 2002). Figure 18.3 summarises the differences and mode of operation in different configurations.

18.2.1.2 Sweeping Gas Membrane Distillation (SGMD)

In a sweeping gas MD, the hydrophobic side of the membrane faces the hot feed, and the vapors generated by the high temperature, after passing through the pores, are swept using an inert gas. Subsequently, the condensation of the gas occurs in the exterior to the membrane setup. The gas barrier here is not static but decreases heat and enhances the mass transfer coefficient (Camacho et al. 2013). Xie et al. (2009) studied the SGMD applications for ammonia removal. PTFE membrane with pore size 0.45 μm, 70% porosity, and 100 and 200 μm thickness, wastewater containing 100 mg/L ammonia at pH 11.5 was used in the experiment (Xie et al. 2009). Distillate flux, the effect of feed temperature, the flow rate of gas and ammonia were examined to analyze the removal of ammonia using SGMD. The experiments show 97% ammonia removal (treated water contained 3.3 mg/L ammonia), and an increase in distillate flux and feed temperature was recorded (Xie et al. 2009). It was concluded that the SGMD process could be effectively used for treating wastewater.

18.2.1.3 Air Gap Membrane Distillation (AGMD)

In AGMD, there is an immobile air gap on the distillate side between the membrane and cold stream. The air gap is crossed by vapor for condensation on the cold surface. Desalination and volatile organic compound removal from aqueous solutions are apt for this design. The drawback in the DCMD and SGMD led to the introduction of the AGMD method. The driving force in this mode is the difference in temperature of the feed liquid and conducting surface (Moulik et al. 2018).

In the Water gap membrane distillation (WGMD) module, the AGMD design is altered by introducing immobile water in place of air (Alawad and Khalifa 2019). For desalination, deionized water is used to avoid contamination. Zarasavand Asadi et al. (2012) used AGMD for the treatment of oily-saline water from a gas refinery (Zarasvand Asadi et al. 2012). Solar energy was used with AGMD apparatus, and the results showed a much decrease in total dissolved solids from 1991 to 91 mg L^{-1}. The objective of this experiment was to obtain drinking water from high salinity oily water, and a sound reduction in the contaminants was observed using AGMD (Zarasvand Asadi et al. 2012).

18.2.1.4 Vacuum Membrane Distillation (VMD)

VMD is done by applying a vacuum on the penetrating side of the membrane, which decreases the pressure radically under the equilibrium pressure of the feed and volatile constituents. VMD process increases permeate flux with less heat loss by evacuating the membrane's pores by deaeration or creating a vacuum on the distillate side using a vacuum pump. VMD gives high flux, and vacuum provides insulation against heat loss by conduction and reduces resistance to mass transfer (Alkhudhiri et al. 2012). One of the applications of VMD was studied in the downstream processing of bioethanol. In converting lignocellulose to bioethanol, hydrolysis produces certain derivatives such as phenolic compounds, furans, and aliphatic acids that act as inhibitors. These by-products reduce the productivity of ethanol. Chen et al. (2013) studied fermentation inhibitor removal from lignocellulose hydrolyzates reaction, and the effect of feed temperature and velocity were investigated (Chen et al. 2013). They concluded that high distillate flux was achieved at high feed temperature due to high heat transfer, and also the removal efficiency increased from 7.26 to 24.79% (Chen et al. 2013). Among all the membrane configurations, DCMD has a widespread application in laboratory research. In commercial applications, AGMD predominates due to its high energy effectiveness and the ability for latent heat retrieval. The air gap in AGMD possesses less heat loss than in DCMD but increases mass transfer resistance.

18.2.1.5 Membrane Characteristics

For efficient MD performance, the membranes are proposed to have specific properties described here with their effect on performance and improvement opportunities. The membrane is a microporous and hydrophobic material acting as a physical obstruction that does not influence separation directly in the MD process. Vapor–liquid equilibrium is established due to water vapor or volatile constituents at the membrane surface. Membranes used in the MD system should have the following properties:

1. High Liquid entry pressure
2. Low thermal conductivity
3. High permeability with Reduced mass transfer resistance
4. Low fouling rate
5. Excellent thermal, mechanical, and chemical stability
6. Long term performance.

High Liquid Entry Pressure

The wetting resistance of a hydrophobic membrane is determined by its liquid entry pressure (LEP) measurement. The LEP is the minimum hydrostatic pressure to be applied to the liquid to penetrate into pores and wet the hydrophobic membrane. For a suitable MD system membrane, high LEP is required, so that entry of the feed solution can be prevented, which is a prerequisite for high flux and quality of the permeate with long-term performance. High LEP can be obtained by low surface energy and high contact angle membrane surfaces. Franken and co-workers related the LEP with liquid surface tension (σ), contact angle (θ) and pore size as following (Franken et al. 1987)

$$LEP = -2B\sigma\frac{\cos\theta}{r_{\max}} \tag{18.2}$$

where B = geometric factor of membrane pore size irregularities (i.e., B = 1 for cylindrical pore); σ = surface tension; θ = contact angle of the liquid on the membrane; r_{max} = maximum pore radius of the membrane.

So, a membrane with a higher contact angle and smaller pore sizes is required for high LEP; surface roughness can also enhance the surface energy for a good MD membrane (Ravi et al. 2020). Introducing super-hydrophobicity in MD membrane creates an air gap between water and membrane surface; this air gap allows the membrane to have larger pores without compromising the LEP requirements. Super-hydrophobic membranes are made by incorporating nanomaterials (NM) in the membrane matrix, which generally imparts surface roughness and lowers the surface energy. Lu et al. (2019) have extensively discussed the membrane-water–vapor interaction characteristics and their improvement approaches (Lu et al. 2019). The nanomaterials have also been shown to affect the contact angle of the membrane. Tijing et al. suggest 5% MWCNT incorporation resulted in a 9.5° change in water contact angle (WCA) of their electrospun composite membrane (Tijing et al. 2016).

Similarly, An et al. (2017) used 2% MWCNT and gained 8.2° WCA increment, Woo et al. (2016) used 5% graphene to enhance WCA by 20.4°. To get some significant change in WCA, we have to mix a sufficient amount of NM, but as the concentration of incorporated NM increases, narrower pores are formed, and flux declines. So, functionalizing the NP with hydrophobic groups like silane or amines can be done. Functionalization not only increases the dispersibility but also increases the wetting resistance of NM at high concentrations without getting agglomerated in the matrix. MD membranes need to exhibit wetting resistance, so surfaces with omniphobic and amphiphobic nature are also gaining attention. Surface modification can be obtained by coating, grafting, and plasma modification. Lu et al. (2016) coated PVDF membrane with silane and SiO_2 by vacuum filtration to get a WCA change of 16.3°. PVDF can also be coated by dip-coating in PDMS to change the surface energy (Jiao et al. 2020). PES membranes were grafted by silane and silicate to increase the WCA to 149° (Rastegarpanah and Mortaheb 2016). Moradi et al. (2016) used PVDF/TiO_2 composite membrane for trace heavy metal removal using MD. An increase in LEP was observed with increasing NP concentration, increasing WCA from 124° to 135°.

High Permeability, Porosity, and Tortuosity

Better results are obtained when a large amount of vapors permeate through the membrane per unit of time. In theory, larger pore size and smaller pore tortuosity in thinner membranes give a high molar flux. A tighter and uniform pore distribution is desirable for MD. But in practice, large pores and thin membranes reduce the LEP and thermal conductivity (Ravi et al. 2020). So optimized parameters help in gaining high LEP and permeability (Wang et al. 2009). Desired values for porosity are 80–90% (Eykens et al. 2016), thickness 0.5–06 μm (Schneider et al. 1988), tortuosity ~0, and thickness 100–200 μm (Tang et al. 2012).

As compared to electrospun membranes, membranes made by phase inversion have high porosity giving a high flux (Eykens et al. 2017). Membrane porosity is characterized as the division of the volume of pores to the mass volume of the layer. It can be estimated by two types of liquids: one that can penetrate the pores (hydrophobic liquid, e.g., IPA) and the other that does not pass (hydrophilic liquid, e.g., water). A highly porous membrane allows larger permeate transport (Khayet and Matsuura 2001). Porosity determined by Smolder and Franken is:

$$\varepsilon = 1 - \frac{\rho_m}{\rho_{pol}} \tag{18.3}$$

where, ε = porosity; ρ_m = membrane density; and ρ_{pol} = polymer density.

The shape of the pores other than cylindrical is referred to as tortuosity (τ). As tortuosity increases, permeate flux decreases (Srisurichan et al. 2006). Macki-Meares formulated tortuosity with porosity as:

$$\tau = \frac{(2 - \varepsilon)^2}{\varepsilon} \tag{18.4}$$

Pore size should be optimum, neither too small to reduce permeate flux nor too large to allow liquid penetration. Hence vapor flux is determined by mean pore size. 100 nm–1 μm pore size membranes are commonly used (El-Bourawi et al. 2006).

Solvent and non-solvent in the case of membrane fabrication play a significant role. So, a weak coagulant solution produces a more porous membrane compared to a strong coagulant. Bonyadi and Chung (2007) and Munirasu et al. (2017) have used alcohols as weak coagulants to get more porous membranes yielding a higher flux. Similarly, Hou et al. (2014) showed flux decline with different solvents like NMP, DMSO, DMAc, and DMF. Few pore-forming additives like PEG (Wang et al. 2008), PVP (Silva et al. 2015), oxalic acid (Zheng et al. 2015), glycerol (Zheng et al. 2015), and Lithium Chloride (Liao et al. 2013; Hemmat et al. 2015) have been administered to get sponge-like structures. These additives are primarily removed in post-treatment on membrane but also alters the membranes WCA (Silva et al. 2015). On increasing the TiO_2 concentration in PVDF composite membrane, Moradi et al. (2016) obtained an increase from 72 to 75% in porosity with decreasing pore size from 300 to 270 nm on 20% blending compared to the pure membrane.

There is an inverse relationship between the permeate flux and the thickness of the membrane. As the thickness increases, flux decreases. A thinner membrane imposes lower mass transfer resistance in terms of thickness, eventually helping in higher flux. Interestingly, increasing the thickness of the upper layer of the dual-layer electrospun membranes help in getting higher porosity and higher flux (Attia et al. 2018).

Fouling Propensity

The growth of undesired materials on the membrane surface or inside the membrane's pores is membrane fouling. Membranes used in MD can be susceptible to a variety of hydrophobic foulants found in feed water. It results in an unfavorable effect on the overall performance of MD, formed due to scaling (salt precipitation), inorganic fouling (collides deposition or salt crystallization), organic fouling (organic compounds accumulation), and bio-fouling (microorganism growth). Factors that affect fouling include (1) foulant characteristics (its' charge, solubility, concentration, diffusivity, molecular size, hydrophobicity, etc.); (2) membrane properties (pore size, hydrophobicity, PSD, surface roughness, surface charge, and functional groups on the surface); (3) operational conditions (solution temperature, flux and flow velocity); and (4) feedwater characteristics (ionic strength, pH, solution chemistry, and presence of inorganic/organic matters) (Tijing et al. 2015). Properties of foulant and their concentration govern the nature of fouling that can happen. The feed's temperature, velocity, and interactions between the membrane surface and foulants affect the fouling significantly. Fouling decreases permeate flux by obstructing the membrane pores and decreasing its effectiveness. However, the fouling strength can be restricted by the flow rate of the feed and its temperature. Fouling on the membrane reduces the temperature gradient available across the membrane. Although MD membranes have low fouling characteristics compared to pressure-driven processes, the elimination of fouling extends the life of the membrane, eventually leading to a reduction in operation costs. Warsinger et al. (2015) and Tijing et al. (2015) have extensively analyzed fouling controlling techniques for MD membranes. Guo et al. (2019) made an innovative photocatalyst BiOBr/Ag-based self-cleaning MD membrane, regenerating under visible light exposure. Electroconductive materials like CNT-PVA /PP can improve the properties by imparting electrical conductivity in composite membranes and improve cleaning abilities (Barbhuiya et al. 2021; Tang et al. 2017). Figure 18.4a–d are SEM images of the fouling in an MD membrane, Fig. 18.4e given below shows the fouling inside the pores and on the membrane surface (Alkhudhiri et al. 2012). Figure 18.4f expresses the extent of flux drop in different fouling scenarios.

Types of Fouling

1. **Inorganic fouling**: Precipitation of hydroxides and accumulation of other inorganic species cause this type of fouling. Due to high bicarbonate concentrations (alkalinity) in the feed water, the most commonly occurring scale in MD membranes is $CaCO_3$, (Gryta 2009) High-temperature strips out CO_2 from the system. So, bicarbonate decomposes to carbonate (Gryta 2005). Scaling is affected by solution pH and temperature, so maintaining pH below 7 hinders the scaling. A non-porous scale significantly blocks the water flow and decreases efficiency. Figure shows the fouling effect of MD performance. Calcium sulfate scaling occurs on supersaturation and causes deformation and cracking on the surface and pores, increasing wetting in MD (Gryta 2005). Similarly, Silica scaling is also significant due to its polymerization making colloidal silica (Milne et al. 2014). Figure 18.4f clearly shows the lowest reduction in flux due to $CaCO_3$ caused fouling compared to other fouling scenarios for the MD system.
2. **Organic fouling**: This type of fouling is caused by soluble Organic particle deposition like debris, hummus, protein and surfactants, and colloidal particles (Abdel-Karim et al. 2021). The organic fouling does not interfere with the membrane matrix but alters its hydrophilicity, eventually leading to pore wetting and reduced performance (Abdel-Karim et al. 2021).
3. **Biofouling**: Biofilm formed by microbes due to the release of polysaccharides causes biofouling. A vicious, slimy, hydrated layer forms on the membrane

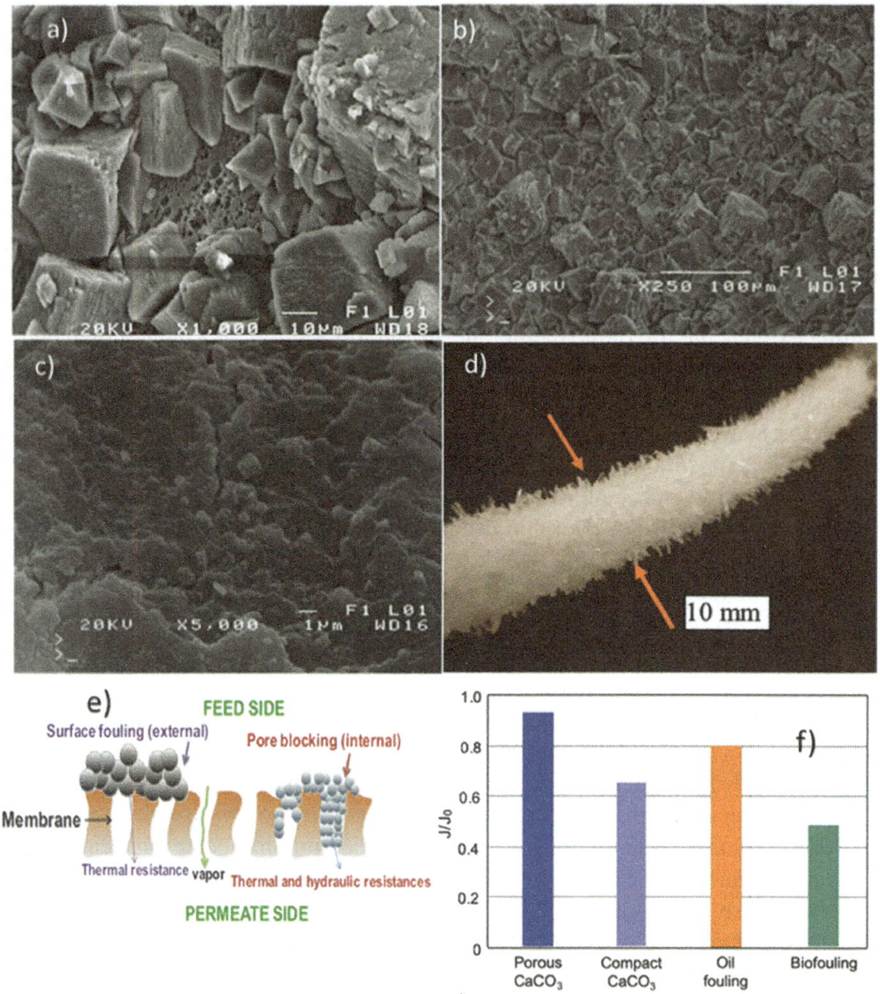

Fig. 18.4 The extent of scaling in MD. **a**, b $CaCO_3$ scaling, **c** biofilm layer, **d** NaCl crystal scaling reprinted with permission from Elsevier from reference (Abdel-Karim et al. 2021), **e** Membrane fouling in the pores and on the surface reprinted with permission from Elsevier from reference (Tijing et al. 2015), **f** Effect of different fouling and the extent of flux drop due to fouling in MD system. Reprinted with permission from Elsevier from reference (Abdel-Karim et al. 2021)

surface, particularly in wastewater applications where feed water contains high quantities of microbes and extracellular substances (EPS) (Bogler and Bar-Zeev 2018). These EPS acts as pore wetting agents, declining the flux and performance of the membranes. Biofouling is affected by the feed temperature (Bogler and Bar-Zeev 2018; Jiang et al. 2020), water recovery (Jiang et al. 2020), cross-flow velocity (Yang et al. 2020), and water composition (Zodrow et al. 2014). Through careful implementation of antibacterials (Jiang et al. 2019) and anti adhesive (Sabri et al. 2019), we can restrict the development of MD. Figure 18.4f shows the highest flux drop caused due to biofouling compared to other fouling scenarios for the MD system.

4. **Colloidal Fouling**: Colloidal fouling can be caused by inorganic and organic substances existing in a system in suspended colloidal form (Xia et al. 2018). They generally have sizes from tens of nanometre to about micron size. Inorganic colloids include silicates, iron, and aluminum oxides, and silts (Tijing et al. 2015). Organic colloids may consist of cell fragments and biopolymers from extracellular and intracellular polymeric substances.

Analytical separation becomes difficult because of fouling as it increases membrane thickness. Therefore softening (Gryta 2010), anti-scaling agent (Al-Anezi et al. 2008), pH modification (Karakulski and Gryta 2005), membrane-based pretreatment (Karakulski and Gryta 2005), and chemical cleaning (Karakulski and Gryta 2005) are done as pre-treatment for feed. Physical cleaning methods like air backwashing, aeration, ultrasonic/vibration, and water flushing are other less explored methods for cleaning.

Thermal Conductivity and Chemical, Thermal and Mechanical Stability

Feed solutions like seawater may contain various reactive chemicals; the cleaning solutions may have abrasive chemicals, so chemical stability of MD membranes is necessary. The operating temperature of the MD may exceed the polymer's melting point, which causes damage to the structure of the membrane. So, Polymers used for MD membranes should be thermally stable. The membrane may face substantial force while being packed in modules or being operated, so sufficient mechanical strength is required for membranes. Thicker membranes provide an excellent mechanical strength but have lower permeability. Additives like $CaCO_3$, cellulose improve the mechanical strength of the membranes (Ravi et al. 2020). Pores of membranes have low conductivity due to the presence of air so, porous membranes have a lower thermal conductivity. Additives like silica aerogel, aluminum fumarate MOF, SiO_2 and PDMS, and MWCNT help improve thermal conductivity (Ravi et al. 2020).

Membrane Materials

Thermally stable, hydrophobic, and strong polymers like Polytetrafluoroethylene (PTFE), polyvinylidene fluoride (PVDF), and polypropylene (PP) are used in the MD process. Among the materials mentioned above, PTFE is highly hydrophobic (the largest contact angle with water), has good chemical and thermal stability and oxidation resistance. PP also shows good thermal and chemical resistance. PVDF membrane can be easily prepared with better hydrophobic character, thermal resistance, and mechanical strength. Fabrication of membranes can be done by stretching, sintering, and phase inversion methods (Camacho et al. 2013). Surface energies and

Table 18.2 Surface energy and thermal conductivity of different materials (Camacho et al. 2013; Gale et al. 2016; Madeira et al. 2018)

Membrane material	Thermal conductivity (W m^{-1} K^{-1})	Surface energy ($\times 10^{-3}$ N/m)
Polytetrafluoroethylene (PTFE)	0.25	9–20
Polypropylene (PP)	0.17	30
Polyvinylidene fluoride (PVDF)	0.19	30.3
Polyethylene (PE)	0.3–0.5	20
Polydimethylsiloxane (PDMS)	0.2	19–21

thermal conductivities of PTFE, PP, and PVDF are mentioned in Table 18.2: Owing to the excellent properties of the ceramics, the hybrid membrane of polymer and ceramic can help in different problems.

18.2.1.6 Performance Evaluation Parameters

The performance of an efficient MD system is characterized by its flux (J), contaminant rejection % (R), and flux recovery ratio (FRR). If W is the mass of distillate obtained (kg), A is the effective area of membrane (m^2), and t is time to obtain this distillate, then cumulative permeate flux (J) of the process can be evaluated by

$$J = \frac{W}{A \times t} \tag{18.5}$$

If C_p and C_f are solute (or pollutant) concentrations in permeate and feed side, respectively, then solute rejection (R) can be computed as

$$R = \left(1 - \frac{C_p}{C_f}\right) \times 100 \tag{18.6}$$

The membrane regeneration can be found by evaluating the flux recovery ratio (FRR), which is the ratio of J_f and J_0, representing the flux of cleaned and pristine membrane, respectively (Fig. 18.5)

$$FRR = \left(\frac{J_f}{J_0}\right) \times 100 \tag{18.7}$$

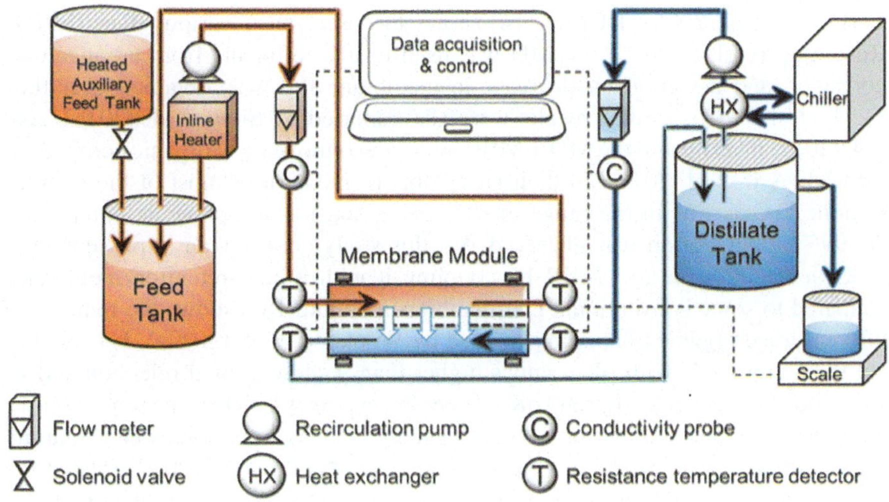

Fig. 18.5 Schematic of a benchtop lab-scale DCMD setup for performance evaluation. Reprinted with permission from Elsevier from reference (McGaughey et al. 2017)

18.3 Emerging Contaminants Removal by Membrane Distillation

MD has been widely investigated for seawater and brackish water desalination due to the high quality and good flux of the produced water. But as the concerns related to emerging contaminants increase, MD can be implemented as a good process for removing chemicals and biological substances. Wastewater treatment through MD has been investigated by Zhu et al. (2014). At the laboratory stage, desalination, groundwater treatment, volatile organics removal, and solar heat utilization have been done for MD (Moradi et al. 2016). There are a limited number of works dealing with emerging contaminants removal using MD (Wijekoon et al. 2014a; Han et al. 2017). Few studies for removal of trace amounts of heavy metals (El-Bourawi et al. 2006; Blanco Gálvez et al. 2009; Zakrzewska-Trznadel et al. 2001), boron (Hou et al. 2010), arsenic (Pal and Manna 2010; Criscuoli et al. 2013) also exist. Wijekoon et al. (2014a) used MD as a post-treatment for membrane bioreactor (MBR) for removing trace organic compounds. The MD-MBR system sufficiently removed the selected 25 TrOCs, but due to salinity buildup during operation, nitrogen and recalcitrant compound removal were negatively affected (Wijekoon et al. 2014b). A pilot-level Air Gap MD was used to remove micropollutants of a WWTP located in Stockholm (Woldemariam et al. 2016).

Song et al. (2018) suggested that integration of MD in an anaerobic Biological Reactor (BR) improves the rejection from 25 to 75% for 26 different pharmaceuticals (Song et al. 2018). MD was also tested for removal of β-estradiol hormone and urea with high water recovery from a life support system used in space missionsc

(Cartinella et al. 2006). DCMD also shows the removal of various dyes (>97%) (Ruiz-Aguirre et al. 2015). DCMD was used to treat effluents from an anaerobic bioreactor, and a significant decrease in membrane flux was obtained over time (Jacob et al. 2015); similar results were also reported by Silva et al. (2018). Asif et al. (2017) investigated the MD-EMBR system to remove trace organic compounds (TrOCs) (Asif et al. 2017). In their study, the degradation of most of the selected phenolic TrOCs was in the range of (87–99%), and for non-phenolic compounds (40–99%), degradation was observed. So, this study confirms an improvement in TrOC degradation in an EMBR by coupling it with a high retention membrane, compared to some conventional UF membranes. Figure 18.6 shows the removal of different trace organic contaminants used by Wijekoon et al. (2014a). It is evident that the compounds with pK_H values higher than 9 show a good rejection and no correlation between rejection and pK_H values lower than 9. Log D value represents the hydrophobicity, and authors suggest that an interplay of hydrophobicity and volatility governs the rejection of these organic compounds. Naidu et al. (2017) treated RO concentrate by DCMD for micropollutants removal, which showed low to moderate rejection for some specific compounds (Naidu et al. 2017).

Silva et al. (2018) used DCMD to remove micropollutants and microorganisms from a real water matrix (Silva et al. 2018). The effect of membrane type and modules, feed temperature, and flow rates were tested. The optimized conditions were then

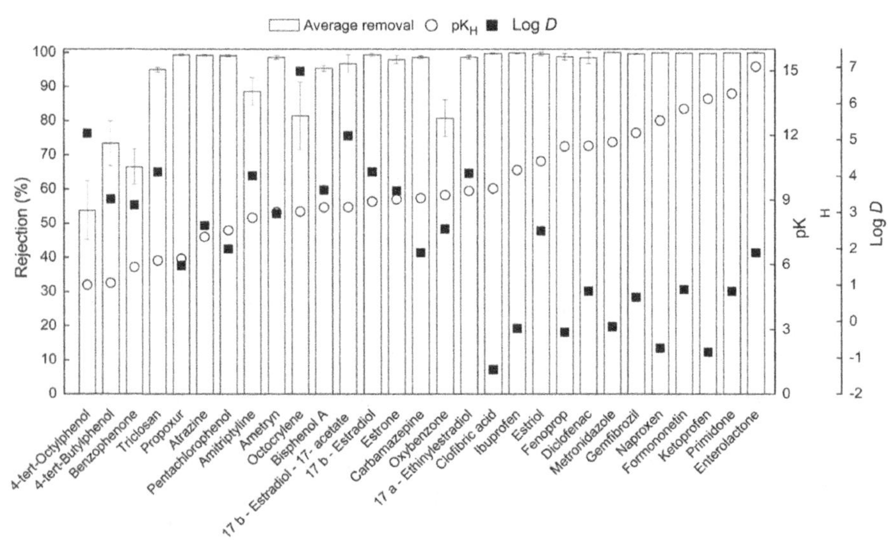

Fig. 18.6 DCMD system rejection for 29 TrOCs by DCMD with respective log D and pK_H values. Log D and pK_H illustrate the values at pH 9. Error bars show a standard deviation from four replicate measurements. A synthetic solution containing approximately 5 mg/L of each TrOC in Milli-Q water was used as the feed. The MD was carried out at feed and distillate temperatures of 40 and 20 °C, respectively. The feed and distillate circulation flow rate was 2 L min^{-1} (corresponding to 11.7 cm s^{-1}). Figure reprinted with permission from Elsevier from reference (Wijekoon et al. 2014a)

used for feed solutions like river water, seawater, and secondary treated municipal wastewater effluent. Diphenhydramine, the first-generation drug, was used as a model organic pollutant and spiked in seawater, and the DCMD system showed a rejection of about 98.7%. Silva et al. (2018) while using a WW matrix, evaluated that out of the three membranes with comparable pore size, a membrane with high porosity, high hydrophobicity, and lowest thickness resulted in the best flux. Won (2017) showed treatment of a set of volatile (V), semi-volatile (SV), non-volatile (NV) organic contaminants, pharmaceuticals, and personal care products (PPCP), and nitrosamine (Won 2017). From all these 32 organics, eight nitrosamines and 22 PPCPs showed acceptable mass recoveries. NV solutes showed a rejection of 90% from the DCMD system. Moradi et al. (2016) investigated the possibility of heavy metal removal using MD (Moradi et al. 2016). Utilizing a PVDF/TiO_2 composite membrane in the VMD setup, they have achieved a 500% concentration increase compared to the initial feed side. Han et al. (2017) investigated ibuprofen removal using MD in the presence of natural organic matter (NOM) and inorganic salts simulating a groundwater matrix (Han et al. 2017). A rejection of c.a. 90% was observed for ibuprofen within the pH range of 2.6–11.

Janowska et al. (2020) proposed an integrated MD and advanced oxidation system using a thermocatalytic membrane generating reactive oxygen species (ROS) (Janowska et al. 2020). These oxygen species are adequate for the degradation of the bisphenol-A. The MD will help in increasing the concentration of these pollutants escalating the degradation process through ROS. This system was able to get a constant water production rate of 1.60 LMH. A thermocatalytic perovskite strontium ferrite ($SrFeO_{3-\delta}$) can achieve 83% mineralization of BPA in 24 h and full decoloration of dye in 60 min, so it was selected for the integrated MD process (Leiw et al. 2013).

FO-MD hybrids are investigated by Lee et al. (2018), Volpin et al. (2019), Li et al. (2020), Arcanjo et al. (2020), Nawaz et al. (2021); for different applications. Nawaz et al. (2021) used FO-MD hybrid to treat high concentrated streams from oil industry-produced water. FO was used to produce a diluted draw solution from a concentrated stream, which subsequently goes through the DCMD for pure water production. Chen et al. (2020) demonstrated a short-chain PFAS Perfluoropentanoic acid (PFPeA) removal using DCMD (Chen et al. 2020). PFAS have a boiling point around 140 °C, so DCMD shows suitability for removal. PTFE flat sheets membranes were used for the demonstration; at a feed temperature of 50 °C, a rejection of PFPeA was obtained. In a 72 h operation, 27.5 kg m^{-2} h^{-1} flux with 82% rejection can be obtained. After some time, due to PFAS fouling layer formation on the membrane, the flux decreased. But this decrease is around 9.1% compared to a 16% decline in the case of NF. Winglee et al. (2017) modeled the DCMD system to treat six representative compounds from oil and gas produced water; they have estimated that the magnification of contaminant concentration in permeate, i.e., the permeate, has concentrations higher than feed (Winglee et al. 2017). Ramlow et al. (2017) have extensively reviewed the textile effluents rejection using the MD process and suggest that MD has shown 80–100% TOC rejection in recent researches (Ramlow et al. 2017). Ruiz-Aguirre et al. (2017)

have used MD for ~ 100% removal of highly resistant spores from wastewater (Ruiz-Aguirre et al. 2017). Table 18.3 lists the different Pollutants with their removal efficiency, MD configuration used and references. As is evident from the table that most of the studies involve the use of the DCMD system because of the simplicity in operation and maintenance. Also, it is clearly visible that the MD system is very efficient in removing most of the pollutants.

18.4 Economics and Energy of MD System

The efficiency of a system is not a sole decider for its practical usability, but its competitiveness is judged based on its cost-effectiveness. For MD based treatment system, its capital cost includes equipment, auxiliary equipment, land, and installation costs, and operating cost includes heating, running, compressor operating costs with maintenance and membrane replacement cost. It can be seen that compared to conventional systems, MD has no significant chemicals required for operation, but the applicability of MD systems is limited by their energy and cost expenditures. Couto et al. (2020) suggests that an NF process, RO process shows an SEC of 0.32 kWh m^{-3} m^{-2}, 1.12 kWh m^{-3} m^{-2} respectively, whereas MD showed an SEC of 0.2 kWh m^{-3} m^{-2} for pumping only, and the heating requirements are significantly higher 41.61 kWh m^{-3} m^{-2}. Couto et al. (2020) found that the MD process has the highest Capex of 1.96 USD m^{-3} compared to 0.50 USD m^{-3} and 0.43 USD m^{-3} for NF and RO, respectively (Couto et al. 2020). Life cycle cost with MD system when operating with waste heat or conventional heat sources varies from 0.26 USD m^{-3} to 18 USD m^{-3} (Saffarini et al. 2012). Contrary to this, MD showed no fouling tendency while NF and RO showed significant flux decline due to fouling. Moreover, cost reduction in the MD process can be realized by reducing the heating energy cost by utilizing low-cost energy such as cogeneration or solar energy (Couto et al. 2020). Figure 18.7 shows that in the case of MD, the energy-related price makes it costlier than NF and RO.

MD process has several opportunities in modern industrial sectors. Desalination and other environmental applications are the promising future of MD. MD technique requires high energy input and long-term operation is prone to membrane wetting and fouling. It also has an uncertain economic cost. The process faces problems in shifting from pilot scale to commercial scale. By using different membrane modules, different permeate fluxes can be obtained. Hence, the permeate flux is affected by module design, MD configuration, and appropriate operating conditions. Reduction in energy consumption is one of the milestones to be achieved. Hybrid MD systems, pressure-driven processes, alternative energy sources such as solar and geothermal energy, and waste heat recovery by installing an MD plant near the nuclear power plant are some of the measures to be taken. This may also bring down the cost of the process. The recent development of the MD solar pilot plant affirms that sustainable industrial growth is possible for MD. However, much more intensive and continuous efforts are needed in basic and applied research, decreasing production costs with

Table 18.3 Different studies on emerging contaminant removal using the MD process

S. No.	Pollutant removed	Removal percentage	Type of configuration used	References
1	Separation of H_2O/HDO and $16H_2O/18H_2O$	~ 100%	Radioactive elements DCMD	Zakrzewaska-Trznadel et al. (2001)
2	Bisphenol A	84%	DCMD	Naidu et al. (2017)
3	Metal plating wastewater	Nickel (~100%)	Modified DCMD	Zoungrana et al. (2016)
4	Hg, Cd, and Pb removal	95–100%	DCMD with FO hybrid	Wu et al. (2017)
5	Cr (VI) removal		DCMD	Bhattacharya et al. (2014)
6	As (III) and As (V)	99.9%	DCMD	Qu et al. (2009)
7	As (III)	> 99.9%	FO-DCMD hybrid	Ge et al. (2016)
8	Pesticide compounds (Atrazine, Clofibric Acid, Dichlorvos, Parathion Methyl, and Phorate)	70–99%	DCMD	Plattner et al. (2018)
9	Sodium salt of ibuprofen	~ 100%	Photocatalysis-DCMD Hybrid	Mozia and Morawski (2012)
10	Boron, Arsenic and Ibuprofen	~ 90%, 100% (B,As)	DCMD	Han et al. (2017)
11	Diclofenac, Ibuprofen, Naproxen	73–100%	Hybrid Photocatalysis–DCMD	Darowna et al. (2014)
12	Diclofenac, Azithromycin, Clarithromycin, Erythromycin, Diphenhydramine	> 99%	DCMD	Silva et al. (2018)
13	Diclofenac, Furosemide, Sulfamethoxazole, Hydrochlorothiazide, Caffeine, Atenolol, Ciprofloxacin, Paracetamol, Trimethoprim, Ranitidine, Metoprolol, Oxazepam, Carbamazepine, Ketoprofen, Propranolol, Citalopram, Bisoprolol etc. (different 37 compounds)	62–100%	AGMD	Woldemariam et al. (2016)
14	Sunset yellow (SY) and rose bengal (RB) dyes and sodium dodecyl sulfate (SDS) surfactant	100%	AGMD	Leaper et al. (2019)
15	Congo red	99%	DCMD	Khumalo et al. (2019)
16	Reactive blue	95.3%	DCMD	Chong et al. (2016)
17	Methylene blue dye, Crystal violet dye, Acid red 18 dye, Acid yellow dye	96–100%	DCMD	An et al. (2016)

(continued)

Table 18.3 (continued)

S. No.	Pollutant removed	Removal percentage	Type of configuration used	References
18	*E. Coli*, *F. Solani* and *Clostridium sp.*	~100%	Liquid gap MD	Ruiz-Aguirre et al. (2017), Ruiz-Aguirre et al. (2015)
19	*B. subtilis* and *Clostridium sp.* spores, from wastewater	~100%	AGMD	Ruiz-Aguirre et al. (2017)
20	29 different selected TrOCs	>95%	thermophilic membrane bioreactor (MBR)-DCMD	Wijekoon et al. (2014a)
21	BPA degradation by thermocatalytic membrane during filtration by MD	83% mineralization in 24 h	DCMD	Janowska et al. (2020)
22	26 different selected TrOCs	76–100%	Anaerobic MBR-DCMD	Song et al. (2018)
23	13 phenolic and 17 non-phenolic trace organic contaminants (TrOCs)	90–99%	DCMD —Enzymatic MBR	Asif et al. (2017)
24	short-chain PFAS Perfluoropentanoic acid (PFPeA)	82% in72 h	DCMD	Chen et al. (2020)
25	Cationic dye Maxilon Blue 5G, Cationic dye Drimarene yellow K-2R, Anionic dye sodium, Fluoresceine	>99.73%	DCMD	Laqbaqbi et al. (2019)
26	Fire retardant Tri(2-chloroethyl)phosphate	>99%	DCMD	Naidu et al. (2017)
27	Antibiotics, pesticides, and Herbicides	84–96%	DCMD	Naidu et al. (2017)

less waste generation, high flexibility, and easy scaling up to construct competitive and innovative pilot plants.

18.5 Summary and Conclusion

Membrane distillation technology is applied for separation processes, such as desalination. In this process, vapor can cross through the membrane, where vapor pressure difference acts as the driving force. This technology can treat wastewater and brines also, but since there is a lack of experimental data for pilot scale and specific membrane and modules, its use is limited at a larger scale. MD is shown to be effective in freshwater production, removal of heavy metals and emerging contaminants. Membrane fouling, membrane wetting are two of the problems faced in this procedure. So, there has been an emphasis on developing membranes with high

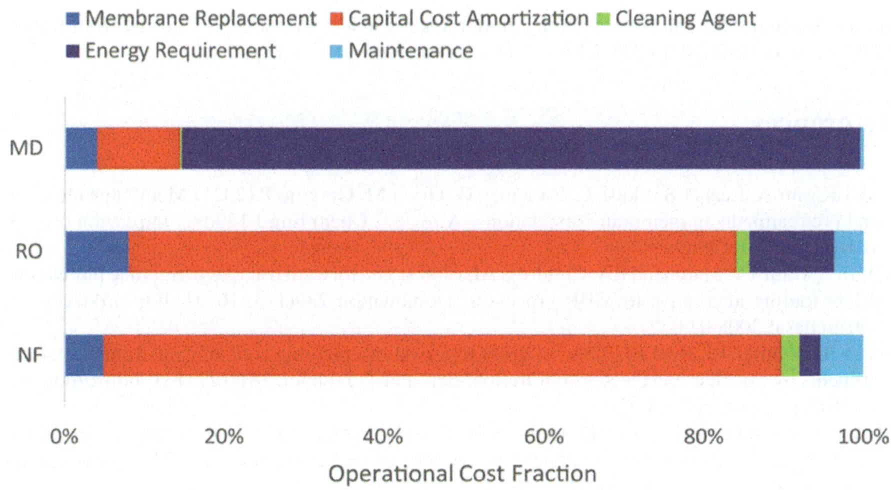

Fig. 18.7 Different cost components and their contribution to treatment cost Reprinted with permission from Elsevier from reference (Couto et al. 2020)

hydrophobic character and anti-fouling characteristics to perform better and purify water. MD is very efficient in the removal of EC. The presence of organic content in the feed tends to alter the hydrophobicity of the membranes, and thus a pre-treatment of the feed is required to preserve the hydrophobicity of the membranes. It is well established that the high specific thermal energy limits MD operation, but the significant advantage lies in its low sensitivity on concentration and quality variation in feed water. A low volume concentrated discharge is created in MD, which can be treated in solar ponds and crystallizers.

The emergence of the water demand reclamation and reuse led to the development of treatment trains fitting the needs and requirements. Hybridization offers new avenues for process improvement and energy saving. The current status shows that the hybrids need to be tested for long-term operations (Abdel-Karim et al. 2021). There has been a surge in the Seawater desalination plants, and the concentrated stream generated needs very careful disposal. MD process has emerged as a significantly good process for such needs. Studies show a slight flux decline after treatment of reject stream of wastewater. Many authors have highlighted RO hybridization with MD to treat the concentrated stream. FO in hybridization with MD uses the draw solution as feed. BR-MD hybrid has also shown a clear performance improvement. In conclusion, the MD systems have a good potential to remove emerging contaminants from the water matrix to almost 100%. MD has been used to treat the concentrate stream for different treatment trains with good efficiency, especially in the presence of emerging and recalcitrant contaminants. So, as the technology develops, more MD systems are employed in hybrid configurations and working with renewable energy of waste heat.

Acknowledgements The corresponding author acknowledges the funding received from ISRO-IIT(B) SPACE TECHNOLOGY CELL (STC) for carrying out this work.

References

Abdel-Karim A, Leaper S, Skuse C, Zaragoza G, Gryta M, Gorgojo P (2021) Membrane cleaning and pretreatments in membrane distillation—A review. Chem Eng J 129696. https://doi.org/10.1016/j.cej.2021.129696

Achilli A, Cath TY, Marchand EA, Childress AE (2009) The forward osmosis membrane bioreactor: a low fouling alternative to MBR processes. Desalination 239(1–3):10–21. https://doi.org/10.1016/j.desal.2008.02.022

Acero JL, Benitez FJ, Real FJ, Teva F (2017) Removal of emerging contaminantsfrom secondary effluents by micellar-enhanced ultrafiltration. Sep. Purif. Technol. 181:123–131. https://doi.org/10.1016/j.seppur.2017.03.021

Al-Anezi K, Somerfield C, Mee D, Hankins N, Hilal N (2008) Effect of anti-scale agents on the solubility of CO_2 in seawater at temperatures of 60–90 °C and pressures of 1–2 bar. Desalination 227(1–3):46–56. https://doi.org/10.1016/j.desal.2007.05.034

Alawad SM, Khalifa AE (2019) Analysis of water gap membrane distillation process for water desalination. Desalination 470:114088. https://doi.org/10.1016/j.desal.2019.114088

Ali I, Basheer AA, Mbianda XY, Burakov A, Galunin E, Burakova I, Mkrtchyan E, Tkachev A, Grachev V (2019) Graphene based adsorbents for remediation of noxious pollutants from wastewater. Environ Int 127:160–180. https://doi.org/10.1016/j.envint.2019.03.029

Alkhudhiri A, Darwish N, Hilal N (2013) Treatment of saline solutions using air gap membrane distillation: experimental study. Desalination 323:2–7. https://doi.org/10.1016/j.desal.2012.09.010

Alkhudhiri A, Hilal N (2018) Membrane distillation-principles, applications, configurations, design, and implementation. In: Emerging technologies for sustainable desalination handbook. Elsevier, King Abdulaziz City for Science and Technology (KACST), Riyadh, Saudi Arabia, pp 55–106. https://doi.org/10.1016/B978-0-12-815818-0.00003-5

Alkhudhiri A, Darwish N, Hilal N (2012) Membrane distillation: a comprehensive review, vol 287, pp 2–18. https://doi.org/10.1016/j.desal.2011.08.027

Alturki A, McDonald J, Khan SJ, Hai FI, Price WE, Nghiem LD (2012) Performance of a novel osmotic membrane bioreactor (OMBR) system: flux stability and removal of trace organics. Bioresour Technol 113:201–206. https://doi.org/10.1016/j.biortech.2012.01.082

Alves VD, Coelhoso IM (2006) Orange juice concentration by osmotic evaporation and membrane distillation: a comparative study. J Food Eng 74(1):125–133. https://doi.org/10.1016/j.jfoodeng.2005.02.019

An AK, Guo J, Jeong S, Lee EJ, Tabatabai SAA, Leiknes TO (2016) High flux and antifouling properties of negatively charged membrane for dyeing wastewater treatment by membrane distillation. Water Res 103:362–371. https://doi.org/10.1016/j.watres.2016.07.060

An AK, Lee EJ, Guo J, Jeong S, Lee JG, Ghaffour N (2017) Enhanced vapor transport in membrane distillation via functionalized carbon nanotubes anchored into electrospun nanofibres. Sci Rep 7. https://doi.org/10.1038/srep41562

Arcanjo GS, Costa FCR, Ricci BC, Mounteer AH, de Melo ENML, Cavalcante BF, Araújo AV, Faria CV, Amaral MCS (2020) Draw solution solute selection for a hybrid forward osmosis-membrane distillation module: effects on trace organic compound rejection, water flux and polarization. Chem Eng J 400. https://doi.org/10.1016/j.cej.2020.125857

Ashoor BB, Mansour S, Giwa A, Dufour V, Hasan SW (2016) Principles and applications of direct contact membrane distillation (DCMD): a comprehensive review. Desalination 398:222–246. https://doi.org/10.1016/j.desal.2016.07.043

Asif MB, Hai FI, Kang J, van de Merwe JP, Leusch FDL, Price WE, Nghiem LD (2018) Biocatalytic degradation of pharmaceuticals, personal care products, industrial chemicals, steroid hormones and pesticides in a membrane distillation-enzymatic bioreactor. Bioresour Technol 247:528–536. https://doi.org/10.1016/j.biortech.2017.09.129

Asif MB, Hai FI, Kang J, van de Merwe JP, Leusch FDL, Yamamoto K, Price WE, Nghiem LD (2017) Degradation of trace organic contaminants by a membrane distillation—Enzymatic bioreactor. Appl Sci (Switzerland) 879. https://doi.org/10.3390/app7090879

Attia H, Johnson DJ, Wright CJ, Hilal N (2018) Comparison between dual-layer (superhydrophobic–hydrophobic) and single superhydrophobic layer electrospun membranes for heavy metal recovery by air-gap membrane distillation. Desalination 439:31–45. https://doi.org/10.1016/j.desal.2018.04.003

Barbhuiya NH, Misra U, Singh SP (2021) Synthesis, fabrication, and mechanism of action of electrically conductive membranes: a review. Environ Sci Water Res Technol. https://doi.org/10.1039/d0ew01070g

Bhattacharya M, Dutta SK, Sikder J, Mandal MK (2014) Computational and experimental study of chromium (VI) removal in direct contact membrane distillation. J. Memb. Sci. 450:447–456. https://doi.org/10.1016/j.memsci.2013.09.037

Blanco Gálvez J, García-Rodríguez L, Martín-Mateos I (2009) Seawater desalination by an innovative solar-powered membrane distillation system: the MEDESOL project. Desalination 246(1–3):567–576. https://doi.org/10.1016/j.desal.2008.12.005

Bogler A, Bar-Zeev E (2018) Membrane Distillation biofouling: impact of feedwater temperature on biofilm characteristics and membrane performance. Environ Sci Technol 52(17):10019–10029. https://doi.org/10.1021/acs.est.8b02744

Bonyadi S, Chung TS (2007) Flux enhancement in membrane distillation by fabrication of dual layer hydrophilic-hydrophobic hollow fiber membranes. J Memb Sci 306(1–2):134–146. https://doi.org/10.1016/j.memsci.2007.08.034

Bound JP, Voulvoulis N (2005) Household disposal of pharmaceuticals as a pathway for aquatic contamination in the United Kingdom. Environ Health Perspect 113(12):1705–1711. https://doi.org/10.1289/ehp.8315

Bradley PM, Journey CA (2014) Assessment of endocrine-disrupting chemicals attenuation in a coastal plain stream prior to wastewater treatment plant closure. J Am Water Resour Assoc 50(2):388–400. https://doi.org/10.1111/jawr.12165

Camacho LM, Dumée L, Zhang J, de Li J, Duke M, Gomez J, Gray S (2013) Advances in membrane distillation for water desalination and purification applications. Water (Switzerland) 5(1):94–196. https://doi.org/10.3390/w5010094

Cartinella JL, Cath TY, Flynn MT, Miller GC, Hunter KW, Childress AE (2006) Removal of natural steroid hormones from wastewater using membrane contactor processes. Environ Sci Technol 40(23):7381–7386. https://doi.org/10.1021/es060550i

Cath TY, Childress AE, Elimelech M (2006) Forward osmosis: principles, applications, and recent developments. J Memb Sci 281(1–2):70–87. https://doi.org/10.1016/j.memsci.2006.05.048

Cath TY, Adams D, Childress AE (2005) Membrane contactor processes for wastewater reclamation in space: II. Combined direct osmosis, osmotic distillation, and membrane distillation for treatment of metabolic wastewater. J Memb Sci 257(1–2):111–119. https://doi.org/10.1016/j.memsci.2004.07.039.

Chen J, Zhang Y, Wang Y, Ji X, Zhang L, Mi X, Huang H (2013) Removal of inhibitors from lignocellulosic hydrolyzates by vacuum membrane distillation. Bioresour Technol 144:680–683. https://doi.org/10.1016/j.biortech.2013.07.021

Chen X, Vanangamudi A, Wang J, Jegatheesan J, Mishra V, Sharma R, Gray SR, Kujawa J, Kujawski W, Wicaksana F, Dumée LF (2020) Direct contact membrane distillation for effective concentration of perfluoroalkyl substances—Impact of surface fouling and material stability. Water Res 182:116010. https://doi.org/10.1016/j.watres.2020.116010

Chen L, Jiang H, Li Y, Zimba BL, Yu X, Chen C, Xiong G, Wu Q (2019) Influences on mechanical properties of chitosan nanofibrous membranes induced by incorporating graphene oxide nanosheets. Mater Res Express 6(7). https://doi.org/10.1088/2053-1591/ab1555

Choi JH, Dockko S, Fukushi K, Yamamoto K (2002) A Novel application of a submerged nanofiltration membrane bioreator (NF MBR) for wastewater treatment. Desalination 146(1–3):413–420. https://doi.org/10.1016/S0011-9164(02)00524-6

Chong KC, Lai SO, Thiam HS, Lee SS, Lau WJ, Mokhtar NM (2016) Reactive blue dye removal by membrane distillation using pvdf membrane. Ind J Sci Technol 9(S1):1–5. https://doi.org/10.17485/ijst/2016/v9is1/110379

Clara M, Strenn B, Gans O, Martinez E, Kreuzinger N, Kroiss H (2005) Removal of selected pharmaceuticals, fragrances and endocrine disrupting compounds in a membrane bioreactor and conventional wastewater treatment plants. Water Res 39(19):4797–4807. https://doi.org/10.1016/j.watres.2005.09.015

Couto CF, Santos AV, Amaral MCS, Lange LC, de Andrade LH, Foureaux AFS, Fernandes BS (2020) Assessing potential of nanofiltration, reverse osmosis and membrane distillation drinking water treatment for pharmaceutically active compounds (PhACs) removal. J Water Process Eng 33:101029. https://doi.org/10.1016/j.jwpe.2019.101029

Criscuoli A, Bafaro P, Drioli E (2013) Vacuum membrane distillation for purifying waters containing arsenic. Desalination 323:17–21. https://doi.org/10.1016/j.desal.2012.08.004

Darowna D, Grondzewska S, Morawski AW, Mozia S (2014) Removal of non-steroidal anti-inflammatory drugs from primary and secondary effluents in a photocatalytic membrane reactor. J Chem Technol Biotechnol 89(8):1265–1273. https://doi.org/10.1002/jctb.4386

Drioli E, Ali A, Macedonio F (2015) Membrane distillation: recent developments and perspectives. Elsevier 356:56–84. https://doi.org/10.1016/j.desal.2014.10.028

El-Abbassi A, Kiai H, Hafidi A, García-Payo MC, Khayet M (2012) Treatment of olive mill wastewater by membrane distillation using polytetrafluoroethylene membranes. Sep Purif Technol 98:55–61. https://doi.org/10.1016/j.seppur.2012.06.026

El-Bourawi MS, Ding Z, Ma R, Khayet M (2006) A framework for better understanding membrane distillation separation process. J Memb Sci 285(1–2):4–29. https://doi.org/10.1016/j.memsci.2006.08.002

Eykens L, De Sitter K, Dotremont C, Pinoy L, Van der Bruggen B (2017) Membrane synthesis for membrane distillation: a review. Sep Purif Technol 182:36–51. https://doi.org/10.1016/j.seppur.2017.03.035

Eykens L, De Sitter K, Dotremont C, Pinoy L, Van Der Bruggen B (2016) How to optimize the membrane properties for membrane distillation: a review. Ind Eng Chem Res. American Chemical Society, pp 9333–9343. https://doi.org/10.1021/acs.iecr.6b02226

Fawell J, Ong CN (2012) Emerging contaminants and the implications for drinking water. Int J Water Resour Dev 28(2):247–263. https://doi.org/10.1080/07900627.2012.672394

Fonseca Couto C, Lange LC, Santos Amaral MC (2018) A critical review on membrane separation processes applied to remove pharmaceutically active compounds from water and wastewater. J Water Process Eng 26:156–175. https://doi.org/10.1016/j.jwpe.2018.10.010

Franken ACM, Nolten JAM, Mulder MHV, Bargeman D, Smolders CA (1987) Wetting criteria for the applicability of membrane distillation. J Memb Sci 33(3):315–328. https://doi.org/10.1016/S0376-7388(00)80288-4

Gale BK, Eddings MA, Sundberg SO, Hatch A, Kim J, Ho T, Karazi SMBT-RM (2016) In: MS, ME (eds). Low-cost MEMS technologies. Elsevier. https://doi.org/10.1016/B978-0-12-803581-8.00530-0

Ge Q, Han G, Chung TS (2016) Effective As(III) removal by a multi-charged hydroacid complex draw solute facilitated forward osmosis-membrane distillation (FO-MD) processes. Environ Sci Technol 50(5):2363–2370. https://doi.org/10.1021/acs.est.5b05402

Goh S, Zhang J, Liu Y, Fane AG (2013) Fouling and Wetting in membrane distillation (MD) and MD-bioreactor (MDBR) for wastewater reclamation. Desalination 323:39–47. https://doi.org/10.1016/j.desal.2012.12.001

Gryta M (2005) Long-term performance of membrane distillation process. J Memb Sci 265(1–2):153–159. https://doi.org/10.1016/j.memsci.2005.04.049

Gryta M (2009) Calcium sulphate scaling in membrane distillation process. Chem Pap 63(2):146–151. https://doi.org/10.2478/s11696-008-0095-y

Gryta M (2010) Desalination of thermally softened water by membrane distillation process. Desalination 257(1–3):30–35. https://doi.org/10.1016/j.desal.2010.03.012

Gryta M, Markowska-Szczupak A, Bastrzyk J, Tomczak W (2013) The study of membrane distillation used for separation of fermenting glycerol solutions. J Memb Sci 431:1–8. https://doi.org/10.1016/j.memsci.2012.12.032

Guo J, Farid MU, Lee EJ, Yan DYS, Jeong S, Kyoungjin An A (2018) Fouling behavior of negatively charged PVDF membrane in membrane distillation for removal of antibiotics from wastewater. J. Memb Sci 551:12–19. https://doi.org/10.1016/j.memsci.2018.01.016

Guo J, Yan DYS, Lam FLY, Deka BJ, Lv X, Ng YH, An AK (2019) Self-cleaning BiOBr/Ag photocatalytic membrane for membrane regeneration under visible light in membrane distillation. Chem Eng J 378. https://doi.org/10.1016/j.cej.2019.122137

Han L, Xiao T, Tan YZ, Fane AG, Chew JW (2017) Contaminant rejection in the presence of humic acid by membrane distillation for surface water treatment. J Memb Sci 541:291–299. https://doi.org/10.1016/j.memsci.2017.07.013

Hausmann A, Sanciolo P, Vasiljevic T, Weeks M, Schroën K, Gray S, Duke M (2013) Fouling of dairy components on hydrophobic polytetrafluoroethylene (PTFE) membranes for membrane distillation. J Memb Sci 442:149–159. https://doi.org/10.1016/j.memsci.2013.03.057

Hemmat A, Ghoreishi SM, Sabet JK (2015) Effect of salt additives on the fabrication of poly (vinylidene fluoride-co-hexafluropropylene) (PVDF-HFP) nanofiber membranes for air gap membrane distillation. Procedia Mater Sci 11:370–375. https://doi.org/10.1016/j.mspro.2015.11.132

Hernández-Maldonado AJ, Blaney L (2019) Contaminants of emerging concern in water and wastewater: advanced treatment processes. Contam Emerg Concern Water Wastewater Adv Treat Process 2019:1–418. https://doi.org/10.1016/C2016-0-05074-X

Hou D, Wang J, Sun X, Luan Z, Zhao C, Ren X (2010) Boron Removal from aqueous solution by direct contact membrane distillation. J Hazard Mater 177(1–3):613–619. https://doi.org/10.1016/j.jhazmat.2009.12.076

Hou D, Fan H, Jiang Q, Wang J, Zhang X (2014) Preparation and characterization of PVDF flat-sheet membranes for direct contact membrane distillation. Sep Purif Technol 135(1):211–222. https://doi.org/10.1016/j.seppur.2014.08.023

Hsu ST, Cheng KT, Chiou JS (2002) Seawater desalination by direct contact membrane distillation. Desalination 143(3):279–287. https://doi.org/10.1016/S0011-9164(02)00266-7

Jacob P, Phungsai P, Fukushi K, Visvanathan C (2015) Direct contact membrane distillation for anaerobic effluent treatment. J. Memb Sci 475:330–339. https://doi.org/10.1016/j.memsci.2014.10.021

Janowska K, Boffa V, Jørgensen MK, Quist-Jensen CA, Hubac F, Deganello F, Coelho FEB, Magnacca G (2020) Thermocatalytic membrane distillation for clean water production. NPJ Clean Water 3(1):34. https://doi.org/10.1038/s41545-020-00082-2

Jia F, Yin Y, Wang J (2018) Removal of cobalt ions from simulated radioactive wastewater by vacuum membrane distillation. Prog Nucl Energy 103:20–27. https://doi.org/10.1016/j.pnucene.2017.11.008

Jiang L, Chen L, Zhu L (2019) Electrically conductive membranes for anti-biofouling in membrane distillation with two novel operation modes: capacitor mode and resistor mode. Water Res 161:297–307. https://doi.org/10.1016/j.watres.2019.06.015

Jiang L, Chen L, Zhu L (2020) Fouling process of membrane distillation for seawater desalination: an especial focus on the thermal-effect and concentrating-effect during biofouling. Desalination 485:114457. https://doi.org/10.1016/j.desal.2020.114457

Jiao L, Yan K, Wang J, Lin S, Li G, Bi F, Zhang L (2020) Low Surface energy nanofibrous membrane for enhanced wetting resistance in membrane distillation process. Desalination 476:114210. https://doi.org/10.1016/j.desal.2019.114210

Junussova LR, Chicherin SV (2020) Wastewater Treatment and application in the advanced nanofiltration system. In: IOP conference series: earth and environmental science, vol 408. https://doi.org/10.1088/1755-1315/408/1/012024

Karakulski K, Gryta M (2005) Water demineralisation by NF/MD integrated processes. Desalination 177(1):109–119. https://doi.org/10.1016/j.desal.2004.11.018

Khaing TH, Li J, Li Y, Wai N, Wong FS (2010) Feasibility study on petrochemical wastewater treatment and reuse using a novel submerged membrane distillation bioreactor. Sep Purif Technol 74(1):138–143. https://doi.org/10.1016/j.seppur.2010.05.016

Khayet M, Matsuura T (2001) Preparation and characterization of polyvinylidene fluoride membranes for membrane distillation. Ind Eng Chem Res 40(24):5710–5718. https://doi.org/10.1021/ie010553y

Khumalo NP, Nthunya LN, De Canck E, Derese S, Verliefde AR, Kuvarega AT, Mamba BB, Mhlanga SD, Dlamini DS (2019) Congo red dye removal by direct membrane distillation using PVDF/PTFE membrane. Sep Purif Technol 211:578–586. https://doi.org/10.1016/j.seppur.2018.10.039

Klaunig JE, Hocevar BA, Kamendulis LM (2012) Mode of action analysis of perfluorooctanoic acid (PFOA) tumorigenicity and human relevance. Reprod Toxicol 33(4):410–418. https://doi.org/10.1016/j.reprotox.2011.10.014

Krause M, Klit A, Blomberg Jensen M, Søeborg T, Frederiksen H, Schlumpf M, Lichtensteiger W, Skakkebaek NE, Drzewiecki KT (2012) Sunscreens: are they beneficial for health? An overview of endocrine disrupting properties of UV-Filters. Int J Androl 35(3):424–436. https://doi.org/10.1111/j.1365-2605.2012.01280.x

Lapara TM, Alleman JE (1999) Thermophilic aerobic biological wastewater treatment. Water Res 33(4):895–908. https://doi.org/10.1016/S0043-1354(98)00282-6

Laqbaqbi M, García-Payo MC, Khayet M, El Kharraz J, Chaouch M (2019) Application of direct contact membrane distillation for textile wastewater treatment and fouling study. Sep Purif Technol 209:815–825. https://doi.org/10.1016/j.seppur.2018.09.031

Leaper S, Abdel-Karim A, Gad-Allah TA, Gorgojo P (2019) Air-gap membrane distillation as a one-step process for textile wastewater treatment. Chem Eng J 360:1330–1340. https://doi.org/10.1016/j.cej.2018.10.209

Lee JG, Jang Y, Fortunato L, Jeong S, Lee S, Leiknes TO, Ghaffour N (2018) An advanced online monitoring approach to study the scaling behavior in direct contact membrane distillation. J Memb Sci 546:50–60. https://doi.org/10.1016/j.memsci.2017.10.009

Lei M, Zhang L, Lei J, Zong L, Li J, Wu Z, Wang Z (2015) Overview of emerging contaminants and associated human health effects. Biomed Res Int 2015:404796. https://doi.org/10.1155/2015/404796

Leiw MY, Guai GH, Wang X, Tse MS, Ng CM, Tan OK (2013) Dark ambient degradation of bisphenol A and acid orange 8 as organic pollutants by perovskite $SrFeO_{3-\delta}$ metal oxide. J Hazard Mater 260:1–8. https://doi.org/10.1016/j.jhazmat.2013.04.031

Li XF, Mitch WA (2018) Drinking water disinfection byproducts (DBPs) and human health effects: multidisciplinary challenges and opportunities. Environ Sci Technol 52(4):1681–1689. https://doi.org/10.1021/acs.est.7b05440

Li N, Wang X, Zhang H, Zhang Z, Ding J, Lu J (2019) Comparing the performance of various nanofiltration membranes in advanced oxidation-nanofiltration treatment of reverse osmosis concentrates. Environ Sci Pollut Res 26(17):17472–17481. https://doi.org/10.1007/s11356-019-05120-2

Li M, Li K, Wang L, Zhang X (2020) Feasibility of Concentrating textile wastewater using a hybrid forward osmosis-membrane distillation (FO-MD) process: performance and economic evaluation. Water Res 172:115488. https://doi.org/10.1016/j.watres.2020.115488

Liao Y, Wang R, Tian M, Qiu C, Fane AG (2013) Fabrication of polyvinylidene fluoride (PVDF) nanofiber membranes by electro-spinning for direct contact membrane distillation. J Memb Sci 425–426:30–39. https://doi.org/10.1016/j.memsci.2012.09.023

Lu X, Peng Y, Ge L, Lin R, Zhu Z, Liu S (2016) Amphiphobic PVDF composite membranes for anti-fouling direct contact membrane distillation. J Memb Sci 505:61–69. https://doi.org/10.1016/j.memsci.2015.12.042

Lu KJ, Chen Y, Chung TS (2019) Design of omniphobic interfaces for membrane distillation—A review. Water Res. Elsevier, pp 64–77. https://doi.org/10.1016/j.watres.2019.06.056

Madeira DMF, Vieira O, Pinheiro LA, de Melo Carvalho B (2018) Correlation between surface energy and adhesion force of polyethylene/paperboard: a predictive tool for quality control in laminated packaging. Int J Chem Eng 2018:2709037. https://doi.org/10.1155/2018/2709037

Mason EA, Mason EA, Malinauskas AP (1983) Gas transportin porous media: the dusty-gas model

McGaughey AL, Gustafson RD, Childress AE (2017) Effect of long-term operation on membrane surface characteristics and performance in membrane distillation. J. Memb. Sci. 543:143–150. https://doi.org/10.1016/j.memsci.2017.08.040

Melin T, Jefferson B, Bixio D, Thoeye C, De Wilde W, De Koning J, van der Graaf J, Wintgens T (2006) Membrane bioreactor technology for wastewater treatment and reuse. Desalination 187(1–3):271–282. https://doi.org/10.1016/j.desal.2005.04.086

Mericq JP, Laborie S, Cabassud C (2011) Evaluation of systems coupling vacuum membrane distillation and solar energy for seawater desalination. Chem Eng J 166(2):596–606. https://doi.org/10.1016/j.cej.2010.11.030

Metcalf, Eddy IAC, Asano T, Burton F, Leverenz H (2007) Water reuse: issues, technologies, and applications, 1st edn. McGraw-Hill Education, New York

Milić N, Milanović M, Letić NG, Sekulić MT, Radonić J, Mihajlović I, Miloradov MV (2013) Occurrence of antibiotics as emerging contaminant substances in aquatic environment. Int J Environ Health Res 23(4):296–310. https://doi.org/10.1080/09603123.2012.733934

Milne NA, O'Reilly T, Sanciolo P, Ostarcevic E, Beighton M, Taylor K, Mullett M, Tarquin AJ, Gray SR (2014) Chemistry of silica scale mitigation for RO desalination with particular reference to remote operations. Water Res 65:107–133. https://doi.org/10.1016/j.watres.2014.07.010

Miraji H, Othman OC, Ngassapa FN, Mureithi EW (2016) Research trends in emerging contaminants on the aquatic environments of Tanzania. Scientifica (Cairo) 2016. https://doi.org/10.1155/2016/3769690

Moradi R, Monfared SM, Amini Y, Dastbaz A (2016) Vacuum enhanced membrane distillation for trace contaminant removal of heavy metals from water by electrospun PVDF/TiO_2 hybrid membranes. Korean J Chem Eng 33(7):2160–2168. https://doi.org/10.1007/s11814-016-0081-y

Morrow CP, Furtaw NM, Murphy JR, Achilli A, Marchand EA, Hiibel SR, Childress AE (2018) Integrating an aerobic/anoxic osmotic membrane bioreactor with membrane distillation for potable reuse. Desalination 432:46–54. https://doi.org/10.1016/j.desal.2017.12.047

Moulik S, Moulik S, Parakala S, Sridhar S, Sridhar S (2018) Membrane distillation: historical perspective and a solution to existing issues in membrane technology. In: Membrane processes. Wiley, Hoboken, NJ, USA, pp 305–356. https://doi.org/10.1002/9781119418399.ch11

Mozia S, Morawski AW (2012) The performance of a hybrid photocatalysis-md system for the treatment of tap water contaminated with ibuprofen. Catal Today 193(1):213–220. https://doi.org/10.1016/j.cattod.2012.03.016

Munirasu S, Banat F, Durrani AA, Haija MA (2017) Intrinsically superhydrophobic PVDF membrane by phase inversion for membrane distillation. Desalination 417:77–86. https://doi.org/10.1016/j.desal.2017.05.019

Naidu G, Jeong S, Choi Y, Vigneswaran S (2017) Membrane distillation for wastewater reverse osmosis concentrate treatment with water reuse potential. J Memb Sci 524:565–575. https://doi.org/10.1016/j.memsci.2016.11.068

Nawaz MS, Son HS, Jin Y, Kim Y, Soukane S, Al-Hajji MA, Abu-Ghdaib M, Ghaffour N (2021) Investigation of flux stability and fouling mechanism during simultaneous treatment of different produced water streams using forward osmosis and membrane distillation. Water Res 198:117157. https://doi.org/10.1016/j.watres.2021.117157

Nikbakht Fini M, Madsen HT, Muff J (2019) The Effect of water matrix, feed concentration and recovery on the rejection of pesticides using NF/RO membranes in water treatment. Sep Purif Technol 215:521–527. https://doi.org/10.1016/j.seppur.2019.01.047

NORMAN (2021) Emerging substances. Norman. https://www.norman-network.net/?q=node/19. Accessed 29 June 2021

Pal P, Manna AK (2010) Removal of arsenic from contaminated groundwater by solar-driven membrane distillation using three different commercial membranes. Water Res 44(19):5750–5760. https://doi.org/10.1016/j.watres.2010.05.031

Pal P, Kumar R, Ghosh AK (2018) Analysis of process intensification and performance assessment for fermentative continuous production of bioethanol in a multi-staged membrane-integrated bioreactor system. Energy Convers Manag 171:371–383. https://doi.org/10.1016/j.enconman.2018.05.099

Phattaranawik J, Fane AG, Pasquier ACS, Bing W (2008) A novel membrane bioreactor based on membrane distillation. Desalination 223(1–3):386–395. https://doi.org/10.1016/j.desal.2007.02.075

Plattner J, Kazner C, Naidu G, Wintgens T, Vigneswaran S (2018) Removal of selected pesticides from groundwater by membrane distillation. Environ Sci Pollut Res 25(21):20336–20347. https://doi.org/10.1007/s11356-017-8929-1

Pordel MA, Maleki A, Ghanbari R, Rezaee R, Khamforoush M, Daraei H, Athar SD, Shahmoradi B, Safari M, Ziaee AH, Lalhmunsiama, Lee SM (2019) Evaluation of the effect of electrospun nanofibrous membrane on removal of diazinon from aqueous solutions. React Funct Polym 139:85–91. https://doi.org/10.1016/j.reactfunctpolym.2019.03.017

Qu D, Wang J, Hou D, Luan Z, Fan B, Zhao C (2009) Experimental study of arsenic removal by direct contact membrane distillation. J Hazard Mater 163(2–3):874–879. https://doi.org/10.1016/j.jhazmat.2008.07.042

Qu D, Sun D, Wang H, Yun Y (2013) Experimental study of ammonia removal from water by modified direct contact membrane distillation. Desalination 326:135–140. https://doi.org/10.1016/j.desal.2013.07.021

Quist-Jensen CA, Ali A, Mondal S, Macedonio F, Drioli E (2016) A study of membrane distillation and crystallization for lithium recovery from high-concentrated aqueous solutions. J Memb Sci 505:167–173. https://doi.org/10.1016/j.memsci.2016.01.033

Radjenović J, Petrović M, Barceló D (2009) Fate and distribution of pharmaceuticals in wastewater and sewage sludge of the conventional activated sludge (CAS) and advanced membrane bioreactor (MBR) treatment. Water Res 43(3):831–841. https://doi.org/10.1016/j.watres.2008.11.043

Rajapaksha PP, Power A, Chandra S, Chapman J (2018) Graphene, electrospun membranes and granular activated carbon for eliminating heavy metals, pesticides and bacteria in water and wastewater treatment processes. Analyst 143(23):5629–5645. https://doi.org/10.1039/c8an00922h

Ramlow H, Machado RAF, Marangoni C (2017) Direct contact membrane distillation for textile wastewater treatment: a state of the art review. Water Sci Technol 76(10):2565–2579. https://doi.org/10.2166/wst.2017.449

Rastegarpanah A, Mortaheb HR (2016) Surface treatment of polyethersulfone membranes for applying in desalination by direct contact membrane distillation. Desalination 377:99–107. https://doi.org/10.1016/j.desal.2015.09.008

Ravi J, Othman MHD, Matsuura T, Ro'il Bilad M, El-badawy TH, Aziz F, Ismail AF, Rahman MA, Jaafar J (2020) Polymeric membranes for desalination using membrane distillation: a review. Desalination 490:114530. https://doi.org/10.1016/j.desal.2020.114530

Richardson SD, Ternes TA (2011) Water Analysis: emerging contaminants and current issues. Anal Chem 83(12):4616–4648. https://doi.org/10.1021/ac200915r

Roman M, Van Dijk LH, Gutierrez L, Vanoppen M, Post JW, Wols BA, Cornelissen ER, Verliefde ARD (2019) Key physicochemical characteristics governing organic micropollutant adsorption and transport in ion-exchange membranes during reverse electrodialysis. Desalination 468. https://doi.org/10.1016/j.desal.2019.114084

Ruiz-Aguirre A, Polo-López MI, Fernández-Ibáñez P, Zaragoza G (2015) Assessing the validity of solar membrane distillation for disinfection of contaminated Water. Desalin Water Treat 55(10):2792–2799. https://doi.org/10.1080/19443994.2014.946717

Ruiz-Aguirre A, Polo-López MI, Fernández-Ibáñez P, Zaragoza G (2017) Integration of membrane distillation with solar photo-fenton for purification of water contaminated with Bacillus Sp. and Clostridium Sp. Spores. Sci Total Environ 595:110–118. https://doi.org/10.1016/j.scitotenv.2017.03.238.

Sabri S, Najjar A, Manawi Y, Eltai NO, Al-Thani A, Atieh MA, Kochkodan V (2019) Antibacterial properties of polysulfone membranes blended with arabic gum. Membranes. https://doi.org/10.3390/membranes9020029

Saffarini RB, Summers EK, Arafat HA, Lienhard VJH (2012) Economic evaluation of stand-alone solar powered membrane distillation systems. Desalination 299:55–62. https://doi.org/10.1016/j.desal.2012.05.017

Sánchez NC, Guzmán-Mar JL, Hinojosa-Reyes L, Palomino GT, Cabello CP (2019) Carbon composite membrane derived from MIL-125-NH2 MOF for the enhanced extraction of emerging pollutants. Chemosphere 231:510–517. https://doi.org/10.1016/j.chemosphere.2019.05.173

Schneider K, Hölz W, Wollbeck R, Ripperger S (1988) Membranes and modules for transmembrane distillation. J Memb Sci 39(1):25–42. https://doi.org/10.1016/S0376-7388(00)80992-8

Schofield RW, Fane AG, Fell CJD (1990) Gas and Vapour transport through microporous membranes II. Membrane distillation. J Memb Sci 53(1–2):173–185. https://doi.org/10.1016/0376-7388(90)80012-B

Siddique MS, Khan SJ, Shahzad MA, Nawaz MS, Hankins NP (2018) Insight into the effect of organic and inorganic draw solutes on the flux stability and sludge characteristics in the osmotic membrane bioreactor. Bioresour Technol 249:758–766. https://doi.org/10.1016/j.biortech.2017.10.064

Silva TLS, Morales-Torres S, Figueiredo JL, Silva AMT (2015) Multi-walled carbon nanotube/pvdf blended membranes with sponge- and finger-like pores for direct contact membrane distillation. Desalination 357:233–245. https://doi.org/10.1016/j.desal.2014.11.025

Silva TLS, Morales-Torres S, Esteves CMP, Ribeiro AR, Nunes OC, Figueiredo JL, Silva AMT (2018) Desalination and removal of organic micropollutants and microorganisms by membrane distillation. Desalination 437:121–132. https://doi.org/10.1016/j.desal.2018.02.027

Sivakumar M, Ramezanianpour M, O'Halloran G (2013) Mine water treatment using a vacuum membrane distillation system. APCBEE Proc 5:157–162. https://doi.org/10.1016/j.apcbee.2013.05.028

Song X, Luo W, McDonald J, Khan SJ, Hai FI, Price WE, Nghiem LD (2018) An anaerobic membrane bioreactor—Membrane distillation hybrid system for energy recovery and water reuse: removal performance of organic carbon, nutrients, and trace organic contaminants. Sci Total Environ 628–629:358–365. https://doi.org/10.1016/j.scitotenv.2018.02.057

Srisurichan S, Jiraratananon R, Fane AG (2006) Mass transfer mechanisms and transport resistances in direct contact membrane distillation process. J Memb Sci 277(1–2):186–194. https://doi.org/10.1016/j.memsci.2005.10.028

Tadkaew N, Hai FI, McDonald JA, Khan SJ, Nghiem LD (2011) Removal of trace organics by MBR treatment: the role of molecular properties. Water Res 45(8):2439–2451. https://doi.org/10.1016/j.watres.2011.01.023

Tang Y, Li N, Liu A, Ding S, Yi C, Liu H (2012) Effect of spinning conditions on the structure and performance of hydrophobic PVDF hollow fiber membranes for membrane distillation. Desalination 287:326–339. https://doi.org/10.1016/j.desal.2011.11.045

Tang L, Iddya A, Zhu X, Dudchenko AV, Duan W, Turchi C, Vanneste J, Cath TY, Jassby D (2017) Enhanced flux and electrochemical cleaning of silicate scaling on carbon nanotube-coated membrane distillation membranes treating geothermal brines. ACS Appl Mater Interfaces 9(44):38594–38605. https://doi.org/10.1021/acsami.7b12615

Thomaidis NS, Asimakopoulos AG, Bletsou AA (2012) Emerging contaminants: a tutorial mini-review. Glob Nest J 14(1):72–79. https://doi.org/10.30955/gnj.000823

Tijing LD, Woo YC, Choi JS, Lee S, Kim SH, Shon HK (2015) Fouling and its control in membrane distillation—A review. J Memb Sci 475:215–244. https://doi.org/10.1016/j.memsci.2014.09.042

Tijing LD, Woo YC, Shim WG, He T, Choi JS, Kim SH, Shon HK (2016) Superhydrophobic nanofiber membrane containing carbon nanotubes for high-performance direct contact membrane distillation. J. Memb. Sci. 502:158–170. https://doi.org/10.1016/j.memsci.2015.12.014

Tomaszewska M, Białończyk L (2016) Ethanol production from whey in a bioreactor coupled with direct contact membrane distillation. Catal Today 268:156–163. https://doi.org/10.1016/j.cattod.2016.01.059

US Enviromental Protection Agency (2009). Contaminant candidate list 4. United State Environmental Protection Agency. US EPA (United States Environmental Protection Agency) Washington, DC, pp 51850–51862

Varma S, Sarode D, Wakale S, Bhanvase BA, Deosarkar MP Removal of nickel from waste water using graphene nanocomposite. Int J Chem Phys Sci 2(Special Issue):132–139

Vasilachi IC, Asiminicesei DM, Fertu DI, Gavrilescu M (2021) Occurrence and Fate of Emerging Pollutants in Water Environment and Options for Their Removal. Water (switzerland). https://doi.org/10.3390/w13020181

Volpin F, Chekli L, Phuntsho S, Ghaffour N, Vrouwenvelder JS, Shon HK (2019) Optimisation of a forward osmosis and membrane distillation hybrid system for the treatment of source-separated urine. Sep Purif Technol 212:368–375. https://doi.org/10.1016/j.seppur.2018.11.003

Wang KY, Chung TS, Gryta M (2008) Hydrophobic PVDF hollow fiber membranes with narrow pore size distribution and ultra-thin skin for the fresh water production through membrane distillation. Chem Eng Sci 63(9):2587–2594. https://doi.org/10.1016/j.ces.2008.02.020

Wang KY, Foo SW, Chung TS (2009) Mixed matrix PVDF hollow fiber membranes with nanoscale pores for desalination through direct contact membrane distillation. Ind Eng Chem Res 48(9):4474–4483. https://doi.org/10.1021/ie8009704

Wang Z, Zhang B, Fang C, Liu Z, Fang J, Zhu L (2019b) Macroporous membranes doped with micro-mesoporous β-cyclodextrin polymers for ultrafast removal of organic micropollutants from water. Carbohydr Polym 222:114970. https://doi.org/10.1016/j.carbpol.2019.114970

Wang CY, Zeng WJ, Jiang TT, Chen X, Zhang XL (2019) Incorporating attapulgite nanorods into graphene oxide nanofiltration membranes for efficient dyes wastewater treatment. Sep Purif Technol 21–30. https://doi.org/10.1016/j.seppur.2018.04.079

Warsinger DM, Servi A, Connors GB, Mavukkandy MO, Arafat HA, Gleason KK, Lienhard JH (2017) Reversing membrane wetting in membrane distillation: comparing dryout to backwashing with pressurized air. Environ Sci Water Res Technol 3(5):930–939. https://doi.org/10.1039/c7ew00085e

Warsinger DM, Chakraborty S, Tow EW, Plumlee MH, Bellona C, Loutatidou S, Karimi L, Mikelonis AM, Achilli A, Ghassemi A, Padhye LP, Snyder SA, Curcio S, Vecitis CD, Arafat HA, Lienhard JH (2018) A Review of polymeric membranes and processes for potable water reuse. Prog Polym Sci 81:209–237. https://doi.org/10.1016/j.progpolymsci.2018.01.004

Warsinger DM, Swaminathan J, Guillen-Burrieza E, Arafat HA, Lienhard VJH (2015) Scaling and fouling in membrane distillation for desalination applications: a review. Desalination 356:294–313. https://doi.org/10.1016/j.desal.2014.06.031

Wijekoon KC, Hai FI, Kang J, Price WE, Guo W, Ngo HH, Nghiem LD (2013) The fate of pharmaceuticals, steroid hormones, phytoestrogens, UV-filters and pesticides during MBR treatment. Bioresour Technol 144:247–254. https://doi.org/10.1016/j.biortech.2013.06.097

Wijekoon KC, Hai FI, Kang J, Price WE, Cath TY, Nghiem LD (2014a) Rejection and fate of trace organic compounds (TrOCs) during membrane distillation. J Memb Sci 453:636–642. https://doi.org/10.1016/j.memsci.2013.12.002

Wijekoon KC, Hai FI, Kang J, Price WE, Guo W, Ngo HH, Cath TY, Nghiem LD (2014b) A novel membrane distillation-thermophilic bioreactor system: biological stability and trace organic compound removal. Bioresour Technol 159:334–341. https://doi.org/10.1016/j.biortech.2014.02.088

Winglee JM, Bossa N, Rosen D, Vardner JT, Wiesner MR (2017) Modeling the concentration of volatile and semivolatile contaminants in direct contact membrane distillation (DCMD) product water. Environ Sci Technol 51(22):13113–13121. https://doi.org/10.1021/acs.est.6b05663

Woldemariam D, Kullab A, Fortkamp U, Magner J, Royen H, Martin A (2016) Membrane distillation pilot plant trials with pharmaceutical residues and energy demand analysis. Chem Eng J 306:471–483. https://doi.org/10.1016/j.cej.2016.07.082

Won D (2017) Treatment performance of direct contact membrane distillation for volatile, semi-volatile and non-volatile organic contaminants in water

Woo YC, Tijing LD, Shim W-G, Choi J-S, Kim S-H, He T, Drioli E, Shon HK (2016) Water desalination using graphene-enhanced electrospun nanofiber membrane via air gap membrane distillation. J Memb Sci 520:99–110. https://doi.org/10.1016/j.memsci.2016.07.049

Wu CY, Chen SS, Zhang DZ, Kobayashi J (2017) Hg Removal and the effects of coexisting metals in forward osmosis and membrane distillation. Water Sci Technol 75(11):2622–2630. https://doi.org/10.2166/wst.2017.143

Wu Y, Kang Y, Zhang L, Qu D, Cheng X, Feng L (2018) Performance and fouling mechanism of direct contact membrane distillation (DCMD) treating fermentation wastewater with high organic concentrations. J Environ Sci (china) 65:253–261. https://doi.org/10.1016/j.jes.2017.01.015

Xia L, Vemuri B, Saptoka S, Shrestha N, Chilkoor G, Kilduff J, Gadhamshetty V (2018) Antifouling membranes for bioelectrochemistry applications. In: Mohan SV, Varjani S, Pandey, ABT-MET (eds) Biomass, biofuels, biochemicals: microbial electrochemical technology: sustainable platform for fuels, chemicals and remediation. Elsevier, pp 195–224. https://doi.org/10.1016/B978-0-444-64052-9.00008-X

Xie Z, Duong T, Hoang M, Nguyen C, Bolto B (2009) Ammonia removal by sweep gas membrane distillation. Water Res 43(6):1693–1699. https://doi.org/10.1016/j.watres.2008.12.052

Yang Y, Bogler A, Ronen Z, Oron G, Herzberg M, Bernstein R (2020) Initial deposition and pioneering colonization on polymeric membranes of anaerobes isolated from an anaerobic membrane bioreactor (AnMBR). Environ Sci Technol 54(9):5832–5842. https://doi.org/10.1021/acs.est.9b06763

Yang G, Fan M, Zhang G (2014) Emerging contaminants in surface waters in China—a short review. Environ Res Lett 9(7). https://doi.org/10.1088/1748-9326/9/7/074018

Zakrzewska-Trznadel G, Harasimowicz M, Chmielewski AG (2001) Membrane processes in nuclear technology-application for liquid radioactive waste treatment. Sep Purif Technol 22–23:617–625. https://doi.org/10.1016/S1383-5866(00)00167-2

Zarasvand Asadi R, Suja F, Tarkian F, Mashhoon F, Rahimi S, Atash Jameh A (2012) Solar desalination of gas refinery wastewater using membrane distillation process. Desalination 291:56–64. https://doi.org/10.1016/j.desal.2012.01.025

Zarebska A, Nieto DR, Christensen KV, Norddahl B (2014) Ammonia recovery from agricultural wastes by membrane distillation: fouling characterization and mechanism. Water Res 56:1–10. https://doi.org/10.1016/j.watres.2014.02.037

Zhang Q, Shuwen G, Zhang J, Fane AG, Kjelleberg S, Rice SA, Mcdougald D (2015) Analysis of microbial community composition in a lab-scale membrane distillation bioreactor. J Appl Microbiol 118(4):940–953. https://doi.org/10.1111/jam.12759

Zhang G, Zhou M, Xu Z, Jiang C, Shen C, Meng Q (2019) Guanidyl-functionalized graphene/polysulfone mixed matrix ultrafiltration membrane with superior permselective, antifouling and antibacterial properties for water treatment. J Colloid Interface Sci 540:295–305. https://doi.org/10.1016/j.jcis.2019.01.050

Zheng L, Wu Z, Zhang Y, Wei Y, Wang J (2015) Effect of Non-solvent additives on the morphology, pore structure, and direct contact membrane distillation performance of pvdf-ctfe hydrophobic membranes. J Environ Sci (china) 45:28–39. https://doi.org/10.1016/j.jes.2015.09.023

Zhu WP, Sun SP, Gao J, Fu FJ, Chung TS (2014) Dual-layer polybenzimidazole/polyethersulfone (PBI/PES) nanofiltration (NF) hollow fiber membranes for heavy metals removal from wastewater. J Memb Sci 456:117–127. https://doi.org/10.1016/j.memsci.2014.01.001

Zodrow KR, Bar-Zeev E, Giannetto MJ, Elimelech M (2014) Biofouling and microbial communities in membrane distillation and reverse osmosis. Environ Sci Technol 48(22):13155–13164. https://doi.org/10.1021/es503051t

Zoungrana A, Cakmakci M, Zengin IH, Inoglu O, Elcik H (2016) Treatment of metal-plating waste water by modified direct contact membrane distillation. Chem Pap 70(9):1185–1195. https://doi.org/10.1515/chempap-2016-0066

Chapter 19
Point-of-Use Drinking Water Treatment Systems and Their Performance in Removal of Emerging Contaminants

Sudharshan Reddy, Najmul Haque Barbhuiya, and Swatantra P. Singh

Abstract In many of the developing countries, people do not have assess to clean water and clean sanitation, and consumption of contaminated water results in the spread of waterborne diseases. Unfortunately, many parts of developing countries do not have centralized water treatment facilities. Therefore, people are practicing some traditional water treatment methods in households. Point-of-use (POU) water treatment systems are one such method of household water treatment systems and are playing an important role in reaching one of the sustainable development goals that are clean water and sanitation. There are various methods available for POU water treatment, mainly the physical removal methods (Biosand filtration, membrane filtration), chemical treatment methods (chlorine disinfection, coagulation/ flocculation disinfection), and light and heat-based methods (boiling, solar disinfection, UV disinfection). All these methods have their own advantages with limitations. However, membrane-based POU water treatment has registered significant importance compared to other methods. Even in developed countries, people are using POU water treatment systems under various circumstances. In addition, in many countries, there is a growing concern about emerging contaminants in drinking water and is important to study POU water filters for the removal of emerging contaminants. Therefore, in this chapter, various types of technologies available for POU water treatment, membrane-based POU water filters, and the performance of household POU water filters in removing emerging contaminants have been discussed.

Keywords Drinking water · Pathogens · Diarrhea · Membrane filtration · Perfluoroalkyl substances · Reverse osmosis

19.1 Introduction

Clean water and sanitation are one of the seventeen Global Goals or Sustainable Development Goals by the UN. Unfortunately, more than 2 billion people are not

---

S. Reddy · N. H. Barbhuiya · S. P. Singh (✉)
Environmental Science and Engineering Department, IIT Bombay, Mumbai, India
e-mail: swatantra@iitb.ac.in

S. P. Singh et al. (eds.), *New Trends in Emerging Environmental Contaminants*, Energy, Environment, and Sustainability,
https://doi.org/10.1007/978-981-16-8367-1_19

having safe drinking water and safe sanitation (World Health Organization (WHO) 2019). Particularly, in developing countries, microbial pollution is the major concern in drinking water even though organic and inorganic pollutants might have their risk contribution. Due to poor sanitation and unhygienic conditions in developing countries, the fecal contaminated water may contact the human body through various means (drinking, bathing, etc.) and spread waterborne diseases. Bacteria, viruses, and protozoa are the main microorganisms responsible for waterborne diseases, and health symptoms include diarrhea, vomiting, and gastroenteritis (Pooi and Ng 2018). Over the long run, damage to the kidney and liver can also occur from waterborne diseases. Diarrhea, polio, hepatitis, typhoid, paratyphoid etc., are some examples of waterborne diseases that originated from microbial contamination (Alsulaili et al. 2020). Other than these, fluorosis and lead poisoning also come under waterborne diseases caused by chemical contamination of drinking water. People may be at risk of these diseases by consuming fecal and chemical contaminated water. It was reported that yearly around 2 million children deaths are happening due to diarrhea and 90% of the deaths are from developing counties (Crump et al. 2004; Venkatesha and Kedare 2014). 80% of diarrhea cases are associated with poor water quality, poor sanitation, and unhygienic practices (Pérez-Vidal et al. 2016). Therefore, the fecal contamination in the drinking water and its distribution systems must be zero.

Almost all the developed countries and major city areas are equipped with centralized water treatment systems because of their affordability. In developing countries and rural areas, the adaptation of a centralized water treatment system is uneconomical. Due to less density of houses and cost of the distribution system, impose more cost on a liter of water, and of course, the average annual income of people living in these areas is low. Even some people living in urban areas do not assess improved drinking water supplies due to their poverty (Berg 2010). Therefore, people have to rely on locally available untreated groundwater and surface water sources like tube wells, dug wells, ponds, lakes, and rivers (Peter-Varbanets et al. 2009). Water from these sources may contaminate by industrial, domestic, and agricultural waste discharges. According to World Health Organization data, around 2 billion people living in developing countries assess drinking water from untreated surface water sources (river, lake, and pond water) (Alsulaili et al. 2020). Almost 70% of surface water in India is chemically and microbiologically contaminated (Water Pollution 2013). Sometimes, even if the treated water is collected for domestic purposes, there are chances of recontamination in the absence of proper storage (Sobsey et al. 2002).

Therefore, it is mandatory to improve the water quality to prevent the spread of waterborne diseases. Hence, decentralized water treatment systems came into existence to treat the contaminated water at household or small community levels, mainly in developing countries. In developed countries, if existing centralized treatment systems fail to treat or distribute the water, then other alternation should be chosen, and decentralized treatment systems play an important role under these circumstances. Generally, decentralized water treatment systems can be described as of following three ways (Peter-Varbanets et al. 2009),

- Point-of-use systems (POU): treats only part of the water needed for drinking and cooking (shown in Fig. 19.1)
- Point-of-entry systems (POE): treats all water supplied to the household (shown in Fig. 19.2)
- Small-scale systems: these are larger than POU and POE systems but smaller than centralized systems.

Point-of-use treatment is a rapidly increasing field in portable water treatment (Ashani 2017). Many studies have reported the capability of POU systems in reducing waterborne diseases by improving the microbial quality of water (Agrawal and Bhalwar 2009; Barna et al. 2014). Other than household applications, people in developed countries are using these POU water filters for traveling, camping, and hiking purposes (Patil et al. 2020). During emergency situations also, POU systems play a crucial role. Immediately after disasters, the supply of drinking water to the victims of vulnerable disasters is important. Usually, many voluntaries, organizations, agencies, and governmental bodies are used to supply water through bottles or packets, which is not sustainable. It is very challenging to treat water on vulnerable

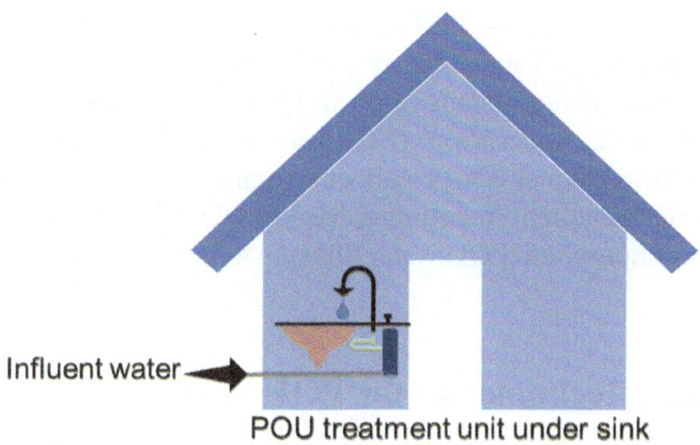

Fig. 19.1 Typical representation of POU treatment unit at the household level

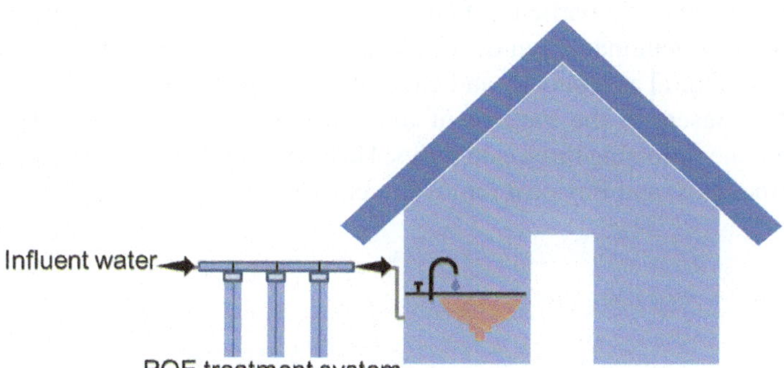

Fig. 19.2 Typical representation of POE treatment system at the household level

sites due to improper infrastructure availability and huge variations in water quality parameters (Loo et al. 2012). So, in order to address this issue, POU water treatment systems can be used (Lantagne and Clasen 2012).

Various methods and technologies for POU water treatment have been developed and practiced in past decades. One can simply classify these methods into the following three ways: (a) Heat and light-based systems (Boiling, Solar disinfection, UV disinfection), (b) Chemical treatment methods (Chlorine disinfection, Coagulation/flocculation), (c) Physical removal methods (biosand filtration, membrane filtration). All these methods have their own advantages with some limitations. Traditional methods like boiling, solar disinfection, chlorine disinfection, and sand filtration give only limited health benefits compared to advanced methods like membrane filtration (Francis et al. 2016). The use of membrane technology for water treatment applications is a growing field of research. Membrane filtration systems are cost-effective, easy to carry, and simple to operate compared to other treatment methods such as sand filtration, chlorination, and flocculation (Shamsuddin et al. 2016). When source water has high fluctuations in water quality parameters, which is the case during disasters like floods, membrane-based systems are very effective in such emergencies. Actually, many centralized treatment systems are designed to remove traditional contaminants (Brown et al. 2017). But, growing concern on water quality and improved developments in sophisticated instruments made identification of various specific and emerging contaminants in drinking water. Therefore, it is important to understand the removal capacities of POU water filters for emerging contaminants. So, the purpose of this chapter is to provide information regarding available technologies for POU water treatment, membrane-based POU water treatment systems, and the performance of household POU water filters in removing emerging contaminants.

19.2 Available Technologies for POU Water Treatment

Similar to the centralized drinking water treatment systems, there are many conventional and advanced technologies available for POU water treatment systems. The overall performance of a particular POU system can be decided by multiple factors, which include contaminants removal efficiency, cost of the treatment system, easy to use, socio-cultural acceptance, and environmental sustainability (Peter-Varbanets et al. 2009). Based on the method of treatment used, one can classify the POU systems into the following three categories: Heat and Light-based systems, Chemical treatment methods, and Physical removal methods.

19.2.1 Heat and Light-Based Systems

Boiling: Boiling is the historically more practiced method of disinfection at the household level of water treatment (Sobsey et al. 2002). Through boiling, it is possible to kill all the microorganisms present in the water. However, this method requires more energy by burning wood or other fuels, which is not encouraged from the environmental and sustainable perspectives (Peter-Varbanets et al. 2009). Another disadvantage with the boiling of water is that it is not safeguarded against future contamination because as the water gets cool again, there are chances of recontamination (Loo et al. 2012). In addition, boiled water alters the taste, and therefore, many people do not use this treatment method.

Solar disinfection (SODIS): It is a simple and cost-effective technique to inactivate the microorganisms present in the water using solar radiation (Pooi and Ng 2018). Water filled in PET or glass bottles allowed sunlight exposure for a certain amount of time so that the pathogens present in the water get inactivated by UV radiation's action and increased water temperature (Peter-Varbanets et al. 2009). Some studies have shown the capability of SODIS in producing microbial-free water and resulting in the reduction of diarrheal episodes (Meyer and Reed 2001; Murinda and Kraemer 2008; Rose et al. 2006; Conroy et al. 1999). But, this method is unsuitable for all weather conditions and low sunlight-intense regions.

UV disinfection: POU systems based on UV disinfection uses certain UV lamps that can produce short wavelength UV light to inactivate the microorganisms (Berg 2010). It was estimated that more than 99.9% killing of protozoa and cysts could be achieved with UV treatment (Peter-Varbanets et al. 2009). But the drawback with this system is the electrical power required to perform the treatment by producing UV light (Venkatesha et al. 2020). Turbidity of the water also affects the UV light transmittance and resulting in low microbial inactivation (Loo et al. 2012). To improve the effectiveness of the system, turbidity has to reduce, and this can achieve by performing pretreatment. Aquaguard Compact is an example of UV based POU system (Venkatesha et al. 2020).

19.2.2 Chemical Treatments

Chlorine disinfection: Inactivating the pathogens in water by adding free chlorine is a simple, effective, and most widely used and accepted method of water treatment. Chlorine may be added into the water in the form of calcium hypochlorite and sodium hypochlorite (Venkatesha and Kedare 2014). At a larger dosage, the microorganisms are destroyed by cell wall oxidation, and at the lower dosage, chlorine penetrates into the cell and then kills the microorganisms (Loo et al. 2012). By adding 2 mg/l of chlorine dosage for about 0.5 h results in entire bacterial removal up to 3 log removal value (LRV) (Gadgil et al. 1997). This treatment method provides a safeguard against future contamination due to residual chlorine. On the other hand, the formation of undesirable by-products from the oxidation of organics by chlorine is the disadvantage of this method.

Coagulation and flocculation: In general, coagulation and flocculation is applying to remove the colloidal particles from the water. During coagulation, particles' destabilization will occur, and at the flocculation stage, floc formation will happen. Then these flocs can remove by settling. Through the coagulation and flocculation process, turbidity and colloids can be effectively removed (Pooi and Ng 2018). Turbidity removal reduces the supporting structure of microorganisms that results in microorganism removal (Pooi and Ng 2018). But, it is impossible to achieve the acceptable level of coliform removal using coagulation and flocculation alone. To improve the removal efficiency, coagulation and flocculation needed to integrate with other treatments.

Coagulation and flocculation coupled with disinfection: The coagulation process has been used in water treatment for more than 100 years and is usually applied to drinking water treatment at community levels and rarely at household levels (Jiang 2015; Pooi and Ng 2018). However, integration of coagulation and flocculation with disinfection is making favorable use at household levels. Here, tablets or powders are used to combine the functioning of coagulation/flocculation and disinfection processes (Kfir et al. 1989; Rodda et al. 1993). By combining flocculation and disinfection using Chlor-floc tablets, the author Rodda et al. (1993) have reported 7-8LRV (Log removal value) for pathogenic bacteria and 6.5LRV for rotavirus (Rodda et al. 1993). PUR water purifier sachet is a product developed by Proctor and Gamble Company is an example of POU water treatment by coupling the flocculation and disinfection (Venkatesha et al. 2020). These products contain calcium hypochlorite as a disinfectant and ferric sulfate as flocculating agent (Pooi and Ng 2018). Using this product, the author Souter et al. (2003) have observed that more than 7LRV for bacteria, 4LRV for viral, and 3LRV for parasites. In the same study, they also detected complete removal of *E.coli* and 99.5% removal of arsenic (Souter et al. 2003). These flocculation-disinfection systems have shown effectiveness in reducing diarrhea episodes among children (Chiller et al. 2006). A combination of flocculation and disinfection methods of POU systems is more suitable when the source water is highly turbid and contaminated (Crump et al. 2004). Taste, odor, and cost are some drawbacks of this method (Lantagne and Clasen 2012).

19.2.3 Physical Removal Methods

Physical removal methods involve filtration processes, and size exclusion is the basic mechanism for the removal of suspended matter, colloidal particles, and pathogens. Different filter media have been used at household levels, and some of those are paper and fabric filters, biosand filters, and membrane filters. Generally, paper and fabric filters are not suggested because they are not effective at removing smaller pathogens like bacteria, viruses, and parasites due to larger filter pores but are capable of removing larger pathogens like swimming larva (Peter-Varbanets et al. 2009).

Any filter material should satisfy some basic criteria: ease of production, physical and chemical stability, effectiveness in removing microorganisms (Jakubczak et al. 2021).

Biosand filters: Biosand filters are similar to slow sand filters and are widely emerging POU household systems where locally available materials like sand are used as filter media (Pooi and Ng 2018; Stauber et al. 2012). This filter media supports the formation of biofilm on its surface that retains the turbidity and larger microbes. Many studies have reported a 44–74% reduction of diarrheal cases by consuming bio-sand filtered drinking water (Liang et al. 2010; Stauber et al. 2012; Tiwari et al. 2009; Duke et al. 2006). Biosand filters remove bacteria in the range of 1-3LRV, viruses in the range of 0.5-3LRV, and protozoa in the range of 2-4LRV (Sobsey et al. 2009; Lantagne and Clasen 2012).

By modifying the filter media, one can achieve better removal efficiencies. For example, Ahammed and Davra reported > 2LRV for *E. coli* by modifying the sand media with iron oxide coated sand (Ahammed and Davra 2011). Similarly, more than 99.99% removal of MS2 and rotavirus was achieved by a modified biosand filter with positively charged iron oxide (Bradley et al. 2011). Removal of contaminants such as nitrate, iron, sulfate, and arsenic was also observed by vinegar-modified anaerobic bio-sand filter (Snyder et al. 2016). The bacterial removal efficiency of biosand filters greatly depends on operating conditions. Increasing the flow rate (daily charge) decreases the bacterial removal efficiency (Ahammed and Davra 2011). Water retention time in the filter improves the removal efficiency (Elliott et al. 2008). Although high removal efficiency for microorganisms can be possible, further to achieve maximum water safety, disinfection of filtered water is recommended (Romero et al. 2020).

The methods mentioned above have their own advantages and drawbacks. However, membrane-based technology is an advanced POU treatment system application, registering its benefits over other treatment methods. Table 19.1 shows the advantages and disadvantages of different methods that are used for POU water treatment. Through membrane filtration, a wide array of contaminants can be removed effectively. In the removal of emerging and specific contaminants, membrane-based POU treatment systems have shown better performance. On the other hand, membrane-based systems have some limitations that need to address. The following section discusses in detail about the membrane-based POU water treatment.

19.3 Membrane-Based POU Systems

19.3.1 Membrane Filtration Technology

For many years, membrane technology has registered its significant importance in water and wastewater treatment systems (Lonsdale 1982; Ezugbe and Rathilal 2020; Ma et al. 2017). Generally, a membrane is a semipermeable barrier between two phases and can transport certain components from one phase to another depending upon the physicochemical characteristics of the membrane and permeates (Mulder 2013). To transport the components, the driving force is acting between the two phases that can be the difference in concentration, pressure, temperature, electrical potential, or osmotic potential (Mulder 2013). If the difference in concentration is the driving force, then this process is dialysis. Similarly, pressure-driven processes are called membrane filtration processes, temperature-driven processes are called membrane

Table 19.1 Advantages and disadvantages of various methods used for POU water treatment (Lantagne and Clasen 2012)

Sl.no	POU treatment	Advantages	Disadvantages
1	Boiling	Simple and easy way of treatment; complete inactivation of all microorganism; socio-culturally practiced traditional method; Independent on source water turbidity	No safeguard against future contamination in the absence of proper storage; not sustainable way of treatment due to fuel consumption for boiling; taste objection
2	Solar disinfection	Simple and easy way of treatment; reduce bacteria, viruses, and protozoa; reduce diarrheal episodes; cost-effective	Source water turbidity impact the removal efficiency; time taking process; requirement of clean plastic bottles; all weather conditions are not favorable
3	UV disinfection	High throughput; fast disinfection	Expensive; turbidity levels of source water alters the system effectiveness
4	Chlorination	Reduce most of bacteria, viruses and protozoa; provide safety against future contamination; simple and cost-effective	Taste and odor problem; lower effectiveness if water is turbid and contaminated with organics and inorganics; chlorine react with organics in water and release undesirable products
5	Coagulation-flocculation/disinfection	Remove bacteria, viruses, protozoa, and some other chemicals and pesticides; suitable for highly turbid water	Multiple steps are required; cost per liter of water is more compared to other household systems

(continued)

Table 19.1 (continued)

Sl.no	POU treatment	Advantages	Disadvantages
6	Biosand filtration	Reported reduction of bacteria and protozoa in water; reported reduction of diarrheal cases; locally available materials can be used for construction of filter; chemical-free treatment; only involves installation cost with minimum maintenance cost; long-lasting life	Low removal efficiency for viruses; no residual protection; regular maintenance is required
7	Household ceramic filters	Low cost; reduced bacteria and protozoa	No safeguard against future contamination; no protection against viruses; needed regular maintenance
8	Microfiltration and ultrafiltration	No chemical requirement for treatment; system unaffected by source water quality	Fouling problem; needed regular backwashing
9	Nano and RO filtration	Complete removal of all most all pathogens; remove specific contaminants like lead, arsenic, etc.; effectively removes emerging contaminants like pesticides, pharmaceutical compounds, and perflouroalkyl substances	Requirement of multiple treatment stages; fouling problem; energy demand; more reject water

distillation, electrically driven processes are called electro-dialysis, and osmotic-driven processes are called forward osmosis (Lonsdale 1982). Most membrane separation processes are based on membrane filtration and are the major process used in POU systems based on membrane separation. Depending upon membrane pore size, the membrane filtration system can be further classified into microfiltration, ultrafiltration, nanofiltration, and reverse osmosis (Fane et al. 2011). Figure 19.3 represents the details of the pore size and particle removal of different membrane filters.

Size exclusion is the main principle of membrane filtration to remove contaminates. Therefore, there is no chemical requirement, the removal process is a single

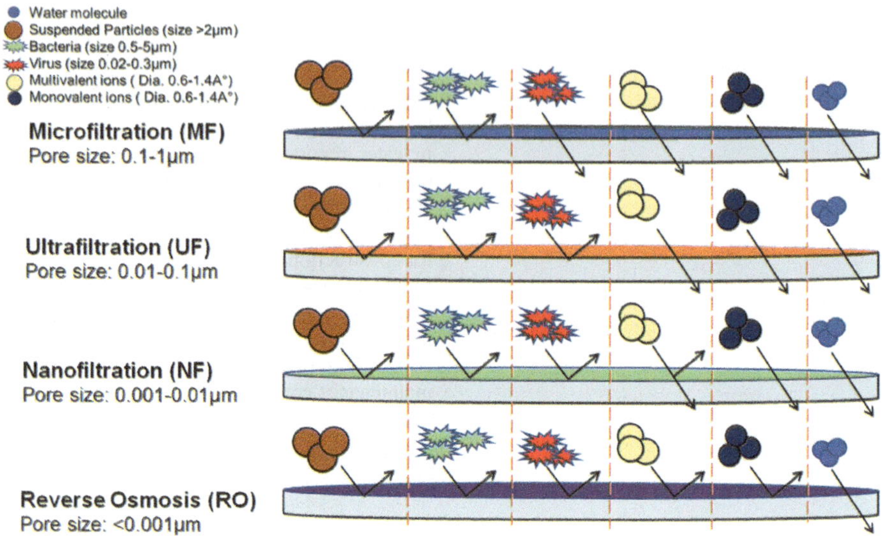

Fig. 19.3 Membrane filtration processes, their size range, and particle removal

step, and less footprint is some positive side of membrane filtration (Peter-Varbanets et al. 2009). Research towards membrane cost minimization and energy requirement has gained must attention from past decades. On the other hand, membrane fouling is the problem that needs to address. In the case of microfiltration and ultrafiltration, backwashing is performed to prevent fouling formation, whereas in nanofiltration and reverse osmosis systems pretreatment is performed (Pooi and Ng 2018).

19.3.2 Microfiltration-Based POU Systems

Ceramic media type and polymer type are two commonly used microfiltration systems (Venkatesha et al. 2020). Ceramic media filters are basically in the shapes of candles or pots with a pore size of about 0.2–3 mm made of porous burnt clay and fillers like sawdust depending upon producer (Sobsey et al. 2002). These filters are capable of retaining the microbes based on size exclusion. Pore size and filter composition are two major factors affecting filter efficiency (Sobsey et al. 2009). Also, many ceramic filters have virus adsorption sites that will improve the virus removal efficiency (Sobsey et al. 2002). It was observed that for bacteria, viruses, and protozoa, the removal efficiencies of ceramic filter systems are 2-6LRV, 0.5-4LRV, and 4-6LRV, respectively (Sobsey et al. 2009). It was also observed that the use of ceramic filters reduces diarrheal episodes (Loo et al. 2012). The effectiveness of these filters can be increased by impregnating them with colloidal silver, which will improve the microbial quality of water by inactivating the microbes and the size exclusion (Peter-Varbanets et al. 2009). Impregnation with silver further helps prevent the formation of a biofilm layer on the filter surface (Sobsey et al. 2002). Katadyn Mini, Potters for Peace filter, and Terafil filters are two examples of ceramic filters impregnated with a silver (Venkatesha et al. 2020). Unlike chlorine disinfection and solar disinfection, the performance of ceramic filters is unaffected by the turbidity of influent water (Clasen et al. 2004). Ceramic filters need regular cleaning to remove the retained materials on the filter to maintain the actual flow conditions, thereby meeting the regular demands.

Other than household applications, filters like Filterpen and Lifestraw family can be used for outdoor purposes, especially during traveling, camping, hiking, and emergency preparedness (Peter-Varbanets et al. 2009). The usage of these filters is very simple. From one side of the filter, water needs to suck by simply putting the other side in the water source (river or stream). Lifestraw personal filters can remove bacteria and protozoa with >6LRV and >3LRV, respectively. More than 3LRV of bacteria and protozoa is possible by using Filterpen and has the capacity to produce 100L clean water over its lifespan (Peter-Varbanets et al. 2009). These filters are not recommended if the water is turbid or virus contaminated.

19.3.3 Ultrafiltration-Based POU Systems

Ultrafiltration (UF) systems require low pressure and are effectively remove turbidity, bacteria and viruses (Peter-Varbanets et al. 2009). POU systems based on UF include emergency applications (disaster relief) and household drinking water treatment (Venkatesha et al. 2020; Loo et al. 2012). The authors Pryor et al. (1998) have conducted three pilot studies on the performance of UF membranes for potable water supply in the developing areas of South Africa. The observations from these studies showed a reduction in the natural organic matter, organic color, and fecal indicators to an acceptable limit (Pryor et al. 1998). These systems require periodic cleaning of filters to eliminate fouling. Sometimes pretreatment stage is needed. The following are some of the most commonly used UF-based POU systems.

Lifestraw® is a product from Vestergaard comprised of hollow fiber membranes of pore size 0.02 μm. Under pressure, feed-water forced into the filter system results in clean water release from pores by retaining bacteria, viruses, and protozoa in hollow fibers. Backwashing is required to remove the filtered material from the system. It can provide safe water for a family of five members for three to five years and has a capacity of filtering 30,000 L of water in its entire lifespan. LifeStraw® family 1.0 is similar to the above, but it provides clean water for up to 5 years for a family of 5 membranes with a lifespan filtration capacity of 18,000 L (Lantagne and Clasen 2012). This product is more suitably helps in emergency preparedness.

A product from Lifesaver, Lifesaver 4000UF bottle is another example of UF-based POU filtration systems. It can produce up to 4000 L of clean and safe water by removing bacteria, viruses, and cysts more than 6LRV, 4LRV, and 3LRV, respectively (Venkatesha et al. 2020). During emergency preparedness, traveling, and camping times, this product is helpful. Skyhydrant™ is another UF-based POU system for community-level having a filtration capacity of 400–1000 l/h with long-lasting life (Butler 2009). It can remove particles of size 2–5 μm with >4LRV (Butler 2009). It can use for short-term and long-term purposes, particularly disaster relief and emergencies.

Some researchers have developed mobile or transportable UF membrane systems. The author He (2009) has developed a system based on UF membranes powered by bicycle and is mainly used during emergencies. Another author, Barbot et al. (2009), has developed a pilot plant with a UF system powered by electricity. BARC has developed a bicycle-mounted water purification unit that is useful when electricity is unavailable and can be easily transportable when people move from one place to another. The authors from the Polytechnic University of Valencia (Spain), under a project the AQUAPOT, have designed a UF-based system that is manually operated and was installed in the rural areas of South America and Africa (Arnal et al. 2001, 2010).

Household water purifiers based on UF system are also available in the market, and some examples of the same include Jaltara (developed by BARC), Moselle (developed by Membrane Filters India pvt. ltd.), and Waterlife little star gold (developed by Waterlife India pvt. ltd.) (Venkatesha et al. 2020).

19.3.4 Nanofiltration and RO-Based POU Systems

RO filtration with membranes of smaller pore size and high pressure removes a wide array of contaminants. Among all the technologies available for POU water treatment, RO-based systems are leading demand in the market. The reason for the huge demand for RO systems is due to their specific features such as easy maintenance and more efficiency in contaminants removal. Complete removal of bacteria, viruses, and protozoa can achieve using RO filters. Specific contaminants like arsenic, lead, uranium, and radium can remove using POU RO systems. Also, these systems effectively remove various emerging contaminants such as pharmaceuticals, pesticides, and perfluoroalkyl substances (Herkert et al. 2020). Accumulation of contaminants on the membrane surface causes fouling problem. In order to handle the fouling issue, pretreatment stages (sedimentation or microfiltration or ultrafiltration or activated carbon) are performed. The efficiency of the filtration system can be improved by replacing the filter element at least once a year (Zhang et al. 2020). Mostly, RO filtration systems are involved multistage treatments such as pre and post-treatments. Integration with other treatment units made the system expensive. Therefore, industrialized countries use RO-based POU filters widely than the transition and developing countries (Peter-Varbanets et al. 2009). The use of RO treatment can remove essential nutrients and minerals from water (Rezaeinia et al. 2018). It is important to retain or replace these essential minerals while using RO-based water treatment systems. There are a variety of products available based on RO systems that are integrated with multiple other treatment technologies in the market. KENT water purifier is one of the examples of RO-based POU water treatment products with good demand in the Indian market.

High-energy demand is another barrier to use RO-based POU systems in rural areas of developing countries because of electricity. The development of a system independent of energy sources is important. Also, some circumstances like uncertainty in water quality and no electricity during emergency situations (disasters) demand RO systems with energy source independence. Solar energy can be used to power the RO system using photovoltaic panels, and this is called ROSI (Reverse Osmosis Solar Installation) (Loo et al. 2012). Schäfer and Richards (2005) did a performance evaluation of the ROSI system for groundwater desalination (using RO/Nano) with ultrafiltration as pretreatment. Unfortunately, no data regarding the feasibility of the ROSI system was available (Schäfer and Richards 2005).

19.4 Performance of Household POU Water Filters in the Removal of Emerging Contaminants

An Increase in concern on environment and public health, improvements in analytical techniques, and development of sophisticated instruments result in identifying various contaminants of emerging concern or emerging contaminants in multiple environmental compartments. Generally, the unregulated newly identified contaminants present in trace amounts and suspected to pose a high risk to human health and ecology refer to emerging contaminants. Pharmaceuticals, personal care products, endocrine disrupting chemicals, nanomaterial, pesticides, surfactants, fire retarders, perfluoroalkyl substances (PFASs), plasticizers, antibiotic-resistant bacteria, hormones all come under contaminants of the emerging concern category (Houtman 2010; Gomes et al. 2020; Kárászová et al. 2020). Due to these contaminants, the actual risk on human health is completely unknown but may cause adverse effects due to their physicochemical nature and longer persistency (Oluwole et al. 2020). These contaminants have their origin from diverse sources, including domestic, industrial, and agricultural practices, and make a pathway into the waste runoff, agricultural runoff, septic tanks, wastewater treatment plants, and landfill leachate (Shahid et al. 2021; Rathi et al. 2021). From there, again, these contaminants get transported and ubiquitously distributed in drinking water sources like surface and groundwater bodies (Ebele et al. 2017; Shahid et al. 2021; Heberer 2002). Also, researchers have identified the presence of emerging contaminants in drinking water treatment plants and tap water at households in many countries (Yim et al. 2009; Takagi et al. 2008; Glassmeyer et al. 2017; Valbonesi et al. 2021). In India, a variety of emerging contaminants with considerable concentration levels in surface-water and groundwater samples from many parts of the country have been reported (Mathew and Kanmani 2020). For example, the groundwater near the Ganga basin contains active pharmaceutical drug compounds such as Naproxen, Caffeine, and Ciprofloxacin in concentrations around 0.002, 0.262, and 0.005 μg/l, respectively

(Sharma et al. 2019). Around 0.002–0.015 μg/l concentration of personal care products and 0.033 μg/l concentration of perfluorinated compounds were also reported in the same basin (Sharma et al. 2016, 2019). In south India, the surface water from the Kaveri river in Tamil Nadu state was reported to have a 13 μg/l concentration of Antiepileptic drug compounds (Ramaswamy et al. 2011). Also, 10.6 μg/l concentration of Antimycotic drug compounds were reported downstream from the Amberpet treatment plant on the Musi river, Hyderabad (Lübbert et al. 2017). In the North-East Indian part, groundwater and surface-water from cities like Nagaon and Dibrugarh in the state of Assam had greater than 5.4 μg/l concentration of DDT (dichlorodiphenyl-trichloroethane) (Mishra and Sharma 2011). In North India, nearly 3.8 and 0.39 μg/l of phthalates, a class of endocrine-disrupting chemicals, were reported for drinking water samples from Okhla industrial region and Jawaharlal Nehru University, New Delhi, respectively (Das et al. 2014). Therefore, emerging contaminants present in drinking water sources are now a global concern, and it is important to provide emerging contaminants-free drinking water to mitigate the suspected risks from these contaminants.

Drinking water treatment plants play a major role in water treatment and are designed to remove traditional contaminants, not for emerging contaminants. However, some studied have investigated the performance of drinking water treatment plants in removing various emerging contaminants from source water (Sorlini et al. 2019). Advanced treatment technologies like adsorption, oxidation processes, and membrane separation processes can remove various emerging contaminants, but still, their concentration levels in treated water are non-ignorable (Gomes et al. 2020; Rathi et al. 2021). In addition, these advanced treatment options for water treatment require more energy and are expensive. Also, establishing a treatment plant with advanced treatment options imposes more cost and is uneconomical in developing countries. Even in developed countries, treating water with advanced techniques at community levels to remove emerging contaminants is not recommended because a very small quantity of treated water is used for drinking and cooking purposes than water used for other household applications (Anumol et al. 2015). Sometimes, there are chances of recontamination of treated water with specific contaminants due to failures in the distribution network systems. Also, there are chances of formation of disinfection by-products in the treated water. Therefore, it is an urgent need and at most, important to have a safe drinking water system at individual household levels and POU water treatment systems can provide an effective solution in removing specific toxic chemicals and emerging contaminants (Brown et al. 2017).

In households, the POU water filter devices are of different types, which include pitcher filters, refrigerator filters, faucet-mounted filters, on-counter filters, and under sink filters (Herkert et al. 2020). These POU treatment systems can remove certain emerging contaminants effectively, but not all the POU filters remove effectively. However, the removal efficiency depends on treatment technology, properties of emerging contaminants, water quality, and operational conditions (Anumol et al. 2015; Patterson et al. 2019). Not many studies have examined POU water treatment systems' effectiveness in removing emerging contaminants, and only limited information and data are available. Mostly, adsorption by activated carbon and RO-based POU systems were tested to evaluate their performance in removing various emerging contaminants.

19.4.1 Activated Carbon-Based POU Systems

Activated carbon-based POU filters are generally made up of different materials like coal and coconut shell. There are two types of activated carbon, granular activated carbon and solid block activated carbon. Compared to granular activated carbon filters, the solid block carbon filters have smaller particles and smaller pore sizes. These smaller pores and smaller particles play a major role in adsorbing the trace organic contaminants due to the high energy of the smaller pores and larger surface area of the smaller particles (Cotruvo et al. 2014; Anumol et al. 2015). Moreover, the chemical properties of organic contaminants also impact the adsorption process (EPA 2006). Figure 19.4 represents the schematic diagram of a household activated carbon-based water treatment system.

Anumol et al. (2015) have studied the activated carbon-based POU filters (Brita filter, PUR filter, Zero Water filter, GE filter and Whirlpool filter) for trace organic contaminants removal (including pharmaceuticals and pesticides) and reported >98% removal efficiency of refrigerator filters (GE filter and Whirlpool filter) and >70% removal efficiency of pitcher filters (Brita filter, PUR filter, and Zero Water filter). The reason for the greater removal efficiency of refrigerator filters than the pitcher filters is due to smaller sizes of carbon particles in refrigerator filters than in pitcher filters. Smaller particles provide a larger surface area, which results in more adsorption of trace organic contaminants. There might also be another factor that affects removal

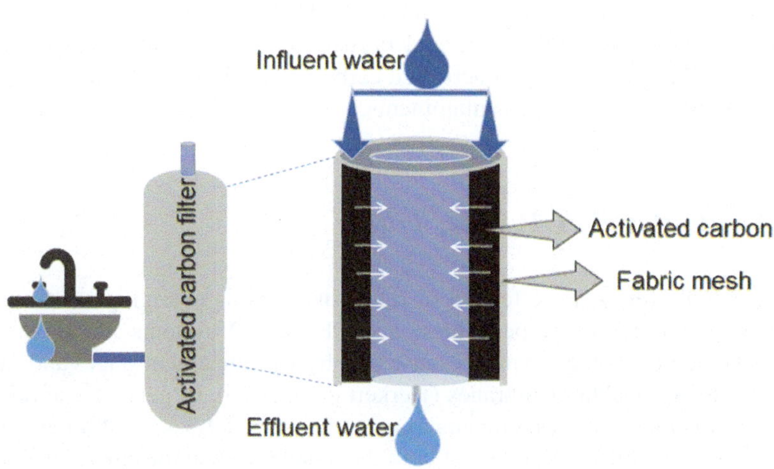

Fig. 19.4 Typical diagram of household activated carbon-based filter

efficiency is type of material used as a carbon source which may not be the same for both refrigerator filters and pitcher filters (Anumol et al. 2015). In the same study, among three pitcher filters, Zero Water filters have shown greater removal efficiency (96.7–97.5%) for PFAS compared to Brita and PUR filters which have shown 52–57% and 79.4–84.8%, respectively. The Zero Water filter consists of anion exchange resins, and the examined PFAS exists mostly in anion form. Therefore, the presence of anion exchange resins in Zero Water filters improves the removal efficiency for PFAS. They have also observed a decrease in removal efficiency for trace organic contaminants with the increase in the expected lifespan of filters (Anumol et al. 2015).

Patterson et al. (2019) also evaluated the performance of two different granular activated carbon-based POU/POE systems for PFAS removal and reported very few or no concentration levels of PFAS in the treated water. Further, they have observed that coconut-based granular activated carbon filters treat more water than coal-based granular activated carbon (Patterson et al. 2019).

In another study by Herkert et al. (2020) observed variations in the performance of POU activated carbon-containing filters in removing PFAS. Only 73% of studied activated carbon filters shown effective removal for PFAS. In the case of activated carbon-based filters, the removal of PFAS was shown to be dependent on the chain length of PFAS and reported around 60–70% long-chain PFAS removal and nearly 40% short-chain PFAS removal (Herkert et al. 2020). Other than chain length, the hydrophobic nature of PFAS also influences their removal. For example, removal efficiency for perfluoroalkyl sulfonic acids (PFSA) was reported more than perfluoroalkyl carboxylic acids (PFCA), which might be due to the greater hydrophobic nature of PESA than PFCA (Herkert et al. 2020). In comparing the filter device types for removal of PFCAs and PFSAs, pitcher filters have shown better performance than refrigerator filters (Herkert et al. 2020). In the same study, almost complete removal of PFAS was observed with dual-stage filters, and the reason for this is unclear. Therefore, research is still needed to understand a better relationship between the activated carbon-based POU filters and PFAS removal efficiency. Further studies can evaluate the performance of activated carbon-based POU filters using different materials to remove emerging contaminants.

19.4.2 RO-Based POU Water Filters

RO-based POU filters are effective in removing various emerging contaminants, including pharmaceuticals, pesticides, and PFAS. RO systems are effective in removing emerging contaminants because they have pretreatment stages and smaller pores of semipermeable membranes (Herkert et al. 2020). In general, the rejection or removal of emerging contaminants from water using RO-based filtration depends on the physicochemical properties of both the membrane and the emerging contaminants. The size exclusion, electrostatic interaction, and hydrophobic interaction are three basic removal mechanisms involved (Fujioka et al. 2013). In the size-exclusion mechanism, the contaminants with a large molecular size than free volume hole-radius can easily retain on the membrane surface. In addition to free volume hole radius, free volume hole distribution and active skin layer thickness can also influence the efficiency of RO membranes (Fujioka et al. 2013). In electrostatic interaction, due to the presence of molecular charge and membrane charge, attraction or repulsion of contaminants occurs. Membranes are mostly negatively charged, and if the contaminants are positively charged, then attraction happens; otherwise, the repulsion happens for negatively charged contaminants (Nghiem and Fujioka 2016). In addition, the hydrophobic nature of the membranes and contaminants affect the removal of emerging contaminants (Nghiem and Fujioka 2016). Figure 19.5 shows the schematic diagram of the RO-based POU household water filter system.

Three RO-based POU systems (iSpring RCS5T, Hydrologic Evolution RO1000, Flexeon LP-700) were studied and noticed a reduction in PFAS in source water below regulatory standard 70 ng/l (Patterson et al. 2019). But, in the same study presence of PFAS in the effluent of the Hydrologic RO system was observed, which might be due to insufficient pre and post-treatments or leakage through membranes. For Hydrologic, after shutdown and startup times also some concentration levels of PFAS were observed in treated water (Patterson et al. 2019).

The authors Herkert et al. (2020) have also studied the RO-based POU filters performance in removing PFAS and observed no relation between chain length of PFAS and removal efficiency, unlike in the case of activated carbon-based filters. They have also reported >90% removal efficiency of RO filters for PFCA and PFSA and >97% for perfluoroalkyl ether acids (PFEA) (Herkert et al. 2020). Further, more

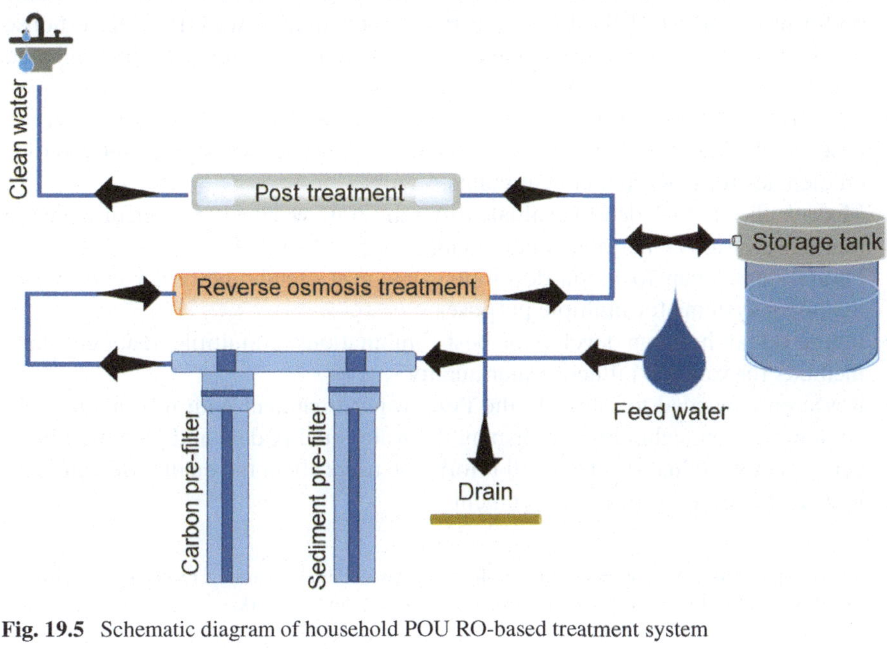

Fig. 19.5 Schematic diagram of household POU RO-based treatment system

studies are needed to evaluate the effectiveness of different RO-based POU systems and check for long-term performance.

19.5 Conclusion

All POU water treatment systems that can improve microbial and chemical water quality have some advantages and disadvantages. Compared to all methods, the preference for membrane-based POU systems is growing rapidly.

- Microfiltration POU systems are effectively removing bacteria and protozoa. These filters have application at the household level, traveling, camping, and emergency situations. These filters can operate without electricity, simply by suction or hand power.
- Ultrafiltration POU systems are effectively removing bacteria, protozoa, and viruses. These filters are the most suitable option for drinking water treatment during emergencies. Various products of this category are available in the market for household, camping, and traveling purposes. Hand power, gravity, and suction energy can operate the system without electricity requirements.
- NF and RO-based POU systems effectively remove a wide array of contaminants with almost complete removal of bacteria, protozoa, and viruses. These filters are effectively removing various emerging and specific contaminants. Energy demand and membrane fouling are two major issues to focus. However, integration of RO treatment with other pre and post-treatment techniques results in better performance. Most of the RO-based POU water treatment systems have house applications and are the most demanded household POU water filters in developed countries.
- Not all POU filters remove emerging contaminants. However, Activated carbon based-POU filters and RO-based POU filters have shown significant removal efficiencies for emerging contaminants.
- More studies are needed to evaluate different POU water filters' performance in removing various emerging contaminants.
- Future studies can focus on developing low-cost membrane-based POU water treatment systems for multiple purposes.
- Future researchers can develop the best combinations of multiple treatment stage methods for variable influent water quality.
- Research is needed to integrate the POU water treatment with information and communication technology to display the water quality data and give the filter's performance evaluation. This will inform the users about the quality of water and indicate the time for filter replacement.

Acknowledgements The corresponding author acknowledges the funding received from ISRO-IIT(B) SPACE TECHNOLOGY CELL (STC) for carrying out this work.

References

Agrawal VK, Bhalwar R (2009) Household water purification: low-cost interventions. Med J Armed Forces India 65:260–263. https://doi.org/10.1016/S0377-1237(09)80019-1

Ahammed MM, Davra K (2011) Performance evaluation of biosand filter modified with iron oxide-coated sand for household treatment of drinking water. Desalination 276:287–293. https://doi.org/10.1016/j.desal.2011.03.065

Alsulaili A, Al-Harbi M, Elsayed K (2020) The influence of household filter types on quality of drinking water. Process Saf Environ Prot 143:204–211. https://doi.org/10.1016/j.psep.2020.06.051

Anumol T, Clarke BO, Merel S, Snyder SA (2015) Point-of-use devices for attenuation of trace organic compounds in water. J Am Water Works Assoc 107:E474–E485. https://doi.org/10.5942/jawwa.2015.107.0129

Arnal Arnal JM, Sancho Fernández M, Martín Verdú G, Lora García J (2001) Design of a membrane facility for water potabilization and its application to third world countries. Desalination 137:63–69. https://doi.org/10.1016/S0011-9164(01)00205-3

Arnal JM, García-Fayos B, Sancho M et al (2010) Design and installation of a decentralized drinking water system based on ultrafiltration in Mozambique. Desalination 250:613–617. https://doi.org/10.1016/j.desal.2009.09.035

Ashani HS (2017) Use of granular activated carbon and carbon block filters at municipal and point of use drinking water treatment for removal of organics, 1–27

Ashok Gadgil et al U (1997) Field-testing UV disinfection of drinking water, 394–396

Barbot E, Carretier E, Wyart Y et al (2009) Transportable membrane process to produce drinking water. Desalination 248:58–63. https://doi.org/10.1016/j.desal.2008.05.038

Barna Z, Antmann K, Pászti J et al (2014) Infection control by point-of-use water filtration in an intensive care unit—a Hungarian case study. J Water Health 12:858–867. https://doi.org/10.2166/wh.2014.052

Berg PA (2010) A new water treatment product for the urban poor in the developing world. In: Proceedings of World environmental and water resources congress 2010 challenges of change, pp 2010–2025. https://doi.org/10.1061/41114(371)210

Bradley I, Straub A, Maraccini P et al (2011) Iron oxide amended biosand filters for virus removal. Water Res 45:4501–4510. https://doi.org/10.1016/j.watres.2011.05.045

Brown KW, Gessesse B, Butler LJ, MacIntosh DL (2017) Potential effectiveness of point-of-use filtration to address risks to drinking water in the United States. Environ Health Insights 11. https://doi.org/10.1177/1178630217746997

Butler R (2009) Skyjuice technology impact on the U.N. MDG outcomes for safe affordable potable water. Desalination 248:622–628. https://doi.org/10.1016/j.desal.2008.05.111

Chiller TM, Mendoza CE, Lopez MB et al (2006) Reducing diarrhoea in Guatemalan children: randomized controlled trial of flocculant-disinfectant for drinking-water. Bull World Health Organ 84:28–35. https://doi.org/10.2471/BLT.04.016980

Clasen T, Brown J, Suntura O, Collin S (2004) Safe household water treatment and storage using ceramic drip filters: a randomised controlled trial in Bolivia. Water Sci Technol 50:111–115. https://doi.org/10.2166/wst.2004.0033

Conroy RM, Elmore Meegan M, Joyce T et al (1999) Solar disinfection of water reduces diarrhoeal disease: an update. Arch Dis Child 81:337–338. https://doi.org/10.1136/adc.81.4.337

Cotruvo JA, Andrew R, Herman R (2014) Point of use and point of entry treatment technologies applicable in the home for controlling chemical, microbial, and aesthetic contaminants in drinking water. Elsevier Ltd

Crump JA, Okoth GO, Slutster L et al (2004) Effect of point-of-use disinfection, flocculation and combined flocculation—disinfection on drinking quality in western Kenya. J Appl Microbiol 97:225–231. https://doi.org/10.1111/j.1365-2672.2004.02309.x

Das MT, Ghosh P, Thakur IS (2014) Intake estimates of phthalate esters for South Delhi population based on exposure media assessment. Environ Pollut 189:118–125. https://doi.org/10.1016/j.envpol.2014.02.021

Duke WF, Nordin RN, Baker D, Mazumder A (2006) The use and performance of BioSand filters in the Artibonite Valley of Haiti: a field study of 107 households. Rural Remote Health 6:570

Ebele AJ, Abou-Elwafa Abdallah M, Harrad S (2017) Pharmaceuticals and personal care products (PPCPs) in the freshwater aquatic environment. Emerg Contam 3:1–16. https://doi.org/10.1016/j.emcon.2016.12.004

Elliott MA, Stauber CE, Koksal F et al (2008) Reductions of *E. coli*, echovirus type 12 and bacteriophages in an intermittently operated household-scale slow sand filter. Water Res 42:2662–2670. https://doi.org/10.1016/j.watres.2008.01.016

EPA (2006) Point-of-use or point-of-entry treatment options for small drinking water systems. US Environ Prot Agency 1–132

Ezugbe EO, Rathilal S (2020) Membrane technologies in wastewater treatment: a review. Membranes (Basel) 10. https://doi.org/10.3390/membranes10050089

Fane AG, Tang CY, Wang R (2011) Membrane technology for water: microfiltration, ultrafiltration, nanofiltration, and reverse osmosis. Treatise Water Sci 4:301–335. https://doi.org/10.1016/B978-0-444-53199-5.00091-9

Francis MR, Sarkar R, Roy S et al (2016) Effectiveness of membrane filtration to improve drinking water: a quasi-experimental study from rural southern India. Am J Trop Med Hyg 95:1192–1200. https://doi.org/10.4269/ajtmh.15-0675

Fujioka T, Oshima N, Suzuki R et al (2013) Rejection of small and uncharged chemicals of emerging concern by reverse osmosis membranes: the role of free volume space within the active skin layer. Sep Purif Technol 116:426–432. https://doi.org/10.1016/j.seppur.2013.06.015

Glassmeyer ST, Furlong ET, Kolpin DW et al (2017) Nationwide reconnaissance of contaminants of emerging concern in source and treated drinking waters of the United States. Sci Total Environ 581–582:909–922. https://doi.org/10.1016/j.scitotenv.2016.12.004

Gomes IB, Maillard JY, Simões LC, Simões M (2020) Emerging contaminants affect the microbiome of water systems—strategies for their mitigation. npj Clean Water 3. https://doi.org/10.1038/s41545-020-00086-y

He Y (2009) Transportable membrane system produces drinking water. Membr Technol 2009:8–9. https://doi.org/10.1016/S0958-2118(09)70163-6

Heberer T (2002) Tracking persistent pharmaceutical residues from municipal sewage to drinking water. J Hydrol 266:175–189. https://doi.org/10.1016/S0022-1694(02)00165-8

Herkert NJ, Merrill J, Peters C et al (2020) Assessing the effectiveness of point-of-use residential drinking water filters for perfluoroalkyl substances (pfass). Environ Sci Technol Lett 7:178–184. https://doi.org/10.1021/acs.estlett.0c00004

Houtman CJ (2010) Emerging contaminants in surface waters and their relevance for the production of drinking water in Europe. J Integr Environ Sci 7:271–295. https://doi.org/10.1080/1943815X.2010.511648

Jakubczak M, Karwowska E, Rozmysłowska-Wojciechowska A et al (2021) Filtration materials modified with 2d nanocomposites—a new perspective for point-of-use water treatment. Materials (basel) 14:1–16. https://doi.org/10.3390/ma14010182

Jiang JQ (2015) The role of coagulation in water treatment. Curr Opin Chem Eng 8:36–44. https://doi.org/10.1016/j.coche.2015.01.008

Kárászová M, Bourassi M, Gaálová J (2020) Membrane removal of emerging contaminants from water: which kind of membranes should we use? Membranes (basel) 10:1–23. https://doi.org/10.3390/membranes10110305

Kfir R, Bateman BW, Pitout BA, Coubrough P (1989) Disinfection of polluted water by chlorine-flocculant tablet. Water Sci Technol 21:207–213. https://doi.org/10.2166/wst.1989.0101

Lantagne D, Clasen T (2012) Point-of-use water treatment in emergency response. Waterlines 31:30–52. https://doi.org/10.3362/1756-3488.2012.005

Liang K, Sobsey M, Stauber C (2010) Use of biosand filters in Cambodia. Water Sanit Progr F Notes

Lonsdale HK (1982) The growth of membrane technology. J Memb Sci 10:81–181. https://doi.org/10.1016/S0376-7388(00)81408-8

Loo SL, Fane AG, Krantz WB, Lim TT (2012) Emergency water supply: a review of potential technologies and selection criteria. Water Res 46:3125–3151. https://doi.org/10.1016/j.watres.2012.03.030

Lübbert C, Baars C, Dayakar A et al (2017) Environmental pollution with antimicrobial agents from bulk drug manufacturing industries in Hyderabad, South India, is associated with dissemination of extended-spectrum beta-lactamase and carbapenemase-producing pathogens. Infection 45:479–491. https://doi.org/10.1007/s15010-017-1007-2

Ma L, Dong X, Chen M, et al (2017) Fabrication and water treatment application of carbon nanotubes (CNTs)-based composite membranes: a review. Membranes (Basel) 7. https://doi.org/10.3390/membranes7010016

Mathew RA, Kanmani S (2020) A review on emerging contaminants in indian waters and their treatment technologies. Nat Environ Pollut Technol 19:549–562. https://doi.org/10.46488/NEPT.2020.V19I02.010

Meyer V, Reed RH (2001) SOLAIR disinfection of coliform bacteria in hand-drawn drinking water. Water SA 27:49–52. https://doi.org/10.4314/wsa.v27i1.5009

Mishra K, Sharma RC (2011) Contamination of aquatic system by chlorinated pesticides and their spatial distribution over North-East India. Toxicol Environ Health Sci 3:144–155. https://doi.org/10.1007/s13530-011-0092-3

Mulder M (2013) Basic principles of membrane technology

Murinda S, Kraemer S (2008) The potential of solar water disinfection as a household water treatment method in peri-urban Zimbabwe. Phys Chem Earth 33:829–832. https://doi.org/10.1016/j.pce.2008.06.025

Nghiem LD, Fujioka T (2016) Removal of emerging contaminants for water reuse by membrane technology. Emerg Membr Technol Sustain Water Treat 217–247. https://doi.org/10.1016/B978-0-444-63312-5.00009-7

Oluwole AO, Omotola EO, Olatunji OS (2020) Pharmaceuticals and personal care products in water and wastewater: a review of treatment processes and use of photocatalyst immobilized on functionalized carbon in AOP degradation. BMC Chem 14:1–29. https://doi.org/10.1186/s13065-020-00714-1

Patil R, Ahmad D, Balkundae P et al (2020) Development of low cost point-of-use (POU) interventions for instant decontamination of drinking water in developing countries. J Water Process Eng 37:101435. https://doi.org/10.1016/j.jwpe.2020.101435

Patterson C, Burkhardt J, Schupp D et al (2019) Effectiveness of point-of-use/point-of-entry systems to remove per- and polyfluoroalkyl substances from drinking water. AWWA Water Sci 1:e1131. https://doi.org/10.1002/aws2.1131

Pérez-Vidal A, Diaz-Gómez J, Castellanos-Rozo J, Usaquen-Perilla OL (2016) Long-term evaluation of the performance of four point-of-use water filters. Water Res 98:176–182. https://doi.org/10.1016/j.watres.2016.04.016

Peter-Varbanets M, Zurbrügg C, Swartz C, Pronk W (2009) Decentralized systems for potable water and the potential of membrane technology. Water Res 43:245–265. https://doi.org/10.1016/j.watres.2008.10.030

Pooi CK, Ng HY (2018) Review of low-cost point-of-use water treatment systems for developing communities. npj Clean Water 1. https://doi.org/10.1038/s41545-018-0011-0

Pryor MJ, Jacobs EP, Botes JP, Pillay VL (1998) A low pressure ultrafiltration membrane system for potable water supply to developing communities in South Africa. Desalination 119:103–111. https://doi.org/10.1016/S0011-9164(98)00126-X

Ramaswamy BR, Shanmugam G, Velu G et al (2011) GC-MS analysis and ecotoxicological risk assessment of triclosan, carbamazepine and parabens in Indian rivers. J Hazard Mater 186:1586–1593. https://doi.org/10.1016/j.jhazmat.2010.12.037

Rathi BS, Kumar PS, Show PL (2021) A review on effective removal of emerging contaminants from aquatic systems: current trends and scope for further research. J Hazard Mater 409:124413. https://doi.org/10.1016/j.jhazmat.2020.124413

Rezaeinia S, Nasseri S, Binesh M et al (2018) Qualitative and health-related evaluation of point-of-use water treatment equipment performance in three cities of Iran. J Environ Heal Sci Eng 16:265–275. https://doi.org/10.1007/s40201-018-0315-5

Rodda N, Bateman B, Kfir R (1993) Removal of *Salmonella typhi*, *Shigella dysenteriae*, *Vibrio cholerae* and rotavirus from water using a water treatment tablet. Water Sci Technol 27:347–350. https://doi.org/10.2166/wst.1993.0373

Romero DAD, de Almeida Silva MC, Chaúque BJM, Benetti AD (2020) Biosand filter as a point-of-usewater treatment technology: influence of turbidity on microorganism removal efficiency. Water (Switzerland) 12. https://doi.org/10.3390/w12082302

Rose A, Roy S, Abraham V et al (2006) Solar disinfection of water for diarrhoeal prevention in southern India. Arch Dis Child 91:139–141. https://doi.org/10.1136/adc.2005.077867

Schäfer AI, Richards BS (2005) Testing of a hybrid membrane system for groundwater desalination in an Australian national park. Desalination 183:55–62. https://doi.org/10.1016/j.desal.2005.05.007

Shahid MK, Kashif A, Fuwad A, Choi Y (2021) Current advances in treatment technologies for removal of emerging contaminants from water—a critical review. Coord Chem Rev 442. https://doi.org/10.1016/j.ccr.2021.213993

Shamsuddin N, Das DB, Starov VM (2016) Membrane-based point-of-use water treatment (PoUWT) system in emergency situations. Sep Purif Rev 45:50–67. https://doi.org/10.1080/15422119.2014.973967

Sharma BM, Bharat GK, Tayal S et al (2016) Perfluoroalkyl substances (PFAS) in river and ground/drinking water of the Ganges River basin: emissions and implications for human exposure. Environ Pollut 208:704–713. https://doi.org/10.1016/j.envpol.2015.10.050

Sharma BM, Bečanová J, Scheringer M et al (2019) Health and ecological risk assessment of emerging contaminants (pharmaceuticals, personal care products, and artificial sweeteners) in surface and groundwater (drinking water) in the Ganges River Basin, India. Sci Total Environ 646:1459–1467. https://doi.org/10.1016/j.scitotenv.2018.07.235

Snyder KV, Webster TM, Upadhyaya G et al (2016) Vinegar-amended anaerobic biosand filter for the removal of arsenic and nitrate from groundwater. J Environ Manage 171:21–28. https://doi.org/10.1016/j.jenvman.2016.02.001

Sobsey MD, World Health Organization. Water S (2002) Managing water in the home : accelerated health gains from improved water supply (prepared by Sobsey MD)

Sobsey MD, Stauber CE, Casanova LM et al (2009) Response to comment on "point of use household drinking water filtration: a practical, effective solution for providing sustained access to safe drinking water in the developing world." Environ Sci Technol 43:970–971. https://doi.org/10.1021/es8026133

Sorlini S, Collivignarelli MC, Miino MC (2019) Technologies for the control of emerging contaminants in drinking water treatment plants. Environ Eng Manag J 18:2203–2216

Souter PF, Cruickshank GD, Tankerville MZ et al (2003) Evaluation of a new water treatment for point-of-use household applications to remove microorganisms and arsenic from drinking water. J Water Health 1:73–84. https://doi.org/10.2166/wh.2003.0009

Stauber CE, Kominek B, Liang KR et al (2012) Evaluation of the impact of the plastic biosand filter on health and drinking water quality in rural tamale, Ghana. Int J Environ Res Public Health 9:3806–3823. https://doi.org/10.3390/ijerph9113806

Takagi S, Adachi F, Miyano K et al (2008) Perfluorooctanesulfonate and perfluorooctanoate in raw and treated tap water from Osaka, Japan. Chemosphere 72:1409–1412. https://doi.org/10.1016/j.chemosphere.2008.05.034

Tiwari SSK, Schmidt WP, Darby J et al (2009) Intermittent slow sand filtration for preventing diarrhoea among children in Kenyan households using unimproved water sources: Randomized controlled trial. Trop Med Int Heal 14:1374–1382. https://doi.org/10.1111/j.1365-3156.2009.02381.x

Valbonesi P, Profita M, Vasumini I, Fabbri E (2021) Contaminants of emerging concern in drinking water: quality assessment by combining chemical and biological analysis. Sci Total Environ 758. https://doi.org/10.1016/j.scitotenv.2020.143624

Venkatesha R, Kedare SB (2014) Seminar report (TD 694) On portable water purifiers. Submitted in partial fulfilment for the Degree of M. Tech. in Technology & Development (Roll No. 143350003) Under the guidance of Prof. Rao AB, Prof. Shireesh B. Kedare Centre for Technology. https://doi.org/10.13140/RG.2.1.3000.3605

Venkatesha R, Rao AB, Kedare SB (2020) Appropriate household point-of-use water purifier selection template considering a rural case study in western India. Appl Water Sci 10:1–15. https://doi.org/10.1007/s13201-020-01207-1

WHO (2019) 1 in 3 people globally do not have access to safe drinking water. UNICEF, WHO. https://www.who.int/news/item/18-06-2019-1-in-3-people-globally-do-not-haveaccess-to-safe-drinking-water-unicef-who

Yim LM, Taniyasu S, Yeung LWY et al (2009) Perfluorinated compounds in tap water from china and several other countries. Environ Sci Technol 43:4824–4829. https://doi.org/10.1021/es900637a

Zhang Z, Zhang W, Hu X, Li K, Luo P, Li X, Xu W, Li S (2020) Evaluating the Efficacy of Point-of-Use Water Treatment Systems Using the Water Quality Index in water 12:1–15

Chapter 20
Electrocoagulation Process for the Removal of Emerging Pollutants in Water and Wastewater

Pramod Kumar, Ashish Kumar, Tabish Nawaz, and Swatantra P. Singh

Abstract The advancement in technology may have created an easy lifestyle, but it has also generated many pollutants that are toxic and cause severe issues for human health called emerging pollutants (EPs). These EPs result from excessive use of chemicals to enhance our daily lives and provide for people's basic needs. Although we may have found several ways to detect some of these EPs by continuous development in technology, the issues of complete removal of them still persist. The chapter seeks electrocoagulation as one of the effective methods to remove the EPs as it is easy to operate technology and requires low cost and maintenance. This method utilizes the combination of electricity and sacrificial metal electrodes to remove pollutants in the form of flocs which either settles in the bottom or float on the reactor's top surface. This chapter further highlights the occurrence, issue, and fate of the EPs. It also describes the electrocoagulation treatment method, its mechanism, and parameters that affect the system operation, along with the applicability of emerging pollutants removal. It is suggested to create and manage a database of EPs that shall help in tracking different classes of EPs in the environment. The results observed from exhaustive literature studies are positive and encouraging for the development of this technique to remove EPs from wastewater efficiently.

Keywords Emerging pollutants · Water · Wastewater · Emerging pollutants · Electrocoagulation · Technology

P. Kumar · A. Kumar · T. Nawaz · S. P. Singh (✉)
Environmental Science and Engineering Department (ESED), Indian Institute of Technology Bombay, Mumbai 400076, India
e-mail: swatantra@iitb.ac.in

T. Nawaz
e-mail: tnawaz@iitb.ac.in

S. P. Singh
Centre for Research in Nanotechnology and Science (CRNTS), Indian Institute of Technology Bombay, Mumbai 400076, India

Interdisciplinary Program in Climate Studies, Indian Institute of Technology Bombay, Mumbai 400076, India

S. P. Singh et al. (eds.), *New Trends in Emerging Environmental Contaminants*, Energy, Environment, and Sustainability,
https://doi.org/10.1007/978-981-16-8367-1_20

20.1 Introduction

Water scarcity is one of the prominent issues faced by humans being in significant parts of the world. We only have 3% freshwater sources available with us, and 2/3 of this is unavailable and is in the form of glaciers and ice caps (Pavithra et al. 2017). One of the prominent causes is the increase in population, and the other is the perpetual contamination of the water resources available to us in the form of rivers, groundwater, lakes, and ponds. The causes of water pollution are the disposal of effluents (without prior treatment) from industries, sewage treatment plants, stormwater carrying waste contaminants, excessive use of pesticides, herbicides, and fertilizers, to the river or on land. These activities cause the release of contaminants to the poisonous, toxic environment and could cause severe health issues to the inhabitants. They also degrade the aquatic life and ecosystem of the point of disposal. The pollutants causing severe issues are metals like chromium, arsenic, nutrients like nitrate and phosphate, oxygen depleting microorganisms, and microorganisms like bacteria, protozoa, and viruses that can cause water-borne diseases. Additional to these, one more category of the pollutants that are new to environmental scientists and could cause more severe issues are emerging pollutants (EPs).

EPs are defined as a compound with severe health effects that have not been studied, and no regulatory measures have been set up for its control. They reach the environment through various anthropogenic sources and are distributed worldwide (Gavrilescu et al. 2015). EPs are the new products of chemicals whose presence has been reported in many studies in the last 30 years, and their effect on the environment and humans is unknown (Deblonde et al. 2011). There are shreds of evidence of EPs in the river near urban areas of Brazil and may be present in the groundwater (Starling et al. 2019). The widespread contamination of the groundwater with the EPs is detected at a concentration level that could be a threat level to the environment (Lapworth et al. 2012). EPs are also observed in the landfill leachate polluting the groundwater sources and ultimately on human life (Kapelewska et al. 2018).

In many cases, these EPs have been present in the environment for a very long time, but due to a lack of methods and techniques to detect them, they are not discoverable yet (Geissen et al. 2015). The compounds found in the wastewater includes phthalates, polychlorobiphenyls (PCBs), polycyclic aromatic compounds (PACs), polycyclic aromatic hydrocarbons (PAHs), Per- and polyfluoroalkyl substances (PFAS), pharmaceuticals, disinfectants, hormones and, many more yet to be discovered (Deblonde et al. 2011). Apart from industrial, one primary source of these EPs is hospital wastewater, which contains pharmaceuticals, radionuclides, solvents, disinfectants, and antibiotics used in a wide range of concentrations for medical research activities (Verlicchi et al. 2010).

Emerging pollutants are synthetic or naturally occurring chemicals that are not commonly monitored in the environment but can enter the environment and cause known or suspected adverse ecological and (or) human health effects. Groundwater and other water sources are primarily the locations of these contaminants, and in some studies, their detection in the groundwater is a tedious task to carry out (Postigo and

Barceló 2015). In some cases, the release of emerging pollutants to the environment has possibly occurred for a long time but may not have been highlighted until new detection methods were developed. In other cases, the synthesis of new chemicals or changes in the use and disposal of existing chemicals can create new sources of emerging pollutants (Geissen et al. 2015). The origin of emerging pollutants can be industrial, agriculture, hospital or laboratory wastewater, or any other source. Emerging pollutants, in large part, are derived from these three categories.

(a) Pharmaceuticals
(b) Personal care products
(c) Endocrine-disrupting compounds

Pharmaceuticals are a set of EPs that are majorly used by the population for medicinal purposes. There are around 2300 active pharmaceuticals ingredients (API) present and used in medicine and are toxic and bioaccumulative. The traces of these compounds and bi-products are reported in many studies in surface water and soil and, in some cases, in groundwater (Pavithra et al. 2017). Industries manufacturing these drugs causing their pollution through unscientific disposal. The use of medication in day-to-day life also leads to its concentration in the municipal wastewaters. It includes antibiotics as significant contributors and, apart from this, analgesics, steroids, beta-blockers, etc. (Rivera-Utrilla et al. 2013). Their mode of action is the reason behind their persistence. Their presence has been recorded in WWTP effluents, sludge, sediments, natural waters, drinking water, and groundwater (Gogoi et al. 2018). Most of these drugs have been found in the aquatic environment in low concentrations of ng/l to μg/l. There is a need for their eco-toxicological impacts on the environment and aquatic life.

Personal care products (PCPs) hold a long list of products that we use daily, like face creams, shampoo, cosmetics, perfumes, steroids, etc. They are sometimes prescribed and not prescribed by the experts yet used by individuals for their personal care. PCPs are generally released in municipal wastewater, as commonly used by the people, but their traces have also been reported in groundwater and aquatic environments. In wastewater treatment plant PCPs after going through various treatments, the likely fate of PCPs is that it converts into CO_2 and water or get attached to the solids if the compound is organic (Gogoi et al. 2018).

Endocrine-disrupting compounds (EDCs) are the chemicals that enter the body either hinder hormones or obstruct the body's normal functioning. These are the category of EPs that alters the body's functioning or mimics the hormonal activity in the body. They are found in the natural environment by human activities and industries through wastewater disposal. Apart from the above, numerous other EPs exist, their existence needs to be detected, and their impacts on the environment and ecology need to be studied.

The primary concern with the EPs is their low concentration in wastewater, and a wide gap exists in the knowledge of their fate, behavior, effects, detection method, and treatment and removal methods (Rosal et al. 2010; Prado et al. 2017; Gogoi et al. 2018; Deblonde et al. 2011). Although many techniques have been used to detect existing or new EPs in the environment, over 150 micropollutants and their transformation were detected using modified liquid chromatography (Hermes et al. 2018), identification of hormonal bioactivities and other contaminants was carried out using bioassays and modified techniques (Houtman et al. 2018). The absence of relevant data on the fate, impact, and detection level makes it more troublesome for governments to control their use in the respective area. It is not easy to manage the levels of persistence in the environment. One more significant issue is that there are no limits on the concentrations of EPs on the wastewater discharge, drinking water, and the environment (Deblonde et al. 2011; Geissen et al. 2015; Gogoi et al. 2018). The wastewater treatment procedure also has adverse effects on the EPs; various biological and chemical degradation processes transform the EPs into more toxic forms than their parent form (Prado et al. 2017; Gogoi et al. 2018). To deal with this issue, we require a treatment method that can remove the pollutants that are known to us, and we have to develop an understanding of the nature of removing the EPs by the technique employed for their removal.

One of the methods that can be implemented for the removal of EPs is electrocoagulation. Electrocoagulation is a complex process that involves various pollutant removal mechanisms operating in synergy (Kabdaşlı et al. 2012). Electrocoagulation is inexpensive compared to chemical coagulation, and also it is a fast pollutant removal process, produces low sludge and a compact process to carried out. It also overcomes the effects of forming additional ions like chloride and sulfide in the solution that are difficult to remove in the later stage (Mollah et al. 2003). Electrocoagulation is an effective method in removing many types of contaminants from wastewater, and it has been used in various industries for wastewater treatment. It is successfully employed to remove chromium, COD, TOC, organic matter, suspended solids, etc., from water and wastewater (Kabdaşlı et al. 2012; Moussa et al. 2017; Islam 2017). It is an effective method in removing active pharmaceutical compounds (Ensano et al. 2019). Electrocoagulation has been proved to be an effective method, Its capacity to remove EPs should be evaluated by implementing desirable changes and modifications in the design and optimizing the method that can provide the best results suited to our purpose.

This chapter discusses a brief introduction about the EPs, their occurrence, and concerns in their regulation. One method of treating EPs, the Electrocoagulaton process, its mechanism, and parameters governing the operation and efficiency of removal have also been discussed. Studies on some of the EPs have mentioned a separate discussion on EPs and the application of electrocoagulation in removing emerging pollutants. The chapter also provides the future challenges in the treatment of EPs from its various sources.

20.2 Electrocoagulation

Electrocoagulation (EC) treatment method involves using electric current to implement a coagulation process to remove suspended particles and other specific pollutants from the water or wastewater; Fig. 20.1 shows the basic setup of an EC process.

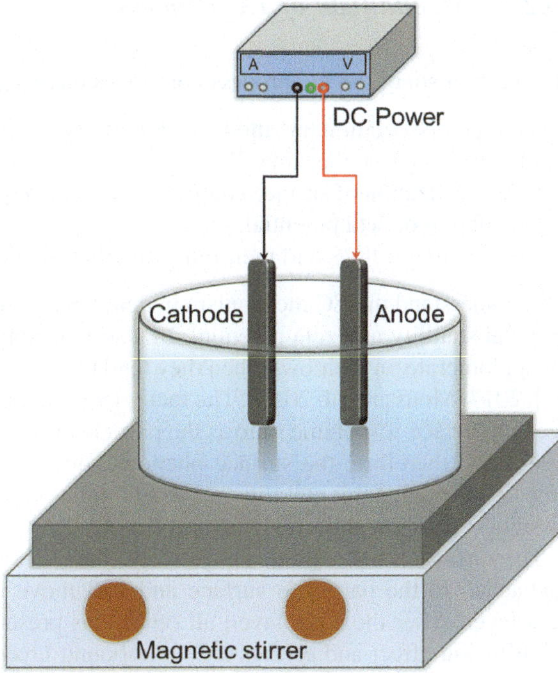

Fig. 20.1 A pictorial representation showing electrocoagulation setup

It uses the concept of electrochemistry to have interactions between electricity, coagulation, and flocculation-causing compounds (Holt et al. 2005).

EC is an electrochemical process that utilizes metal electrodes, also named sacrificial electrodes, and electricity to remove pollutants from water and wastewater. It causes the production of coagulant form metal dissolution due to the passage of electricity that induces coagulation in the electrolyte (Mollah et al. 2001; Sahu et al. 2014; Garcia-Segura et al. 2017; Islam 2017). It is used in the broad-spectrum process and has been used to remove total suspended solids, heavy metals, emulsified oil, bacteria, and many other pollutants from water. The use of electricity to treat water is not a new concept; this concept was first used in the nineteenth century. Electrocoagulation was used to treat drinking water in England in the nineteenth century, and wastewater treatment with EC was started in the US. On a large scale, EC was used to treat drinking water by the US in 1946 (Vik et al. 1984; Wang et al. 2007). Later on, due to high operation costs and significant investment, EC was no longer used to treat water and wastewater. Another reason for neglecting EC was introducing chemicals for the coagulation process, majorly aluminum and iron salts (Emamjomeh and Sivakumar 2009; Moussa et al. 2017; Islam 2017). In the last three decades, concern over water availability and its overuse has directed attention towards the effective treatment of water that can be reused for different processes and strict water quality standards and environmental regulations.

20.2.1 Mechanism of EC Process

The mechanism of the EC process comprises three significant steps:

1. Metal dissociation at anode and hydrolysis of water to hydroxyl ion and hydrogen gas at the cathode.
2. Destabilization of surface charge on suspended particles and other pollutants or reduction of zeta potential.
3. Formation of flocs and their removal either by floatation or precipitation.

To understand the EC mechanism in detail, we need to understand the concept of colloidal stability and zeta potential. In a suspension, when suspended particles do not agglomerate on their own, then they tend to be in a stable condition (Hakizimana et al. 2017; Moussa et al. 2017). The factor that primarily supports this stability is the sizeable surface-to-volume ratio as the particle in suspension is too tiny and has very little mass; therefore, the surface phenomenon dominates the mass phenomenon. Now to understand this stability, we should focus on the surface of the particle. The possible surface condition is shown in Fig. 20.2.

The surface of the colloid is negatively charged, and the positive layer of cations is attached to the particle's surface and will move it; this layer is known as the stern layer. After the stern layer, all other ions present move over the layer due to the diffusion effect and creates a double-bound layer, as shown in the figure. The potential at the end of the double-layer is known as Zeta potential. Other particles also arrange themselves in the same manner. When these two-particle come close to each other, the repulsive force between them resists them to agglomerate. To destabilize the particle and make it agglomerate with other particles, we need to overcome this potential. The process is done with the help of metal ions induced in the system

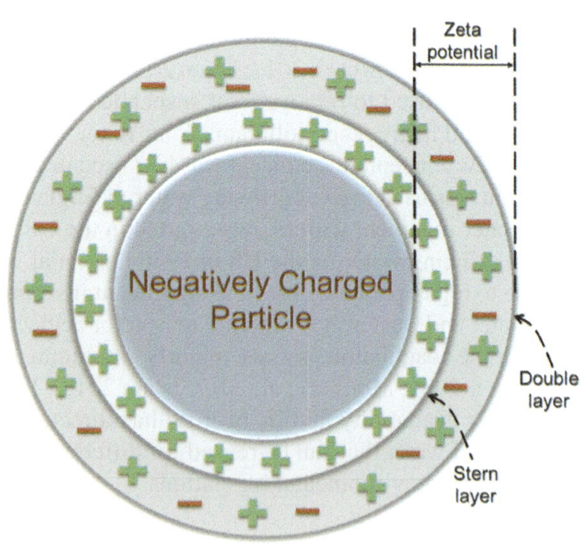

Fig. 20.2 Graphical representation of zeta potential around negatively charged particle

by chemical coagulation dissociation of metal electrodes through the EC process. The role of coagulant is to compress the double and reduce the repulsive action and causing agglomeration of particles and thus particle after agglomeration particle size increases and settlement of particle take place.

In the EC process, the metal species generated through metal dissociation at anode and formation of hydroxyl ion and H_2 gas at the cathode as shown in Fig. 20.30.3:

- At anode metal oxidation takes place, metal cations are formed:

$$M \rightarrow M^{+} + Ze^{-} \tag{20.1}$$

  In the above equation, Z is the number of electrons transferred in the dissociation process; it depends on the mole of the metal (Mollah et al. 2001; Moussa et al. 2017; Islam 2017).
- At the cathode, water is converted into hydroxyl ion and hydrogen gas:

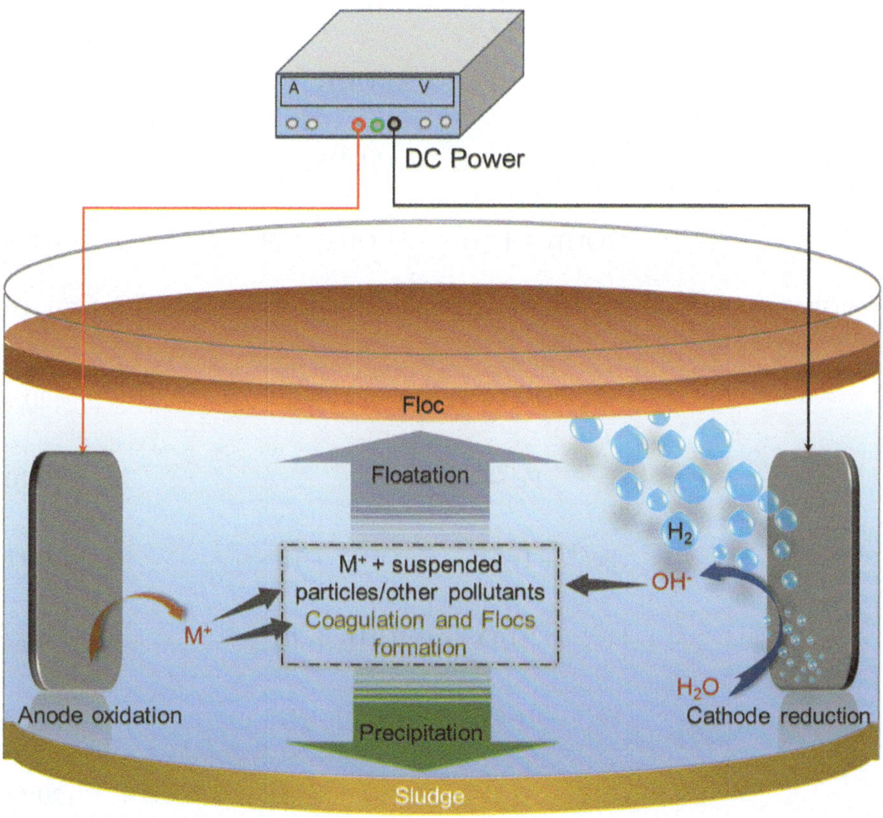

Fig. 20.3 Schematic diagram showing processes in the electrocoagulation reactor (adopted and modified from Hakizimana et al. 2017)

$$3H_2O \rightarrow \frac{3}{2}H_2 + 3OH^- \tag{20.2}$$

If aluminum and iron electrodes are used, the metal ions formed from metal dissociation will undergo rigorous reactions to produce aluminum and iron metal hydroxides or polyhydroxides. All the reactions depend on the pH of the solution.

In the case of Aluminium electrode following reaction occurs (Hakizimana et al. 2017):

$$Al^{3+} + H_2O \rightarrow Al(OH)^{2+} + H^+ \tag{20.3}$$

$$Al^{3+} + 2H_2O \rightarrow Al(OH)^{2+} + 2H^+ \tag{20.4}$$

$$Al^{3+} + 3H_2O \rightarrow Al(OH)^{2+} + 3H^+ \tag{20.5}$$

$$Al(OH)^{2+} + H_2O \rightarrow Al(OH)_{2+} + H^+ \tag{20.6}$$

$$Al(OH)_{2+} + H_2O \rightarrow Al(OH)_3 + H^+ \tag{20.7}$$

$$Al(OH)_3 + H_2O \rightarrow Al(OH)_4^- + H^+ \tag{20.8}$$

In the case of Iron electrodes, during anode oxidation can causes either ferrous or ferric cations Hakizimana et al. 2017):

$$Fe \rightarrow Fe^{2+} + 2e^- \tag{20.9}$$

$$Fe \rightarrow Fe^{3+} + 3e^- \tag{20.10}$$

$$Fe^{2+} + H_2O \rightarrow Fe(OH)^+ + H^+ \tag{20.11}$$

$$Fe^{2+} + 2H_2O \rightarrow Fe(OH)_2 + 2H^+ \tag{20.12}$$

$$Fe^{3+} + H_2O \rightarrow Fe(OH)^{2+} + H^+ \tag{20.13}$$

$$Fe^{3+} + 2H_2O \rightarrow Fe(OH)^+ + 2H^+ \tag{20.14}$$

The generation of the enormous amount of metal ions formed by dissociation of metal electrodes causes compression of the double layer around the surface of the suspended particle, and there is a decrease in zeta potential, causing stability of the suspended particle. In continuation of this process, the repulsive force decreases, the particle starts combining, and the formation of flocs occurs in the system.

With the decrease in the repulsive force and neutralization of the surface charge, agglomeration of the particles takes place, and that leads to the formation of flocs which either get precipitated in the bottom or get floated up with the hydrogen gas and forms a flocs layer on the top (Moussa et al. 2017; Hakizimana et al. 2017; Islam 2017) as shown in the figure.

20.2.2 Parameters Governing EC Process

In the EC process, several parameters are there which affect the operation and efficiency of the process. Operational, water and wastewater features and the geometry of the reactor and electrode are a factor that affects the process. The factors affecting the process are as follows:

1. **Electrode arrangement**: The electrode in the EC process is generally arranged in parallel or series. The energy consumption in series is generally higher than in parallel arrangement due to more resistance offered by the series connections. In the EC cell, the electrodes can be arranged in any manner; based on the system's need, researchers either use one pair of electrodes or more than one pair of electrodes. The pair of electrodes are then arranged in three manners, as shown in Fig. 20.4.

   There is a low potential difference in the MP-P connection as the current is divided between the electrode compared to the MP-S connection in which the current is added up and has more potential difference. In the third arrangement, the outer electrode is connected to the power supply, and the inner electrode is placed between the electrodes. In the inner electrode, the surface gets polarised as per the outermost electrode charge, and there is an opposite charge on the two sides of the electrode (Mollah et al. 2004; Moussa et al. 2017; Hakizimana et al. 2017). It creates a mini cell, in which the inner electrode surface act as both anode and cathode, but in this, more initial energy is required to develop the surface charge and cause metal dissociation. Most of the studies used either MP-P or BP-P connection in the EC process to remove color, turbidity, metals, non-metals, and organic compounds. Both provide almost the same removal efficiency; however, MP-P proves to be more cost-effective and less utilizes less energy for operation (Moussa et al. 2017).
2. **Type of power supply**: In the EC process, we can use both DC and AC power supplies to the electrode. DC power supply leads to the oxidation of the anode and its continuous deterioration by its consumption and formation of oxide layer on the cathode called cathode passivation. To resolve the passivation issue, adding chloride ions reduces the passivation if used insufficiently or using alternate pulse current (Moussa et al. 2017; Hakizimana et al. 2017). AC supply

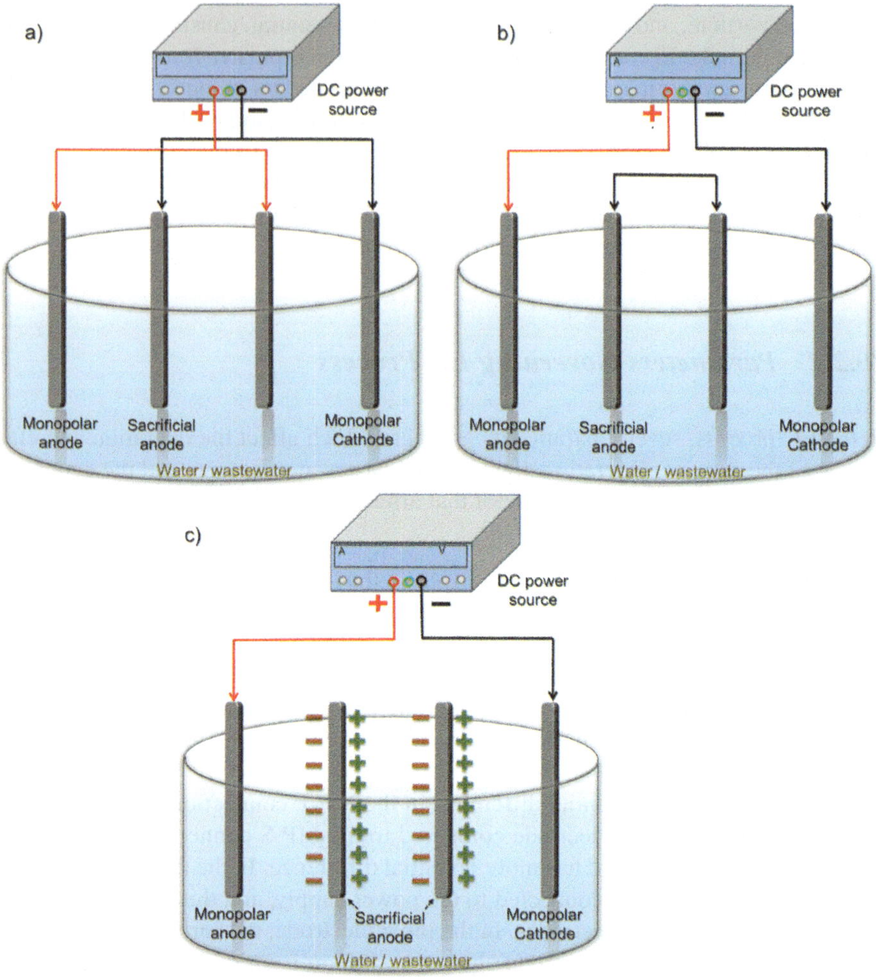

Fig. 20.4 Electrocoagulation reactor in **a** monopolar parallel; **b** monopolar series; **c** bipolar parallel arrangement (adopted and modified from Hakizimana et al. 2017)

reduces the passivation effect and provides less energy consumption, better efficiency, reduced consumption of single electrodes, and sometimes lesser retention time. However, it disturbs the stability of the dipole structures formed in the system as a continuous change in the polarity will hinder the electrophoretic transfer of the charged particle that leads to dipole–dipole interaction, and the system will contain non-spherical particle that does not settle with ease (Mollah et al. 2001, 2004; Sahu et al. 2014).

3. **Current density (CD)**: Current density is defined as the ratio of applied current to the effective electrode area. It can be calculated as:

$$CD = \frac{I}{A}$$

where CD is the current density in mA/cm^2, I is the applied current, and A is the effective area of the electrode in cm^2. The dissociation of metal ions in the EC reactor is directly proportional to the current density. It has been mentioned in various studies that after a specific value of current density, the quality of the treated water does not change; that value is called critical current density (Moussa et al. 2001; Sahu et al. 2014; Hakizimana et al. 2017; Islam 2017). Other parameters are also affected by selecting an optimum value for current density such as pH, temperature, and water flow rate as it controls the formation of metal ions and hydroxyl and hydrogen gas that causes a change in pH and hence changes in the rest of the parameters.

4. **Concentrations of ions**: The presence of various anions affects the destabilization properties of metal particles. Sulfate particles are known to hinder the deterioration/metal disintegration from the electrode, and thus they decline the destabilization of colloids and current effectiveness. Then again, chloride and nitrate particles forestall the hindrance of sulfate particles by separating the latent layer framed. The presence of chloride particles also diminishes the unfavorable impact of sulphate particles, leading to the precipitation of salts on the anodes when the salt fixation is adequately high. The above process structures a protecting layer that diminishes the current effectiveness by expanding the potential between cathodes (Hakizimana et al. 2017; Moussa et al. 2017). The presence of other ions in the solution also hinders the adsorption process due to the preferential adsorption of other ions. (Vasudevan and Laxmi 2011)
5. **Initial pH of the solution**: pH is one of the parameters which affects conductivity, dissociation of metal, hydrolyzed species, and reactions that take place in the reactor. The pH plays a vital role in the EC process; it governs many processes and sometimes directs the operation and working of EC. In studies, it is observed that the initially acidic pH of the solution changes to alkaline and vice versa after the EC process. The reactions 20.3–20.14 mentioned above are all dependent on the pH. They all occur at different pH and proceed as the pH of the solution changes. Controlling pH value in the EC process helps control the EC process, and sometimes it is needed to get a specified result. Al and Fe hydroxides and polyhydroxides are also pH-dependent, and their formation affects and directs the process (Moussa et al. 2001; Vasudevan and Laxmi 2011; Eyvaz et al. 2009; Sahu et al. 2014; Hakizimana et al. 2017; Islam 2017). In studies, people maintain pH at around 4–8 to get the desired results and maximize the removal efficiency.
6. **Inter-electrode distance (IED)**: The distance between the electrodes is also one of the essential parameters. The ohmic potential, which is a function of the solution conductivity, increases by increasing the IED. The increase in IED affects the electrolysis energy consumption; in low conductivity, it will increase the energy consumption. The low value of IED produces more bubbles leading to high mass transfer and a high reaction rate between the coagulant species and

pollutant (Mollah et al. 2001; Sahu et al. 2014; Hakizamana et al. 2017). The IED also governs the time of treatment in the batch reactor to reach a specified efficiency and reaction time in a continuous reactor. In studies, the value of IED is taken from 0.2 to 3 cm and sometimes even more than that value.

7. **Design of reactor**: To reach the desired output for an EC processor to maximize the efficiency of the process, we need to ensure minimized potential drop, adequate mass transfer between coagulant species and pollutant, least passivation of electrodes, and to achieve all these, we required and effective reactor design. EC reactor configuration is vital since it influences the general performance of the EC interaction through its impact on the working boundaries, to be specific, flow regime, flocs development, evacuation yield, and buoyancy/settling attributes. EC reactors have been planned after some fundamental rules, essentially the operating mode and the objectives to be reached. EC reactor configuration can be arranged based on three essential qualifications. The first is whether a reactor is designed as a batch or a continuous system, for example, the feed mode. For a nonstop framework, reactors are ceaselessly taken care of in water/wastewater and work, while the activity is done with a fixed wastewater volume for every treatment cycle in a bunch interaction (Mollah et al. 2001; Holt et al. 2005; Sahu et al. 2014; Hakizamana et al. 2017). The subsequent qualification is the strategy used to isolate the collected contaminations. The last one is the plan to calculate the electrode arrangement that characterizes the current dissemination in the cell.

   Apart from the feed mode and electrode arrangement, EC design also affects reactor working volume used to determine the area to volume ratio (A/V) through the geometry of the reactor. EC A/V ratio is one of the scale-up parameters in plant design that allows developing a full-scale EC reactor.

8. **Conductivity**: Conductivity of the water/wastewater that needs treatment will directly affect energy consumption. More conductivity implies less consumption of energy and vice versa. Therefore the conductivity of the solution is enhanced by adding various electrolytes like NaCl, KCl, KNO_3, NH_4Cl, etc.
9. **Temperature**: In electrochemical processes, pollutant removal efficiency depends on the temperature of the solution, and the increasing temperature results in the pollutant removal efficiency increases. However, at high temperatures, an increase in the solubility of the precipitates leads to the decreasing of the pollutant removal efficiency (Eyvaz et al. 2009; Vasudevan and Laxmi 2011; Sahu et al. 2014)

20.2.3 Ecomomic Aspect of EC Process

The cost effectiveness of any process is the significant factor for its use in the industry. In case of EC, its economic analysis would take fixed operation and capital cost. The fixed cost will comprise the cost of total equipment cost including electrolytic cell, pumps and cost of their set up. The operation cost includes the cost of chemicals, electrodes and energy consumption (Kabdaşlı et al. 2012; Segura et al. 2016; Hakizimana et al. 2017). Apart from this, cost pertaining to labour, maintenance, sludge dewatering and disposal should also be considered for accurate cost analysis. Studies have compared the cost of chemical coagulation and electrocoagulation. Moussa et al. 2017 compared the cost analysis of chemical coagulation and electrocoagulation of studies on wastewater treatment like textile, metallic, can manufacturing waste, dye house, tannery wastewater and cadmium removal. The comparison concluded that the cost of electrocoagulation was 1.3–1.5 times less than chemical coagulation. Various models have also been developed for cost estimation. Hakizimana et al. 2017 used a model to calculate the electrode material and energy consumption using chemical used, sludge generated per m^3 of wastewater treated, unit price of electrode, chemicals used, electrical energy used and sludge and handling and disposal charges. The conclusion form the model is that the cost generally vary depending on the unit price of material used for electrode and its arrangement. It was also concluded that EC was cheaper than the chemical coagulation depending upon the current density variation, but under low current density value it's more economical than chemical coagulation.

20.2.4 Advantage and Disadvantages of EC Process

The advantage of the EC process is as follows:

- EC is a simple and easy to operate method and handles most of the problems during running.
- It produces palatable, clear, colorless, and odorless water after wastewater treatment.
- The sludge formed by EC tends to be readily settable and easy to de-water because it mainly comprises metallic oxides/hydroxides. Above all, it is a common sludge-producing technique.
- Flocs formed by EC tend to be much larger, contain less bound water, are acid-resistant, and more stable than chemical floc, and therefore, can be separated faster by filtration.
- EC produces effluent with less total dissolved solids (TDS) content than chemical treatments, contributing to a low water recovery cost.
- The EC process generated no secondary pollution compared to chemical coagulation, and therefore, no issues of neutralizing excess chemicals.
- The gas bubbles produced during electrolysis can carry the pollutant to the top of the solution, where it can be more easily concentrated, collected, and removed.
- It requires less maintenance as electricity is provided in a controlled manner without any moving part. It can be used with the help of solar energy and hence can be conveniently used in rural areas.

Disadvantages of the EC process.

- It requires periodic removal of the sacrificial electrode due to its oxidation during operation.
- The use of electricity in many places may be expensive.
- It causes the passivation of the electrode due to the formation of an oxide layer on the cathode. However, this process does not cause rapid and may take sufficient time depending on the characteristics of the wastewater.
- High conductivity of the wastewater is needed for the smooth working of the process,

20.3 Application of EC for the Removal of Emerging Pollutants

EC has a broad application in domestic as well as in industrial wastewater treatment. It has been used in various industries like dye and textile, electroplating, pharmaceutical, food processing, tannery, petroleum, pulp and paper industry, etc. EC process removes not only the suspended particles and colloidal particles but is also proven effective in removing oil, heavy metals, and complex compounds present in the wastewater (Fig. 20.5).

Emerging pollutants are complex compounds that are not easily removed by conventional treatment methods. The EC process carries out some studies on the removal of EPs, such as commonly used pharmaceuticals. Table 20.1 describes the reactor design, parameters varied, electrodes used, and individual results by the various studies. In most studies, batch mode is preferred for the pollutant treatment

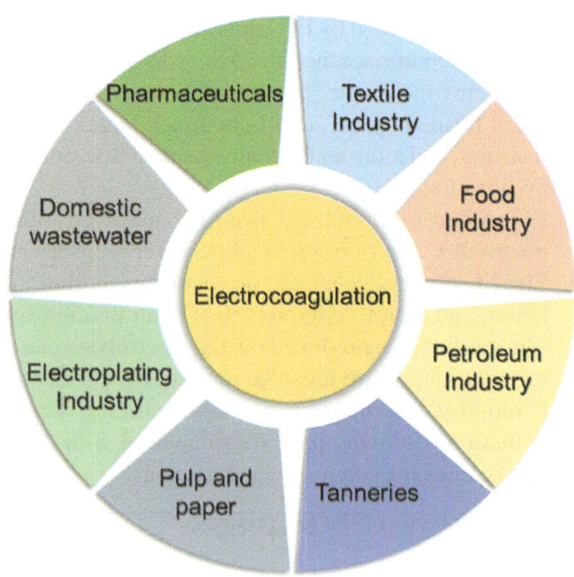

Fig. 20.5 Potential application of electrocoagulation process in various industries

Table 20.1 Studies on emerging pollutants removal by EC process

Pollutants	Reactor design	Electrodes used and size	Result	Remark	Author
Diclofenac (DCF), carbamazepine (CBZ) and amoxicillin (AMX)	Real municipal wastewater Batch mode Continuos supplt (CS), intermittent power supply (IPS)	Al 5 × 16.5 × 0.2 cm (anode) SS 6 × 16.5 × 0.2 cm (cathode) 300 rpm	CD- 1.8 mA/cm^2 Time- 19 h DCF- 44% (CS) 28% (IPS) CBZ- 40%(CS) 24% (IPS) AMX- 43% (CS) 28% (IPS)	An increase in initial concentrations reduces the removal efficiency	Ensano et al. (2019)
AMX	Batch reactor EC-NF	Al 11.9 × 4.7 × 0.4 cm	The sequential EC–NF model catered the best efficiency of 98.2 and 97.5% at the pH values of 2.5 and 10. EC (68.46% at pH 2.5)	Calcium containing electrolyte enhances the removal	Oulebsir et al. (2020)
(Diclofenac), a sulfonamide antibiotic (sulfamethoxazole) and a β-blocker (atenolol) All with the concentration of 100 mM	Batch reactor	Al Graphite Both with area 125 cm^2 and inter-electrode distance of 100 mm in a single cell and 24 mm in multiple cells	Removal efficiency (one cell system) SMX-75 DCF-68 ATE-55 Removal efficiency (5-cell system) SMX-99.8 DCF-100 ATE-85	When tested in a real wastewater sample, the results do not replicate the results obtained in the lab	Liu et al. (2020)
AMX	Batch reactor	(a) Iron as anode Ti as cathode (b) Graphite as anode (c) Reticulated vitreous carbon as anode	(a) 85% in 4 h (b) 10% in 60 min (c) 95% in 75 min	AMX with 30 ppm is used to create the solution Ultrasound waves seem to restrain the EC process	Padilla-robles et al. (2015)

(continued)

Table 20.1 (continued)

Pollutants	Reactor design	Electrodes used and size	Result	Remark	Author
Ciprofloxacin (CIP)	Batch	Al electrode 3 × 2 × 0.2 cm	CIP removal- 90.34% NaCl (0.07–0.1 M) 0.07 M IED—1 cm CD- 12.5 mA/cm^2 CIP- 32.5 mg/l pH- 7.78 reaction time – 20 min	The obtained results concludes sweep flocculation as a determinant step for CIP removal	Saeid et al. (2017)
CIP	Continuous	Two SS cathode CIP = 10 mg/l Al anode 6 × 2.7 cm	CIP – 88% Initial pH- 5.6 Temp- 25.5 Time- 100 min CD- 5.6 mA/cm^2 Flow rate- 25.9 mL/min	The second order kinetic model provides best fit and suggested that chemical adsorption mechanism controlled the CIP adsorption process	Mehdi et al. (2017)
CIP	Batch	Pure iron 10 × 5 × 0.2 IED 2 mm	CIP removal = 86.6 CD = 4.325 mA/cm^2 Process time = 10 min Initial CIP = 5 mg/l pH = 4	RSM seems to be an excellent step to predict the removal efficiency	Barısci and Turkay (2016)
Bisphenol A	Batch Artificial wastewater	Ti electrode	40% with nitrate 42% with bicarbonate 80% with bicarbonate in 20 min reaction time	In the absence of nitrate, 100% removal is observed	Jo et al. (2016)

as all the studies are under the test phase, and their results will be used to further the development of the treatment process.

Ensano et al. (2019) carried out the study for the removal of three pharmaceuticals, namely, Diclofenac (DCF), carbamazepine (CBZ), and amoxicillin (AMX). The study's objective is to analyze the viability of the electrocoagulation process in removing pharmaceuticals compounds from natural wastewater. The study is conducted in batch experimental runs using a simple laboratory-scale electrochemical reactor with aluminum and stainless steel or iron as anode and cathode. The effects of variation in current density (0.3, 0.5, 1.15, and 1.8 mA cm^{-2}), initial pharmaceutical concentration (0.01, 4, and 10 mg/l), electrolysis time (3, 6, and 19 h), and current application mode on the removal of pharmaceuticals were observed. The application of the current supply is carried out in continuous and intermittent mode (5 min on and 20 min off). The continuous mode gives better results than intermittent mode but uses 96% more energy than intermittent, making it costly. The increase in initial concentration (1000 times) also reduces the removal efficiency; the possible reasons for this would be low hydrophobicity of the selected pharmaceuticals or the competition for active sites on the coagulant surface between the impurities present and pharmaceuticals compounds. The variation in current density and electrolysis time was also observed; no significant changes are observed at high current density values (1.15 and 1.8 mA cm^{-2}) and high electrolysis duration (6 and 19 h). The reason explained is the formation of an oxidized layer on the electrodes, causing the reduction in the release of coagulant species and bubbles and thereby reducing pharmaceuticals removal efficiencies. The study concludes that the charge neutralization and electro flotation was the reason for AMX and DCF removal, while CBZ was removed due to its greater tendency in the coagulant species.

Another study by Liu et al. (2020), aimed to investigate the removal of the nonsteroidal anti-inflammatory drug (diclofenac, DCF), a sulfonamide antibiotic (sulfamethoxazole, SMX), and a β-blocker (atenolol, ATE) in synthetic wastewater by different types of electrodes, aluminum, and graphite. The removal of the pharmaceuticals was more significant in the graphite electrode system as compared to aluminum. It gives 100, 99.8, and 85% DCF, SMX, and ATE removal. In the graphite electrode system, the removal takes place through the electrooxidation process, and it follows two steps, one is direct oxidation of the pharmaceuticals on the surface electrode, and the second one was the indirect oxidation of the pharmaceuticals by the strong oxidants produced by the electrolysis of compounds like chloride and water. These strong oxidants help in degrading the pharmaceuticals. The study concludes that the removal efficiency was affected by the electrodes system and the nature of pharmaceuticals present. It also concludes that the degradation of the pharmaceuticals was effective as compare to the separation of medicines.

In some studies, more than 80% removal efficiency is obtained by applying a combination of current density, inter-electrode distance, pH, and initial pollutant combination. A series of experiments is carried out by applying various combinations to get the maximum efficiency possible from the EC process. In studies, initial concentration and current density play a vital role in the removal of the pollutant. One controls the amount of metal dissociates in the system and the other the amount adsorbed on the surface of flocs formed. A possible combination of both satisfying their amount is tried to obtain from these series experiments.

Effect of pH also plays a crucial role, some of the pollutants are only able to adsorb at a particular pH value, and sometimes the combination of pH and like AMX is removed at a pH value of 8–10, and as in this range, presence of OH^-

ions provides a competitive nature of the AMX ions and assist in its removal. It was also observed that the combination of pollutants affects the removal behavior of the pollutants, as seen in the case of SMX, DCF, and ATE removal. They have higher removal efficiency if treated alone, while in combination, their removal efficiency decreases as competitive nature between them and different practical pH values of removal. Another parameter that influences the removal of the pollutant is the electrode material. Mostly aluminum and the iron electrode are used in the process, and removal efficiency in the 20–80% is obtained. The use of titanium, graphite, and carbon electrodes has enhanced the removal process. The use of graphite and carbon electrode seems to be most effective in removing CIP from the system.

Another point that requires attention is that applying other treatment methods in combination with the EC process can be effective in EPs degradation. A study that uses the EC-NF model to treat AMX from the system gives good results. Other possible combinations could also provide an area to be studied. The negative point observed from these studies is that most of the studies cannot scale up the process, and the effect of EC in the natural wastewater has not been seen in any of the studies. The above indicates a gap in the area that needs to be fulfilled.

20.4 Future Challenges

Many EPs in the environment are not detected or reported due to lack of knowledge or low concentrations. The biggest challenge with EPs is their detection, identification, quantification, transformation during treatment, and presence in various environmental matrices; therefore, it is essential to gain knowledge on their occurrence and fate (Geission et al. 2015). The knowledge of the fate and occurrence of EPs in the lab and natural environment is still a significant concern and challenge, and the absence of known EPs in wastewater does not guarantee the absence of its transformative pollutants. The system's need is to build protocols for ecotoxicity determination under different organisms and different effects by using suitable endpoints (Gogoi et al. 2018). Future research should be more focused on the treatment of the EPs and their fate in the wastewater treatment plants and the natural environment (Deblonde et al. 2011).

The current state of the art methods focus on specific EPs classes and by far do not cover all the EPs of potential concern; some of the highly hazardous EPs that are monitored have meager detection limits to carry out the risk assessment (Geission et al. 2015). These hurdles can be taken care of by creating and managing a combined database for all EPs known to us and using software to analyze the results obtained. In the future, the focus should be made on the behavioral aspect of EPs and the development of a risk-based model for screening EPs.

20.5 Summary and conclusion

Water scarcity has led us to a situation where we are trying to reuse every drop of water we have. One of the prominent causes of this situation is the disposal of pollutants in the water bodies and land without any prior treatment. This has led to an increase in wastewater by disposing it on freshwater resources such as rivers, ponds, lakes, and groundwater. This causes the release of pollutants like heavy metals, organic compounds, microorganisms like bacteria, protozoa, and viruses in the environment that severely affect the ecosystem and, hence, the environment. Emerging pollutants are the new set of contaminants observed in this list. They are present in low concentrations and in such a form that is not easily detected by the various methods. Various studies focus on developing new methods of detecting this kind of pollutants and treating these pollutants.

One of the methods of treating such kinds of pollutants is electrocoagulation. Electrocoagulation is a technique that is similar to chemical coagulation but more effective than that, and it utilizes electricity and electrodes to do coagulation and floatation in the system. The mechanism of the process involves the generation of coagulant by metal dissociation, destabilization of colloid and floc formation, and removal of the pollutant from the system either by floatation or precipitation. The process has various operational parameters to optimize the result like electrode arrangements, current density, type of power supply, interelectrode distance, and other parameters like temperature, the concentration of ions, conductivity, and pH of the wastewater also influence the result. The process has advantages like low sludge production, ease of operation, bigger and fewer water flocs, no secondary pollutants, and less maintenance. It has some disadvantages like passivation of the electrode, replacement of electrodes, and costly treatment in areas of high electricity cost.

Studies were carried out on the removal of EPs by using electrocoagulation. The studies presented the applicability of electrocoagulation for the treatment of wastewater containing pharmaceuticals. The removal of the pharmaceuticals was attributed to the physicochemical characteristics of the pharmaceuticals. Various operational parameters were uses to remove the pharmaceuticals from the system, like current density, initial pharmaceutical concentration, electrolysis time, and electrode material. The results observed are positive and encourage the development of this technique to remove EPs from wastewater efficiently. There is still room for improvement in the process that can bring more effective results and bring more understanding of the removal nature of the EPs.

Acknowledgements The corresponding author acknowledges the funding received from Centre of Excellence in Oil, Gas & Energy, IIT Bombay for carrying out this work.

References

Barısci S, Turkay O (2016) Optimization and modelling using the response surface methodology (RSM) for ciprofloxacin removal by electrocoagulation. Water Science and Technology 73. https://doi.org/10.2166/wst.2015.649

Deblonde T, Cossu-Leguille C, Hartemann P (2011) Emerging pollutants in wastewater: a review of the literature. Int J Hyg Environ Health 214(6):442–448. https://doi.org/10.1016/j.ijheh.2011.08.002

Emamjomeh MM, Sivakumar M (2009) Review of pollutants removed by electrocoagulation and electrocoagulation/flotation processes. J Environ Manage 90(5):1663–1679. https://doi.org/10.1016/j.jenvman.2008.12.011

Ensano BMB, Borea L, Naddeo V, Belgiorno V, de Luna MDG, Balakrishnan M, Ballesteros FC (2019) Applicability of the electrocoagulation process in treating rea l municipal wastewater containing active pharmaceutical compounds. J Hazard Mater 361(August 2018):367–373. https://doi.org/10.1016/j.jhazmat.2018.07.093

Eyvaz M, Kirlaroglu M, Aktas TS, Yuksel E (2009) The effects of alternating current electrocoagulation on dye removal from aqueous solutions. Chem Eng J 153(1–3):16–22. https://doi.org/10.1016/j.cej.2009.05.028

Garcia-Segura S, Eiband MMSG, de Melo JV, Martínez-Huitle CA (2017) Electrocoagulation and advanced electrocoagulation processes: a general review about the fundamentals, emerging applications, and its association with other technologies. J Electroanal Chem 801:267–299. https://doi.org/10.1016/j.jelechem.2017.07.047

Gavrilescu M, Demnerová K, Aamand J, Agathos S, Fava F (2015) Emerging pollutants in the environment: present and future challenges in biomonitoring, ecological risks, and bioremediation. New Biotechnol 32(1):147–156. https://doi.org/10.1016/j.nbt.2014.01.001

Geissen V, Mol H, Klumpp E, Umlauf G, Nadal M, van der Ploeg M, van de Zee SEATM, Ritsema CJ (2015) Emerging pollutants in the environment: a challenge for water resource management. Int Soil Water Conserv Res 3(1):57–65. https://doi.org/10.1016/j.iswcr.2015.03.002

Gogoi A, Mazumderb P, Tyagic VK, Chamindad GGT, Ane AK, Kumar M (2018) Occurrence and fate of emerging contaminants in water environment: A review. Groundwater for Sustainable Development 6:169–180. https://doi.org/10.1016/j.gsd.2017.12.009

Hakizimana JN, Gourich B, Chafi M, Stiriba Y, Vial C, Drogui P, Naja J (2017) Electrocoagulation process in water treatment: a review of electrocoagulation modeling approaches. Desalination 404:1–21. https://doi.org/10.1016/j.desal.2016.10.011

Hermes N, Jewell KS, Wick A, Ternes TA (2018) Quantification of more than 150 micropollutants including transformation products in aqueous samples by liquid chromatography-tandem mass spectrometry using scheduled multiple reaction monitoring. J Chromatogr A 1531:64–73. https://doi.org/10.1016/j.chroma.2017.11.020

Holt PK, Barton GW, Mitchell CA (2005) The future for electrocoagulation as a localised water treatment technology. Chemosphere 59(3):355–367. https://doi.org/10.1016/j.chemosphere.2004.10.023

Houtman CJ, ten Broek R, Brouwer A (2018) Steroid hormonal bioactivities, culprit natural and synthetic hormones and other emerging contaminants in wastewater measured using bioassays and UPLC-tQ-MS. Sci Total Environ 630:1492–1501. https://doi.org/10.1016/j.scitotenv.2018.02.273

Islam SMDU (2017) Electrocoagulation (EC) technology for wastewater treatment and pollutants removal. Sustain Water Resour Manage 5(1):359–380. https://doi.org/10.1007/s40899-017-0152-1

Jo E-Y, Lee T– K, Kim Y, Park CG (2016) Effect of anions on the removal of bisphenol A in wastewater by electro-oxidation process. Desalination and water treatment, 57:29500–29508. https://doi.org/10.1080/19443994.2016.1197156

Kabdaşlı I, Arslan-Alaton I, Ölmez-Hancı T, Tünay O (2012) Electrocoagulation applications for industrial wastewaters: a critical review. Environ Technol Rev 1(1):2–45. https://doi.org/10.1080/21622515.2012.715390

Kapelewska J, Kotowska U, Karpińska J, Kowalczuk D, Arciszewska A, Świrydo A (2018) Occurrence, removal, mass loading and environmental risk assessment of emerging organic contaminants in leachates, groundwaters and wastewaters. Microchem J 137:292–301. https://doi.org/10.1016/j.microc.2017.11.008

Lapworth DJ, Baran N, Stuart ME, Ward RS (2012) Emerging organic contaminants in groundwater: a review of sources, fate and occurrence. Environ Pollut 163:287–303. https://doi.org/10.1016/j.envpol.2011.12.034

Liu YJ, Huang YL, Lo SL, Hu CY (2020). Comparing the effects of types of electrodes on the removal of multiple pharmaceuticals from water by electrochemical methods. Water (Switzerland):12(9). https://doi.org/10.3390/W12092332

Mehdi Y, Saeid A, Aghasi M, Dolatabadi M (2017). Optimization of electrocoagulation process for efficient removal of ciprofloxacin antibiotic using iron electrode; kinetic and isotherm studies of adsorption. J Mol Liq 225:544–543. http://dx.doi.org/10.1016/j.molliq.2016.11.093

Mollah M, Yousuf A, Schennach R, Parga JR, Cocke DL (2001) Electrocoagulation (EC). science and applications. J Hazard Mater 84, 29–41

Mollah MYA, Morkovsky P, Gomes JAG, Kesmez M, Parga J, Cocke DL (2004) Fundamentals, present and future perspectives of electrocoagulation. J Hazard Mater 114(1–3):199–210. https://doi.org/10.1016/j.jhazmat.2004.08.009

Moussa DT, El-Naas MH, Nasser M, Al-Marri MJ (2017) A comprehensive review of electrocoagulation for water treatment: potentials and challenges. J Environ Manage 186(1):114886. https://doi.org/10.1016/j.desal.2020.114886

Oulebsir A, Chaabane T, Tounsi H, Omine K, Sivasankar V, Flilissa A, Darchen A (2020) Treatment of artificial pharmaceutical wastewater containing amoxicillin by a sequential electrocoagulation with calcium salt followed by nanofiltration. J Environ Chem Eng 8:104597. https://doi.org/10.1016/j.jece.2020.104597

Padilla-robles BG, Alonso A, Martínez-delgadillo SA, González- M (2015) Chemical engineering and processing : process intensification electrochemical degradation of amoxicillin in aqueous media. Chem Eng Process 94:93–98. https://doi.org/10.1016/j.cep.2014.12.007

Pavithra KG, Senthil Kumar P, Sundar Rajan P, Saravanan A, Naushad M (2017) Sources and impacts of pharmaceutical components in wastewater and its treatment process: a review. Korean J Chem Eng 34(11):2787–2805. https://doi.org/10.1007/s11814-017-0255-2

Postigo C, Barceló D (2015) Synthetic organic compounds and their transformation products in groundwater: occurrence, fate and mitigation. Sci Total Environ 503–504:32–47. https://doi.org/10.1016/j.scitotenv.2014.06.019

Prado M, Borea L, Cesaro A, Liu H, Naddeo V, Belgiorno V, Ballesteros F (2017) Removal of emerging contaminant and fouling control in membrane bioreactors by combined ozonation and sonolysis. Int Biodeterior Biodegradation 119:577–586. https://doi.org/10.1016/j.ibiod.2016.10.044

Rivera-Utrilla J, Sánchez-Polo M, Ferro-García MÁ, Prados-Joya G, Ocampo-Pérez R (2013) Pharmaceuticals as emerging contaminants and their removal from water. A review. Chemosphere 93(7):1268–1287. https://doi.org/10.1016/j.chemosphere.2013.07.059

Rosal R, Rodríguez A, Perdigón-Melón JA, Petre A, García-Calvo E, Gómez MJ, Agüera A, Fernández-Alba AR (2010) Occurrence of emerging pollutants in urban wastewater and their removal through biological treatment followed by ozonation. Water Res 44(2):578–588. https://doi.org/10.1016/j.watres.2009.07.004

Saeid A, Asadipourc A, Pournamdaric, Behnamb B, Rahimid HR, Dolatabadie M (2017) Removal of ciprofloxacin from hospital wastewater using electrocoagulation technique by aluminum electrode: Optimization and modelling through response surface methodology. Process Saf Environ Prot 109:538–547. http://dx.doi.org/10.1016/j.psep.2017.04.026

Sahu O, Mazumdar B, Chaudhari PK (2014) Treatment of wastewater by electrocoagulation: a review. Environ Sci Pollut Res 21(4):2397–2413. https://doi.org/10.1007/s11356-013-2208-6

Starling MCVM, Amorim CC, Leão MMD (2019) Occurrence, control and fate of contaminants of emerging concern in environmental compartments in Brazil. J Hazard Mater 17–36. https://doi.org/10.1016/j.jhazmat.2018.04.043

Vasudevan S, Lakshmi J (2011) Effects of alternating and direct current in electrocoagulation process on the removal of cadmium from water—a novel approach. Sep Purif Technol 80(3):643–651. https://doi.org/10.1016/j.seppur.2011.06.027

Verlicchi P, Galletti A, Petrovic M, BarcelÓ D (2010) Hospital effluents as a source of emerging pollutants: an overview of micropollutants and sustainable treatment options. J Hydrol 389(3–4):416–428. https://doi.org/10.1016/j.jhydrol.2010.06.005

Vik EA, Carlson DA, Eikum AS, Gjessing ET (1984) Electrocoagulation of potable water. Water Res 18(11):1355–1360. https://doi.org/10.1016/0043-1354(84)90003-4

Chapter 21
Application of Microbial Fuel Cells for the Treatment of Emerging Contaminants from Wastewater: An Overview

Bhavana Kanwar, Najmul Haque Barbhuiya, Indrajeet Vijay Kharade, Amritanshu Shriwastav, and Swatantra P. Singh

Abstract With the growing energy crisis, industrialization, rapid urbanization, and increased population, it is necessary to sort out environmental pollution and energy crisis issues. The detection of newly identified or emerging contaminants (ECs) into our aquatic environment is of serious concern for the health and safety of the whole ecosystem, and the existing conventional wastewater (WW) treatment processes are not designed to treat these unidentified contaminants. In this context, Bio-electrochemical systems such as microbial fuel cells (MFCs) are considered a prospective technology in removing these ECs and electricity generation from WW through microbial metabolisms. The MFC unit includes an anode, cathode, cation-sensitive membrane, and an external wire. MFCs have shown significant advantages, including converting substrate into energy, operating at wide ranges of temperature and pH with diverse biomass, generating a meager amount of activated sludge from WW, and zero energy requirement for aeration. Additionally, MFCs have been seen as a resolution for water and energy issues due to their capability to treat WW and electricity generation. However, MFCs have numerous challenges in-field applications, such as turbulence in each compartment, membrane resistance in the proton transportation process, etc. This chapter focuses on the application of MFCs towards the removal of various ECs from the aqueous environment. Applications of MFC technology can also be worked in its power generation ability and sustainable energy generation. Although the current applications of MFC technology are still at the laboratory level, it has a great potential for commercial applications in the near future.

Keywords Microbial fuel cell · Activated sludge · Emerging contaminants · Microbial metabolisms · Antibiotics · Wastewater

---

B. Kanwar · N. H. Barbhuiya · I. V. Kharade · A. Shriwastav · S. P. Singh (✉)
Environmental Science and Engineering Department, IIT Bombay, Mumbai, India
e-mail: swatantra@iitb.ac.in

S. P. Singh et al. (eds.), *New Trends in Emerging Environmental Contaminants*, Energy, Environment, and Sustainability,
https://doi.org/10.1007/978-981-16-8367-1_21

21.1 Introduction

Population growth with rapid urbanization and industrialization are the active reasons for the rapid generation of wastewater (WW). The WW is in general, is the water supply after it has been used for a variety of purposes (Bose et al. 2018c). With industrialization worldwide, WW of enormous volumes is being pumped into rivers, streams, and the sea itself. The effect of dumping this untreated/partially treated WW is extreme besides the harm it can cause to the marine environment and fisheries. The WW includes mostly water (99.9%) together with smaller amounts of suspended and dissolved organic and inorganic solids (Bose et al. 2020). The basic organic substances that are present in WW are fats, proteins, starches, soaps, cleansers, with their disintegrations and various other environmental pollutants viz. synthetic dyes, phenolic compounds, antibiotics, nitrogen-based compounds, pesticide, ECs, trace organic compounds etc. (Mandal and Das 2018; Bruggen et al. 2003). In order to prevent the damaging effects of WW on the environment and aquatic ecosystem, treatment of all the contaminants to meet defined standards is a necessity (Bose et al. 2020; Kumar and Bishnoi 2017; Logan et al. 2006).

WW treatment plants must ensure that the WW is treated properly before the subsequently treated water is released/discharged to land or water. A WW treatment process comprises a series of unit operations designed to degrade/remove WW contaminants to an acceptable level as permitted in the standards. Several configurations and combinations of unit operation are possible to make up for a treatment process, yet many standard technologies have been developed. Mainly the conventional WW treatment process involves primary, secondary, and tertiary stages. Primary treatment involves the removal of inorganic particles at the beginning of the process, while secondary treatment involves the degradation of organic waste via biological conversion. The final tertiary treatment removes suspended impurities, nutrients, or disinfectants before discharging the WW (Bose et al. 2020). Among the available technologies of WW treatment in the field, membrane separation has proven effective for treating these effluents, and filtration processes such as ultrafiltration and nanofiltration in conjunction with reverse osmosis have increased importance (Bose et al. 2020; Gao et al. 2017). However, these processes can be successfully integrated with biological methods in a novel way using MFCs and may contribute to environmental protection while supporting the water infrastructure (Gao et al. 2017).

Over the last few decades, MFCs have achieved immense universal significance. In the bioremediation process, MFCs utilize bacteria to oxidize organic and inorganic substances at the anode along with the bioelectricity generation (He et al. 2015). The main advantages of MFC technology include conversion of substrate energy directly into electricity, it can work in a wide range of operating temperatures, less activated sludge is produced in the process, no energy is required for aeration, and doesn't require gas treatment (He et al. 2015; Mandal and Das 2018). Generally, the domestic WW has high organic matter dissolved in sewage ranging from 4.92 to 7.97 kWh/kg COD, which is a potential source of energy generation (Du et al.

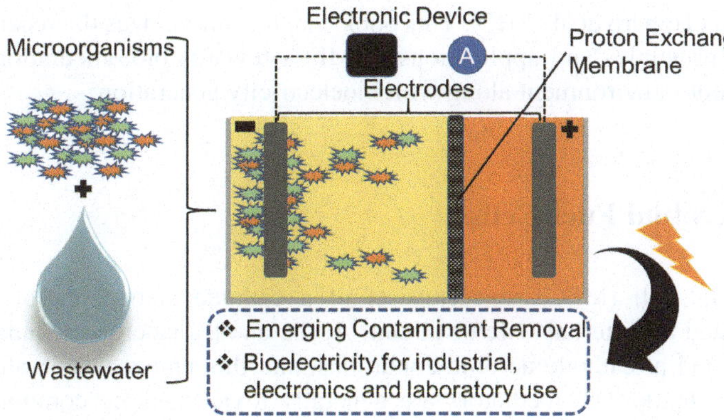

Fig. 21.1 Schematic of MFCs

2007). Figure 21.1 shows various applications of MFCs with the microbes and WW substrates for bioelectricity production.

MFCs improve the growth of bio-electrochemically active microbes during the treatment process, demonstrating good operating system stability. In some cases, up to 90% of the COD removal with columbic efficiency of 80% have been reported (Kim et al. 2005). The efficiency in removal rate of pollutants achieved in the MFC is similar to those with industrial WW treatment systems without large energy consumption. MFCs result in reduced sludge handling/disposal costs as these remove up to 90% BOD and COD with 50–90% lesser sludge production as compared to other bioprocesses (Du et al. 2007). Liu et al. (2004); in their single prototype chamber, microbial fuel cell reactor removed 80% of the COD of the wastewater and generated electrical power up to a maximum of 26 mWm^{-2}. COD removal and power generation were a function of hydraulic retention time (HRT), and results showed that COD removal increased from 40% at HRT 3 h to a maximum of 80% at HRT of 33 h and generated power up to a maximum of 26 mWm^{-2}. The buildup of volatile fatty acids does not negatively affect working since MFCs use fatty acids like acetate and butyrate as feedstock. The MFC technology has an eco-friendly approach when compared to conventional technologies in controlling environmental pollution and can be a potential substitute for the traditional activated sludge process (Do et al. 2018).

Based on the contaminants to be removed, various kinds of WW can be used as feed for the cathode/anode chambers, including municipal WW, food processing factories' WW, dairy industry WW, palm oil mill effluent, recalcitrant pharmaceutical industrial effluent, petrochemical industry WW, etc. (Boghani et al. 2017; Khan et al. 2017; Mansoorian et al. 2016; Morris and Jin 2012; Nor et al. 2015; Zhang et al. 2015). MFC technology has also shown substantial improvements in terms of various aspects associated with the electrodes, including high specific surface area, biocompatibility, surface roughness, and electrons transfer capacity. Full-scale implementation of MFC depends on ambient temperature, atmospheric pressure, and near-neutral pH condition (Trapero et al. 2017). This book chapter summarizes the recent studies on working principles and applications of MFCs towards bioremediation of ECs from an aqueous environment along with bioelectricity generation.

21.2 Microbial Fuel Cells

Microbial fuel cell (MFC) unit consists of a cathode, anode, cation-selective membrane, and an external wire or circuit. At the anode, anaerobic conditions are maintained, and at cathode aerobic conditions, and this unit may operate with or without a mediator. The mediators are generally toxic phenolic compounds that exchange electrons with fuels or oxidants at the reaction sites of the biocatalysts, diffuse to the surface of the electrode, and exchange electrons there. The mediated fuel cells are the addition of external bacteria to oxidize the substrate (Saeed et al. 2015). In the anodic chamber, microorganisms are supplied with a suitable substrate that is anaerobically degraded and releases electrons. The external circuit transport these released electrons from anode to cathode and the protons generated are passed selectively across the membrane. To regulate the flow of proton from anode to cathode, a proton exchange membrane (PEM) is used. For the adhesion of PEM to the cathode terminal, hydrogels or ion exchange resins are used. The rich substrate fed to the anode chamber function as a fuel to the biocatalysts presented inside the chamber. Bacteria decompose the organic compounds, electrons generated are transported directly or indirectly to the electrodes by mediator's help and then travel to the cathode chamber with the outer circuit. This process of flow of electrons from the external circuit provides the basis for bioelectricity generation (Fig. 21.2). Simultaneously in the anodic chamber, protons are generated during the organic

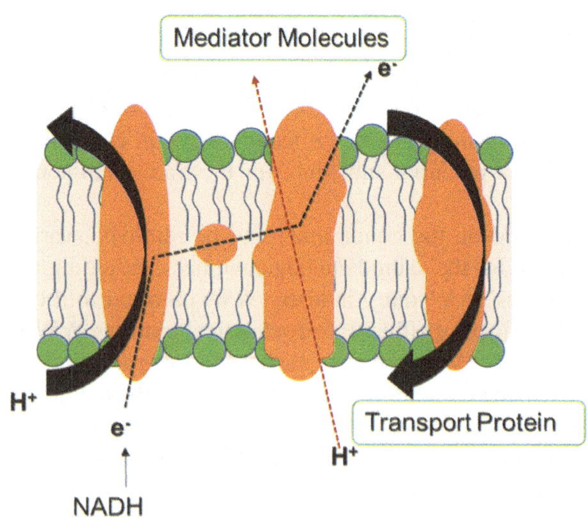

Fig. 21.2 Electron transfer chain in bacterial cell wall for energy productions in MFC

matter degradation and pass through the PEM to the cathode chamber. Lastly, this transferred proton reacts with O_2, ferric chloride, potassium dichromate, potassium ferricyanide, etc., to produce water (Bose et al. 2020; Kumar et al. 2017; Mandal and Das 2018). The following equations can give the reactions taking place in an MFC:

$$\text{Reactions at the anode}: (CH_2O)_n + nH_2O \rightarrow nCO_2 + 4ne^- + 4nH^+ \quad (21.1)$$

$$\text{Reactions at the cathode}: O_2 + 4ne^- + 4nh^+ \rightarrow 2H_2O \quad (21.2)$$

The main working principle of MFC's is reliant on the use of microorganisms as biocatalysts at cathode and anode chambers. For achieving the desired results, various configurations (as shown in Fig. 21.3) of the reactors and materials are used. These systems are susceptible to the environment and operating conditions too. Some of these conditions are substrate type, electrode material, reactor configuration, and temperature. The change in working temperature directly affects the hydrogen production and COD removal in MFCs; therefore, temperatures between 30 and 40 °C are ideal for MFCs. It has been found that the highest COD removal is achieved in the range of 25–35 °C (Li and Chen 2018). When the increase in temperature is to the extreme, the overall microbial activity decreases. One of the advantages of MFCs is that they can withstand temperatures change without interfering with much power generation and WW treatment. Studies have shown that a temperature range between 20 and 35 °C can change power generation and hydrogen production in MFCs only by 10% (Li and Chen 2018). This temperature factor helps in the real-world manufacturing of MFCs. In the last few decades, MFC technology has been substantially improved. However, it has met numerous challenges in practical scale-up and commercial application, such as turbulence in each compartment and membrane resistance in the proton transportation process (Sharma and Li 2010). The other constraint of MFCs is that they cannot function at very low temperatures owing to the fact that microbial reactions are slow at low temperatures (Chen et al. 2008).

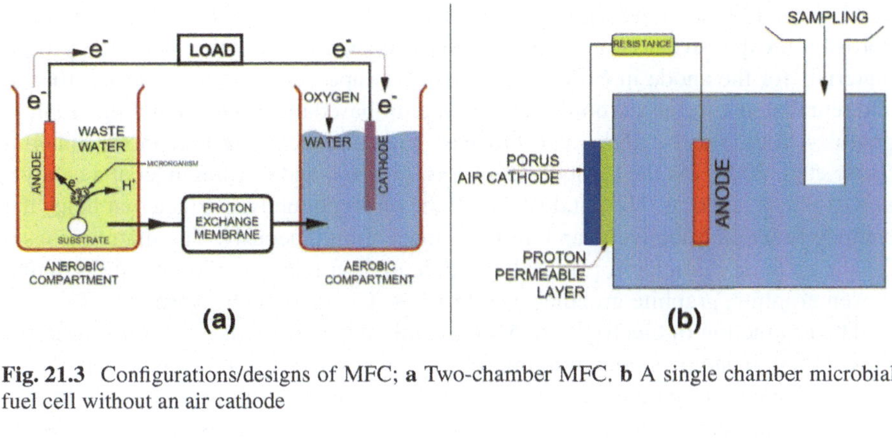

Fig. 21.3 Configurations/designs of MFC; **a** Two-chamber MFC. **b** A single chamber microbial fuel cell without an air cathode

The microorganism's capacity to degrade the complex substrate depends on the constituents of the microbial community and its adaptation. Various kinds of microbial communities have been used for bioelectricity generation through the MFC. As a catalyst, exoelectrogens are used for organic biomass degradation and electricity generation. Around 35 pure cultures have been reported as exoelectrogens in MFCs, for example, *Geobacter*, *Rhodoferax, Shewanella*, etc. (Koók et al. 2018). Various other microorganisms from archaea (methanogens), fungi, and bacteria have been used in MFCs (Koók et al. 2018). It is observed that pure cultures are useful in MFCs, and in these pure cultures electrogenic microorganisms could be easily monitored and analyzed (Kim et al. 2002). Most of the MFC studies are done with pure bacterial cultures due to their high electron transfer efficiency as compared to mixed cultures. However, pure cultures are confined to the low growth rate, have low energy transfer rate, need high substrate specifications, and there is always a continuous risk of contamination of unwanted microbes (Bond and Lovley 2003; Kim et al. 2002).

Microbial colonies are also used in MFCs, and these microbial colonies are biofilms consists of a distinct or ideal group of microbes. Mixed cultures are appropriate for bioelectricity generation due to the interaction between microorganisms (Kumar et al. 2017). Mixed microbial community members assist as well as compete for carbon and energy. That helps in the biodegradation of a complex raw influent. In mixed culture, each species can perform its specific function. The function performed by one species of microorganisms may be crucial for the other species. To overcome this problem, pure as well as mixed cultures are used in the MFCs. This way purpose of MFCs for bioelectricity generation and effluent treatment is fulfilled (Kumar et al. 2017).

Various types of electrodes have been developed and examined in recent years for MFCs. In MFCs, the electrodes are conductors as well as habitats for electrogenic microbes. Electrodes used in MFCs have features like high specific surface area, biocompatibility, considerable surface roughness, and capacity to effectively transfer electrons between microbes and electrodes (Kumar et al. 2019). Transfer of electrons to the anode by microbes is via direct contact or chemical mediator. The anodic material is selected based on its electrically conductive, ease of availability, scalability to large sizes, non-corrosive nature, high specific surface area and porosity, inexpensiveness, and less fouling propensity. The most commonly used materials for the anode in MFC include carbon paper, carbon cloth, carbon foams, and reticulated vitreous carbon (RVC). Graphite-based materials, including graphite granules, rods, sheets, fibers, and brushes, have also been used as anode material (Bose et al. 2020). At the cathode, electrons, protons, and oxygen meet at a catalyst in a tri-phase reaction. The catalyst should be on a conductive surface that helps the cathode function effectively, and all must meet simultaneously. The materials used for the cathode fabrication primarily include carbon paper, carbon cloth, graphite, woven graphite, graphite granules, and brushes (Chaturvedi and Verma 2016).

The production of electricity in MFC mainly depends on the substrates used for the cultivation and metabolism of microbes. The organic substrates fed to MFCs are pure substances like glucose, lactose, acetate, etc., chemical substances like cellulose, fatty acids, protein, sugar, etc. The organic substrate also consists of different WWs generated from domestic, industrial, and agricultural activities. Substrates contain various pollutants, such as biodegradable organic matter, nitrogen, phosphorous, suspended solids, detergents and soaps, fertilizers, pesticides, heavy metals, and other inorganic compounds. Microbes generate electricity by using these substrates as feed (Xiao and He 2014).

In MFCs, the energy conversion is from chemical energy to electrical energy through the oxidation of organic matter by electrochemically active bacteria. Different designs for the construction of MFCs are mainly based on the chambers viz. single-chambered MFC, stacked MFC, and two-chamber MFC (Tharali et al. 2016). In a single-chamber MFC, the anodic chamber is coupled with an air cathode upon which the protons and electrons are transferred. The dual-chamber design of MFC consists of chambers partitioned by an ion-exchange membrane giving it a classical design for MFC. In contrast, a stacked MFC has connections in series or in parallel with each other for a combination of MFCs (Tharali et al. 2016). MFCs use biological organisms to produce electricity from the feed WW along with pollutants removal. Therefore, this process brings down COD and biological oxygen demand (BOD) of WW along with the production of bioelectricity (Bose et al. 2018a). MFC can successfully be integrated with standard as well as advanced WW treatment technology to attain desired treatment instead of processing a single technology (Bose et al. 2018a, 2018b).

The performance of MFCs is mainly influenced by several factors such as subsequent supply and consumption of oxygen in the cathode chamber as MFCs require a continuous supply of organic matter. Oxidation of substrates in anode chamber as electrons are generated when substrate oxidation takes place by bacteria, affecting the reaction rate in MFCs and the growth of bacteria. There should be a proper electron shuttle from anode compartment to anode surface and permeability of proton exchange membrane (Chaturvedi and Verma, 2016; Rahimnejad et al. 2011). The most used PEM is Nafion due to its highly selective permeability of protons (Sharma and Li, 2010). The MFCs appear to be encouraging as a possible approach to treating contaminants and energy recovery in the environment.

21.3 Emerging Contaminants

The newly identified ECs are rapidly occurring in our water resources, and these are not considered to be good for the health and safety of the consuming environment (Sun et al. 2012; Wang et al. 2017; Yu and Chu 2009). These ECs consist of various pharmaceutical compounds, surfactants, personal care products, industrial additives, and numerous chemicals claimed to be endocrine disrupters. The current conventional WW treatment plants are not appropriately designed to remove these ECs, posing a serious threat to the water supply network. The existing WW treatment system is inefficient in removing the ECs as these new pollutants have not been monitored due to a lack of stringent regulations. These pollutants are being dumped, knowingly or unknowingly, in the aquatic environment that affect the living organism and the ecosystem with it (Bolong et al. 2009; Yu and Chu 2009; Sun et al. 2012). MFCs application in the bioremediation of various kinds of ECs present in the WW has been discussed below.

Antibiotics in pharmaceutical compounds as ECs are unique medicines that fight bacterial infections in people and animals. They act selectively on pathogenic bacteria without affecting the human cells and tissues. The antibiotics into the natural water bodies mainly come from the treated or poorly treated effluents of municipal sewage treatment plants (STPs) and pharmaceutical manufacturing industries. As per Michael et al. (2013), urban sewage or industrial treatment plants are more likely to be hotspots for the discharge of antibiotics and antibiotic-resistant genes (ARGs) in the environment (Michael et al. 2013). Antibiotics in pharmaceutical chemicals are considered toxic, hazardous, and recalcitrant bio-accumulative compounds and acknowledged as ECs (Chen and Zhou 2014; Michael et al. 2013).

Chloramphenicol (CAP), a broad-spectrum nitroaromatic antibiotic, is present in the medical practice on a larger scale as it was the first synthetic antibiotic introduced (Feder et al. 1981). CAP is cost-effective, and since its introduction, this antibiotic has served global application in the animal breeding industry. But in several developed countries, CAP has been banned for its use in animals foods and production because of aplastic anemia and its severe toxicity in animals and humans (Feder et al. 1981; Martelli et al. 1991). However, in many developing countries, CAP is still widely used due to its affordable price and is often detected in the effluent WW of breeding farms. Zhang et al. (2017) investigated the MFC performance in terms of CAP removal with 50 mg/L and showed 84% degradation within 12 h (Zhang et al. 2017). In this system, acetate significantly removed CAP as the electron donor and improved the CAP removal rate. In addition, the antibacterial activity test showed reduced biotoxicity of CAP towards *Escherichia coli* (ATCC 25,922) and *Shewanella oneidensis* (MR-1). MFC promoted the growth and multiplication of electrogenic bacteria as depicted by microbial community composition, and the electrons supplied by electrogenic bacteria are important for the increased CAP removal rate (Zhang et al. 2017).

Sulphamethoxazole (SMX) is an antibiotic largely consumed in the breeding industry due to its cheap and effective behavior. SMX is not easily biodegraded and is broadly present in surface water and the effluent of WW treatment plants (Batt et al. 2006; Clara et al. 2005; Larcher and Yargeau 2012; Wei et al. 2011). Wang et al. (2016) showed the successful removal of SMX and its degradation product, 3-amino-5-methylisoxazole (3A5MI), in MFC reactors (Wang et al. 2016). The process removed around 85% of 20 ppm SMX within 12 h, and this biodegradation rate is rapid than any other, as reported previously in the literature (Avisar et al. 2010; Gómez-Ramos et al. 2011). The reduction in biotoxicity has been observed in the antibacterial activity test of SMX towards *Shewanella oneidensis* MR-1 and *Escherichia coli* DH5α after MFC treatment, while these microorganisms have been severely inhibited by SMX (Fig. 21.4). Wang et al. (2016) concluded that MFCs enhance the removal of refractory contaminants from the environment by providing a long acclimation period and enhancing microbes' degradation ability towards recalcitrant chemicals (Wang et al. 2016).

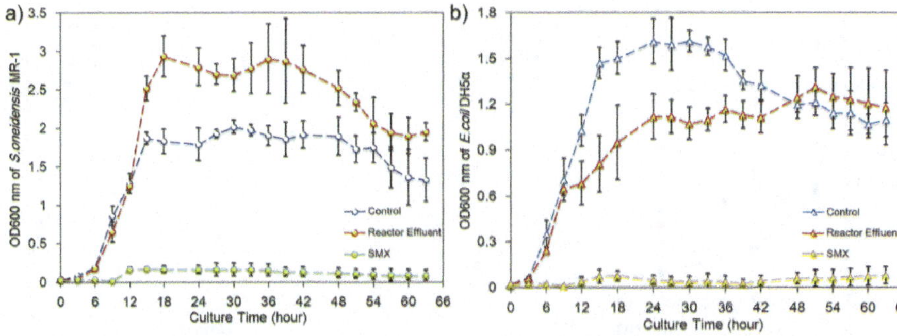

Fig. 21.4 Growth curve of *S. oneidensis* MR-1 (**a**) and *E. coli* DH5α (**b**) in the control group, MFC effluent group and SMX group. Adapted from (Wang et al. 2016), with permission from Elsevier

Acetaminophen (ACTP) or acetyl para-aminophenol (APAP), popularly named as Paracetamol (PAM), is a common over-the-counter painkiller and antipyretic. PAM is safe for use at the prescribed dose, but even small overdoes can do a fatal liver breakdown (Joss et al. 2005). With this common over-the-counter strategy and its general application in pharmaceutical industries and daily life, paracetamol has established itself as an important WW contaminant. Zhang et al. (2015); used an MFC-Fenton system with dual-chamber MFC reactors for the bio-electrochemical degradation of PAM, and Fenton reactions have been introduced to MFCs in the absence of an external power supply (Zhang et al. 2015). In the cathode chamber of an MFC, hydroxyl radicals mainly mineralized PAM in the presence of iron and oxygen. In the anode chamber, in-situ generation of free hydroxyl radicals got enhanced by bio-electrons generated from pollutant oxidation. When compared to the conventional Fenton reactions system, no requirement of continuous addition of reagent has been observed in the MFC-Fenton system. The degradation efficiency of PAM was found to be 70% within 9 h at initial pH 2.0, external resistance of 20Ω, and a total iron concentration of 5 mg/L. In addition, complete mineralization of 25% PAM has been observed, and the majority has been mainly converted to less toxic intermediate metabolites of p-nitrophenol via p-aminophenol and dicarboxylic/carboxylic acids. All these findings suggested that the MFC-Fenton system could be applied as an energy-saving and effective treatment for PAM-containing WW (Zhang et al. 2015).

For the application of MFCs in the ECs removal, Wang et al. (2017) studied the removal efficiency of four ECs, including bisphenol A (BPA), estrone (El), sulfamethazine (SM2), and triclocarban (TCC) in the MFC-Fenton system, owing to their regular detection in WW. Wang et al. (2017) used glucose as the substrate and graphite rod with stacked graphite granules as the electrode, as these increase cathodic H_2O_2 synthesis and improve the current density of the MFC system (Wang et al. 2017). This MFC-Fenton integrated system showed efficient removal of the selected ECs within 24 h in batch as well as continuous modes for open and closed circuits (Fig. 21.5). An open circuit is an electric circuit in which current does not flow, whereas a closed-circuit means a complete electrical connection around which currently circulates. In

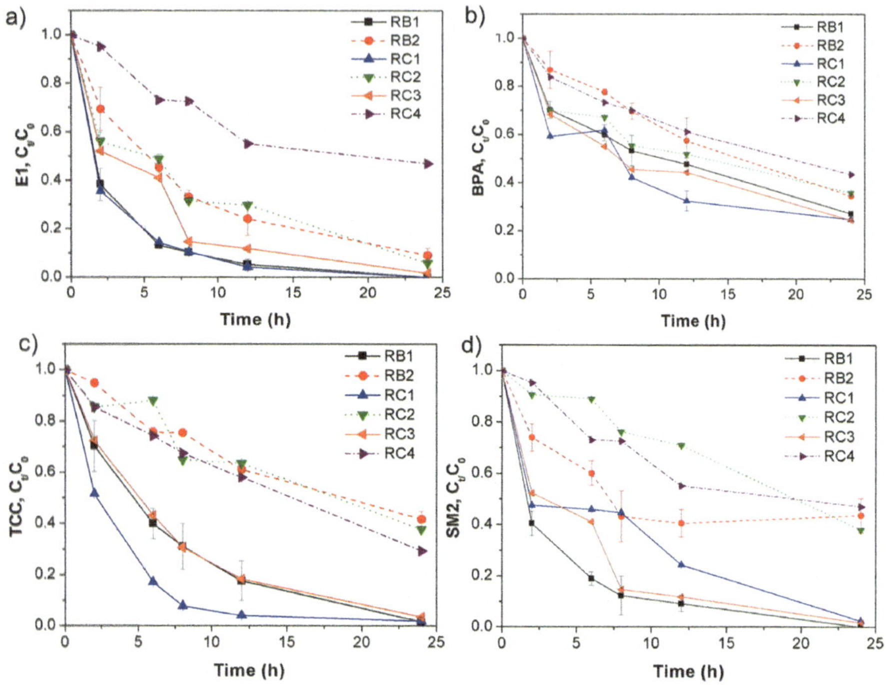

Fig. 21.5 The degradation of four ECs ($C_0 = 1$ mg/L) **a** E1, **b** BPA, **c** TCC and **d** SM2 in batch and continuous operation MFCs. RB1: batch MFC operated with close circuit; RB2: batch MFC operated with open circuit; RC1: continuous MFC operated with close circuit; RC2: continuous MFC operated with open circuit; RC3: continuous MFC operated with close circuit with control of pH in the cathodic electrolyte at 3.0; RC4: continuous MFC operated with open circuit with control of pH in the cathodic electrolyte at 3.0. Adapted from (Wang et al. 2017), with permission from Elsevier

this system, closed circuits resulted in greater potential for bioremediation of ECs when compared to the open circuit, as shown in Fig. 21.5. The ECs' decrease was attributed to the Fenton reaction of in-situ generated H_2O_2 and the addition of Fe^{2+} in the cathodic chamber to form OH radicals that is effective oxidizers to degrade organic pollutants and the adsorption onto the graphite. Wang et al. (2017); found that pseudo-first-order kinetics (Eq. 21.3) has been followed for ECs removal. In comparison with control groups with open circuits, the removal rate constants of most ECs with the close circuit as RBI, RC1, and RC3 were significantly on the higher side regardless of the operation modes or pH control. These findings suggest that the Fenton process significantly improved ECs removal in closed circuits. The study stated that with the involvement of the two-removal mechanism viz. adsorption and Fenton process, the overall reduction might differ depending on the various ECs and their reactivity in Fenton reactions. Also, pH-controlled continuous flow MFC with cathodic electrolyte exhibited improvement on in-situ production of H_2O_2 and should be taken into consideration in the prospective application of the MFC-Fenton system (Wang et al. 2017).

The literature says that different ECs have different reactivity towards chemical oxidation (Marie Deborde et al. 2005); therefore, the involvement of the MFC-Fenton process in the removal of EC will vary with different ECs dependent on the ECs reactivity in the Fenton reaction. To some extent, it reflects the capacity of the MFC-Fenton system for the removal of ECs under different conditions. The first-order logarithmic decay mode is as:

$$\mathrm{Ln}(C_t/C_0) = -K_{app} \times t \tag{21.3}$$

where C_t and C_0 are the concentration of ECs at time t and time zero, respectively, and t is the reaction time, represents different emerging contaminants. Kapp is the apparent rate constant.

High electrical energy is necessary for the electro-Fenton process in the removal of pharmaceuticals and personal care products (PPCPs) from WW. Wang et al. (2018) developed a microbial electro-Fenton cell process for treating PPCPs in WW and it proved to be a novel and more cost-effective process as compared to the stand-alone Fenton process (Wang et al. 2018). The carbamazepine (CBZ) removal in this microbial electro-Fenton cell process has been observed significantly higher (90%) than that in the conventional electro-Fenton process (62%) within 24 h operation time. Also, when comparing with other biological treatment processes, the CBZ removal rate in the microbial electro-Fenton cell process has been 10–100 times faster than that in other biological treatment processes. In the future, the MFC-Fenton process can be used as a new successful, cost-effective strategy for the effective removal of ECs from WW (Wang et al. 2018).

Dyes that contain one or more azo bonds (—N = N—) are used extensively in plastic, textile, cosmetic, leather, and food industries (O'neill et al. 1999). Due to the azo bond, these dyes show color and related chromophores; so, when these dyes are discharged into the surface water, they affects the aesthetics, their breakdown products are carcinogenic, or toxic to humans and ultimately to the whole environment (Ganesh et al. 1994). An anthraquinone-based vinylsulphone dye Reactive Blue-19 (RB19) is hard to decompose by chemical oxidation as its anthraquinone structure is stabilized by resonance. RB19 has a fixation efficiency of 75–80% on the cotton textiles, and the unfixed dye remains for a very long period in the environment, signaling that WW containing RB19 needs to be adequately treated (Holkar et al. 2014; Weber and Stickney 1993). Holkar et al. (2018) prepared synthetic textile WW containing RB19 dye of concentrations from 50 mg/L to 800 mg/L to produce electricity with electrogenic bacterium *Klebsiella sp. C.* The power density of 84 mW/m^2 have been generated with a current density of 533 mA/m^2 in this MFC process with the decolorization efficiency of 90% within 24 h under optimized conditions i.e. temperature of 37 °C, pH 7, glucose to microbe ratio of 60 and 6 g/L glucose concentration (Holkar et al. 2018).

Jayaprakash et al. (2016) used Pseudomonas catalyzed MFC for decolourizing synthetic WW containing some azo dyes viz. methyl orange (MO), congo red, RB172, reactive yellow 145, and reactive red 2. Decolorization efficiency between 74 to 94% has been reported with azo dyes in the anodic chamber (Jayaprakash et al. 2016). Narayanasamy and Jayaprakash (2018) achieved the best performance in MFC with the highest power density of 1575 ± 223.26 pW/m^2 and decolorization of 89% for synthetic WW containing azo dye MO (Narayanasamy & Jayaprakash, 2018). Gomaa et al. (2019) used genetically modified Escherichia coli K-12 with MtrA, MtrC and MtrCAB to evaluate the active involvement of each of these genes in energy generation and decolorization of Congo red in MFC using a double chamber configuration (Gomaa et al. 2019). The highest decolorization of 80% in 36 h with MtrCAB insert showed under MFC conditions and MtrC for the power density of 59 mW/m^2 (Gomaa et al. 2019). Table 21.1 shows list of various studies using MFC for the treatment of ECs.

Table 21.1 List of various studies using MFC for treatment of ECs

Emerging contaminants	Conc. (mg/L)	Anode and Cathode	MFC configuration	Reference
Chloramphenicol (CAP)	50	Active-carbon felt	Dual chamber MFC	Zhang et al. (2017)
Sulphamethoxazole (SMX)	20	Carbon felt with titanium wire	Two-chamber MFC	Wang et al. (2016)
Paracetamol (PAM)	5	Graphite felt and single graphite plate	MFC-Fenton system with dual-chamber MFC	Zhang et al. (2015)
Bisphenol A (BPA),	1	Graphite rod with stacked graphite granules	Dual-chamber MFC	Sun et al. (2012); Wang et al. (2017)
Estrone (El)	1	Graphite rod with stacked graphite granules	Dual-chamber MFC	Wang et al. (2017)
Sulfamethazine (SM2)	1	Graphite rod with stacked graphite granules	Dual-chamber MFC	Wang et al. (2017)
Triclocarban (TCC)	1	Graphite rod with stacked graphite granules	Dual-chamber MFC	Wang et al. (2017)
Carbamazepine (CBZ)	10	Carbon brush and Gas diffusion cathode	Single-chamber air–cathode MFC	Wang et al. (2018)
Dyes	50–800	Graphite electrode	Double chamber MFC	Gomaa et al. (2019); Holkar et al. (2018); Jayaprakash et al. (2016); Narayanasamy and Jayaprakash (2018)

21.4 Summary and Conclusion

The continuous occurrence and detection of ECs in water bodies should be monitored continuously and not overlooked. The negative impacts of these ECs on aquatic life and humans have been highlighted in many previous studies. These impacts have urged the removal of ECs not only from drinking water but also from WW. To address the issue of ECs, many researchers have studied the potential of MFCs for their removal. MFC is the newest and most novel eco-friendly approach of bioelectricity generation along with contaminant degradation, and the bioremediation is achieved using microorganisms for the oxidation of the organic and inorganic substances. MFCs can convert substrate energy to electricity directly, operate in a wide range of temperatures, produce less sludge, and require neither energy for the aeration nor any gas treatment. A wide variety of WW can be used as feed for the MFCs, including municipal, food processing, dairy, palm oil mill, and pharmaceutical WW. The high organic content in the WW makes it a suitable source for energy generation using MFCs. Thus, this chapter summarizes the recent studies on working principles and applications of MFCs in tackling the ECs, including antibiotics, microplastics, and synthetic dyes.

MFCs generally consist of cathode, anode, external circuit wire, and a cation-selective membrane. Anaerobic and aerobic conditions are maintained at the anode and cathode sides, respectively. Microorganisms supplied at the anode side degrade the suitable substrates and release electrons which are transported via the external circuit wire to the anode. The protons generated are then passed through the PEM to regulate the proton transport. This electron transport is the basis for the generation of bioelectricity. Microorganisms used in the MFC are the important biocatalysts for the systems susceptible to electrode material, substrate type, operating temperature, and reactor configuration. The use of microbial colonies of the distinct or ideal microbial group has also been demonstrated to be effective in MFC applications. A more extended acclimation period could greatly enhance the degradation ability of microbes towards recalcitrant chemicals. The other important component of the MFCs is the electrodes: anode and cathode, which act as the conductors as well as the home for the microbes. The preferred features of the electrodes include biocompatibility, high specific surface area, effective electron transfer capacity, and considerable surface roughness. The selection of electrode material also depends on the factors like ease of availability, non-corrosive nature, scalability to large sizes, porosity, inexpensiveness, and antifouling propensity. The materials that have been commonly used as electrode materials are carbon cloth, carbon paper, carbon foam, RVC graphite, woven graphite, graphite granules, and brushes. The bioelectricity generation in the MFCs depends on the substrates used for the microorganisms. The substrates used can be pure substances like glucose or can consist of different WWs from municipal, industrial or agricultural activities. The MFCs are generally of three designs based on the electrode chambers: single-chamber, double-chamber, and stacked MFCs.

The MFCs have been demonstrated to effectively deal with various ECs, including antibiotics, microplastics, and synthetic dyes. The pharmaceutical and micropollutants like CAP, SMX, PAM, BPA, El, SM2, and TCC have been shown to be effectively removed (more than 70%) within 12 to 24 h of treatment time in MFC and integrated MFC-Fenton systems. Most of the studies showed reduced biotoxicity of the MFC treated effluent towards *Escherichia coli* and *Shewanella oneidensis* with respect to the controls. Bacterial growth, electron transport via electrogenic bacteria, long acclimation period, adsorption on electrodes, and degradation by the radicals generated (MFC-Fenton) contributed to the enhanced removal of these ECs. Even some ECs like CBZ can be removed 10–100 times faster by the microbial electro-Fenton MFCs than biological treatment processes. WW containing chemically hard to degrade dyes like RB19 and azo dyes have also been reported to be decolorized with ~80–90% decolorization efficiency within 24–32 h under the respective optimized conditions.

Although MFCs have shown promising results in degrading ECs, more work is required for studying the degradation kinetics, the initial formation of biofilm, their stability, and its effect on MFCs performance. MFC technology could also be aimed at future space technology applications due to its power generation capability, biosensor application, and sustainable energy generation. Also, some challenges need to be overcome for the more comprehensive application and acceptance of MFCs. These challenges include turbulence in the electrode chambers, membrane resistance in the proton transportation process, inability to function at very low temperatures, and the use of MFCs in large-scale commercial applications. Overall, the MFCs can treat ECs that are a challenge for conventional treatment plants and provide energy.

Acknowledgements The corresponding author acknowledges the funding received from SERB, Department of Science and Technology, Government of India for carrying out this work.

References

Avisar D, Lester Y, Mamane H (2010) pH induced polychromatic UV treatment for the removal of a mixture of SMX, OTC and CIP from water. J Hazard Mater 175(1–3):1068–1074. https://doi.org/10.1016/J.JHAZMAT.2009.10.122

Batt AL, Bruce IB, Aga DS (2006) Evaluating the vulnerability of surface waters to antibiotic contamination from varying wastewater treatment plant discharges. Environ Pollut 142(2):295–302. https://doi.org/10.1016/j.envpol.2005.10.010

Boghani H, Kim JR, Dinsdale RM, Guwy AJ, Premier GC (2017) Reducing the burden of food processing washdown wastewaters using microbial fuel cells. Biochem Eng J 117:210–217. https://doi.org/10.1016/j.bej.2016.10.017

Bolong N, Ismail AF, Salim MR, Matsuura T (2009) A review of the effects of emerging contaminants in wastewater and options for their removal. Desalination 239(1–3):229–246. https://doi.org/10.1016/j.desal.2008.03.020

Bond DR, Lovley DR (2003) Electricity production by Geobacter sulfurreducens attached to electrodes. Appl Environ Microbiol 69(3):1548–1555. https://doi.org/10.1128/AEM.69.3.1548-1555.2003

Bose D, Dhawan H, Kandpal V, Vijay P, Gopinath M (2018a) Sustainable power generation from sewage and energy recovery from wastewater with variable resistance using microbial fuel cell. Enzyme Microb Technol 118:92–101. https://doi.org/10.1016/j.enzmictec.2018.07.007
Bose D, Gopinath M, Vijay P (2018b) Sustainable power generation from wastewater sources using microbial fuel cell. Biofuels Bioprod Biorefin 12(4):559–576. https://doi.org/10.1002/bbb.1892
Bose D, Kandpal V, Dhawan H, Vijay P, Gopinath M (2018c) Energy recovery with microbial fuel cells: bioremediation and bioelectricity. Energy Environ Sustain 7–33. https://doi.org/10.1007/978-981-10-7413-4_2
Bose D, Rawat R, Bahuguna R, Vijay P, Gopinath M (2020) Sustainable approach to wastewater treatment and bioelectricity generation using microbial fuel cells. In: Current developments in biotechnology and bioengineering: resource recovery from wastes. Elsevier. https://doi.org/10.1016/B978-0-444-64321-6.00003-3
Chaturvedi V, Vema P (2016) Microbial fuel cell: a green approach for the utilization of waste for the generation of bioelectricity. Bioresour Bioprocess 3(1):1–14. https://doi.org/10.1186/s40643-016-0116-6
Chen K, Zhou JL (2014) Occurrence and behavior of antibiotics in water and sediments from the Huangpu River, Shanghai, China. Chemosphere 95:604–612. https://doi.org/10.1016/j.chemosphere.2013.09.119
Chen G-W, Choi S-J, Lee T-H, Lee G-Y, Cha J-H, Kim C-W (2008) Application of biocathode in microbial fuel cells: cell performance and microbial community. Appl Microbiol Biotechnol 79(3):379–388. https://doi.org/10.1007/S00253-008-1451-0
Clara M, Strenn B, Gans O, Martinez E, Kreuzinger N, Kroiss H (2005) Removal of selected pharmaceuticals, fragrances and endocrine disrupting compounds in a membrane bioreactor and conventional wastewater treatment plants. Water Res 39(19):4797–4807. https://doi.org/10.1016/j.watres.2005.09.015
Deborde M, Rabouan S, Duguet J-P, Legube B (2005) Kinetics of aqueous ozone-induced oxidation of some endocrine disruptors. Environ Sci Technol 39(16):6086–6092. https://doi.org/10.1021/ES0501619
Do MH, Ngo HH, Guo WS, Liu Y, Chang SW, Nguyen DD, Nghiem LD, Ni BJ (2018) Challenges in the application of microbial fuel cells to wastewater treatment and energy production: a mini review. Sci Total Environ 639:910–920. https://doi.org/10.1016/j.scitotenv.2018.05.136
Du Z, Li H, Gu T (2007) A state of the art review on microbial fuel cells: A promising technology for wastewater treatment and bioenergy. Biotechnol Adv 25(5):464–482. https://doi.org/10.1016/j.biotechadv.2007.05.004
Feder HM, Osier C, Maderazo EG (1981) Chloramphenicol: a review of its use in clinical practice. In: Reviews of infectious diseases, vol. 3, issue 3. Oxford Academic, pp 479–491. https://doi.org/10.1093/clinids/3.3.479
Ganesh R, Boardman GD, Michelsen D (1994) Fate of azo dyes in sludges. Water Res 28(6):1367–1376. https://doi.org/10.1016/0043-1354(94)90303-4
Gao C, Liu L, Yang F (2017) Development of a novel proton exchange membrane-free integrated MFC system with electric membrane bioreactor and air contact oxidation bed for efficient and energy-saving wastewater treatment. Biores Technol 238:472–483. https://doi.org/10.1016/J.BIORTECH.2017.04.086
Gomaa OM, Fapetu S, Kyazze G, Keshavarz T (2019) Probing the mechanism of simultaneous bioenergy production and biodegradation process of Congo red in microbial fuel cells. J Chem Technol Biotechnol 94(7):2092–2097. https://doi.org/10.1002/jctb.5892
del Mar Gómez-Ramos M, Mezcua M, Agüera A, Fernández-Alba AR, Gonzalo S, Rodríguez A, Rosal R (2011) Chemical and toxicological evolution of the antibiotic sulfamethoxazole under ozone treatment in water solution. J Hazard Mater 192(1):18–25. https://doi.org/10.1016/J.JHAZMAT.2011.04.072
He CS, Mu ZX, Yang HY, Wang YZ, Mu Y, Yu HQ (2015) Electron acceptors for energy generation in microbial fuel cells fed with wastewaters: a mini-review. Chemosphere 140:99–105. https://doi.org/10.1016/j.chemosphere.2015.03.059

Holkar CR, Pandit AB, Pinjari DV (2014) Kinetics of biological decolorisation of anthraquinone based reactive blue 19 using an isolated strain of Enterobacter sp.F NCIM 5545. Biores Technol 173:342–351. https://doi.org/10.1016/J.BIORTECH.2014.09.108

Holkar CR, Arora H, Halder D, Pinjari DV (2018) Biodegradation of reactive blue 19 with simultaneous electricity generation by the newly isolated electrogenic Klebsiella sp. C NCIM 5546 bacterium in a microbial fuel cell. Int Biodeterior Biodegrad 133:194–201. https://doi.org/10.1016/j.ibiod.2018.07.011

Jayaprakash J, Parthasarathy A, Viraraghavan R (2016) Decolorization and degradation of monoazo and diazo dyes in Pseudomonas catalyzed microbial fuel cell. Environ Prog Sustain Energy 35(6):1623–1628. https://doi.org/10.1002/ep.12397

Joss A, Keller E, Alder AC, Göbel A, McArdell CS, Ternes T, Siegrist H (2005) Removal of pharmaceuticals and fragrances in biological wastewater treatment. Water Res 39(14):3139–3152. https://doi.org/10.1016/j.watres.2005.05.031

Khan MZ, Nizami AS, Rehan M, Ouda OKM, Sultana S, Ismail IM, Shahzad K (2017) Microbial electrolysis cells for hydrogen production and urban wastewater treatment: a case study of Saudi Arabia. Appl Energy 185:410–420. https://doi.org/10.1016/j.apenergy.2016.11.005

Kim HJ, Park HS, Hyun MS, Chang IS, Kim M, Kim BH (2002) A mediator-less microbial fuel cell using a metal reducing bacterium, Shewanella Putrefaciens. Enzyme Microb Technol 30(2):145–152. https://doi.org/10.1016/S0141-0229(01)00478-1

Kim JR, Min B, Logan BE (2005) Evaluation of procedures to acclimate a microbial fuel cell for electricity production. Appl Microbiol Biotechnol 68(1):23–30. https://doi.org/10.1007/S00253-004-1845-6

Koók L, Kanyó N, Dévényi F, Bakonyi P, Rózsenberszki T, Bélafi-Bakó K, Nemestóthy N (2018) Improvement of waste-fed bioelectrochemical system performance by selected electro-active microbes: process evaluation and a kinetic study. Biochem Eng J 137:100–107. https://doi.org/10.1016/j.bej.2018.05.020

Kumar SS, Bishnoi NR (2017) Coagulation of landfill leachate by $FeCl_3$: process optimization using Box–Behnken design (RSM). Appl Water Sci 7(4):1943–1953. https://doi.org/10.1007/s13201-015-0372-1

Kumar SS, Malyan SK, Basu S, Bishnoi NR (2017) Syntrophic association and performance of Clostridium, Desulfovibrio, Aeromonas and Tetrathiobacter as anodic biocatalysts for bioelectricity generation in dual chamber microbial fuel cell. Environ Sci Pollut Res 24(19):16019–16030. https://doi.org/10.1007/s11356-017-9112-4

Kumar SS, Kumar V, Malyan SK, Sharma J, Mathimani T, Maskarenj MS, Ghosh PC, Pugazhendhi A (2019) Microbial fuel cells (MFCs) for bioelectrochemical treatment of different wastewater streams. Fuel 254(May):115526. https://doi.org/10.1016/j.fuel.2019.05.109

Larcher S, Yargeau V (2012) Biodegradation of sulfamethoxazole: current knowledge and perspectives. Appl Microbiol Biotechnol 96(2):309–318. https://doi.org/10.1007/s00253-012-4326-3

Li S, Chen G (2018) Factors affecting the effectiveness of bioelectrochemical system applications: data synthesis and meta-analysis. Batteries 4(3):1–18. https://doi.org/10.3390/batteries4030034

Liu H, Ramnarayanan R, Logan BE (2004) Production of electricity during wastewater treatment using a single chamber microbial fuel cell. Environ Sci Technol 38(7):2281–2285. https://doi.org/10.1021/ES034923G

Logan BE, Hamelers B, Rozendal R, Schröder U, Keller J, Freguia S, Aelterman P, Verstraete W, Rabaey K (2006) Microbial fuel cells: methodology and technology. Environ Sci Technol 40(17):5181–5192. https://doi.org/10.1021/es0605016

Mandal SK, Das N (2018) Application of microbial fuel cells for bioremediation of environmental pollutants: an overview. J Microbiol Biotechnol Food Sci 7(4):437–444. https://doi.org/10.15414/jmbfs.2018.7.4.437-444

Mansoorian HJ, Mahvi AH, Jafari AJ, Khanjani N (2016) Evaluation of dairy industry wastewater treatment and simultaneous bioelectricity generation in a catalyst-less and mediator-less membrane microbial fuel cell. J Saudi Chem Soc 20(1):88–100. https://doi.org/10.1016/j.jscs.2014.08.002

Martelli A, Mattioli F, Pastorino G, Robbiano L, Allavena A, Brambilla G (1991) Genotoxicity testing of chloramphenicol in rodent and human cells. Mutat Res Genet Toxic 260(1):65–72. https://doi.org/10.1016/0165-1218(91)90081-V

Michael I, Rizzo L, McArdell CS, Manaia CM, Merlin C, Schwartz T, Dagot C, Fatta-Kassinos D (2013) Urban wastewater treatment plants as hotspots for the release of antibiotics in the environment: a review. Water Res 47(3):957–995. https://doi.org/10.1016/j.watres.2012.11.027

Morris JM, Jin S (2012) Enhanced biodegradation of hydrocarbon-contaminated sediments using microbial fuel cells. J Hazard Mater 213–214:474–477. https://doi.org/10.1016/j.jhazmat.2012.02.029

Narayanasamy S, Jayaprakash J (2018) Improved performance of Pseudomonas aeruginosa catalyzed MFCs with graphite/polyester composite electrodes doped with metal ions for azo dye degradation. Chem Eng J 343:258–269. https://doi.org/10.1016/j.cej.2018.02.123

Nor MHM, Mubarak MFM, Elmi HSA, Ibrahim N, Wahab MFA, Ibrahim Z (2015) Bioelectricity generation in microbial fuel cell using natural microflora and isolated pure culture bacteria from anaerobic palm oil mill effluent sludge. Biores Technol 190:458–465. https://doi.org/10.1016/j.biortech.2015.02.103

O'neill C, Hawkes FR, Hawkes DL, Lourenço ND, Pinheiro HM, Delée W (1999) Review colour in textile effluents-sources, measurement, discharge consents and simulation: a review. https://doi.org/10.1002/(SICI)1097-4660(199911)74:11

Rahimnejad M, Ghoreyshi AA, Najafpour G, Jafary T (2011) Power generation from organic substrate in batch and continuous flow microbial fuel cell operations. Appl Energy 88(11):3999–4004. https://doi.org/10.1016/J.APENERGY.2011.04.017

Saeed HM, Husseini GA, Yousef S, Saif J, Al-Asheh S, Abu Fara A, Azzam S, Khawaga R, Aidan A (2015) Microbial desalination cell technology: a review and a case study. Desalination 359:1–13. https://doi.org/10.1016/j.desal.2014.12.024

Sharma Y, Li B (2010) The variation of power generation with organic substrates in single-chamber microbial fuel cells (SCMFCs). Biores Technol 101(6):1844–1850. https://doi.org/10.1016/J.BIORTECH.2009.10.040

Sun Q, Li Y, Chou PH, Peng PY, Yu CP (2012) Transformation of bisphenol A and alkylphenols by ammonia-oxidizing bacteria through nitration. Environ Sci Technol 46(8):4442–4448. https://doi.org/10.1021/es204424t

Tharali AD, Sain N, Osborne WJ (2016) Microbial fuel cells in bioelectricity production. Frontiers in Life Science 9(4):252–266. https://doi.org/10.1080/21553769.2016.1230787

Trapero JR, Horcajada L, Linares JJ, Lobato J (2017) Is microbial fuel cell technology ready? An economic answer towards industrial commercialization. Appl Energy 185:698–707. https://doi.org/10.1016/j.apenergy.2016.10.109

Van Der Bruggen B, Vandecasteele C, Van Gestel T, Doyen W, Leysen R (2003) A review of pressure-driven membrane processes in wastewater treatment and drinking water production. Environ Prog 22(1):46–56. https://doi.org/10.1002/ep.670220116

Wang L, Liu Y, Ma J, Zhao F (2016) Rapid degradation of sulphamethoxazole and the further transformation of 3-amino-5-methylisoxazole in a microbial fuel cell. Water Res 88:322–328. https://doi.org/10.1016/j.watres.2015.10.030

Wang Y, Feng C, Li Y, Gao J, Yu CP (2017) Enhancement of emerging contaminants removal using Fenton reaction driven by H_2O_2-producing microbial fuel cells. Chem Eng J 307:679–686. https://doi.org/10.1016/j.cej.2016.08.094

Wang W, Lu Y, Luo H, Liu G, Zhang R, Jin S (2018) A microbial electro-fenton cell for removing carbamazepine in wastewater with electricity output. Water Res 139:58–65. https://doi.org/10.1016/j.watres.2018.03.066

Weber EJ, Stickney VC (1993) Hydrolysis kinetics of Reactive Blue 19-Vinyl Sulfone. Water Res 27(1):63–67. https://doi.org/10.1016/0043-1354(93)90195-N

Wei R, Ge F, Huang S, Chen M, Wang R (2011) Occurrence of veterinary antibiotics in animal wastewater and surface water around farms in Jiangsu Province, China. Chemosphere 82(10):1408–1414. https://doi.org/10.1016/j.chemosphere.2010.11.067

Xiao L, He Z (2014) Applications and perspectives of phototrophic microorganisms for electricity generation from organic compounds in microbial fuel cells. Renew Sustain Energy Rev 37:550–559. https://doi.org/10.1016/J.RSER.2014.05.066

Yu CP, Chu KH (2009) Occurrence of pharmaceuticals and personal care products along the West Prong Little Pigeon River in east Tennessee, USA. Chemosphere 75(10):1281–1286. https://doi.org/10.1016/j.chemosphere.2009.03.043

Zhang L, Yin X, Li SFY (2015) Bio-electrochemical degradation of paracetamol in a microbial fuel cell-Fenton system. Chem Eng J 276:185–192. https://doi.org/10.1016/j.cej.2015.04.065

Zhang Q, Zhang Y, Li D (2017) Cometabolic degradation of chloramphenicol via a meta-cleavage pathway in a microbial fuel cell and its microbial community. Bioresour Technol 229:104–110. https://doi.org/10.1016/j.biortech.2017.01.026

Chapter 22
Material and Process-Related Contaminants in Solar Photovoltaics: Key Issues, and Future Prospects

Venkata S. N. Chava and Sreeprasad T. Sreenivasan

Abstract The limited availability of fossil fuel sources coupled with the health and environmental risks associated with their use lead to the increased focus on renewable energy resources such as solar photovoltaics (PV) as a potential energy source for the future. Currently, significant research is focused on improving the efficiency (i.e., reducing the cost per watt power) and long-term reliability of solar cells to make PV cells competent with fossil fuels. On the other hand, little attention is given to understanding and assessing long-term environmental impacts associated with the contaminants produced during the manufacturing and application of solar cells. Hence, it is imperative to review and evaluate the critical environmental issues relevant to solar PV, especially in emerging PV technologies. This chapter will introduce different PV technologies, including silicon PV, thin-film PV, and perovskite solar cells, and outline the materials and the processes used in PV technologies. A review of the health and environmental impact of Sn- and Pb- based PV technologies and the need for alternative technologies such as Sn- and Pb-free perovskite PV will be presented. The potential environmental, energy, and health impacts and a review of possible mitigation strategies related to perovskite solar cells-induced hazards are also presented.

Keywords Sustainable energy · CO_2 emissions · Photovoltaics · Solar panel · Module · Silicon · CdTe · CIGS · Perovskite · Recycling · Toxicity · Health and safety · Environmental impact

22.1 Introduction

Land, energy, and water are among the most precious and critical resources for human survival. The mismanagement of these resources has, however, contributed negatively to climate change. Currently, fossil fuels are the primary energy sources from which

---

V. S. N. Chava · S. T. Sreenivasan (✉)
Department of Chemistry and Biochemistry, The University of Texas at El Paso, 500 W. University Avenue, El Paso, TX 79968, USA
e-mail: sreenivasan@utep.edu

S. P. Singh et al. (eds.), *New Trends in Emerging Environmental Contaminants*, Energy, Environment, and Sustainability,
https://doi.org/10.1007/978-981-16-8367-1_22

we generate energy for our daily needs, including industries, transportation, etc. In the last century, due to huge scientific and technological advancements, there has been a significant rise in the consumption of energy derived from fossil fuels (Smalley 2005). In this process, the adverse effects of fossil fuel emissions on climate are largely ignored, which led to a significant rise in global temperatures. Now, there is a consensus (Carlton et al. 2015), after decades of arguments (Oreskes 1686), about the negative consequences of the emissions, including severe human health and safety concerns, due to the production and use of these fossil fuels. Addressing climate change has been identified as one of the top 10 global challenges for the twenty-first century (Smalley 2005). Two major initiatives, including the Kyoto protocol (1995) and the Paris climate agreement (2015), are launched to address the climate challenge. However, the progress made is negligible when compared to the threat we are facing (Climate Change 2019). In the last couple of decades, there is a shift towards the use of alternative energy sources. These alternative energy sources include solar, wind, biomass, hydrogen, geothermal, etc. There is a non-uniform shift across the globe when it comes to adopting the renewable energy sources for public and industrial applications (Fig. 22.1) (Xu et al. 2019). A major reason for this disparity could be the lack of government support in the form of incentives to the companies and customers that can lead to the cost per watt of these alternative energy sources comparable to the fossil fuel sources (Lund 2009; Koseoglu et al. 2013). Moreover, the shift toward alternative sources has taken place at much slower rates than the increased CO_2 emissions from fossil fuels. In this scenario, an important question arises: why do countries across the globe rely heavily on fossil-fuel-based energy sources? One important reason for this could be the lack of alternative energy sources that can meet the existing energy demand (Smalley 2005; Armaroli and Balzani 2007; Dorian et al. 2006).

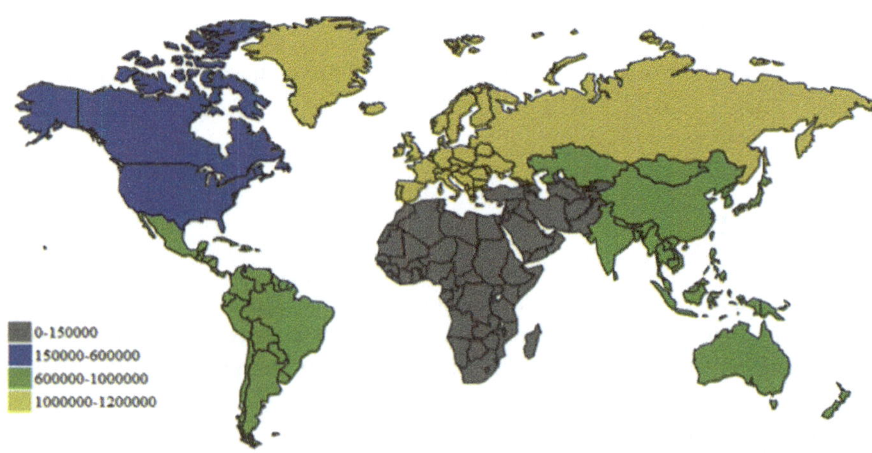

Fig. 22.1 Renewable energy generation by region in 2016 Unit: Gwh *Sources* International Energy Agency, the World Bank). Reproduced with permission from Elsevier (Xu et al. 2019)

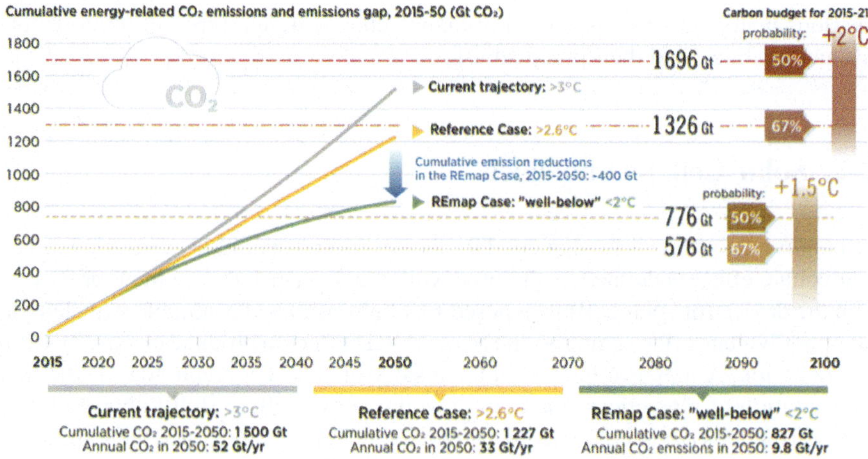

Fig. 22.2 A pathway for a well-below 2 °C climate target, towards 1.5 °C. Cumulative energy-related CO_2 emissions for the period 2015–2050 and emission budgets for 2015–2100 for 1.5 and 2 °C scenarios (Gt CO_2) (Global energy transformation: a roadmap to 2050 2019).

It is predicted that if renewable energy resources can provide two-thirds of the total global energy demand, this can help restrict the global temperature rise under 2 °C (Fig. 22.2) (Gielen et al. 2019). Therefore, immediate action for a fast transition towards renewable and clean energy sources is needed for addressing climate change problems. Such transition in global energy use is multi-faceted that is constantly progressing due to evolving technologies, socioeconomics, and government policies. Figure 22.2 summarizes a trajectory of cumulative energy-related CO_2 emissions with corresponding changes in global temperatures. To restrict the increase in temperature below 2 °C, the CO_2 emissions should drop below 9.8 GT/Year by 2050 from the current 52 GT/Year rate.

Electric power generation using solar photovoltaics has always been considered the primary choice for a clean and renewable energy economy. The belief in PV technologies is due to the significant availability of solar radiation that can be harnessed to generate power across the world. Given the increasing demand for renewable energy production, considerable investments are made in this sector, including research and manufacturing. However, the health and safety concerns arising due to the generation and consumption of solar PV are often overlooked. Therefore, it is imperative for both scientists and engineers to seriously consider and evaluate all possible scenarios of solar PV manufacturing and applications that could lead to undesirable health and safety consequences. Finally, both the government and policymakers must encourage new R&D innovations, manufacturing, and installations in solar PV. They should also consider imposing strict regulations on PV technologies, especially on those that may pose serious health and safety concerns. Hence, this chapter aims to provide a comprehensive overview of the safety and health concerns related to various materials, processes used to manufacture different types of photovoltaic cells and modules, and their practical use.

22.2 Solar Cell Technologies

The photovoltaic effect is fundamental for the design and operation of solar cells. The photovoltaic effect is defined as the process that generates either voltage or current when the device (or solar cell) is exposed to a light source of a suitable wavelength. Solar photovoltaics (PV) employs the photovoltaic effect to produce electricity from solar radiation. A major milestone in the history of solar PV technology is the first demonstration of a practical silicon photovoltaic (PV) cell, at Bell Laboratories in 1953 (Perlin 2004), that converted solar energy into electricity. Since then, different PV technologies have emerged, including silicon PV, thin-film PV, dye-sensitized PV, perovskite PV, etc. (Amin et al. 2017). The classification of these technologies (or devices) is based on the absorbing film (or active layer) used in the PV cell. Among these technologies, Si and thin-film (CdTe) based solar panels have been used for practical applications for almost > 2 decades. Figure 22.3 shows a comparison of the share of the total PV installations worldwide by technology.

22.2.1 Overview of the Material Properties of Absorbing Materials Used in Different PV Technologies

In a solar cell, the absorbing material (or active layer) is the key component that absorbs light and generates e–h pairs and a photovoltage using the photovoltaic effect.

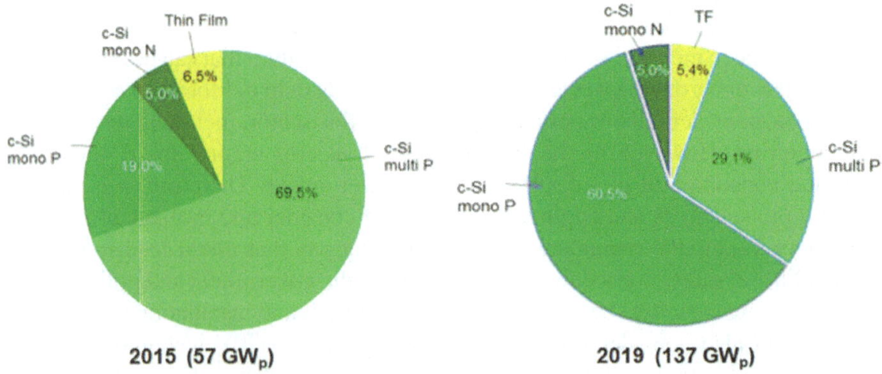

Fig. 22.3 Comparison of market share of different PV technologies. Reproduced with permission from Elsevier (Benda and Černá 2020)

Conventionally, the solar cells are designed and fabricated by forming suitable junctions (e.g., p-n, Schottky) using inorganic semiconductors as absorbing films. While there exist several semiconductors with a suitable bandgap to absorb incoming solar radiation and produce a measurable current, only a few of them are used in PV cells due to constraints related to material quality, cost, and availability that are critical for commercialization. Semiconducting materials with a direct bandgap of 1.5 eV, high optical absorption coefficient (10^5 cm^{-1}), high quantum efficiency are considered as an ideal choice for terrestrial solar cell applications (Wadia et al. 2009). In addition, the semiconducting absorber should be able to form a good electronic junction (homo/hetero/Schottky) with suitably compatible materials. While elemental semiconductors are preferable from the manufacturing point of view, there are no suitable materials available with a direct bandgap close to 1.5 eV. Interestingly, even though silicon (indirect bandgap of $E_g = 1.1$ eV), is not an ideal material for PV, it still dominates the PV market due to various reasons including advancements in silicon microelectronics and reliability, as will be discussed later in other sections. Due to indirect bandgap nature of Si, the Si wafers used in PV must be at least 50 mm thick for effective solar absorption, unless the optical absorption is enhanced using some additional design and process steps. Besides silicon, other elemental semiconductor materials that showed some promise in laboratory research are diamond-like carbon (Ma and Liu 2001), and fullerene films (Narayanan and Yamaguchi 2003). However, the synthesis and formation of electrical contacts with these materials is a challenging task. There also exists a big library of compound semiconductor or alloys that are suitable for PV applications. A metastable film of a-Si:H, with advantages including large tailorable bandgap, and high optical absorption coefficient, is also a suitable semiconductor material for PV applications (Rech and Wagner 1999). In the early days, the a-Si:H-based solar PV technology was considered as a major contender to challenge the crystalline Si photovoltaics dominance. Despite of its promise, the a-Si:H technology is not a commercial success. become a reality. A few other compound semiconductors based thin-film solar cells, including GaAs, CdTe, Cu_2S, Cu_2O, InP, Zn_3P_2, etc., are also reported. Of these, GaAs and InP with direct bandgaps of 1.4 eV and 1.34 eV, respectively, are ideal for photovoltaic applications, but the high wafer costs together with device processing challenges severely restricted their commercialization. Besides the active semiconductor material, careful consideration of other materials required for device fabrication is paramount. With increasing number of components, the number of possible materials increases in geometrical proportion. On the other hand, the phase diagram of the compound semiconductors and alloys in bulk form is quite complex, and therefore poses a significant challenge in the quality and control of the material growth (Altamura and Vidal 2016; Bosson et al. 2017). In addition to binary semiconductors, there are various ternary semiconductor based PV cells. Cu_2ZnSnS_4 (popularly referred to as CZTS) is a quaternary absorber materials with an optimal bandgap (1.5 eV) and high absorption coefficient (>10^4 cm^{-1}) (Katagiri et al. 2009). Even though inorganic semiconductors can offer high structural stability critical for long term reliability, organic semiconductors (e.g., fullerenes), which combine superior optoelectronic properties and processing properties have attracted significant research attention (Brabec and Durrant 2008).

The absorption of incoming photons by organic semiconductors, creates an exciton, which is bound pair of electron and hole, with a binding energy of approximately 0.5 eV. Each excitons when dissociated will generate an unbound electron–hole pair at the dissociation sites, and these charge carriers (electron and holes) are subsequently separated and transported to the suitable contact electrodes. A majority of semiconducting polymers are not optimal absorber of incoming solar radiation as they have $E_g > 2.0$ eV (Hoppe and Sariciftci 2004). In addition, owing to low charge-carrier and exciton mobility, the thickness of the polymer absorbing film cannot be above 100 nm. In these materials, Similar to silicon, light-trapping strategies can be employed to enhance the optical absorption. Despite of these limitations, organic semiconductors provide unique benefits such as solution based device processing at room temperature on flexible substrates. Further, the device fabrication by employing low-energy deposition methods such as spin or blade coating, substantially reduces the device costs (Brabec and Durrant 2008; Zhao et al. 2018). Perovskite solar cells and quantum dot solar cells are a recent addition to the family of solar cells that have shown tremendous research progress in the last decade (Park 2015; Kamat 2013). Perovskites have unique material properties, including direct bandgap, high charge-carrier mobilities, low exciton binding energies, wide-wavelength absorption, which makes them good candidates for solar cell applications. There are significant scientific developments in these new device technologies that exploited low dimensional quantum confinement effects for achieving enhanced device performance (Wright et al. 2020). Figure 22.4 summarizes the historical progress of the different solar cell technologies with corresponding best cell efficiencies certified by NREL.

While Si and thin-film PV technologies have shown tremendous growth in terms of their installations, owing to their cost advantages, emerging PV technologies such as DSSC and perovskite solar cells (PSC) have the potential to reach the commercial

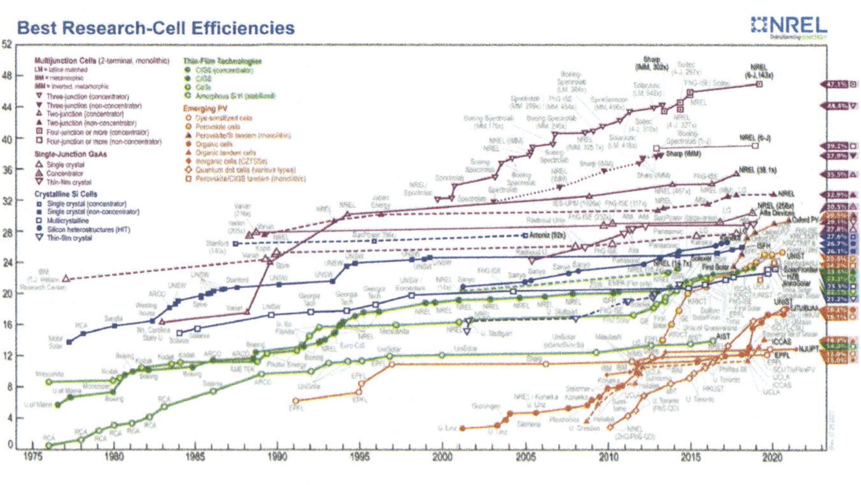

Fig. 22.4 A plot showing the historical data of highest certified cell efficiencies of different PV cells (Best research-cell efficiency chart 1976)

market and compete with Si and thin-film PV technologies. Hence, considerable research efforts have been devoted in the last decade to improve the efficiency and stability of the PSC technology.

The solar cells based on silicon wafers are generally referred to as the first-generation PV technologies. Currently, about 90% of the commercial solar cells are manufactured using silicon wafers, either single crystalline or multi-crystalline (Fig. 22.3). The dominance of silicon solar cells stems from the high efficiencies of Si PV modules with demonstrated long-term stability. Starting from wafer manufacturing to solar cell fabrication technology, the early success in the field of silicon PV was driven by R&D advances in silicon microelectronics. The major share of Si PV technology cost lies in the silicon wafer, which is more than fifty-percent of the total module manufacturing cost. Alternatively, the thin-film solar cells are expected to lower the PV module costs due to the relatively inexpensive absorbing films used in them. The thin film based photovoltaics are commonly referred to as second-generation PV technologies. In addition to the their lower module cost, the silicon supply shortage has also led contributed for the rapid commercialization of thin-film photovoltaics modules. Several distinct thin-film technologies are now available, based on a-Si, poly-crystalline Si, mixed phases, or chalcogenides. There is also a third generation of solar cell technologies that include dye-sensitized solar cells, organic solar cells, perovskite solar cells, etc. The primary difference between the third generation differs from the first and second generation is in terms of materials used for cell formation or in terms of the operation principle of the device. The third-generation PV solar cells have entered the commercial market recently, and the 3rd generation PV technologies are yet to prove their full potential in the PV market (Kim et al. 2018a).

22.2.2 First-Generation (or Silicon) Solar Cells

Silicon solar cells are classified into three categories based on the silicon used. These include monocrystalline (m-Si), multi-crystalline (mc-Si), and amorphous silicon (a-Si). Silicon is an indirect bandgap material that is successfully used to make commercial solar cell modules for almost 4 decades. Several different silicon solar cell structures are designed and optimized for achieving high efficiency are emerged in the last 20 years. These structures are presented in Fig. 22.5.

22.2.2.1 Silicon Solar Cell Fabrication Process Steps

While there are different types of Si solar cell technologies (Fig. 22.6), the process steps discussed below are common and applicable in most Si solar cell fabrication with minor changes in the cell design and process. The fabrication of silicon solar

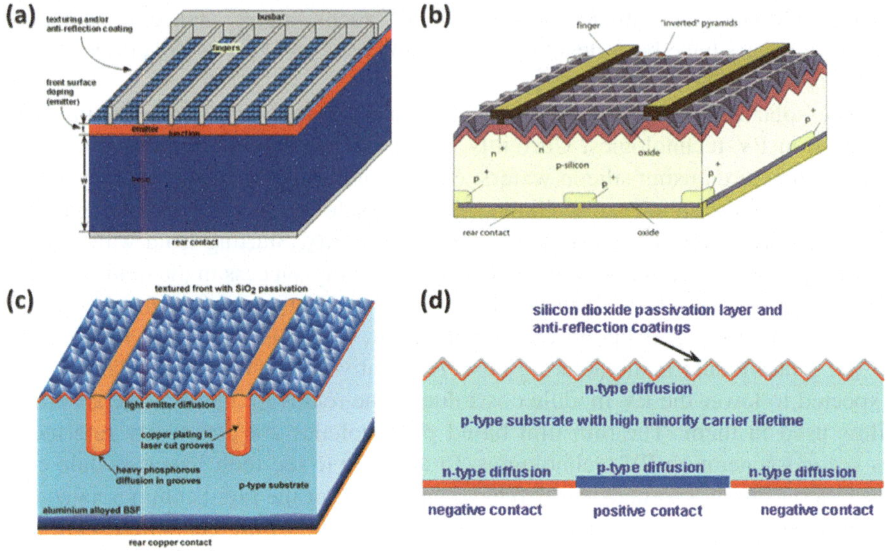

Fig. 22.5 Schematic showing various types of silicon solar cell structures. **a** Standard screen-printed silicon solar cell **b** passivated emitter rear localized (PERL), **c** buried-contact cell structure **d** back contact, back junction (BC-BJ). Reproduced with permission from www.pveducation.org Honsberg and Bowden (2019)

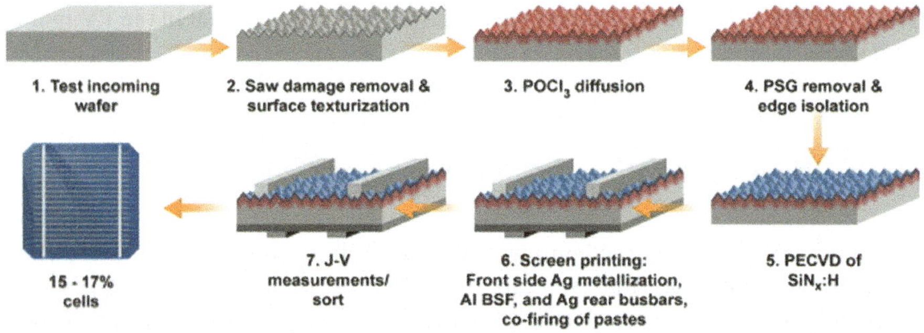

Fig. 22.6 Process flow for fabricating a standard c-Si solar cell. Reproduced with permission from Elsevier (Goodrich et al. 2013)

cells begins with a monocrystalline or multi-crystalline silicon wafer substrate (p-type, ~ 300–500 μm thick, 6 inch wafer). The cell fabrication process steps are discussed below.

Saw damage removal: The silicon wafers used in PV manufacturing are obtained by cutting ingots using either slurry based sawing or diamond wire sawing. In general, the sawing process results in a small mechanical damage and a surface contamination layer on each face of the wafer. Therefore, as a first step of the cell manufacturing process, this damage and surface contamination should be removed. Typically, 10–20 μm of silicon is etched from both top and bottom faces of wafers. The etch depth is based on the type of sawing used i.e. diamond wire sawing or slurry based sawing. The saw damage removal is typically done using acidic (mixture of HNO_3, HF and CH_3COOH) or hot alkaline (NaOH or KOH in water) solution (Park et al. 2009).

Surface Texturization: After the saw damage removal etching step, the silicon wafer surface reflects more than 30% of the incident light. To minimize the optical absorption losses due to surface reflection, the wafer surface undergoes a process called "texturing" (Abdullah et al. 2016). The reflection losses in commercial silicon solar cells are reduced mainly by random chemical texturing. The texturization of mono-crystalline silicon wafers with randomly distributed pyramids on the surface can be achieved by treating the wafer in a mixture of diluted NaOH and KOH with isopropanol (IPA) at 80 °C. This random texturization process is not so effective in case of multi-crystalline silicon wafers compared to mono-crystalline silicon wafers. Through this texturization process, the optical reflection losses from the silicon wafer surface can be lowered down to 10% from 30% seen at the end of the saw damage removal step.

Emitter diffusion (junction formation): As described previously, the photovoltaic cell operation requires a semiconductor junction to separate the photogenerated electron–hole pairs and develop photovoltage. Therefore, an n-type impurity (e.g. phosphorus) is introduced into the starting p-type silicon wafers in this step to form the junction. $POCl_3$ is the commonly used source for phosphorous doping in silicon solar cell manufacturing (Wolf et al. 2015). Typically the diffusion of $POCl_3$ is done inside a furnace at a temperature of ~ 900 °C. The p–n junction is formed about 0.5 μm below the top surface when the phosphorous diffusion is carried for 30 min at 900 °C.

There exist alternatives technologies to $POCl_3$ furnace diffusion for dopant deposition, including screen printing (Kwon et al. 2011), spin-on dopants (SOD) (Yadav et al. 2015), or chemical vapor deposition (CVD) (Schmich et al. 2007).

PSG Removal and Edge isolation: In the previous step, the diffusion of phosphorous dopant source happens via the gas phase, and as a result, the emitter layer is usually formed on the entire surface of the wafer. This results in the formation of shunts between the n-type and the p-type region of the device. To address this issue or isolate the front and the rear side emitter, edge isolation is done after the emitter diffusion step. Typical edge isolation methods (Comparison of different techniques for edge isolation 2001) include mechanical (Arumughan et al. 2005), laser cutting (Abbott et al. 2007; Kray et al. 2007; Zhang et al. 2010), or plasma etching. Plasma etching is the most common and popular method used in industrial silicon solar cell manufacturing. By using fluoride or oxide plasma environment, electrical isolation of the front and the rear emitter is achieved by removing ~ 2–5 μm of silicon from the edges.

Antireflection coating (ARC): As discussed earlier, the top surface of silicon reflects ~ 10% of the incident light after the surface texturing step, and this results in a significant loss in terms of the device efficiency. Therefore, further reduction of surface reflectance from silicon solar cell can help improve the cell efficiency. This can be achieved by depositing an antireflection coating (ARC) on the top surface. A critical material requirement for ARC is the suitable refractive index of the ARC film. In general, materials with refractive index between 1.4 and 2.7 can be used as an antireflective coating (ARC) for silicon solar cells. Silicon nitride (SiN_x) is the popular choice as ARC for silicon solar cells due to its optimal refractive index and bulk passivation of surface states (Sopori 2003), and as a result enhancement in the short circuit current density (J_{sc}) and open circuit voltage (V_{OC}) can be observed. Typically, the SiN_x is deposited using plasma enhanced chemical vapor deposition (PECVD).

Double layer ARC coatings [e.g., SiN_x/SiO_2 (Kim et al. 2013), MgF_2/SiN_x (Chu 2011)] have also been shown to mitigate the surface reflection losses and improve the passivation. In the last decade, plasmonic solar cells (Matheu et al. 2008) that leveraged plasmonic scattering of metal nanoparticles has emerged as a new paradigm in solar cells. Owing to silicon's indirect bandgap, plasmonic scattering can enhance the absorption in conventional silicon solar cells (Green and Pillai 2012) Plasmonic silicon solar cells with improved efficiencies are reported by leveraging the surface plasmon resonance (SPR) of metal nanoparticles [e.g., Ag NPs (Sardana et al. 2015; Sardana 2014)] deposited on the top surface of silicon.

Metal contact formation (metallization): The process of contact formation is a vital solar cell processing step because it strongly affects various properties of the silicon solar cell, such as short circuit current, open-circuit voltage, series resistance, shunt resistance, and the fill factor. For high-efficiency solar cells, the front electrodes should have low series resistance and low area coverage (Schroder and Meier 1984) The most widely used metal contacting technique for silicon solar cells is screen-printing (Hoenig et al. 2012). The front side contact is achieved by a screen-printed silver paste (Hilali et al. 2006). In contrast, the rear side electrode formation and surface passivation are attained by alloying a screen-printed aluminum paste with silicon (Gatz et al. 2011). The pastes are subsequently dried in an oven at a temperature of approximately 300 °C.

Contact firing: The screen-printed contacts initially lie on top of the insulating antireflection coating. The silicon solar cells undergo a short heat treatment at temperatures up to 900 °C, using an IR belt furnace (Cooper et al. 2010). During the firing process, the antireflection coating layer experiences selective dissolution such that the front Ag contacts penetrate through the antireflection coating onto the emitter while avoiding deep penetration into bulk silicon. Simultaneously, the back surface field (BSF) formation with aluminum is achieved.

22.2.3 Second-Generation (or Thin-Film) PV Cells

Second-generation photovoltaics primarily include thin-film solar cells such as a-Si or Cadmium Telluride (CdTe) or (CIGS) (Chopra et al. 2004; Lee and Ebong 2017). These technologies are already matured and have been in the market for almost 2 decades.

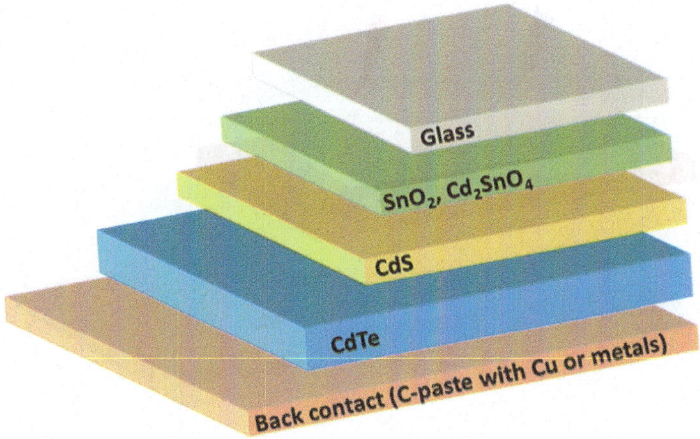

Fig. 22.7 Schematic of a CdTe solar cell device structure

22.2.3.1 CdTe Solar Cells

The CdTe is the leading contender to the Si PV and comprises ~ 5% of the total installation worldwide in 2019 (Fig. 22.3). Also, this technology entered the commercial market almost 2 decades ago.

CdTe Solar Cell Fabrication Process

The most commonly used CdTe solar cell is fabricated by forming a heterojunction between CdS and CdTe. Figure 22.7 shows the cross-sectional image of CdTe solar cell structure fabricated on glass substrate with a junction formed between CdTe and CdS. The first step in the CdTe solar cell device fabrication is the substrate cleaning. The commonly used substrate in CdTe solar cells is SnO_2-coated glass that can be obtained commercially. After cleaning the SnO_2-glass substrate, CdS film is deposited using chemical-bath deposition (CBD). This is followed by the deposition of absorbing CdTe thin-film. Different techniques including close-spaced sublimation, vapor-transport deposition, or electrodeposition for CdTe deposition. After CdTe deposition, CdTe is prepared (chemical etching) for back contact formation. The back contact can be carbon paste containing Cu_xTe and HgTe, or other metals (e.g. Cu). The inclusion Cu in the back contact improves both the performance and stability of the CdTe solar cells.

22.2.3.2 CIGS Solar Cells

$CuIn_xGa_{1-x}Se_2$ (or CIGS) solar cell is another prominent example for thin-film solar cells. A schematic of the device is shown in Fig. 22.8. In these devices, the absorbing

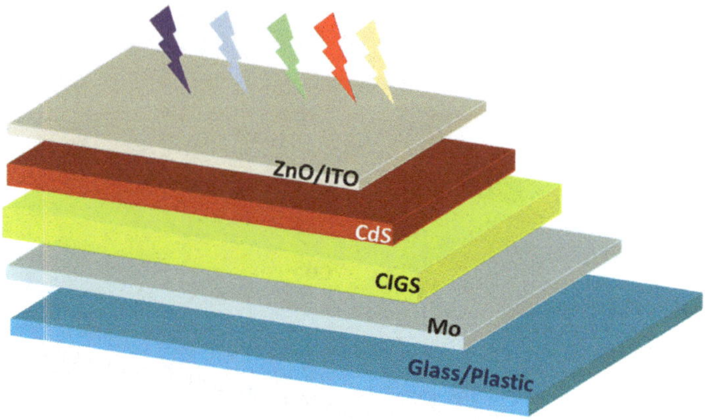

Fig. 22.8 Schematic of a CIGS solar cell device structure

film is formed by alloying $CuInSe_2$ and $CuGaSe_2$, and the bandgap of the absorbing film can be tuned by modifying the composition of In and Ga. Both $CuInSe_2$ and $CuGaSe_2$ have a direct bandgap, very high absorption coefficient (10^5 cm^{-1}),and typically 1–2 μm thick active absorbing films are used in the devices.

CIGS Solar Cell Fabrication Process

The substrates used for CIGS cell fabrication includes either glass, metal foil, or plastic. The first step in the device fabrication is coating of 0.5–1 μm Mo film on glass (i.e., substrate) using sputtering process. The next step after Mo film coating is the formation of active CIGS absorber film. The CIGS film can be deposited on top of Mo film using different methods such as evaporation of elements, sputtering of metals followed by selenization with H_2Se, reactive sputtering of metals with Se vapor, or printing of metals from ink precursors followed by selenization. The growth of CIGS is assisted by Na atoms that are diffused through Mo film into the CIGS. After this CdS layer is formed on top of CIGS film by using CBD technique. This is followed by deposition of a bilayer consisting of intrinsic and conducting ZnO. ZnO can be deposited using either sputtering or CVD. The absence of Na in the device may lower the device efficiency by 2–3%.Therefore, when non-Na-containing substrates are used, alternative steps are required to improve the efficiency such as introducing NaF during the CIGS deposition or forming a 6–12 nm NaF layer on Mo back contact. These cells can be further processed to make CIGS modules for practical applications. The schematic for the CIGS module fabrication process flow is shown in Fig. 22.9.

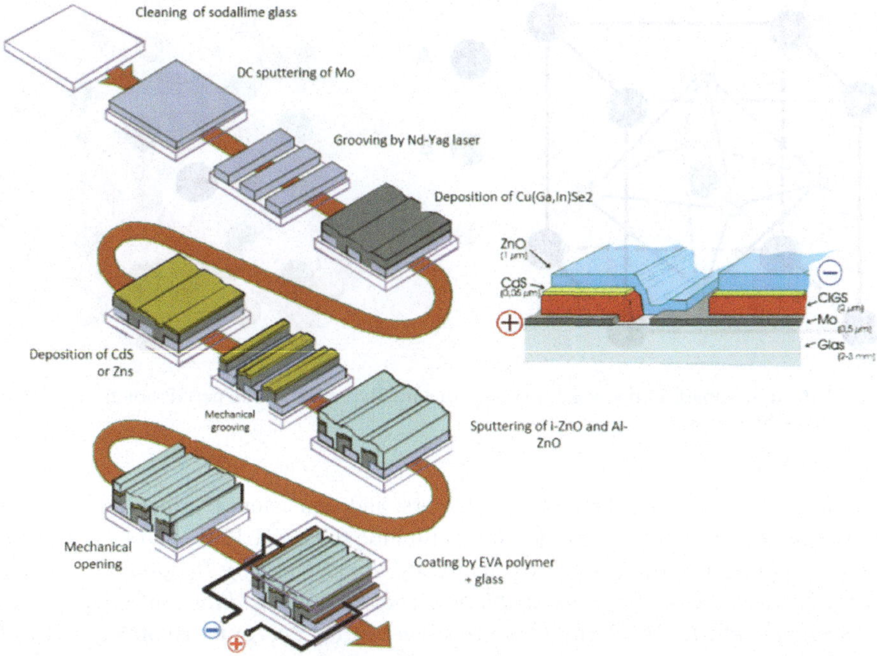

Fig. 22.9 (Left) A schematic showing the CIGS photovoltaic module fabrication process steps starting from cleaning of the glass substrate. (Right) A cross-section view of CIGS solar cell depicting different layers in the device. Reproduced with permission from EDP Sciences-SIF (Slaoui 2017)

22.2.4 Third Generation (or Emerging) Solar PV Cells

Third-generation solar cell technology includes dye-sensitized solar cells, perovskite solar cells (PSCs), quantum dot (QD) solar cells, etc. In this section, we limit our discussion to PSC-based solar cells owing to their potential for commercialization and multiples environmental issues related to the processing and use of PSCs.

22.2.4.1 Perovskite Solar Cells (PSCs)

Perovskite solar cells are the devices in which the light-harvesting active layer has a perovskite structure. The perovskites have a general formula of ABX_3 with a cubic structure. In ABX_3, A is a monovalent cation (cesium (Cs), methylammonium (MA), or formamidinium (FA)); B is a divalent cation (Pb or Sn); and X is a halogen (e.g., Cl, Br, or I) (Fig. 22.10). Several perovskite compounds are reported in the literature, and only a few are actively pursued for PSCs applications. Among them, organic–inorganic hybrid perovskites (OIHP) have been the leading candidates. The OIHP compounds have shown great promise in applications, including solar cells,

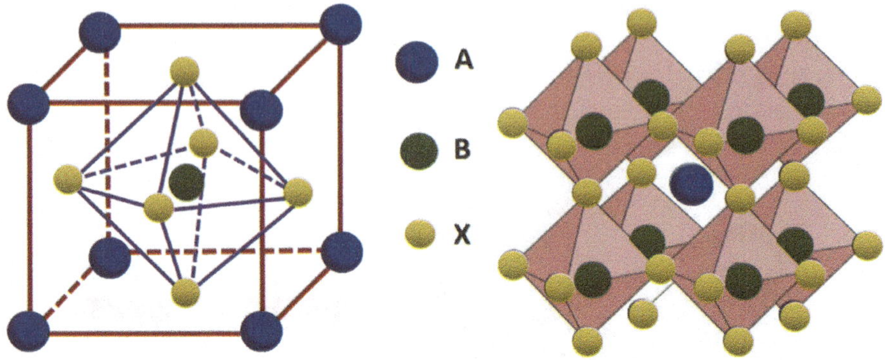

Fig. 22.10 **a** Schematic of perovskite crystal structure. Reproduced with permission from Elsevier (Zhao and Wang 2018)

light-emitting diodes (LEDs), photodetectors, and transistors (Kojima et al. 2009). In particular, the Methyl Ammonium Lead Iodide ($CH_3NH_3PbI_3$) perovskite owing to its excellent material properties, including *direct bandgap, tunable visible and near-infrared light absorption, ambipolar transport, long diffusion lengths, high absorption coefficient (*10^4 cm^{-1}*), low exciton binding energy (*~ 10 meV*), and high charge carrier mobilities* combinedly contributed significantly to the progress of perovskite solar cells (PSCs) (Stranks et al. 2013; Herz 2017; Tan et al. 2014; Xing et al. 2013).

22.2.4.2 Perovskite Solar Cell (PSC) Fabrication Process

Perovskite solar cell (PSC) fabrication predominately uses solution methods for depositing different films that are part of the device structure. Besides their exciting optoelectronic properties, the feasibility of solution-based synthesis and fabrication is also a major reason for tremendous progress of PSCs through university laboratory research. The fabrication technology of PSCs has evolved significantly (Park and Zhu 2020). Optimization of each process step in the device fabrication is critical for the enhancing the respective film properties and also the overall device efficiency and stability. Beside the absorbing perovskite film, the PSC device structure may require two additional layers, that are deposited on either side of the perovskite film. These layers are often referred to as charge transport layers, and one of these two layers function as electron transport layer (ETL) and the other one as hole transport layer (HTL). In literature, a huge library of materials are reported that can function as either ETL or HTL.

In this section, we introduce general steps involved in the $CH_3NH_3PbI_3$ perovskite device fabrication (Dharmadasa et al. 2019). The first step in the PSC device fabrication process is the preparation of materials and solutions required for depositing each film in the device structure. In the device structure presented below, the PCBM and PEDOT:PSS are used as ETL and HTL, respectively.

Perovskite Solar Cell (PSC) device fabrication steps:

The device process flow is summarized in Fig. 22.11. Pre-cleaned conducting ITO-glass or FTO-glass substrates are used as substrates for the device fabrication. Note that, typically, all the process steps, except substrate preparation, related to device fabrication are done inside a glove box. The substrate is first pre-cleaned using commercial soap solution (e.g. Hellmannex). The pre-cleaning step is followed by solvent cleaning which includes cleaning in an ultrasonic bath with DI water,

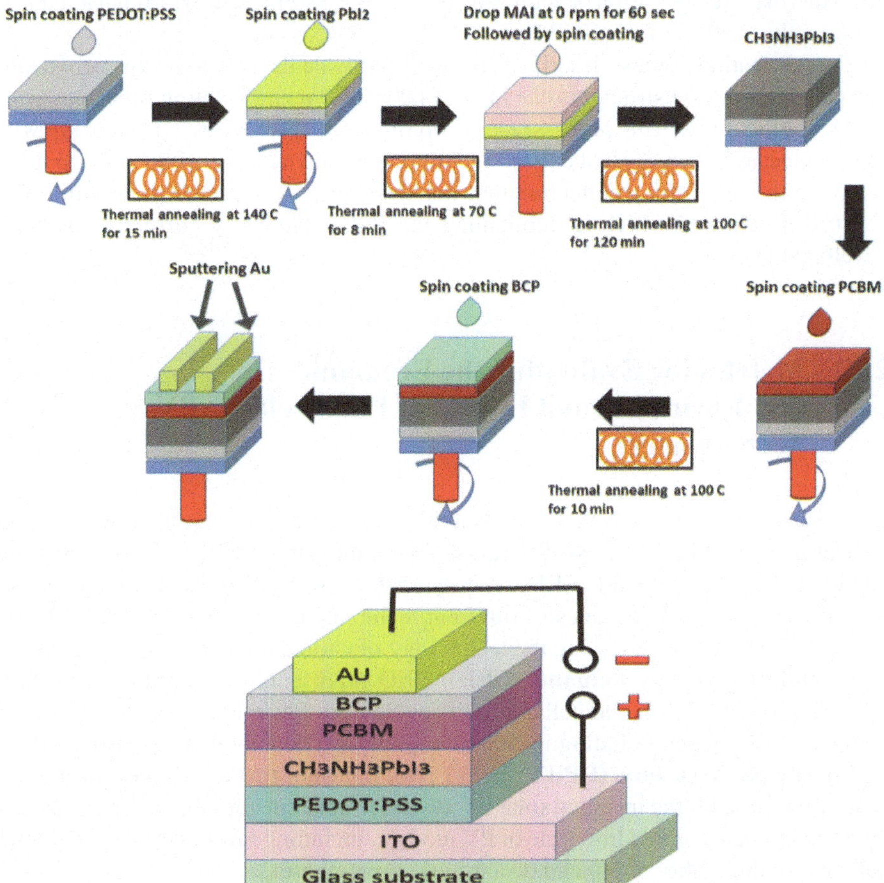

Fig. 22.11 Sample device process fabrication steps. (Top) Schematic diagram depicting different process steps involved in the fabrication of a typical PSC device with $CH_3NH_3PbI_3$ perovskite absorbing film. (Bottom) Schematic of completed PSC device with PCBM and PEDOT:PSS as electron transport layer (ETL) and hole transport layer (HTL), respectively. Reproduced with permission from Springer Science (Dharmadasa et al. 2019)

acetone, and Isopropanol, in that order, for 20 min each. The cleaned substrates can be subjected to ultraviolet-ozone treatment for 30 min to improve the surface adhesion. A thin film of PEDOT:PSS is uniformly coated onto the cleaned ITO substrate. The film is then annealed at 140 °C for 15 min. After this, the $CH_3NH_3PbI_3$ (1.2 M in 0.7 ml DMF and 0.3 ml DMSO) solution prepared overnight is spin-coated on the sample with spinner programmed to 1000 rpm for 5 s and 6000 rpm for 45 s. After 10 s, in the second step, 500 μL of toluene can be dripped on the sample. The spin-coated films are annealed at 100 °C for 10 min. The PCBM dissolved in chlorobenzene (20 mg/mL) is spin-coated onto the perovskite layer at 1000 rpm for 60 s and annealed at 80 °C for 15 min. Then, 1 mg/mL of BCP solution in ethanol is spin coated at 4000 rpm for 30 s. Finally, 100 nm thick Ag electrodes are deposited using thermally evaporation or sputtering (~ 1×10^{-6} Torr pressure) through a shadow mask on the surface of the BCP layer.

The fabrication process described above primarily deals with p-i-n type perovskite device. Other device structures can also be fabricated using solution processes with slight modifications in the process steps to optimize film properties. Different types of PSC device architectures that exist in literature are shown in Fig. 22.12. Mesoporous is among the first PSC architecture that was developed. The planar perovskite solar cell (p-i-n or n-i-p) which is dominating the PSC technology currently has been developed later.

22.3 Metrics for Evaluating the Economic and Environmental Impact of Photovoltaic (PV) Technologies

Before discussing the environmental, health, and safety aspects, we summarize some critical metrics used to assess different PV technologies. Table 22.1 compares the cumulative energy demand (CED), energy payback time (EPBT), and Greenhouse as emission rate (GHG) values of different technologies (Ludin et al. 2018) These are key metrics while assessing the technology of choice for future energy needs.

Cumulative energy demand (CED): CED represents the amount of primary energy used within the solar cell entire life cycle from raw materials extraction until the end-of-life stage, including installation, transportation, and energy-mix used.

Energy payback time (EPBT): EPBT is an indicator that represents the amount of time required for the installed solar PV system to generate the equivalent amount of energy used in the whole life cycle of PV module, including raw materials extraction, cell and module fabrication, and decommissioning phase.

Green House Gas (GHG) emissions rate: A key goal in moving towards the renewable energy technologies such as solar is to reduce the GHG emissions. Therefore, it is essential to develop an metric to understand the respective trends. In this

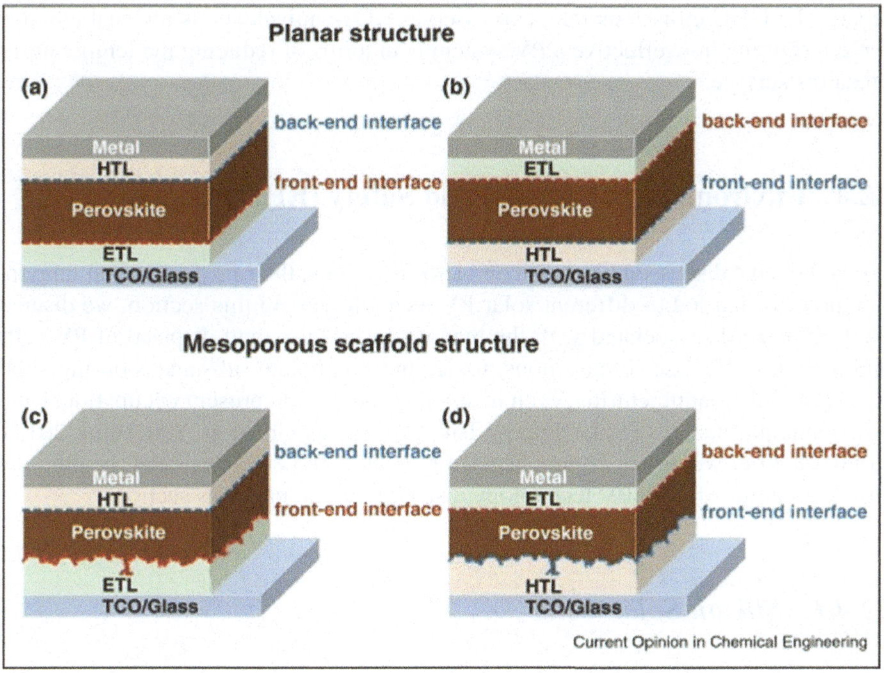

Fig. 22.12 Schematics of PSC and module architecture. **a** planar n-i-p PSCs. **b** planar p-i-n PSCs, **c** mesoporous n-i-p, and **d** mesoporous p-i-n device structures. Reproduced with permission from Elsevier (Manspeaker et al. 2017)

Table 22.1 A comparison of different PV technogies using CED, EPBT, and range of GHG emissions based on literature reports from the year 2000

Solar PV technology	Range of CED	Range of EPBT	Range of GHG emissions (g CO_2 eq/kWh)
Mono-Si	1123–8050	1.4–7.3	29–671
Multi-Si	1034–5150	0.8–4.17	12.1–569
a-Si	862–1731	1.1–3.2	8.1–57
CdTe	811–1803	0.79–2.7	8.9–66
CIS	1105–1684	1.3–2.8	33–95
DSSC	277–365	0.6–1.8	9.8–25
Perovskite	379–821	0.2–5.4	56.65–497.2
Quantum dot	370–1030	0.9– 1.51	2.89–5

Reproduced with permission from Elsevier (Ludin et al. 2018) regard, the GHG emissions rate, expressed as CO_2-equivalents, is a valuable index for determining how effective a PV system is in terms of reducing the temperatures on earths surface.

22.4 Environmental, Health, and Safety (EHS) Impacts

So far we have discussed key semiconductor materials, their properties and fabrication process that led to different solar PV technologies. In this section, we discuss the EHS hazards associated with the manufacturing, use, and disposal of PV cells and modules. The use of hazardous, toxic, and flammable substances during solar cell or module manufacturing, even in small amounts, can present occupational and environmental hazards (Solar Energy Isn't Always as Green as You Think 2014). In this section, we give a comprehensive review of the environmental, health, and safety concerns of each PV technology discussed in the previous sections.

22.4.1 Silicon Solar Cells

In this section, we review the toxic effects of silicon-PV cells. The first life cycle assessment (LCA) research was on silicon-solar cells was reported 15 years ago (Koroneos et al. 2006). Despite being the oldest PV technology with more than 3 decades of field performance studies, the number of studies investigating different types toxic effects at different stages of silicon solar cell life cycle is limited in the literature.

22.4.1.1 Occupational Health Issues

As discussed in the silicon solar cell process steps, multiple acids including hydrofluoric acid (HF), nitric acid (HNO_3) as well as alkalis (e.g., NaOH) are used at different stages in the silicon solar cell processing. The potential hazards from using these acids and solvents include acid burns and fumes inhalation (Fthenakis 2018). The inhalation of dopant gases and vapors, e.g., $POCl_3$ and B_2H_6 that are commonly used as dopant sources for silicon, is hazardous. Even though $POCl_3$ is in liquid form, the generation of toxic P_2O_5 and Cl_2 gaseous effluents in the diffusion furnaces is a significant health concern for the safety of the workers. Therefore, adequately designed ventilation should be in place in the process area to avoid any possible inhalation hazards. The flammability of silane (SiH_4) used in SiN_x deposition using PECVD poses significant safety hazard. Further, the by products from SiN_x deposition are also considered occupational health hazards (Moskowitz and Fthenakis 1991). There were reports of workers being affected by the silane leak in plants (Biello 2010). One pathway to avoid silane transport and frequent installations is storing locally and avoid frequent transport and installation. Engineering and safety protocols such as proper gas-handling systems are used to minimize the risks of fire and explosion from explosive and hazardous gases and their by products in PV manufacturing (Fthenakis 2018; Moskowitz and Fthenakis 1991).

22.4.1.2 Environmental Issues

The environmental issues in silicon solar cell technology primarily arise in liquid and solid waste generation. The first step in silicon solar cell manufacturing is obtaining polysilicon wafers from metallurgical grade silicon (Zulehner 2000). In this step, metallurgical-grade silicon is treated with hydrochloric acid to form trichlorosilane ($SiHCl_3$). Then, this $SiHCl_3$ reacts with hydrogen to form polysilicon. A by-product of this refinement process is liquid silicon tetrachloride ($SiCl_4$). The $SiCl_4$ can form corrosive and toxic hydrogen chloride (HCl) has upon its reaction with water or air On average the production of every ton of polysilicon involves producing three to four tons of $SiCl_4$ (Wu et al. 2010). In addition, there is solid and liquid waste generated during wafer slicing, cleaning, and etching process steps. There is also solid waste generated during module assembly (e.g., stainless-steel wire) and the wafers slicing from ingots (abrasive slurry), and liquid slurry waste (Li 2021).

The fabrication of crystalline silicon panels requires various materials such as silicon, glass, polymers, silver, copper, boron, phosphorous, tin, tin oxide, lead (Sica et al. 2018) Even though there may not be any environmental concern during the silicon PV module operation, the pH dependent leaching of heavy metals such as Al, Cu, Ni, and Pb from module pieces is a significant concern. Therefore, it is critical to recycle the solar panels safely to avoid any problems related to drinking water, aquatic life, soil, agriculture, etc.

22.4.1.3 Solutions for Safer and Environmentally Friendly Silicon PV Technology

Since the waste produced by using oil-based slurry during silicon wafer slicing can have significant negative environmental impact, extensive cleaning of the wafers prior to silicon solar cell fabrication is required. Therefore, slurries based on glycol or water should be preferred as an alternative to oil-based slurry. The generation of toxic effluents from the belt-furnaces or tube-furnaces during the $POCl_3$ diffusion step demands regular cleaning of the diffusion chamber with acids such as HF. The use of belt-furnaces, on the other hand, can offer an environmentally friendly process technology because of the water-soluble and non-toxic materials or gases used for the doping step (Yadav et al. 2015). Typically, laser cutting or plasma etching are used for edge isolation after $POCl_3$ diffusion step. Also, the use of CF_4 in plasma etching should be avoided as it is a dangerous greenhouse gas. Hence, safer alternatives to CF_4 should be considered in future for plasma etching step. Also, a safer non-pyrophoric alternative to SiH_4 for ARC coating step should be considered. The soldering residues formed while making the interconnections in a PV module requires cleaning with chlorofluorocarbons (CFC) compounds, which can cause atmospheric ozone depletion. The CFC usage should be avoided, and instead, low-residue fluxes should be used in PV manufacturing facilities (Fthenakis and Moskowitz 2000). If the residues or fluxes on the cells cannot be avoided, materials that produces water soluble fluxes or residues should be given priority for safer practices (Evans 2007; Tsuo et al. 1998).

22.4.2 Thin-Film Solar Cells

In the case of thin-film solar cells, including a-Si, CdTe, CIGS, etc., the environmental, health, and safety concerns are more severe compared to silicon solar cells as discussed below. Table 22.2 shows the total mass of different elements used to manufacture crystalline and thin-film solar cells.

22.4.2.1 Occupational Health and Safety Issues

The degree of occupational health hazards related to Cd and Te compounds can vary. This is because each process step in the device fabrication can produce compounds with different toxicities. In addition, the physical state and the mode of exposure can

Table 22.2 Total mass of elements in one module piece for c-Si, a-Si, CdTe, and CIGS

Element	Total mass (in mg) per 1 module piece ($5 \times 5\ cm^2$)			
	c-Si	a-Si	CdTe	CIGS
Ag	7.8 ± 0.9	2.2 ± 0.3	0.05 ± 0.005	1.2 ± 0.4
Sn	21.3 ± 1.1	31.0 ± 1.7	12.5 ± 3.9	19.1 ± 0.4
Zn				16.1 ± 1.6
Cd			14.9 ± 1.6	0.2 ± 0.001
Te			15.9 ± 1.1	
In				19.2 ± 0.7
Ga				0.7 ± 0.2
Se				8.2 ± 0.8
Al	167.2 ± 49.9			
Mo			13.0 ± 1.8	5.0 ± 0.2
Cu	7.8 ± 0.9	130.4 ± 16.7	74.5 ± 4.7	146.2 ± 5.7
Ni		1.0 ± 0.2		
Pb	15.9 ± 1.2			

Reproduced with permission from IOP Publishing Ltd (Nover et al. 2017) Copyright (2017) The Japan Society of Applied Physics

Table 22.3 Summary of the hazardous materials used in silicon and thin film solar cell manufacturing

PV technology	Material/precursor	Purpose	Safety concerns
Si	HF	Etchant	Noxious, corrosive
	Chlorosilane	Si deposition	Toxic fumes upon decomposition
a-Si	Chlorosilane	a-Si deposition	Toxic fumes upon decomposition
	Diborane	a-Si deposition	Highly toxic
	Hydrogen	a-Si deposition	Fire hazard
	Phosphine	a-Si deposition	Toxic and flammable
	Silane	a-Si deposition	Fire and explosive
CdTe	Cd compounds	CIS film deposition	Carcinogenic
CIS	Cd Compounds	CIS film deposition	Carcinogenic
CIS	Hydrogen selenide	CIS film deposition	Highly toxic and flammable
CIS	Hydrogen sulfide	CIS film deposition	Highly toxic and flammable

lead to health hazards of varying severities (Table 22.3). The Occupational Safety and Health Administration (OSHA) agency of the US has regulated the use of Cd (including CdTe) due to reasons related to its toxicity and potential for causing lung carcinogen (Occupational Safety and Health Administration (1971)). The accidental ingestion of Cd compounds by personnel in the research laboratories and production environments is possible when proper and safe handling practices of these toxic and hazardous gases are not in place (Elinder and Järup 1996; Genchi et al. 2020). The environmental risks can vary depending on the physical state of the Cd compound used or Cd products released to the environment. For instance, as with many metal particulate matter, fabrication processes in which fine particles of Cd compounds are used or produced present more significant health risks. Hence, the laboratories and industrial facilities that uses volatile or soluble Cd compounds must take significant care during their use, and must discharge volatile by products into the environment only after careful treatment.

CdTe is typically deposited using spray pyrolysis. Given that spray pyrolysis is not an efficient process, a high amount of by products and fine particles are generated during the CdTe deposition step. Furthermore, these by products/particles can also get deposited on the spray process chamber wall. As a consequence, periodic cleaning (either scraping or chemical removal) of the reactors is required. More importantly, spray pyrolysis deposition of CdTe can produce dangerous by products such as Cd and CdO. Therefore, proper care must be taken while operating the spray pyrolysis reactors to avoid accidental exposure to Cd and CdO fumes (Kolesnikov et al. 2017).

Currently, there is only a limited information on the toxicity of process related emissions for the CIS and CGS solar cells. One known major hazard in CIS processing is the use of hydrogen selenide feedstock which is highly toxic. While mechanical scribing and deposition are known to produce airborne hazards (e.g., Cu, In, and Cd) in CIS and CdS processing, existing data, which is not extensive, suggest concentrations of these elements are below their threshold levels.

22.4.2.2 Environmental Issues

Chronic exposure to Cd compounds poses serious health concerns to the public (Järup and Åkesson 2009). The Cd compounds may be released into the environment as a by product of different manufacturing steps or as a waste from the uncontrolled disposal of used photovoltaic modules at the end of their life. However, the magnitude of potential emission from different manufacturing and laboratory facilities can vary depending on the fabrication processes used and the implementation of emission controls.

As mentioned before, the techniques used for the deposition of CdTe also play a key role in determining the amount of effluents generated in CdTe cell manufacturing. For example, electrodeposition is considered as a less hazardous process. On the other hand, a less efficient deposition methods such as processes spray pyrolysis (with an estimated efficiency of 10%) would generate significant amount of hazardous (e.g. $CdCl_2$, CdO, HCl, H_2S, and thiourea) by products (Fthenakis and Moskowitz 2000). Even though some of these by products will be deposited on the spray pyrolysis reactor's walls, most of them will escape through exhaust in gaseous form. Previous studies indicates significantly less CO_2 emissions (20 g /kWh) from CdTe photovoltaics during the lifecycle compared to that emitted from fossil fuel plants (500–1000 g CO_2/kWh). Also, emission of other priority pollutants such as SO_2, NO_X, particulates, etc., amounts to only 2–4% of that emitted from fossil-fuel plants (Fthenakis et al. 2008).

Beyond the traditional pollutants, the heavy-metal emissions (e.g., Cd, Se) are also important in case of CdTe and CdS, CIGS solar cells. Experimental investigations also confirmed that the Cd can be safely encapsulated in the module using molten glass which makes it less fire-proof for residential use (Fthenakis 2004). In addition, the direct cadmium emissions in the life cycle of CdTe solar cells amount to 0.02 g per GWh of energy produced under average US conditions (Fthenakis et al. 2005). It is worth noting that the magnitude of cadmium emission from CdTe PV cells is two orders less compared to an average coal-burning plant that emits 2 g Cd per GWh (Fthenakis 2004). First Solar, a leading manufacturer of CdTe solar modules, expressed strong confidence about the green nature of this PV technology. Their studies confirmed there is no occupational exposure hazard to Cd based on tests conducted over 15 years. The sample size for this assessment is approximately 1100 Cd exposure tests performed on the workers. These results indicate thin-film PV can be benign to the environment compared to fossil-fuel-based energy sources.

It is also essential to consider the potential risks involved with thin-film technology at the end of their lifetime. In spite of serious concerns about the cadmium in CdTe, and CIGS/CdS solar cells, the hazardous level is not directly related to the amount of toxic compounds used in a module but it mainly depends on leaching out potential of cadmium (Nover et al. 2017). Moreover, there is a possibility for cadmium release into atmosphere if the PV modules are disposed in landfills. The simulation studies using Toxic Characteristic Leaching Procedure (TCLP) can help understand the potential for leaching of the toxic elements. All three cadmium based photovoltaic modules i.e., CdTe, CIGS and CdS have passed TCLP studies and thus confirming that these thinfilm PV modules are safe to use under normal conditions. In contrast, recent investigation suggested that long-term leaching studies can provide better insights on potential harms due to leaching out of toxic elements by considering different conditions (e. g. pH, environment) (Nover et al. 2017). Even though existing literature did not provide any direct evidence of significant hazards under normal scenarios, the recycling of the heavy metal based solar panels should be done with carefully at the end of their life. Hence, the recollection of used modules should be a major priority for both manufacturers and the governments. Toxic emissions are much lower in the life cycle of thin-film photovoltaics than in the life cycles of alternative photovoltaic- and conventional-power systems (Fthenakis et al. 2008).

22.4.3 Perovskite Solar Cell (PSCs)

As mentioned in the first section of this chapter, the PSCs offer unique advantages compared to Silicon or thin-film (CdTe or CIGS) based PV technologies, such as the ease of chemical tunability, solution processability. However, these same advantages can be the downside with significant negative environmental impact. An important downside of the PSC technology is that the presence of heavy metals in the perovskite absorbing film. A schematic of the perovskite solar cell life cycle is shown in Fig. 22.13.

22.4.3.1 Occupational Safety and Health Hazards

While the Pb presents a significant health hazard, perovskite solar cell device processing is typically conducted in a glove box (Salvador et al. 2020). This can potentially minimize the exposure to Pb-containing compounds. Even though the perovskite device fabrication takes place inside glove box, the unintentional or accidental exposure to Pb present in perovskite can occur at other times, including testing in ambient conditions. Upon contact with Pb-containing materials or devices, the ingestion and uptake can possibly happen via different routes (e.g., gastrointestinal, respiratory, and dermal). Besides the intoxication by the perovskites, the use of the toxic solvents such as dimethylformamide (DMF) and dimethylsulfoxide (DMSO) in the solution processing of the PSCs presents significant occupational safety and health challenges. In addition, the high miscibility of these solvents with water can increases their bioavailability. As a result, there is a significant risk of absorbing these toxic solvents by oral ingestion, dermal contact or permeable cell membrane (Ménorval et al. 2012). Due to these reasons, sufficient personal protection should be in place when working with these solvents. (e.g. 15 mil latex gloves, dust masks

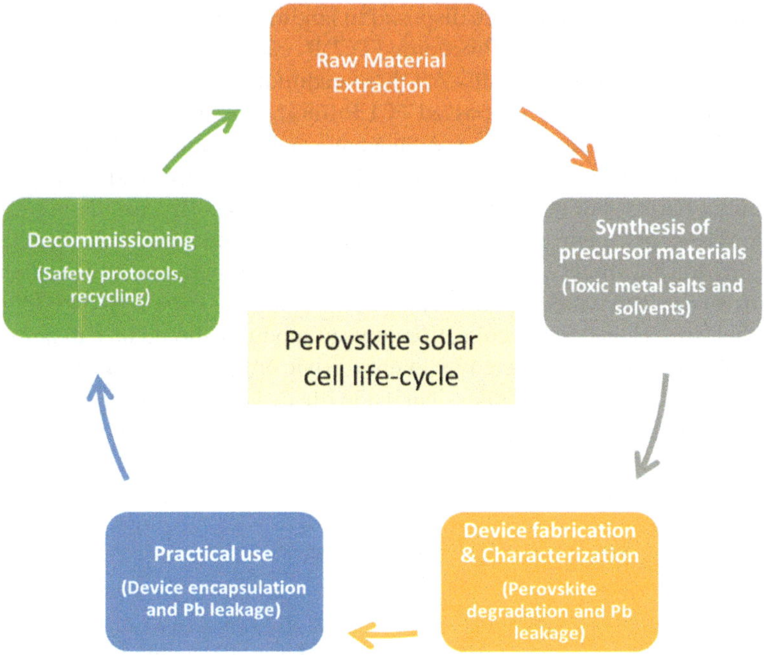

Fig. 22.13 Schematic of the life cycle of perovskite solar cells and associated hazards

and glasses). Further, the safe storage of waste products in the perovskite solar cell fabrication laboratories is paramount. Owing to the toxicity of DMSO and DMF, alternative solutions should be pursued in PSC processing. Recent studies show the feasibility of using tetraethyl orthocarbonate (with perovskite precursors) and anisole (with PCBM) as potential alternatives that can lead to avoiding the use of DMSO or DMF in PSC processing in future (Gardner et al. 2016; Zhang et al. 2020; Tian et al. 2019; Wang et al. 2017).

The Sn based perovskites also poses some toxic effect (Babayigit et al. 2016a). Owing to its lower reduction potential Sn(II) can be converted into Sn(IV) easily in ambient. As a result, the operators can be easily exposed to Sn(II) and Sn(IV) during the perovskite synthesis. This is a serious health hazard as the net gastrointestinal absorption of Sn depends on the oxidation state, and, Sn(II) and Sn(IV) are not readily interconverted when being transported in the human body. The conversion to Sn(IV) can be slowed down in laboratories where low-oxygen and dry environment is maintained. A summary of health hazards that may arise due to exposure to Pb and Sn are shown in Fig. 22.14.

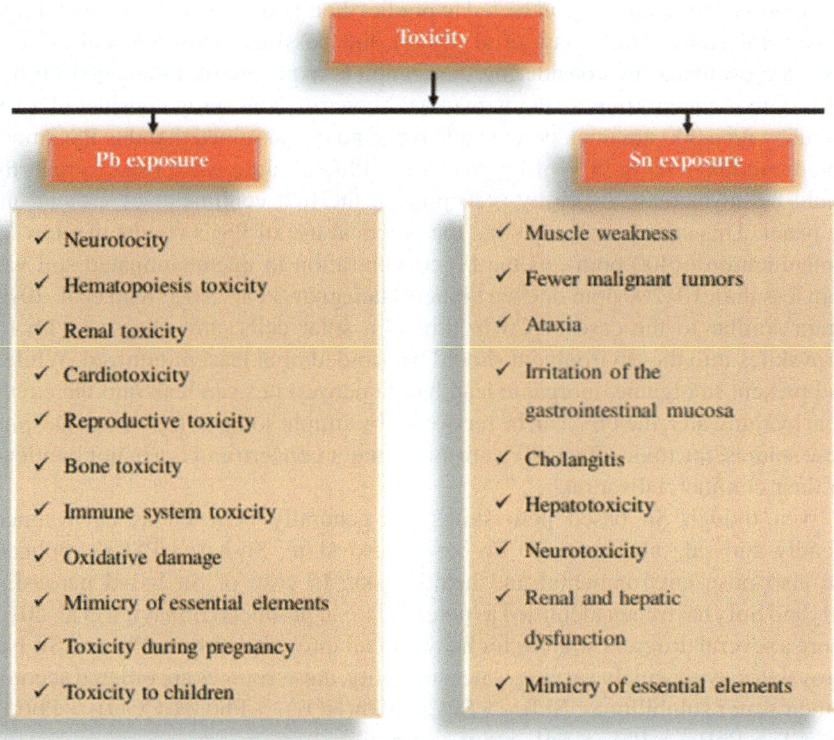

Fig. 22.14 Schematic diagram showing the health hazards for living beings upon exposure to Pb and Sn, which are the popular metal constituents present in perovskite films. Reproduced with permission from Wiley (Debnath et al. 2020)

22.4.3.2 Environmental Impact

A major challenge with $CH_3NH_3PbI_3$ based PSCs is the presence of heavy metal Pb. Lead is known to damage the nervous system and cause brain disorders. The solubility product (K_{sp}) of PbI_2 in water is on the order of 10^{-8}, which is much higher than those of other common heavy metal compounds used in solar cells (Fabini 2015). For instance, the K_{sp} values for CdS, PbS, and CdTe lie in the range of 10^{-27}–10^{-34}. Furthermore, unlike PbI_2 with three-atoms, CdS, PbS and CdTe are binary compounds, with up to 14 orders of smaller solubilities than PbI_2 in water (https://pubs.acs.org/doi/full/10.1021/acs.jpclett.5b00504). it can decompose into CH_3NH_3I–HI–CH_3NH_2 (methylamine). Hence, soil, rivers, and groundwater may be contaminated by Pb when the installed PSC modules are disposed without proper care. Also, the PbI_2 (lead iodide) formed upon decomposition of perovskite, can be further decomposed into Pb. This Pb can find its way into the groundwater via the effluent in significant quantities that can cause severe health risks. There is very little history avilable on the perovskite module life cycle analysis, and hence simulation studies must be conducted to predict their potential environmental impact. In a recent study (Hailegnaw et al. 2015), the possible environmental effects of PSCs are evaluated by considering the complete exposure of a damaged Pb-based perovskite device to rain as the worst-case scenario. The study considered a PSCs module with a 300 nm thick perovskite layer, and it is assumed that this PSC module would contain ~ 0.4 g m^{-2} of Pb. When the PSC module is exposed to rain, there would be an increase of Pb by ≈70 ppm in the first centimeter of ground under the panel. This study concluded that the practical use of PSCs results in a low level contamination (<400 ppm), as the Pb concentration in uncontaminated soil varies from less than 10–200 ppm or even higher (Hailegnaw et al. 2015; Kim et al. 2018b). Again similar to the case of Cd in thin film solar cells, any release of Pb from perovskites into the environment should be avoided or at least minimized. While the lead present in organic–inorganic lead halide perovskites can leak into the environment by rainwater, the Pb^{2+} can be recovered by simple solution treatment. Despite of these studies, the toxic effect of Pb are still a major concern and is a major bootleneck for their commercialization.

Even though Sn based perovskites are generally believed as environmental friendly and safe alternative to Pb based perovskite, Sn-based PSCs' application can also pose environmental and health risks. In case of Sn based perovskites, SnI_2 and SnI_4 have been identified as potential toxic products (Babayigit et al. 2016b). There a several different sources for heavy metal intoxication from Pb- and Sn-based perovskites. with widely varying water solubility, these sources are ordered according to their water solubility as: $SnBr_2 > SnCl_2 > Pb(NO_3)_2 > Pb(CH_3COO)_2 > Pb(OH)_2 > PbCl_2 > PbBr_2 > PbI_2 > SnI_4 > SnI_2 > SnO_2$.

Moderate exposure to different environments including humidity, oxygen, elevated temperature, etc., can degrade perovskites into harmful heavy metal compounds. Furthermore, the structural failure of a photovoltaic module can lead to leaching of these compounds into the environment. Like CdTe modules, even if the PSC modules are encapsulated there is potentially a higher environmental risk when these panels are exposed to rain. Especially during their practical use, the PSC modules need to be installed in outdoors or rooftops, and there is a potential for these panels to get damaged and consequently exposed to moisture, rain, etc.

Even though recent studies indicated that the amount of Pb spilled from defective PSCs would be orders of magnitude smaller than that emitted from other energy-related activities, we need to remember that this conclusion based on a limited data. Hence, careful toxicological investigation is highly required to further understand the dangers associated with local accumulation of perovskites (Fabini 2015). It is also possible, in a worst-case scenario, the perovskite degradation products could be transported by soil water and eventually accumulate in unpredictable locations (Nelson and Campbell 1991). The composition of Pb penetrated into soil is highly dependent on soil composition and acidity (McBride et al. 1997).

One may think that the total metal content per square meter of a solar panel is only a few hundred milligrams which is less risky. However, the large scale implementation of the PSC technologies may cause severe safety and health hazards. At this point, due to limited field data, it is better not to conclude the environmental impacts of PSCs as either positive or negative. Similar to thin film solar cells, the safe deployment of PSC technology relies entirely on adopting precautionary measures against contamination at each stage of the device's life, from fabrication to disposal/recycling. In this direction, dedicated EHS standards should be developed as part of academic research during their development. Such standards can be beneficial especially when multiple personnel work on PSC device fabrication at the same time. Finally, the safety protocols developed in academic research labs can be easily implemented in industrial settings where the expected risks can be further mitigated by the automation of production processes.

22.5 Summary and Conclusion

Solar PV technologies have tremendous potential to offer sustainable and cleaner energy resources for future energy demands. While Si and thin-film solar cell technologies currently lead the worldwide PV installations, emerging technologies such as DSSC and perovskite solar cells can potentially enter the commercial market in the next decade. Especially considering the material shortage issues with thin-film technologies such as CdTe, there is a strong need for these emerging PV technologies to be commercially available and compete with Si PV in terms of price and long-term durability. Currently, significant research efforts are underway to address the stability issues, especially for perovskite solar cells. But, in perovskite solar cells, the researcher community should also look for alternative materials to the currently used Pb and Sn containing perovskites. Simultaneously, the PSC research community should consider using safe chemical processing technologies to minimize health and safety impacts on the researchers and workforce. Still, solar PV technology is deemed a green technology and can significantly contribute to addressing climate challenges. However, more studies are required to understand the emerging PV technologies' environmental impacts and life cycle, mainly based on PSCs. Considering the relatively less data available on the LCA of PSC technologies, simulation studies should be conducted to understand the potential future challenges in this area. Furthermore, the ecotoxicity of leached chemicals from different types of solar cell devices should be investigated further as currently there is only limited literature exists in this area. in future studies, significant attention should be given to accumulating more ecotoxicological data. Moreover, the governments and industries across the globe should work together to develop sustainable practices to address the occupational safety and health concerns of the personnel working in PV manufacturing, waste management, environmental impacts. Finally, a greater emphasis on recycling PV panels should be given to minimize the hazardous waste from the PV cells and minimize their environmental footprint.

References

Abbott MD, Trupke T, Hartmann HP, Gupta R, Breitenstein O (2007) Laser isolation of shunted regions in industrial solar cells. Prog Photovoltaics Res Appl 15:613–620

Abdullah MF et al (2016) Research and development efforts on texturization to reduce the optical losses at front surface of silicon solar cell. Renew Sustain Energy Rev 66:380–398

Act An (1970) Occupational safety and health act of 1970. Public Law 91:596

Altamura G, Vidal J (2016) Impact of minor phases on the performances of CZTSSe thin-film solar cells. Chem Mater 28:3540–3563

Amin N et al (2017) Solar photovoltaic technologies: from inception toward the most reliable energy resource. In: Encyclopedia of sustainable technologies, pp 11–26. https://doi.org/10.1016/B978-0-12-409548-9.10092-2

Armaroli N, Balzani V (2007) The future of energy supply: challenges and opportunities. Angew Chemie Int Ed 46:52–66

Arumughan J, Pernau T, Hauser A, Melnyk I (2005) Simplified edge isolation of buried contact solar cells. Sol Energy Mater Sol Cells 87:705–714

Babayigit A, Ethirajan A, Muller M, Conings B (2016a) Toxicity of organometal halide perovskite solar cells. Nat Mater 15:247–251

Babayigit A et al (2016b) Assessing the toxicity of Pb-and Sn-based perovskite solar cells in model organism Danio rerio. Sci Rep 6:18721

Benda V, Černá L (2020) PV cells and modules—State of the art, limits and trends. Heliyon 6

Best research-cell efficiency chart (1976). https://www.nrel.gov/pv/cell-efficiency.html

Biello D (2010) Explosive silicon gas casts shadow on solar power industry. Scientific American https://www.scientificamerican.com/article/explosive-gas-silane-used-to-make-photovoltaics/

Bosson CJ et al (2017) Cation disorder and phase transitions in the structurally complex solar cell material Cu2ZnSnS4. J Mater Chem A 5:16672–16680

Brabec CJ, Durrant JR (2008) Solution-processed organic solar cells. MRS Bull 33:670–675

Carlton JS, Perry-Hill R, Huber M, Prokopy LS (2015) The climate change consensus extends beyond climate scientists. Environ Res Lett 10:094025

Chopra KL, Paulson PD, Dutta V (2004) Thin-film solar cells: an overview. Prog Photovoltaics Res Appl 12:69–92

Chu J-Y (2011) 17.2% efficiency multicrystalline solar cells by optimizing structure of the MgF_2/SiNx double antireflection layer. J Photonics Energy 1:017001

Climate Change (2019). https://www.un.org/en/global-issues/climate-change

Cooper IB et al (2010) Understanding and use of IR belt furnace for rapid thermal firing of screen-printed contacts to Si solar cells. IEEE Electron Device Lett 31:461–463

Debnath T, Kim E, Lee K, Nath NCD (2020) Halide perovskite solar cells with biocompatibility. Adv Energy Sustain Res 1:2000028

Dharmadasa IM, Rahaq Y, Ojo AA, Alanazi TI (2019) Perovskite solar cells: a deep analysis using current–voltage and capacitance–voltage techniques. J Mater Sci Mater Electron 30:1227–1235

Dorian JP, Franssen HT, Simbeck DR (2006) Global challenges in energy. Energy Policy 34:1984–1991

Elinder CG, Järup L (1996) Cadmium exposure and health risks: recent findings. Ambio 25:370–373

Evans JW (2007) A guide to lead-free solders: physical metallurgy and reliability. Springer Science & Business Media

Fabini D (2015) Quantifying the potential for lead pollution from halide perovskite photovoltaics. J Phys Chem Lett 6:3536–3548

Fthenakis VM (2004) Life cycle impact analysis of cadmium in CdTe PV production. Renew Sustain Energy Rev 8:303–334

Fthenakis VM, Moskowitz PD (2000) Photovoltaics: environmental, health and safety issues and perspectives. Prog Photovoltaics Res Appl 8:27–38

Fthenakis VM et al (2005) Emissions and encapsulation of cadmium in CdTe PV modules during fires. Prog Photovoltaics Res Appl 13:713–723

Fthenakis VM, Hyung CK, Alsema E (2008) Emissions from photovoltaic life cycles. Environ Sci Technol 42:2168–2174

Fthenakis VM (2018) Overview of potential hazards. In: McEvoy's handbook of photovoltaics: fundamentals and applications (2018). https://doi.org/10.1016/B978-0-12-809921-6.00035-5

Gardner KL et al (2016) Nonhazardous solvent systems for processing perovskite photovoltaics. Adv Energy Mater 6:1600386

Gatz S et al (2011) 19.4%-efficient large-area fully screen-printed silicon solar cells. Phys Status Solidi Rapid Res Lett 5:147–149

Genchi G, Sinicropi MS, Lauria G, Carocci A, Catalano A (2020) The effects of cadmium toxicity. Int J Environ Res Public Health 17:3782

Gielen D, Gorini R, Wagner N, Leme R, Gutierrez L, Prakash G, Asmelash E et al (2019) Global energy transformation: a roadmap to 2050

National Renewable Energy Laboratory (NREL) (2021) Best research-cell efficiencies. https://www.nrel.gov/pv/cell-efficiency.html. Accessed June 15 2021

Goodrich A et al (2013) A wafer-based monocrystalline silicon photovoltaics road map: utilizing known technology improvement opportunities for further reductions in manufacturing costs. Sol Energy Mater Sol Cells 114:110–135

Green MA, Pillai S (2012) Harnessing plasmonics for solar cells. Nat Photonics 6:130–132

Hailegnaw B, Kirmayer S, Edri E, Hodes G, Cahen D (2015) Rain on methylammonium lead iodide based perovskites: possible environmental effects of perovskite solar cells. J Phys Chem Lett 6:1543–1547

Hauser A, Hahn G, Spiegel M, Fath P, Bucher E, Feist H, Breitenstein O, Rakotoniaina JP (2001) Comparison of different techniques for edge isolation. In: 17th European photovoltaic solar energy conference, pp 1739–1742

Herz LM (2017) Charge-carrier mobilities in metal halide perovskites: fundamental mechanisms and limits. ACS Energy Lett 2:1539–1548

Hilali MM et al (2006) Effect of Ag Particle Size in Thick-Film Ag Paste on the Electrical and Physical Properties of Screen Printed Contacts and Silicon Solar Cells. J Electrochem Soc 153:A5

Hoenig R et al (2012) Impact of screen printing silver paste components on the space charge region recombination losses of industrial silicon solar cells. Sol Energy Mater Sol Cells 106:7–10

Honsberg CB, Bowden SG (2019) Photovoltaics Education Website. www.pveducation.org.www.pveducation.org

Hoppe H, Sariciftci NS (2004) Organic solar cells: an overview. J Mater Res 19:1924–1945

Järup L, Åkesson A (2009) Current status of cadmium as an environmental health problem. Toxicol Appl Pharmacol 238:201–208

Kamat PV (2013) Quantum dot solar cells. The next big thing in photovoltaics. J Phys Chem Lett 4:908–918

Katagiri H et al (2009) Development of CZTS-based thin film solar cells. Thin Solid Films 517:2455–2460

Kim J et al (2013) Double antireflection coating layer with silicon nitride and silicon oxide for crystalline silicon solar cell. J. Electroceramics 30:41–45

Kim DH, Whitaker JB, Li Z, van Hest MFAM, Zhu K (2018a) Outlook and challenges of perovskite solar cells toward terawatt-scale photovoltaic module technology. Joule 2:1437–1451

Kim BJ, Lee S, Jung HS (2018b) Recent progressive efforts in perovskite solar cells toward commercialization. J Mater Chem A 6:12215–12236

Kojima A, Teshima K, Shirai Y, Miyasaka T (2009) Organometal halide perovskites as visible-light sensitizers for photovoltaic cells. J Am Chem Soc 131:6050–6051

Kolesnikov E et al (2017) Investigation of discharged aerosol nanoparticles during chemical precipitation and spray pyrolysis for developing safety measures in the nano research laboratory. Ecotoxicol Environ Saf 139:116–123

Koseoglu NM, van den Bergh JCJM, Lacerda JS (2013) Allocating subsidies to R&D or to market applications of renewable energy? Balance and geographical relevance. Energy Sustain Dev 17:536–545

Kray D, Hopman S, Spiegel A, Richerzhagen B, Willeke GP (2007) Study on the edge isolation of industrial silicon solar cells with waterjet-guided laser. Sol Energy Mater Sol Cells 91:1638–1644
Kwon TY et al (2011) Screen printed phosphorus diffusion for low-cost and simplified industrial mono-crystalline silicon solar cells. Sol Energy Mater Sol Cells 95:14–17
Lee TD, Ebong AU (2017) A review of thin film solar cell technologies and challenges. Renew Sustain Energy Rev 70:1286–1297
Li J et al (2021) Progress in recovery and recycling of kerf loss silicon waste in photovoltaic industry. Sep Purif Technol 254:117581
Ludin NA et al (2018) Prospects of life cycle assessment of renewable energy from solar photovoltaic technologies: A review. Renew Sustain Energy Rev 96:11–28
Lund PD (2009) Effects of energy policies on industry expansion in renewable energy. Renew Energy 34:53–64
Ma ZQ, Liu BX (2001) Boron-doped diamond-like amorphous carbon as photovoltaic films in solar cell. Sol Energy Mater Sol Cells 69:339–344
Manspeaker C, Venkatesan S, Zakhidov A, Martirosyan KS (2017) Role of interface in stability of perovskite solar cells. Curr Opin Chem Eng 15:1–7
Matheu P, Lim SH, Derkacs D, McPheeters C, Yu ET (2008) Metal and dielectric nanoparticle scattering for improved optical absorption in photovoltaic devices. Appl Phys Lett 93:113108
McBride M, Sauvé S, Hendershot W (1997) Solubility control of Cu, Zn, Cd and Pb in contaminated soils. Eur J Soil Sci 48:337–346
de Ménorval MA, Mir LM, Fernández ML, Reigada R (2012) Effects of dimethyl sulfoxide in cholesterol-containing lipid membranes: a comparative study of experiments in silico and with cells. PLoS One 7:e41733
Moskowitz PD, Fthenakis VM (1991) A checklist of suggested safe practices for the storage, distribution, use and disposal of toxic and hazardous gases in photovoltaic cell production. Sol Cells 31:513–525
Narayanan KL, Yamaguchi M (2003) Photovoltaic effects of a:C/C60/Si (p-i-n) solar cell structures. Sol Energy Mater Sol Cells 75:345–350
Nelson WO, Campbell PGC (1991) The effects of acidification on the geochemistry of Al, Cd, Pb and Hg in freshwater environments: a literature review. Environ Pollut 71:91–130
Nover J et al (2017) Long-term leaching of photovoltaic modules. Jpn J Appl Phys 56:08MD02
Occupational Safety and Health Administration (1971)
Oreskes N (2004) The scientific consensus on climate change. Science (80):306, 1686
Park NG (2015) Perovskite solar cells: an emerging photovoltaic technology. Mater Today 18:65–72
Park NG, Zhu K (2020) Scalable fabrication and coating methods for perovskite solar cells and solar modules. Nat Rev Mater 5:333–350
Park H et al (2009) Improvement on surface texturing of single crystalline silicon for solar cells by saw-damage etching using an acidic solution. Sol Energy Mater Sol Cells 93:1773–1778
Perlin J (2004) Silicon solar cell turns 50. NREL/BR-520–33947. National Renewable Energy Lab., Golden, CO. (US)
Koroneos C, Stylos N, Moussiopoulos N (2006) LCA of multicrystalline silicon photovoltaic systems. Part 1: present situation and future perspectives. Int J Life Cycle Assess 11:129–136
Rech B, Wagner H (1999) Potential of amorphous silicon for solar cells. Appl Phys A Mater Sci Process 69:155–167
Salvador M, Motter CE, McCulloch I (2020) Hidden perils of lead in the lab: guidelines for containing, monitoring, and decontaminating lead in the context of perovskite research. Chem Mater 32:7141–7149
Sardana SK et al (204) Influence of surface plasmon resonances of silver nanoparticles on optical and electrical properties of textured silicon solar cell. Appl Phys Lett 104:073903
Sardana SK, Chava VSN, Komarala VK (2015) Morphology and optical properties of sputter deposited silver nanoparticles on plain, textured and antireflection layer coated textured silicon. Appl Surf Sci 347:651–656

Schmich E, Schillinger N, Reber S (2007) Silicon CVD deposition for low cost applications in photovoltaics. Surf Coat Technol 201:9325–9329
Schroder DK, Meier DL (1984) Solar cell contact resistance—a review. IEEE Trans Electron Devices 31:637–647
Sica D, Malandrino O, Supino S, Testa M, Lucchetti MC (2018) Management of end-of-life photovoltaic panels as a step towards a circular economy. Renew Sustain Energy Rev 82
Slaoui A (2017) Inorganic materials for photovoltaics: status and futures challenges. In: EPJ web of conferences, vol 148. EDP Sciences, p 00007. https://www.epj-conferences.org/articles/epjconf/abs/2017/17/epjconf_epssif2017_00007/epjconf_eps-sif2017_00007.html
Smalley RE (2005) Future global energy prosperity: the terawatt challenge. MRS Bull 30:412–417
Solar Energy Isn't Always as Green as You Think (2014). https://spectrum.ieee.org/green-tech/solar/solar-energy-isnt-always-as-green-as-you-think
Sopori B (2003) Silicon nitride processing for control of optical and electronic properties of silicon solar cells. J Electron Mater 32:1034–1042
Stranks SD et al (2013) Electron-hole diffusion lengths exceeding 1 micrometer in an organometal trihalide perovskite absorber. Science 342(80):341–344
Tan ZK et al (2014) Bright light-emitting diodes based on organometal halide perovskite. Nat Nanotechnol 9:687–692
Tian S et al (2019) A facile green solvent engineering for up-scaling perovskite solar cell modules. Sol Energy 183:386–391
Tsuo Y, Gee J, Menna P, Strebkov D (1998) Environmentally benign silicon solar cell manufacturing. In: Photovoltaic solar energy conversion (1998)
Wadia C, Alivisatos AP, Kammen DM (2009) Materials availability expands the opportunity for large-scale photovoltaics deployment. Environ Sci Technol 43:2072–2077
Wang J et al (2017) Highly efficient perovskite solar cells using non-toxic industry compatible solvent system. Sol RRL 1:1700091
Wolf A et al (2015) Status and perspective of emitter formation by POCL3-diffusion. In: 31st European PV solar energy conference and exhibition, pp 414–419
Wright AD et al (2020) Intrinsic quantum confinement in formamidinium lead triiodide perovskite. Nat Mater 19:1201–1206
Wu L, Ma Z, He A, Wang J (2010) Studies on destruction of silicon tetrachloride using microwave plasma jet. J Hazard Mater 173:305–309
Xing G et al (2013) Long-range balanced electron-and hole-transport lengths in organic-inorganic CH3NH3PbI3. Science 342(80):344–347
Xu X, Wei Z, Ji Q, Wang C, Gao G (201) Global renewable energy development: influencing factors, trend predictions and countermeasures. Resour Policy 63:101470
Yadav A, Singh G, Nekovei R, Jeyakumar R (2015) c-Si solar cells formed from spin-on phosphoric acid and boric acid. Renew Energy 80:80–84
Zhang L, Shen H, Yang Z, Jin J (2010) Shunt removal and patching for crystalline silicon solar cells using infrared imaging and laser cutting. Prog Photovoltaics Res Appl 18:54–60
Zhang M, Xin D, Zheng X, Chen Q, Zhang WH (2020) Toward greener solution processing of perovskite solar cells. ACS Sustain Chem Eng 8:13126–13138
Zhao X, Wang M (2018) Organic hole-transporting materials for efficient perovskite solar cells. Mater Today Energy 7:208–220
Zhao W et al (2018) Environmentally friendly solvent-processed organic solar cells that are highly efficient and adaptable for the blade-coating method. Adv Mater 30:1704837
Zulehner W (2000) Historical overview of silicon crystal pulling development. Mater Sci Eng B Solid-State Mater Adv Technol 73:7–15

Chapter 23
Colloid Transport in Porous Media at Multiple Length Scales

Mahima John Horta, Yerramilli Sai Rama Krishna, and N. Seetha

Abstract Colloidal contaminants such as pathogenic microorganisms and engineered nanoparticles enter subsurface from various sources such as land application of wastewater, reuse of untreated sewage for irrigation and sanitary landfills. Understanding colloid transport in the subsurface is essential for assessing the safe distance of drinking water wells from the source of contamination, bioremediation of contaminated sites, and degree of treatment required for land application of wastewater. There is a large disparity in the length scales associated with colloid transport in soil. This includes micrometre, centimetre, tens of centimetres, metre and kilometre scales which are representative of a single soil capillary, representative elementary volume scale, 1D lab scale, 3D lab scale, and the field scale, respectively. Colloid deposition mechanisms are scale dependent and are governed by the heterogeneity at that scale. Hence, the observed transport processes and the estimated parameters at a smaller scale may not simulate the observations at a larger scale. It is imperative to link the processes and the associated parameters across scales to better predict transport behavior at larger scales. This chapter discusses the mechanisms of colloid retention in porous media at various scales, the effect of heterogeneity on colloid transport at each scale, and upscaling of transport processes.

Keywords Colloids · Heterogeneity · Scale · Upscaling

23.1 Introduction

Colloids are particles with sizes ranging between 10 nm and 10 μm and are ubiquitous in the subsurface. Colloids exist in groundwater in as mineral precipitates of iron, aluminium and manganese, natural organic matter such as humic acid and fulvic acid, bacteria, viruses, clays and engineered nanoparticles (McCarthy and McKay 2004). Colloidal contaminants such as pathogenic microorganisms and engineered

M. J. Horta · Y. S. R. Krishna · N. Seetha (✉)
Department of Civil Engineering, Indian Institute of Technology Hyderabad, Hyderabad, Telangana 502285, India
e-mail: seetha@ce.iith.ac.in

S. P. Singh et al. (eds.), *New Trends in Emerging Environmental Contaminants*, Energy, Environment, and Sustainability,
https://doi.org/10.1007/978-981-16-8367-1_23

nanoparticles enter the subsurface zone through various sources such as landfills, industrial sludge, leakage from sewers, and accidental spills. The vadose zone acts as a barrier to protect groundwater from contamination of infiltrated colloids by retaining a portion of them. The uncaptured colloids in the vadose zone move down and reach the groundwater table, thereby contaminating it. In addition to their contamination potential, colloids may also affect the transport behaviour of other contaminants in the subsurface (Grolimund and Borkovec 2005). The various mechanisms involved in the retention of the colloids in soil include attachment to the grain surface or solid-water interface (SWI), air–water interface (AWI) and the contact region or air–water-solid interface (AWS), straining, and ripening. The transport behaviour of colloids in the soil is controlled by various physico-chemical factors such as the flow velocity, water chemistry, soil type, colloid type, and degree of saturation (Ryan and Elimelech 1996; Bradford et al. 2007; Han et al. 2006). Understanding colloid transport in the subsurface is essential for assessing the safe distance of drinking water wells from the source of contamination, bioremediation of contaminated sites, and degree of treatment required for land application of wastewater.

Flow and transport processes in the subsurface are scale dependent, with the scales spanning from micrometres to kilometres (Vogel and Roth 2003). Colloid transport in soils has been studied experimentally and numerically at various scales including micrometre, centimetre, tens of centimetres, metre and kilometre scales which are representative of a single soil capillary, representative elementary volume (REV) scale, 1D lab scale, 3D lab scale, and the field scale, respectively. The above scales are studied through microchannels (Lazouskaya and Jin 2008), micromodels (Wan and Wilson 1994a; Zhang et al. 2013) and flow cells packed with soil grains (Gao et al. 2006; Morales et al. 2009), laboratory soil columns (Bradford et al. 2004; Schijven et al. 2000a; Keller et al. 2004), laboratory sand-box setup (Schijven et al. 2000a; Phenrat et al. 2010) and field experiments (Vilks et al. 1997; Becker et al. 1999), respectively. Figure 23.1 shows the schematic of various scales associated with colloid transport in porous media. Studies at each of the above scales not only provide valuable information on the mechanisms and factors controlling colloid transport at that scale, but also help to infer the mechanisms behind the observations at larger scales.

The most common and comfortable scale for colloid transport studies is the laboratory column scale. However, soil is heterogeneous at all scales (Wheatcraft and Cushman 1991) and exhibits a high degree of spatial variability of flow and transport properties at various scales (Vogel and Roth 2003; Hopmans and Schoups 2005). Heterogeneities exist in various forms at each scale of observation. Heterogeneity is broadly classified into two types: chemical heterogeneity and physical heterogeneity. Chemical heterogeneity manifests in the form of spatially variable surface properties of the soil grains caused by the distribution of minerals such as oxides of iron, magnesium and aluminium, hydroxyl groups, and organic matter. The patches of oxides of iron, magnesium and aluminium present on the grain surface are positively charged, whereas the primary minerals such as quartz and feldspar are negatively charged at the typical environmental conditions (Johnson et al. 1996). Humic acid is the most widely present form of organic matter in soils and has a negative surface charge under

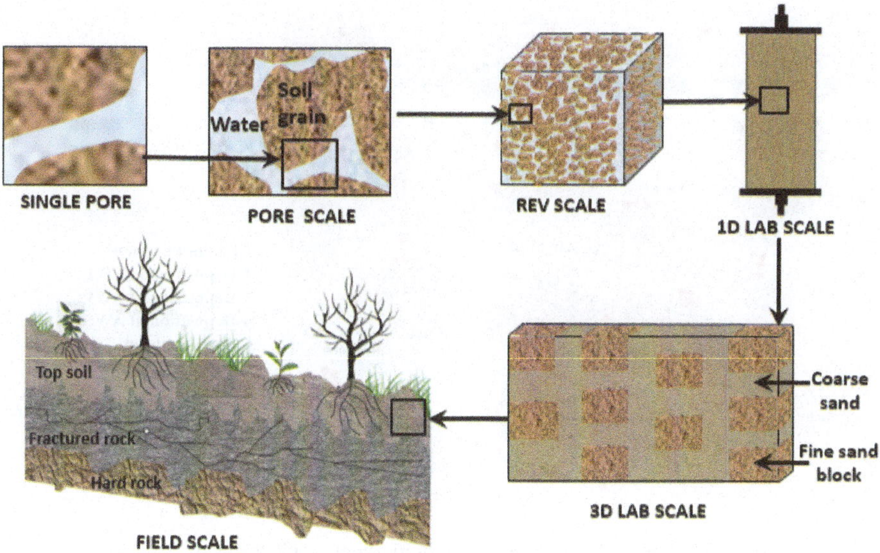

Fig. 23.1 Schematic of various scales associated with colloid transport in porous media

the environmental conditions in the subsurface (Chen et al. 2020). Physical heterogeneity is due to the surface roughness of soil grains, distribution in the grain sizes and shapes, and spatial variation in the hydraulic properties of soil (Morales et al. 2009; Zhang et al. 2010; Johnson et al. 1996; Saiers et al. 1994a). Surface roughness on the grain surface typically spans from nanometers to micrometers (Argent et al. 2015; Shen et al. 2015), whereas spatial variation in the hydraulic properties of soil exists at all scales (Vogel and Roth 2003; Wheatcraft and Cushman 1991). Hence, the transport processes observed and the parameters obtained at a given scale may not be representative of the phenomenon at the larger scales. For example, heterogeneity at the field scale introduces additional complexities such as preferential flow paths and retention by vegetation, which may not be present in the lab scale experiments which are usually conducted using a homogeneous soil without vegetation (Vilks et al. 1997; Burkhardt et al. 2008). Hence, it is important not only to study the processes at various scales but also to link the processes across various scales to understand the effect of small-scale processes on large-scale mechanisms. This chapter discusses the mechanisms of colloid retention in porous media at various scales, the effect of heterogeneity at each scale on colloid transport, and the various scaling approaches.

23.2 Colloid Retention Mechanisms

This section briefly discusses various colloid retention mechanisms in soil such as deposition to various interfaces including SWI, AWI, and AWS, blocking, straining,

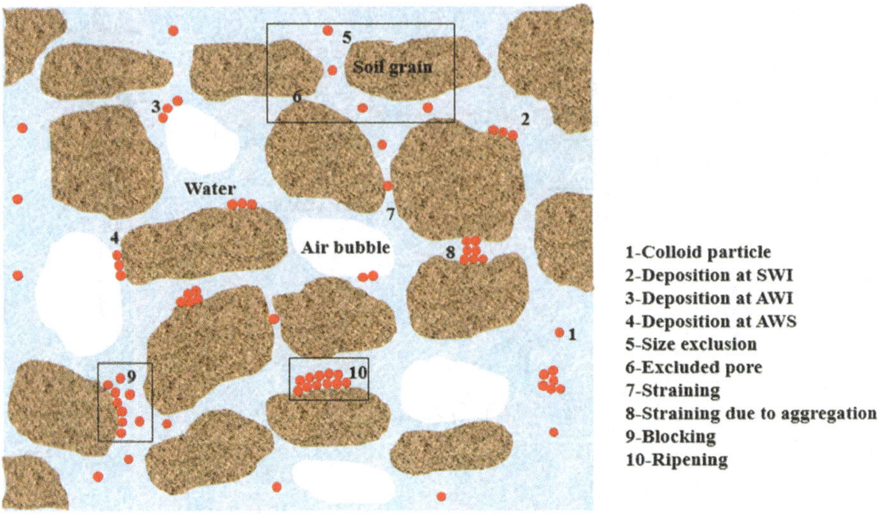

Fig. 23.2 Schematic representation of various colloid retention mechanisms

ripening, and size exclusion. A schematic representation of colloid retention mechanisms is shown in Fig. 23.2. More detailed information about different mechanisms can be found in Liu et al. (2019), Babakhani et al. (2017), Bradford et al. (2014).

23.2.1 Deposition at Interfaces

Colloid deposition in porous media occurs at the SWI under saturated conditions. Under unsaturated conditions, colloids can also get deposited at additional retention sites such as AWI and AWS. This leads to greater retention of colloids under unsaturated conditions than saturated conditions (Wan and Wilson 1994a; Crist et al. 2004). Colloid deposition to the interfaces depends on the interaction energy between them, having a length scale of hundreds of nanometres, which can be calculated using Derjaguin-Landau-Vervey-Overbeek (DLVO) theory (Deraguin and Landau 1941; Verwey 1947). For SWI, the interaction energy is the sum of electrostatic double-layer energy, London van der Waals energy, and Born repulsion energy (Bradford and Torkzaban 2008), whereas for the AWI it is the sum of electrostatic double layer energy, London van der Waals energy, and the hydrophobic energy (Lazouskaya and Jin 2008). Electrostatic double-layer energy can be attractive or repulsive depending on the dissimilarity or similarity of surface charges on the colloid and the interface. Under the typical environmental conditions, the colloids, the grains, and the AWI are negatively charged. This makes the electrostatic energy between colloid and SWI as well as between colloid and AWI repulsive. London van der Waals energy is attractive at SWI and repulsive at AWI (Bradford and Torkzaban 2008). Hydrophobic energy is attractive, and the magnitude of the energy depends on the contact angle of the colloid with AWI (Zevi et al. 2012). The conditions for the deposition of colloids to an interface is said to be favourable if the interaction energy profile is devoid of an energy barrier (zone of repulsion), and the conditions for deposition are said to be unfavourable in the presence of an energy barrier in the energy profile (Seetha et al. 2015). Colloid deposition onto the grain surface is irreversible and reversible under favourable and unfavourable conditions, respectively. The attachment rate coefficient of colloid to the grain surface under favourable conditions in saturated porous media can be calculated using colloid filtration theory (CFT) (Yao et al. 1971). Colloid retention at AWI is found to be irreversible as the colloids are held strongly by capillary forces (Wan and Wilson 1994a). Deposition to SWI has been observed for carboxylate-modified latex microspheres (Bradford et al. 2003, 2004; Torkzaban et al. 2008; Wan and Wilson 1994b; Li and Johnson 2005; Tong and Johnson 2006; Lazouskaya and Jin 2006), colloidal silica (Saiers et al. 1994a, b), viruses (Han et al. 2006; Schijven et al. 2000a; Torkzaban et al. 2006a, b, 2008; Jin et al. 2000), bacteria (Bai et al. 2016; Hornberger et al. 1992), protozoa (Tufenkji et al. 2004), and different types of nanoparticles including silica nanoparticles (Wang et al. 2012a), silver nanoparticles (Yecheskel et al. 2018; Flory et al. 2013; Kanel et al. 2015; Kumahor et al. 2015), and iron nanoparticles (Tosco and Sethi 2010; Raychoudhury et al. 2014). Different types of colloids such as carboxylate-modified latex microspheres (Torkzaban et al. 2008; Lazouskaya and Jin 2006), silica colloids (Lenhart and Saiers 2002; Saiers and Lenhart 2003), clay colloids (Gao et al. 2004), viruses (Han et al. 2006; Torkzaban et al. 2006a, b, 2008; Jin et al. 2000), TiO_2 nanoparticles (Fang et al. 2013), and silver nanoparticles (Kumahor et al. 2015) have been found to get deposited at AWI and AWS.

23.2.2 Blocking

If the soil surface has limited capacity for particle deposition, then the available surface area for deposition decreases with time because of particle deposition, and the deposited particles block further deposition leading to a decrease in the deposition rate. This mechanism is termed as blocking (Ryan and Elimelech 1996; Saiers et al. 1994b). Two common approaches to model blocking are Langmurian approach and random sequential adsorption approach (Ryan and Elimelech 1996; Deshpande and Shonnard 1999). Both the approaches introduce a dynamic blocking function which accounts for the fraction of the sites on the grain surface available for colloid deposition. Blocking has been found to occur for various colloidal particles including carboxylate-modified latex beads (Torkzaban et al. 2008, 2012), amidine-modified polystyrene latex microspheres (Liu et al. 1995; Johnson and Elimelech 1995), colloidal silica (Johnson et al. 1996; Saiers et al. 1994b), bacteria (Deshpande and Shonnard 1999; Torkzaban et al. 2008b; Rijnaarts et al. 1996), viruses (Sasidharan et al. 2017; Bhattacharjee et al. 2002), TiO_2 nanoparticles (Toloni et al. 2014; Chowdhury et al. 2011; Chen et al. 2011), silver nanoparticles (Yecheskel et al. 2018;

Neukum et al. 2014), graphene oxide nanoparticles (Sun et al. 2015a; Feriancikova and Xu 2012; Qi et al. 2014), and fullerene nanoparticles (Li et al. 2008; Wang et al. 2008).

23.2.3 Straining

Straining is a physical mechanism in which particles get trapped in the pore throats which are smaller than the size of the particles. It is a depth-dependent phenomenon and is more prominent at the inlet of the porous medium. It is characterized by a hyper-exponential retention profile of colloids with distance. Though straining mainly occurs for micron-sized particles, it has also been observed to occur for nanoparticles due to the formation of aggregates which are larger in size than the individual nanoparticles (Bradford et al. 2004; Fang et al. 2013; Qi et al. 2014; Rastghalam et al. 2019; Fan et al. 2015). The main criterion for straining to happen is the threshold value for the ratio of particle diameter to grain diameter (d_p/d_g). Straining occurs when this criterion is exceeded. The threshold value depends on the type and size of colloid and soil grains, aggregation kinetics, gradation of soil (Bradford et al. 2005) and flow rate (Xu et al. 2006). Various studies have reported different threshold values of d_p/d_g for various colloidal types. For example, the threshold values for carboxylate-modified latex microspheres reported are 0.008 (Xu et al. 2006, 2008), 0.005 (Bradford et al. 2004), 0.0016 (Shen et al. 2008), and 0.0017 (Bradford et al. 2002). For *E. coli*, the reported values are 0.002 (Foppen et al. 2005) and 0.005 (Bradford et al. 2006). The threshold values for ZnO nanoparticles in the literature are 0.003 (Sun et al. 2015b) and 0.0093 (Jiang et al. 2012). The threshold values for *Pseudomonas* strain, TiO_2 nanoparticles, and nano-zero valent iron (nZVI) particles are 0.0005 (Tan et al. 1994), 0.001 (Fang et al. 2013), and 0.0002 (Raychoudhury et al. 2014), respectively. Straining can also occur due to the formation of heteroaggregates when two or more colloids get transported together. The threshold value of d_p/d_g for straining to occur is different for heteroaggregates and homoaggregates. For example, the threshold values of d_p/d_g for heteroaggregates of hydroxyapatite and goethite nanoparticles (Wang et al. 2015a) and heteroaggregates of hydroxyapatite and haematite nanoparticles (Wang et al. 2015b) are 0.0046 and 0.0027, respectively. Film straining is observed in unsaturated porous medium wherein the colloids get trapped at thin water films around soil grains when the thickness of the water film is smaller than the diameter of the colloid (Wan and Tokunaga 1997).

23.2.4 Ripening

Ripening is a mechanism where the previously deposited particles on the surface of the soil grains act as favourable sites for the deposition of particles arriving at the grain surfaces (Camesano et al. 1999). Ripening is dominated by particle–particle interactions rather than particle-grain interactions. As a result, the deposition rate increases with time (Wang et al. 2014a). Ripening is observed prominently in nanoparticles such as nano zero valent iron particles (Tosco and Sethi 2010; Hosseini and Tosco 2013; Basnet et al. 2015), hydroxyapatite nanoparticles (Wang et al. 2012b, 2014a, 2015a, b), TiO_2 nanoparticles (Chen et al. 2011; Wang et al. 2014b), and fullerene nanoparticles (Cheng et al. 2005), due to favourable particle–particle interactions (Wang et al. 2014a; Dagtepe and Chikan 2010). In addition, ripening was also observed in bacteria (Nascimento et al. 2006; Camesano and Logan 1998; Kim et al. 2009) and viruses (Gitis et al. 2002).

23.2.5 Size Exclusion

Size exclusion is described as the phenomenon in which colloidal particles travel faster than the conservative tracer (Ginn 2002). Size exclusion is influenced by the size of the colloids, and is significant for larger colloids that cannot travel through small pores and tend to travel along larger pores (Sirivithayapakorn and Keller 2003a). Size exclusion results in larger velocity and lesser dispersivity for larger colloids. Size exclusion has been observed for different colloid types including latex beads (Keller et al. 2004; Bradford et al. 2003; Sirivithayapakorn and Keller 2003a; Auset and Keller 2004), viruses (Keller et al. 2004; Anders and Chrysikopoulos 2005; Sinton et al. 2000), bacteria (Sinton et al. 2000), silver nanoparticles (Braun et al. 2015; Cornelis et al. 2013; Sagee et al. 2012), and TiO_2 nanoparticles (Solovitch et al. 2010).

23.3 Scales Associated with Colloid Transport

In this section, we discuss the various scales at which the transport of colloids has been studied, including the pore scale, the column scale, the 3D sand-box scale, and the field scale. The transport processes at various scales are interlinked, and the observed mechanism at a given scale is due to the collective effect of the various processes occurring at smaller scales.

23.3.1 Pore Scale

The pore scale, ranging from micrometres to millimetres, is the smallest scale at which colloid transport is studied. Pore scale studies offer the advantage of visualizing the mechanisms behind the observations at larger scales. At this scale, transport of colloids is studied in a single soil capillary, a 2D network of pores, and a 3D packing of soil grains by performing experiments through microchannels, micromodels and flow cells packed with solid grains, respectively. Micromodel represents the complex structure of soil using an interconnected network of pore bodies and pore throats having pore dimensions in the range of 10–100 μm (Wan and Wilson 1994a; Zhang et al. 2013). Flow cell is at the REV scale and is the smallest scale at which both microscopic and macroscopic processes can be studied (Smith et al. 2008). Various mechanisms such as attachment to SWI (Argent et al. 2015; Chen and Flury 2005), AWI (Lazouskaya and Jin 2006, 2008; Wan and Wilson 1994a; Wan et al. 1994) and the contact line (Zhang et al. 2013; Crist et al. 2004; Zevi et al. 2012), retention at grain to grain contacts (Gao et al. 2006; Johnson et al. 2010), size exclusion (Sirivithayapakorn and Keller 2003a; Auset and Keller 2004), and remobilization of deposited colloids from SWI, AWI and AWS due to transients in flow or chemistry (Gao et al. 2006; Lazouskaya et al. 2011) are identified through visualization studies. Moreover, pore-scale studies also help to understand the effect of pore geometry, pore connectivity, nanoscale/microscale heterogeneity, and grain packing on colloid transport and retention. Detailed information about the different methods of imaging the fluid distribution and transport of colloids in porous media can be found in the review papers (Lazouskaya et al. 2011; Keller and Auset 2007; Ochiai et al. 2006).

Microchannel experiments, with the channel width being 10's to 100's of micrometres, help to clearly differentiate colloid retention at SWI, AWI and AWS, and provide information on the dynamics of colloid movement between the interfaces. Table 23.1 lists various mechanisms that have been identified at the pore scale through experiments in microchannels. Figure 23.3 shows the retention of colloids at SWI, AWI and AWS, imaged by performing microchannel experiments (Lazouskaya and Jin 2008). Colloids attached to AWI are found to be in random motion due to Brownian diffusion and hydrodynamic effects and are observed to slide along AWI and deposit at AWS (Lazouskaya and Jin 2008). The greater the colloid concentration at AWI, the greater is the concentration at AWS (Lazouskaya and Jin 2006). Hydrophobic force is responsible for the attraction of colloids to AWI, and the colloids are held at the interface due to the capillary forces (Lazouskaya and Jin 2006, 2008; Flury and Aramrak 2017). The colloids attached to AWS appear to be stationary and irreversible (Lazouskaya and Jin 2008) and is attributed to the low flow velocities at the collector surface and retention of the colloids near thin water films at AWS. Lazouskaya and Jin (2008) observed greater retention of hydrophobic than hydrophilic colloids at AWI. Zevi et al. (2012) observed from experiments in an open microchannel that colloid retention at AWS is larger for small contact angles due to counter flow, and the flow vortices that formed closer to AWS helped deliver the colloids to SWI below the contact line. Drainage and imbibition events cause the mobilization of deposited colloids in soil. The movement of the contact line due to drainage or imbibition mobilizes the attached colloids from SWI to AWS, AWS to AWI, AWI to AWS, and SWI into bulk solution (Lazouskaya et al. 2011). The mobilization of colloids from interfaces during the passage of the front depends on the magnitudes of various forces between the colloids and the interfaces (Lazouskaya et al. 2011, 2013).

Table 23.1 Mechanisms identified through microchannel studies

Mechanism	References	Colloid	Key finding
Sorption at SWI, AWI, and AWS	Lazouskaya and Jin (2006)	Carboxylate-modified latex colloids	Colloids preferentially attached to and accumulated at AWI
	Lazouskaya and Jin (2008)	Sulfate- and carboxylate-modified latex colloids	Colloids attached to AWI translate along and move towards the contact line, where they get attached. Contact line serves as a preferential location for colloid retention compared to AWI
	Zevi et al. (2012)	Synthetic polystyrene microspheres	Colloid retention at the AWS is negatively correlated with contact angles
Remobilization of attached colloids	Lazouskaya et al. (2011)	Yellow-green fluorescent carboxylate-modified polystyrene microspheres	Moving contact line mobilized the deposited colloids at the channel wall
	Lazouskaya et al. (2013)	Carboxylate modified and sulfate colloids	Colloid mobilization in hydrophilic substrate is due to imbibition and in hydrophobic substrate the mobilization is due to drainage
	Noordmans et al. (1997)	Polystyrene particles	Detachment of colloids from the collector surfaces is due to high surface tensions forces at the colloid-collector interface

Micromodel studies predominantly help to understand the colloid travel pathways as affected by the distribution in the sizes of pore throats and pore bodies, pore connectivity, and sorption at different interfaces in a collection of pores. Table 23.2 lists the various micromodel studies and the identified mechanisms. Network topology has been found to have a significant effect on colloid dispersion, with a heterogeneous or zig-zag network resulting in more dispersion than a regular network due to longer travel paths and detouring (Auset and Keller 2004; Baumann et al. 2010). The different pore topologies are shown in Fig. 23.4 (Baumann et al. 2010). Auset and Keller (2004) observed from micromodel studies that dispersion decreases with increasing colloid size as larger colloids travel along central streamlines resulting in less detours and faster velocities. The phenomenon of size exclusion depends

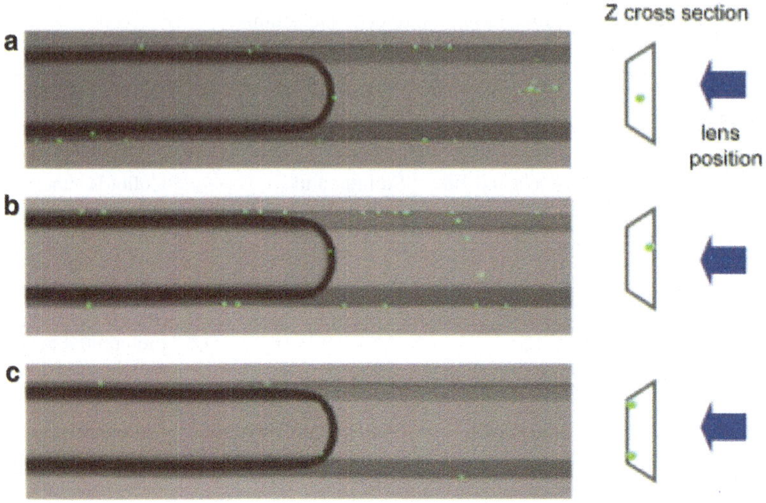

Fig. 23.3 Colloid (green dots) retention at **a** AWI, **b** contact line at the front wall, and **c** contact line at the back wall (Lazouskaya and Jin 2008). The flow direction is from right to left. Reprinted from Lazouskaya and Jin (2008), copyright 2008, permission obtained from Elsevier

on the pore size distribution of the network, pore connectivity and colloid size. A colloid can enter a pore throat if the ratio of the throat to colloid diameter is greater than 1.5 (Sirivithayapakorn and Keller 2003a). Large colloids are observed to move through preferential flow paths showing size exclusion effect, whereas the small colloids travelled through longer pathways. Colloids can also be entrapped in the pore throats smaller than the colloid size resulting in straining. Straining was found to occur when the ratio of pore throat to colloid diameter is less than 1.8, and was prominent at the inlet of the micromodel (Auset and Keller 2006). Colloid retention was found to be significantly influenced by surface roughness as observed from micromodel studies (Argent et al. 2015; Auset and Keller 2006). Auset and Keller (2006) observed that a rough micromodel retained more colloids than a smooth micromodel. Also, nanoscale roughness reduces the depth of primary minimum, and hence only a fraction of the surface is found to be available for colloid retention at SWI even at high ionic strengths (Argent et al. 2015). Also, surface roughness has been found to enhance colloid straining (Bradford and Bettahar 2006). Colloid attachment to AWI is observed to be irreversible due to strong capillary forces that hold the colloids at AWI (Wan and Wilson 1994a; Wan et al. 1994). Particle attachment to AWI depends on particle hydrophobicity, ionic strength, and surface charge of the particle. Hydrophobic colloids showed greater retention at AWI than the hydrophilic colloids, and retention was larger for positively-charged particles than negatively-charged particles (Wan and Wilson 1994a). Also, retention increased with increasing ionic strength (Wan and Wilson 1994a). The retention of colloids at AWI and glass-water interface observed in the micromodel study of Wan and Wilson (1994a) is shown in Fig. 23.5. Colloid retention increases with decreasing water saturation due

Table 23.2 Mechanisms observed through micromodel studies

Mechanism	References	Colloid	Key finding
Dispersion of colloids	Auset and Keller (2004)	Carboxylate modified latex polystyrene microspheres	Dispersion depends on the pore geometry and colloid size
	Baumann et al. (2010)	Fluorescent carboxylate polystyrene microspheres	Dispersion is higher in heterogeneous network and increases with ionic strength
Size exclusion	Sirivithayapakorn and Keller (2003a)	Carboxylate-modified latex colloids	Ratio of pore throat to colloid diameter for colloids to enter new pores $\geq$1.5
Straining	Auset and Keller (2006)	Polystyrene latex particles	Straining is significant at the inlet of the micromodel. Colloids retained at wall cervices and irregularities
Film straining	Auset et al. (2005)	*E. coli*	Infiltration front causes mobilization of the colloids at thin film and drainage front traps colloids in thin water films
Sorption at AWI, SWI and AWS	Wan and Wilson (1994a)	Latex microspheres, Na-montmorillonite, P. cepia Arthrobacter	Sorption at AWI is irreversible due to strong capillary forces
	Wan et al. (1994)	P. cepia, Arthrobacter	Sorption of hydrophobic colloids to AWI is more than hydrophilic colloids
	Zhang et al. (2013)	Carboxylated fluorescent microspheres	Prominent attachment at contact line than liquid–liquid interface and solid–liquid interfaces
	Argent et al. (2015)	Fluoresbrite carboxyl latex and borosilicate glass beads	Attachment to AWS observed at high ionic strength. Attachment to SWI decreases with increasing ionic strength due to surface roughness

(continued)

Table 23.2 (continued)

Mechanism	References	Colloid	Key finding
Remobilization of deposited colloids	Sirivithayapakorn and Keller (2003b)	Carboxylate-modified latex colloids	Colloids attached to AWI gets mobilized during the dissolution of air bubbles
	Zhang et al. (2013)	Carboxylated Fluorescent microspheres	Remobilization is seen at moving contact line and the fluid–fluid interface
	Nishad and Al-Raoush (2021)	Polystyrene and carboxylate-modified polystyrene colloids	Mobilized hydrophilic colloids at gas–water interface get redeposited in the gas–water-solid interface due to strong capillary forces

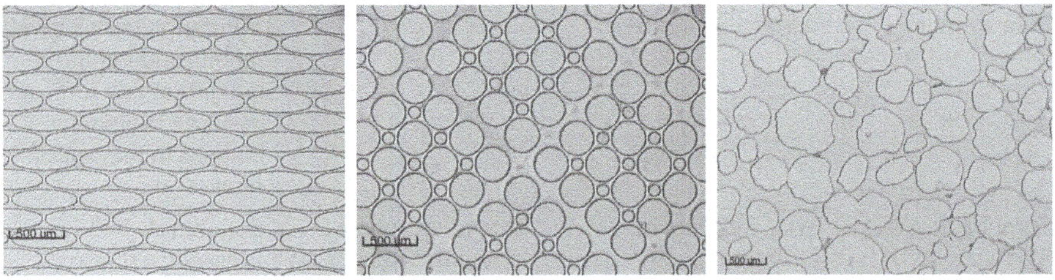

Fig. 23.4 Pore structure in the ellipsoid, diamond and heterogeneous micromodel (Baumann et al. 2010). Reprinted from Baumann et al. (2010), copyright 2010, permission obtained from Elsevier

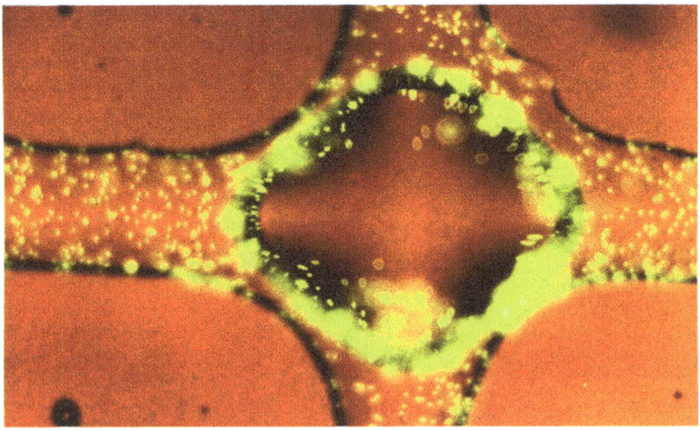

Fig. 23.5 Retention of hydrophobic colloids (yellow green) at air–water and glass-water interfaces (Wan and Wilson 1994a). Reprinted from Wan and Wilson (1994a), copyright 2010, permission obtained from John Wiley and sons

to increase in the area of liquid–liquid interface and liquid–liquid-solid contact line (Zhang et al. 2013).

Transients in flow remobilize the deposited colloids from various interfaces. Zhang et al. (2013) observed remobilization of attached colloids from the liquid–liquid interface and contact line during imbibition as shown in Fig. 23.6. Also, the moving liquid–liquid interface and contact line scoured the colloids attached to the solid–liquid interface. However, no remobilization was observed during drainage as the remobilized colloids were re-deposited downstream at the liquid–liquid interface and contact line. A similar observation was made by Nishad and Al-Raough (2021) that the moving gas–water interface mobilized both hydrophilic and hydrophobic colloids from SWI during drainage. However, the mobilized hydrophilic colloids at gas–water interface get re-deposited onto the gas–water–solid interface or thin water films due to their greater capillary potential as the micromodel considered in their study was hydrophilic. The amount of remobilization of the deposited

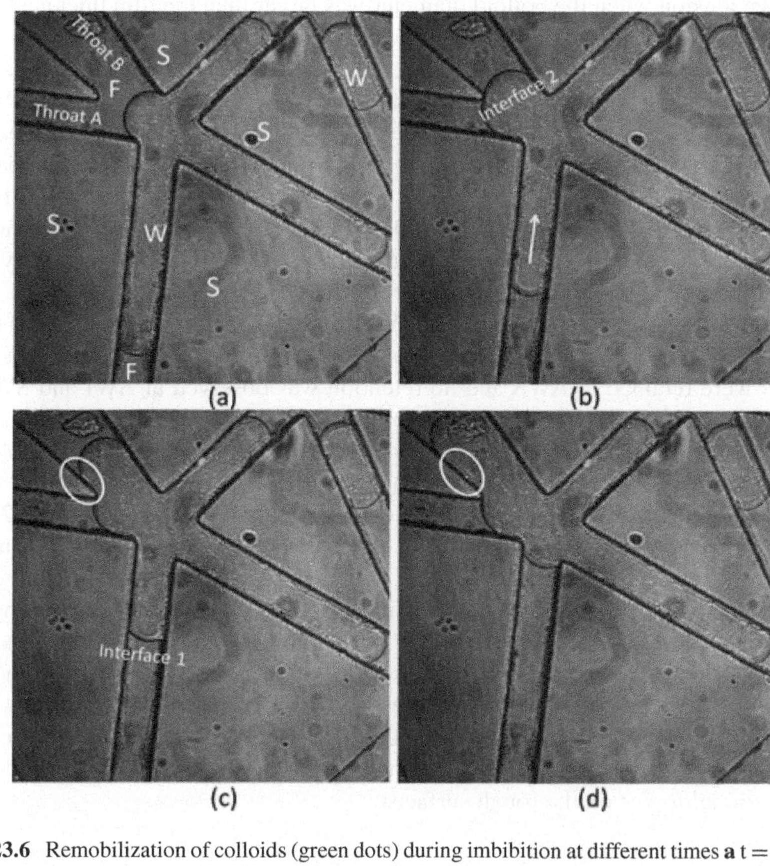

Fig. 23.6 Remobilization of colloids (green dots) during imbibition at different times **a** t = 45 min, **b** t = 45 min 16 s, **c** 45 min 31 s, and **d** 46 min 22 s (Zhang et al. 2013). Reprinted from Zhang et al. (2013), copyright 2013, permission obtained from John Wiley and sons

colloids depends on the change in the degree of water saturation (Zhang et al. 2013). In another study, Sirivithyapakorn and Keller (2003b) observed remobilization of colloids caused by the dissolution of air bubbles during imbibition. As dissolution progressed, colloids attached to the AWI were brought closer to each other and adhered to form clusters. When the air bubble disappeared, the colloidal clusters were transported down gradient where they attached to AWI or trapped in pore throats. At low water saturation, colloids tend to attach to immobile thin water films (Auset et al. 2005). Imbibition causes the immobile water zones to get mobilized, thereby releasing the colloids trapped in the thin films.

Flow cells with a length scale of centimetres help to visualize colloid transport mechanisms in 3D packed porous media (Crist et al. 2004). Table 23.3 lists the different flow cell studies and the key findings. Gao et al. (2006) studied the retention and remobilization of colloids in a sand-packed flow cell. Colloid retention occurred at AWI, thin films and immobile water zones. Film straining was observed in low moisture content experiments where the colloids were trapped in thin water films due to surface tension when the colloid diameter was larger than the film thickness. The zones branching out from larger flows to form dead-end distributaries with low pore water velocity istermed as immobile zones. The colloids enter the immobile zones by diffusion and slow advection and accumulate at AWI. Straining of the colloids in the thin films and accumulation at the AWS of the pendular ring is shown in Fig. 23.7 (Gao et al. 2006). These entrapped colloids were mobilized by increasing the flow rate tenfold which induces shear stresses causing some of the colloids in the thin films to get dislodged (Gao et al. 2006). The retained colloids in the immobile water zones were mobilized as they converted to mobile water zones due to physical perturbation, as shown in Fig. 23.8 (Gao et al. 2006).

In contrast to the findings of Wan and Wilson (1994a) and Wan et al. (1994) from micromodel studies, Crist et al. (2004) observed from flow cell experiments that colloids were retained at AWS and no retention was observed at AWI and SWI as shown in Fig. 23.8. Zevi et al. (2005) observed that hydrophilic colloids retained mainly on AWS, whereas hydrophobic colloids attached to AWI, SWI and AWS. The smaller hydrophilic colloids were found to attach throughout the water film. Colloid retention at grain to grain contact is shown in Fig. 23.9. Johnson et al. (2010) observed colloid retention at grain to grain contact when the ratio of colloid diameter to collector diameter is greater than 0.005, and the colloids retained at grain to grain contacts could not be detached by decreasing the ionic strength or by increasing the velocity. Bradford et al. (2005) observed that straining increased with the textural contrast of the sand grains. Smaller colloids were found to significantly attach to the open surfaces due to surface roughness (Johnson et al. 2010). From flow cell experiments, Morales et al. (2009) observed a greater retention of colloids on rough sand than on smooth ones under both saturated and unsaturated conditions due to high frictional forces on the rough surfaces.

Table 23.3 Mechanisms observed through flow cell experiments

Mechanism	References	Porous media	Colloid	Key finding
Sorption at AWS, AWI and SWI	Zevi et al. (2005)	Silica sand	Carboxylate polystyrene microspheres, blue polystyrene microspheres	Colloids retained at AWS when film thickness is approximately equal to colloid diameter
	Crist et al. (2004)	Quartz sand	Blue dyed polystyrene latex microspheres	Drainage causes higher retention of colloids at the AWS
	Smith et al. (2008)	Quartz sand	Latex microspheres, bacteria, Poly lactic acid microspheres	Colloids attached to grain surfaces. Rate constants estimated at pore scale able to simulate Darcy scale observations
	Morales et al. (2009)	Quartz sand	Polystyrene dyed colloids	Colloid deposition is more on rough than on smooth sand
Remobilization	Gao et al. (2006)	Quartz sand	Carboxylate-modified latex microspheres	Transient flow mobilizes colloids retained in thin water films and immobile zones
Straining	Johnson et al. (2010)	Glass beads	Fluorescent carboxylate-modified polystyrene latex microspheres	Straining is prominent for colloid to collector ratio greater than 0.005
	Bradford et al. (2005)	Quartz sand-homogenous and layered	Yellow-green fluorescent latex microspheres	Increase in textural contrast increases straining

(continued)

Table 23.3 (continued)

Mechanism	References	Porous media	Colloid	Key finding
Film straining	Gao et al. (2006)	Quartz sand	Carboxylate-modified latex microspheres	Low moisture experiments trap colloids in thin films due to surface tension

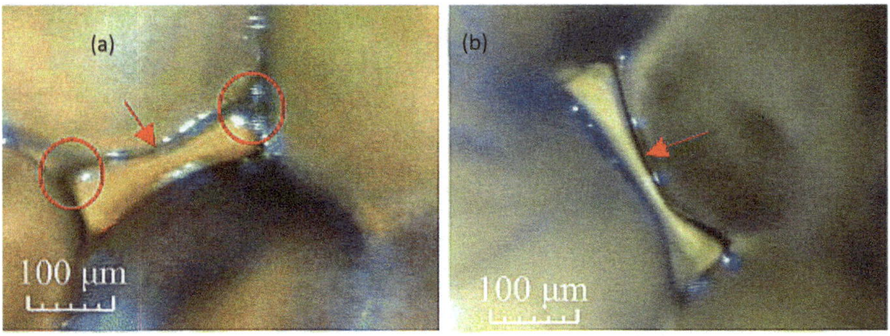

Fig. 23.7 Retention of colloids (blue dyed) **a** within the films (shown with red arrow) and **b** held at AWI of pendular rings (red circles) (Gao et al. 2006). Reprinted from Gao et al. (2006), copyright 2006, permission obtained from John Wiley and sons

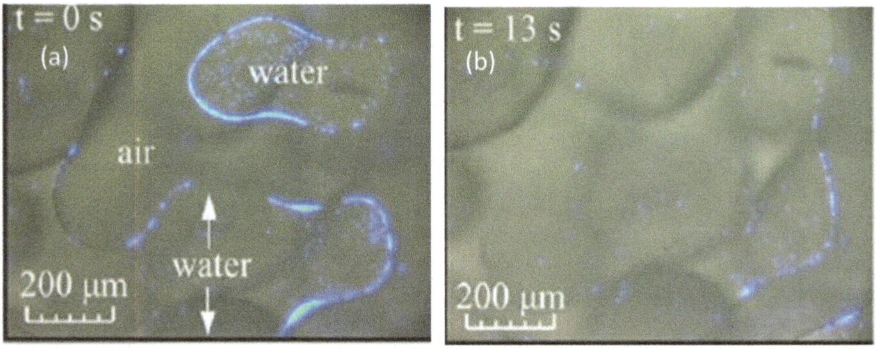

Fig. 23.8 Conversion of **a** immobile water zones to **b** mobile water zones in transient conditions (Gao et al. 2006). The blue fluorescent dyed colloids are trapped at the AWI in immobile zones and a faint bluish line of colloids is seen on remobilization. Reprinted from Gao et al. (2006), copyright 2006, permission obtained from John Wiley and sons

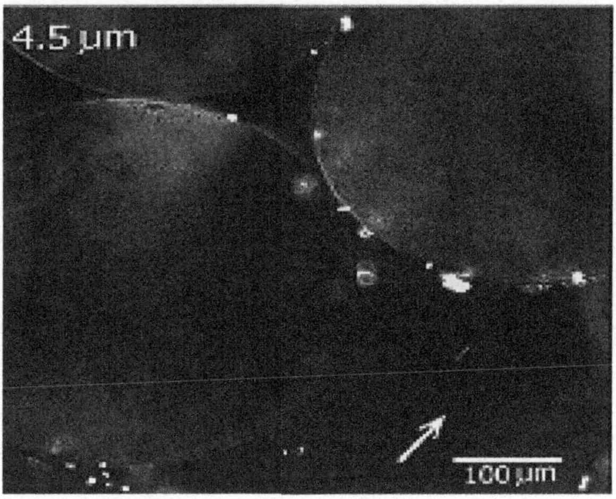

Fig. 23.9 Retention of colloids (white) at the grain to grain contacts (Johnson et al. 2010). Reprinted from Johnson et al. (2010), copyright 2010, permission obtained from Elsevier

23.3.2 Column Scale

The column scale represents one-dimensional flow and transport. It has a length scale of tens of centimeters, and is much larger than the REV scale. Colloid transport studies are usually carried out at the column scale by performing laboratory experiments. The various underlying mechanisms of colloid retention, such as deposition to SWI and AWI, blocking, straining, and ripening, are inferred from the observed breakthrough curves and the retention profiles (Bradford et al. 2004; Keller et al. 2004; Torkzaban et al. 2008; Wan and Wilson 1994b; Saiers et al. 1994b; Wang et al. 2012a; Wan and Tokunaga 1997; Camesano et al. 1999). Colloid transport at this scale is modelled mathematically using a 1D advection–dispersion equation modified to account for one or more retention mechanisms such as one-site or two-site deposition, blocking, straining, ripening, and aggregation. One-site and two-site deposition represent the different types of deposition sites available on the soil surface, including equilibrium site (Schijven et al. 2000a; Jin et al. 2000; Torkzaban et al. 2006a, b), and reversible- and irreversible-kinetic sites (Li and Johnson 2005; Bai et al. 2016; Hornberger et al. 1992; Tufenkji et al. 2004; Wang et al. 2012a; Shen et al. 2008; Hosseini and Tosco 2013; Ma et al. 2017). The values of the macroscopic parameters representing various retention mechanisms are estimated by fitting the experimental breakthrough curves or the retention profiles with the advection–dispersion-deposition equation. Some of the column studies where a linear kinetic model describes the deposition of colloids to the grain surface in a homogeneous porous medium under saturated conditions are given in Table 23.4. Table 23.5 lists some of the column experimental studies where straining mechanism was observed along with deposition to the grain surface in a homogeneous porous medium. Table 23.6 lists some of the column experimental studies where blocking or size exclusion occurred along with deposition to the grain surface in a homogeneous porous medium. Some of the studies where deposition to both SWI and AWI occurred in a homogeneous porous medium under

Table 23.4 List of column experimental studies with deposition to the grain surface under saturated conditions

References	Porous medium	Colloid	Mathematical model	Site-1		Site-2	
				Attachment rate coefficient (min^{-1})	Detachment rate coefficient (min^{-1})	Attachment rate coefficient (min^{-1})	Detachment rate coefficient (min^{-1})
Tufenkji et al. (2004)	Ultrapure quartz sand	Cryptosporidium oocysts	One-site irreversible deposition	0.72–2.58	0	–	–
Wang et al. (2012a)	Accusand	Silica nanoparticles	One-site irreversible deposition	6.83×10^{-5} to 0.0096	0	–	–
Bai et al. (2016)	Sand	*E. coli* and *Klesbiella oxytoca*	One-site irreversible deposition	0.012–0.211	0	–	–
Syngouna and Chrysikopoulos (2012)	Quartz sand	ϕX174 and MS2	One-site irreversible deposition and inactivation	0.00002–0.018	0	–	–
Sadeghi et al. (2011)	Quartz sand	PRD1	One-site reversible deposition with inactivation	0.000075–0.0346	0.000015–0.0033	–	–
Hornberger et al. (1992)	Quartz sand	Bacterial strains	One-site reversible deposition	0.005–0.126	0.0002–0.0015	–	–
Hosseini and Tosco (2013)	Natural coarse sand	nano-Fe/Cu	Two-site reversible deposition	0.36–0.78	0.96–2.34	1.44–2.52	0.12–0.192

(continued)

Table 23.4 (continued)

References	Porous medium	Colloid	Mathematical model	Site-1		Site-2	
				Attachment rate coefficient (min^{-1})	Detachment rate coefficient (min^{-1})	Attachment rate coefficient (min^{-1})	Detachment rate coefficient (min^{-1})
Li and Johnson (2005)	Quartz sand	Fluorescent carboxylate-modified latex beads	One-site reversible deposition	0.001–0.18	0.00166–0.009	–	–

Table 23.5 List of column experimental studies with colloid deposition to the grain surface and straining under saturated conditions

References	Porous medium	Colloid	Straining ratio (d_c/d_p)	Mathematical model	Attachment rate coefficient (min^{-1})	Detachment rate coefficient (min^{-1})	Straining coefficient (min^{-1})
Fang et al. (2013)	River sand	TiO_2 nanoparticles	0.0012–0.0097	Two-site kinetic model with one site reversible and the other site irreversible with straining	0.13–1.04		
Foppen et al. (2005)	Coarse and fine quartz sand	E.coli ATCC25922	0.007, 0.00286	Two-site kinetic model with one site reversible and the other site irreversible with straining	0.001–0.0139	0.00007–0.0055	17–152
Bradford et al. (2004)	Quartz sand	Fluorescent carboxylate-modified latex beads	0.0015–0.0213	One-site irreversible kinetic model with straining			0.065–2
Bradford et al. (2003)	Quartz sand	Fluorescent carboxylate-modified latex beads	0.0006–0.0213	One-site irreversible kinetic model with straining			0.0041–2.160

Table 23.6 List of column experimental studies with colloid deposition to the grain surface and blocking or size exclusion under saturated conditions

Mechanism	References	Porous medium	Colloid	Mathematical model	Attachment rate coefficient (min^{-1})	Detachment rate coefficient (min^{-1})
Blocking	Toloni et al. (2014)	Quartz sand	TiO_2 nanoparticles	One-site irreversible kinetic model with blocking	0.03–0.09	0
	Torkzaban et al. (2008b)	Ultrapure quartz sand	*E. coli D21g*	One-site irreversible kinetic model with blocking	0.04–0.91	0
	Sasidharan et al. (2017)	Quartz sand	MS2, PRD1 and $\phi \times$ 174	One-site reversible kinetic model with blocking and inactivation	0.071–1.836	0.00017–0.065
	Torkzaban et al. (2008)	Ottawa quartz sand	Carboxylate-modified latex beads	One-site reversible kinetic model with blocking	0.01–0.26	0.001–0.29
Size exclusion	Keller et al. (2004)	Quartz sand	MS2, Carboxylate-modified latex beads		5.55E-5 to 0.00216	

Table 23.7 List of column experimental studies with colloid deposition to the grain surface and AWI under unsaturated conditions

References	Porous medium	Colloid	Mathematical model	Attachment rate coefficient (min^{-1})	Detachment rate coefficient (min^{-1})	Straining coefficient (min^{-1})
Lenhart and Saiers (2002)	Quartz sand	Silica colloids	Dual rate law irreversible kinetic model	0.0013–0.0081		
Saiers and Lenhart (2003)	Quartz sand	Silica colloids	Two site irreversible attachment model with straining	1.14×10^{-4} to 3.21×10^{-2} (SWI); 1.14×10^{-3} to 1.66×10^{-2} (AWI)	2.04×10^{-4} to 3.83×10^{-3}	4.25×10^{-4}
Gao et al. (2004)	Quartz sand	Kaolinite and illite clay	Two site irreversible attachment model with straining	0.0032–0.012 (AWI and SWI)		0.0028–0.003
Torkzaban et al. (2006a)	Quartz sand	MS2 and $\phi \times 174$	Two site reversible kinetic model with attachment to AWI and SWI	0.00005–0.01 (SWI); 0.0006–0.0097 (AWI)	0.0001–0.0095 (SWI); 5×10^{-6} to 0.00002 (AWI)	

unsaturated conditions are listed in Table 23.7. The range of values of rate coefficients for attachment, detachment and straining estimated from column experiments are 6.83×10^{-5} to 2.58 min^{-1}, 5×10^{-6} to 2.34 min^{-1}, 6.76×10^{-6} to 152 min^{-1}, respectively. However, these macroscopic 1D parameters are averaged quantities obtained by averaging the pore-scale processes and depend on the physico-chemical conditions of the system studied, including the flow velocity, ionic strength, pH, type of colloid and soil, size of colloid and soil grains, and degree of saturation. These estimated parameters may not be applicable to predict colloid transport under a different physico-chemical condition and may not be able to predict colloid transport at larger scales due to spatial variability in the soil properties. Hence, it is essential to link the pore-scale processes and parameters quantitatively to column scale processes and parameters to predict colloid transport under a wide range of physico-chemical conditions. Most commonly, pore-scale visualization studies explain the observed mechanisms at the column scale only qualitatively (Bradford et al. 2005; Auset et al. 2005).

The effect of geochemical heterogeneity on colloid retention in porous media at the column scale has been investigated by many researchers and is given in Table 23.8. For example, Johnson et al. (1996) and Chen et al. (2020) performed column experiments to study the transport of silica and natural soil colloids, respectively, through a chemically heterogeneous porous medium made up of clean and iron

Table 23.8 Effect of chemical heterogeneity on colloid transport in column studies

References	Porous media	Colloid	Percentage of heterogeneity	Mathematical model	Key finding
Johnson et al. (1996)	A mixture of ferric oxyhydroxide coated and uncoated sand	Silica microspheres	0–16%	Two site irreversible kinetic model with blocking	Colloid retention increased with increasing degree of heterogeneity
Chen et al. (2020)	A mixture of (1) pure quartz sand and sand coated with goethite, and (2) pure quartz sand and sand coated with goethite and humic acid	Soil colloids	0.5% Goethite, 0.4% Humic acid	–	Colloid retention increased in the presence of goethite and decreased in the presence of humic acid
Han et al. (2006)	Glass beads	MS2 and ϕ × 174	1:1mixture of hydrophilic and hydrophobic glass beads	–	Higher retention in chemically heterogeneous medium. Greater retention in saturated medium
Hopmans and Schoups (2005)	Quartz sand mixed with amino silane	Carboxyl latex	10%	Two-site irreversible model	Colloid transport is affected by the fraction of chemical heterogeneity and not its spatial variability

oxide-coated quartz sand, thereby creating unfavourable and favourable sites for deposition. The column is homogeneous, with the degree of chemical heterogeneity being uniform throughout the porous medium. They found that even a small fraction of favourable sites significantly increased colloid deposition, and the deposition rate decreased with time due to blocking. The experimental results of Johnson et al. (1996) were simulated using a patch-wise charge heterogeneity model (or a two-site kinetic model) with separate deposition kinetics for favourable and unfavourable sites. The deposition rate coefficients to favourable and unfavourable sites were estimated independently using the transport data from experiments performed with only iron-oxide coated sand and clean sand, respectively. In a column packed with clean quartz sand, Hopmans and Schoups (2005) studied the effect of layered heterogeneity created by aminosilane-modified quartz sand at different depths. They found that the spatial distribution of the chemical heterogeneity of the porous medium does not affect the deposition rate and the transport behaviour of carboxyl latex particles for a given fraction of heterogeneity. The presence of organic matter on the soil surface decrease the deposition of colloids as they offer unfavourable conditions for deposition (Chen et al. 2020). In contrast to the above studies, virus retention has been found to be lesser in a chemically heterogeneous porous medium composed of hydrophilic and hydrophobic glass beads than a homogeneous porous medium of hydrophilic glass beads (Han et al. 2006). This is because of flow bypassing the hydrophobic regions in heterogeneous media, thereby a fraction of favourable sites were not accessible to viruses. Also, virus retention has been found to be greater under saturated than unsaturated conditions in a chemically heterogeneous porous medium composed of hydrophilic and hydrophobic glass beads (Han et al. 2006).

At the column scale, the various forms of physical heterogeneity studied are grain surface roughness (Shen et al. 2015), layered heterogeneity (Johnson et al. 1996; Rod et al. 2018), embedding a tubule of soil in a larger matrix of another soil (Bradford et al. 2004; Saiers et al. 1994a; Lv et al. 2016; Mishurov et al. 2008), and a mixture of soil grains of various sizes (Rod et al. 2018). A detailed review of the effect of physical and chemical heterogeneity on colloid transport in the subsurface can be found in Ryan and Elimelech (1996) and McCarthy and McKay (2004). Table 23.9 describes the effect of physical heterogeneity on colloid transport at the column scale.

Surface roughness caused greater attachment and detachment of fullerene nanoparticles onto sand than glass beads in laboratory columns (Shen et al. 2015). The effect of structural heterogeneity on the transport of silica colloids was studied by embedding a tubule of coarse-grained sand in a matrix of fine-grained sand throughout the length of the column (Saiers et al. 1994a). Both the tracer and colloid breakthrough curves from heterogeneous columns were characterized by multiple inflections, indicative of the preferred flow path through coarse sand as shown in Fig. 23.10, whereas those from a homogeneous column packed with either coarse-grained or fine-grained sand had only a single peak. The experimental results from heterogeneous columns were successfully simulated using a 2D advection–dispersion equation with first-order reversible kinetic deposition, in which the kinetic rate coefficients for fine- and coarse-grained soil were estimated by fitting the breakthrough curves of homogeneous columns of fine- or coarse-grained soil with 1D advection–dispersion-deposition equation. It was found that about 60% of colloidal mass transmitted through the preferred flow path. Similar kind of breakthrough curves characterized by early arrival and multiple inflections were also observed by Lv et al. (2016) for the transport of TiO_2 nanoparticles in a structurally heterogeneous soil composed of fine sand embedded in the centre of a column packed with coarse sand and both the sand types extending throughout the column length. It was also found that the TiO_2 nanoparticle retention in heterogeneous soil is lesser than homogeneous

Table 23.9 Effect of physical heterogeneity on colloid transport at column scale

References	Porous media	Colloid	Form of heterogeneity	Mathematical model	Key finding
Shen et al. (2015)	Sand, glass beads	Fullerene	Surface roughness		Sand retained more colloids than glass beads
Saiers et al. (1994a)	Quartz sand (fine and coarse)	Colloidal silica	Tubule of coarse sand embedded in the matrix of fine sand	2D advection–dispersion equation with first-order reversible kinetic deposition	Preferential flow paths cause a non-sigmoidal breakthrough curve
Bradford et al. (2004)	Quartz sand of varying sizes	Yellow-green fluorescent latex microsphere	Coarse or fine-textured soil lens embedded in fine or coarse textured matrix soil	One-site reversible kinetic model with straining	Flow bypassed fine-textured soil, and colloids retained at textural interfaces
Lv et al. (2016)	Quartz Sand	nTiO_2	Fine sand embedded in coarse sand	One-site irreversible deposition with blocking	Preferential flow paths through coarse sand
Mishurov et al. (2008)	Quartz sand	Carboxylate-modified microspheres	Fine sand tubes within a column of coarse sand	–	Higher recovery of colloid due to exchange between the fine and coarse sand

soil under the same experimental conditions, indicating that structural heterogeneity can facilitate colloid transport. Further, the length of the embedded fine sand had minimal effect on TiO_2 nanoparticle transport, pointing to the dominance of the preferential flow path created by the coarse sand on the colloid transport behaviour. The experimental results were simulated using a two-domain model accounting for the exchange of mass between fine and coarse sands and irreversible deposition to the sand types with blocking.

In another study, Bradford et al. (2004) investigated the transport of fluorescent latex microspheres in a physically heterogeneous system consisting of various combinations of coarse or fine-textured cylindrical soil lens embedded in the centre of a larger cylinder of fine or coarse textured matrix soil. Physical heterogeneity led to flow bypassing the fine-textured soil, thereby causing early arrival of both tracer and colloids due to size exclusion and long tailing in the breakthrough curves, as shown in Fig. 23.11. Also, colloids were retained at the textural interfaces, and retention increased with increasing particle size and decreasing grain size due to straining.

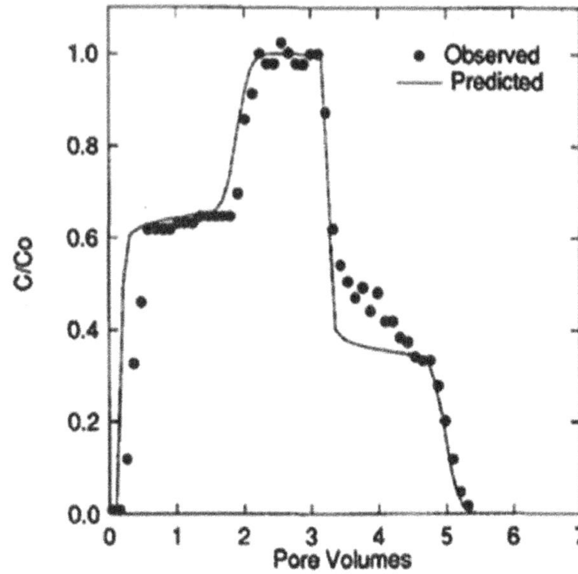

Fig. 23.10 Breakthrough curve of colloidal silica from a physically heterogeneous porous media (Saiers et al. 1994a). Reprinted from Saiers et al. 1994a, copyright 1994, permission obtained from Elsevier

Mixed physical heterogeneity created by mixing glass beads of two different sizes was found to retain more SiO_2 colloids than sequential layered heterogeneity of coarse and fine glass beads due to smaller pore size in the mixed heterogeneous porous media (Rod et al. 2018).

23.3.3 Sand-Box Scale

One dimensional column experiments cannot account for the spatial variability in hydraulic conductivity and the complex three-dimensional flow and transport processes existing at the field scale. For example, spatial variability of rainfall, soil heterogeneity, and vegetation have been found to play a significant role in colloid transport in the subsurface at the field scale (Loveland et al. 2003; Wang et al. 2019; Yu et al. 2012) and cannot be studied at the column scale. Two- or three-dimensional laboratory sand-box experiments having a length scale of meters can provide a close representation of the field conditions in the lab and can be used to obtain a better understanding of the processes occurring in the field. A typical experimental set-up of a sand-box study is shown in Fig. 23.12. The experimental setup consists of a chamber made up of acrylic or glass to hold the porous medium, pumps, constant head reservoirs at the inlet and outlet to maintain flow through the porous medium, screens to distribute the flow along the entire crosssection of the tank, colloid injection well, and sampling ports at multiple locations along the longitudinal and transverse directions to collect the samples for analysis (Schijven et al. 2000a). The sand-box studies

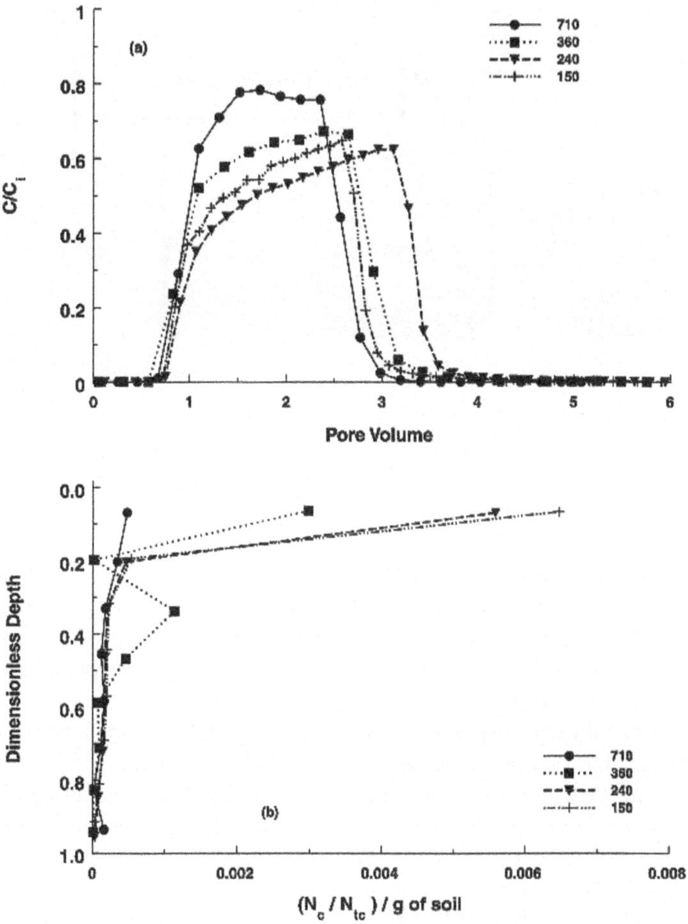

Fig. 23.11 Breakthrough curves (**a**) and spatial distribution of colloids (**b**) in a physically heterogeneous porous media (Bradford et al. 2004). Reprinted from Bradford et al. (2004), copyright 2005, permission obtained from Elsevier

reported in the literature along with the associated mechanisms and mathematical models are listed in Table 23.10.

Transport of *Pseudomonas putida* bacteria through saturated quartz sand was studied by Chrysikopoulos et al. (2012) by performing 3D sand-box experiments. The experimental results were simulated using a 3D advection–dispersion equation in a uniform flow field, accounting for first-order reversible deposition and first-order decay in liquid and solid phases. The effect of layered chemical heterogeneity on the transport of silica coated zirconia colloids through sand was studied by Loveland et al. (2003) in a two dimensional acrylic tank. A mixture of ferric oxyhydroxide coated (favourable site) and uncoated (unfavourable site) sand was filled in three layers parallel to the direction of the flow in the tank, with each layer having a

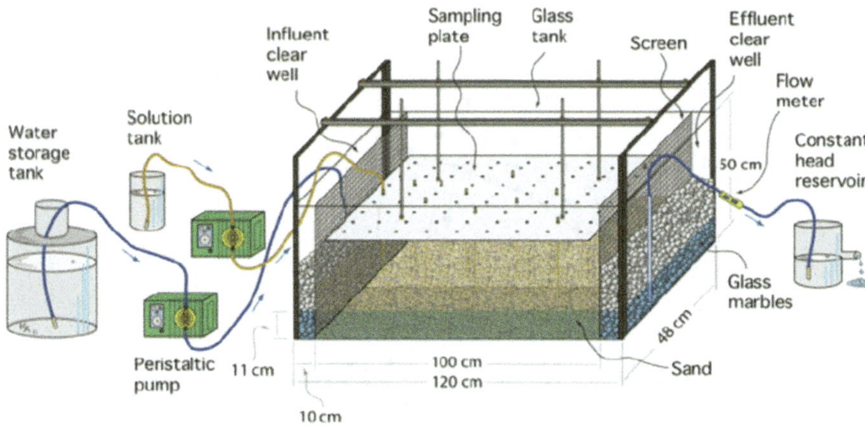

Fig. 23.12 Typical set-up of sand-box experiments (Chrysikopoulos et al. 2012). Reprinted by permission from springer (Chrysikopoulos et al. 2012), copyright 2012

different fraction of coated sand: 0.001, 0.1 and 0.05 in the top, middle and bottom layers, respectively. Figure 23.13 presents the experimental breakthrough curves from Loveland et al. (2003). It can be seen that as the fraction of favourable sites for deposition increases, the time to breakthrough increases and the peak breakthrough concentration decreases. Since transverse dispersion of tracer and colloids was found to be negligible, colloid transport in each layer was modelled using a one dimensional two-site irreversible kinetic model with blocking (Fig. 23.13).

Phenrat et al. (2010) studied the effect of layered physical heterogeneity of soil on the transport of nZVI particles using a 2D flow cell filled with five layers of fine, medium and coarse sand (Fig. 23.14). The nZVI suspensions were injected through an injection well into the third layer from the top, and the background solution flowed from left to right of the flow cell. The velocity distribution in the layers simulated using a two dimensional groundwater flow equation showed that flow happened predominantly through the medium and coarse sand, with negligible flow through the fine sand (Fig. 23.14). Phenrat et al. (2010) observed from experiments that nZVI particles were transported through preferential flow paths through coarse and medium sand, and accumulated significantly at the point of injection (Fig. 23.3; X = 0–7.5 cm and Y = 7–12 cm) in the fine sand layer due to its low permeability, and in the low velocity regions (Fig. 23.3; X = 0–8 cm and Y = 1–4 cm, X = 17–30 cm and Y = 4–8 cm) (Fig. 23.15).

The role of surface runoff on the transport of colloids in the subsurface was studied by Yu et al. (2011) by performing sand-box experiments in the presence of a simulated rainfall of uniform intensity. The tracer and kaolinite clay colloids applied to one end of the sand box appeared predominantly in the surface runoff, and their mass recovered from surface flow was higher than that of subsurface flow. This is due to the exchange of tracer and colloids from the subsurface to the surface runoff caused by film diffusion, raindrop induced exchange, and pumping exchange. Both the

Table 23.10 List of studies on transport of colloids performed at the sand-box scale

References	Colloid source type	Dimensions (L * W * H) in m	Porous media	Colloid type	Mathematical model	Attachment rate coefficient (min^{-1})	Detachment rate coefficient (min^{-1})	Inference
Loveland et al. (2003)	Distributed source at the inlet	5 × 0.1 × 1	Layered heterogeneity: Mixture of uncoated and ferric oxyhydroxide coated sand	Silica Zirconia colloids	1D advection–dispersion equation with irreversible deposition in each layer	1.069 (favourable site), 0.00031 (unfavorable site)	0	Deposition increased with an increase in favourable sites or fraction of coated sand
Phenrat et al. (2010)	Point source injection into a layer of fine sand	0.3 × 0.18 × 0.025	Layered heterogeneity: fine, medium and coarse sand	Fe nanoparticles	–	–	-	Particles transported through preferential flow paths through coarse and medium sand
Yu et al. (2011)	Distribute source at inlet using a flow spreader	1.53 × 0.402 × 0.1	Quartz sand	Kaolinite clay colloids	–	–	–	Transport of tracer and colloids was similar in surface runoff

(continued)

Table 23.10 (continued)

References	Colloid source type	Dimensions (L * W * H) in m	Porous media	Colloid type	Mathematical model	Attachment rate coefficient (min^{-1})	Detachment rate coefficient (min^{-1})	Inference
Schijven et al. (2000a)	Distributed source at the inlet	1.2 × 0.48 × 0.5	Quartz sand	*P. putida*	3D advection–dispersion equation with reversible sorption and decay	0.013	0.0043	Reversible deposition and decay of bacteria in the aquifer
Yu et al. (2013)	Distribute source at inlet using a flow spreader	0.2 × 0.2 × 0.1	Quartz sand with vegetation and no drainage	Carboxylate-modified latex microspheres	First order irreversible deposition and diffusive mass exchange between overland flow and sand			Colloids get trapped in the soil present at the roots of dense vegetation
Bradford et al. (2015)	Sprinkler source at inlet	2.25 × 0.15 × 0.16	Ultrapure quartz sand	*E.Coli* bacteria	Transient storage model with a term for exchange between surface and subsurface layers	–	–	Tailing of E.coli in the surface runoff and exchange between surface and subsurface layers increased with an increase in runoff slope

(continued)

Table 23.10 (continued)

References	Colloid source type	Dimensions (L * W * H) in m	Porous media	Colloid type	Mathematical model	Attachment rate coefficient (min^{-1})	Detachment rate coefficient (min^{-1})	Inference
Mondal et al. (2018)	Distributed source at the inlet	0.7 × 0.025 × 0.18	Quartz sand	CMC-nZVI particles	–	–	–	Effect of sand grain, concentration and flow velocity
Li et al. (2016)	Distributed source at the inlet	0.55 × 0.014 × 0.45	Silica sand	CMC-nZVI particles	–	–	–	Viscous fingering was observed when CMC-nZVI particles were suspended in the sand
Yu et al. (2012)	Distribute source at inlet using a flow spreader	1.53 × 0.402 × 0.1	Quartz sand	Carboxylate-modified latex microspheres	–	–	–	Dense vegetation helps in preventing the contaminants from entering surface runoff

(continued)

Table 23.10 (continued)

References	Colloid source type	Dimensions (L * W * H) in m	Porous media	Colloid type	Mathematical model	Attachment rate coefficient (min^{-1})	Detachment rate coefficient (min^{-1})	Inference
Wang et al. (2019)	Spraying with spray bottles on the surface	0.2 × 0.5 × 0.1	Loamy sand	DNA labelled particle tracers				Attachment, straining and size exclusion affect the transport of DNA labelled particle tracers

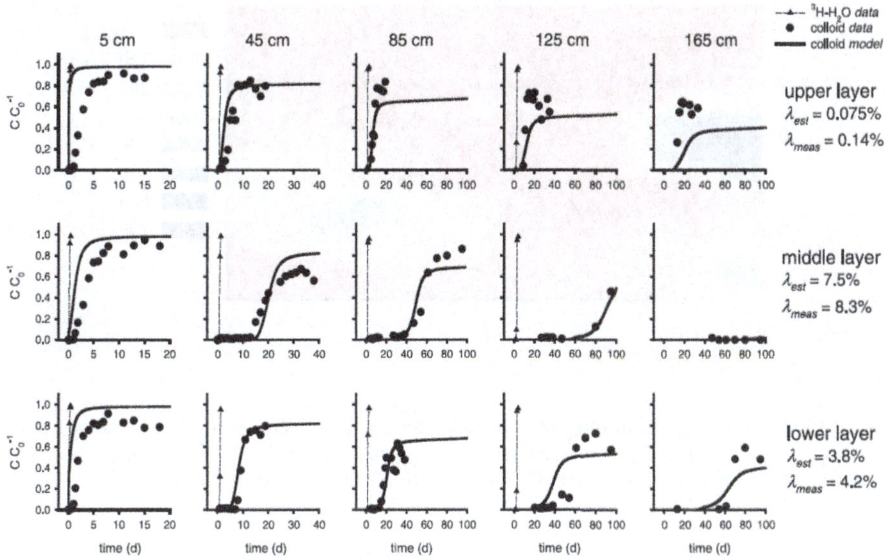

Fig. 23.13 Tracer and colloid concentrations as a function of time. Best fits produced by the colloid transport model are shown as solid curves. Travel distance is listed along the top of the graphs. λ_{est} and λ_{meas} are the measured and estimated values of favourable surface fraction (Loveland et al. 2003). Reprinted from Loveland et al. 2003, copyright 2003, permission obtained from Elsevier

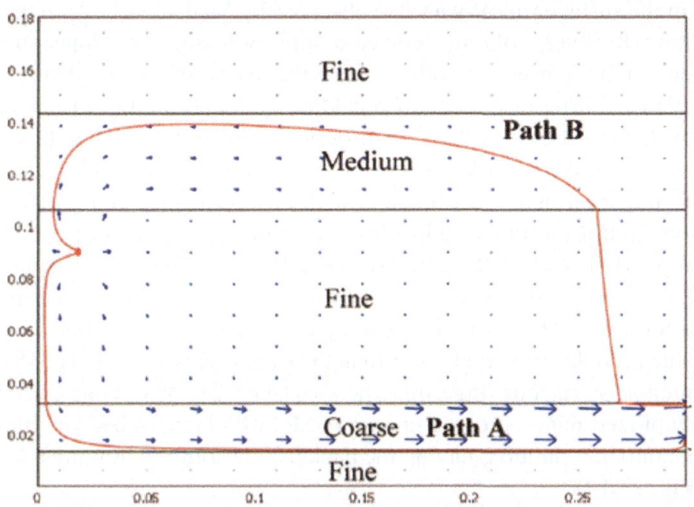

Fig. 23.14 Schematic of 2D flow cell with heterogeneous layered packing and simulated flow paths (Phenrat et al. 2010). Reprinted with permission from Phenrat et al. (2010). Copyright 2010 American Chemical Society

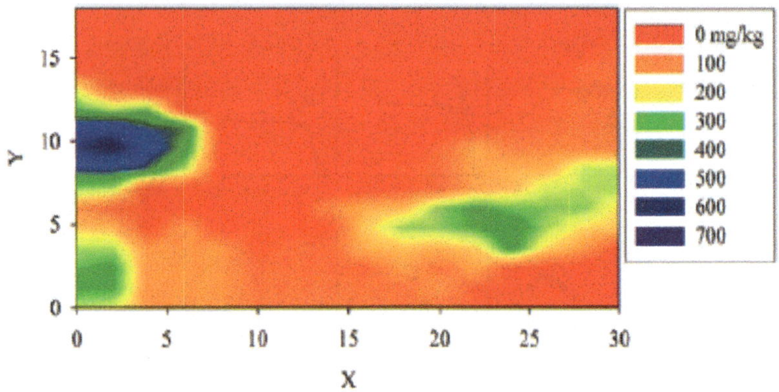

Fig. 23.15 Contour map illustrating nZVI particles deposited onto sand grains (mg of nanoparticles per kg of sand. X and Y axis represent distances in cm (Phenrat et al. 2010). Reprinted with permission from Phenrat et al. (2010). Copyright 2010 American Chemical Society

kaolinite and tracer breakthrough curves in the surface runoff showed extended tailing indicating their slow release from the soil to the surface runoff (Fig. 23.16). Also, the transport of kaolinite resembled tracer transport in the surface runoff, whereas the mobility of kaolinite was lower than that of the tracer in the subsurface due to its deposition onto the grain surface. Similarity in the transport behavior of colloids and tracer in the surface runoff was also observed by Yu et al. (2013) in the presence of vegetation. However, colloids showed a slightly lesser peak concentration than the tracer due to the removal of colloids from the overland flow by dense vegetation (Yu et al. 2013). Tailing of the *E. coli* breakthrough curve in the surface runoff was observed by Bradford et al. (2015), with lesser peak and greater tailing for larger runoff slopes when the *E. coli* suspension was applied at the soil surface. This is because of greater amounts of exchange with the subsurface and greater extent of physical nonequilibrium in the runoff layer with increasing slope. In another study, Wang et al. (2019) studied the effect of rainfall events on the transport of DNA labelled particle tracers through loamy sand in a sloped lysimeter receiving periodic rainfall for ten days. They found that the particle recovery rate was less than the tracer, and the particles travelled faster than the tracer. Mondal et al. (2018) observed the phenomenon of viscous fingering when a suspension of viscous carboxymethyl cellulose-stabilized nano zero valent iron (CMC-nZVI) particles was injected as a point source into the sandbox, having the background solution flowing from the inlet to outlet (Fig. 23.17).

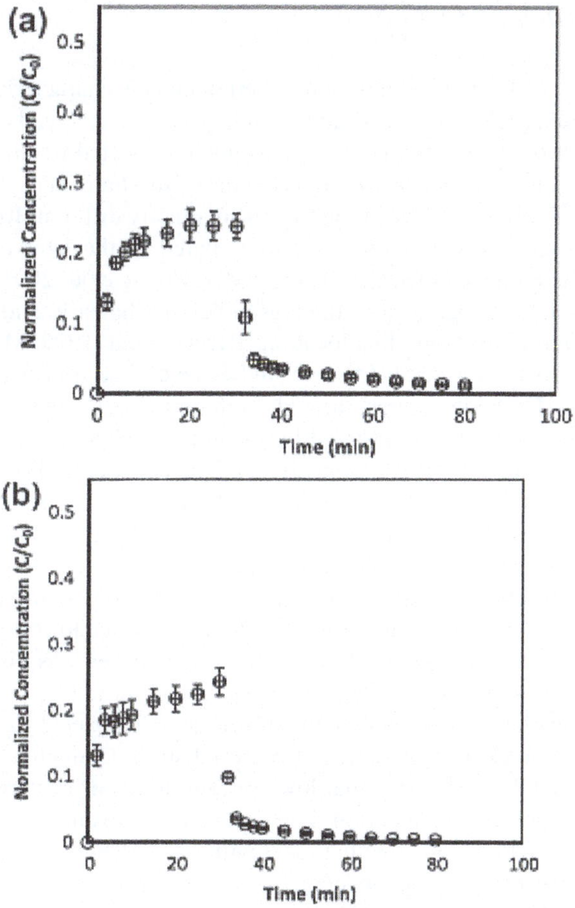

Fig. 23.16 Transport of **a** bromide and **b** kaolinite in surface flow (Yu et al. 2011). Reprinted from Yu et al. (2011), copyright 2011, permission obtained from Elsevier

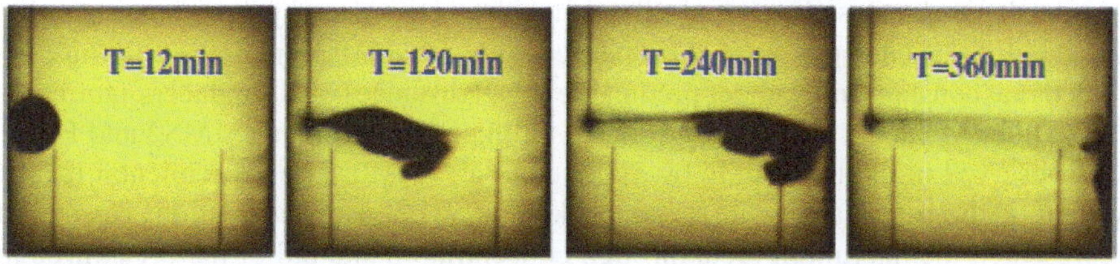

Fig. 23.17 CMC-nZVI plume movement in the sandbox during point injection (Mondal et al. 2018). Reprinted from Mondal et al. (2018), copyright 2018, permission obtained from Elsevier

23.3.4 Field Scale

Field having a length scale of kilometers is characterized by a high degree of spatial heterogeneity in hydraulic conductivity, which leads to colloid transport occurring through preferential flow paths, interconnected fractures and macropores formed by earthworm burrows or root channels (Burkhardt et al. 2008). The deposition behavior of colloids has been found to be distinctly different for the first few meters from the source of contamination as compared to the later distance in the aquifer due to heterogeneity (Schijven et al. 1999; Zhang et al. 2001). The spatial heterogeneity at the field scale is also the reason behind the early and late breakthrough of colloids at different sampling locations (Becker et al. 1999; McKay et al. 2000). Table 23.11 lists the various field-scale studies on colloid transport reported in the literature.

Colloid transport through preferential pathways caused by fractures and macropores has been observed by many researchers at the field scale (Vilks et al. 1997; Becker et al. 1999; Göppert and Goldscheider 2008). Vilks et al. (1997) observed in a granite formation that silica colloids were transported over a distance of 17 m, and the breakthrough curves of both tracer and silica colloids were characterized by sharp initial peaks and long tails, indicative of the effect of fracture geometry and hydraulic conditions within the fracture zone, and is shown in Fig. 23.18. The peak concentration of colloids eluted was less than tracer due to colloid deposition onto the fracture surface. Similar kind of breakthrough curves was observed by Higgo et al. (1993) for the transport of fluvic acid particles in field conditions and is attributed to the spatial heterogeneity of the subsurface. Zhang et al. (2001) observed extended tailing of breakthrough curves of bacterial strain Comamonas DA001 in the field for four months, indicating that low concentrations of bacteria were present in the groundwater for a long period. The experimental results were modelled using a one dimensional two-site kinetic model with one site being reversible and the other site being irreversible. The rate coefficient of attachment of bacteria to the soil has been found to be spatially variable, with its value decreasing with increasing travel distance. This may be due to decreased bacterial adhesion with increased transport distance because of alteration of bacterial surface properties with transport. Hence, the column estimated rate constants may under predict the travel distances of colloidal contaminants in the field (Zhang et al. 2001). Extended tailing of bacteriophages MS2 and PRD1 was observed in the field study by Schijven et al. (1999). The experimental results were modelled using a one-site reversible kinetic model with inactivation. The model represented the maximum breakthrough concentration and tail of the breakthrough curves satisfactorily, but was unable to simulate the end of the rising limb and the onset of the falling limb. The bacteriophage removal was larger for the first few meters than later with a reduction of $3\log_{10}$ and $8\log_{10}$ in 2.4 and 30 m of travel distance, and the estimated attachment rate coefficient decreased with the travel distance. Similarly, in another study, Schijven et al. (Schijven and Šimůnek 2002) observed a nonlinear reduction in the concentration of MS2, PRD1, *Clostridium bifermentans*, and *E. coli* in a sandy aquifer, with the removal being largest in the first 8 m than

Table 23.11 List of colloid transport experiments conducted in the field

References	Field location	Formation	Colloid type	Source of colloid	Mathematical Model	Attachment rate coefficient (min^{-1})	Detachment rate coefficient (min^{-1})	Inference
Göppert and Goldscheider 2008)	Holloch cave, Astro-German Alps	Karst aquifer	Yellow-green carboxylate-modified latex microspheres and bright blue microspheres					Colloids travel at higher velocities than solutes due to preferential flow in fractures
Becker et al. (1999)	New Hampshire	Fractured granite/Schists	Carboxylate-modified latex microspheres	Well injection	–	–	–	Early arrival of colloids due to size exclusion and also greater retention due to aggregation and settling
Zhang et al. (2015)	Jieliu catchment in southwest China	Layered heterogeneity: Purple soil, mudrock and sandstone in layers one below the other	Natural colloids	Naturally occurring subsurface colloids	–	–	–	Infiltration due to rainfall events results in remobilization of colloids

(continued)

Table 23.11 (continued)

References	Field location	Formation	Colloid type	Source of colloid	Mathematical Model	Attachment rate coefficient (min^{-1})	Detachment rate coefficient (min^{-1})	Inference
Vilks et al. (1997)	Underground research laboratory, South Manitoba	Granite fractures	Silica colloids	Bore hole injection	–	–	–	Sharp peaks and long tailing in breakthrough curves suggesting preferential flow due to fractures
McKay et al. (2000)	Oak ridge national laboratory	Fractured shale saprolite	MS2, PRD1, carboxylate-modified latex microspheres and INA	Injection wells	–	–	–	Colloidal tracer remained in the wells for hundreds of days and are mobilized due to changes in flow and chemistry of the infiltrating water

(continued)

Table 23.11 (continued)

References	Field location	Formation	Colloid type	Source of colloid	Mathematical Model	Attachment rate coefficient (min^{-1})	Detachment rate coefficient (min^{-1})	Inference
Masciopinto et al. (2008)	Nardo freshwater aquifer, Salento, Italy	Layered sandstone, limestone and dolomite deposits	E. coli, *Clostridium spores*, somatic coliphages, faecal coliform, fecal streptococci and total bacteria	Groundwater recharge through a sink hole	3D reversible model with inactivation and blocking			Microbes travelled hundreds of kilometers downstream from the point of injection
Bennett et al. (2010)	San Francisco Bay, Palo Alto, California	Alluvial coarse grained sediments with silt and clays	CMC-nZVI particles	Injection during push pull tests	–	–	–	Increase in residence time of colloids during remediation results in immobilization of colloids
Ryan and Gschwend (1990)	Central Delaware, New Jersey	Pine barren lands with muck and peat soils	Natural colloids	Well injection			–	Depletion of iron in the absence of oxygen increased the mobilization of clay colloids

(continued)

Table 23.11 (continued)

References	Field location	Formation	Colloid type	Source of colloid	Mathematical Model	Attachment rate coefficient (min^{-1})	Detachment rate coefficient (min^{-1})	Inference
Higgo et al. (1993)	Glacial sand aquifer	Layered heterogeneity in field with layers of clay, angular and well-rounded sand grains	Fulvic acid, sodium polystyrene p-sulphonate, Silica and carboxylate modified latex microsp heres	Boreholes	–	–	–	Comparison of field and column studies revealed that heterogeneity in field conditions affected the transport of colloids
Zhang et al. (2001)	Oyster site	Silt loam layer with sediments comprised of quartz, feldpsar, Fe and Al hydroxides	Comamonas DA001	Injection wells	One site reversible kinetic model	1.66×10^{-5} to 0.00066	8.33×10^{-7} to 0.0004	Extended tailing of bacteria observed under natural gradient conditions
Schijven et al. (1999), Schijven and Šimůnek (2002)	Castricum, The Netherlands	Dune sand	MS2 and PRD1 bacteriophages	Dune recharge	2D two-site reversible model with inactivation	6.94×10^{-8} to 0.0031	3.61×10^{-7} to 0.036	Extended tailing of bacteriophages observed
Schijven and Šimůnek (2002), Schijven et al. (2000b)	Someren in southeast Netherlands	Sandy aquifer	MS2 and PRD1, spores of Clostridium bifermentans (R5) and Escherichia coli (WR1)	Deep well injection	2D two-site reversible model with inactivation	6.94×10^{-8} to 0.0035	3.61×10^{-7} to 0.036	Extended tailing of bacteriophages observed

(continued)

Table 23.11 (continued)

References	Field location	Formation	Colloid type	Source of colloid	Mathematical Model	Attachment rate coefficient (min^{-1})	Detachment rate coefficient (min^{-1})	Inference
Burkhardt et al. (2008)	Arable land near Merzenhausen, Germany	Orthic Luvisol with silt (80%) and clay (20%)	Synthetic polymer resin microspheres	Spraying on field soil surface with pesticide sprayer	–	–	–	Rainfall caused remobilization of colloids, and the colloids were transported into the subsurface mainly through macro pore structures

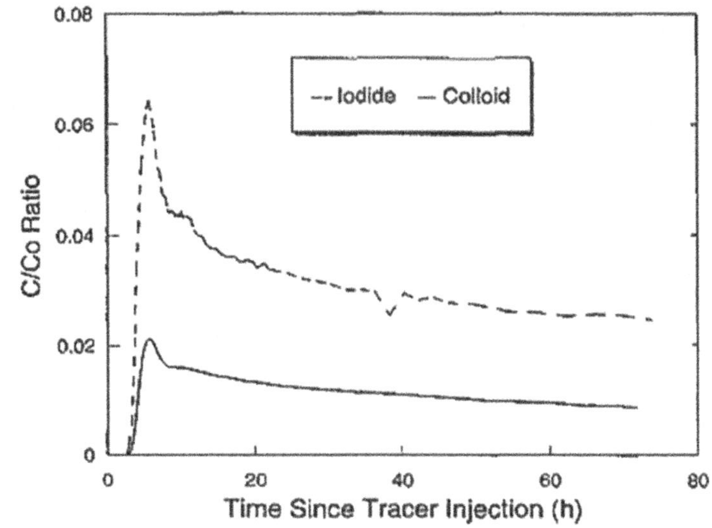

Fig. 23.18 Tracer (iodide) and colloid elution profiles in granite formation (Vilks et al. 1997). Reprinted from Vilks et al. (1997), copyright 1997, permission obtained from Elsevier

thereafter: $6\log_{10}$, $5\log_{10}$ and $7.5\log_{10}$ for MS2 and PRD1, *Clostridium bifermentans*, and *E. coli*, respectively. This nonlinear reduction with distance is attributed to the geochemical heterogeneity caused by the presence of ferric oxyhydroxides in the first 8 m from the injection well, which provide a favourable site for the attachment of microorganisms. Later, Schijven and Simunek (2002) reanalysed the field data of Schijven et al. (1999, 2000b) and found that a two dimensional two-site kinetic model with inactivation was capable of simulating the observed breakthrough curves from both the field sites (Fig. 23.19). Becker et al. (1999) observed early arrival of fluorescent carboxylate-modified latex colloids compared to solutes in fractured granite/schist field site in central New Hampshire and attributed this effect to size exclusion. Also, the retention was found to be greater for larger than smaller colloids due to aggregation and settling. Carboxylate polystyrene microspheres were detected in a spring 2.5 km downstream of the cave stream in a karst aquifer where they were injected (Göppert and Goldscheider 2008). The microspheres were found to travel faster than the solutes during low flow following preferential flow paths.

Transient flow due to rainfall has been found to remobilize the deposited colloids from the soil. McKay et al. (2000) observed colloids such as fluorescent latex microspheres, bacteriophages PRD1 and MS2, INA (a strain of *Pseudomonas syringae*) injected in a fractured shale saprolite appeared in the downstream observation wells for several days after the day of injection, as shown in Fig. 23.20. This is due to the remobilization of the deposited colloids by changes in flow and chemistry caused by the infiltrating rainwater. In another study, Masciopinto et al. (2008) detected pathogenic microorganisms such as *E. coli*, *clostridium spores*, *somatic coliphages*,

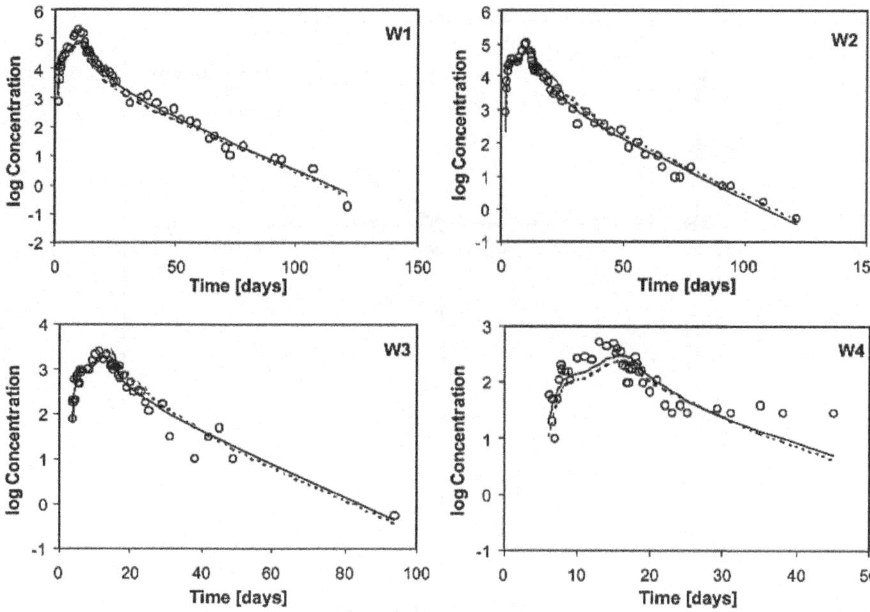

Fig. 23.19 Measured and optimized MS2 breakthrough curves for dune recharge experiment. Dashed lines represent one site model and full lines represent a two site model (Schijven and Šimůnek 2002) Reprinted from Schijven and Šimůnek (2002), copyright 2002, permission obtained from Elsevier

faecal coliform, *faecal streptococci* and total bacteria for several kilometers downstream of the point of injection of treated municipal wastewater in Nardo fractured aquifer, Salento, Italy. They found that a safe setback distance of 8000 ± 4800 m was needed between the point of wastewater injection and the point of water use. The results from this study were also modeled using a three dimensional kinetic model accounting for the inactivation of microorganisms. Figure 23.21 presents the observed and simulated breakthrough of pathogens at different wells considered in this study. Rainfall caused the deposited synthetic microspheres retained on the soil surface to move to larger depths in an arable land in Germany (Burkhardt et al. 2008). Microspheres were transported to the subsurface mainly by the hydrologically active macropore structures, thereby bypassing large parts of the soil matrix. The critical role of antecedent moisture content and rainfall intensity on the mobilization of natural colloids and its transport through surface flow, interflow in soil having macropores, and subsurface flow in the underlying fractured mudrock in a sloping farmland plot in Jieliu catchment in southwest China was studied by Zhang et al. (2015). Hydrodynamic shear caused by the overland flow broke and dispersed the soil aggregates, resulting in the release of soil colloids in the surface runoff. Infiltration from rainfall resulted in the mobilization of colloids from the soil layer and underlying rock due to the movement of the air–water interface. Colloid concentration in surface runoff has been found to be directly proportional to the rainfall intensity, whereas colloid

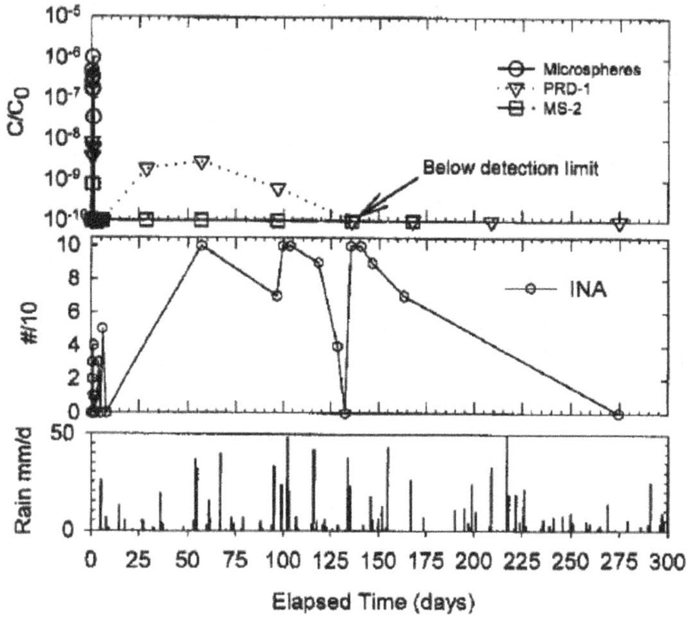

Fig. 23.20 Concentration of colloids in an observation well for a period of 0–300 days after injection (McKay et al. 2000). Reprinted from McKay et al. (2000), copyright 2005, permission obtained from John Wiley and Sons

concentration in the fracture flow peaked during the rising stage of the fracture flow when the overlying soil was not saturated. However, when the overlying soil was saturated, the maximum colloid concentration in the fracture coincided with the peak discharge as shown in Fig. 23.22. The concentration of colloids in the surface runoff was always 1–2 orders higher than in the subsurface flow. Colloidal particles of smaller diameter ($< 10\ \mu m$) dominated the surface runoff as larger ones might have been removed by settling and entrapment by surface vegetation. In contrast, larger colloids dominated the lateral and subsurface flows, with the smaller particles being deposited to the grain or the fracture surface. The effect of geochemical conditions on the mobilization of natural colloids from subsurface was investigated by Ryan and Gschwend (1990). The suspended colloid concentrations were higher in anoxic groundwater than in oxic groundwater. Anoxic conditions led to depletion of iron present on soil surfaces, causing the mobilization of clay colloids attached to the soil particles, whereas under oxic conditions, the oxidized iron created heterogeneous patches on soil surface creating favorable conditions for colloid attachment.

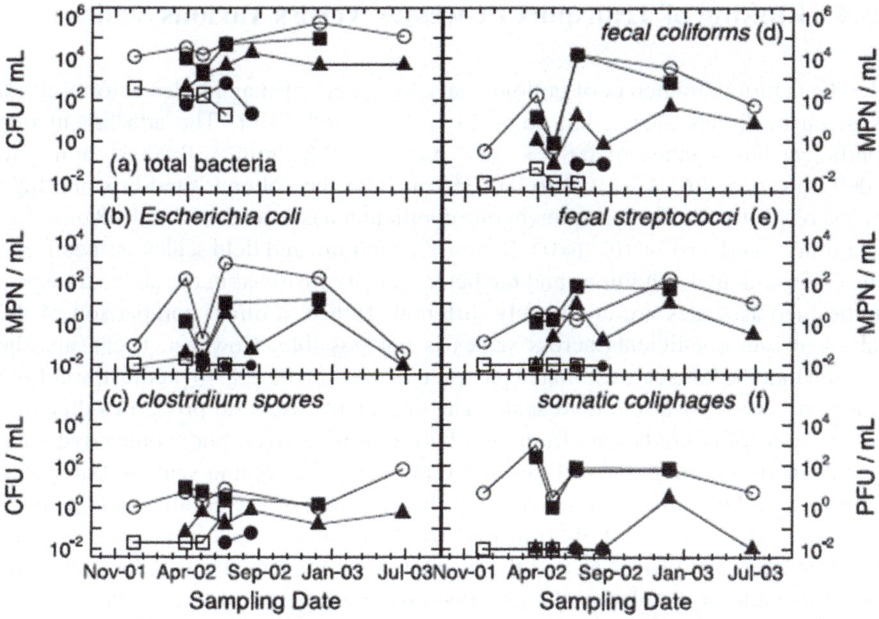

Fig. 23.21 Observed **a** total bacterial count at 37 °C; **b** *E. coli*, **c** *Clostridium spores*, **d** faecal coliforms, **e** faecal streptococci, and **f** *somatic coliphages* at the sinkhole (open circles), Well #1 (solid squares), Well #2 (solid triangles), Well #5 (open squares) and Well #4 (solid circles, background values) (Masciopinto et al. 2008). Reprinted from Masciopinto et al. (2008), copyright 2008, permission obtained from John Wiley and Sons

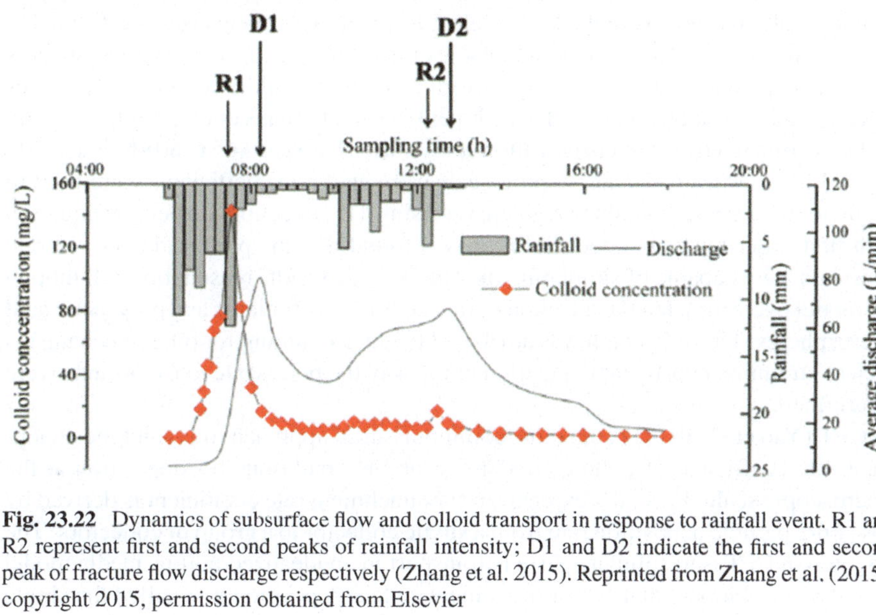

Fig. 23.22 Dynamics of subsurface flow and colloid transport in response to rainfall event. R1 and R2 represent first and second peaks of rainfall intensity; D1 and D2 indicate the first and second peak of fracture flow discharge respectively (Zhang et al. 2015). Reprinted from Zhang et al. (2015), copyright 2015, permission obtained from Elsevier

23.4 Linking of Transport Processes Across Various Scales

The deposition parameters of colloids vary by orders of magnitudes across scales, as given in Tables 23.4, 23.5, 23.6, 23.7, 23.10 and 23.11. The attachment rate coefficient has a range of 6.83×10^{-5} min^{-1} to 2.58 min^{-1}, 0.00031 min^{-1} to 1.069 min^{-1}, and 6.94×10^{-8} to 0.0035 min^{-1} at the column, sand-box and field scales, respectively. The detachment rate coefficient has a range of 5×10^{-6} min^{-1} to 2.34 min^{-1}, and 3.61×10^{-7} to 0.036 min^{-1} at column and field scales, respectively. The environmental conditions and the heterogeneity involved in the above transport studies across scales are completely different. Hence, a direct comparison of the values of rate coefficients across scales is not possible. However, it can also be inferred that the lab-scale estimated parameters may not be able to predict the colloid transport behaviour at the field scale. The dominant structural properties dictating the transport of colloids span from distribution of pore sizes, pore connectivity and heterogeneity on the grain surfaces at the pore and the column scale to the spatial variability in hydraulic conductivity and the presence of preferential flow paths at the field scale. Hence, colloid transport mechanisms are scale dependent and the observed transport processes and the estimated parameters at a smaller scale may not be capable of simulating the observations at a larger scale. It is imperative to link the processes and the associated parameters across scales to understand the effect of small-scale processes on large-scale processes. Though a number of studies have been reported in the literature at each pore, column, sand-box and field scales independently, studies that link the colloid transport processes across scales are rare (Smith et al. 2008; Zhang et al. 2015; Velimirovic et al. 2020).

Smith et al. (2008) found that pore-scale derived rate constants can be used to simulate colloid transport at the REV scale. Flow cell experiments were performed to visualize the transport and deposition of latex microspheres, *E. coli* and microspheres made of poly lactic acid in silica sand, as well as to obtain colloid breakthrough curves at REV scale. The attachment and detachment rate coefficients of colloids to the grain surface estimated from the visualization study varied between 465 to 808 h^{-1}, and 303 to 1193 h^{-1}, respectively. These pore-scale estimated rate coefficients were used to simulate the observed breakthrough curves using 1D advection–dispersion equation with first-order reversible deposition. It was found that the pore-scale derived rate constants were capable of simulating the Darcy-scale results reasonably well though the model overestimated the attachment rate coefficients for latex and poly lactic acid microspheres. This mismatch was attributed to not accounting for other mechanisms such as straining, non-linear deposition or the way the pore-scale rate constants were determined.

CFT (Yao et al. 1971) is the most commonly used approach to predict the attachment rate coefficient of colloids under favourable conditions for deposition at the macroscopic scale. Here, the expression for attachment rate coefficient is derived by upscaling particle deposition rate from a single collector to a group of collectors. The first step in CFT is to simulate particle transport in an idealized porous media model to find a correlation equation for the dimensionless deposition rate called the single collector contact efficiency. A number of correlation equations for single collector contact efficiency are available in the literature, which differs in the model geometry, range of parameter values, and the mechanisms considered for particle attachment to the collector surface (Tufenkji et al. 2004; Velimirovic et al. 2020; Kamai et al. 2015; Long and Hilpert 2009; Ma et al. 2013; Nelson and Ginn 2011). The control volume at the Darcy scale is assumed to be made up of identical collectors, and upscaling to the Darcy scale is performed by applying mass balance at the Darcy scale based on particle removal by a single collector (Nelson and Ginn 2011; Logan et al. 1995; Molnar et al. 2015). CFT has been found to perform really well for micrometer-sized particles under favourable conditions for deposition (Tufenkji et al. 2004; Nelson and Ginn 2011; Molnar et al. 2015; Elimelech and O'Melia 1990; Tufenkji and Elimelech 2005). However, CFT has been found to overestimate the deposition rate at low flow velocities, which are typical of groundwater flow conditions, and for nanometer-sized particles (Ma et al. 2013; Nelson and Ginn 2011; Petosa et al. 2010).

Seetha et al. (2017) upscaled the transport of nanoparticles under unfavourable conditions from pore to the Darcy scale in a homogeneous porous medium using pore-network modeling. Pore-network modeling offers the advantage of studying the macroscopic behaviour by accounting for the physics at the underlying pore scale. The complex structure of the porous medium can be represented mathematically as an interconnected network of pore throats and pore bodies (Raoof et al. 2010). The first step in upscaling is simulating nanoparticle transport in a single pore for a range of physicochemical conditions, and thereby developing correlation equations for nanoparticle deposition parameters in a single pore in terms of various pore-scale parameters (Seetha et al. 2015). The pore-scale derived expressions for deposition parameters were then incorporated into a pore-network model to upscale nanoparticle transport to the Darcy scale. Pore-network simulations were performed for a wide range of parameter values to obtain the breakthrough curves of nanoparticle concentration. The latter was then fitted with a macroscopic 1D advection–dispersion-deposition equation to estimate the values of Darcy-scale rate deposition coefficients. Correlation equations for the Darcy-scale deposition rate coefficients, under unfavorable conditions, were provided as a function of measurable Darcy-scale parameters, including: porosity, mean pore throat radius, mean pore water velocity, nanoparticle radius, ionic strength, dielectric constant, viscosity, temperature, and surface potentials of the particle and grain surfaces. However, the effect of heterogeneity of the grain surface was not accounted for in the studies of Seetha et al. (2015, 2017), which resulted in a multiplier in their correlation equations whose value needed to be determined from calibration.

Velimirovic et al. (2020) upscaled the transport of humic acid-coated goethite nanoparticles from 1D column scale to 3D lab scale by performing experiments in a large-scale 3D container (9 m $\times$6 m $\times$ 4.5 m) in the lab having randomly distributed fine and coarse sand blocks in three layers representing a heterogeneous aquifer (Fig. 23.23). The experimental results were simulated using a 3D advection–dispersion equation with irreversible deposition. The attachment rate coefficient for the 3D modelling was obtained by fitting the breakthrough curves from 1D column experiments with 1D advection–dispersion-deposition equation. The model results

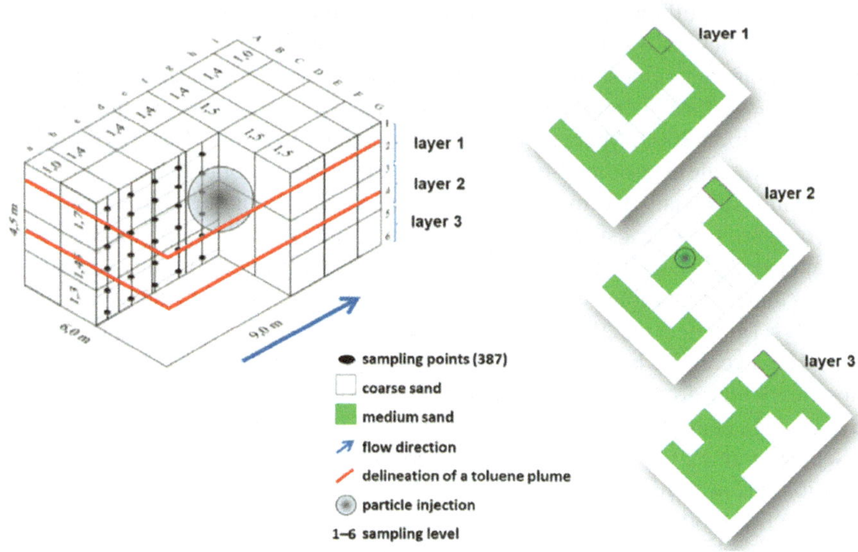

Fig. 23.23 Schematic of the large scale 3D container (Velimirovic et al. 2020)

showed that the particles tend to travel through the high permeable coarse sand blocks. The simulated results were found to correctly reproduce the arrival time and the shape of the nanoparticle breakthrough curve. However, the model results were found to either overestimate or underestimate the concentration in some monitoring wells. This discrepancy was attributed to settling and aggregation of the nanoparticles which was not accounted for in the model. The other theoretical scaling approaches available in the literature that can be adopted for upscaling colloid transport processes are the scale-way approach (Vogel and Roth 2003; Roth et al. 1999) and power averaging approach (Scheibe and Yabusaki 1998).

From the above studies, it can be seen that upscaling of colloid transport processes beyond the REV scale has not been done. Though some studies have shown that the column-scale estimated deposition parameters through homogeneous soils can predict (reasonably well) colloid transport behaviour through a heterogeneous soil at column (Johnson et al. 1996; Saiers et al. 1994a) or sand-box scales (Loveland et al. 2003; Velimirovic et al. 2020), the heterogeneity at the lab scale is an artificially created structured one. However, the spatial variability of heterogeneity at the field scale is very large and it is not possible to exactly resolve the heterogeneity variation. Also, it is not yet known what should be the scale of resolution of field-scale heterogeneity. Hence, more studies need to be conducted to link the transport processes and the associated parameters across scales, and to accurately predict colloid transport behaviour at the field scale.

23.5 Conclusions

Colloid transport in porous media is a multi-scale problem with the length scales spanning from micrometres at the pore scale to kilometres at the field scale. Colloid transport studies at each scale provide valuable information about the factors and mechanisms controlling the behaviour at each scale. The colloid deposition mechanisms are scale dependent, and are governed by the spatial heterogeneity at each scale. Spatial heterogeneity manifests in the form of roughness at the pore-scale, distribution in grain sizes and pore connectivity at the REV scale, structured heterogeneity at the column and sand-box scale, and high degree of spatial variability in the hydraulic conductivity and the presence of macropores and vegetation at the field scale. This results in the colloid breakthrough curve at the field scale being characterized by a sharp peak and extended tailing, unlike the breakthrough curve at the lab scale. Hence, the observed transport processes and the estimated parameters at a smaller scale may not be capable of simulating the observations at a larger scale. Thus it is imperative to link the processes and the associated parameters across scales to better predict the transport behaviour at larger scales. Colloid deposition onto the grain surface have been upscaled from pore to the REV scale in saturated homogeneous porous media, but not beyond the REV scale. Also, the scale effects on the parameters describing other transport mechanisms such as blocking, straining, deposition to SWI, AWI and AWS under partially saturated conditions, and the scale effect of heterogeneity on the various transport mechanisms is not known yet. Though some studies have shown that the colloid transport behaviour through a structurally heterogeneous soil at column or sand-box scales can be predicted by knowing the deposition parameters for each type of soil in the heterogeneous system, the heterogeneity at the lab scale is an artificially created structured one. However, the spatial variability of heterogeneity at the field scale is very large and the scale of resolution of spatial heterogeneity information needed at the field scale to predict colloid transport behaviour is not known yet. Hence, more experimental and numerical studies need to be conducted to upscale colloid transport processes and the associated parameters from the lab scale to the field scale, and thereby accurately predict colloid transport behaviour at the field scale.

Acknowledgements The corresponding author acknowledges the funding received from Department of Science and Technology, Government of India (Sanction number: DST/INSPIRE/04/2016/000013) for carrying out this work.

References

Anders R, Chrysikopoulos CV (2005) Virus fate and transport during artificial recharge with recycled water. Water Resour Res 41(10)

Argent J, Torkzaban S, Hubbard S, Le H, Amirianshoja T, Haghighi M (2015) Visualization of micro-particle retention on a heterogeneous surface using micro-models: influence of nanoscale surface roughness. Transp Porous Media 109(2):239–253
Auset M, Keller AA (2004) Pore-scale processes that control dispersion of colloids in saturated porous media. Water Resour Res 40(3)
Auset M, Keller AA (2006) Pore-scale visualization of colloid straining and filtration in saturated porous media using micromodels. Water Resour Res 42(12)
Auset M, Keller AA, Brissaud F, Lazarova V (2005) Intermittent filtration of bacteria and colloids in porous media. Water Resour Res 41(9)
Babakhani P, Bridge J, Doong RA, Phenrat T (2017) Continuum-based models and concepts for the transport of nanoparticles in saturated porous media: a state-of-the-science review. Adv Coll Interface Sci 246:75–104
Bai H, Cochet N, Drelich A, Pauss A, Lamy E (2016) Comparison of transport between two bacteria in saturated porous media with distinct pore size distribution. RSC Adv 6(18):14602–14614
Basnet M, Di Tommaso C, Ghoshal S, Tufenkji N (2015) Reduced transport potential of a palladium-doped zero valent iron nanoparticle in a water saturated loamy sand. Water Res 68:354–363
Baumann T, Toops L, Niessner R (2010) Colloid dispersion on the pore scale. Water Res 44(4):1246–1254
Becker MW, Reimus PW, Vilks P (1999) Transport and attenuation of carboxylate-modified latex microspheres in fractured rock laboratory and field tracer tests. Groundwater 37(3):387–395
Bennett P, He F, Zhao D, Aiken B, Feldman L (2010) In situ testing of metallic iron nanoparticle mobility and reactivity in a shallow granular aquifer. J Contam Hydrol 116(1–4):35–46
Bhattacharjee S, Ryan JN, Elimelech M (2002) Virus transport in physically and geochemically heterogeneous subsurface porous media. J Contam Hydrol 57(3–4):161–187
Bradford SA, Bettahar M (2006) Concentration dependent transport of colloids in saturated porous media. J Contam Hydrol 82(1–2):99–117
Bradford SA, Torkzaban S (2008) Colloid transport and retention in unsaturated porous media: A review of interface-, collector-, and pore-scale processes and models. Vadose Zone J 7(2):667–681
Bradford SA, Yates SR, Bettahar M, Simunek J (2002) Physical factors affecting the transport and fate of colloids in saturated porous media. Water Resour Res 38(12):63–71
Bradford SA, Simunek J, Bettahar M, Van Genuchten MT, Yates SR (2003) Modeling colloid attachment, straining, and exclusion in saturated porous media. Environ Sci Technol 37(10):2242–2250
Bradford SA, Bettahar M, Simunek J, Van Genuchten MT (2004) Straining and attachment of colloids in physically heterogeneous porous media. Vadose Zone J 3(2):384–394
Bradford SA, Torkzaban S, Walker SL (2007) Coupling of physical and chemical mechanisms of colloid straining in saturated porous media. Water Res 41(13):3012–3024
Bradford SA, Kim HN, Haznedaroglu BZ, Torkzaban S, Walker SL (2009) Coupled factors influencing concentration-dependent colloid transport and retention in saturated porous media. Environ Sci Technol 43(18):6996–7002
Bradford SA, Wang Y, Kim H, Torkzaban S, Šimůnek J (2014) Modeling microorganism transport and survival in the subsurface. J Environ Qual 43(2):421–440
Bradford SA, Headd B, Arye G, Šimůnek J (2015) Transport of *E. coli* D21g with runoff water under different solution chemistry conditions and surface slopes. J Hydrol 525:760–768
Bradford SA, Simunek J, Bettahar M, Tadassa YF, van Genuchten MT, Yates SR (2005). Straining of colloids at textural interfaces. Water Resour Res 41(10)
Bradford SA, Simunek J, Walker SL (2006) Transport and straining of *E. coli* O157: H7 in saturated porous media. Water Resour Res 42(12)
Braun A, Klumpp E, Azzam R, Neukum C (2015) Transport and deposition of stabilized engineered silver nanoparticles in water saturated loamy sand and silty loam. Sci Total Environ 535:102–112
Burkhardt M, Kasteel R, Vanderborght J, Vereecken H (2008) Field study on colloid transport using fluorescent microspheres. Eur J Soil Sci 59(1):82–93

Camesano TA, Logan BE (1998) Influence of fluid velocity and cell concentration on the transport of motile and nonmotile bacteria in porous media. Environ Sci Technol 32(11):1699–1708
Camesano TA, Unice KM, Logan BE (1999) Blocking and ripening of colloids in porous media and their implications for bacterial transport. Colloids Surf A 160(3):291–307
Chen G, Flury M (2005) Retention of mineral colloids in unsaturated porous media as related to their surface properties. Colloids Surf A 256(2–3):207–216
Chen G, Liu X, Su C (2011) Transport and retention of TiO2 rutile nanoparticles in saturated porous media under low-ionic-strength conditions: measurements and mechanisms. Langmuir 27(9):5393–5402
Chen Y, Ma J, Wu X, Weng L, Li Y (2020) Sedimentation and transport of different soil colloids: effects of Goethite and humic acid. Water 12(4):980
Cheng X, Kan AT, Tomson MB (2005) Study of C60 transport in porous media and the effect of sorbed C60 on naphthalene transport. J Mater Res 20(12):3244–3254
Chowdhury I, Hong Y, Honda RJ, Walker SL (2011) Mechanisms of TiO_2 nanoparticle transport in porous media: Role of solution chemistry, nanoparticle concentration, and flowrate. J Colloid Interface Sci 360(2):548–555
Chrysikopoulos CV, Syngouna VI, Vasiliadou IA, Katzourakis VE (2012) Transport of Pseudomonas putida in a 3-D bench scale experimental aquifer. Transp Porous Media 94(3):617–642
Cornelis G, Pang L, Doolette C, Kirby JK, McLaughlin MJ (2013) Transport of silver nanoparticles in saturated columns of natural soils. Sci Total Environ 463:120–130
Crist JT, McCarthy JF, Zevi Y, Baveye P, Throop JA, Steenhuis TS (2004) Pore-scale visualization of colloid transport and retention in partly saturated porous media. Vadose Zone J 3(2):444–450
Dagtepe P, Chikan V (2010) Quantized Ostwald ripening of colloidal nanoparticles. J Phys Chem C 114(39):16263–16269
Deraguin BV, Landau L (1941) Theory of the stability of strongly charged lyophobic sols and of the adhesion of strongly charged particles in solution of electrolytes. Acta Physicochim: USSR 14:633–662
Deshpande PA, Shonnard DR (1999) Modeling the effects of systematic variation in ionic strength on the attachment kinetics of Pseudomonas fluorescens UPER-1 in saturated sand columns. Water Resour Res 35(5):1619–1627
Elimelech M, O'Melia CR (1990) Effect of particle size on collision efficiency in the deposition of Brownian particles with electrostatic energy barriers. Langmuir 6(6):1153–1163
Fan W, Jiang XH, Yang W, Geng Z, Huo MX, Liu ZM, Zhou H (2015) Transport of graphene oxide in saturated porous media: effect of cation composition in mixed Na–Ca electrolyte systems. Sci Total Environ 511:509–515
Fang J, Xu MJ, Wang DJ, Wen B, Han JY (2013) Modeling the transport of TiO_2 nanoparticle aggregates in saturated and unsaturated granular media: effects of ionic strength and pH. Water Res 47(3):1399–1408
Feriancikova L, Xu S (2012) Deposition and remobilization of graphene oxide within saturated sand packs. J Hazard Mater 235:194–200
Flory J, Kanel SR, Racz L, Impellitteri CA, Silva RG, Goltz MN (2013) Influence of pH on the transport of silver nanoparticles in saturated porous media: laboratory experiments and modeling. J Nanopart Res 15(3):1–11
Flury M, Aramrak S (2017) Role of air-water interfaces in colloid transport in porous media: a review. Water Resour Res 53(7):5247–5275
Foppen JWA, Mporokoso A, Schijven JF (2005) Determining straining of Escherichia coli from breakthrough curves. J Contam Hydrol 76(3–4):191–210
Gao B, Saiers JE, Ryan JN (2004) Deposition and mobilization of clay colloids in unsaturated porous media. Water Resour Res 40(8)
Gao B, Saiers JE, Ryan J (2006) Pore-scale mechanisms of colloid deposition and mobilization during steady and transient flow through unsaturated granular media. Water Resour Res 42(1)
Ginn TR (2002) A travel time approach to exclusion on transport in porous media. Water Resour Res 38(4):12–21

Gitis V, Adin A, Nasser A, Gun J, Lev O (2002) Fluorescent dye labeled bacteriophages—A new tracer for the investigation of viral transport in porous media: 2. Studies of deep-bed filtration. Water Res 36(17):4235–4242

Göppert N, Goldscheider N (2008) Solute and colloid transport in karst conduits under low-and high-flow conditions. Groundwater 46(1):61–68

Grolimund D, Borkovec M (2005) Colloid-facilitated transport of strongly sorbing contaminants in natural porous media: mathematical modeling and laboratory column experiments. Environ Sci Technol 39(17):6378–6386

Han J, Jin Y, Willson CS (2006) Virus retention and transport in chemically heterogeneous porous media under saturated and unsaturated flow conditions. Environ Sci Technol 40(5):1547–1555

Higgo JJW, Williams GM, Harrison I, Warwick P, Gardiner MP, Longworth G (1993) Colloid transport in a glacial sand aquifer. Laboratory and field studies. In: Colloids in the aquatic environment. Elsevier, pp 179–200

Hopmans, J. W., & Schoups, G. H. (2005). 74: Soil Water Flow at Different Spatial Scales.

Hornberger GM, Mills AL, Herman JS (1992) Bacterial transport in porous media: evaluation of a model using laboratory observations. Water Resour Res 28(3):915–923

Hosseini SM, Tosco T (2013) Transport and retention of high concentrated nano-Fe/Cu particles through highly flow-rated packed sand column. Water Res 47(1):326–338

Jiang X, Tong M, Lu R, Kim H (2012) Transport and deposition of ZnO nanoparticles in saturated porous media. Colloids Surf A 401:29–37

Jin Y, Chu Y, Li Y (2000) Virus removal and transport in saturated and unsaturated sand columns. J Contam Hydrol 43(2):111–128

Johnson PR, Sun N, Elimelech M (1996) Colloid transport in geochemically heterogeneous porous media: modeling and measurements. Environ Sci Technol 30(11):3284–3293

Johnson PR, Elimelech M (1995) Dynamics of colloid deposition in porous media: Blocking based on random sequential adsorption. Langmuir 11(3):801–812

Johnson WP, Pazmino E, Ma H (2010) Direct observations of colloid retention in granular media in the presence of energy barriers, and implications for inferred mechanisms from indirect observations. Water Res 44(4):1158–1169

Kamai T, Nassar MK, Nelson KE, Ginn TR (2015) Colloid filtration prediction by mapping the correlation-equation parameters from transport experiments in porous media. Water Resour Res 51(11):8995–9012

Kanel SR, Flory J, Meyerhoefer A, Fraley JL, Sizemore IE, Goltz MN (2015) Influence of natural organic matter on fate and transport of silver nanoparticles in saturated porous media: laboratory experiments and modeling. J Nanopart Res 17(3):1–13

Keller AA, Auset M (2007) A review of visualization techniques of biocolloid transport processes at the pore scale under saturated and unsaturated conditions. Adv Water Resour 30(6–7):1392–1407

Keller AA, Sirivithayapakorn S, Chrysikopoulos CV (2004) Early breakthrough of colloids and bacteriophage MS2 in a water-saturated sand column. Water Resour Res 40(8)

Kim HN, Bradford SA, Walker SL (2009) *Escherichia coli* O157: H7 transport in saturated porous media: role of solution chemistry and surface macromolecules. Environ Sci Technol 43(12):4340–4347

Kumahor SK, Hron P, Metreveli G, Schaumann GE, Vogel HJ (2015) Transport of citrate-coated silver nanoparticles in unsaturated sand. Sci Total Environ 535:113–121

Lazouskaya V, Jin Y (2006) Interfacial interactions and colloid retention under steady flows in a capillary channel. J Colloid Interface Sci 303(1):171–184

Lazouskaya V, Jin Y (2008) Colloid retention at air–water interface in a capillary channel. Colloids Surf A 325(3):141–151

Lazouskaya V, Wang LP, Gao H, Shi X, Czymmek K, Jin Y (2011) Pore-scale investigation of colloid retention and mobilization in the presence of a moving air–water interface. Vadose Zone J 10(4):1250–1260

Lazouskaya V, Wang LP, Or D, Wang G, Caplan JL, Jin Y (2013). Colloid mobilization by fluid displacement fronts in channels. J Colloid Interface Sci 406:44–50

Lenhart JJ, Saiers JE (2002) Transport of silica colloids through unsaturated porous media: Experimental results and model comparisons. Environ Sci Technol 36(4):769–777
Li X, Johnson WP (2005) Nonmonotonic variations in deposition rate coefficients of microspheres in porous media under unfavorable deposition conditions. Environ Sci Technol 39(6):1658–1665
Li Y, Wang Y, Pennell KD, Abriola LM (2008) Investigation of the transport and deposition of fullerene (C60) nanoparticles in quartz sands under varying flow conditions. Environ Sci Technol 42(19):7174–7180
Li J, Rajajayavel SRC, Ghoshal S (2016) Transport of carboxymethyl cellulose-coated zerovalent iron nanoparticles in a sand tank: effects of sand grain size, nanoparticle concentration and injection velocity. Chemosphere 150:8–16
Liu D, Johnson PR, Elimelech M (1995) Colloid deposition dynamics in flow-through porous media: role of electrolyte concentration. Environ Sci Technol 29(12):2963–2973
Liu G, Xue W, Wang J, Liu X (2019) Transport behavior of variable charge soil particle size fractions and their influence on cadmium transport in saturated porous media. Geoderma 337:945–955
Logan BE, Jewett DG, Arnold RG, Bouwer EJ, O'Melia CR (1995) Clarification of clean-bed filtration models. J Environ Eng 121(12):869–873
Long W, Hilpert M (2009) A correlation for the collector efficiency of Brownian particles in clean-bed filtration in sphere packings by a Lattice-Boltzmann method. Environ Sci Technol 43(12):4419–4424
Loveland JP, Bhattacharjee S, Ryan JN, Elimelech M (2003) Colloid transport in a geochemically heterogeneous porous medium: aquifer tank experiment and modeling. J Contam Hydrol 65(3–4):161–182
Lv X, Gao B, Sun Y, Dong S, Wu J, Jiang B, Shi X (2016) Effects of grain size and structural heterogeneity on the transport and retention of nano-TiO_2 in saturated porous media. Sci Total Environ 563:987–995
Ma H, Hradisky M, Johnson WP (2013) Extending applicability of correlation equations to predict colloidal retention in porous media at low fluid velocity. Environ Sci Technol 47(5):2272–2278
Ma E, Ouahbi T, Wang H, Ahfir ND, Alem A, Hammadi A (2017) Modeling of retention and re-entrainment of mono-and poly-disperse particles: effects of hydrodynamics, particle size and interplay of different-sized particles retention. Sci Total Environ 596:222–229
Masciopinto C, La Mantia R, Chrysikopoulos CV (2008) Fate and transport of pathogens in a fractured aquifer in the Salento area, Italy. Water Res Res 44(1)
McCarthy JF, McKay LD (2004) Colloid transport in the subsurface: past, present, and future challenges. Vadose Zone J 3(2):326–337
McKay LD, Sanford WE, Strong JM (2000) Field-scale migration of colloidal tracers in a fractured shale saprolite. Groundwater 38(1):139–147
Mishurov M, Yakirevich A, Weisbrod N (2008) Colloid transport in a heterogeneous partially saturated sand column. Environ Sci Technol 42(4):1066–1071
Molnar IL, Johnson WP, Gerhard JI, Willson CS, O'carroll DM (2015) Predicting colloid transport through saturated porous media: a critical review. Water Resour Res 51(9):6804–6845
Mondal PK, Furbacher PD, Cui Z, Krol MM, Sleep BE (2018) Transport of polymer stabilized nano-scale zero-valent iron in porous media. J Contam Hydrol 212:65–77
Morales VL, Gao B, Steenhuis TS (2009) Grain surface-roughness effects on colloidal retention in the vadose zone. Vadose Zone J 8(1):11–20
Nascimento AG, Totola MR, Souza CS, Borges MT, Borges AC (2006) Temporal and spatial dynamics of blocking and ripening effects on bacterial transport through a porous system: a possible explanation for CFT deviation. Colloids Surf, B 53(2):241–244
Nelson KE, Ginn TR (2011) New collector efficiency equation for colloid filtration in both natural and engineered flow conditions. Water Resour Res 47(5)
Neukum C, Braun A, Azzam R (2014) Transport of stabilized engineered silver (Ag) nanoparticles through porous sandstones. J Contam Hydrol 158:1–13
Nishad S, Al-Raoush RI (2021) Colloid retention and mobilization mechanisms under different physicochemical conditions in porous media: a micromodel study. Powder Technol 377:163–173

Noordmans J, Wit PJ, Van Der Mei HC, Busscher HJ (1997) Detachment of polystyrene particles from collector surfaces by surface tension forces induced by air-bubble passage through a parallel plate flow chamber. J Adhes Sci Technol 11(7):957–969
Ochiai N, Kraft EL, Selker JS (2006) Methods for colloid transport visualization in pore networks. Water Resour Res 42(12)
Petosa AR, Jaisi DP, Quevedo IR, Elimelech M, Tufenkji N (2010) Aggregation and deposition of engineered nanomaterials in aquatic environments: role of physicochemical interactions. Environ Sci Technol 44(17):6532–6549
Phenrat T, Cihan A, Kim HJ, Mital M, Illangasekare T, Lowry GV (2010) Transport and deposition of polymer-modified Fe0 nanoparticles in 2-D heterogeneous porous media: effects of particle concentration, FeO content, and coatings. Environ Sci Technol 44(23):9086–9093
Qi Z, Zhang L, Chen W (2014) Transport of graphene oxide nanoparticles in saturated sandy soil. Environ Sci Process Impacts 16(10):2268–2277
Rajagopalan R, Tien C (1976) Trajectory analysis of deep-bed filtration with the sphere-in-cell porous media model. AIChE J 22(3):523–533
Raoof A, Hassanizadeh SM, Leijnse A (2010) Upscaling transport of adsorbing solutes in porous media: pore-network modeling. In: First international conference on frontiers in shallow subsurface technology. European Association of Geoscientists & Engineers, pp cp-150
Rastghalam ZS, Yan C, Shang J, Cheng T (2019) Nanoscale titanium dioxide ($nTiO_2$) aggregation and transport in the co-presence of dissolved phosphate, illite colloid, and Fe oxyhydroxide coating. Colloids Surf A: Physicochem Eng Aspects 578:123560
Raychoudhury T, Tufenkji N, Ghoshal S (2014) Straining of polyelectrolyte-stabilized nanoscale zero valent iron particles during transport through granular porous media. Water Res 50:80–89
Rijnaarts HH, Norde W, Bouwer EJ, Lyklema J, Zehnder AJ (1996) Bacterial deposition in porous media related to the clean bed collision efficiency and to substratum blocking by attached cells. Environ Sci Technol 30(10):2869–2876
Rod K, Um W, Chun J, Wu N, Yin X, Wang G, Neeves K (2018) Effect of chemical and physical heterogeneities on colloid-facilitated cesium transport. J Contam Hydrol 213:22–27
Roth K, Vogel HJ, Kasteel R (1999) The scaleway: a conceptual framework for upscaling soil properties. Modelling of transport processes in soils at various scales in time and space. 24–26
Ryan JN, Elimelech M (1996) Colloid mobilization and transport in groundwater. Colloids Surf A 107:1–56
Ryan JN, Gschwend PM (1990) Colloid mobilization in two Atlantic Coastal Plain aquifers: field studies. Water Resour Res 26(2):307–322
Sadeghi G, Schijven JF, Behrends T, Hassanizadeh SM, Gerritse J, Kleingeld PJ (2011) Systematic study of effects of pH and ionic strength on attachment of phage PRD1. Groundwater 49(1):12–19
Sagee O, Dror I, Berkowitz B (2012) Transport of silver nanoparticles (AgNPs) in soil. Chemosphere 88(5):670–675
Saiers JE, Hornberger GM, Harvey C (1994a) Colloidal silica transport through structured, heterogeneous porous media. J Hydrol 163(3–4):271–288
Saiers JE, Hornberger GM, Liang L (1994b) First-and second-order kinetics approaches for modeling the transport of colloidal particles in porous media. Water Resour Res 30(9):2499–2506
Saiers JE, Lenhart JJ (2003). Colloid mobilization and transport within unsaturated porous media under transient-flow conditions. Water Resour Res 39(1)
Sasidharan S, Torkzaban S, Bradford SA, Cook PG, Gupta VV (2017) Temperature dependency of virus and nanoparticle transport and retention in saturated porous media. J Contam Hydrol 196:10–20
Scheibe T, Yabusaki S (1998) Scaling of flow and transport behavior in heterogeneous groundwater systems. Adv Water Resour 22(3):223–238
Schijven JF, Šimůnek J (2002) Kinetic modeling of virus transport at the field scale. J Contam Hydrol 55(1–2):113–135

Schijven JF, Hoogenboezem W, Hassanizadeh M, Peters JH (1999) Modeling removal of bacteriophages MS2 and PRD1 by dune recharge at Castricum, Netherlands. Water Resour Res 35(4):1101–1111
Schijven JF, Hassanizadeh SM, Dowd SE, Pillai SD (2000a) Modeling virus adsorption in batch and column experiments. Quant Microbiol 2(1):5–20
Schijven JF, Medema G, Vogelaar AJ, Hassanizadeh SM (2000b) Removal of microorganisms by deep well injection. J Contam Hydrol 44(3–4):301–327
Seetha N, Majid Hassanizadeh S, Mohan Kumar MS, Raoof A (2015) Correlation equations for average deposition rate coefficients of nanoparticles in a cylindrical pore. Water Resour Res 51(10):8034–8059
Seetha N, Raoof A, Kumar MM, Hassanizadeh SM (2017) Upscaling of nanoparticle transport in porous media under unfavorable conditions: pore scale to Darcy scale. J Contam Hydrol 200:1–14
Shen C, Zhang M, Zhang S, Wang Z, Zhang H, Li B, Huang Y (2015) Influence of surface heterogeneities on reversibility of fullerene (nC60) nanoparticle attachment in saturated porous media. J Hazard Mater 290:60–68
Shen C, Huang Y, Li B, Jin Y (2008) Effects of solution chemistry on straining of colloids in porous media under unfavorable conditions. Water Resour Res 44(5)
Sinton LW, Noonan MJ, Finlay RK, Pang L, Close ME (2000) Transport and attenuation of bacteria and bacteriophages in an alluvial gravel aquifer. NZ J Mar Freshwat Res 34(1):175–186
Sirivithayapakorn S, Keller A (2003a) Transport of colloids in saturated porous media: A pore-scale observation of the size exclusion effect and colloid acceleration. Water Resour Res 39(4)
Sirivithayapakorn S, Keller A (2003b) Transport of colloids in unsaturated porous media: a pore-scale observation of processes during the dissolution of air-water interface. Water Resour Res 39(12)
Smith J, Gao B, Funabashi H, Tran TN, Luo D, Ahner BA, Steenhuis TS, Hay AG, Walter MT (2008) Pore-scale quantification of colloid transport in saturated porous media. Environ Sci Technol 42(2):517–523
Solovitch N, Labille J, Rose J, Chaurand P, Borschneck D, Wiesner MR, Bottero JY (2010) Concurrent aggregation and deposition of TiO_2 nanoparticles in a sandy porous media. Environ Sci Technol 44(13):4897–4902
Sun Y, Gao B, Bradford SA, Wu L, Chen H, Shi X, Wu J (2015a) Transport, retention, and size perturbation of graphene oxide in saturated porous media: effects of input concentration and grain size. Water Res 68:24–33
Sun P, Shijirbaatar A, Fang J, Owens G, Lin D, Zhang K (2015b) Distinguishable transport behavior of zinc oxide nanoparticles in silica sand and soil columns. Sci Total Environ 505:189–198
Syngouna VI, Chrysikopoulos CV (2012) Transport of biocolloids in water saturated columns packed with sand: effect of grain size and pore water velocity. J Contam Hydrol 129:11–24
Tan Y, Gannon JT, Baveye P, Alexander M (1994) Transport of bacteria in an aquifer sand: experiments and model simulations. Water Resour Res 30(12):3243–3252
Toloni I, Lehmann F, Ackerer P (2014) Modeling the effects of water velocity on TiO_2 nanoparticles transport in saturated porous media. J Contam Hydrol 171:42–48
Tong M, Johnson WP (2006) Excess colloid retention in porous media as a function of colloid size, fluid velocity, and grain angularity. Environ Sci Technol 40(24):7725–7731
Torkzaban S, Hassanizadeh SM, Schijven JF, De Bruin HAM, de Roda Husman AM (2006b) Virus transport in saturated and unsaturated sand columns. Vadose Zone J 5(3):877–885
Torkzaban S, Bradford SA, van Genuchten MT, Walker SL (2008) Colloid transport in unsaturated porous media: the role of water content and ionic strength on particle straining. J Contam Hydrol 96(1–4):113–127
Torkzaban S, Wan J, Tokunaga TK, Bradford SA (2012) Impacts of bridging complexation on the transport of surface-modified nanoparticles in saturated sand. J Contam Hydrol 136:86–95
Torkzaban S, Hassanizadeh SM, Schijven JF, Van Den Berg HHJL (2006a). Role of air-water interfaces on retention of viruses under unsaturated conditions. Water Resour Res 42(12)

Torkzaban S, Tazehkand SS, Walker SL, Bradford SA (2008b) Transport and fate of bacteria in porous media: coupled effects of chemical conditions and pore space geometry. Water Resour Res 44(4)

Tosco T, Sethi R (2010) Transport of non-Newtonian suspensions of highly concentrated micro-and nanoscale iron particles in porous media: a modeling approach. Environ Sci Technol 44(23):9062–9068

Tufenkji N, Elimelech M (2005) Breakdown of colloid filtration theory: role of the secondary energy minimum and surface charge heterogeneities. Langmuir 21(3):841–852

Tufenkji N, Miller GF, Ryan JN, Harvey RW, Elimelech M (2004) Transport of cryptosporidium oocysts in porous media: role of straining and physicochemical filtration. Environ Sci Technol 38(22):5932–5938

Velimirovic M, Bianco C, Ferrantello N, Tosco T, Casasso A, Sethi R, Wagner D, Miyajima K, Klaas N, Meckenstock RU, Hofmann T (2020) A large-scale 3D study on transport of humic acid-coated goethite nanoparticles for aquifer remediation. Water 12(4):1207

Verwey EJW (1947) Theory of the stability of lyophobic colloids. J Phys Chem 51(3):631–636

Vilks P, Frost LH, Bachinski DB (1997) Field-scale colloid migration experiments in a granite fracture. J Contam Hydrol 26(1–4):203–214

Vogel HJ, Roth K (2003) Moving through scales of flow and transport in soil. J Hydrol 272(1–4):95–106

Wan J, Tokunaga TK (1997) Film straining of colloids in unsaturated porous media: Conceptual model and experimental testing. Environ Sci Technol 31(8):2413–2420

Wan J, Wilson JL (1994a) Visualization of the role of the gas-water interface on the fate and transport of colloids in porous media. Water Resour Res 30(1):11–23

Wan J, Wilson JL (1994b) Colloid transport in unsaturated porous media. Water Resour Res 30(4):857–864

Wan J, Wilson JL, Kieft TL (1994) Influence of the gas-water interface on transport of microorganisms through unsaturated porous media. Appl Environ Microbiol 60(2):509–516

Wang Y, Li Y, Fortner JD, Hughes JB, Abriola LM, Pennell KD (2008) Transport and retention of nanoscale C60 aggregates in water-saturated porous media. Environ Sci Technol 42(10):3588–3594

Wang C, Bobba AD, Attinti R, Shen C, Lazouskaya V, Wang LP, Jin Y (2012a) Retention and transport of silica nanoparticles in saturated porous media: effect of concentration and particle size. Environ Sci Technol 46(13):7151–7158

Wang D, Bradford SA, Harvey RW, Gao B, Cang L, Zhou D (2012b) Humic acid facilitates the transport of ARS-labeled hydroxyapatite nanoparticles in iron oxyhydroxide-coated sand. Environ Sci Technol 46(5):2738–2745

Wang D, Su C, Liu C, Zhou D (2014a) Transport of fluorescently labeled hydroxyapatite nanoparticles in saturated granular media at environmentally relevant concentrations of surfactants. Colloids Surf A 457:58–66

Wang X, Cai L, Han P, Lin D, Kim H, Tong M (2014b) Cotransport of multi-walled carbon nanotubes and titanium dioxide nanoparticles in saturated porous media. Environ Pollut 195:31–38

Wang D, Jin Y, Jaisi DP (2015a) Effect of size-selective retention on the cotransport of hydroxyapatite and goethite nanoparticles in saturated porous media. Environ Sci Technol 49(14):8461–8470

Wang D, Jin Y, Jaisi DP (2015b) Cotransport of hydroxyapatite nanoparticles and hematite colloids in saturated porous media: mechanistic insights from mathematical modeling and phosphate oxygen isotope fractionation. J Contam Hydrol 182:194–209

Wang C, McNew CP, Lyon SW, Walter MT, Volkman TH, Abramson N, Sengupta A, Wang Y, Neto AA, Pangle L, Troch PA, Dahlke HE (2019) Particle tracer transport in a sloping soil lysimeter under periodic, steady state conditions. J Hydrol 569:61–76

Werth CJ, Zhang C, Brusseau ML, Oostrom M, Baumann T (2010) A review of non-invasive imaging methods and applications in contaminant hydrogeology research. J Contam Hydrol 113(1–4):1–24

Wheatcraft SW, Cushman JH (1991) Hierarchical approaches to transport in heterogeneous porous media. Rev Geophys 29(S1):263–269
Xu S, Liao Q, Saiers JE (2008) Straining of nonspherical colloids in saturated porous media. Environ Sci Technol 42(3):771–778
Xu S, Gao B, Saiers JE (2006) Straining of colloidal particles in saturated porous media. Water Resour Res 42(12)
Yao KM, Habibian MT, O'Melia CR (1971) Water and waste water filtration. Concepts and applications. Environ Sci Technol 5(11):1105–1112
Yecheskel Y, Dror I, Berkowitz B (2018) Silver nanoparticle (Ag-NP) retention and release in partially saturated soil: column experiments and modelling. Environ Sci Nano 5(2):422–435
Yu C, Gao B, Muñoz-Carpena R, Tian Y, Wu L, Perez-Ovilla O (2011) A laboratory study of colloid and solute transport in surface runoff on saturated soil. J Hydrol 402(1–2):159–164
Yu C, Gao B, Muñoz-Carpena R (2012) Effect of dense vegetation on colloid transport and removal in surface runoff. J Hydrol 434:1–6
Yu C, Muñoz-Carpena R, Gao B, Perez-Ovilla O (2013) Effects of ionic strength, particle size, flow rate, and vegetation type on colloid transport through a dense vegetation saturated soil system: experiments and modeling. J Hydrol 499:316–323
Zevi Y, Dathe A, McCarthy JF, Richards BK, Steenhuis TS (2005) Distribution of colloid particles onto interfaces in partially saturated sand. Environ Sci Technol 39(18):7055–7064
Zevi Y, Gao B, Zhang W, Morales VL, Cakmak ME, Medrano EA, Sang W, Steenhuis TS (2012) Colloid retention at the meniscus-wall contact line in an open microchannel. Water Res 46(2):295–306
Zhang P, Johnson WP, Scheibe TD, Choi KH, Dobbs FC, Mailloux BJ (2001) Extended tailing of bacteria following breakthrough at the Narrow Channel Focus Area, Oyster Virginia. Water Resour Res 37(11):2687–2698
Zhang C, Kang Q, Wang X, Zilles JL, Müller RH, Werth CJ (2010) Effects of pore-scale heterogeneity and transverse mixing on bacterial growth in porous media. Environ Sci Technol 44(8):3085–3092
Zhang Q, Hassanizadeh SM, Karadimitriou NK, Raoof A, Liu B, Kleingeld PJ, Imhof A (2013) Retention and remobilization of colloids during steady-state and transient two-phase flow. Water Resour Res 49(12):8005–8016
Zhang W, Tang XY, Weisbrod N, Zhao P, Reid BJ (2015) A coupled field study of subsurface fracture flow and colloid transport. J Hydrol 524:476–488

The manufacturer's authorised representative in the EU is Springer Nature Customer Service Centre GmbH, Europaplatz 3, 69115 Heidelberg, Germany. If you have any concerns regarding our products, please contact ProductSafety@springernature.com

Printed and bound by CPI Group (UK) Ltd, Croydon, CR0 4YY
30/09/2026
02225238-0017